TA 365 PET

WITHDRAWN
FROM STOCK
QMUL LIBRARY

Acoustics and Noise Control

PEARSON

We work with leading authors to develop the strongest educational materials in acoustics, bringing cutting-edge thinking and best learning practice to a global market.

Under a range of well-known imprints, including Prentice Hall we craft high quality print and electronic publications which help readers to understand and apply their content, whether studying or at work.

To find out more about the complete range of our publishing, please visit us on the World Wide Web at:
www.pearsoned.co.uk

Acoustics and Noise Control

Third edition

R J Peters PhD, MSc, FIOA, CEng, CPhys, BSc, BA, DIC, ARCS
Principal Acoustics Consultant at Applied Acoustics Design

Visiting Research Fellow, London South Bank University
Lecturer in Acoustics and Noise Control (freelance)
formerly Head of School of Environmental Engineering at NESCOT
(North East Surrey College of Technology)

B J Smith BSc, PhD, FIOA, FCIBSE, FCIOB, FIMA, MInstP, CMath, CPhys, CEng
formerly Director of Studies and Head of School of Building,
Ulster Polytechnic (University of Ulster)

Margaret Hollins LLB, LLM, Solicitor
Deputy Head of the Department of Urban, Environment and Leisure Studies,
London South Bank University

Prentice Hall
is an imprint of

Harlow, England • London • New York • Boston • San Francisco • Toronto
Sydney • Tokyo • Singapore • Hong Kong • Seoul • Taipei • New Delhi
Cape Town • Madrid • Mexico City • Amsterdam • Munich • Paris • Milan

Pearson Education Limited
Edinburgh Gate
Harlow
Essex CM20 2JE
England

and Associated Companies throughout the world

Visit us on the World Wide Web at:
www.pearsoned.co.uk

First published 1982
Second edition published 1996
Third edition published 2011

© Pearson Education Limited 2011

The rights of Robert J Peters, Brian J Smith, Margaret Hollins and Stephanie Owen to be identified as authors of this work have been asserted by them in accordance with the Copyright, Designs and Patents Act 1988.

All rights reserved. No part of this publication may be reproduced, stored in a retrieval system, or transmitted in any form or by any means, electronic, mechanical, photocopying, recording or otherwise, without either the prior written permission of the publisher or a licence permitting restricted copying in the United Kingdom issued by the Copyright Licensing Agency Ltd, Saffron House, 6–10 Kirby Street, London EC1N 8TS.

QM LIBRARY
(MILE END)

All trademarks used herein are the property of their respective owners. The use of any trademark in this text does not vest in the author or publisher any trademark ownership rights in such trademarks, nor does the use of such trademarks imply any affiliation with or endorsement of this book by such owners.

Pearson Education is not responsible for the content of third party internet sites.

ISBN: 978-0-273-72468-1

British Library Cataloguing-in-Publication Data
A catalogue record for this book is available from the British Library

Library of Congress Cataloging-in-Publication Data
A catalogue record for this book is available from the Library of Congress

10 9 8 7 6 5 4 3 2 1
15 14 13 12 11

Typeset in 9.5/11.5 pt Minion by 73
Printed and bound by Ashford Colour Press Ltd., Gosport

Brief contents

Contents

Preface

Although the underlying physical laws and principles of acoustics do not change there have been very many changes and additions to standards, laws and regulations, codes of practice relating to noise, and in noise measurement techniques and noise control technology since the last edition of *Acoustics and Noise Control* was published in 1995. Therefore much of the material relating to these aspects of the second edition is now severely out of date. The fact that, despite this, the book is still being used and found to be of value by many people has provided much of the inspiration for the creation of this third edition.

This new edition has been almost completely re-written and re-structured. Chapters 1 to 6 have been completely reorganized and now contain some of the material formerly contained in the later chapters 7, 8 and 9. Each chapter starts with a brief introduction indicating what is to be covered and making links with other chapters. An important feature of the original book, written many years ago by Brian Smith, was its student-centred approach, with many worked examples, and list of questions (with answers to numeric parts) at the end of each chapter. This approach has been retained. The excellent chapter on law written originally by Stephanie Owen has been comprehensively and expertly revised and updated by Margaret Hollins.

The book is an introductory text for those completely new to the subject but it also aims to serve the needs of the practitioner working in local government environmental health departments as well as at junior consultant level, and is a suitable introduction to more advanced texts, a list of which is given in the bibliography. Through worked examples the book illustrates the application of prediction and design calculations routinely used in noise control practice, with an emphasis on the assumptions and limitations that apply to these calculations, and, new to this edition, with derivations added in appendices at end of chapters, for the benefit of readers who may wish to delve deeper into the theoretical background to the subject. A bibliography contains lists of more advanced texts, of reports, standards and codes of practice, and a complete list of formulae and equations is given in an appendix. The glossary was a much valued feature of previous editions and has been completely revised and the number of entries considerably increased.

This new edition covers much of the Institute of Acoustics Diploma syllabus, and those of various MSc courses in acoustics, as well as the acoustics and noise control components of degree and higher level technician courses in architecture, construction, engineering, environmental health, environmental science, health and safety and occupational hygiene. The chapter on law relating to noise will be of use to lawyers, barristers and other professionals dealing with noise cases, but also of interest to many non-expert readers with an interest in the legal framework relating to noise control in Europe and the UK.

Doing, seeing and listening will always enhance understanding and enjoyment of any subject and this is particularly so in the study of acoustics. Therefore an appendix provides a list of possible experiments, exercises and observations for the reader to carry out. A wide variety of audio and animated visual demonstrations are available via the internet. These can enhance explanations given in the text, and readers are referred to these and encouraged to investigate for themselves.

It is hoped that readers of this new edition will find it useful and informative.

Bob Peters
October 2010

Authors' acknowledgements

The authors wish to acknowledge the important contribution of Stephanie Owen, who wrote the law chapter in the second edition, which is the basis of the revision in this edition. They also wish to thank the Institute of Acoustics for giving permission to use material from the Institute's Distance Learning material and examination questions.

Publisher's acknowledgements

We are grateful to the following for permission to reproduce copyright material:

Figures
Figure 1.16 from The Open Door Web Site IB Physics Waves Diffraction files, http://www.saburchill.com/physics/chapters2/0008.htm; Figure 1.22 from http://hyperphysics.phy-astr.gsu.edu/hphys.html; Figures 2.5, 2.8, 2.9, 2.10, 2.11, 2.12, 2.13, 2.17, 2.18 from Distance Learning notes for the General Principles of Acoustics Module of the Institute of Acoustics Diploma in Acoustics and Noise Control; Figure 2.16 from ISO9613-2 (1996) General Method of Calculation, Table A.1; Figure 3.9 from ISO 532:1975, Acoustics – Expression of the subjective magnitude of sound or noise, Part 2: Method for calculating loudness level; Figure 3.10 from ISO recommendation R507, 1966; Figure 5.15 from BS EN ISO 10534-2001; Figure 6.15 from BS8233; Figure 7.14 from BS7385:part 2 1993. Permission to reproduce extracts from British Standards is granted by the British Standards Institution (BSI). No other use of this material is permitted. British Standards can be obtained in PDF or hard copy formats from the BSI online shop: can be obtained in PDF or hard copy formats from the BSI online shop: http://shop.bsigroup.com or by contacting BSI Customer Servcies for hard copies only: Tel: +44 (0)20 8996 9001, Email: cservices@bsigroup.com; Figure 3.1 from http://hyperphysics.phy-astr.gsu.edu/hbase/sound/ear.html; Figure 3.4 from http://www.gpnotebook.co.uk/simplepage.cfm?ID=845873165, © Oxbridge Solutions; Figures 3.5, 3.6, 3.7 adapted from *Examples of Industrial Audiograms* (M.E. Bryan and W. Tempest 1978); Figure 3.8 from http://www.aist.go.jp/aist_e/latest_research/2003/20031114/20031114.html, Reproduced courtesy of AIST, http://aist.go.jp; Figure 4.5 Adapted from an article in *Acoustics Bulletin*, Vol. 33 No. 4, July/August 2008m, pp. 36–40 (P. Brooker); Figures 6.2, 8.4, 8.5, 8.6 from Bruel & Kjaer, www.bksv.com, Courtesy of Brüel and Kjaer; Figures 6.11, 6.12 from Building Regulations Approved Document E, 2003, Crown Copyright material is reproduced with permission under the terms of the Click-Use License; Figures 6.13, 6.14, 6.16, 6.17 from Building Regulations Approved Document E, Crown Copyright material is reproduced with permission under the terms of the Click-Use License; Figure 7.13 from *Mechanical Vibration and Shock Measurements*, Bruel &

Kjaer, Courtesey of Brüel & Kjær; Figures 6.2, 8.4, 8.5, 8.6 from Bruel & Kjaer, www.bksv.com, Courtesy of Brüel & Kjær; Figure 8.7 from Bruel and Kjaer, www.bksv.com, Courtesy of Brüel & Kjær; Figure 8.19a from Bruel & Kjaer product data sheet, Courtesy of Brüel & Kjær; Figure 8.19b from Bruel & Kjaer product data sheet, Courtesy of Brüel & Kjær; Figures 9.10, 9.17, 9.18 from *Controlling Noise at Work*, Health and Safety Executive, Contains public sector information published by the Health and Safety Executive and licensed under the Open Government Licence v.1.0; Figure 9.12a courtesy of Sheldon Waters; Figure 9.12b from CIBSE.

Tables

Table 1.1 in Chapter 1 Appendix from BS EN ISO 266:1997 Preferred Frequencies; Table 5.2 from BS8233; Table 6.4 from ISO 717-2:2004; Table 6.6 from BS EN 717-1: 1990; Table on page 186 from Table B.2 of BS5228:2009 part 2; Table on page 190 from Table 1 from BS6472; Table 8.1 from BS EN 61672-1:2003; Table 8.2 from BS EN 61672-1:2003, Permission to reproduce British Standards is granted by the British Standards Institution (BSI). No other use of this material is permitted. British Standards can be obtained in PDF or hard copy formats from the BSI online shop: http://shop.bsigroup.com or by contacting BSI Customer Services for hard copies only: Tel: +44 (0)20 8996 9001, Email: cservices@bsigroup.com; Table 3.2 from Guidelines for Community Noise, 1999, World Health Organization, http://www.who.int/docstore/peh/noise/guidelines2.html; Table 5.3 from Building Bulletin 93 (BB93); Tables on page 159 and page 160, from Building Regulations Approved Document E, 2003, Crown Copyright material is reproduced with permission under the terms of the Click-Use License.

Text

Questions from IOA Diplomas reproduced by permission of the Institute of Acoustics; Extract on pages 325–6 from Civil Aviation Act 1982, Crown Copyright material is reproduced with permission under the terms of the Click-Use License; Extract on pages 289–90 from Statutory Nuisance (Appeals) Regulations 1995 (S.I. 1995/2644) Regulation 2(2), Crown Copyright material is reproduced with permission under the terms of the Click-Use License; Extract on pages 310–11 from Social Security (Industrial Injuries) (Prescribed Diseases) Regulations 1985 (SI 1985 No 967), Crown Copyright material is reproduced with permission under the terms of the Click-Use License.

We are grateful to the Institute of Acoustics for permission to include material from the General Principles of Acoustics Distance Learning Notes, and also some 'end of chapter questions' from Diploma Examination Papers (denoted by (IOA) after the question).

In some instances we have been unable to trace the owners of copyright material, and we would appreciate any information that would enable us to do so.

Terminology and notation

It is current practice in the literature on the subject to represent all decibel quantities by L, with appropriate subscripts such as L_p, L_I, L_{Aeq}, $L_{EP,d}$, L_{AE} (see Glossary for fuller list). Again in line with common usage, pascals (Pa) have been used instead of N/m^2.

Symbols

A list of the main symbols used in the book is given below. The list is not exhaustive and other symbols are defined in the text.

a acceleration
A frequency weighting
A attenuation, acoustic absorption, acceleration amplitude
B bandwidth
c velocity of sound
C frequency weighting, spectral adaptation terms
d depth or thickness
D frequency weighting, source dimensions
D level difference
e the exponential number (= 2.718 . . .)
f frequency
F fast time weighting, force amplitude
g gram
g acceleration due to gravity
h height, hour
Hz hertz
I impulse time weighting
I sound intensity
J joule
j the complex operator, i.e. $\sqrt{(-1)}$
k kilo (e.g. kg, km, kHz)
k wave number, $k = 2\pi/\lambda$, stiffness
K elastic modulus
l length
L level in decibels
m metres
m mass, surface density, minutes
n standing wave ratio
N newtons, number of decibels, number of items, noys
p sound pressure
P atmospheric pressure
Pa pascals
Q dynamic magnification factor, directivity factor
r distance, damping constant, reflection coefficient
R sound reduction index/transmission loss, real part of a complex impedance, electrical resistance, universal gas constant
R_C Room constant
S slow time weighting, area (m^2), Sones, microphone sensitivity
t time, thickness, transmission coefficient
T reverberation time, transmissibility, period, duration, temperature
u acoustic particle velocity
v vibration velocity
V vibration velocity amplitude, vibration dose value, volume
w vibration frequency weighting
W watt
x distance, displacement
X displacement amplitude, reactance i.e. imaginary part of a complex impedance, static deflection
z acoustic impedance
Z frequency weighting
α acoustic absorption coefficient
γ ratio of specific heats of a gas
δ logarithmic decrement (vibration), path difference (barriers)
Δ small increment
ε energy density
η isolation efficiency
θ angle
λ wavelength $\lambda = 2\pi/k$
ξ damping ratio
π ratio of circumference to diameter of a circle, pi = 3.142 . . .
ρ density
σ radiation efficiency
φ phase difference
ω angular frequency, $\omega = 2\pi f$

Where the same symbol has more than one meaning, multiple representation has been retained to comply with common usage. The meaning will be clear from the context and is made clear in the text.

Chapter 1 The nature and behaviour of sound

1.1 A qualitative picture of wave motion

Acoustics is the science of sound, and sound is a wave motion. In a wave a change or disturbance in some physical property of a medium is transmitted through that medium. For example when a sound occurs in air the sound wave causes the particles in the air to move to and fro (i.e. to vibrate), and because the particles are elastically connected (air being an elastic medium) this vibration is transmitted through the air. The vibrating layers of air contain energy and so another feature of all waves is that they contain energy. The essential features of a medium which is able to transmit sound waves are that it must possess elasticity and inertia (mass); sound waves can travel through solids, liquids and gases but not through a vacuum. In any real medium there will always be some frictional processes at work so that some of the energy of the vibrating particles of the medium will be lost to the sound wave and turned into heat, a process known as sound absorption.

The two simplest types of sound waves are **spherical waves** and **plane waves** (see Figure 1.1), and it helps to understand them if we consider the analogy of waves on the surface of water. If we drop a small object such as a stone into water we see circular ripples travelling outwards. The invisible spherical sound waves in air are their three-dimensional counterparts. If the stretch of water is linear, e.g. a canal, and the stone is replaced by a long plank of wood we would see plane ripples move along the water surface.

The wavefront represents the leading edge of the wave, i.e. it tells us how far the wave has travelled and the rays, always perpendicular to the wavefronts, indicate the direction in which the wave is travelling.

These two forms of wave are idealized models of wave propagation and are useful because waves for sound sources can often approximate to one of these models. Sound from a loudspeaker tends to radiate equally in all directions at low frequencies (i.e. like spherical waves) but be much more directional (i.e. more like plane waves) at high frequencies.

Plane waves travelling in one direction only are the simplest form of waves and can be used to explain frequency and wavelength.

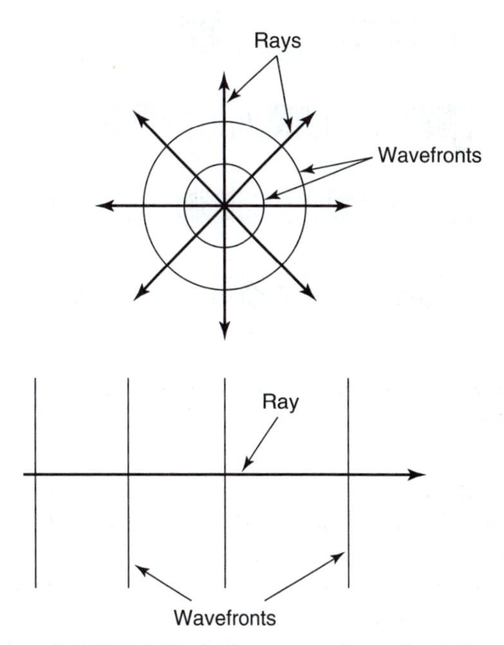

Figure 1.1 Sketch illustrating rays and wavefronts for spherical and plane waves

In a sound wave in air, as a result of the to and fro motion, sometimes the air particles are bunched together, causing a very slight increase in pressure in the atmospheric pressure (a compression) and sometimes causing them to be spaced further apart, causing a very slight reduction in pressure (a rarefaction). This is shown in Figure 1.2 where compressions and rarefactions from the vibrations of a tuning fork are shown travelling in one dimension (down a tube or pipe for example). These very small fluctuations in pressure in the tube constitute the sound pressure caused by the passage of the sound wave down the tube.

The disturbance caused by the sound waves could be described in terms of the vibrations of the air particles, either as a displacement, as a velocity or as an acceleration, and these alternatives will be discussed in more detail in Chapter 7 on vibration. However, since these movements cannot be seen, and since our human ears and our microphones respond to the changes in

Figure 1.2 Propagation of a one-dimensional sound wave

pressure caused by sound waves it is more usual to measure and describe sound waves in terms of sound pressure, in pascals (Pa.).

The simplest form of plane wave occurs when the vibration of the air particles causes a sinusoidal variation in sound pressure with time and can be used to explain frequency and wavelength. The sound pressure in such a plane wave varies with both distance and time, as shown in Figures 1.3 and 1.4. This sinusoidal variation of sound pressure with time represents a sound with a single frequency, called a pure tone.

1.2 Frequency and wavelength and sound speed

After a certain amount of time, called the period, T, of the motion, the cycle repeats itself (Figure 1.3). The frequency, f, of the vibration and of the wave is the number of cycles of the motion which occur in one second:

$$f = 1/T$$

Thus frequency, f, is measured in cycles per second or hertz (abbreviation Hz).

The wavelength, λ, is the minimum distance between points on the wave where the air particles are vibrating in step or in phase as shown in Figures 1.1 and 1.3.

The relationship between sound speed, frequency and wavelength

In order for air particles, which are one wavelength apart, to be in phase, it must be the case that the wave travels one wavelength in the time that it takes for any one of the particles to complete one cycle of motion. Since the number of such cycles completed in one second corresponds to the frequency of the wave, and since the wave velocity is the distance travelled by the wave in one second, it follows that frequency, f, wavelength, λ,

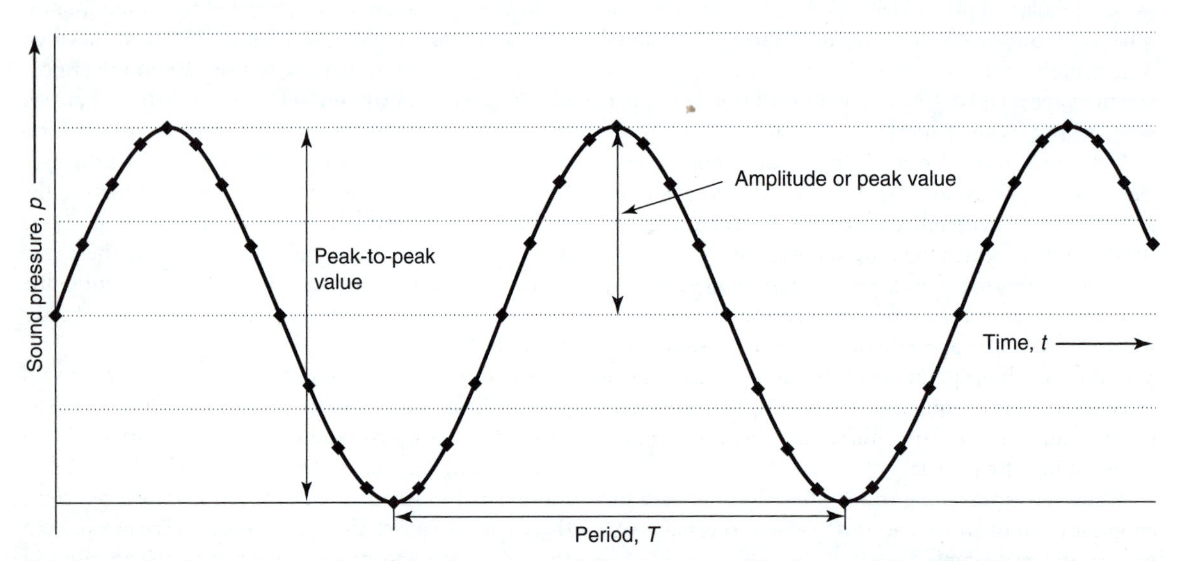

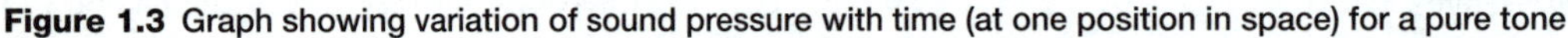

Figure 1.3 Graph showing variation of sound pressure with time (at one position in space) for a pure tone

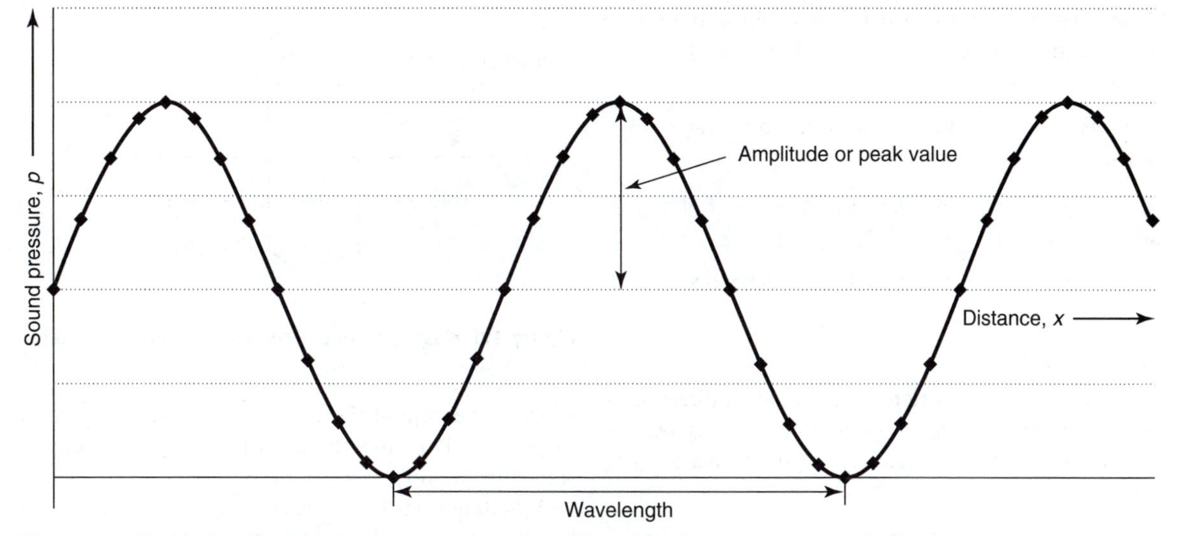

Figure 1.4 Graph showing variation of sound pressure with position (at one moment of time) for a pure tone

and wave velocity, c, are related by the well known equation:

$$c = f\lambda$$

For sound waves in air the speed of sound ranges, approximately, between 330 and 340 metres per second, depending upon air temperature. Thus for a frequency of 100 Hz, at the lower end of the audio range the wavelength will be about 3.3 metres, whereas at the much higher frequency of 1000 Hz it is about 0.33 metres, i.e. the lower the frequency the greater the wavelength, and vice versa.

Note that the frequency of the wave is determined only by the source of the sound. The sound velocity depends on the medium through which the wave is travelling. These two factors then determine the wavelength in the medium, according to the equation $c = f\lambda$. If the sound moves from one medium to another, from air into water, for example, the frequency will remain the same in both media, but because of the difference in sound velocities the wavelengths in the two media will be different.

Can you sketch the curve for positions one quarter, one half and three quarters of a wavelength away?

Can you sketch the curve for positions one quarter, one half and three quarters of a cycle later?

Example 1.1

A plane sound wave in air has a single frequency of 660 Hz. Taking the velocity of sound in air as 330 m/s, what is the phase difference:

(a) between two points in the wave at distance of 0.125 metres apart, at the same moment?

(b) at the same position, but separated by 0.0001 seconds?

Solution

Wavelength = sound velocity/frequency = 330/660 = 0.5 metres.

(a) In terms of phase, a distance of one wavelength, 0.5 metres, corresponds to a phase difference of 360 degrees. Therefore a distance of 0.125 metres corresponds to a phase difference of $360 \times (0.125/0.5) = 90°$.

(b) In terms of phase, a time period of one cycle of the vibration, i.e. 1/660 of a second = 0.00152 seconds, corresponds to a phase difference of 360 degrees. Therefore a time separation of 0.0001 seconds corresponds to a phase difference of $360 \times (0.0001/0.00152) = 23.7°$.

1.3 A mathematical description of a plane progressive wave

The variation of sound pressure with time (Figure 1.3) may be represented by the equation:

$p = A\sin(\omega t)$, where $\omega = 2\pi f$ = angular frequency, and A = sound pressure amplitude.

The variation of sound pressure with distance (Figure 1.4) may be represented by the equation:

$p = A\sin(kx)$, where $k = 2\pi/\lambda$ = wave number, and A = sound pressure amplitude.

Note: These two equations may be combined to produce one equation which gives the sound pressure p for any position, x, at any time, t:

$p = A\sin(\omega t - kx)$ for the case where $p = 0$ at $x = 0$ when $t = 0$

$p = B\cos(\omega t - kx)$ for the case where $p = B$ at $x = 0$ when $t = 0$

$p = A\sin(\omega t - kx) + B\cos(\omega t - kx)$ generally

Example 1.2

A plane progressive wave travelling in the x direction is presented by the following equation which gives the instantaneous sound pressure at any distance x and any time, t:

$p = 0.9 \sin (3142t - 9.25x)$ Pa

Calculate the frequency, wavelength, peak sound pressure of the sound, and the speed of sound in the medium through which the sound is travelling.

Solution

By comparison with the general equation $p = A\sin(\omega t - kx)$:

A = amplitude or peak sound pressure = 0.9 Pa

$\omega = 2\pi f = 3142$, therefore frequency $f = 3142/2\pi = 500$ Hz

$k = 2\pi/\lambda = 9.25$, therefore wavelength $\lambda = 2\pi/9.25 = 0.68$ m

velocity of sound $c = f\lambda = 500 \times 0.68 = 340$ m/s

1.4 The audible range of sound pressures and frequencies

The audible range of sound pressures is from about 2×10^{-5} Pa (or 20×10^{-6} Pa, i.e. 20 μPa) to about 20 Pa. The audible range of sound frequencies is from about 20 Hz to about 20,000 Hz (or 20 kHz). Acoustic waves with frequencies above the audible range are called ultrasonic and those with frequencies below are called infrasonic.

1.5 Sound pressure, sound power, sound intensity and acoustic impedance

The sound pressure is related to the motion of the particles in the medium which cause it, and is most easily related to the particle velocity. These two quantities are related by the specific acoustic impedance of the wave, z, which is the ratio of the acoustic pressure, p (measured in pascals (Pa)) and the acoustic particle velocity, v (measured in m/s).

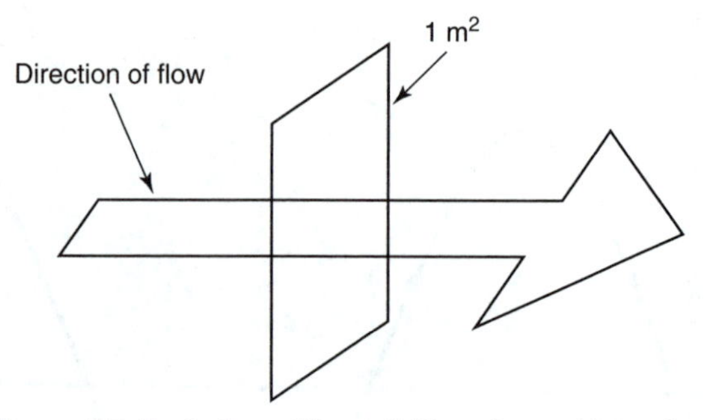

Figure 1.5 Illustration of the definition of sound intensity

Vibrating particles possess mechanical (potential and kinetic) energy, and the transmission of a disturbance through a medium involves the flow of energy. Sound power is the rate of flow of energy, i.e. energy transmitted per unit time is measured in watts (or joules (J) per second). The sound intensity at any point in the medium in any given direction is defined as the rate of flow of energy per unit area in that direction (Figure 1.5) and measured in W/m^2.

Exercise

If the sound intensity being transmitted through an open window of area 0.5 m^2 is 0.1 W/m^2 what is the sound power being transmitted, and how much sound energy will be transmitted through the window in 1 hour?

Answer

0.05 W and $0.05 \times 3600 = 180$ joules or 0.05×10^{-3} kW hour.

In any point in a sound wave the relationship between the acoustic pressure, the acoustic particle velocity and the sound intensity at that point are related by the following set of equations which also involve a quantity called the specific acoustic impedance, z, of the wave:

$$p = zv$$
$$I = pv$$
$$I = p^2/z$$
$$I = zv^2$$

For a wave travelling in one direction, called a plane wave, the specific acoustic impedance, z, depends only upon the nature of the medium, and has the value of the product ρc where ρ is the density of the medium, in kg/m^3, and c is the velocity of sound in the medium, in m/s. For air the value of ρc varies with temperature and with atmospheric pressure but it typically varies between about 410 and 420 kg m/s^2 for typical atmospheric conditions. For water it is much larger, 1.5×10^6 kgm/s^2. The quantity ρc is also known as the characteristic acoustic impedance of the medium.

Note: In terms of the fundamental units m, kg and s the units of specific acoustic impedance are kgm/s^2 as above, but sometimes the quantity is expressed in terms of newtons (N) or pascals (Pa), as either Nsm^{-3} or as Pa s/m, and also as the rayl (after Lord Rayleigh, one of the pioneers of acoustics).

Example 1.3

Calculate the acoustic particle velocity and acoustic intensity at a point in a plane wave where the sound pressure is 0.001 Pa. Take the value of ρc as 415 kgm/s^2.

Solution

Particle velocity, $v = p/\rho c = 0.001/415$
$= 2.41 \times 10^{-6}$ m/s
Acoustic intensity, $I = p^2/\rho c = (0.001)^2/415$
$= 2.41 \times 10^{-9}$ W/m^2

Example 1.4

Calculate the sound pressure and the acoustic particle velocity at a point in a plane wave where the sound intensity is 1×10^{-6} W/m^2. Take the value of ρc as 415 kgm/s^2.

Solution

$p^2 = I\rho c = (1 \times 10^{-6})415 = 4.15 \times 10^{-4}$

therefore $p = \sqrt{(4.15 \times 10^{-4})} = 0.020$ Pa

$V = I/p = 1 \times 10^{-6}/0.020 = 5.0 \times 10^{-5}$ m/s

Sound power

Sound sources have a tremendous range of sound powers, varying from about 10^{-9} W for the human voice when whispering, to millions of watts radiated by a space rocket during launching. The human voice radiates about 20×10^{-6} W during conversation, and this could increase to about 10^{-3} W when shouting. A pneumatic drill used for road breaking radiates about 1 W, and a typical figure for the noise output of a jet airliner is about 50,000 W. Audible sounds can have a very wide range of intensities, from 10^{-12} W/m^2, i.e. a millionth of a millionth of a watt per square metre (the threshold of hearing for the average person) to more than 100 W/m^2 (approaching the threshold of pain) – a range of more than a million million to one. Sound intensity is a useful quantity because it can be related to the sound power of the noise source, and is one of the important factors in the subjectively assessed 'loudness' of the sound.

1.6 More complex waveforms – frequency analysis, peak and RMS values

The pure tone is the simplest sort of sound. It is produced, for example, by a tuning fork or by a loudspeaker fed with a sinusoidal voltage signal. The pure tone is a single frequency sound. Most sounds have a waveform which is more complicated than the simple sine wave and they can be considered to contain more than one frequency. The next simplest type of sound is that produced by a musical instrument, playing one single note. The waveform is harmonic, that is it repeats itself, but in a more complicated way than the sine wave of the pure tone waveform. The French mathematician Fourier showed that such a waveform could be 'built up' or 'synthesized' by combining together a number of simple sinusoidal waveforms. The frequencies of these components are the fundamental frequency (the repeating frequency of the complex waveform) and multiples of it, called harmonics. Figure 1.6 shows a harmonic waveform and illustrates how it may be produced from a fundamental and two harmonics.

The reverse process to Fourier synthesis is Fourier analysis. This means the 'breaking down' or analysing of complex waveforms into their component frequencies. Mathematical techniques can be used to do this theoretically for relatively simple repeating waveforms such as in Figure 1.6, and more complicated waveforms can also be analysed numerically using computers. The analysis can also be performed using frequency analysis instrumentation attached to the sound measuring equipment. This consists of a series of electronic filters which allow only sounds within a particular range of frequencies to be measured.

The frequency spectrum of sounds

A frequency analysis may be displayed on a graph called the frequency spectrum showing the amplitude of the different frequencies (or frequency bands). The frequency spectrum and the waveform can be considered as two alternative ways of describing a sound (see Figure 1.7). For a pure tone, the frequency spectrum is exceedingly simple – it consists of a single line indicating the amplitude of the single frequency. A harmonic waveform also has a relatively simple spectrum, consisting of a line

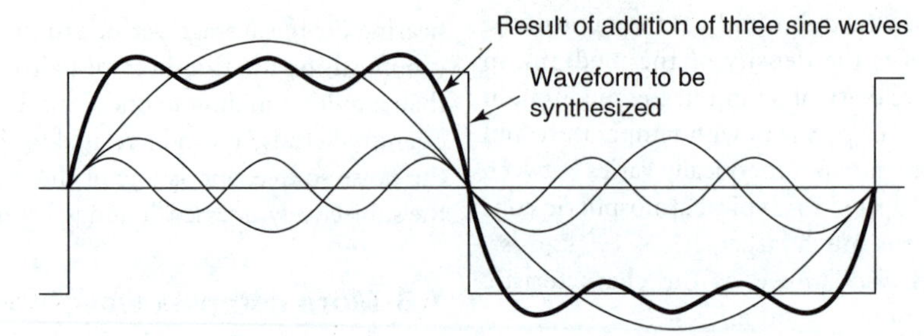

Figure 1.6 Synthesis of a harmonic waveform, showing how a square wave can be built up from a series of sine waves (only the first three harmonics are shown)

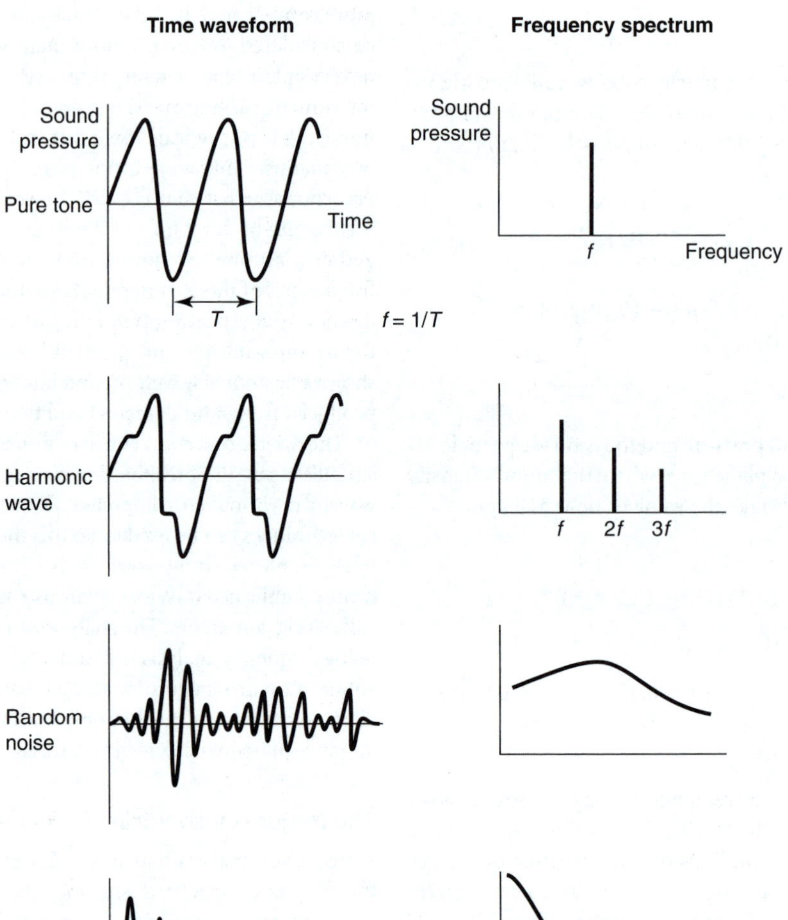

Figure 1.7 Some different types of waveform and their frequency spectra

spectrum of the fundamental and the harmonics. The spectrum envelope shows the relative amplitudes of the different harmonics and the fundamental, and thus the spectra of the same note played on two different musical instruments would be different.

It is worth mentioning two more types of waveform. Random waveforms describe sounds caused by processes which are random, and so never exactly repeat themselves. Hence the waveform is not repetitive. Examples are noise produced by traffic, by wind and by many sorts of machine. Even when it seems that the noise is produced by some repeating event such as the rotation of an engine, or of gear teeth etc., the noise often has a random component, since the machine process never repeats exactly from cycle to cycle due to slight changes in speed load and other conditions.

Transient waveforms die away to zero after the passage of a period of time. They arise from a source which provides only a short transient burst of acoustic energy such as a plucked violin string, or the impact of a hammer blow onto a rivet. The sound energy is dissipated by frictional processes. The simplest transient waveform resembles a sine wave but with an amplitude which decays with time. Examples are the noise produced by impulses such as those from punches, presses, hammering and mechanical handling of goods.

Frequency spectra can be assigned to transient and random sounds, but they are more complicated than periodic sounds. A random noise contains a little of all frequencies and so its frequency spectrum is a continuous curve, called a broadband spectrum. The frequency spectra of transient sounds are also complicated. In the case of the plucked violin string the spectrum will obviously contain the fundamental frequency and its harmonics. In the case of a repeated transient, such as the impacts between teeth in a gear mechanism, the repetition rate of the impacts and its harmonics will also be important.

Figure 1.7 shows some examples of different types of waveform and their frequency spectra.

Frequency analysis and frequency spectra

In most cases noise has a broadband spectrum, i.e. it contains a mixture of all frequencies, but some more than others. If we wish to wish to investigate the frequency content of the noise in more detail we have to split the frequency range into bands, and measure the sound pressure level in each band. This process is called frequency analysis, and the graph showing how the sound pressure level varies with the frequency of each band is called the frequency spectrum of the noise.

There are two ways of splitting the frequency range into bands: either on a constant bandwidth basis, or on a constant percentage bandwidth.

Constant bandwidth

In the constant bandwidth method each band has the same width, so that if for example the bandwidth is 100 Hz then one of the bands might be from 100 Hz to 200 Hz, and others might be, for example from 200 to 300 Hz, or from 1100 to 1200 Hz or from 10,000 to 10,100 Hz. The centre frequency of each band would be half way between its upper and lower cut-off points, so that for the band between 100 and 200 Hz the centre frequency would be 150 Hz.

The constant bandwidth approach is usually only used for what is called 'narrow band frequency analysis' in which the main purpose is to identify with precision a particular frequency component in the spectrum, usually a pure tone which might be causing a disturbance (an annoying hum, whine or whistle), and to aid the diagnosis of the cause of such a component, e.g. to link it to a particular fan, motor or turbine via the rotational speed and number of rotors.

Typically the constant bandwidths might be 300 Hz, 100 Hz, 10 Hz, 3 Hz or 1 Hz, and they might be arranged in contiguous bands, e.g. 100 to 200, 200 to 300, 300 to 400 Hz etc. but it is also possible that the centre frequency will be continuously variable allowing a 'frequency sweep' to take place.

Constant percentage bandwidth

Covering the entire audio range (20 to 20,000 Hz) using constant bandwidth filters requires too many bands, and so, for convenience (but also because of the way the ear responds to broadband sound), constant percentage bandwidth frequency analysis is most frequently used for routine assessment of broad band noise. The most commonly used methods use octave and third octave bands, although one sixth, octave, one twelfth and one twenty fourth octaves are also available.

In musical terms an octave is a range for a particular frequency, say 256 Hz (which is 'middle C' on the piano) to double that frequency, i.e. 512 Hz. The internationally defined system of octave bands for sound measurement has a series of contiguous bands named by their nominal centre frequencies at 16 Hz, 31.5 Hz, 63 Hz, 125 Hz, 250 Hz, 500 Hz, 1000 Hz (or 1 kHz), 2 kHz, 4 kHz, 8 kHz and 16 kHz. In much noise measurement and assessment work the more limited range of seven bands, from 63 Hz to 4 kHz, is often used, with the 31.5 Hz and 8000 Hz bands sometimes being included.

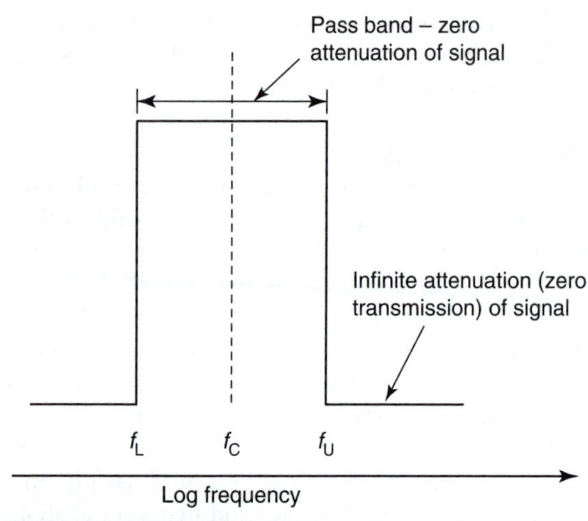

Figure 1.8 Characteristic features of a constant percentage bandwidth filter

In the one third octave system each octave is split into three bands. The standard range used for most building acoustics measurements consists of 16 bands, with nominal centre frequencies from 100 Hz to 3150 Hz: 100, 125, 160, 200, 250, 315, 400, 500, 630, 800, 1000, 1250, 1600, 2000, 2500 and 3150 Hz.

Sometimes an extended range, with five extra bands, is used, from 50 Hz to 5000 Hz. What are the five extra bands?[1]

Log and linear frequency scales

Unlike the constant bandwidth system the constant percentage bands get wider (in absolute terms) as the frequency increases, so that, for example the bandwidth of the 1000 Hz octave band is 10 times that of the 100 Hz band. When we plot an octave band spectrum, with sound pressure level in dB on a vertical scale, and octave band frequency on the horizontal scale we assign equal intervals of space to each octave along the horizontal axis. In doing this we are, in effect, plotting frequency on a logarithmic (log for short) frequency scale.

Using a logarithmic frequency scale the centre frequency f_C is half way between upper and lower cut-off frequencies (as shown in Figure 1.8); it is also known as the geometric mean of the upper (f_U) and lower (f_L) cut-off frequencies:

$$f_C = \sqrt{(f_L \times f_U)}$$

The exact centre frequencies of the standardized acoustic bands are slightly different from the nominal centre frequencies (except for the 1000 Hz band): for example the exact centre frequencies of the nominal 500, 2000, 800 and 1250 bands are 501.2, 1995.3 Hz, 794.3, 1258.9 Hz.

[1] Answer: 50 Hz, 63 Hz, 80 Hz, 4000 Hz and 5000 Hz.

The relationship between the upper, lower and centre frequencies of the various bands is defined by mathematical series based either on base 2 or on base 10.

For the base 10 series, considered be the more accurate of the two, the relationships between f_L, f_C and f_U are:

For octaves: $f_L = f_C/10^{0.15}$ and $f_U = f_C \times 10^{0.15}$

And for one third octaves:

$f_L = f_C/10^{0.05}$ and $f_U = f_C \times 10^{0.05}$

In the base 2 series the multiplying or dividing factor of $10^{0.15}$ for octaves is replaced by $2^{0.5}$ and for one third octaves the factor of $10^{0.05}$ is replaced by $(2)^{1/6}$.

Since $10^{0.15} = 1.4124$ and $2^{0.5} = 1.4142$ the difference between the two methods is very small, and the same is true of the third octave multiplying and dividing factors, since $10^{0.05} = 1.122018$ and $2^{1/6} = 1.122462$.

Example 1.5

For the 1000 Hz octave band:

$f_L = 1000/1.4124 = 708.0$ Hz and

$f_U = 1000 \times 1.4124 = 1412.4$ Hz

And for the 2000 Hz one third octave band:

$f_L = 2000/1.122018 = 1782.5$ Hz and

$f_U = 2000 \times 1.122018 = 2244.0$ Hz

The various frequencies for all the bands are specified in International Standard ISO 266:1997 *Acoustics – Preferred frequencies.* Appendix 1.1 gives a full list of octave and one third octave frequencies and bandwidths.

Recognizing pure tones

Sometimes a broadband noise from machinery contains a tonal component which is audibly recognizable as a whine, a hum or a whistle. The tonal quality makes the noise more annoying, and some noise assessment methods (such as BS 4142 for example) impose a penalty of 5 dB on such a noise (i.e. they rate it as being 5 dB higher than its measured value). It can therefore be important to recognize and agree on when a tone is present in a noise. In some cases identification and agreement can be reached simply by listening, but in less clear cut cases an objective measurement method is required. A fairly common 'rule of thumb' approach is to look at the third octave spectrum. A pure tone is indicated in any band which is several decibels higher than its immediate neighbouring bands. A 5 dB increase is a fairly strong indication of tonality and a 10 dB step between neighbouring bands is conclusive. There are more sophisticated techniques involving 1, 1/6, 1/12 and 1/24 octave bands,

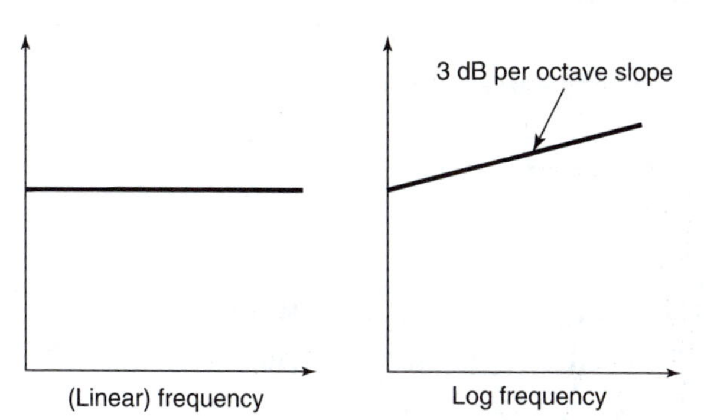

Figure 1.9 Characteristics of white noise

but they are beyond the scope of this chapter. Methods for the detection of pure tones in signals are described in ISO 1996-2:2007.

White and pink noise

These are two special types of broadband noise that both have flat frequency spectra.

In the case of white noise it is the spectrum plotted on a linear frequency scale which would be flat (i.e. a horizontal line graph), because white noise is defined as having equal sound energy per constant bandwidth, or equal energy per Hz. If the white noise were plotted on a log frequency scale (e.g. measured in octaves or third octaves) then the spectrum would slope upwards at 3 dB increase per octave, because the bandwidth increases with frequency. (See Figure 1.9.)

In the case of pink noise there is equal energy per percentage bandwidth, i.e. equal energy in each octave or each one third octave band.

Can you plot the corresponding graphs for pink noise? (See Figure 1.10.)

Frequency weightings

Sometimes the additional detail of a frequency spectrum is not needed, but it is required to describe the noise by a single number which still in some way takes the broad spectrum characteristic of the noise into account, i.e. that it is predominantly high or low frequency in character. A number of single figure frequency weightings have been devised for this purpose, the best known being the A and C frequency weightings.

These are described in more detail in section 1.10.

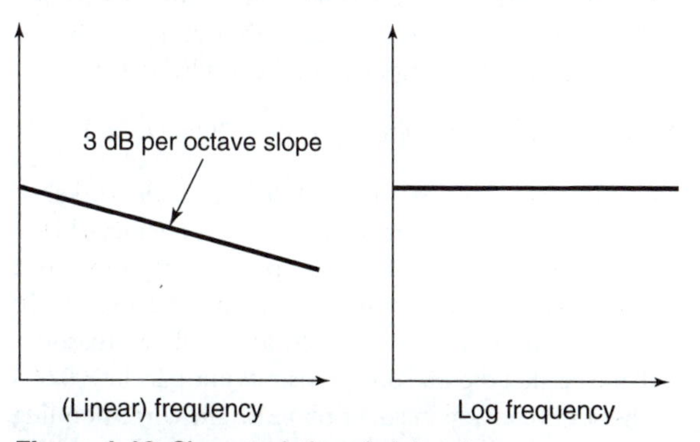

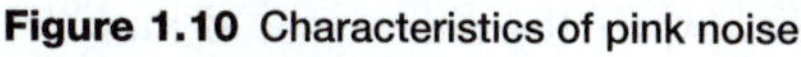

Figure 1.10 Characteristics of pink noise

The magnitude of sound pressures – RMS and peak values

If a pure tone was being played over the radio and the volume was turned up, the amplitude of the sound pressure would be increased – the sound would become louder. The amplitude is thus a convenient measure of the magnitude of the sound and can be related to its intensity and loudness, which will be discussed later.

With a more complicated waveform, however, it is not so easy. One might think that the magnitude of the peak pressure of the waveform would be the value which would be most useful. However, the sound pressure might be near to the peak value for only a small fraction of the duration of the sound, and might not be very closely related to the subjective impression of the sound. Perhaps an 'average' sound pressure would be a better measure of the 'size' of a sound? However, if we look at the sinusoidal waveform of a pure tone, we see that, taken over a complete cycle, the average sound pressure, including rarefactions and compressions, is zero. This is true for all waveforms, not just a pure tone. We need an 'average' which takes into account the magnitude of the sound pressure fluctuations but not their direction (positive and negative) so that the compressions and rarefactions do not average out. There are various possible ways of obtaining a 'non-zero' average sound pressure, but the one most commonly used is the root mean square (or RMS) sound pressure. This can best be described by looking at the waveform shown in Figure 1.11. In effect the sound level meter first 'squares' the signal, that is multiplies it by itself. This has the effect of producing a pressure-squared waveform, which is always positive. (Remember that in algebra minus one times minus one gives plus one.) The next stage is to take the average (or mean value) of this pressure-squared waveform – called the 'mean pressure squared'. Finally, by taking the square root of this value, we get back to a pressure – the root mean square pressure (strictly the square root of the mean pressure squared). The process is illustrated in Figure 1.11.

Most sound level meters have electronic circuits which convert the microphone signal into an RMS value corresponding to the RMS sound pressure. The RMS pressure is used because it can be related to the average intensity of the sound and to the loudness of the sound.

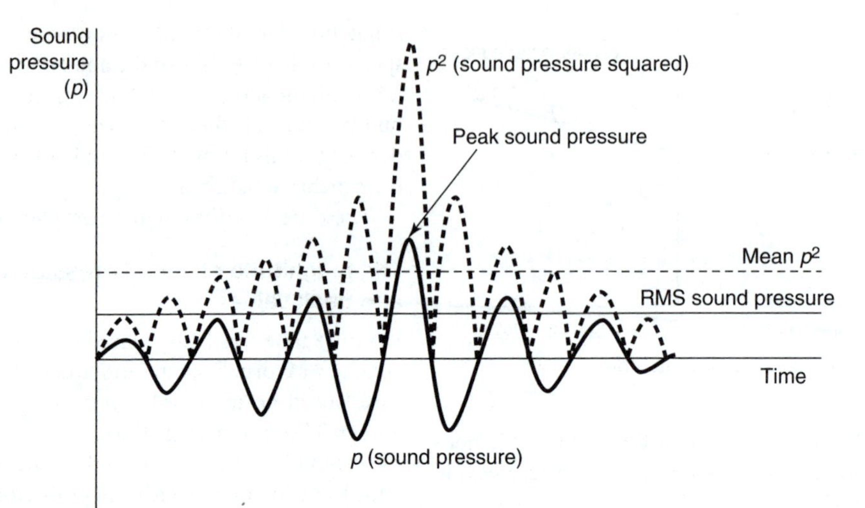

Figure 1.11 Illustrating peak and RMS values of a waveform

For a pure tone it can be shown that the peak pressure and the RMS pressure are simply related:

$$p_{RMS} = p_{peak}/\sqrt{2} = 0.707 \times p_{peak}$$

For more complex signals there is no simple relationship between the two.

There are occasions when it is important to measure the peak value of a complex sound waveform, or the peak to peak value, Examples could be the sound produced by impulsive noise, such as gunfire, explosions or punch presses. Some specialist impulse sound level meters can measure peak as well as RMS sound pressures.

In decibel terms the peak level of a pure tone is 3 dB higher than the RMS value, but the difference is higher for impulsive noise. The ratio of the peak value of a signal to its RMS value is called the **crest factor** of the signal. The more impulsive the signal the higher will be its crest factor.

1.7 The decibel scale

The decibel scale is used for comparing and measuring powers (electrical as well as acoustic), and related quantities such as sound intensity and sound pressure, so it is necessary to explain the relationships between these quantities.

However, although sound intensity is important as the basis of many prediction calculations, sound pressure is a more useful quantity in practical terms, and is the quantity which is always measured using the microphone of the sound level meter. The intensity (I) of sound at a point is proportional to the square of the sound pressure (p) at that point:

$$I \propto p^2$$

Thus for example if the sound pressure is doubled then the sound intensity increases fourfold. It is because of this important relationship that the RMS value of pressure is often measured, since the average value of 'p^2' will be proportional to the mean or average intensity of the sound over the measurement period.

The decibel scale

The decibel scale is a logarithmic scale for measuring or comparing energies or powers, or related quantities such as sound intensity.

If two sounds have intensities I_1 and I_2 then on a decibel scale I_2 may be said to be N dB above I_1, where:

$$N = 10\log(I_2/I_1)$$

A similar scale may be used to compare the sound power outputs (W_2 and W_1) from two noise sources, i.e.

$$N = 10\log(W_2/W_1)$$

Sound pressure is related to the sound intensity as discussed above, and so two sounds with sound pressures p_1 and p_2 may also be compared on the decibel scale:

$$N = 10\log(I_2/I_1) = 10\log(p_2/p_1)^2 = 20\log(p_2/p_1)$$

Table 1.1 gives some examples of how the scale works.

Thus a 20 dB noise reduction, typically achieved by a single glazed window for example, corresponds to a 100-fold reduction in sound intensity, and the 50 dB sound insulation which is typically achieved by a masonry wall means that the wall only transmits one part in 100,000 of the sound energy incident upon it. The corresponding sound pressure ratios are the square root of those for

Table 1.1 How the scale works

Intensity ratio I_2/I_1	Decibel difference N dB
1	0 dB
2	3 dB
3	4.8, i.e. approx. 5 dB
4	6 dB
5	7 dB
6	7.8, i.e. approx. 8 dB
7	8.5 dB
8	9 dB
9	9.5 dB
$10 = 10^1$	10 dB
$100 = 10^2$	20 dB
$1000 = 10^3$	30 dB
$10{,}000 = 10^4$	40 dB
$100{,}000 = 10^5$	50 dB
$1{,}000{,}000 = 10^6$	60 dB

intensity, so that a 20 dB reduction is equivalent to a pressure ratio of 10:1 ($= \sqrt{100}$). The table may be extended by adding or subtracting values from the right-hand column and multipyling or dividing the corresponding values from the left-hand column, or vice versa. Thus a decibel reduction of 26 dB (= 20 + 6) corresponds to a sound intensity ratio of 400 ($= 100 \times 4$), and a sound pressure ratio of 20 ($= \sqrt{400}$). [Alternatively we could use arithmetic because if $26 = 10\log(W_2/W_1)$ then $W_2/W_1 = 10^{2.6} = 398$ (not exactly 400 because, more accurately, an intensity ratio of 400 gives 10log400 = 26.02 dB).]

Reasons for using the decibel scale

The first reason is one of convenience, because by using a logarithmic scale the very large range of audible sound pressures (5 million:1, corresponding to an even larger range of sound intensities of 25 million million:1) is compressed into a much more manageable range of about 120 dB). The second reason is that the human response to sound is also logarithmic, with each tenfold increase (i.e. 10 dB) in sound intensity being judged, on average, to double the loudness of the sound, so that a 100-fold increase, i.e. 20 dB, would produce a fourfold increase in loudness and a 1000-fold increase (30 dB) will increase the loudness by a factor of 8. A 3 dB increase, which corresponds to a doubling of sound intensity, produces a small but noticeable subjective increase of loudness in typical situations, but a 1 dB increase is only just noticeable under the most favourable listening conditions. The third reason is again one of convenience, for it is easier to deal with decibel values which are added to or subtracted from each other (e.g. 20 dB + 30 dB = 50 dB) than with very small ratios which would have to be multiplied or divided by each other (e.g. $0.01 \times 0.001 = 0.00001$).

Reference values: sound pressure level, sound intensity level and sound power level

The statement that machine A produces a noise level which is 10 dB higher than machine B uses the decibel scale in a relative way, without assigning an absolute value to either of the two levels.

The use of internationally agreed reference levels gives an absolute value to quantities measured on a decibel scale. The reference value of sound pressure, p_0, is 2×10^{-5} Pa, or 20×10^{-6} or 20 micropascals, which represents the threshold of hearing for the average young person with normal hearing and corresponds to 0 dB. Values measured on a decibel scale relative to this value are called sound pressure levels, and denoted by the symbols SPL or L_p. The reference value, W_0, for the sound power level scale (L_W) is 10^{-12} W, and for sound intensity level scale (L_I) the reference value (I_0) is 10^{-12} W/m^2. Thus:

$$L_p = 20\log(p/p_0)$$

$$L_I = 10\log(I/I_0)$$

$$L_W = 10\log(W/W_0)$$

These three different scales represent three different physical quantities (i.e. sound pressure, power and intensity) although in each case the measure is in dB. Which of the three is being referred to should be made clear by the context, but occasionally this may be made specific, or given emphasis by quoting the reference value, e.g. 120 dB re. 10^{-12} W, which makes it clear that this a sound power level.

The relationship between the reference values I_0 and p_0 are such that for plane waves (when $I = p^2/\rho c$) the sound intensity level has (approximately) the same numerical value as the sound pressure level. This is because, approximately, $I_0 = p_0^2/\rho c$, depending upon the value of ρc used. Sound level meters, which measure sound pressure levels, may therefore be used to give an approximate indication of the magnitude of sound intensity levels, but unlike more specialist sound intensity meters, described in Chapter 9, they cannot give any indication of the direction of flow of sound energy.

Example 1.6

Calculate the sound pressure level at a point where the sound pressure is 5.0×10^{-3} Pa.

Solution

$L_p = 20\log(p/p_0) = 20\log(5.0 \times 10^{-3}/2.0 \times 10^{-5})$
$= 48 \text{ dB re. } 2.0 \times 10^{-6} \text{ Pa}$

Example 1.7

Calculate the sound intensity level at a point where the sound intensity is 8.5×10^{-7} W/m^2.

Solution

$L_I = 10\log(I/I_0) = 10\log(8.5 \times 10^{-7}/1 \times 10^{-12})$
$= 59.3 \text{ dB re. } 1 \times 10^{-12} \text{ W/m}^2$

Example 1.8

Calculate the sound pressure, sound intensity and sound intensity level at a point in a plane wave at which the sound pressure level is 75 dB. Take the specific acoustic impedance of air as 415 Nsm^{-3}.

Solution

$L_p = 20\log(p/p_0)$

from which $p = p_0 \times 10^{(Lp/20)} = 2.0 \times 10^{-5} \times 10^{(75/20)}$
$= 0.112$ Pa

$I = p^2/\rho c = (0.112)^2/415 = 3.0 \times 10^{-5} \text{ W/m}^2$

$L_i = 10\log(I/I_0) = 10\log(3.0 \times 10^{-5}/1 \times 10^{-12})$

$= 75 \text{ dB re. } 1 \times 10^{-12}$

Example 1.9

Calculate the sound intensity, sound pressure and sound pressure level at a point in a plane wave at which the sound intensity level is 90 dB. Take the specific acoustic impedance of air as 415 Nsm^{-3}.

Solution

$L_I = 10\log(I/I_0)$

From which $I = I_0 \times 10^{(L_I/10)} = 1 \times 10^{-12} \times 10^{9.0}$
$= 1 \times 10^{-3} \text{ W/m}^2$

$I = p^2/\rho c$, from which $p = \sqrt{(I \times \rho c)}$
$= \sqrt{(1 \times 10^{-3} \times 415)} = 0.64$ Pa

$L_p = 20\log(0.64/2.0 \times 10^{-5})$
$= 90 \text{ dB re. } 2.0 \times 10^{-5} \text{ Pa}$

Combining sound pressure levels

When more than one noise source is operating at once it becomes necessary to consider how the individual sound pressure levels combine. Since the decibel values are based on logarithms we should not expect them to obey the rules of ordinary arithmetic. We learnt earlier that a doubling of sound energy, power or intensity corresponds to an increase of 3 dB, and so if two machines each individually produce a level of, say, 90 dB at a certain point, then when both are operating together we should expect the combined sound pressure level to increase to 93 dB, but certainly not to 180 dB!

Table 1.2 Combining decibels

Add to the higher level	Difference between levels
3	0
3	1
2	2
2	3
1	4
1	5
1	6
1	7
1	8
1	9
0	10

Table 1.2 gives a method for combining levels in pairs, based on adding to the higher level a correction which depends upon the difference between the two levels. Although this is only an approximate method it should give results which are accurate to the nearest dB, which is satisfactory for most purposes.

Example 1.10

As an example of the use of Table 1.2 consider the combination of four decibel levels: 82 dB, 84 dB, 86 dB and 88 dB.

The levels are combined in pairs using Table 1.2. The first two levels in the series, 82 and 84, are combined to give 86 dB. This 'running total' of 86 dB is then combined with the next in the list, 86 dB, to give a new running total of 89 dB, which is then combined with the final value, 88 dB, to give a total combined level of 92 dB.

Note that according to Table 1.2 differences of 10 dB or more are negligible, so that the lower of the two levels may be ignored. Although the levels may be taken in any order, it is convenient to take them in ascending order, as in this example, so that lower values can be combined first, and so may make a significant contribution when combined with the higher levels.

The combined value, L_T, of several levels, $L_1, L_2, L_3, \ldots, L_N$ may also be calculated, accurately, using the formula:

$$L_T = 10\log[10^{L_1/10} + 10^{L_2/10} + 10^{L_3/10} + 10^{L_4/10} + \cdots + 10^{L_N/10}]$$

Note that:

$$10^{L_1/10} = p_1^2/p_0^2 = (p_1/p_0)^2$$

and

$$10^{L_2/10} = p_2^2/p_0^2 = (p_2/p_0)^2 \text{ etc.}$$

Therefore in this formula each term inside the square brackets is related to the value of p^2, i.e. is related to the intensity of each of the component noise levels, and therefore the sum of the terms inside the square brackets relates to the total intensity of all the noises. In effect the formula is combining three steps into one calculation:

1. Turn each level back into a sound intensity.
2. Add (arithmetically) the intensities to find the total intensity.
3. Turn this total sound intensity back into a sound pressure level.

Applying this to Example 1.10:

$$L_T = 10\log[10^{8.2} + 10^{8.4} + 10^{8.6} + 10^{8.8}] = 91.6 \text{ dB}$$

which agrees with the earlier result of 92 dB using the chart method.

Subtracting decibels

A similar approach to that for combining decibels may be used to 'subtract' a component sound level from a total level. A common use for this technique is to correct a measured noise level for the effects of background. The result, L_{A-B}, of subtracting level L_B from a higher level L_A is given by:

$$L_{A-B} = 10\log[10^{L_A/10} - 10^{L_B}]$$

As an example, suppose that the noise from a machine is measured (including the contribution of background noise) and found to be 87 dBA but when the machine is switched off the background noise alone is measured as 83 dBA. A more accurate value for the machine noise may be obtained by 'subtracting' the 83 dBA background noise from the combined level of 87 dBA; i.e. $10\log[10^{8.7} - 10^{8.3}] = 84.8$ dBA. Note that if the measured noise level is more than 10 dB above background the correction is less than 0.5 dB and is usually considered to be negligible.

Averaging sound pressure levels

Sometimes it is necessary to find the average value of a number of sound level measurements. A good example would be in building acoustics where in order to find a representative value of the sound level in a room a number of measurements are taken at different positions within the room, and an average value is calculated.

The appropriate average value is that which corresponds to the average sound intensity, or the average value of p^2. The average value, L_{AVGE}, of several levels, $L_1, L_2, L_3, \ldots, L_N$ may also be calculated using the formula:

$$L_{AVGE} = 10\log[(10^{L_1/10} + 10^{L_2/10} + 10^{L_3/10} + 10^{L_4/10} + \cdots + 10^{L_N/10}) \times (1/N)]$$

In this formula the value inside the square brackets [. . .] corresponds to the average sound intensity.

Example 1.11

Calculate the average of four decibel levels: 82 dB, 84 dB, 86 dB, 88 dB.

Solution

$$L_{AVGE} = 10\log[(10^{8.2} + 10^{8.4} + 10^{8.6} + 10^{8.8})/4] = 85.6 \text{ dB}$$

Note that this average value, sometimes called the logarithmic average, is different from the arithmetic average of the four levels, which is 85.0 dB. In this case the difference is only 0.6 dB, but it will increase when the range of levels to be averaged is greater. The logarithmic average will always be higher than the arithmetic average.

Exercise

Compare the logarithmic and arithmetic averages of the following pairs of levels:
(i) 80 dB and 90 dB
(ii) 70 dB and 90 dB.

Answer

(i) Logarithmic average = 87.4 dB; arithmetic average = 85 dB.
(ii) Logarithmic average = 87.0 dB; arithmetic average = 80 dB.

In these two examples the lower level is making a negligible contribution to the total intensity and so the average of the two values is half that of the larger one, or in decibel terms 3 dB lower.

Time weighted average sound levels

In the above example, the four different sound levels have been given equal weighting when calculating the average level, because they represent sound levels which are not varying with time but measured at different positions. In a different situation it might be the case that the sound level varies during different periods of the day. In this case it will be necessary to take into account the duration of each sound level when calculating the average over the entire period.

Suppose for example that over an eight-hour period at a particular location the sound levels in Example 1.11 had been measured over different periods, let us say 82 dB for 4 hours, 84 dB for 2 hours, 86 dB for 1.5 hours and 88 dB for 0.5 hours.

In this case the time weighted average level would be obtained by weighting each of the sound intensity values by the appropriate duration, and then dividing by the total duration (8 hours) as follows:

$$L_{\text{AVGE}} = 10\log[(4 \times 10^{8.2} + 2 \times 10^{8.4} + 1.5 \times 10^{8.6} + 0.5 \times 10^{8.8})/8] = 84.1 \text{ dB}$$

This value is the **time average level**, also widely known as the **continuous equivalent noise level, L_{eq}**. L_{eq} is discussed in much more detail in Chapter 4, together with other methods of measuring time varying noise,

Exercise

How will the time weighted average change if the durations of the 82 and 88 dB levels are interchanged, i.e. 82 dB for 0.5 hours and 88 dB for 4 hours (the other two components remaining the same)?

Answer
86.7 dB.

1.8 Equal loudness contours and the A-weighting network, dBA

A-weighted decibels – dBA

The vast majority of noise measurements are of A-weighted decibels, dBA. The A-weighting is the result of an electronic frequency weighting network in the sound level meter which attempts to build the human response to different frequencies into the reading indicated by a sound level meter, so that it will relate to the loudness of the noise. The relationship between the A-weighting scale and loudness is explained in more detail below.

Loudness

Loudness is a measure of the subjective impression of the magnitude of a sound. It is mainly related not only to the intensity of the sound but also to its frequency. Intensity and frequency are measures of the physical characteristics of the noise, independent of human response, whereas loudness is a measure of human response. Frequency weighting networks were first introduced into sound level meters in the 1950s in an attempt to simulate the equal loudness contours which show how the loudness of pure tones is related to sound pressure level and frequency. The contours were developed from the results of experiments in which listeners were asked to judge between the loudness of a reference tone at 1000 Hz, set at a constant level (e.g. 60 dB for the 60 dB contour), and a test tone, at another frequency. The level of the test tone is adjusted until the subject judges that the two tones are equally loud. The level and frequency are then plotted, as one point on the contour, and the experiment is then repeated at another frequency. The contours (Figure 1.12) show that human hearing is most sensitive to frequencies in the 1 to 4 kHz range, with a reduced response at the low and very high frequency ends of the spectrum. Thus, for example, a pure tone of say 60 dB at 1 kHz will sound louder than a tone of the same level at say 100 Hz, a measure of the difference being the increase in level of the lower frequency tone needed for it to sound equally as loud as the 1 kHz note (approximately 9 dB in this case).

The shape of the equal loudness contours (Figure 1.12) indicate that human frequency response to sound varies with the sound level, with contours of higher level being 'flatter' than lower ones. Therefore three different frequency weightings, A, B and C, were originally devised, for use with sounds of low, medium and high levels, with a fourth weighting, D, being added later, specifically for use with aircraft noise (Figure 1.13).

Although the correlation of dBA with loudness is only approximate, and there are more accurate methods of determining loudness, the A-weighted SPL has, despite some limitations, become universally accepted as the simplest way of measuring a noise which does give some correlation with human response. The B- and D-weighting networks are no longer commonly used, but the C weighting is sometimes used, as explained below.

The value of the frequency weightings are specified by British and International Standard BS EN 61672-1: 2003 *Electroacoustics – Sound level meters – Part 1: Specification.*

Low frequency noise – the C-weighting

The value of the A-weighting is largest at low frequencies, and one of the most common criticisms of the use of dBA is that it may undervalue the disturbing effects of low frequency noise, such as that from commercial and industrial fans, which sometimes only becomes noticeable at night-time, when other sources of noise have ceased to operate. A good indication of the low frequency content in a noise may be obtained by comparing the A-weighted and the unweighted values of sound pressure level. If the difference between these is small then the sound contains mainly medium and high frequencies, whereas a low frequency noise will have a dBA value which is well below the unweighted value. The C-weighting is the 'flattest' of the weightings and so dBC

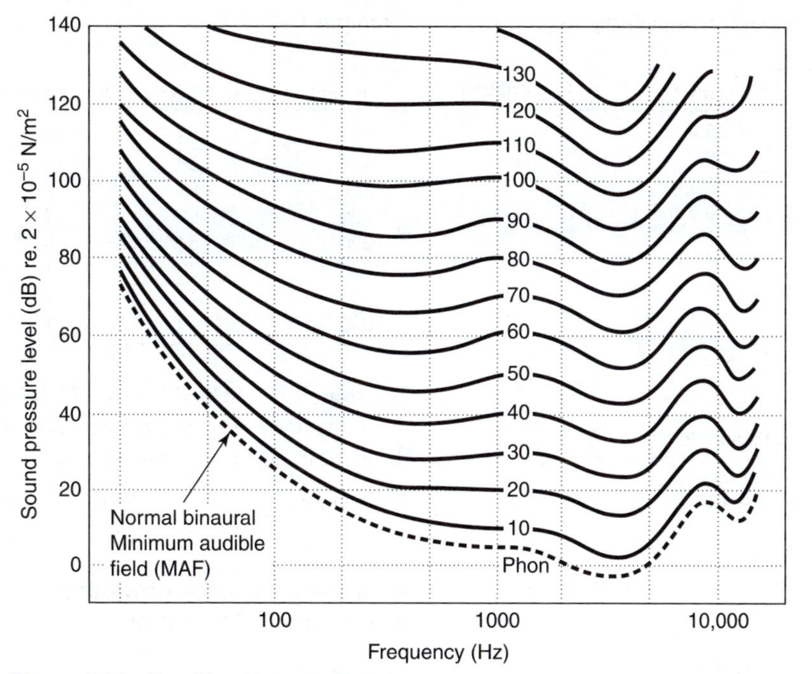

Figure 1.12 Equal loudness contours

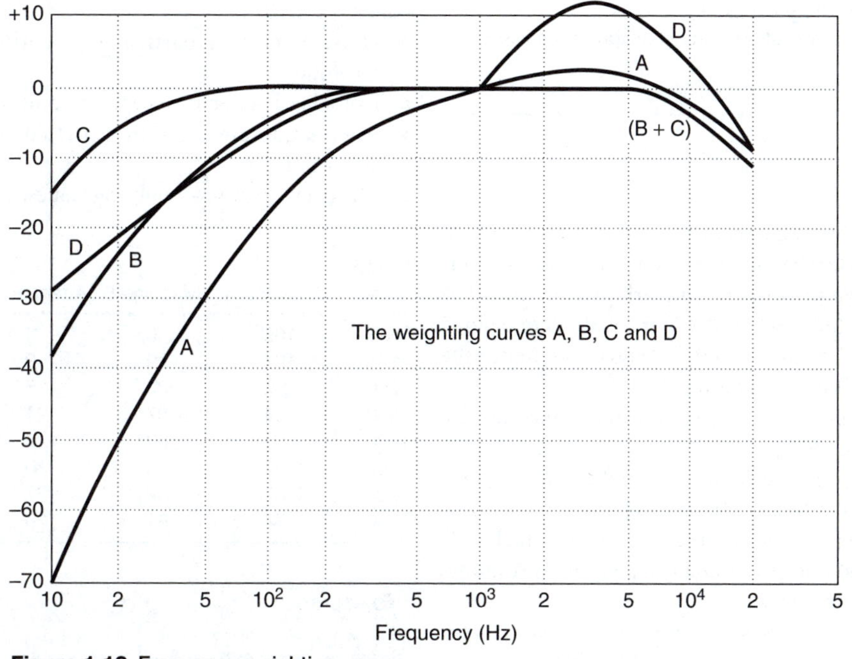

Figure 1.13 Frequency weighting curves

is sometimes used as an approximation to the unweighted SPL, and the difference between dBA and dBC used as an indication of low frequency content of a noise. The C-weighting has also become increasingly used for measuring the peak sound pressure of impacts and impulsive noise, for which the A-weighting is inappropriate because of its effect on the low frequency components of the noise.

In 2003 a new Z (for zero) frequency weighting was introduced into BS EN 61672-1 which defines

Table 1.3 A and C octave band weighting (to nearest decibel)

Octave band (Hz)	A-weighting	C-weighting
63	−39.4	−1
125	−26.2	0
250	−16.1	0
500	−8.6	0
1000	0	0
2000	1.2	0
4000	1.0	−1
80,000	−1.1	−3

Table 1.4 Some typical sound levels in dBA

140 dBA	Threshold of pain
120 dBA	Jet aircraft at 100 m
110 dBA	
100 dBA	Road drill, loud disco
90 dBA	DIY drill (close to ear), lorry (roadside)
80 dBA	Traffic at a busy roadside
70 dBA	Hair dryer
60 dBA	Washing machine
50 dBA	TV in lounge
40 dBA	Quiet office
30 dBA	Bedroom at night
20 dBA	Broadcasting studio (background noise level)
10 dBA	
0 dBA	Threshold of hearing

performance of sound level meters. Prior to this different sound level meters had their own ways of measuring unweighted sound pressure levels, variously described as 'flat' or 'unweighted' or 'linear' but these were not defined in any national or international standards – hence the growing use of the C-weighting.

The A-weighted sound pressure level may also be written as L_{pA} or as L_A.

Table 1.3 shows the A- and C-weightings, in octave bands, taken from BS EN 61672-1:2003.

Also shown is a table of typical dBA levels (Table 1.4) and the A-weighting curves and equal loudness contours (Figures 1.12 and 1.13).

The study of loudness and how it is measured and estimated is continued in more detail in Chapter 3.

Calculation of dBA value from octave band sound pressure levels

Octave band sound pressure levels may be A-weighted and combined to give the A-weighted sound pressure level, in dBA.

Example 1.12

Octave band frequency	Octave band SPL	A-weighting	A-weighted SPL
63	103	−26	77
125	96	−16	80
250	89	−9	80
500	82	−3	79
1000	84	0	84
2000	79	+1	80
4000	73	+1	74
8000	69	−1	68

The overall A-weighted level is obtained by combining the individual A-weighted band values, i.e. by combining the levels in the final column above:

$$L_A = 10\log[10^{7.7} + 10^{8.0} + 10^{8.0} + 10^{7.9} + 10^{8.4} + 10^{8.0} + 10^{7.4} + 10^{6.8}] = 88.5 \text{ dB}$$

Attenuation, in dBA, produced by a noise control measure

Example 1.13

The following information is given below, in octave bands (all in dB):

- Noise level at a particular reception point from machine A
- Noise level at a same reception point from machine B
- Attenuation produced by a particular type of noise enclosure
- The octave band A-weighting values.

Octave band	Machine A	Machine B	Attenuation	A-weighting
63	105	68	4	−26
125	107	79	9	−16
250	99	82	15	−9
500	94	87	21	−3
1000	91	92	24	0
2000	87	96	30	+1
4000	82	89	27	+1
8000	79	81	26	−1

Exercise

Calculate:

- the overall noise levels for machines A and B, in both dBA and dB(LIN)
- the noise levels in each case after the machines have been enclosed
- the attenuation, in dBA produced by the enclosure for each machine.

What conclusions can you draw about the difference between the dBA and dB(LIN) values, and between the amount of attenuation provided, in each case?

Answer

- Machine A without attenuation: 110 dB(LIN) and 98 dBA; after attenuation: 84 dBA.
- Machine B without attenuation: 99 dB(LIN) and 99 dBA; after attenuation: 72 dBA.
- Attenuation of enclosure: for machine A: 14 dBA, for machine B: 27 dBA

Machine A produces mostly low frequency noise, and therefore the dB(LIN) value is much higher than the dBA, whereas for machine B which produces mostly high frequency noise the two values are similar. Because the enclosure produces more attenuation at higher than at lower frequencies the dBA reduction it provides is greater for machine B. The conclusion is that it is not possible to give a single figure dBA reduction for an enclosure (or any other noise control device) because the reduction will depend on the spectrum of the noise being treated.

1.9 Types of elastic waves in solids and fluids

There are a wide variety of different types of elastic waves which can be transmitted through matter. In an unbounded, i.e. infinite, expanse of solid two types of waves are possible, **compressive waves** (sometimes called **P waves**) and **shear waves** (called **S-waves**). These two types of waves are important in transmitting vibration through the ground, e.g. from trains road traffic, construction and demolition activities, and are also responsible for transmitting shocks from earthquakes. In a compressive wave the to and fro motion of the particles of the medium are in the same direction as the direction of travel of the wave itself – such waves are called **longitudinal waves.** In the case of shear waves the particle vibration is perpendicular to the direction of wave travel – this type of wave is an example of a **transverse wave.**

If the solid is bounded, or finite, as in the case of a beam, rod or plate for example, then several other types of waves are possible, such as **flexural or bending waves**, or **torsional waves.** These types of waves are a combination of compressive and shear waves. Yet another type of elastic wave, called **surface waves**, can occur only on and close to the surfaces of solids and liquids. The ripples on the surface of a pond and waves on the ocean or at the seaside are obvious examples, but a type of surface wave called a **Rayleigh wave** is also important in transmitting vibration through the ground over relatively short distances, and very high frequency ultrasonic surface waves in solids are important in the electronics and telecommunications industries. Bending, or flexural, waves in plates are examples of transverse waves, as are elastic waves on a wire or string, and all types of surface waves.

Shear forces applied to a fluid, i.e. a liquid or a gas, cause flow to occur. Thus unlike solids, liquids and gases cannot transmit shear waves, and so the only type of elastic waves that can travel through the bulk of a fluid are compressive waves, and therefore sound waves in air are of this type, and are longitudinal waves.

The velocity of elastic waves in fluids and solids

An outcome of solving the wave equation for elastic waves travelling in an elastic medium is that the speed or velocity at which such waves travel through the medium is given by the formula:

$$C = \sqrt{(K/\rho)}$$

where K is the elastic modulus of the medium (in N/m^2 or Nm^{-2}), and ρ is the density of the medium (in kg/m^3 or kgm^{-3}).

Elastic modulus = stress/strain
= (force per unit area)/
(fractional change in deformation)

Note that strain is a dimensionless quantity (i.e. length/length), and so the dimensions of the elastic modulus are the same as those of stress, or pressure, i.e. N/m^2 or Nm^{-2}.

There are many different elastic moduli, for various types of elastic deformation, e.g. compression, shear, torsion, bending etc. and so a corresponding number of different types of elastic waves – compressional, shear, bending etc.

A solid is able to withstand both shear and compressional forces and so all possible types of waves may be transmitted. A fluid (liquid or gas) can only withstand compressive forces, but not shear forces.

Therefore the only type of elastic waves that can travel in a fluid are compressional waves – hence sound waves in water and air are compressional waves and are **longitudinal.** Solids can, in addition to compressive waves, also transmit shear and bending waves – these are **transverse** waves.

For compression waves in a fluid the appropriate elastic modulus is the **bulk modulus**, where the appropriate form of strain is the fractional change in volume of a fluid element subjected to a uniform fluid pressure (acting equally in all directions).

In a gas, such as air, there are in principle two possible values for the bulk modulus, depending upon whether or not the heat flow changes caused by the compressions and rarefactions of the sound wave can take place quickly enough to follow (i.e. keep in step with) these pressure fluctuations, i.e. whether or not the changes are isothermal or adiabatic.

In fact the heat flow cannot keep pace with the pressure fluctuations, i.e. the compressions and rarefactions take place under adiabatic conditions, and for this situation the bulk modulus = γP, where γ is a constant for a particular gas (1.4 for air) and P is the atmospheric pressure.

Hence speed of sound in air:

$$c = \sqrt{(\gamma P/\rho)}$$

From this it can be seen that at a given temperature the velocity of sound in gases with very low density, e.g. helium and hydrogen, is higher than for denser gases such as carbon dioxide. In fact, a change of a factor of 4 in the density would lead to a doubling of the velocity of sound. The velocity of sound for some different gases is shown below:

Gas	Sound velocity at 0°C, m/s
Oxygen	317.2
Air	331.0
Hydrogen	1269.0
Carbon dioxide	258.0

Example 1.14

Calculate the velocity of sound at 20° Celsius.

Solution

At 20° Celsius the density of air = 1.2 kgm^{-3} and atmospheric pressure = 1.01 × 10^5 Pa, or 101 kPa.

Therefore:

$$c = \sqrt{(1.4 \times 101{,}000/1.2)}$$
$$= 343.3 \text{ m/s or ms}^{-1}$$

For an ideal gas, pressure P, temperature T and volume V are related by the ideal gas equation $PV = \mathrm{R}T$ (where R = universal gas constant).

Therefore because of the factor P/ρ in the above formula, the speed of sound is independent of atmospheric pressure.

The speed of sound in a gas does, however, depend upon temperature according to:

$$C = \sqrt{(\gamma \mathrm{R}T)}$$

And so the sound speed is proportional to the absolute temperature, in kelvins (K).

Example 1.15

The speed of sound in air is 343.3 ms^{-1} at a temperature of 20° Celsius. What will it be at 0° Celsius?

Solution

0° Celsius = 273 + 0 = 273 kelvins
20° Celsius = 273 + 20 = 293 kelvins

Now since c is proportional to $\sqrt{T}$, if c_1 and c_2 are the sound speeds at temperatures T_1 and T_2 respectively then:

$$c_2/c_1 = \sqrt{(T_2/T_1)}$$

In this example:

$c_1 = 343.3$ ms^{-1}, $T_1 = 293$ K and $T_2 = 273$ K.

Therefore at 0° Celsius:

$$c_2 = 343.3 \times \sqrt{(273/293)} = 331.4 \text{ ms}^{-1}.$$

Dispersive and non-dispersive waves

The velocity of sound in air does not vary with the frequency of the sound – a fact of importance to theatre and concert goers – since it means that audience members in both the front and rear seats will all receive the same mixture of different frequencies issuing from the performers on stage. In this respect air is a non-dispersive medium for sound waves. Some types of elastic waves are, however, dispersive, most notably flexural or bending waves in solid plates in beams, where the wave speed increases with frequency. These types of waves are important for the transmission of sound in buildings and other structures and will be discussed again in Chapter 6.

Note also that the speed of sound in air is independent of the amplitude of the sound wave, i.e. of the sound pressure level, for small amplitudes only. (This is called the linear acoustics approximation.)

1.10 Absorption and attenuation of sound

Sound absorption is a process whereby sound energy is lost from a sound wave and converted to heat as a result of some sort of frictional process. Since the energy content of sound waves is usually very small the actual temperature rises which result from sound absorption processes are usually negligible. The sound absorption can usually be thought of in terms of some frictional process which occurs between vibrating molecules which are transmitting the sound. Frictional processes also occur in vibrating bodies (e.g. vibrating panels) but in these cases the energy loss is usually referred to as 'damping'. These two terms (damping and absorption) have much in common.

Absorption processes can occur within the sound transmitting medium, or at the interface with another medium, during reflection and scattering. Absorption which occurs during reflection at surfaces is discussed in detail in Chapter 5. All of the sound absorption mechanisms which occur in air are frequency dependent, the amount of absorption being proportional to the square of the frequency of the sound.

One of the causes of sound absorption in air, molecular absorption, is due to the vibration and rotation of the oxygen and water vapour molecules in the air, and is therefore dependent on the relative humidity of the air as well as on the air temperature.

Absorption and attenuation

It is important to distinguish between the terms absorption and attenuation. Attenuation simply means the reduction of sound level by whatever means, and not just by the process of absorption, i.e. conversion to heat through frictional processes. Attenuation may arise because of the sound spreading with distance, or as a result of the effect of scattering, diffraction, interference or refraction, and also as a result of sound insulation, isolation, the use of enclosures, barriers, silencers, as well as because of sound absorption.

Diffraction, reflection, interference and refraction

All waves, including sound waves demonstrate the following behaviour: **diffraction**, **reflection**, **interference** and **refraction**. Some of this material will be revision of schooldays for some, and new for others. Alternative and additional explanations may be found in physics textbooks and there is also much helpful information available on many websites.

1.11 Diffraction

Diffraction is about the interaction between sound waves and solid objects. There are two different aspects to consider:

- What happens when a sound wave meets an obstacle in its path, i.e. to what extent is the sound scattered by, or bent around, or reflected by the object?
- What happens when the object is vibrating? How are the resulting sound waves radiated by the vibrating object? In other words, what is the resulting pattern of the sound radiated? How directional is the radiation?

Therefore diffraction of sound is important in determining:

- the effectiveness of noise barriers (limited by the extent to which sound 'bends' around the edges of the barrier)
- the directionality of noise sources, including the human voice
- the directionality of microphones, and of human hearing
- scattering of sound by objects, including by microphones, sound level meters, and by the human head and body.

Of overriding importance in all diffraction issues is the ratio of the size of the obstacle (D) to the wavelength of the sound waves (λ): D/λ.

It is easiest to consider first of all the two extreme cases (see Figure 1.14):

- If $D/\lambda \gg 1$, i.e. obstacle much larger than wavelength, then the object casts a sharp shadow, regions

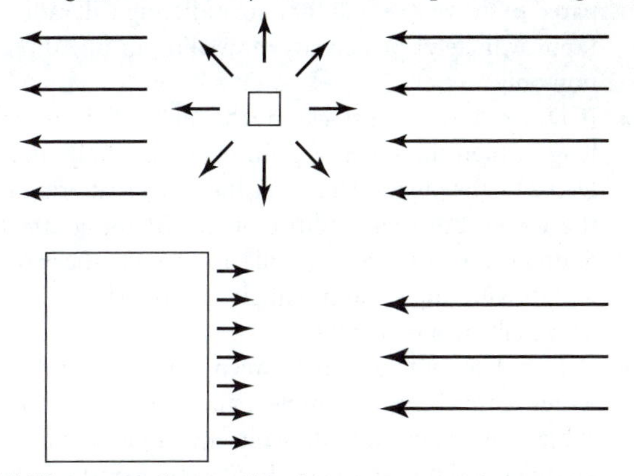

(a) Waves meeting an object

Upper: $\lambda \gg D$, i.e. large $\boldsymbol{\lambda}$ (low frequency) waves and/or small object: the waves are scattered by the object (more or less equally in all directions) but otherwise the waves travel on undisturbed by the object.

Lower: $D \gg \lambda$, i.e. large object and small $\boldsymbol{\lambda}$ (high frequencies). The waves are reflected at the surface of the object. There is no bending of the waves around the object and there is a quiet shadow zone behind the object.

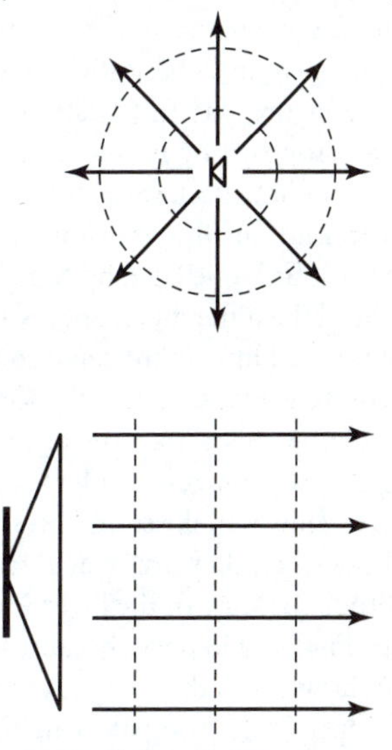

(b) Vibrating source radiating waves

Upper: $\lambda \gg D$, i.e. small loudspeaker radiating large $\boldsymbol{\lambda}$ (low frequency) waves: the waves are radiated equally in all directions.

Lower: $D \gg \lambda$, i.e. large loudspeaker radiating small $\boldsymbol{\lambda}$ (high frequencies). The sound radiation is highly directional (approximating to plane waves).

Figure 1.14 The two extreme cases of diffraction

'behind' the object are effectively shielded from the sound, and sound is reflected from the surface of the object facing the sound waves. This is what happens with light waves. Sound sources which are large compared to the wavelength they are radiating will radiate (approximately) plane waves travelling in one direction only.

- If $D/\lambda \ll 1$, i.e. obstacle much smaller than wavelength, then the waves are almost completely unaffected by the object. There may be some scattering of the waves, but there is little or no shielding effect. Sound sources which are small compared to the wavelength will radiate (approximately) spherical waves, i.e. equally in all directions.
- If $D \approx 1$ the interaction between the object and the waves is much more complex, and the theory is the subject of many quite difficult equations in undergraduate and postgraduate physics and acoustics textbooks.

The Huygens-Fresnel theory of secondary wavelets

Why does diffraction occur? For example why does sound bend around corners? An insight can be gained by considering a theory of secondary wavelets, put forward by the Dutch physicist Huygens. He proposed that every point on a wavefront acts as a point source of new identical secondary wavelets radiating spherically in the direction the wave travel (see Figure 1.15a). (A later modification to this theory was proposed by Fresnel to explain why the wavelets did not radiate backwards towards the original source.) We can find out the position and shape of the new wavefront, a little while later, by combining the contributions from all the different secondary wavelets.

To explain this in a little more detail consider just two sources of secondary waves, S_1 and S_2 (see Figure 1.15b) radiating waves towards a receiver at position R. The waves leaving the two sources are identical in amplitude and are in phase. However, the sound pressures they produce (p_1 and p_2) when they arrive at R will be different both in amplitude (because of the inverse square law and spherical spreading) and in phase because they have travelled different distances to the receiver position.

If the difference in the two path lengths is d then the phase difference is given by:

$$\text{Phase difference, } \varphi = 2\pi d/\lambda$$

(on the basis that a path difference of one wavelength will result in a phase change of 2π).

The total sound pressure at R from the wavelets from the two sources will be, according to the principle of superposition,

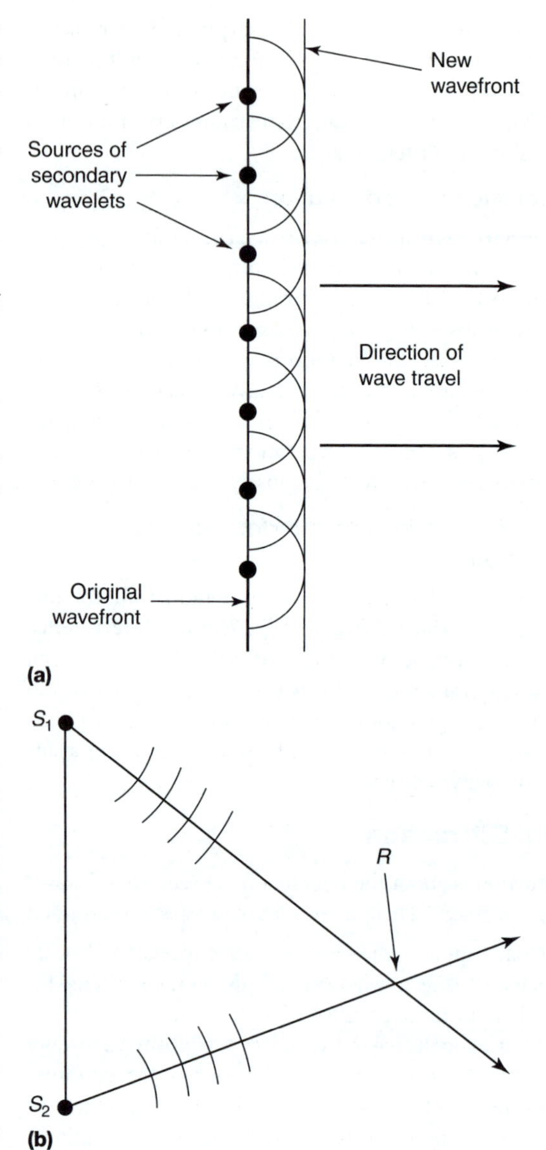

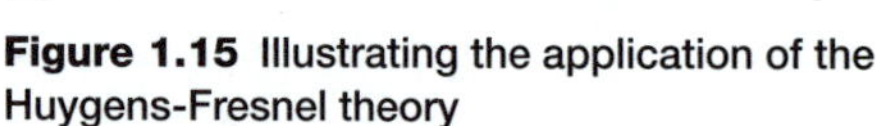
Figure 1.15 Illustrating the application of the Huygens-Fresnel theory

$$p_{\text{TOTAL}} = p_1 + p_2$$

This will be an equation rather like:

$$p_{\text{TOTAL}} = (A/r_1)\sin(\omega t - kr_1) + (A/r_2)\sin(\omega t - kr_2)$$

where A is the pressure amplitude of the wavelets at the sources, r_1 and r_2 their path lengths to the receive point R, and ω the angular frequency ($= 2\pi f$).

To find out the new wavefront we have to add in the contributions from all the other point sources on the existing wavefront (in theory an infinite number of them), and then we have to repeat the exercise for all other (infinite number of) possible reception points – quite a lot of mathematics. If we were to do all this maths we would come up with the unsurprising result that the new wavefront is another plane wavefront, just a little bit further forward than the original one, and so on – a result we know intuitively by watching the movement of ripples on the surface of water. We could go through an exactly similar analysis to show that the wavefronts on the surface of a spherical wave will combine to form a new slightly advanced spherical wave, again a result that we know from everyday experience.

What this tells us is that in order to make a completely new plane wavefront contributions are needed from all of the entire original wavefront. In cases where the wave encounters an obstacle, or passes through an opening, it is the wavelets at the edge of the wavefront that spread out into the shadow area, causing the effect of the sound 'spreading around the corner' (see Figure 1.16).

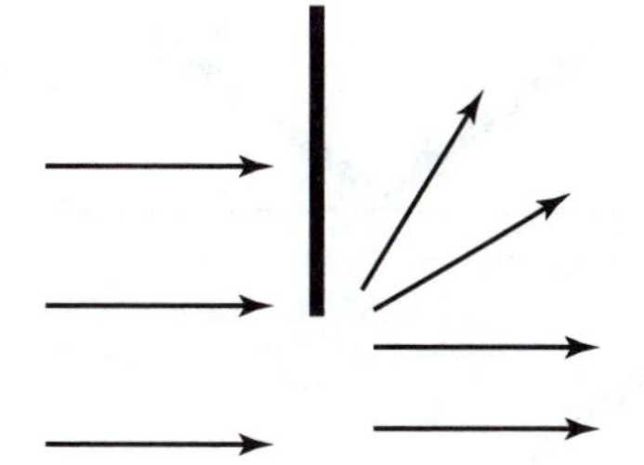

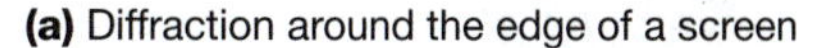

(a) Diffraction around the edge of a screen

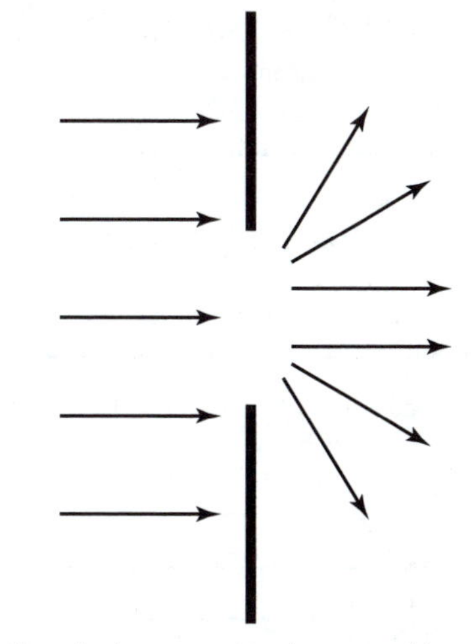

(b) Diffraction at a large aperture (compared to wavelength)

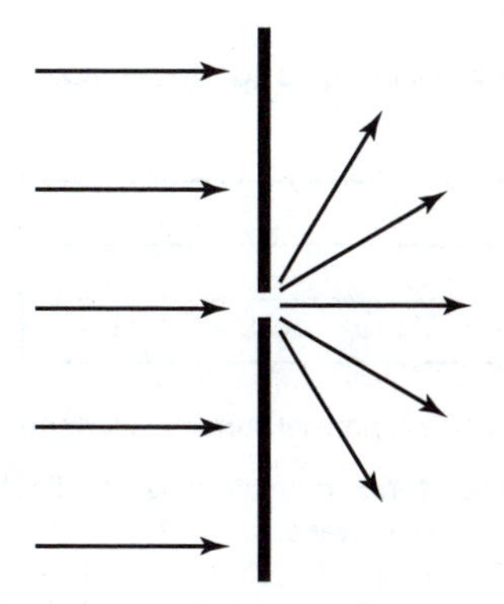

(c) Diffraction by a narrow aperture (compared to wavelength)

Figure 1.16 Diffraction around edges and through openings

1.12 Reflection

This is really a special case of diffraction. When sound waves meet a surface (of different acoustic impedance) which is large compared to the wavelength then sound is reflected (see Figure 1.17). If the reflecting surface is plane and smooth then **specular reflection** occurs, i.e. angle of incidence equals angle of reflection (as with light and a glass mirror, and balls on a billiard table). If the surface is rough then **diffuse reflection** occurs, with sound being scattered in all directions (as with light and frosted glass).

As with the optical case, concave surfaces can cause sound to be focused and convex surfaces cause sound to be dispersed.

A nearby wall or floor can reflect sound and increase sound pressure levels compared to positions at similar distances from the source but well away from reflecting surfaces. Very close to the surface (just a few centimetres) differences of up to 6 dB may be observed, because the incident and reflected waves may be considered to be coherent. At distances of about 1 m away, when the incident and reflected waves are considered to be incoherent (i.e. uncorrelated), increases of about 3 dB compared to the free field situation may be expected; and it is considered that sound levels further than about 3 m from the wall may not be significantly influenced by the reflections.

When we are holding a sound level meter, reflections from our body can affect the sound level at the microphone and so we should always hold the meter at arm's length away from the body, or use a tripod, in order to minimize these effects.

For the same reasons (reflections from the head and body) readings from a dosemeter worn by an employee working close to a noisy machine may never quite agree with those from a sound level meter at similar distances from the machine: difference of about 2 dB are sometimes observed.

Note that smaller surfaces and obstacles will scatter (or diffract) the sound in a more complex manner.

An interesting example of the application of sound reflection, prior to the development of radar in

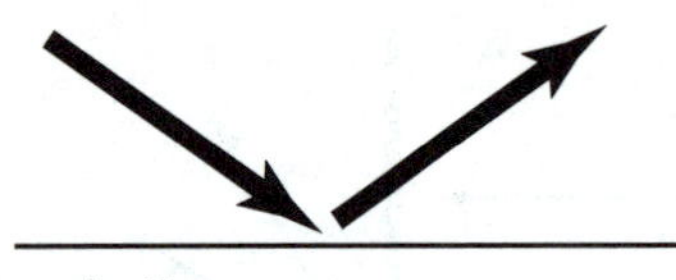

(a) Specular reflection

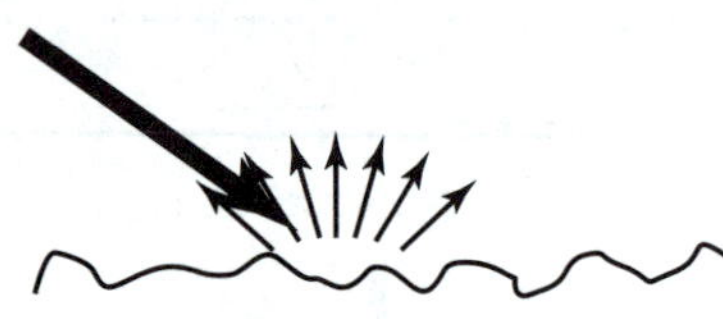

(b) Diffuse reflection (diffusion)

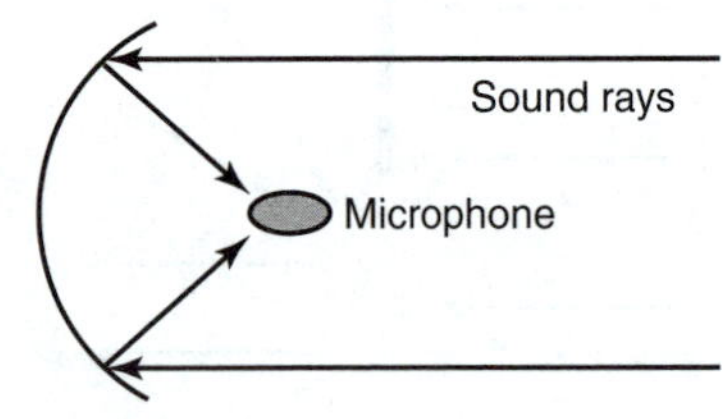

(c) Reflection (focusing) by a concave surface

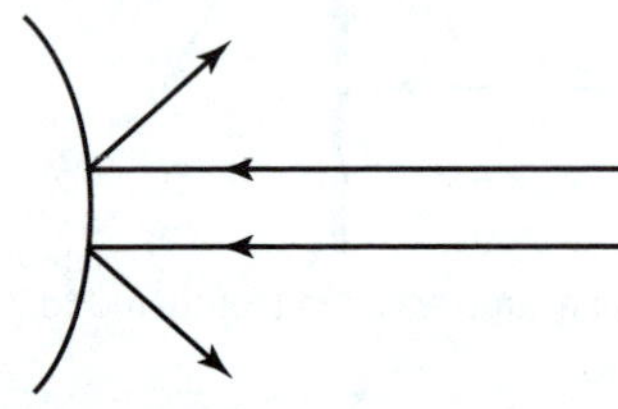

(d) Reflection (dispersion) by a convex surface

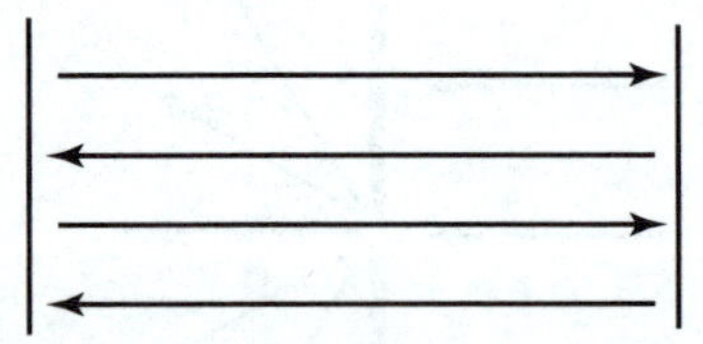

(e) Reflection between parallel surfaces (flutter echoes)

Figure 1.17 Illustrating some aspects of reflection of sound waves from surfaces

the Second World War, was the use of large concrete 'sound mirrors' at Dungeness in Kent and other places along the UK coast, to try to give advance warning of the approach of enemy aircraft. A number of parabolic reflectors, tens of metres in diameter were constructed, with microphones placed at their focal points. (Further information is available on several websites in response to a search under 'sound mirrors at Dungeness' or similar.)

Partial reflection

Sound waves are reflected when they arrive at an interface with a medium with a different acoustic impedance, but some sound energy will also be transmitted into the second medium. The greater the change in specific acoustic impedance, the greater is the fraction of sound energy which is reflected. Air has a specific acoustic impedance of about 415 kgm/s^2 at room temperature and water has a value of about 1.5 million kgm/s^2. Therefore, in view of this very large difference, we should expect that almost all of the sound energy would be reflected at an air–water interface, and only a small proportion of energy will be transmitted.

The fraction of sound energy, R, which is reflected at an interface between two media with acoustic impedances z_1 and z_2 is given by:

$$R = [(z_1 - z_2)/(z_1 + z_2)]^2$$

For an air–water interface the fraction of sound energy reflected will be 0.998899272. Therefore the fraction which is transmitted is:

$$(1 - 0.99899272)$$

In decibels this means that when a sound wave in air arrives at an interface with water the level of the sound wave transmitted into the water will be almost 30 dB below the level in air (and vice versa if the direction of wave travel is reversed). This is relevant to the role of the middle ear in facilitating the transmission of sound from the outer ear to the inner ear, as will be discussed in Chapter 3.

1.13 Interference and standing waves

The phenomenon of interference describes what happens when two or more sound waves meet (see Figure 1.18).

The situation is governed by the **principle of superposition**, which states that the resulting total disturbance at any point is the algebraic sum of the disturbance caused by each of the waves at that point at that moment in time. (The 'disturbance' may be specified as an acoustic pressure, particle velocity, displacement or acceleration, and the term algebraic recognizes that theses disturbances may be positive or negative.)

In order that waves may display interference they must be **coherent**, i.e. their waveforms must be identical in shape (i.e. in time profile), although they may be different in amplitude.

In the vast majority of situations, e.g. in the case of waves from two different noise sources, the waves will not have similar waveforms, and will not be coherent.

The most likely circumstance in which the waves would be coherent is if they are both pure tones derived from the same source so that they have the same sinusoidal waveform and the same frequency. If they also have the same or similar amplitudes it is possible that at a particular moment at a particular position the two waves could either cancel each other out (called **destructive**

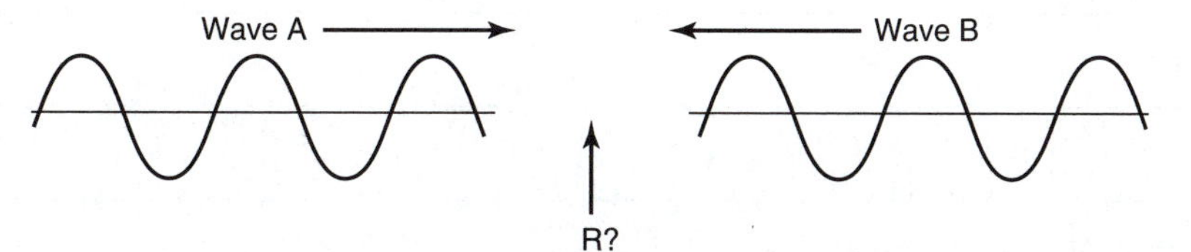

(a) What happens when waves A and B meet at point R?

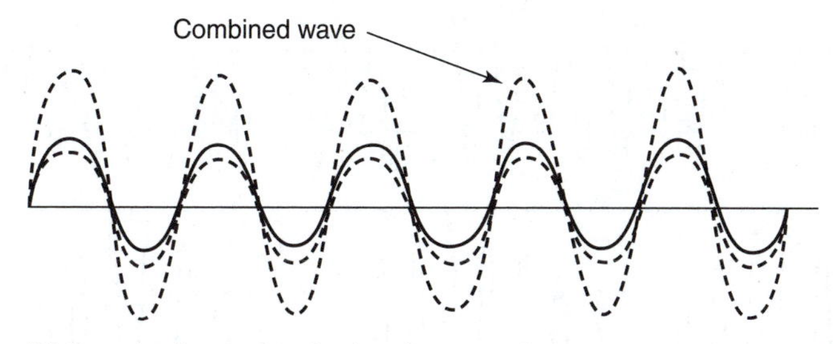

(b) Completely constructive interference, when waves A and B are in phase

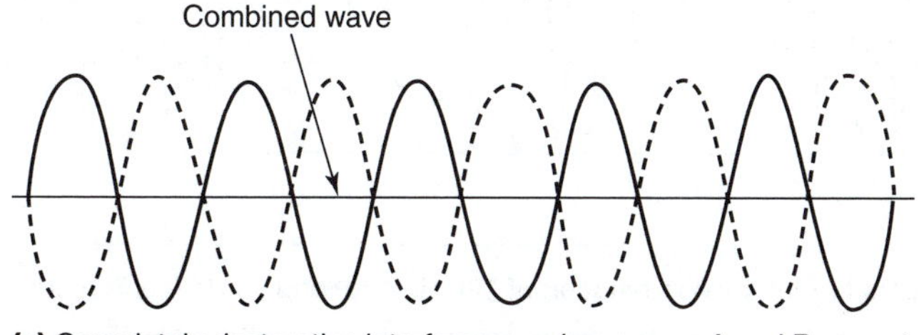

(c) Completely destructive interference, when waves A and B are out of phase

Figure 1.18 Illustrating the problem of interference

interference) or reinforce each other with doubled amplitude (**constructive interference**). This could happen, for example, at a point between two loudspeakers facing each other, with both being supplied by exactly the same pure tone signal. Under these circumstances, however, the net effect would simply be a 3 dB increase in level because as the waves from each source travel onwards the pattern of interference at a particular point would be continually changing from constructive to destructive and back again.

Stationary or standing waves

Reflection of sound produces waves which are coherent and which can interfere constructively or destructively with the original incident wave. Under certain circumstances **stationary** or **standing waves** are produced. These are steady patterns of interference characterized by large variations in amplitude with position, so that there will be alternately positions where the amplitude is a minimum, called **nodes**, as a result of destructive interference, and positions of maximum amplitude (called **antinodes**) resulting from a constructive interference.

Beats

Beats are amplitude modulated pure tones where the amplitude of a higher frequency appears to be modulated by a lower frequency, for example a frequency of 100 Hz whose amplitude is modulated at a frequency 10 Hz. This can arise from a combination of two tones of slightly different frequency but equal or nearly equal similar amplitude. Consider for example two tones of 100 Hz and 90 Hz with equal amplitudes. They start off in phase but after five cycles they are out of phase, and cancel each other out (destructive interference), and after a further five cycles are back in phase, then out of phase five cycles later, and so on. This is shown in Figure 1.19.

As shown earlier the sound pressure variation with time of two pure tones of the same amplitude A and frequencies f_1 and f_2 may be described by the equations $p_1 = A\sin(2\pi f_1 t)$ and $p_2 = A\sin(2\pi f_2 t)$. Their combination, $p_1 + p_2$, is given by:

$$\begin{aligned} p_1 + p_2 &= A\sin(2\pi f_1 t) + A\sin(2\pi f_2 t) \\ &= 2A\cos[2\pi\{(f_1 - f_2)/2\}t]\,\sin[2\pi\{(f_1 + f_2)/2\}t] \end{aligned}$$

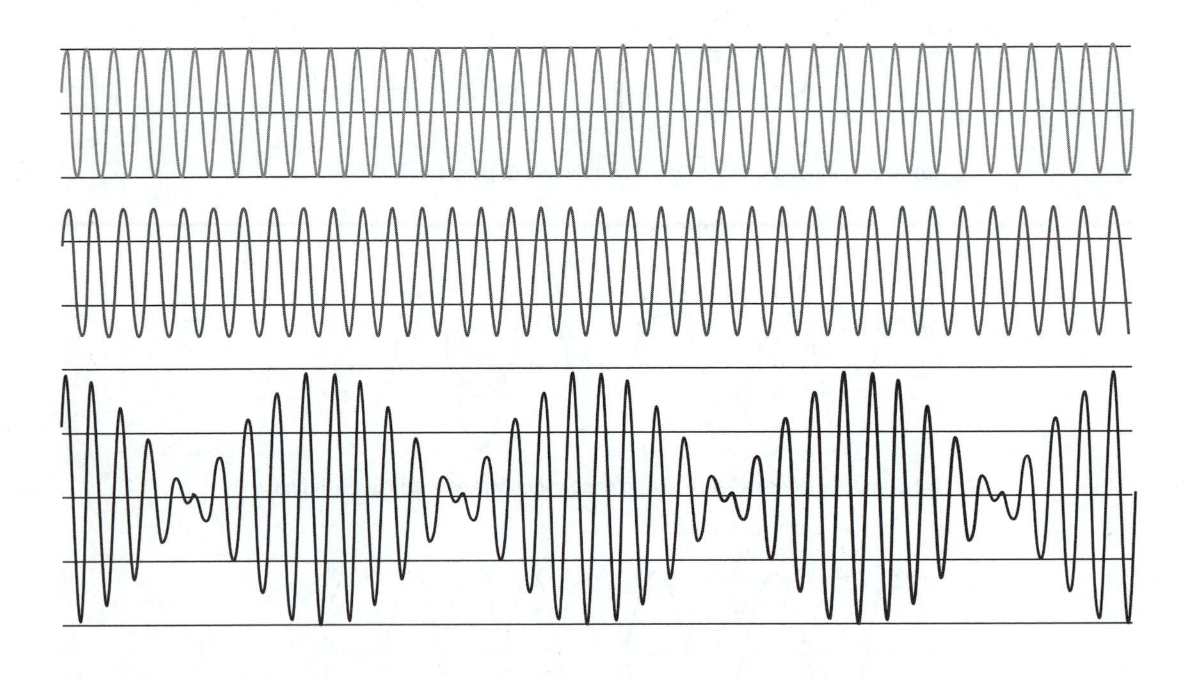

Figure 1.19 Illustrating beats formed by combination of 100 Hz (top) and 90 Hz (centre) pure tone over 40 cycles of the 100 Hz tone

The expression on the second line, obtained using a standard trigonometric relationship, represents a pure tone of frequency $(f_1 + f_2)$ modulated by a tone of frequency $(f_1 - f_2)$.

This shows that the beat frequency f_B is the difference between the two component frequencies, f_1 and f_2:

$$f_B = f_1 - f_2$$

Beats can occur between different notes in some musical compositions and are used in the tuning of musical instruments. They may also be produced by rotating machines such as fans or motors which produce tones related to their rotation speed. If there are two such machines, nominally identical and rotating at the same speeds, beats can occur if in practice the two machines are running at slightly different speeds.

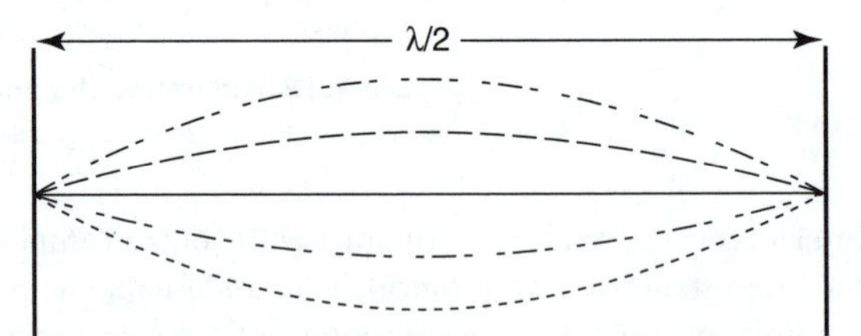

(a) Position of the string during one cycle of vibration performing its first mode at a frequency at which the length of the string equals one half a wavelength.

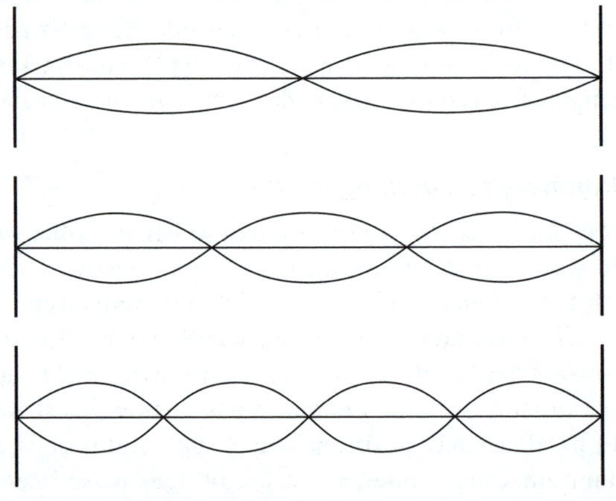

(b) Second, third and fourth modes of vibration of a string

Figure 1.20 Mode shapes of standing waves in strings

One-dimensional standing waves

Standing waves in strings, pipes and rods are important in the operation of many musical instruments.

Standing waves in strings

The simplest case to consider is that of a string stretched between two fixed ends (see Figure 1.20). When a wave is created in the string it vibrates between the ends as a result of reflection. Interference will occur continuously between the waves travelling in the opposite directions, and in general these patterns of interference will be continually changing. But there are certain special frequencies at which steady, or standing, waves are set up. The lowest frequency, called the **fundamental**, or **first harmonic**, occurs at the frequency at which the length of the string is equal to one half of a wavelength. Under these circumstance every point on the string is vibrating in phase, but the amplitude increases from zero at the ends to a maximum in the middle. This is also called the first mode of vibration of the string, and, like every other mode, is characterized by its frequency and also by its mode shape which describes the shape of the vibrating string. (Note that you may also come across the alternative terms **eigenfunction** (mode shape) and **eigenfrequency** (mode frequency) in technical literature.)

The next mode of vibration occurs at a frequency for which the wavelength in the string will be one wavelength (or two half wavelengths). The mode shape for this mode contains two maxima (or antinodes) ¼ and ¾ along the length of the string, and three minima, or nodes, one at each end and one in the middle. The frequency of this mode, called the second harmonic, is twice that of that of the first mode.

Note that there will always be nodes at the two ends, because the string is fixed or clamped here, and so no motion is possible. This is an example of what is called a 'boundary condition' which acts as a constraint on what standing wave patterns (modes) are allowed for a string of this particular length and wave speed.

There are an infinite number of modes of vibration of such a string. The common link is that in each case the mode frequency is such that the length of the string is equal to a whole number of half wavelengths. For a string of length L along which waves travel at speed c and wavelength λ (remembering from earlier in the chapter that $c = f\lambda$):

Mode number	Wavelength (λ)	Frequency ($= c/\lambda$)
1	$1(2L)$	$1(c/2L)$
2	$(1/2)(2L)$	$2(c/2L)$
3	$(1/3)(2L)$	$3(c/2L)$
4	$(1/4)(2L)$	$4(c/2L)$
⋮	⋮	⋮
n	$(1/n)(2L)$	$n(c/2L)$

It can be seen that in this case the frequencies are in the ratio of 1:2:3: . . . :n, etc.

Exercise

Compare the vibration (amplitudes and phases) at the same positions along a string: (a) when progressive waves only are travelling along the string (assume for example that the string is very long so that there are no reflected waves); and (b) when there are standing waves (as described above) present. Assume that the vibration in both cases is of a single frequency (sinusoidal waves) and that there is no damping or attenuation of the waves.

Answer

In the case of progressive waves the amplitude is the same at all positions on the string (because there is no attenuation) but there is a phase difference between the vibration cycles of adjacent parts of the string (such that the phase difference is 180°, or half a cycle) for positions half a wavelength apart and 360°, or one whole cycle, for positions one wavelength apart (see Figures 1.3 and 1.4 earlier).

For the standing waves all points on the string which lie in between two nodes always vibrate in phase with each other (but with different amplitudes), and they are out of phase, by half a cycle, with the points which are located between the adjacent pair of nodes. Unlike the progressive wave case the amplitude of vibration varies with position according to the mode shape from zero at nodes to a maximum at antinodes (see Figure 1.20).

When we pluck the string the resulting pattern of vibration will be made up of a combination of these different modes, and the frequency content of the resulting sound will be a mixture of these special natural frequencies, and of these frequencies only. In other words other frequencies are not allowed or do not occur.

Why is this?

The answer is that it is only at these frequencies (i.e. when an exact number of half wavelengths fit in exactly between the ends of the string) that the pattern of interference that occurs at different points along the string is constant, leading to the observed standing wave patterns of vibration. At any other frequency the pattern of interference at any point on the string is constantly changing, and averaged over time so no interference effects are observed.

Although the note arising from the plucking of the string will always contain the same mixture of frequencies (the line frequency spectrum discussed briefly earlier in the chapter, see Figure 1.7) the exact nature of the mix, i.e. the relative amounts of different frequencies (and the shape of the frequency spectrum) will change, depending on how and where the note is plucked. In the case of the same note played on different musical instruments such as a violin, guitar or piano for example, the mix of frequencies and the quality of the note reaching the ear (called the 'timbre') will also depend on other factors such as how effectively the different frequencies of vibration are converted to airborne sound, depending on the mechanical structure of the instrument. This is why the same musical note will have a different quality when played on different instruments.

Exercise

If the string is held, or clamped, at its centre point (i.e. as well as at both ends) what difference will this make to the vibration of the string when it is plucked?

Answer

The answer is that those modes which have a maximum amplitude at the centre will be suppressed, i.e. will not occur (because movement of the string at its centre point is now no longer allowed). Therefore the note will contain a mixture of only the even numbered harmonics (2, 4, 6 etc.) for which there is a node at the centre point of the string, and which are therefore unaffected by the central point being clamped.

What happens if instead of being plucked the string is subjected to a continuous vibration?

This obviously depends on the frequency or frequencies in the applied vibration. The string will respond very selectively producing large amplitudes that occur at its own natural frequencies and very little response at other frequencies. This is known as the phenomenon of resonance, and the natural frequencies are also known as the resonance frequencies of the string.

Standing waves in pipes

Sound waves in a pipe are reflected at the end of the pipe if it is terminated by a rigid cap at the end, called a closed end. Reflection also occurs if the end of the pipe is open. This is because the acoustic impedance presented to the sound waves by the air inside the pipe is different from that of the open air just outside the end of the pipe. The change of impedance causes the waves to be reflected back down the pipe.

The difference in acoustic impedance depends on the diameter of the pipe compared to the wavelength of the sound, so that for a given wavelength the change in impedance (compared with open air), and therefore the fraction of sound reflected at the open end increases as the pipe diameter gets smaller.

Exercise

How might you modify the end of the pipe in order to reduce reflection and increase the amount of sound radiated from the end of the pipe?

Answer

The answer is to flare (i.e. gradually increase the diameter towards) the end of the pipe (like a trumpet).

Change of phase on reflection

There is an important difference between the reflection at the open and closed ends of a pipe. At the closed end a sound wave is reflected without any change of phase, so that a compression is reflected as a compression (and a rarefaction as a rarefaction), but a change of phase of half a cycle occurs at the open end, where a compression will be reflected as a rarefaction, and vice versa.

Acoustic particle displacement and acoustic pressure amplitudes

The standing waves in strings were described in terms of the amplitude of vibration of the string at various points along its length. Standing waves of sound in pipes may be described either in terms of the amplitudes of vibration of the air particles in the sound wave (i.e. similar to the string case) or in terms of the acoustic pressure at different positions along the pipe. These two quantities will be half a cycle out of phase with each other. This means that at a closed end of a pipe the particle displacement will always be a minimum (zero, as for the string) but the acoustic pressure amplitude will be a maximum, and conversely at an open end the particle movement will be a maximum but the acoustic pressure will be zero.

The two cases (open–open and open–closed pipes)

There are two possible cases to consider (i.e. two sets of boundary conditions): pipes which are open at both ends, and pipes which are open at one end but closed at the other (see Figure 1.21).

For a pipe open at both ends the standing wave pattern must meet the boundary condition that the particle displacement is a maximum at both ends. The lowest frequency at which this can occur is when the length of the pipe is equal to one half wavelength, and the mode shape has a particle displacement minimum (node) half way along the pipe. The next mode is when the length of the pipe equals one wavelength (two half wavelengths) and the frequency is double that of the first mode. It can be seen that for the open–open pipe the natural frequencies are in the same 1:2:3 ratio as for the string, but the mode shapes are different.

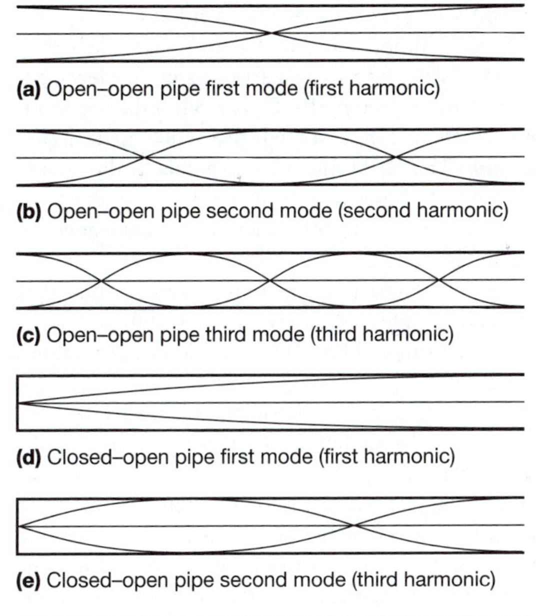

(a) Open–open pipe first mode (first harmonic)

(b) Open–open pipe second mode (second harmonic)

(c) Open–open pipe third mode (third harmonic)

(d) Closed–open pipe first mode (first harmonic)

(e) Closed–open pipe second mode (third harmonic)

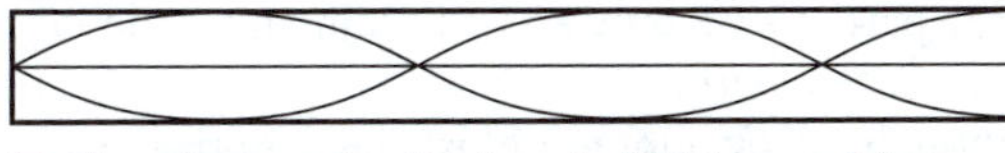

(f) Closed–open pipe third mode (fifth harmonic)

Note that these sketches indicate amplitudes of particle displacement, which is half a cycle out of phase with acoustic pressure, so that at an open end particle displacement is a maximum but acoustic pressure in zero. The converse is true at the closed end of a pipe, i.e. particle displacement is zero but acoustic pressure is a maximum.

Figure 1.21 Mode shapes of standing waves in pipes

In the case of the open–closed pipe the standing wave pattern must meet the boundary condition that the particle displacement is a maximum at one end and a minimum at the other end. The lowest frequency at which this can occur is when the length of the pipe is equal to one quarter of a wavelength. The next frequency will occur when the length of the pipe is equal to three quarters of a wavelength, and the mode shape has a displacement minimum (node) two thirds of the way along the length of the pipe from the closed end, as well as at the closed end.

Exercise

What will be the frequency of the next mode, and how many nodes and antinodes will there be?

Answer

The frequency for which the length of the pipe is equal to one and a quarter wavelengths ($5c/4L$), 3 nodes and 3 antinodes.

For a pipe of length L along which sound travels at velocity c (and wavelength λ, remembering that $c = f\lambda$):

Mode number	Wavelength (λ)	Frequency ($= c/\lambda$)
1	$4L/1$	$1(c/4L)$
2	$4L/3$	$3(c/4L)$
3	$4L/5$	$5(c/4L)$
4	$4L/7$	$7(c/4L)$
⋮	⋮	⋮
n	$4L/(2n - 1)$	$(2n - 1)(c/4L)$

It can be seen that in this case the frequencies are in the ratio of 1:3:5:7:n etc. The first, and all of the other 'odd' harmonics, are missing.

Exercise

Assuming that the human ear canal can be modelled as a pipe of 30 mm length which is closed at one end and open at the other end, what will be the frequencies of the first two modes of the pipe (assuming the velocity of sound in air is 330 m/s)?

Answer

2750 Hz and 8250 Hz.

A mathematical description of one-dimensional standing waves

We learnt earlier that a one-dimensional progressive wave travelling in the (positive) x direction may be described by:

$$p_1 = A\sin(\omega t - kx),$$

where $\omega = 2\pi f$ and $k = 2\pi/\lambda$

So a wave travelling in the opposite (i.e. negative x) direction will be represented by:

$$p_2 = A\sin(\omega t + kx)$$

It therefore follows that when these two waves meet, which is what happens in standing waves, the total sound pressure $p_1 + p_2$ is given by:

$$p = p_1 + p_2 = A\sin(\omega t - kx) + A\sin(\omega t + kx)$$

and it can be shown that using some well known relationships in trigonometry that:

$$p = 2A\sin(kx)\cos(\omega t)$$

Let us examine each of the three parts in this expression in turn. Since the highest value both the sine and cosine terms can have is 1, then '$2A$' represents the maximum value or amplitude of the combined wave. If we now consider a particular moment in time, then we see that the value of sound pressure varies sinusoidally with position x, and at any given position x the sound pressure varies sinusoidally with time. This is the behaviour of a standing wave as illustrated in Figure 1.20 for a string or Figure 1.21 for a pipe.

Example 1.16

A standing wave in a pipe is described by the formula:

$$p = 0.1 \times \sin(9.46X)\cos(3217.4t)$$

Calculate the maximum amplitude, frequency, wavelength and sound speed.

Solution

By comparison with the formula:

$$p = 2A\sin(kx)\cos(\omega t)$$

The maximum amplitude of the standing wave = 0.1 Pa

$\omega = 3217.4 = 2\pi f$, from which $f = 3217.4/2\pi = 512\,\text{Hz}$

$k = 9.46 = 2\pi/\lambda$, from which $\lambda = 2\pi/9.46 = 0.66$ m

and $c = f\lambda = 512 \times 0.66 = 340$ m/s

Two-dimensional standing waves in membranes and plates

A membrane (such as a drumskin) is rather like a two-dimensional version of a stretched string. In both cases these devices derive their elasticity from being stretched, i.e. from being in a state of tension. When not stretched the string or membrane has no intrinsic stiffness of its own. When the string is plucked or the drum is struck a local change in displacement is produced which travels as a wave along the string or across the drumskin.

A plate has its own built in stiffness, a resistance to bending, and waves in plates are called bending waves or flexural waves.

In both cases the boundaries of the plate or membrane act as reflectors of the waves travelling across them. The case of the membrane is much simpler because in order for the membrane to be in tension it must be fixed or clamped at its edges, i.e. there will be always be a boundary condition of zero displacement at the edges, as for strings. For a plate there may be a variety of different boundary conditions. The edges of a plate may be completely free (able to move and to rotate) or they may be hinged (free to rotate but not to move) or clamped (can neither move nor rotate).

Specification of two-dimensional modes

Two integer numbers are needed to specify each two-dimensional standing wave pattern, or mode. In a rectangular membrane for example the (1, 1) mode is the one which is a combination of the first mode that would occur in each dimension. The (2, 1) mode corresponds to the second mode in one direction, and the first in the other direction.

Some examples of mode shapes of membranes are shown in Figure 1.22.

The mode shapes of a square plate can be amazingly diverse and complicated, as a result of the different possible combinations of boundary conditions and methods of excitation.

In 1787 the German physicist and musician Ernst Chladni devised a method for visualizing the modes of a vibrating plate, by sprinkling the plate with a fine powder. When the plate was set into vibration by stroking an edge with a violin bow the powder particles are set into motion and move away from the antinodes and migrate towards and settle in the nodes. If a different mode is excited the particles become excited again and settle into the new nodal pattern.

Although many more sophisticated optical techniques, such as laser holography, are available for visualization and research into modes of vibration of structures, Chladni's simple method remains a very popular and impressive way of demonstrating the wide variety of mode shapes that can be exhibited by even the simplest structure, such as a square metal plate.

There are many websites which provide excellent demonstrations of Chladni's technique and of resonance in plates (search under 'Chladni's figures' or similar).

Knowledge of the mode shapes of structures such as sound radiating panels of machinery is of great importance to the noise control engineer in devising measures to reduce vibration and noise radiation. The position of nodal and antinodal lines tells the engineer the optimum positions to stiffen or support a structure in order to minimize the effect of important noise radiating modes of vibration.

In this chapter we have considered one- and two-dimensional standing waves. In Chapter 5 three-dimensional standing waves in spaces, called room modes, will be described.

Interference of ground reflections

Partial interference can also occur outdoors between sound reaching a receiver directly from the source and that which arrives after the reflection from the ground. This results in a reduction in sound level at the receiver, called **ground attenuation**. This will be discussed in more detail in Chapter 2.

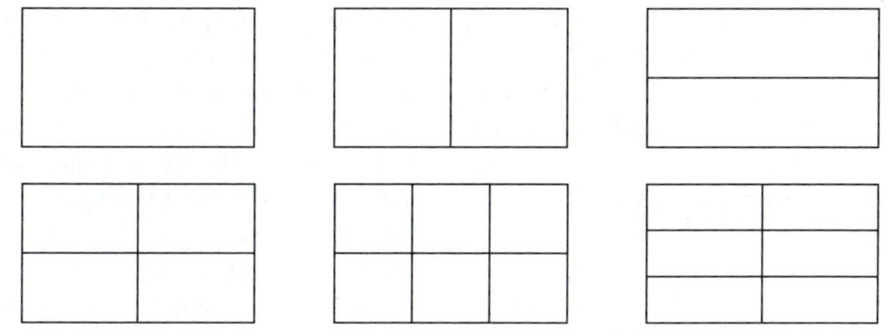

The dividing lines separate adjacent areas which are half a cycle out of phase. The modes are, from left to right: upper: (1, 1), (2, 1), (1, 2), lower: (2, 2), (3, 2), (2, 3).

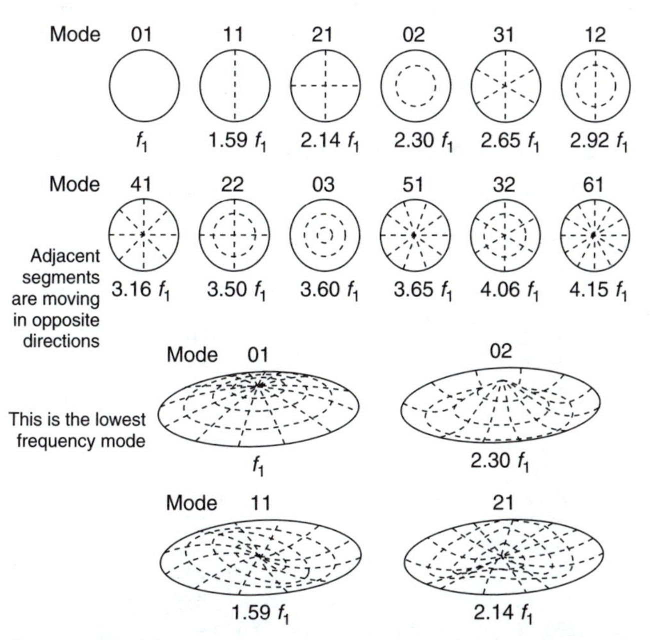

Figure 1.22 Some modes of vibration of rectangular and circular membranes

1.14 Refraction

Refraction is the change of direction of a wave that occurs when the wave travels from a medium with a wave speed C_1 to a different medium with a wave speed C_2 (see Figure 1.23).

The change of direction in Figure 1.23 is best described with reference to the dotted line which is perpendicular to the (horizontal) boundary between the two media, and called the 'normal'. The rule is that when the wave travels from a medium where the wave speed is faster, to one where it is travelling more slowly (as in the diagram on the left) then the change in direction involves a bending 'towards the normal'. Conversely, as in the diagram on the right, when the wave travels from a 'slower' to a 'faster' medium then the wave direction is bent 'away from the normal'.

The angles which the incident and refracted rays make with the normal are known as the angle of incidence and angle of refraction (i and r in the sketches) respectively, and the 'law of refraction' is that $\sin(i)/\sin(r) = C_1/C_2$.

A well known application of refraction, based on the fact that the speed of light is very different in air and in

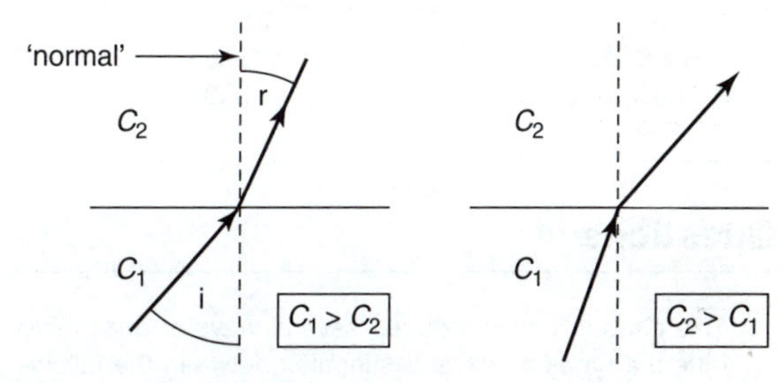

Figure 1.23 Illustrating refraction

glass, is in the design of optical lenses used in spectacles, and in optical instruments such as microscope and telescopes. A balloon filled with carbon dioxide (which has a different sound speed from that of air) would behave in a similar way, as an acoustic lens. (This was demonstrated in an experiment by K. F. J. Sondhauss in 1852.)

The speed of sound changes slightly with air temperature from about 330 to 340 metres a second between 0° and 20° Celsius (see earlier, section 1.9). It also changes with wind speed depending upon whether it is travelling with or against the wind, which may vary its speed from 0 to 10 metres a second. Both temperature and the wind speed vary with height above the ground, therefore so does the speed of sound. This results in refraction effects similar in principle to those which occur when light enters a glass lens, except that in this case the changes in the sound speed are far smaller and more gradual. They result in sound rays being bent either towards the sky or towards the ground. These effects only become noticeable over considerable distances, more than a kilometre, but are responsible for our observations that noise from a distant motorway or factory carries better on some days than others, depending on wind direction and weather conditions. These effects exist and make prediction of sound levels over very long distances notoriously difficult and can cause day to day variation in sound level at these distances of many decibels (10 to 20 dB). These effects are discussed in more detail in Chapter 2.

Appendix 1.1 Octave and one third octave band centre bands and bandwidths

(Exact centre band frequencies taken from BS EN ISO 266:1997 *Preferred frequencies*)

Nominal centre band frequency (Hz)	Exact centre band frequency (Hz)	1/3 octave band (Hz)	Octave band (Hz)
25.0	25.119		
31.5	31.623	28 to 36	
40.0	39.811	35 to 45	22 to 45
50	50.119	45 to 56	
63	63.096	56 to 71	45 to 89
80	79.433	71 to 89	
100	100.000	89 to 112	
125	125.89	112 to 141	89 to 178
160	158.49	141 to 178	
200	199.53	178 to 242	
250	251.19	242 to 282	178 to 355
315	316.23	288 to 355	
400	398.11	355 to 447	
500	501.19	447 to 563	354 to 709
630	630.96	562 to 708	
800	794.33	708 to 892	
1000	1000.00	891 to 1123	707 to 1414
1250	1258.9	1122 to 1413	
1600	1584.9	1413 to 1779	
2000	1995.3	1778 to 2240	1411 to 2822
2500	2511.9	2238 to 2819	
3150	3162.3	2917 to 3549	
4000	3981.1	3547 to 4469	2815 to 5630
5000	5011.9	4464 to 5625	
6300	6309.6	5621 to 7082	
8000	7943.3	7077 to 8916	5617 to 11,234
10,000	10,000.0	8909 to 11,225	

Questions

1 This chapter has introduced several ways of describing different types of wave. Distinguish between the following: progressive and standing waves; transverse and longitudinal waves; plane and spherical waves; dispersive and non-dispersive waves; elastic and electromagnetic waves; compressive and shear waves.

2 Determine the sound power level range (re. 10^{-12} W) of the human voice which is from about 10 to 50 microwatts.

3 The voice level of a lecturer teaching on the Diploma in Acoustics course corresponds to an average sound pressure level over a one hour lecture of 60 dB at a distance of 1 m in free field conditions. Estimate the sound power in watts and hence the amount of hot air (sound energy) produced by the lecturer if he lectures for 100 hours over the duration of the course (you may ignore hot air generated by students).

4 Estimate the acoustic energy generated by (a) an airliner during a three-hour flight and (b) a pop concert. Compare these with the amount of energy needed to boil a 1 kW kettle of water, which takes 2 minutes.

5 A plane sound wave in air has a single frequency of 500 Hz. Taking the velocity of sound in air as 340 m/s, what is the phase difference:

(a) Between two points in the wave at distance of 0.51 metres apart, at the same moment?
(b) At the same position, but separated by 0.001 seconds?

6 Either the sound pressure level p, or the acoustic particle velocity u, or the acoustic intensity I, are tabulated below for five plane sound waves. Fill in the gaps in the table taking the acoustic impedance of air as 415 kgm²/s.

	p pascals	U m/s	I W/m²
1	0.01 Pa		
2		0.001 m/s	
3			0.0001 W/m²
4	0.2 Pa		
5		0.002 m/s	
6			0.00002 W/m²

7 By how much will the sound pressure, sound intensity, sound pressure level and sound intensity level produced by a sound source at a certain point increase if its sound power is increased by (a) 5 times, (b) 15 times and (c) 25 times.

8 By how much will the sound intensity, sound pressure level and sound intensity level produced by a sound source at a certain point increase if its sound pressure is increased by (a) 5 times, (b) 15 times and (c) 25 times.

9 What will be the corresponding increases in sound pressure (as a ratio) if the sound pressure level is increased by (a) 10 dB, (b) 20 dB, (c) 30 dB?

10 Calculate the sound pressure levels (re. 2×10^{-5} Pa] corresponding to the following sound pressures: (i) 1 Pa; (ii) 10 Pa; 8.0×10^{-3} Pa; (iv) 4.0×10^{-4} Pa; (v) 2.0×10^{-5} Pa; (vi) 1.0×10^{-5} Pa.

11 Calculate the sound pressure, in pascals, corresponding to the following sound pressure levels: (i) 40 dB; (ii) 37 dB; (iii) 34 dB; (iv) 100 dB; (v) 97 dB; (vi) 80 dB.

12 In a certain factory space the noise level with all machines running is 101 dB. One machine alone produces a level of 99 dB. What would be the level in the factory with all machines running except this one?

13 The sound pressure level of a machine is measured in octave band in the presence of a background noise, which is also measured, with the machine turned off. The results of the measurements are shown below.

Using the process of decibel subtraction calculate the octave band levels corrected for the influence of background noise, and then, using these corrected levels, calculate the A-weighted and unweighted (also called Z-weighted) sound pressure levels produced by the machine.

Octave band (Hz)	Machine on (dB)	Machine off (dB)
125	92	88
250	88	82
500	85	77
1000	78	66
2000	76	63
4000	75	55
8000	70	52

14 The sound pressure levels from five noise sources are measured individually at a point in a workshop and their (logarithmic) average value is 80 dBA. What will the sound pressure level be when all five sources are in operation? If four of the levels were 72 dBA, 74 dBA, 76 dBA and 80 dBA what was the level produced by the fifth source?

15 In underwater acoustic a reference sound pressure of 1×10^{-6} (one micropascal) is used rather than the reference of 20×10^{-6} which is used in water. For any given sound pressure (e.g. 1 Pa), what is the difference in sound pressure levels, in dB, measured relative to these two different values?

16 The constant sound pressure levels from a certain sound source measured in the 500 Hz octave band is 85 dB, and in the 400 Hz and 500 Hz one third octave bands are 78 and 80 dB. What is the sound pressure level in the 630 Hz one third octave band?

17 The constant sound pressure levels from a certain sound source measured in the 1000 Hz octave band is 80 dB. Assuming that the sound energy is spread uniformly throughout the band, what is the sound pressure level in the 800, 1000 and 1250 Hz one third octave bands?

18 Two identical machines are operating in a noisy workshop. The measured A-weighted reverberant sound pressure level is 91.1 dB. When one machine is switched off, the reverberant level drops to 89.5 dB. What would the reverberant level be when both of these machines are switched off? (IOA 2008 (part question))

19 The noise level from a factory with 10 identical machines measured near some residential property was found to

be 54 dB. The maximum permitted is 50 dB at night. How many machines could be used during the night?

20 The A-weighted reverberant sound level in a factory is currently 83 dB. When a new machine was tested in the factory the spectrum below was recorded (with the rest of the factory not operating). Calculate the maximum number of these machines which can be installed in the factory (in addition to the existing plant) if the A-weighted sound level is not to exceed 85 dB.

Octave band (Hz)	63	125	250	500	1000	2000	4000	8000
Octave band SPL (dB)	64	76	73	69	66	64	59	50
A-weighting (dB)	−26	−16	−9	−3	0	1	1	−1

(IOA 1996 (part question))

21 Explain the meaning of the terms reflection diffraction and superposition of sound waves. Discuss the action of:

- reflection in the development of Eyring's theory of reverberation time
- diffraction in noise reduction by barriers
- superposition in resonance in a quarter wave tube.

(IOA 1997)

22 Flue gases from an exhaust fan are delivered to atmosphere via a 5 m long metal chimney. Calculate the first three modes of the chimney assuming (a) that the pipe behaves as an open–closed pipe and (b) as a closed–open pipe. Take the velocity of sound in the gases as 340 m/s at a temperature of 20° Celsius. How will the frequencies change when the gas temperature increases to 100° Celsius?

23 Calculate the frequency of the first three modes of a vibrating string of length 5.0 m, if the velocity of transverse waves moving along the string is 200 m/s. The velocity of waves travelling along the string is proportional to the square root of the tension in the string (assuming that all other variables remain constant). How would the frequencies change if the string tension was doubled?

24 How long must an open–open pipe be in order that its lowest note will be 261 Hz, if the speed of sound in air in the pipe is 340 m/s?

25 What are the upper and lower frequencies of the 500 Hz octave band, and of the 400 Hz, 500 Hz and 630 Hz one third octave bands?

26 A violinist can change the pitch of a string when it is played on an instrument either by turning the appropriate 'tuning' peg or by pressing the string against the finger board, i.e. by 'stopping' the string. Give a brief explanation of the physical mechanisms responsible for changing the pitch in both of these actions. (IOA 2009)

27 (a) For the following explain, with an illustrative example, each term:
(i) refraction and diffraction
(ii) longitudinal and transverse waves
(iii) plane and spherical waves
(iv) progressive and standing waves
(v) coherent and incoherent sources.

(b) (i) Explain how two progressive acoustic waves can combine to produce beats.
(ii) How is the beat frequency related to the frequencies of the individual waves?

(c) The equation of an acoustic wave is $p = p_0 \sin(1000\,\pi t - 9.15x)$ Pa.
(i) What type of wave is it?
(ii) What are the frequency, wavelength and velocity of the wave?
(iii) If the wave strikes an absorbent boundary and 50% of the energy is absorbed and 50% reflected what is the pressure amplitude of the reflected wave as a fraction of the amplitude of the incident wave? (IOA 2006)

28 (a) The pressure variation of a plane wave in air is represented by the equation:

$$p = 0.9\cos(1608t - 4.73x)\ \text{Pa}$$

where the symbols have their usual meaning.
Determine the frequency, wavelength and the L_p (re. 20 μPa) of this wave.

(b) The sound pressure level when two machines A and B operate simultaneously in a noisy factory is 87.1 dB (re. 20 μPa).

If B is turned off, the level falls to 83.1 dB. With B on and A off, the level in the factory is 86.0 dB. Determine the level generated by each machine alone and the level of the background noise. (IOA 1992)

29 (a) (i) Explain the difference between spherical waves and plane waves in acoustics. For each type of wave, give an example of circumstances which approximate to it.
(ii) What is meant by the terms sound pressure, sound power and sound intensity.
(iii) Show how the relation between intensity I and sound power W for a point source on a reflecting plane, $I = W/2\pi r^2$, where r is the distance from the source, leads to

$$L_p = L_W - 20\log r - 8$$

where L_p and L_W are sound pressure level and sound power level respectively.

(b) The pressure variation in pascals at position x m and at time t s for a plane wave in air is given by the expression $p = 0.15\cos(314t - 15.2x)$.
(i) Sketch the corresponding variation in pressure with position at a fixed time. On your sketch

indicate the wavelength and amplitude of the pressure variation. Determine the values of these quantities.

(ii) Sketch the corresponding variation in pressure at a fixed position. Indicate the period of the pressure variation on your sketch. Determine values for the period and frequency of the pressure variation. (IOA 2008)

30 For each of the following pairs of quantities give their definitions and explain any relationships between them:

(a) sound pressure and acoustic particle velocity
(b) sound intensity and sound pressure
(c) sound pressure level and sound intensity level
(d) sound reduction index and level difference
(e) sound pressure level and loudness level.

(IOA 2005 (part (d) relates to Chapter 6))

31 (a) A progressive sound wave is represented by the equation:

$$y = 5 \times 10^{-6}\sin(2000\pi t - (\pi x/0.17))$$

Where x and y are in metres and t is in seconds. Determine the following with respect to the wave:

(i) amplitude;
(ii) frequency;
(iii) velocity.

(b) Another type of wave is represented by:

$$y = 10^{-6}\cos(\pi x/0.17) \times \sin(2000\pi t)$$

(i) Explain how such a wave is produced.
(ii) Show how this equation gives rise to the presence of nodes and antinodes along the wave.

(c) Describe an experimental procedure used to determine the absorption coefficient of a sample in a standing wave tube. (IOA 2009)

Chapter 2 Sound propagation

2.1 Introduction

This chapter builds on those properties of sound waves that determine how sound travels – diffraction, reflection, interference and refraction – which were described in Chapter 1 but starting with the geometric spreading of sound leading to the inverse square law of sound propagation and an introduction to the absorption of sound. These ideas are then applied to sound propagation outdoors. Some of the important features of sound sources and how they radiate sound are also considered. Throughout this chapter free field conditions are assumed, i.e. the effect of reflection from room surfaces indoors are not considered. This will be dealt with in Chapter 5. This chapter concludes with a discussion of the Doppler effect.

2.2 The geometric spreading of sound

Reminder: Sound intensity is sound power per unit area. Therefore for sound power, W, passing through an areas S, the sound intensity, I, is given by:

$$I = W/S$$

If we assume an idealized model of a sound source, i.e. a simple point source which radiates sound energy equally in all directions (i.e. is omni-directional) and also assume free field conditions (i.e. that there are no reflecting or absorbing surfaces nearby), the wavefronts radiated by such a source will be spherical.

The surface area of a sphere of radius, r, is $4\pi r^2$. Therefore using $S = 4\pi r^2$ in the above equation allows us to find the sound intensity at a distance, r, from a point source of power W:

$$I = W/4\pi r^2$$

Example 2.1

What is the acoustic intensity at a distance of 10 metres from an idealized point source of sound power of 1 watt under free field conditions? And what are the values of the corresponding sound intensity level, sound pressure and sound pressure level?

Solution

$$I = W/4\pi r^2 = 1.0/4\pi(10)^2 = 7.95 \times 10^{-4}\ \text{W/m}^2$$

Sound intensity level:

$$L_I = 10\log(1/I_0)$$
$$= 10\log(7.95 \times 10^{-4}/1 \times 10^{-12})$$
$$= 89\ \text{dB re. } 1 \times 10^{-12}\ \text{W/m}^2$$

In order to obtain the sound pressure we shall have to assume that at a distance of 10 m from the source the spherical waves will have flattened out sufficiently to be considered as plane waves, so that (from earlier):

$$I = p^2/z = p^2/(\rho c)$$

Taking the value of ρc (the specific acoustic impedance of air) to be 420 rayls:

$$p^2 = I(\rho c) = 7.95 \times 10^{-4} \times 420 = 0.3339\ \text{Pa}^2$$

from which:

$$p = \sqrt{(0.3339)} = 0.578\ \text{Pa}$$

Sound pressure level:

$$L_p = 20\log(p/p_0) = 20\log(0.578/2.0 \times 10^{-5})$$
$$= 89\ \text{dB re. } 2.0 \times 10^{-5}\ \text{Pa}$$

(This checks out – remember that in a plane wave the numerical values of sound intensity level and sound pressure level are the same, arising from the relationships between the chosen values of I_0 and p_0.)

Inverse square law

The equation $I = W/4\pi r^2$ tells us that the acoustic intensity, I, is inversely proportional to the distance, r, from the source:

$$I \propto 1/r^2$$

Therefore if the distance from the source doubles, the sound intensity reduces to one quarter of its value (because the sound power has been spread over a sphere of four times the area). This is known as the inverse square law, relating sound intensity and distance from source.

This relationship can be used to find the intensity (I_2) at any distance (r_2) provided that the intensity (I_1) at one particular distance (r_1) is known:

$$(I_2/I_1) = (r_1/r_2)^2$$

Example 2.2

Given that the sound intensity at a distance of 10 m from the source in the above example is 7.95×10^{-4} W/m^2, what is the sound intensity at a distance of 25 m?

Solution

$$I_2/(7.95 \times 10^{-4}) = (10/25)^2$$

from which: $I_2 = 1.27 \times 10^{-4}$ W/m^2

How far will it be before the intensity drops to 5×10^{-5} W/m^2?

$$(5.0 \times 10^{-5}/7.95 \times 10^{-4}) = (10/r_2)^2$$

$$0.0629 = 100/(r_2)^2$$

$$(r_2)^2 = 100/0.0629$$

from which: $r_2 = 39.9$ m from the source

The variation of sound pressure with distance from the source

Since $I \propto p^2$ (intensity proportional to pressure squared)

and $I \propto 1/r^2$ (intensity inversely proportional to distance squared)

it follows that $p^2 \propto 1/r^2$

and therefore: $p \propto 1/r$ (sound pressure inversely proportional to distance)

This relationship can be used to find the sound pressure (p_2) at any distance (r_2) provided that the sound pressure (p_1) at one particular distance (r_1) is known:

$$(p_2/p_1) = (r_1/r_2)$$

Example 2.3

Given that the sound pressure at a distance of 10 m from the source in the above example is 0.578 Pa, what is the sound pressure at a distance of 25 m?

Solution

$$p_2/0.578 = 10/25$$

from which: $p_2 = 0.23$ Pa

The '6 dB per doubling of distance' rule

In decibel terms a doubling (or halving) of sound intensity corresponds to an increase (or a reduction) of 6 dB, so that in decibels the inverse square law corresponds to the '6 dB per doubling of distance' rule.

For example, applying this rule to the source in the above example (89 dB at 10 m) we can deduce the sound pressure level at further distances as follows:

89 dB at 10 m
83 dB at 20 m
77 dB at 40 m
71 dB at 80 m
65 dB at 160 m

and so on, until the level falls below that of background noise.

Expressing the inverse square law in decibels

Since on a day to day basis we deal with sound pressure levels, in decibels, rather than sound pressures in Pa it is more convenient to express the above equations in decibel form.

The decibel version of $I = W/4\pi r^2$ is:

$$L_p = L_W - 20\log r - 11$$

Where L_p is the sound pressure level at a distance of r metres from the sound source of sound power level, L_W.

The decibel version of $(p_2/p_1) = (r_1/r_2)$ is:

$$L_2 = L_1 - 20\log(r_2/r_1)$$

Example 2.4

Calculate the sound pressure level generated at a distance of 10 metres from a sound source of sound power level of 120 dB, under free fields condition.

$$L_p = L_W - 20\log r - 11$$

Solution

In this case: $L_W = 120$ dB and $r = 10$

so that: $L_p = 120 - 20\log(10) - 11$

$= 120 - 20 - 11$

$= 89$ dB

Example 2.5

What will be the sound pressure level at a distance of 160 m from the source?

$$L_2 = L_1 - 20\log(r_2/r_1)$$

Solution

In this case: $L_1 = 89$ dB, $r_1 = 10$ m, $r_2 = 160$ m

so that: $L_2 = 89 - 20\log(160/10)$
$= 89 - 24$
$= 65$ dB

which agrees with the earlier result obtained using the '6 dB per doubling of distance' rule.

Derivation of formula

From earlier $I = W/4\pi r^2$

and also $I = p^2/\rho c$

therefore $p^2/\rho c = W/4\pi r^2$

and so $p^2 = W \times (1/r^2) \times [\rho c/4\pi]$

taking $\rho c = 420$ for air: $\rho c/4\pi = 33.4$

and so: $p^2 = W \times (1/4\pi r^2) \times [33.4]$

Note: In what follows always remember that we can do anything to one side of an equation provided that we do exactly the same to the other side.

Multiply both sides by $(p_0)^2$ where $p_0 = 2.0 \times 10^{-5}$ so that $(p_0)^2 = 4 \times 10^{-10}$

$(p/p_0)^2 = W \times (1/r^2) \times [33.4/4 \times 10^{-10}]$

Multiply top and bottom of the right-hand side by 10^{-12}:

$(p/p_0)^2 = (W/10^{-12}) \times (1/r^2)$
$\times [33.4 \times 10^{-12}/4 \times 10^{-10}]$

But $10^{-12} = W_0$

Therefore:

$(p/p_0)^2 = (W/W_0) \times (1/r^2) \times [0.0835]$

Take the log of both sides and multiply by 10:

$10\log(p/p_0)^2 = 10\log(W/W_0) + 10\log(1/r^2)$
$+ 10\log(0.0835)$

Taking each of these terms in turn:

$10\log(p/p_0)^2 = L_p =$ sound pressure level at distance r from source

$10\log(W/W_0) = L_W =$ sound power level of source

$10\log(1/r^2) = -20\log(r)$

$10\log(0.0835) = -10.7 = 11$ (to nearest decibel)

Therefore:

$L_p = L_W - 20\log r - 11$

Note: a slightly simpler derivation, based on sound intensity (I) and sound intensity level (L_I), starts with: $I = W/4\pi r^2$ The next stage is to divide both sides by 10^{-12}, and the end point is the equation: $L_I = L_W - 20\log r - 11$. We can then invoke the assumption that for plane waves $L_I = L_p$, and we have same formula as above. You may wish to fill in the few lines of algebra for yourself.

Further developments of inverse square relationships

So far we have assumed an ideal point source which radiates spherical waves in a free field environment. The application of these ideas to real sound sources involves the introduction of the concepts of near and far fields, and of source directionality. These ideas are developed later in this chapter.

Of course there are other mechanisms, such as air absorption, and ground and atmospheric attenuation which cause the sound level to reduce, as well as wavefront spreading, and these will be discussed later in this chapter.

Rate of change of sound level with distance

The way in which the rate of change of acoustic intensity and sound level with distance is described may be different for different mechanisms.

In some cases the sound pressure level reduces linearly with distance when plotted on a logarithmic scale (see Figure 2.1). The obvious example is the effect of waveform spreading discussed earlier from a point source in a free field. In these cases it is most convenient to express the rate of attenuation with distance in terms of the number of dB reduction per doubling of distance (or alternatively in dB per decade of distance). In the case of the simple theory of a point source in a free field this is a 6 dB per doubling of distance, or 20 dB per decade of distance. In practice the attenuation rate can be greater than this (e.g. 7 or 8 dB per double distance) as a result of the additional contribution from ground attenuation, or it

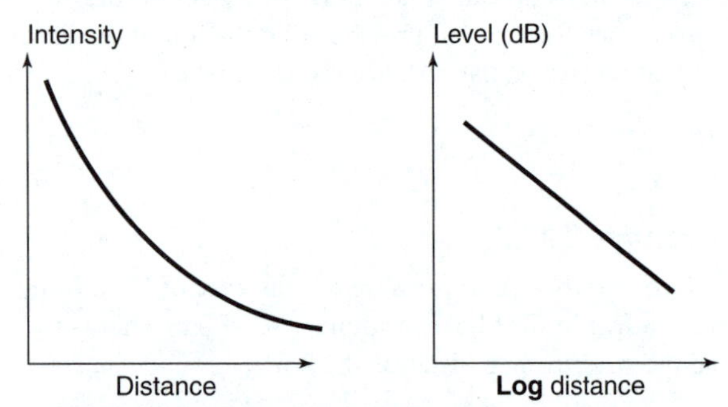

Figure 2.1 Illustrating change in sound intensity and sound level with distance arising from geometrical spreading, i.e. 'inverse square law'. Note linear distance scale (left), and log distance (right). (Curve on the left is of the form Intensity, I, proportional to $(1/r^2)$.)

may be less than 6 dB per double distance, as a result of reflection down a street for example.

In such a case the inverse square law formula may be adapted:

$$L_1 - L_2 = N\log(r_2/r_1)$$

where $N = 20$ represents the theoretical inverse square law case, as above. If the levels L_1 and L_2 at the two distance r_1 and r_2 are known then N may be found, and the levels at other distances predicted, assuming the rate of change with distance remains the same.

Example 2.6

The sound level at distances of 10 and 20 metres from a sound source outdoors is measured and found to be 75 dBA and 66 dBA respectively. Estimate the level at a distance of 60 metres from the source, assuming a constant rate of decrease of sound level with distance.

Solution

In this case $L_1 - L_2 = 75 - 66 = 9$ dB when $r_2/r_1 = 2$, i.e. 9 dB per doubling of distance.

Therefore: $9 = N\log(2/1) = N \times 0.3$ from which $N = 30$.

Therefore $L_1 - L_2 = 30\log(r_2/r_1)$, and so: $75 - L_2 = 30\log(60/10)$, from which $L_2 = 51.7$ dB.

In other cases, such as when sound is passing through a sound absorbing medium, the reduction in sound pressure level is proportional to the distance travelled through the medium, and it is then more convenient to describe the attenuation in terms of dB per metre, or dB per 100 m, or dB per km (see Figure 2.2).

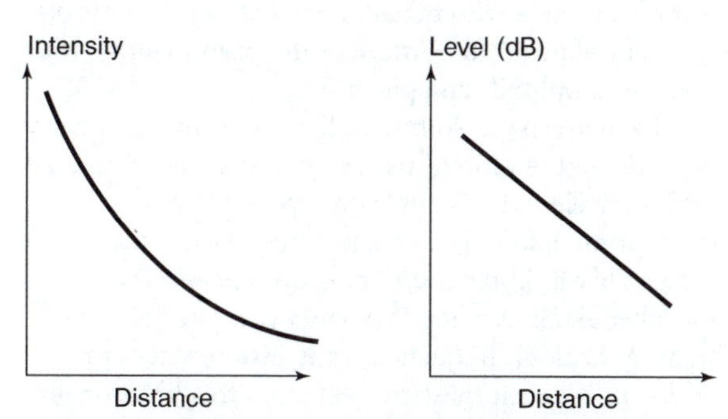

Figure 2.2 Illustrating how sound intensity and sound level vary with distance as a result of absorption, i.e. from sound travelling through a sound absorbing medium. Note that distance is plotted on a linear scale in both cases. (The curve on the left indicates an exponential decay of Intensity (I) with distance (x) travelled into the sound absorbing medium, of the form $I = I_0e^{-kx}$ where k is the attenuation constant for the medium and I_0 is the initial intensity when $x = 0$. When intensity is converted to level in decibels a straight line relationship occurs (right-hand side))

Because of this, care is needed when attempting to estimate variation of sound pressure level with distance when both of these two phenomena (distance attenuation and sound absorption) are involved.

2.3 Monopole, dipole and vibrating piston sources

So far we have met two idealized sorts of waves: plane waves which transmit sound energy only in one direction and spherical waves which transmit sound equally in all directions. What sound sources produce these waves? And what waves are produced by real, everyday, sound sources?

Spherical wave radiator

Imagine a spherical surface which is vibrating radially, i.e. while always maintaining its spherical shape its radius is periodically increasing and decreasing, rather like a pulsating spherical balloon. Such a source would radiate spherical waves into the medium surrounding it, and is called, variously, a monopole source, a simple source or a point source.

The action of the pulsating sphere on the surrounding medium is to create local density and pressure fluctuations (i.e. compressions and rarefactions) that give rise to a spherically symmetrical propagating sound wave.

Although, as explained below, a single cabinet loudspeaker may approximate to a spherical source over a certain frequency range, dodecahedron sources are available involving 12 identical loudspeakers arranged in a spherical array which give a better approximation to an ideal monopole source. If the monopole sound source is in a free field with no reflecting surfaces, the sound pressure level falls at 6 dB per doubling of distance.

Plane wave radiator

Plane waves are radiated by vibrating plane surfaces provided that they are large enough and vibrate in a uniform manner. Each point on the vibrating surface acts as a source of spherical waves (rather like Huygens secondary wavelets theory described in Chapter 1, when discussing diffraction). If all points on the vibrating surface are vibrating with the same amplitude and phase then the waves radiated from each point on the surface will combine to form plane waves radiating from the surface. Strictly speaking, in order for perfectly plane waves to be produced the vibrating surface must be infinite in extent and also rigid. If it is not infinite in extent then spherical waves will be produced at the edges of the vibrating surface and if it is not rigid the vibration amplitude may vary in amplitude and phase at different points on the

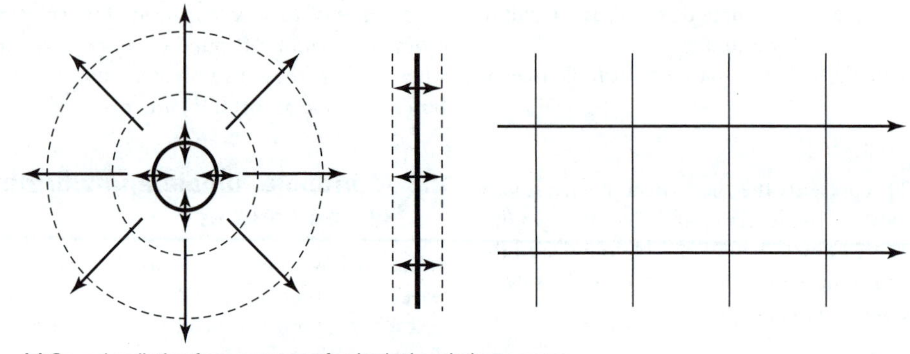

(a) Sound radiation from sources of spherical and plane waves

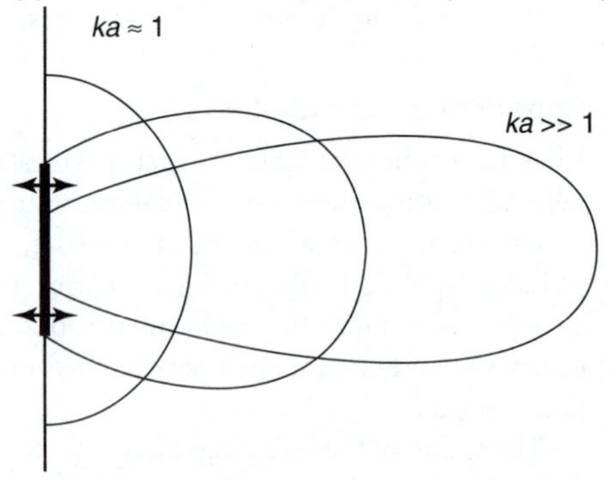

(b) Sound radiation from a piston source of radius a for values of ka of approximately 1, 5 and 10

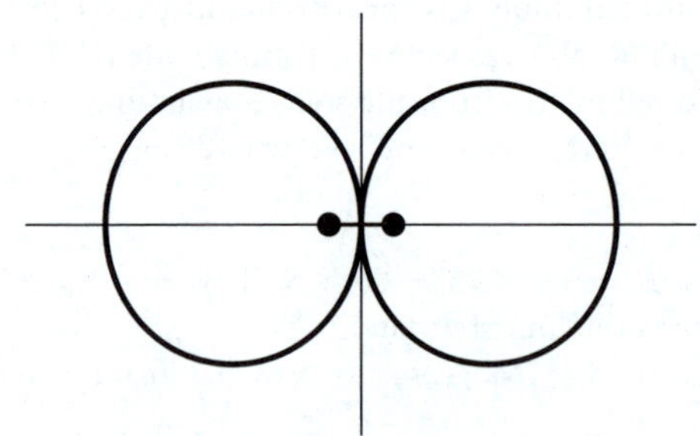

(c) Sound radiation from a dipole source

Figure 2.3 Radiation from sound sources

surface, and the individual wavelets will not combine to produce plane waves.

In practice, if the surface is large compared to the wavelength of sound then it is likely to approximate to a plane wave source, and will radiate sound predominantly in one direction.

Real sources

Perfectly plane waves and perfectly spherical waves are idealized models of sound wave propagation, and so are the sources which produce them (Figure 2.3). In practice the laws of diffraction tell us that any vibrating surface of dimensions, much less than the wavelength it is radiating, will approximate to a simple source of sound, and radiate approximately spherical waves; and far enough away from such a source the waves will behave like plane waves. Conversely any vibrating surface that is large compared to the wavelength will radiate predominantly in one direction and will approximate to a plane sound wave radiator.

Piston source

A source model which gives a closer approximation to the behaviour of real sources, and is intermediate between the simple source of spherical waves and the plane wave generator is the plane vibrating piston source (see again Figure 2.3). The usual version of this model is a rigid circular piston vibrating in a direction along its axis. The piston is set in a sound reflecting wall (called an infinite rigid baffle in some textbooks) so that the sound from the piston can only be radiated in the forward direction, i.e. there is no radiation from the back of the piston. All points on the surface of the piston vibrate with the same amplitude and phase.

The pattern of sound radiation from the piston depends on the ratio of its size (diameter, D) compared to the wavelength (i.e. on D/λ). When $D/\lambda \ll 1$ (e.g. small piston and large λ, i.e. low frequencies) the piston behaves like a simple source radiating spherical waves. At the other extreme, when $D/\lambda \gg 1$ (e.g. large piston and small λ, i.e. high frequencies) the piston radiates plane waves. In between these two extremes the directionality of the piston sources increases with frequency from nearly omni-directional at low frequencies to highly directional at high frequencies.

Dipole source

A dipole source consists of two identical simple sources separated by a short distance and operating half a cycle out of phase with each other, so that when one of them is sending out compressions the other is sending out rarefactions. The line joining the two sources defines the axis of the dipole (see again Figure 2.3).

The sound pressure at any position surrounding the dipole is obtained by summation of the individual sound pressures produced by each source, according to the principle of superposition. Generally the resulting sound field from the dipole is highly directional and there is no sound radiation in the direction at right angles to the axis, because of destructive interference. Overall the dipole source is a less efficient sound radiator than the monopole and the sound pressure levels fall off faster than the 6 dB per doubling of distance rule for the ideal monopole source.

A loudspeaker cone, i.e. a loudspeaker taken out of its cabinet, is a good approximation to a dipole source because when the front of the cone is sending out compressions the back of the cone is radiating rarefactions. The piston source discussed above, without its wall or baffle, would also be a dipole. The main purpose of the loudspeaker cabinet is to enclose the loudspeakers so that radiation from the back of the cone is prevented from interfering with the radiation from the front of the cone, so that with the enclosure the overall directionality and efficiency of sound radiation is improved.

There are also quadrupole sources and other higher n-pole sources, which are more directional and less efficient radiators than the dipole sources. These types of sources can feature in sound radiation from turbulent fluid flow (i.e. aerodynamic or hydrodynamic noise).

2.4 Near and far acoustic fields of sound sources

We can apply the inverses square law to predict the sound intensity, I, and sound pressure level, L_p, at different distances, r, from a point source, of sound power W (or sound power level L_W) in a free field (i.e. no reflecting surfaces) using one of the following:

$$I = W/(4\pi r^2)$$

$$L_p = L_W - 20\log r - 11$$

If the sound power or sound power level of the source was unknown we could predict the sound intensity and sound pressure level at some remote distance from the source, r_2, by using measured values taken closer to the source (at distance r_1) and again, applying the inverse square law:

$$I_2 = I_1(r_2/r_1)^2$$

$$L_2 = L_1 - 20\log(r_2/r_1)$$

Note: the '6 dB per doubling of distance' rule is embodied in these equations.

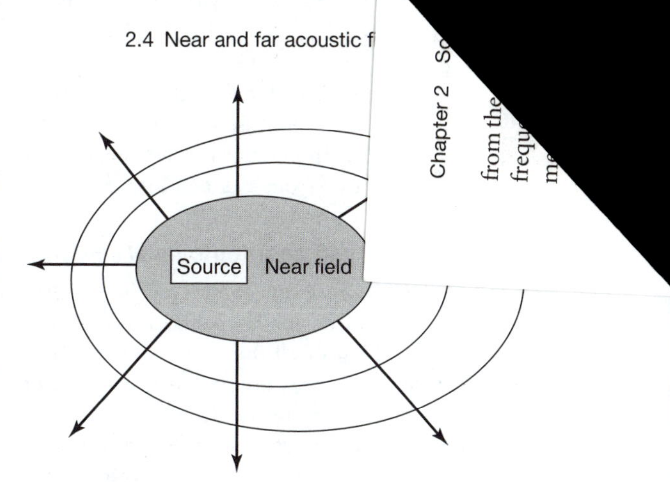

Figure 2.4 Near and far field regions of a sound source

Can these useful predictive methods be applied to real sources and not just to idealized point sources? The answer is yes, provided that the distance from the source is far enough away to be in the 'far field' of the sound source, and provided also that we take into account any possible directionality of the source.

Near and far field of sound sources

Close to the source the sound waves originating from all points on the source can combine to produce complicated patterns of sound. The sound pressure may vary with position and distance from the source in a complicated way and be difficult to predict. This area is called the near field of the source.

On moving away a sufficient distance from the source the sound field tends to settle down into a succession of smooth wavefronts and the intensity falls off in accordance with the inverse square law (and the '6 dB per doubling of distance' rule). This is called the far field of the sound source (see Figure 2.4).

The important practical consequences of this are that:

- The simple prediction methods based on the inverse square may be extended to real sources, but only for far field positions.
- Only sound pressure levels measured in the far field may be used to predict levels at other far field distances.
- Sound power levels may only be used to predict sound pressure levels at far field positions, but not in the near field.
- It will be necessary to take into account the possible directivity of the sound source when making predictions.

The extent of the near field region

In general, the extent of the near field cannot be defined precisely, but ideally, to be in the far field the distance

...source should be several wavelengths of the lowest ...ency sound considered and also several times the di...ension of the source. (In practice at least one or two wavelengths and one or two source dimensions away.)

The hydrodynamic and geometric near fields

There are two reasons which can account for the complex variations of sound pressure levels which may occur close to a sound source. First of all the sound pressure and acoustical particle velocity may not be in phase, as they in the far field. This gives rise to the hydrodynamic near field. Secondly, interference may occur between sound waves arriving from different parts of the source. This gives rise to the geometric near field. Further details of these effects maybe found in texts such as that by Bies and Hansen, if required.

2.5 Directivity of sound sources

The directivity pattern of an ideal point source is circular, i.e. representing spherical sound radiation, equal in all directions. Any real i.e. non-ideal sound source will have a non-circular directivity pattern associated with its radiation. In practice this means that the sound intensity and sound pressure level at the same distance may be different in different directions.

The directivity (also called directionality) of the radiation of a source may be quantified in two ways, in terms of either a directivity factor or a directivity index.

The directivity factor

The directivity factor, Q, is the ratio of the sound intensity in a given direction measured at a certain distance from the source divided by the average sound intensity (i.e. averaged over all directions) at the same distance and is defined as:

$$Q = I/I_{avge} = (p/p_{avge})^2$$

The average values I_{avge} and p^2_{avge} are the values which would be obtained from an imaginary omni-directional source emitting the same sound power.

The directivity factor is a ratio, so that a Q of 0.5 in a certain direction means that the intensity (or p^2 value) in that direction is half of the average value, i.e. half of what would occur in that direction from an omni-directional source of the same sound power. Similarly a Q value of 2 would indicate twice the sound intensity in a certain direction as compared to the average. It follows that over all directions the various Q values must average to one.

The directivity index

The second approach to quantifying directivity involves the introduction of the directivity index D. This is expressed in decibels and is the difference between the sound pressure level, L_θ in a certain direction (defined by the angle θ) measured at a certain distance from the source and the average sound pressure level L_{avge} which would be produced at the same distance by a notional omni-directional source of the same sound power:

$$D = L_\theta - L_{avge}$$

The value of D may be positive or negative so for example a value of $D = 3$ dB means that in a certain direction the sound pressure level is 3 dB higher than it is on average over all directions at the same distance, whereas a value of $D = -2$ dB means 2 dB less in a certain direction than on average. Note that L_{avge} must be obtained from the logarithmic average over levels over all directions.

The two terms are simply related:

$$D = 10\log Q \quad \text{and} \quad Q = 10^{D/10}$$

So that for example a value of $Q = 2$ corresponds to a value of $D = 3$ dB, and $Q = 0.25$ to a value of $D = -6$ dB.

Example 2.7

Measurements are made of the sound pressure levels at positions around an electric drill in an anechoic room (a special test room with completely sound absorbing surfaces, which therefore provides free field test conditions). Six measurement positions are selected, each at 2 m from the drill: four in the horizontal plane of the drill (positioned in the centre of the room) at the compass points (N, S, E and W) and directly above and below the drill (A and B).

The sound pressure levels (all in the same octave band) are: N 80 dB, S 78 dB, E 76 dB, W 82 dB, A 74 dB and B 84 dB.

Calculate the directivity index in each direction. Also calculate the sound power level of the machine.

Solution

First it is necessary to calculate the logarithmic average of the six sound pressure levels:

$$L_{avge} = 10\log[(10^{7.4} + 10^{7.6} + 10^{7.8} + 10^{8.0} + 10^{8.2} + 10^{8.4})/6] = 80.3 \text{ dB}$$

In the direction N: $D = 80 - 80.3 = -0.3$ dB, and $Q = 10^{-0.3/10} = 0.93$.

Similarly for all the other directions:

	D	*Q*
N	−0.3 dB	0.93
S	−2.3 dB	0.59
E	−4.3 dB	0.37
W	1.7 dB	1.48
A	−6.3 dB	0.23
B	3.7 dB	2.3

The sound power level may be found using the average sound pressure at 2 m and the inverse square law formula:

$80.3 = L_W - 20\log(2) - 11$

from which: $L_W = 97.3$ dB re. 10^{-12} W

Note: The choice of an electric drill in the above example was consistent with the requirement that ideally the sound pressure level measurements should be in the far field of the source. The predominant octave bands for such as source are usually the high frequency ones of 2000 Hz and 4000 Hz, and the sound wavelengths in air at these frequencies are small enough for the measurement distance of 2 m to meet the far field conditions of being both several wavelengths and several machine dimensions away from the source.

The example has illustrated how both sound power levels and directivity indices of small sound sources may be determined.

The standard methods for determining sound power levels will be discussed in later chapters on measurement and on noise control, but the above example illustrates the principle of one of the methods, the free field method described in ISO 3745:2003.

The standard requires that spatial averaging of sound pressure levels around the source should be done in such a way as to ensure that each measurement point represents a constant area on the measurement sphere. For this purpose recommended sets of Cartesian (i.e. *x*, *y*, *z*) co-ordinates are available of the points for such measurements. An example set of relative co-ordinates is given in Table 2.1.

Table 2.1 Example of co-ordinates for measurements around a sphere to achieve spatial averaging, extracted from ISO 3745:2003

No.	$\frac{x}{r}$	$\frac{y}{r}$	$\frac{z}{r}$
1	−0.99	0	0.15
2	0.50	−0.86	0.15
3	0.50	0.86	0.15
4	−0.45	0.77	0.45
5	−0.45	−0.77	0.45
6	0.89	0	0.45
7	0.33	0.57	0.75
8	0.66	0	0.75
9	0.33	−0.57	0.75
10	0	0	1.0
11	0.99	0	−0.15
12	−0.50	0.86	−0.15
13	−0.50	−0.86	−0.15
14	0.45	−0.77	−0.45
15	0.45	0.77	−0.45
16	−0.89	0	−0.45
17	−0.33	−0.57	−0.75
18	0.66	0	−0.75
19	−0.33	0.57	−0.75
20	0	0	−1.0

An omni-directional source near reflecting planes

A simple directional source can be created by placing an omni-directional source on a hard flat surface which acts as a reflecting plane (see Figure 2.5).

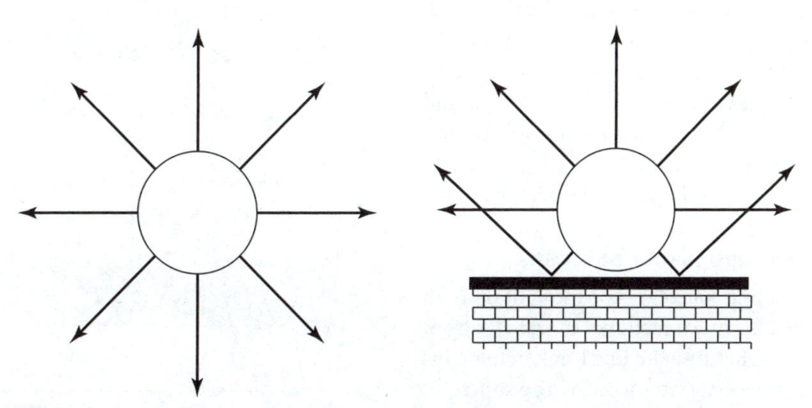

Figure 2.5 Sound sources in the free field and above a reflecting surface

Table 2.2 Directivity associated with source position

Source position outdoors	Source position indoors	Q	D
Free field	Source at centre of room	1	0
Source adjacent to one reflecting plane (ground or vertical wall)	Centre of wall, floor or ceiling	2	3 dB
Source at junction of two reflecting planes (wall and ground)	Half way along junction of wall and floor (or wall and ceiling, or wall and wall)	4	6 dB
Source at junction of three reflecting planes (e.g. corner of courtyard)	Corner of room	8	9 dB

Clearly the energy that would normally propagate downwards is reflected upwards resulting in doubling of the sound intensity in the upper hemisphere. Hence the directivity factor for the source in this location is 2 in all directions above the plane and it follows that the directivity index is 3 dB. This means the sound pressure level for the source will be 3 dB higher when it is close to the reflecting surface than it would be if it was in an open space with no reflecting surfaces nearby.

For a similar source placed at the junction of a wall and a floor, the directivity factor becomes 4 and the directivity index is 6 dB. In the corner formed between two walls and a floor, the values become 8 and 9 dB respectively.

The effect of nearby reflecting surfaces on source directivity is summarized in Table 2.2.

2.6 The prediction of sound levels from real sound sources

Having considered the effects of near and far field of real sound sources and their directionality, it is easily possible to adapt the simple point source model of prediction based on the inverse square law to real sound sources. This is achieved simply by including the directivity index of the source in the equation:

$$L_p = L_W - 20\log r - 11 + D$$

and remembering that only far field sound levels may be predicted in this way.

Later in this chapter in the section on outdoor sound propagation this formula will be extended to include sound propagation effects other than attenuation due to distance, but still assuming free field conditions. The effect of reflection on sound indoors will be discussed in Chapter 5.

Extension to linear and planar sources

The simplest form of line source can be thought of as an infinitely long chain of omni-directional sources far from any reflecting surfaces. Because the line is taken to be infinite in length, it is necessary to describe the source in terms of the sound power per unit length. The wavefronts radiated in the far field of such a source will clearly be cylindrical (see Figure 2.6) rather than the spherical waves from a point source.

Since a cylinder of radius r metres will have an area of $2\pi r$ for each metre of length, and considering how the wavefront from a line source spreads (i.e. increases in area and therefore decreases in intensity) with distance, it follows that a doubling of distance leads to the sound intensity being reduced by a factor of 2 (compared with 4 in the case of spherical spreading), which corresponds to a decrease of 3 dB per doubling of distance (compared with 6 dB per doubling of distance for spherical waves). The variation of sound intensity and sound pressure level with distance are given by the following equations where W_L and L_{WL} represent sound power per unit length and sound power level per unit length:

$$I = W/2\pi r \quad \text{and} \quad Lp = L_{WL} - 10\log r - 8$$

$$\text{and} \quad L_2 - L_1 = 10\log(r_2/r_1)$$

The sound radiating from freely flowing traffic moving on a road (i.e. on an acoustically reflecting plane) produces a good approximation to cylindrical spreading.

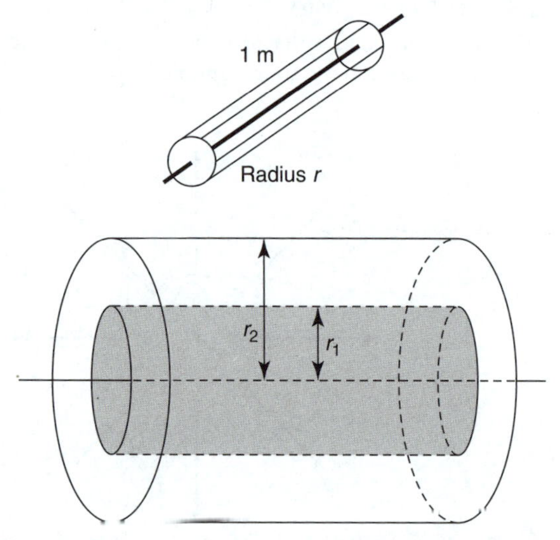

Figure 2.6 Line source with cylindrical spreading

Sound radiation from plane surfaces

A plane source such as a large open aperture in the wall of a factory produces plane waves in its immediate vicinity. Over the region close to the source there is no beam spreading and the radiated sound level remains more or less constant. At great distances the wavefronts become spherical and thus the sound pressure level falls at 6 dB per doubling of distance.

In between these two situations, and determined by the relationship between the dimensions of the plane and the wavelength of the sound, there is a region where the reduction may approximate to the 3 dB per doubling of distance of a line source.

A fuller treatment, sometimes attributed to Rathe, of a finite planar source of height a and length $b\,(b > a)$ identifies three zones with a rate of fall 0, 3 and 6 dB per doubling of distance respectively as shown in Figure 2.7.

The variation at great distance is at least understandable, since from such distances the source looks small. The middle section also makes sense in that diffraction is already having an effect by spreading the beam over an expanding surface from a/π onwards where a is the smaller of the two dimensions.

Example 2.8

Noise from inside a factory building is being radiated through its façade. The average sound pressure level measured externally and close to the façade, of dimensions 15 m × 4 m is 70 dB in a certain octave band. Estimate the sound pressure level in that band radiated from the façade at positions opposite the centre of the façade, at distances of 1 m, 2 m, 4 m, 8 m and 100 m away.

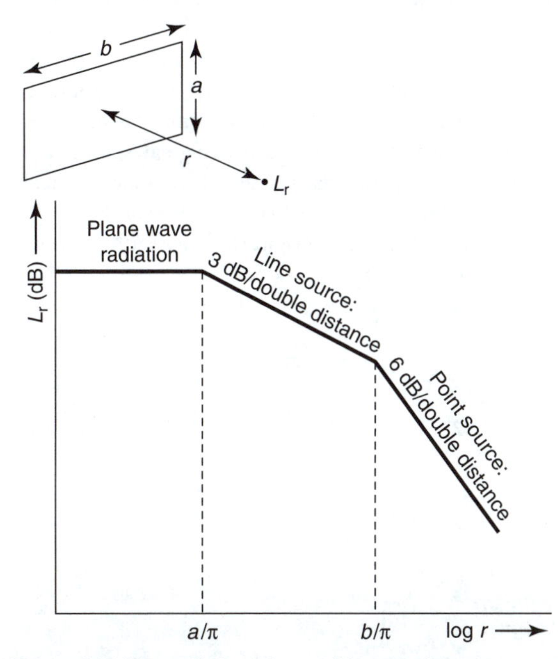

Figure 2.7 The radiation of sound from a plane area: variation of sound pressure level with distance (Rathe method)

Solution

$a = 4/\pi = 1.3$ m and $b = 15/\pi = 4.8$ m

Therefore, according to the Rathe method, at distances up to 1.3 m there is plane wave radiation with no attenuation with distance; between 1.3 m and 4.8 m the level reduces at a rate of 3 dB per doubling of distance; and beyond 4.8 m the level reduces at a rate of 6 dB per double distance:

- Level at 1 m = 70 dB
- Level at 2 m = 70 − 10log(2/1.3) = 68.1 dB
- Level at 4.8 m = 70 − 10log(4.8/1.3) = 64.3 dB

Beyond 4.8 m there is point source radiation:

- Level at 8 m = 64.3 − 20log(8/4.8) = 59.9 dB
- Level at 100 m = 64.3 − 20log(100/4.8) = 37.9 dB

Alternative method

An alternative method is available but only for distances $r \gg b$. The method assumes that at large distances from the façade the façade can be considered to behave like a point source radiator. This is only valid for distances r very much greater than the largest façade dimension, i.e. $r \gg b$.

The calculation has two stages.

Stage 1

Estimate the sound power level radiated by the façade using:

$$L_W = L_p + 10\log S$$

where L_p is the sound pressure level at the façade and S is the surface area of the façade.

$$L_W = 70 + 10\log(15 \times 4) = 87.8 \text{ dB}$$

Stage 2

Calculate the sound pressure level at distance r assuming point source and inverse square law propagation:

At 100 m:

$$L_p = 87.8 - 20\log(100) - 11 = 36.8 \text{ dB}$$

This compares with 37.9 dB, about a 1 dB difference, using the Rathe method; i.e. an acceptable level of agreement bearing in mind that both methods are approximate ones.

Summary: types of sound sources

We have discussed two important features of sound sources; their near field and far field regions, and their directivity. Another important characteristic, their acoustic radiation efficiency, will be discussed in Chapter 7, on vibration.

2.7 Outdoor sound propagation

There are several effects which modify propagation in an outdoor situation and they are often collectively referred to as aspects of excess attenuation – that is to say excess over beam spreading or geometrical attenuation. These can be classified into those associated with absorption either by the body of air though which the sound passes or as a result of interaction of the sound waves with the terrain over which the sound travels, with obstacles such as topographical or man-made barriers, those associated with meteorological effects (i.e. temperature and wind velocity gradients) and those associated with vegetation.

All the effects of the above influences are summed into a single excess attenuation term to be added to the general equation for the far field sound pressure level from a directional source under conditions of unobstructed propagation:

$$L_p = L_W - 20\log r - 11 + D - A_{excess}$$

where A_{excess} is the total effect of atmospheric and environmental propagation effects.

The algorithms for the various propagation effects can be quite complex and are continually being refined in environmental noise modelling packages. The following sections give general guidance of the effects.

2.8 Absorption of sound in air

Under normal circumstances the absorption of sound by air is not considered very important when compared with other factors involved in sound propagation such as the spreading of the sound wavefront, ground attenuation, refraction and scattering in the atmosphere, diffraction by barriers etc. Air absorption only becomes significant when considering propagation outdoors over long distances (more than a few hundred metres), and particularly when high frequency sound is involved. Air absorption is much more important at ultrasonic frequencies and is the reason why many technical devices (burglar alarms etc.) employing ultrasonic beams have a limited range in air. Air absorption of sound energy only becomes significant in sound propagation indoors at the very highest frequencies and for the very largest spaces such as theatres and concert halls. Usually it is negligible compared with the absorption which occurs when the sound strikes a surface such as a wall or floor.

An approximate expression for the excess attenuation given by air absorption is given in ISO 9613 *Acoustics – Attenuation of sound during propagation outdoors – Part 2: A general method of calculation.* At a temperature of 20° Celsius and a relative humidity of 50% the predicted attenuation is:

- 0.5 dB/km at 500 Hz
- 1.5 dB/km at 1 kHz
- 6 dB/km at 4 kHz.

2.9 Attenuation from propagation close to the ground

When source and receiver are fairly close to the ground there is an interaction between the sound wave travelling directly from source to receiver and the sound wave arriving at the receiver after reflection with the ground (see Figure 2.8). One way of analysing sound propagation situations like this is by introducing an image source S_i. The strength of this image source is determined by the reflection coefficient of the surface which is dependent upon the angle of incidence and the impedance of the ground. Depending on the type of ground surface and sound frequency a change of phase change may occur on reflection, and attenuation can arise as a result of partial interference between the direct and ground reflected waves. Ground attenuation is usually higher for soft ground (grassland, vegetation etc.) than for hard ground, and greatest for a range of mid-frequencies of a few hundred hertz than for higher and lower frequencies. The situation is also complicated by interaction with atmospheric turbulence with the direct and reflected sound propagation.

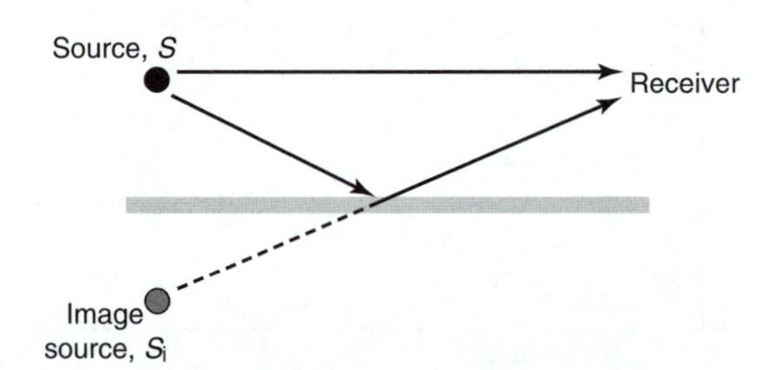

Figure 2.8 Direct and ground reflected sound transmission paths for a sound source outdoors

2.10 Refraction of sound in the atmosphere

Refraction is the change of direction of a sound wave which occurs when sound moves from one medium to another medium with a different sound velocity. In the atmosphere there are sound velocity gradients, i.e. changes in sound speed with height above the ground caused by changes in air temperature (temperature gradients) and wind speed gradients, and by combinations of both.

Temperature gradients

When sound travels from one medium to another medium with a slower sound speed, a sound ray (denoting the direction of travel of the sound wave) bends towards the 'normal' (the perpendicular to the interface between the two media). This corresponds to the case of sound moving from a layer of warm air in the atmosphere to a layer of cooler air (see Figure 2.9). (This is the same direction as the bending of light when moving from air into glass.)

In the atmosphere the change in temperature with height is gradual but continuous rather than occurring in discrete layers, with the result that sound rays are gradually bent into a curved path (see Figure 2.10).

Normally, during daytime, air temperatures fall with increasing height above the surface of the earth: this is called temperature lapse. The changes in the temperature of the layers in the atmosphere cause sound rays from an omni-directional source to bend upwards creating a partial acoustic shadow zone (see Figure 2.11). During the night-time and occasionally during the day, temperature inversions occur with the air near the ground colder and the propagation paths bend towards the ground (see Figure 2.12). Over large distances, this can give rise to variations in sound level of up to 20 dBA.

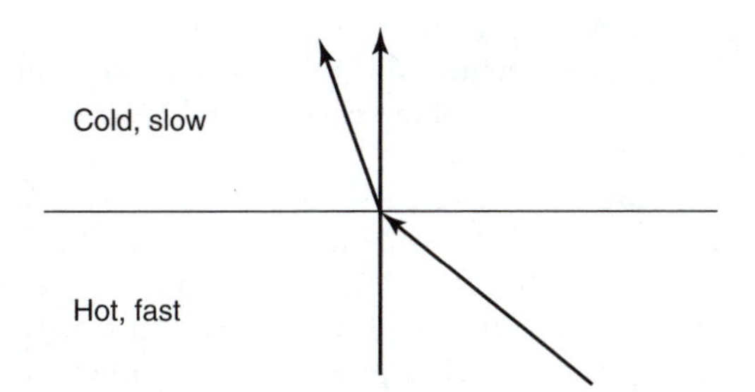

Figure 2.9 Change in direction of sound wave between warm and cold air due to the change in speed of sound

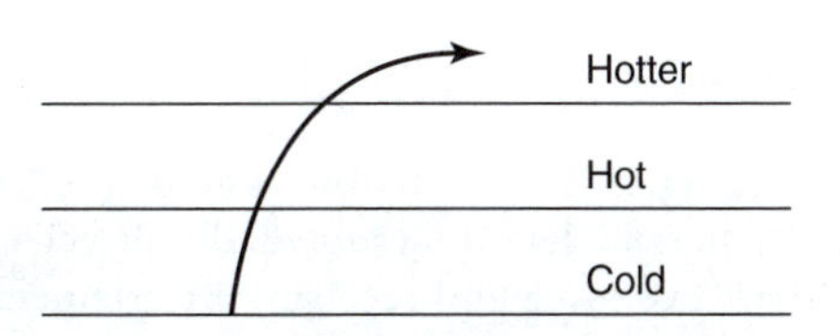

Figure 2.10 Bending of sound wave in a layered medium due to the change in speed of sound

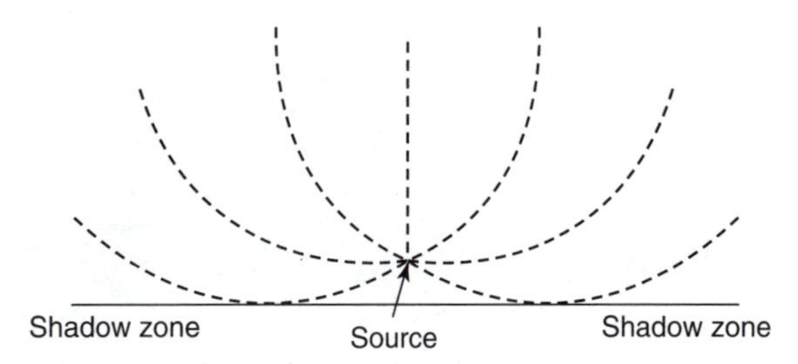

Figure 2.11 Bending of sound wave under daytime conditions of temperature variation above ground

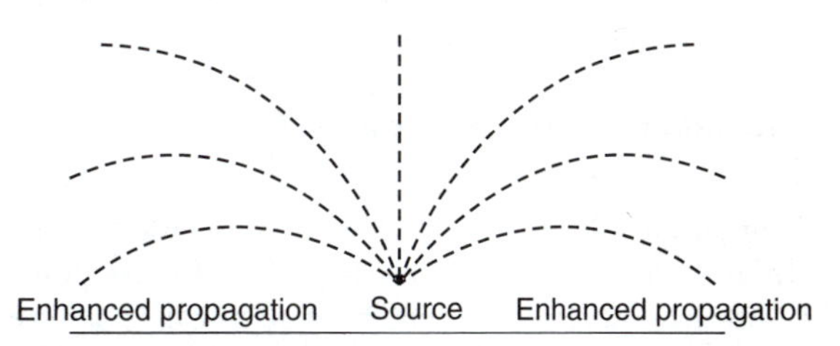

Figure 2.12 Bending of sound wave under night-time or temperature inversion conditions

Wind effects

It is well known that sound levels are usually greater upwind than downwind from the source. This effect is not because the sound is blown along by the wind but is a consequence of the gradient of wind speed above the ground surface. The velocity of the wind increases with distance from the ground. As with temperature gradients discussed above, this wind speed gradient leads to a change in direction of the sound wavefronts with the apparent effect of a 'bending' of the wavefronts (see Figure 2.13). In the direction of the wind the sound waves from the source are 'bent downwards, towards the ground' leading to higher noise levels in that direction (see Figure 2.13). In the opposite direction the sound waves are 'bent upwards, towards the sky' and lower noise levels are observed.

Under some circumstances a partial acoustic shadow zone can be observed in the upwind direction. When the

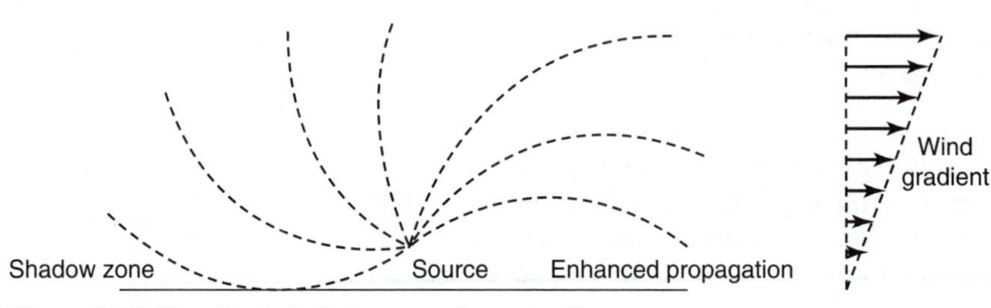

Figure 2.13 The effect of wind on sound propagation

wind velocity exceeds about 3 m/s, wind effects usually dominate over temperature effects.

Anomalous sound propagation in the atmosphere

A simple introduction to refraction of sound waves in the atmosphere has been given above, but these various effects can occasionally combine to form what are sometimes known as 'anomalous sound propagation conditions'. Under these circumstances sound may become trapped in atmospheric refraction channels and travel for very long distances with very little attenuation before returning to earth, while the sound may not be heard at much shorter intermediate distances from the source. There are many famous examples in history of such occurrences featuring single very loud events such as from explosions and from military battles.

One such incident which occurred in 1666 was recorded in the diaries of John Evelyn and Samuel Pepys. During a military engagement of the English and Dutch fleets in the English Channel on 1 June the sounds of the guns were heard in London, but not on the South Downs, Deal or Dover, nor at any points between the battle and London. An internet search under 'anomalous sound propagation in the atmosphere' (or similar) will reveal many more recent examples, such as explosions at munitions factories. The American acoustician Charles D. Ross has written on how the effects of atmospheric refraction of sound may have influenced many battles in the American civil war.

Refraction of sound in the oceans

Refraction of sound waves in the oceans also occurs and can result in sound propagation channels between certain ocean depths and for certain sound frequencies. These channels are much more stable (and often almost permanent) than those which occur in the atmosphere. Sound which is trapped in such channels does not suffer the normal dispersion with distance which usually occurs, and this can result in sound being carried for very long distances with relatively little attenuation. Such channels are used by whales in the oceans to communicate over long distances.

2.11 Barriers

The effectiveness of a barrier in reducing sound transmission between source and receiver is limited by diffraction of sound over the top and around the sides of the barrier, which is determined by the size of the barrier compared with wavelength of the sound. The attenuation provided by the barrier (also known as the insertion loss) is the reduction in noise level at the receiver arising from the noise source as a result of the presence of the barrier. The usual basis of prediction is that the sound transmission through the barrier is negligible and can be ignored compared with diffraction around the edges of the barrier. There are a variety of methods for predicting the barrier attenuation arising from diffraction, but one of the simplest, described below for a long thin barrier, developed by Maekawa, is based on path difference.

The path difference, δ, i.e. the additional sound path for the rays travelling over a simple barrier (see Figure 2.14), is given by $\delta = (a + b) - c$.

The wavelength, λ, of the incident sound is taken into consideration using the concept of the Fresnel Number which is defined by $N = 2\delta/\lambda$. There are two versions of the formula for the predicted barrier attenuation depending on the receiver position. If the receiver is in the so called 'illuminated zone' so that there is direct line of sight of the sound source from the receiver position then the barrier attenuation, A_{barrier}, is:

$$\text{Attenuation} = 10\log(3 - 20N)\ \text{dB}$$

If the receiver is in the shadow zone of the barrier, i.e. if the barrier has interrupted the direct line of sight between source and receiver, the attenuation is given by:

$$\text{Attenuation} = 10\log(3 + 20N)\ \text{dB}$$

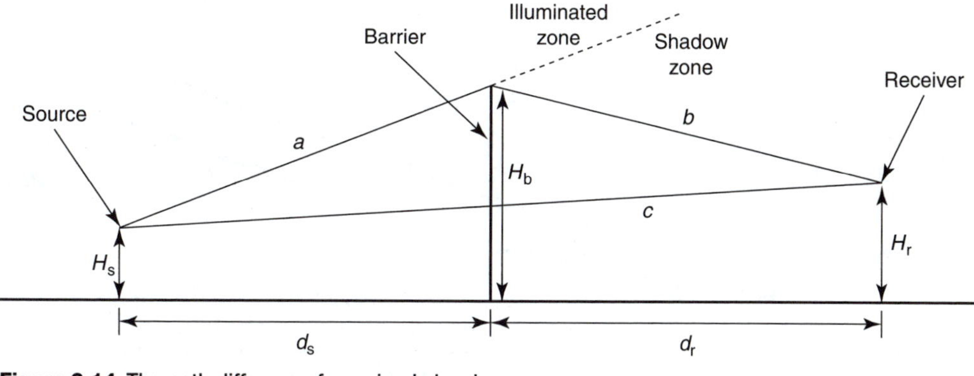

Figure 2.14 The path difference for a simple barrier

For the simple case shown in Figure 2.14 the demarcation between shadow and illuminated zones is a continuation of a line joining source and the top of the barrier, so that the receiver in Figure 2.14 is in the shadow zone. If the receiver position is on the line of sight, so that δ, and therefore N in the above formulae, are both zero then the predicted barrier attenuation is 5 dB, with predicted attenuations being lower than this for receiver positions in the illuminated zone, and higher in the shadow zone.

These equations are consistent with the prediction from diffraction theory in that the greatest attenuation is achieved for a given barrier with sound components of the highest frequency, that is to say the smaller wavelength, i.e. the higher frequency.

In practice the attenuation achieved with a barrier will be less than that determined from the above equations due to a combination of all the effects associated with propagation of sound as discussed above.

It is important to appreciate that the maximum attenuation obtained with a practical barrier is about 15 to 20 dBA for very high barriers or barriers very close to either the source or receiver. This means that the attenuation of sound through the barrier should be at least 10 dB better than this, in order that the direct transmission should be negligible. Sound transmission through barriers will be discussed in Chapter 6, but based on the 'mass law' of sound insulation it is generally considered that a barrier of surface density of around 15 kg/m^2 will be sufficient to achieve this, and that making the barrier heavier than this will not achieve any further attenuation (unless the barrier is also made longer and higher).

Example 2.9

Estimate the reduction for a source with the octave band spectrum shown below, if a barrier is installed:

- height of the source is 0.5 m
- height of receiver is 1.2 m
- height of the barrier is 3 m
- source to barrier is 10 m
- barrier to receiver is 20 m.

Solution

The first step is to determine the path length difference from the dimensions given, by using Pythagoras' theorem for three right-angled triangles with a, b and c as hypoteneuse (see Figure 2.14):

$$a = [(H_b - H_s)^2 + (d_s)^2]^{0.5} = [(3 - 0.5)^2 + 10^2]^{0.5}$$

$$b = [(H_b - H_r)^2 + (d_r)^2]^{0.5} = [(3 - 1.2)^2 + 20^2]^{0.5}$$

$$c = [(H_r - H_s)^2 + (d_s + d_r)^2]^{0.5}$$
$$= [(1.2 - 0.5)^2 + (10 + 20)^2]^{0.5}$$

This leads to a = 10.3 m, b = 20.1 m and c = 30 m.

Therefore path difference $\delta = (a + b) - c = 0.38$ m.

The next step is to determine the attenuation from the barrier, as tabulated below.

Freq. (Hz)	63	125	250	500	1 K	2 K	4 K	8 K
$N = 2\delta/\lambda$	0.14	0.27	0.54	1.08	2.16	4.32	8.64	17.29
$10\log(3 + 20N)$	7.6	9.2	11.4	13.9	16.7	19.5	22.5	25.4

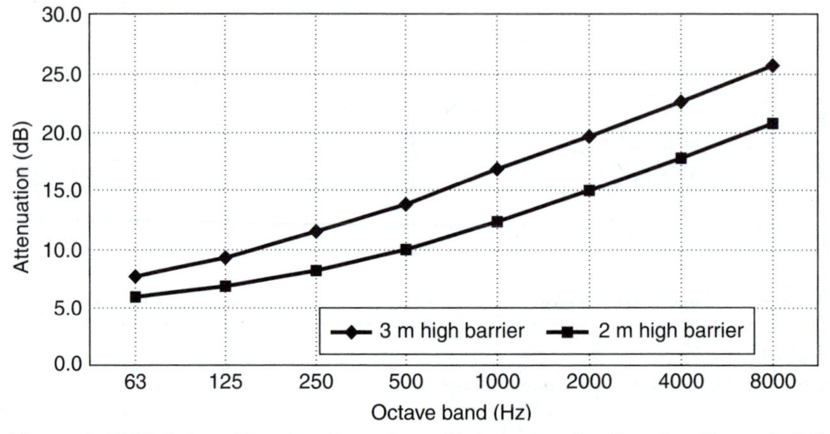

Figure 2.15 Variation of barrier attenuation with frequency (re. Exercise, Example 2.9)

Note how the predicted attenuation of the barrier varies considerably across the frequency spectrum, increasing with frequency from 7.6 dB at 63 Hz to 25.4 dB at 8 kHz – see Figure 2.15.

Exercise

Determine the attenuation if the height of the barrier in the example is reduced to 2 m.

Answer

This is shown graphically on Figure 2.15.

More on barriers

The above section sets out the basic principles. A significant amount of research effort is applied to investigating variations on the basic theme. Absorbing surfaces facing the traffic, inclined surfaces and barriers with wide, flat, absorbing tops are among modifications of interest. The goal is often to achieve a higher noise reduction while limiting the obtrusive nature of a barrier.

Limitations to the performance of barriers

Reflections from nearby surfaces can partly fill in the acoustic shadow. Turbulence near the top of the barrier and propagation through the barrier results if its mass is not adequate or if there are penetrations arising from poorly fitting gates, gaps between the components or even decorative features with holes.

If the barrier has a non-absorbing face, reflections from that face may increase the levels on the opposite side of the road. In this case other, more complicated, propagation paths could exist involving, for example, the side of a large truck and, perhaps, a barrier on the opposite side of the road.

It is also important to note that if there was attenuation by soft ground prior to installation of a barrier, the observed reduction from the barrier will be less than if there was hard ground. Much of attenuation previously provided by the soft ground will be negated by the installation of the barrier.

2.12 Other sound propagation effects

Vegetation

A row or trees or bushes will have an insignificant physical effect on sound proportion (although it can give a psychological effect). If sufficiently dense to completely block the view along the noise propagation path, the foliage of trees and shrubs will provide a small amount of noise attenuation. This attenuation is dependent on the frequency of the sound source and the distance from the dense foliage, as shown below.

Figure 2.16 shows the octave band attenuation due to propagation with distance through dense foliage (from ISO 9613-2:1996 *Acoustics – Attenuation of sound during propagation outdoors*) for distances between 10 m and 200 m. At distance greater than 200 m the attenuation should be taken as that for 200 m.

Other effects

Mist and fog have negligible effects on sound propagation through the air, but of course they are often indicative or certain type of atmospheric propagation, such as

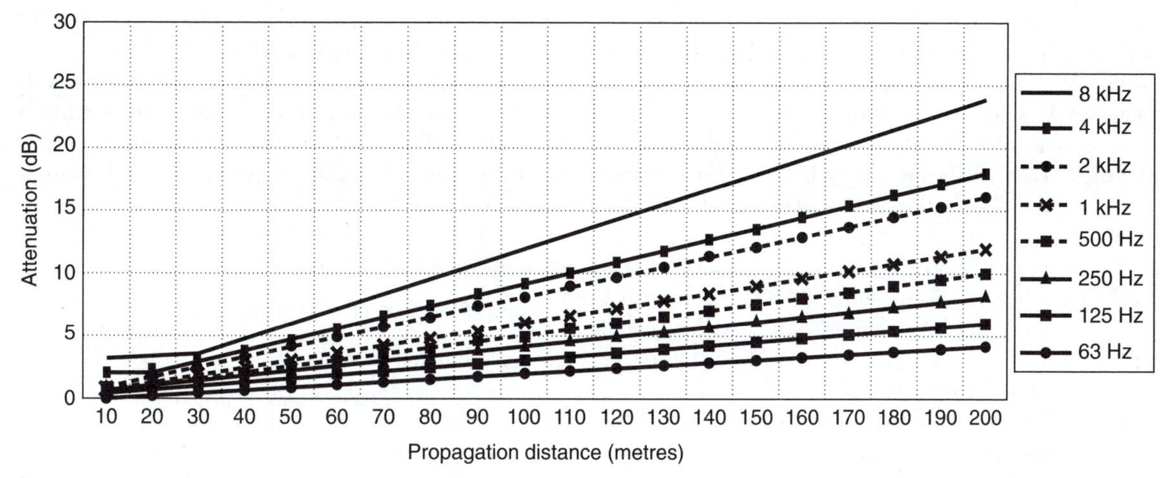

Figure 2.16 Attenuation of noise with propagation distance through dense foliage. Taken from ISO 9613:1966, Table 1

temperature inversion conditions, which do have a significant influence.

As indicated above, thin layers of trees and bushes also have a negligible physical effect on sound propagation but the possible psychological effect, for example in hiding the noise source from the view of the receiver, can be significant. The noise of wind in trees can also provide an acceptable masking of unwanted sounds.

The various propagation effects described above can combine together in several ways, so that, for example, atmospheric refraction and turbulence can affect barrier attenuation, which can also be affected by ground attenuation, which in turn can be affected by weather conditions. Scattering and reflection of sound by topographical features are also factors to be considered.

2.13 Prediction of outdoor sound levels and noise mapping

Each of the factors considered above (air absorption, ground attenuation, atmospheric scattering, barriers etc.) produces additional sound attenuation to be added to that produces by the geometric dispersion of the sound wave (i.e. the distance attenuation). If these various attenuations are predicted in decibels they simply add onto the sound prediction equation:

$$L_p = L_W - 20\log r - 11 + D - A_{air} - A_{ground} - A_{turbulence} - A_{refraction} - A_{barrier} - \cdots$$

Since these effects will be frequency dependent it will be necessary to repeat this calculation in each octave band. The prediction of these different attenuation factors is very difficult and complex, and subject to much uncertainty.

The need for accurate and reliable prediction methods plays a very important in planning to minimize the impact of potentially noisy developments, and this requirement has received additional impetus with the EC requirement for noise mapping of areas throughout Europe. Many different prediction models have been developed in different countries for different types of noise (e.g. aircraft noise, road traffic noise, train noise and industrial noise). Examples include the Nordic models, CRTN and CRN, and CONCAWE and ISO 9631, but there are many others. Many different software platforms have been produced to implement the calculation methods of these models including SoundPlan, CADNA and many others. Much research has been carried out and continues to improve prediction methods and, in Europe, to produce a single unified prediction model, to include all types of noise source, called Harmonoise, for the purpose of the continuing noise mapping exercise.

2.14 The Doppler effect

This is the apparent change in frequency of a sound that occurs when there is relative motion between source and receiver. A common example is observed when an emergency vehicle sounding a siren approaches, passes and then recedes from an observer.

The effect was first proposed by the Austrian scientist Christian Doppler in 1842. In the most general case the medium can move as well, but we shall only consider below the two cases of the moving source and moving receiver.

Before doing so it is important to emphasize that the frequency of a sound derives only from it source, and that the sound speed depends only on the medium (e.g. air and water). The apparent change of frequency which occurs in the Doppler effect arises from an apparent change of wavelength or of sound speed.

Also remember that the wavelength λ is the distance travelled by a wave during one cycle (i.e. in time $1/f$ seconds, where f is the frequency), so that distance travelled = speed × time = $c(1/f) = \lambda$ (since $c = f\lambda$).

If an observer is moving towards a sound source it is evident that he/she will encounter more sound waves per second than if he/she remains stationary, hence the apparent increase in frequency. Similarly if the source approaches the observer.

As will be seen below, the two cases do however differ in certain respects.

Source moving towards a listener

When the source moves towards the observer, successive peaks and troughs arrive more frequently as compared to when the source is stationary, and they are more closely spaced, as shown in Figure 2.17, and this results in an apparent reduction in wavelength and consequent increase in frequency (since sound speed remains constant). The opposite effect occurs if the source moves away from the observer in which case the wavefronts reaching the observer are more widely spaced, and there is an apparent increase in wavelength and reduction in frequency.

Returning to the case where the source moves towards the observer, the distance between wavefronts arriving at the observer will be reduced from $c(1/f)$ in the case of the stationary source, by the distance which the source moves towards the observer in one wave cycle, i.e. by $v(1/f)$.

Therefore since distance between successive peaks = apparent wavelength, λ', is given by:

$$\lambda' = c(1/f) - v(1/f) = (c - v)(1/f)$$

and since apparent frequency, $f' = c/\lambda'$, then

$$f' = c/[(c - v)(1/f)] = [c/(c - v)]f$$

$$f' = [c/(c - v)]f$$

Example 2.10

If a train moves at 300 km/h = 80 m/s and the whistle is at 1 kHz find the apparent frequency.

Solution

$$\begin{aligned}\text{Apparent frequency} &= \{c/(c - v)\} \times f' \\ &= \{340/(340 - 80)\} \times 1000 \\ &= 1308 \text{ Hz}\end{aligned}$$

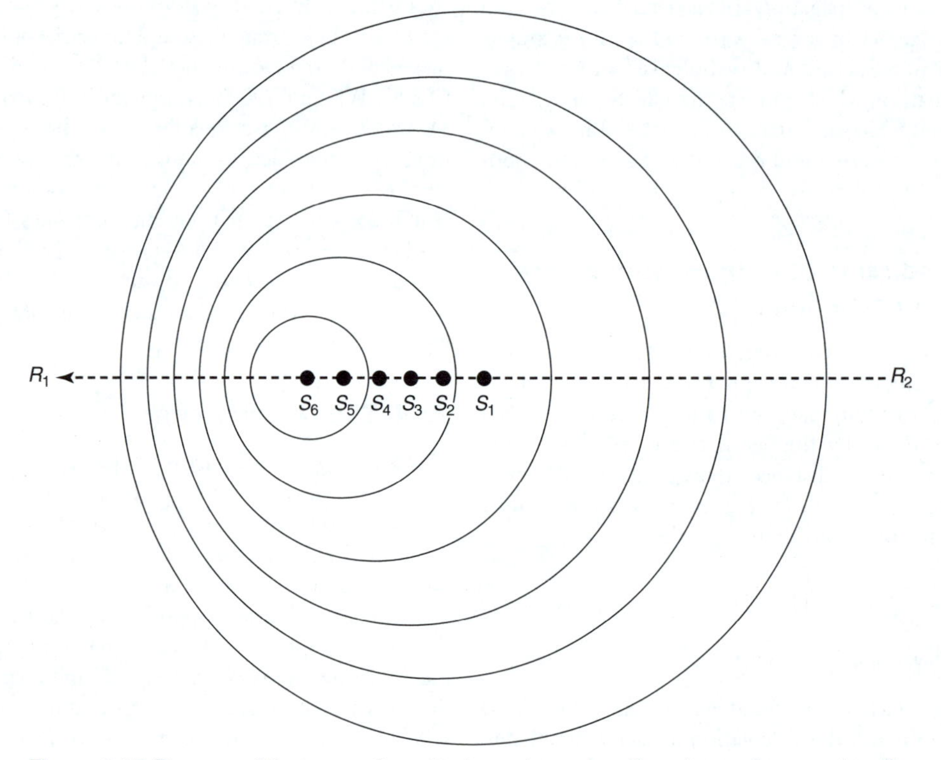

Figure 2.17 The case of the source S moving towards receiver R_1 and away from receiver R_2

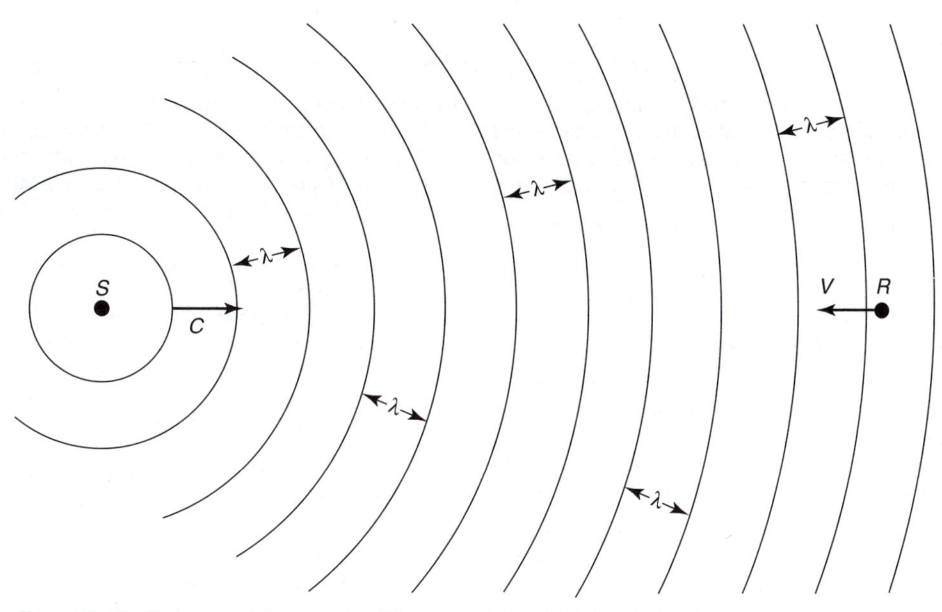

Figure 2.18 The case of the receiver (R) moving towards the source (S)

Observer moving towards a source

In this case the spacing between successive peaks reaching the observer is unchanged, i.e. the wavelength is unchanged, but because of the motion of the observer the frequency with which the observer receives successive peaks is increased, i.e. the effective speed of sound is increased from c to $c + v$. (See Figure 2.18.)

Since the wave length (= wave speed/frequency) is unchanged:

$$c/f = (c + v)/f'$$

and so effective frequency, $f' = [(c + v)/c]f$.

The same result can be obtained by considering how many wavefronts the observer moving towards the source meets per second. If the observer remained stationary he/she would receive f wavefronts per second. By moving towards the source at a speed of v m/s he/she encounters an additional v/λ wavefronts.

Therefore the number of waves crossed per second by this moving observer, which is the apparent frequency of the sound (f') is:

$$\begin{aligned} f' &= f + (V/\lambda) \\ &= f + [v/(c/f)] \\ &= f + vf/c \\ &= f[1 + (v/c)] \end{aligned}$$

Apparent frequency $= f(1 + v/c) = [(c + v)/c]f$ (as above).

Comparing this with the formula for the moving source case (above) it can be seen that the two formulae have different forms, so that it is not simply a case of considering the relative velocity between source and receiver. A more extensive theory which takes into account the effect of the medium as well is needed to fully explain this in detail. The following example (which uses the same speed as in the previous example) confirms the difference between the two cases.

Example 2.11

If a receiver moves at 300 km/h = 80 m/s towards a whistle with a frequency of 1 kHz, find the apparent frequency.

Solution

$$\begin{aligned} \text{Apparent frequency} &= \{(c + v_0)/c\} \times f_s \\ &= \{(340 + 80)/340\} \times 1000 \\ &= 1235 \text{ Hz} \end{aligned}$$

Note that this is a different result from the previous case of the moving source.

Applications of the Doppler effect

There are a number of medical ultrasonic applications of the Doppler effect including in echocardiology for visualizing the heart structure, and for detection of blood flow. In the audio field the Doppler effect has been used to provide special sound effects in modern music (search websites under 'Leslie loudspeaker' for more details).

Questions

1 Explain the following properties of sound waves – reflection, diffraction, interference and diffraction – and discuss their relevance to the measurement and prediction of sound levels outdoors.

2 Describe and discuss the following factors affecting sound propagation outdoors:

- air absorption
- ground attenuation
- refraction due to sound velocity gradients.

3 Discuss the variation of sound pressure level with distance from point, line and area sources for unobstructed propagation conditions. Draw a sketch graph showing how the sound pressure level from a sound radiating plane surface varies with distance from the surface.

4 Explain the terms near field and far field, used to describe the sound field surrounding a noise source and their significance with respect to prediction of noise levels from real noise sources such as machines. What are the factors which determine the extent of these regions?

5 What does diffraction theory tell you (in broad general terms) about:

(a) the directivity (or directionality) of noise sources
(b) the shielding effects of screens and barriers
(c) the directionality of microphones.

6 Explain briefly why the performance of a noise barrier depends on frequency. Why should a barrier be located either close to the source or close to the receiver? (IOA 2009 (part question))

7 The noise level from a machine outdoors was measured at a distance of 10 metres and 20 metres from the source, and found to be 77 dBA and 68 dBA respectively. Assuming that the reduction with distance follows the same rate at larger distances, estimate the noise level at 120 metres from the source.

8 Using sketches where appropriate:

(a) Describe the influences of height, thickness, construction, mass and distance to source/receiver on the performance of outdoor noise barriers.
(b) Describe how reflection from buildings, other barriers and vehicles can affect the performance of a barrier and give ways in which those effects can be controlled.
(c) Describe two ways in which weather conditions can affect the performance of a barrier. (IOA 2007)

9 (a) Define the terms 'near field' 'far field' and 'directivity index'.
(b) Distinguish between the 'inherent' directivity of a source and directivity induced by source location, giving an example of each.
(c) How could the inherent directivity of a source be determined? (IOA 2007 (part question))

10 (a) Outdoor sound propagation is influenced by meteorological conditions. Give brief descriptions of the physical effects associated with the meteorological conditions (excluding rain) and discuss how these effects are likely to vary with frequency and with range. What dry meteorological conditions give rise to the greatest sound attenuation at a given range?
(b) Identify two other environmental factors and one property of the source that could influence the propagation of sound from a fixed source outdoors. For each factor describe how they influence the sound level as a function of range.
(c) During tests of sound propagation at a particular flat grassland site with a fairly steady background level of 54 dB at 1 kHz, it is established that the ground effect at 1 kHz and at a range of 1 km is 10 dB, the maximum atmospheric absorption is 0.0015 dB m^{-1} and the maximum attenuation due to meteorological conditions at this range is 10 dB. What should be the sound power level of a 1 kHz siren, with a directivity index of 3 dB in the direction of interest, to ensure audibility in all dry conditions at a distance of 1 km? State any assumptions that you make. (IOA 2006)

11 (a) Show that Makaewa's simple theory of the attenuation offered by a thin, infinitely long barrier leads to the conclusion that a listener placed at such a height that a source can just be seen above such a barrier will benefit from some attenuation. How much attenuation will be achieved?
(b) A thin partial barrier in the form of a vertical rectangle measuring 6 m (long) × 4 m (high) is placed symmetrically between a source of frequency 500 Hz and a microphone each of height 2 m above the ground and at a distance of 5 m from the barrier. Calculate the attenuation offered by the barrier assuming that both the propagation through the barrier and ground attenuation are negligible. (IOA 2004)

12 (a) Explain what you understand by the 'near field' and the 'far field' of an acoustic source. Define the 'directivity index' of an acoustic source in terms of intensity. How is the directivity of a source taken into account during the determination of power levels in anechoic and semi-anechoic enclosures?
(b) What is meant by the 'cut-off frequency' of an anechoic enclosure? How would you set about qualifying such an enclosure for sound power level measurements?
(c) Derive the relationship between sound pressure level and sound power level due to a directional source in a semi-anechoic environment and hence, by rearrangement, an expression for the directivity index.
(d) A source is situated at the centre of the floor of a semi-anechoic enclosure. The sound pressure levels

(L_a) measured at floor level and 8 metres from the centre of the source at various angles (θ) are:

Angle (θ)	0	30	60	90
SPL (dB)	64	75	62	49

Compute the directivity index of the source at these angles, given that the sound power level of the source is 95 dB. (IOA 2001)

13 (a) Why is a wall or fence only effective for reducing noise when its height is large compared with the wavelength of the noise? State another important construction requirement for a wall or fence to perform effectively as a noise barrier.

(b) The owner of a café has installed an air conditioning unit on the ground, at the foot of the café wall adjacent to the café's car park, in a position which is 50 m from the nearest house. The unit has a rated sound power level of 90 dB and emits noise predominantly in the 500 Hz octave band. What sound pressure levels might be expected 2 m from the unit and at the façade of the nearest house? State any assumptions made in deriving this value.

(c) Complaints from the owner of the nearest house are likely unless this noise can be reduced by 12 dB at the level of the first floor, 4.2 m above the ground. Would a fence of height 2 m installed 5 m from the air conditioning unit achieve the required noise reduction?

(d) After the noise barrier is installed a further noise survey shows that the noise level 2 m from the unit has increased. What is the most likely reason for this? (IOA 2008)

14 (a) Define the terms sound power, sound intensity and sound pressure, and state their units.

(b) Show that the sound pressure level falls by 3 dB and 6 dB per doubling of distance for a line and point source, respectively (IOA 1998 (part question))

15 (a) Describe the principal factors that should be taken into account when determining the attenuation of noise with distance from a noise source. Indicate which of these factors are only significant close to the source, and which are only significant at longer distances.

(b) A portable generator of sound power level 95 dBA is located on the ground close to a large reflecting wall. What noise level would be expected 1 in from the façade of a house 75 m from the generator position?

(c) If a noise barrier is interposed between the generator and the house, such that the top of the barrier is on the sight line between the generator and the monitoring position at the house, what reduction in noise level at the house would be expected? Assume barrier attenuation, A_B, $= 10\log(3 + 20N)$ dB where $N = 2\delta/\lambda$. (IOA 1995)

16 Explain the meaning of the terms reflection, diffraction and superposition of sound waves.

Discuss the action of:

(a) Reflection in the development of Eyring's theory of Reverberation Time.

(b) Diffraction in noise reduction by barriers.

(c) Superposition in resonance in a quarter wave tube. (IOA 1997)

17 (a) Define the terms sound power and sound intensity. Hence derive the simplest equation relating these quantities for an omni-directional sound source placed on a hard flat surface outdoors. List the assumptions made in your derivation.

(b) Show that for such a source the sound pressure level decreases by 6 dB if the distance from the source is doubled. (IOA 1996 (part question))

18 Given that the intensity of a plane wave can be related to the pressure by the equation:

$$I = p^2_{\text{avge}}/\rho c$$

Derive an expression for the sound pressure level as a function of the acoustic power output of a point source radiating into a hemisphere at a distance r. ($\rho c = 410$ kgm^{2}s^{-1}) (IOA 1994 (part question))

19 (a) Deduce an expression relating the observed frequency of a signal from a moving source by a stationary observer to the actual frequency.

(b) A rotating disc has a signal attached to its perimeter. The diameter of the disc is 1 m and it rotates at 2 Hz. What is the ratio of the maximum to minimum frequency received by a stationary observer situated beyond but within the plane of the disc? (Take the velocity of sound, c, to be 332 ms^{-1}.)

(c) Discuss, giving an estimation of numerical values, the importance of the effect mentioned in sections (a) and (b) in the field of environmental noise. (IOA 1992)

20 Sound waves may display: (a) reflection (b) refraction (c) diffraction (d) interference. Explain the meanings of these terms.

Discuss what effect, if any, these properties might play in:

(i) the design and resultant quality of a concert hall.

(ii) the environmental noise produced by an open-air rock concert. (IOA 1991)

21 (a) Explain with reference to physical phenomena why the noise reduction by a barrier depends on its location and frequency.

(b) A small compressor is to be installed on the ground near the external façade of a factory and at a position 50 m from the nearest residential housing. The measured noise levels at 10 m from a similar installation are tabulated below:

Frequency (Hz) Octave band	63	125	250	500	1000	2000	4000
SPL (dB)	102	96	96	95	77	65	58
A-weighting (dB)	−26	−16	−8	−30	0	+1	+1
Barrier attenuation (dB)	7	8	10		15	17	20

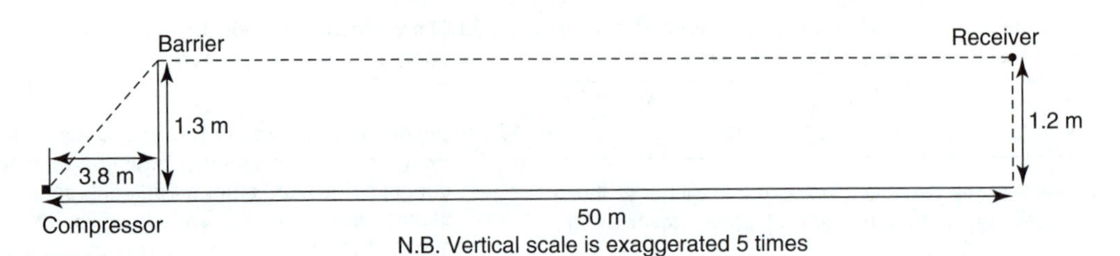

In order to reduce the noise from the compressor at nearby properties, it is proposed to erect a long noise barrier 1.3 m high at a distance of 3.8 m from the source, as shown in the figure above; which also indicates the standard height for a receiver of 1.2 m. The expected barrier attenuation at various octave band frequencies is also tabulated, except at 500 Hz.

(i) Calculate the barrier attenuation at 500 Hz in dB. Assume the speed of sound in air to be 330 ms^{-1}.

(ii) Calculate the expected A-weighted noise level in dB from the compressor at the specified position, with and without the barrier, and hence the noise reduction in dB expected from the barrier.

(c) Explain why measured noise reduction achieved by the barrier might be different from that predicted above. (IOA 2006)

22 (a) (i) List three materials that are typically used in the construction of roadside screens.

(ii) Describe the advantages and disadvantages of two of these types of material.

(iii) In what circumstances is it particularly advantageous for barrier materials to have sound-absorbent characteristics?

(b) Suggest, by reference to acoustic principles, which of the following circumstances are likely to increase or reduce the effectiveness of a roadside barrier, relative to a site which is flat and level, stating any assumptions you make:

(i) the site is level and the receptor is close to the road

(ii) the site is level and the receptor is far from the road

(iii) the site slopes downwards towards the road which is level with the local ground

(iv) the site slopes upwards towards the road which is level with the local ground

(v) the site is level and the road is in a cutting

(vi) the site is level and the road is on an embankment.

(c) A long, straight road crosses a flat, level site. Fifty metres from the road is a receptor point 1.5 m above ground level. You wish to screen the receptor point with a 3 m high barrier. Suggest the best place to put the barrier, setting out the factors you considered in reaching your decision.

(d) Suppose there is already a long barrier 3 metres high running alongside the road at a distance of 3 m from the edge of the carriageway. You wish to increase the shielding of the receptor by installing another 3 m barrier.

(i) What position would give the greatest additional insertion loss?

(ii) Assuming that the path difference over each barrier is the same, approximately what is the maximum insertion loss that the additional barrier can give? (IOA 2003)

23 (a) List five factors other than space or visual intrusion to be considered when barriers are constructed to protect people from the effects of environmental noise.

(b) Discuss the benefits of noise barriers compared with other noise control options for road traffic, railway and aircraft noise reduction.

(c) Describe the ways in which the audible features of road traffic, railway and aircraft noise change with distance from the source. How is noise at long distances from these sources affected by meteorological conditions and topographical features?

(d) Describe the extent to which the factors described in (c) above are taken into account in the calculations for road traffic noise (CRTN) and railway noise (CRN). (IOA 2004)

Chapter 3 Human response to noise

3.1 Introduction

This chapter discusses human hearing and the subjective attributes of sounds, in particular the concept of noisiness and the relevant indices. Other aspects of human response to noise are also briefly discussed.

There are many websites that provide useful and interesting supplements to these notes, on topics such as the details of the anatomy of the ear and hearing mechanisms, defects of hearing (including audio demonstrations), audiometry, and the effects of noise on health. Some examples of web addresses are given in the text but there are many others which may be accessed via the use of a search engine such as Google or similar.

3.2 The ear

Anatomy of the ear

A diagrammatic section of the ear is shown in Figure 3.1, with a further explanatory diagram illustrating the operation of the ossicles shown in Figure 3.2. Another interactive diagram of the ear can be found at *http://www.hearingcenteronline.com/ear2.shtml.* Many images of the components of the ear can be seen at *http://www.augie.edu/perry/ear/hearmech.htm.* The ear is divided into three main zones, each with a distinct function.

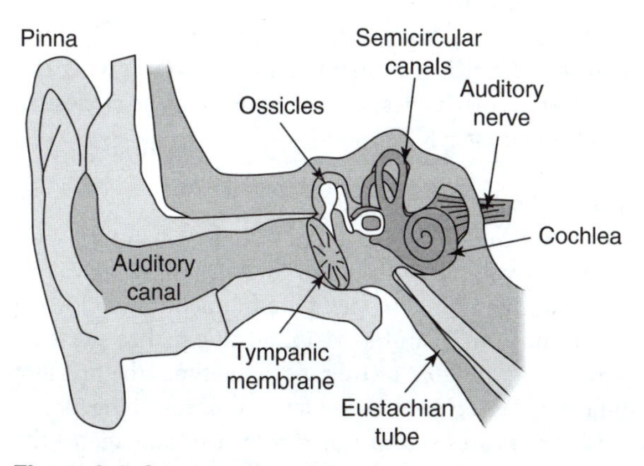

Figure 3.1 Cross-section of the ear

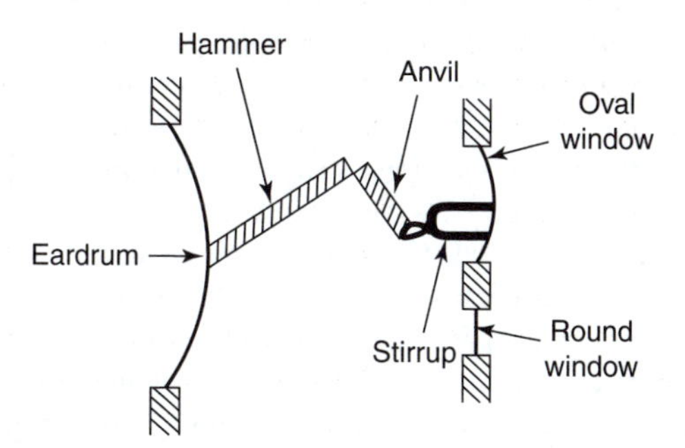

Figure 3.2 How the ossicles in the middle ear work

The **external ear** consists of:

- the pinna – the fleshy outside portion and ear lobe
- the external meatus – the ear (or auditory) canal
- the tympanic membrane – the eardrum.

The sound wave is 'funnelled' down the canal and the pressure impinges on the membrane causing it to vibrate. Under the conditions of maximum sensitivity an eardrum moves a distance approximately the diameter of a hydrogen atom at the threshold for a young person with no history of hearing disorders. This is indeed a small movement, yet we are to consider this later as relatively large in relation to other movements in the operating mechanism of the ear.

The **middle ear** is the second component and is an air filled cavity beyond the tympanic membrane containing the **ossicles**, which are the three smallest bones in the body.

These bones are called the malleus, incus and stapes, commonly known as the hammer, the anvil and the stirrup, these names arising from their shapes. These three bones form a conductive chain from the tympanic membrane to the oval window, which is a membrane at the barrier to the internal ear. The ossicles act as a mechanical system which transforms the relatively large displacement but small force provided by the movement of the eardrum into a small displacement but involving a larger force at the oval window.

The semicircular canals contain sensors that provide information about the spatial attitude of the head in three dimensions.

The **inner ear** is the third component consisting of the **cochlea** which is a small tube, coiled like a snail, and set within the protection of the hardest bone in the body after the teeth. The tube of the cochlear is broadly separated into two main parts, the upper or scala vestibuli and the lower or scala tympani. These interconnect through a gap at the apical end which is called the helicotrema. The two main parts are separated and supported by a complex structure, the most important part of which is the basilar membrane. This is about 35 mm long, 2 mm wide at the near, or basal, end enlarging to about 3 mm wide at the apical end. The oval window, connected to the ossicles, forms the end of the upper part and a membrane called the round window seals the lower part.

The cochlea is filled with liquid and the movement of the oval window causes an increase in hydrostatic pressure in the upper part which effectively pushes the basilar membrane down into the lower space. The liquid in the cochlea is effectively incompressible so the movement of the basilar membrane is accommodated by a contrary movement of the round window. The basilar membrane includes ranks of hair like structures which move with the fluid and the membrane. Embedded at the base of these hairs are nerve cells which detect the movement and generate electrical signals for transmission to the brain.

As an acoustic medium, the cochlea is very small so, even though the fluid has a velocity of sound exceeding that of air, the pressure increase caused by the inward movement of the oval window (and vice versa) takes effect more or less instantaneously along the length of the cochlea, i.e. in the time it takes for a sound wave travelling at approximately 1500 m/s to travel the length of the cochlea, i.e. about 25 microseconds.

The acoustic propagation path between the exterior of the ear and the inner ear contains a complex variety of interactive elements, ending in the signal processing which occurs within the cochlea and the brain. Each part of the path responds differently to different frequencies, i.e. each has its own frequency response, so that the resulting overall frequency response is the combination of these different parts.

The outcome is that the hearing response varies with frequency and has maximum sensitivity between around 3 and 5 kHz.

More detail on the function of the ear's components

The acoustic function of the outer ear (or pinna)

The shape of the pinna appears to impose distortions on an incident sound in the frequency range above 5000 Hz which are interpreted by the ear in such a way as to infer the distance and general direction of the source. For example the jangling of keys used as a source generally allows location of the source in space with one ear, without moving the head. It is not known exactly how the inner ear interprets the signals provided by the pinna.

The ear canal

The ear (or auditory) canal acts as a form of tuned musical instrument giving about a 20 dB increase in pressure at the tympanic membrane over the sound pressure level at the pinna at certain frequencies.

Acoustic reflex

The tensor tympani muscles are attached to the ossicles which themselves communicate with the eardrum. Any tendency of the eardrum to exceed certain excursion limits causes a tightening of these muscles in an attempt to limit further movement and thus protect the chain of the hearing process.

This tensioning of these muscles is accompanied by a sensation of pain. If exposure to excessive noise is a possible cause, this sensation of pain must be taken as evidence of conditions likely to lead to hearing loss.

Impedance matching and signal amplification measures

When sound in air arrives at the boundary of a body of water, because of the vast difference in acoustic impedance almost all the sound energy will be reflected at the interface and only about 0.1% of the incident sound energy, representing about a 30 dB reduction, will be transmitted into the water. The purpose of the middle ear is to act as an impedance matching device between the air in the outer ear and the cochlear fluid of the inner ear.

The 'lever action' of ossicles provide a mechanical advantage of about 3:1, while the relative sizes of the larger eardrum and the smaller oval window result in an overall mechanical advantage of about 15:1.

The average length of the ear canal is about 35 mm, and so the canal is resonant at about 4 kHz, giving rise to a further mechanical advantage at frequencies close to 4 kHz of about 3:1.

The middle ear

The middle ear is air-filled and the eustachian tube connects this volume to the nasal cavity. Swallowing causes the eustachian tube to open for a moment thus allowing air to move in or out as required to equalize pressure across the tympanic membrane. Sometimes the pressure differential is too great for voluntary opening of the tube to solve the problem and this can lead to damage to the inner ear, either temporary or permanent.

3.3 Aspects of hearing

Pitch discrimination and theories of cochlear mechanics

The human ear can readily discern the change in pitch generated by varying the frequency of a pure tone test signal by a few Hz.

An early theory of how the ear–brain system managed to do this was proposed by Helmholtz (1821–94), who suggested that the basilar membrane and any other membranes that may be attached to it, acted rather like an array of individual transducing components along its length, rather in the manner of a host of tuned microphones each responding to a certain pure tone component. It was imagined that each tuned element was connected via biological wires to the brain and that the brain assembled a total sensation by summing the signals along each of the wires.

The basis for this lay in imagining that the pieces of the basilar membrane acted like the wires of a piano, since if a tuning fork is struck and applied to the frame of the piano, the strings that are, by virtue of their mass and tension, closest in natural frequency to the tuning fork vibrate in a form of sympathetic or resonant oscillation.

A later theory was developed in the 1950s by Bekesy and called the 'place theory' of the cochlea. According to this theory the cochlea was divided into regions along its length with the hair cells in each region being most sensitive to sound in a certain range or band of frequencies, very approximately like one third octave bands, with the hair cells close to the apex (the helicotrema) being most sensitive to low frequencies and those at the base of the cochlea to high frequencies, rather like a bank of third octave band filters in an audio frequency analyser.

This theory was supported to a degree by the knowledge that accidental damage to the apical end of the cochlea resulted in a loss of ability to hear low notes, and the reverse is true in respect of damage towards the basal end.

Attempts were made to relate these ideas to the mechanical structure of the cochlea, with the mass and stiffness of the basilar membrane of the cochlea (which contains the hair cells) varying along its length so that different parts act rather like individual mass–spring systems each with a different resonance frequency. (You will learn more about vibration response and the natural frequencies of mass–spring systems in Chapter 7 on vibration.)

However, experiments on the mechanical vibration response of the cochlea of cadavers (using guinea pigs for example) indicated that because of the effects of damping by the fluid in the cochlea the resonant response of the different parts of the basilar membrane was far too broad to account for the highly frequency selective response of the ear.

A possible explanation for the discrepancy was that rather than the basilar membrane acting as a passive mechanical system (i.e. like a bank of different mass-spring systems) there was an active feedback system in operation between the hairs cells in the cochlea and the signal processing that occurred in the brain. Such a feedback system would be capable of tuning the cochlea frequency response sufficiently to account for human frequency selectivity.

Within the last few decades evidence has been found of such active physiological processes occurring within the cochlea, including the existence of otoacoustic emissions. These are acoustic signals originating within the cochlea and emanating from the ear in response to sound stimuli presented at the outer ear. They have been loosely called 'cochlea echoes' but they are not true echoes in the acoustic sense. Spontaneous emissions, in the absence of acoustic stimuli, have also been observed.

Otoacoustic emission techniques are now routinely used for testing the hearing of newborn infants, and are being developed as a method of giving early indication of the presence of noise induced hearing loss.

Binaural localization

This simply refers to using our two ears to localize sources and is an interesting example of how the ear–brain system weighs two pieces of information and comes up with the most probable solution.

A sound which arrives at the human ear from an oblique direction (i.e. not from straight ahead or from directly behind the head) will arrive at one ear before the other, the difference being known as the ITD (the inter-aural time difference). If the sound is a continuous note such as pure tone the time difference will be detected as a difference in phase between the signal at the two ears. There will also be a difference in the sound pressure at the two ears as a result of the slightly different distances between the source and each ear, and more importantly in most cases, as a result of diffraction of the sound by the head. This difference is known as the inter-aural intensity difference (IID) and will depend on the frequency and wavelength of the sound, the size and shape of the head, and the direction from which the sound approaches the ears. All of this information is related to the IID by the head transfer function, governed by the laws of diffraction. (There is more about diffraction in Chapter 1.)

The hearing mechanism detects the IID and the ITD (or the corresponding phase difference) of an incoming sound wave and the brain then analyses this information, together with its stored knowledge of the head transfer function to make a judgement about the direction from which the sound is coming. Additional information about the direction in the vertical plane is obtained by tilting the head.

Although both the ITD and IID information is used making this judgement, it is the ITD which is the more useful at low frequencies and the IID is used more at higher frequencies. The use of hearing protectors, while necessary to reduce the risk of damage from high levels of noise, can interfere with this processing and reduce the ability to detect the direction that a sound is coming from.

Combination tones and harmonics

A system displaying non-linearity (for example an over-driven loudspeaker with the cone hitting the endstops) with a single pure tone of frequency, f, as input will produce in the sound output an array of distortion components at twice, three times, four times etc., the original frequency. These are collectively known as harmonic distortion components and hi-fi designers go to great lengths to minimize this effect. Moreover, if two pure tones of frequencies f_1 and f_2 are presented to the input, the output comprises signal components at the two original frequencies, various harmonics of these frequencies and some combination frequencies at, for example, $f_1 + f_2, f_1 - f_2, 2 \times f_1 + f_2, f_1 + 2 \times f_2$, etc.

It is pertinent to ask if similar frequency series are generated by our exceedingly non-linear ear–brain system? The answer is yes; they can be detected electronically within the voltage signals being passed to the brain. They can also be made audible by various techniques using, for example, different detection tones presented at the same time for the purpose. Yet, however many distortion products are generated en route to the brain, we still derive the sensation that we are listening to a pure tone or a musical instrument and so forth.

The ability to detect signals in noise

The ability of a listener to understand his name spoken at a distance in a noisy room at perhaps 20 dB below the general background level is really a remarkable illustration of the ear's signal processing system. The only inputs to this system are via two eardrums, one on each side of the head. Each of which, of course, can only be in one position at one time, and yet from these moving devices originates all the decoding and extraction of signals from noise which is of course remarkably in advance of the achievements of present-day electronic techniques. In relation to environmental noise this ability may be a disadvantage as a resident may hear and become annoyed by a sound well within the overall noise. This may lead to difficulties when there is a need to quantify, i.e. measure, this noise in order to assess if the complaint is justified and to develop appropriate noise reduction.

Masking

Masking is the phenomenon of one sound interfering with the perception of another sound.

According to Bekesy's place theory a single pure tone will cause a maximum displacement of the basilar membrane at a position along the cochlea which depends upon the frequency of the tone. The response of the basilar membrane is not, however, symmetrical about the position of maximum excitation, but rather it falls off much more slowly towards the high frequency end of the cochlea (base of the cochlea) than it does towards the low frequency end (apex of cochlea, or helicotrema). This has the effect that low frequency sounds will raise the threshold of detection of (i.e. will mask) high frequency sounds much more easily than high frequency sounds will mask low frequency sounds.

When the masking sound is broadband rather than a tone, the situation has been summarized (Kryter, 1985– see Bibliography at the end of the book) as follows:

1. Narrowband noise causes greater masking around its centre frequency than does a pure tone of that frequency. This should be evident, since a larger portion of the basilar membrane is excited by the noise.
2. Narrowband noise is more effective than pure tones in masking frequencies above the band centre frequency.
3. A noise bandwidth is ultimately reached, above which any further increase in bandwidth has no further influence on the masking of a pure tone at its centre frequency. This implies that the ear recognizes certain critical bandwidths, associated with the regions of activity on the basilar membrane.
4. The threshold of a masked tone is normally raised to the level of the masking noise only in the critical bandwidth centred on that frequency.
5. A tone which is a few decibels above the masking noise seems about as loud as it would sound if the masking noise were not present.

Masking sound can be used to advantage. For example a general background sound can be introduced into a large open office to mask out some of the noise from other workers in the area. Masking noise, like background music, can also be used to provide some privacy between a consulting room and an adjacent waiting room. It is important in these applications to ensure that the masking noise itself does not cause annoyance or disturbance.

Critical bands

The cochlea of the inner ear acts rather like a mechanical frequency spectrum analyser, as though it were made up of overlapping filters having bandwidths equal to the critical bandwidth. The critical bandwidth varies from slightly less than 100 Hz at low frequency to about one third octave at high frequencies. The audible range of frequencies comprises about 24 critical bands.

The pioneering work of Fletcher in the 1940s, on the masking of pure tones by bands of noise led to the discovery of critical bands in the cochlear response. Fletcher found that only a narrow band of noise surrounding the tone was responsible for the masking.

Critical bands are important in understanding many auditory phenomena: the perception of loudness, and of phase, pitch and timbre, the phenomenon of masking, and even in our understanding of human appreciation of music.

For example, a listener can be presented with a narrow band of noise centred on, say, 1 kHz and the bandwidth of the noise increased while the total energy is maintained constant. It is found that initially the sensation of loudness also remains constant. When the bandwidth reaches a certain value (the critical bandwidth) the loudness appears to progressively increase.

The ear will only detect two separate tones if they are separated in frequency by an amount dependent on the frequency. This may be interpreted as meaning that a tone excites a small region of the basilar membrane, called a critical band. If the critical bands of the two tones do not overlap then two separate tones are heard, but if the critical bands do overlap then two separate tones cannot be heard, but rather a single tone is heard, with a modulated amplitude.

Harmonic restoration

A superficial discussion of hearing sometimes states that the pitch of a musical signal is determined by the fundamental, i.e. lowest component frequency, and that the harmonic mix determines the quality of the sound. That this is incorrect is clear when we consider a small loudspeaker in a radio or set of ear buds for an MP3 player. Theory shows that a small radiating surface is unable to generate significant powers of low frequency sound unless impossibly large excursions of the surface are postulated. Hence the very large movements of a woofer, i.e. a large radiating area, when generating low frequency sounds. Since a small loudspeaker cannot produce much energy at low frequencies, this implies that the lower-harmonics will be absent from the reproduced musical sound. Yet this is not observed so the simple theory is therefore invalid. In fact what appears to happen is that the ear discerns pitch on the basis of the separation between the remaining harmonics (this will often be the same as the frequency of the missing fundamental, which cannot be reproduced by the small loudspeaker). Thus although, from a tiny loudspeaker, the quality will not be great, the sensation of pitch does not change. It is as though the ear–brain system uses prior experience to replace the missing harmonics.

Speech intelligibility and noise

The bulk of the information content in speech lies between 250 Hz and 5 kHz. The dynamic range in any band is about 30 dB. There is a variety of approaches to quantifying speech intelligibility, and the converse, namely speech privacy, and these are treated in a later section. A simple measure that is useful for field use (rather than research work) is speech interference level, SIL. This is the arithmetic average of the noise levels in the 500 Hz, 1 kHz and 2 kHz octave bands. It is considered that, for most noise spectra, the SIL is 9 units less than the A-weighted sound pressure level, to an accuracy of about 4 units. A guide to voice effort needed for face to face communication at a distance r metres can be gauged from:

$$VL_A \geq 1.33\,(\mathrm{SIL} + 20\log r) - 36$$

where VL_A is the A-weighted voice level and this is taken as: 57 dB for normal voice; 65 dB for raised voice; 74 dB for a very loud voice; and 82 dB for a shout.

References

Some of the information in this section has been taken from the sources listed below, from which further information may be obtained (see also the Bibliography):

Bies, D. A. and Hansen, C. H. (1996) *Engineering Noise Control*, 2nd edition, E & F N Spon, Chapter 2.

Broch, J. T. (1973) *Acoustic Noise Measurements*, Bruel and Kjaer.

Moore, D. R. and McAlpine, D. (1991) Biology of Hearing and Noise-induced Hearing Loss, *Acoustics Bulletin*, Vol. 16, No. 2, April.

Acoustical Society of America (1987) booklet accompanying disc of Auditory Demonstrations.

3.4 Audiometry

Types of audiometry

Air-conduction audiometry is the most commonly used form of measuring acuity of hearing. Earphones are placed on the subject and tone bursts over a series of frequencies are presented together with a means of indicating whether the subject can hear the signal at that particular level. Traditionally, audiometric data is presented as illustrated in Figure 3.3. The zero line, called audiometric zero, is the threshold of hearing. If, for the subject to hear the sound, the sound level has to be increased above the value on this threshold curve then the subject is said to have a raised threshold. This of course indicates a reduction in acuity and is evidence of some deafness. The magnitude of the threshold shift shows the extent of that deafness. On the audiogram chart, the change in threshold is plotted below the audiometric zero. Separate audiograms are presented for each ear.

Bone conduction audiometry involves the vibration of the skull by direct contact with an oscillating device (held against the mastoid bone, just behind the ear) which bypasses the external and middle ear and so tests the performance of the inner ear.

Otoacoustic emissions are sounds measured in the external ear canal that are a characteristic of the working cochlea. Probe and click stimuli are utilized in the performance of this test. Otoacoustic emission is used in the screening as well as the diagnosis of hearing impairment in neonates and young children.

Techniques for audiometry

Manual audiometry requires a trained person to operate the instrument. The subject indicates by pressing a button when he/she can hear the pitch at the level at which it is being presented. Some uncertainty is involved in respect of the judgement of the technician or other person operating the instrument and this is particularly so when the subject shows uncertainty about a particular response, raising the question of whether a second attempt should be offered.

Automatic or Bekesy audiometry involves the automatic presentation of tone bursts with an array of frequencies and levels. If the subject indicates audibility the level presented is changed to a lower value until the subject's response reverses. At this point, the level is changed upwards and the process continued throughout the frequency range. Level changes are automatically coupled to a pen recorder or printer which traces the subject's response automatically. There is clearly a benefit in speed, but more importantly in consistency of operations since variations in judgement of the operator are taken out of the equation. There is another benefit of this form of audiometry for patients exhibiting what is called loudness recruitment. Such a person is unable to hear quiet sounds but can hear intense

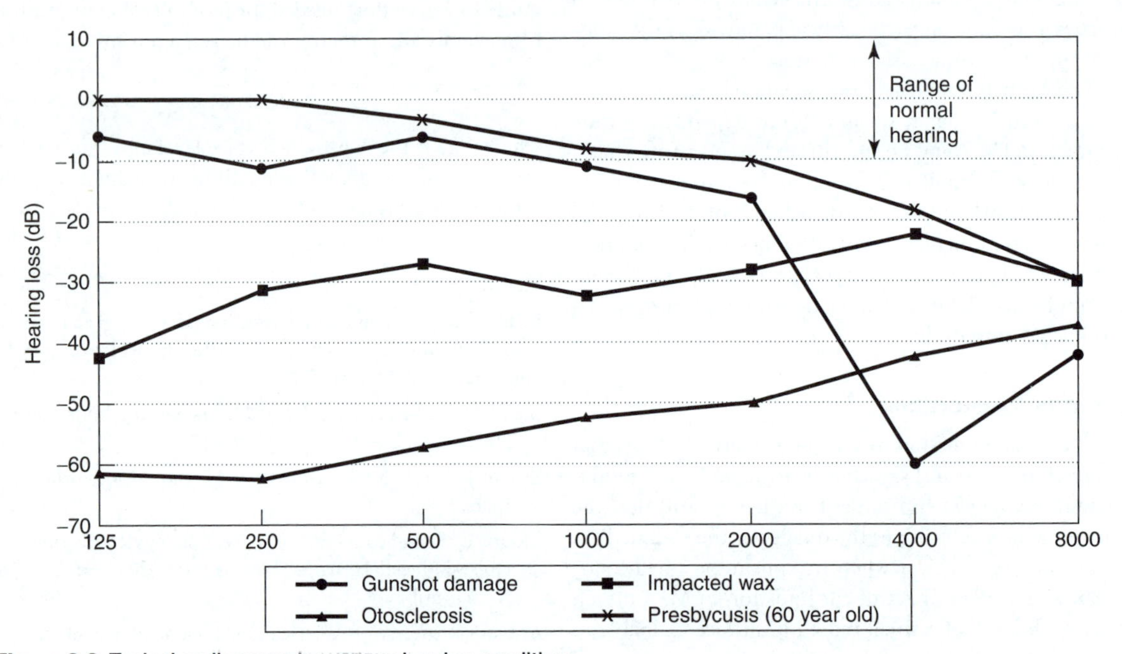

Figure 3.3 Typical audiograms for various hearing conditions

sounds very loudly, perhaps, for example, reacting with startle when a door slams. Recruitment is evidenced by the pen making very small excursions between the 'can't hear' and 'can hear' positions. This information is often of value to a medical practitioner looking to make a diagnosis of the origin of the hearing loss.

Errors in audiometry

Proper training is required for those undertaking audiometry. The types of errors that can occur in the audiograms if sufficient care is not taken include:

- Set and reset errors for headphones. Repeatedly placing them on a subject and making a measurement will normally produce some scatter in the results.
- Excessive noise level in the booth; this noise level should be more than 10 dB below the hearing threshold when the attenuation of the cups of the earpieces has been allowed for.
- Lack of care in the case of profound unilateral deafness. The intracranial attenuation is only about 40 dB. This means that a person with virtually no residual hearing in one ear can perceive the tone being presented at a high level by virtue of the level of hearing available in the other ear. In this case a band of masking noise is applied to the contralateral ear to overcome this.
- Calibration drift. Performance checks need to be made regularly to ensure there are no calibration problems with the equipment.
- Malingering, although this usually shows up as a lack of consistency in responses. The training for an audiometrist includes techniques to deal with this.

3.5 Types and sources of hearing loss

There are many forms of hearing loss and a few of the more common types are discussed below. You can listen to examples of hearing loss at a number of sites on the web including *http://www.utdallas.edu/~thib/rehabinfo/tohl.htm* and *http://www.hearingcenteronline.com/sound.shtml*

Conductive hearing loss

This is the result of defects in the conductive parts of the hearing mechanism, i.e. of the outer and middle ear. Examples are beads or wax in the ear canal, otosclerosis, otitis media and perforated or partially immobilized eardrum etc.

The effect of conductive hearing loss is generally like turning down the amplification of a radio so that the attenuation is more or less the same across the frequency range (see Figure 3.4). As the inner ear is still functioning, there is some hope of medical relief where hearing loss arises in this way, via hearing aid amplification or implants.

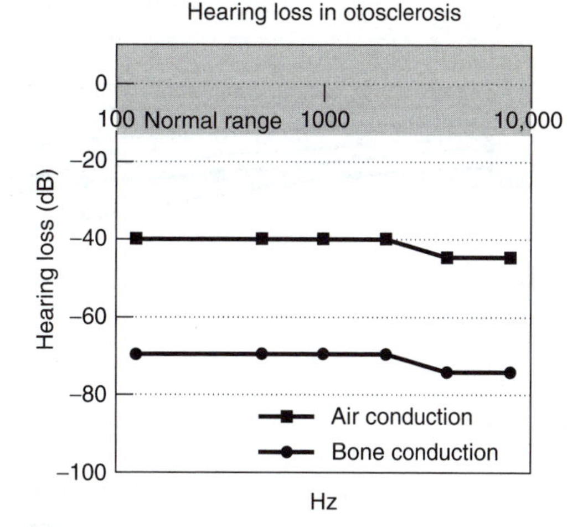

Figure 3.4 Example of an audiogram showing conductive hearing loss

Otitis media is an infection causing inflammation of the middle ear which can cause the eardrum to be sucked in towards the middle ear through a reduction of pressure in the middle ear.

Glue ear is an infection, fairly common among children, and causes blockage of the eustachian tube.

Otosclerosis is the development of hard or bony deposits at the junction of ossicles, impairing their usual range of movement.

Sensorineural hearing loss

This is associated with damage to the inner ear and the mechanisms implicated in generating and interpreting the nerve impulses. The effect therefore varies across the frequency spectrum rather like adjusting the individual levels on a graphic equalizer. The main causes include presbycusis, excessive noise exposure, medications and other factors which affect nerves.

Presbycusis is also referred to as 'old age hearing loss' (see Figure 3.5). There is variation in the incidence of this condition: some people reach the age of 70 with virtually zero hearing loss, others may be profoundly deaf. The reduction in acuity commences with the higher frequencies and progresses gradually with age down through the frequency spectrum. This process is irreversible and hearing aids can only provide some relief.

Note that in Figure 3.5 the loss increases with frequency. Further information on presbycusis is given in

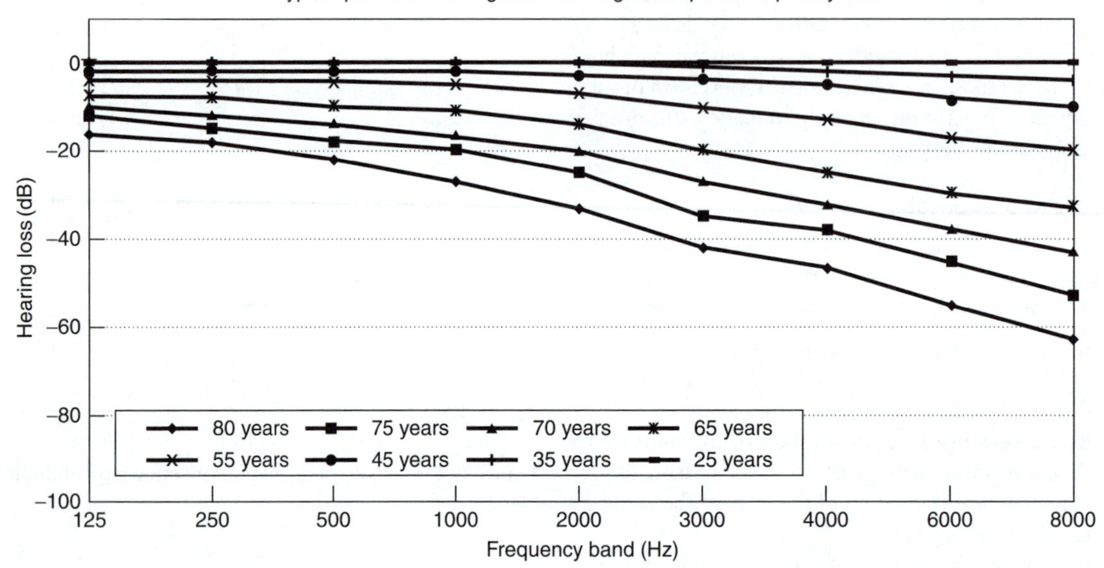

Figure 3.5 Example of an audiogram showing sensorineural hearing loss from presbycusis

standard BS EN ISO 7029:2000 *Acoustics – The statistical distribution of hearing thresholds as a function of age.*

Noise induced hearing loss

An audiogram taken before and after spending only a short time in a high noise environment will probably show a threshold shift, with the greatest difference being in the 4 to 6 kHz region. This is a **temporary threshold shift (TTS)** and after a period in relative quiet a repeat audiogram should show a return to normal. Repeated exposure to excess noise leads to more episodes of the temporary threshold shift, each of which takes a little longer to return to normal. Further exposure to the noise leads to a **permanent threshold shift (PTS)**. See Figures 3.6 and 3.7.

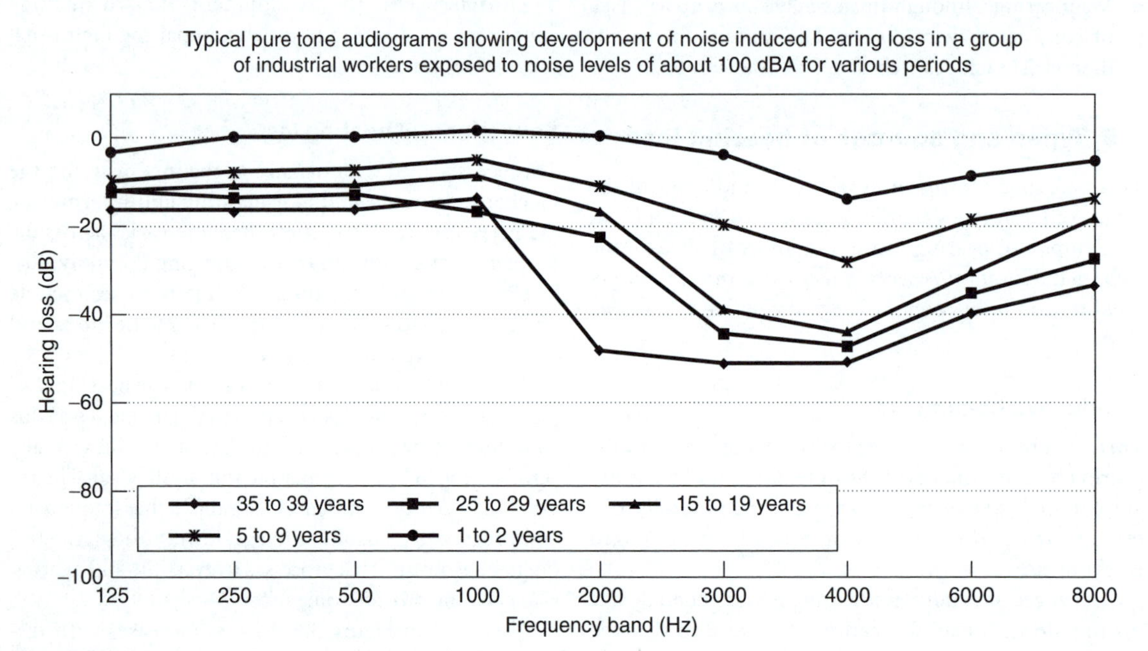

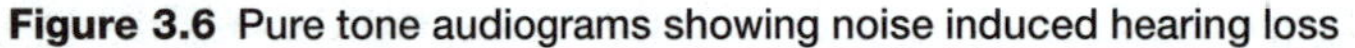
Figure 3.6 Pure tone audiograms showing noise induced hearing loss

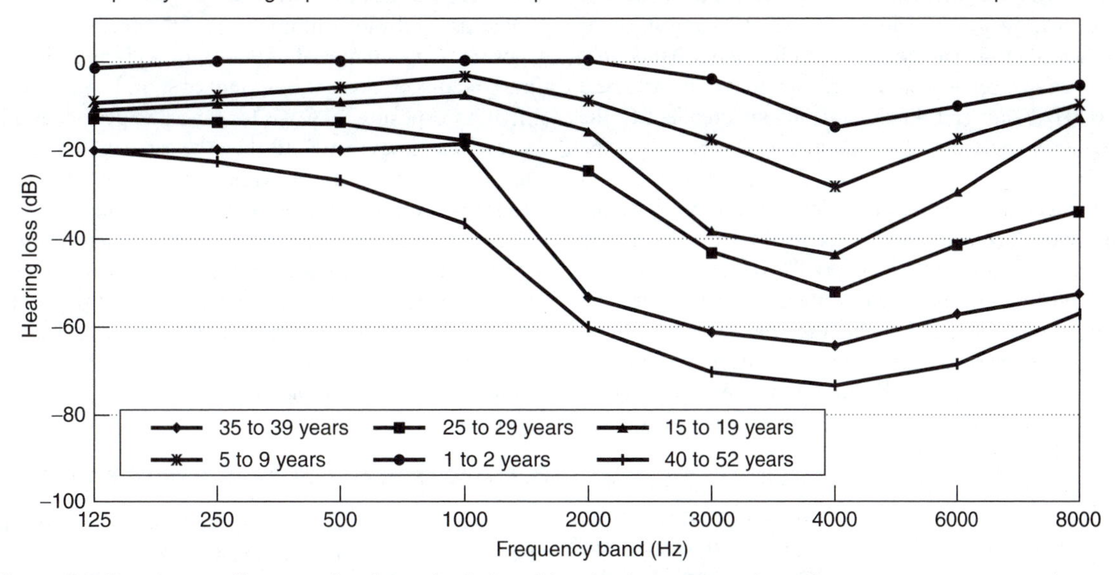

Figure 3.7 Pure tone audiograms showing noise induced hearing loss and presbycusis

Acoustic trauma is the immediate onset of hearing damage as a result of exposure to very loud noises such as explosions or from gunfire.

Note that the loss usually shows as a notch in the audiogram around 4 kHz (see Figure 3.6). Hearing loss can also occur as a result of defects in the transmission of the nerve impulses from the cochlea to the brain via the auditory nerve and as a result of impairment of that part of the brain which interprets these signals (cortical deafness).

Hearing impairment may also be caused by diseases such as measles, mumps, Ménière's disease, and by the effect of certain (ototoxic) drugs.

Loudness recruitment

An impaired ear may suffer an effect called recruitment, in which the apparent dynamic range of sufferers is greatly compressed. While the sufferer's threshold has been raised, tolerance for loud sounds has been sharply reduced, and tolerable listening is thus confined to a narrow range of 'loud enough but not too loud'.

As an example: a normal ear may be able to hear extremely quiet sounds (say between 0 and 20 dB) but can also tolerate very loud sounds, up to about 110 dB. An ear with loudness recruitment may well be unable to hear sounds, particularly at high frequencies, below 50 dB, but finds that any sounds above 80 dB are uncomfortable to listen to, and may be distorted.

It is interesting to note that a person suffering from loudness recruitment arising from hearing damage can detect smaller changes in sound level than people with normal hearing.

Tinnitus

Tinnitus is the sensation of sound (e.g. a 'ringing' in the ears) for which there is no acoustic source. It arises from a malfunction of hair cells in the cochlea which can arise from a number of reasons including physical disease and the effect of certain ototoxic drugs. Tinnitus is a conditions that almost everyone experiences on a temporary basis in a mild form (e.g. as the result of a cold). However for some people severe and continuous tinnitus can be a disabling condition. Tinnitus may arise from and occur together with noise induced hearing loss (NIHL), but noise induced hearing loss may occur without the symptoms of tinnitus, and conversely tinnitus may occur for reasons other than exposure to noise. Nevertheless the association with NIHL is sufficiently strong that questions about the onset of tinnitus are regularly used in routine medical checks on employees exposed to noise as an indicator of risk of NIHL.

3.6 Estimation of risk of noise induced hearing loss

Noise induced hearing loss is a major component of claims for injury at work in the UK and in most other industrialized countries. Susceptibility to hearing damage from high levels of noise exposure varies widely from person to person with some people suffering only minimal hearing loss when exposed at a certain level whereas others exposed to the same noise experiencing substantial hearing loss. The hearing level frequency bands that are usually most affected by occupational noise are the 3 to 6 kHz bands and this is generally independent of the frequency spectrum of the damaging noise. This is the frequency range where the ear is most sensitive to sound and where the consonants which provide much of the information content and are essential for understanding speech, have most of their energy. This often means that a person with noise induced hearing loss may be aware that someone is talking to them because they can hear the low frequency vowel sounds but may not be able to understand what is being said because they cannot distinguish the consonants that often occur at the beginning and end of words.

Apart from exposure to individual bursts of very high levels of noise (such as from explosions, gunfire etc.), where hearing damage is immediately obvious, occupational hearing loss usually occurs gradually, as a result of continuous accumulated noise exposure over weeks, months and years. The damage is insidious, and may not at first be noticeable to the subject, which is why it is important that those exposed to high levels of noise either at work or as a result of noisy leisure activities should undergo regular audiometric tests to check their hearing level. It is also the case that the noise induced damage to the hair cells in the cochlea adds to the increasing loss of these cells due to ageing (presbycusis). The early stages of loss may affect only the highest frequencies, and the hearing loss may not be noticeable until the damage extends to the speech carrying frequencies and the ability to hear conversational speech becomes impaired.

Research by Burns and Robinson in the UK in the 1960s, and others, indicated that on average the degree of hearing damage of employees over a working lifetime could be best correlated with the total amount of A-weighted sound energy received by the ear, and this led to the development of the continuous equivalent sound level $L_{Aeq,T}$, and the publication of a guideline limit for daily noise exposure at work of 90 dBA over an eight-hour period, in the Code of Practice for Reducing the Exposure of Employed Persons to Noise, published by the UK Health and Safety Executive in 1972.

The continuous equivalent noise level, L_{Aeq} (also now called the average sound level), has remained the basis for estimating the risk of hearing loss due to noise exposure, and the guideline limits eventually became the basis of limits set out in the 1990 Noise at Work Regulations, subsequently replaced by the 2005 Control of Noise at Work Regulations. L_{Aeq} is explained further in Chapter 4 and the regulations discussed in Chapter 10.

Because of the logarithmic basis of the decibel scale an exposure to a sound level of 90 dB for a given period of time contains 10 times the amount of sound energy content as an exposure to 80 dB for the same period, and if the level increases to 100 dB then the exposure increases 100-fold (compared to the same period at 80 dB). This means that exposure to even short bursts of high levels of noise may become a dominant factor in determining overall daily noise exposure and risk of hearing damage.

The probability of hearing damage arising from a particular noise exposure history may be estimated using ISO 1999:1990 *Determination of occupational noise exposure and of noise induced hearing impairment.*

3.7 Assessment of loudness of sound

Loudness is the subjective judgement of the magnitude of a sound, is probably the most important subjective characteristic of a sound.

The concepts of loudness and the equal loudness contours were introduced in Chapter 1, mainly as an introduction to and basis for the A-weighting network, and A-weighted sound pressure level, dBA. The original equal loudness contours were developed by Fletcher and Munson in the 1930s and by Robinson and Dadson in the 1950s, and eventually embodied in ISO 226. As a result of more recent work ISO 226 was updated in 2003. (This work is described in a report by Suzuki *et al.* which may be found at: *www.nedo.go.jp/itd/grant-e/report/00pdf/is-01e.pdf*)

A comparison of the 1987 and 2003 curves in the ISO standard is shown in Figure 3.8. It is interesting to note that there are some large differences of up to 15 dB at low frequencies.

Loudness level (Phons) and loudness (Sones)

While in practice the use of the A-weighting is adequate to indicate the approximate loudness of sounds, there are situations when more precise estimates of loudness are necessary and this is when the Phon and Sone scales are used.

The Phon scale is also called the loudness level scale, and because of the way it is based on the equal loudness contours, which are plotted in decibels, it is essentially a decibel scale. The Sone scale was developed to give a measure of loudness which is not in decibels. The Sone is a unit of loudness (or sometimes called a loudness index). The Sone scale is based on the fact that on average an increase of 10 dB represents a doubling of loudness, as does each subsequent 10 dB increase.

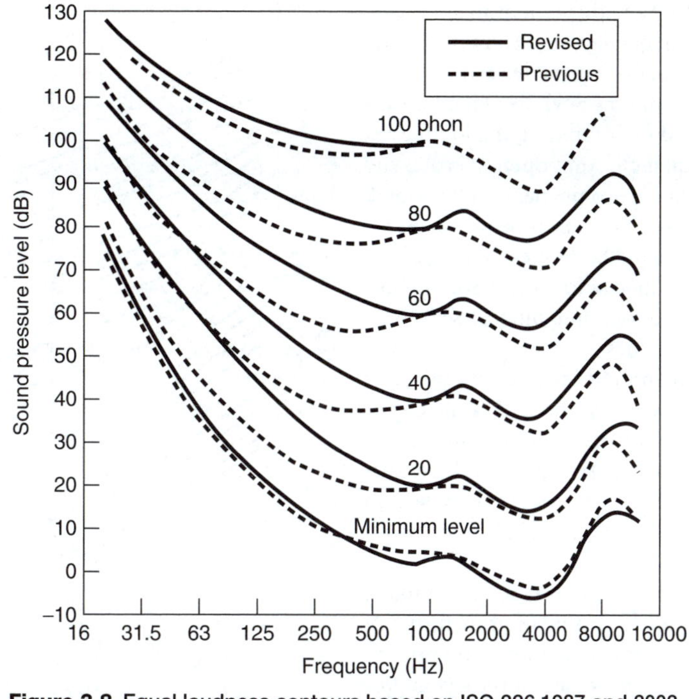

Figure 3.8 Equal loudness contours based on ISO 226 1987 and 2003

Taking 40 Phons as an arbitrary starting point for the scale, we have 40 Phons being equivalent to 1 Sone, and therefore a 10 dB increase to 50 Phons gives 2 Sones, and so on:

40 Phons = 1 Sone
50 Phons = 2 Sones
60 Phons = 4 Sones
70 Phons = 8 Sones
80 Phons = 16 Sones, etc.

But also:

40 Phons = 1 Sone
30 Phons = 0.5 Sones
20 Phons = 0.25 Sones

This simple numerical relationship between the Phon value and the Sone value can be expressed by the following equations:

$$\text{Sones} = 2^{(\text{Phons}-40)/10}$$

$$\text{Phons} = 40 + 33\log(\text{Sones})$$

The equal loudness curves are easy to use but they only apply for pure tones. How do we estimate the loudness of more complex, broadband sounds? There are two approaches that can be used; either to measure the loudness directly using a panel of listeners, or to develop methods for estimating the loudness from objective sound pressure level measurements (in either octave or third octave bands)

Loudness determined by subjective means

A panel of listeners are asked to compare the loudness of the broadband sound with that of a pure tone at 1000 Hz. The level of the pure tone is adjusted until it is judged to be equally as loud as the broadband sound under test.

This is not an easy task, particularly if the sound in question has a distinctive quality such as that from a motorcycle exhaust, but one that can be accomplished with practice. If the measured level of the reference tone is, say, 50 dB then it follows that the complex sound also has a loudness level of 50 Phons and a loudness of 2 Sones. In stating such a result it should be added that it has been obtained by subjective means.

Loudness calculated from spectral data

The above method for determination of loudness is very time consuming and expensive and therefore much research has been carried out over many years (and still continues) into developing a method for estimating the loudness of broadband sounds from objective measured data, i.e. from their sound pressure level frequency spectrum.

The method is based on the theory that the brain assigns a loudness to each frequency band component and then combines them in some way to arrive at a sensation of the loudness of the sound as a whole. There is also evidence that the method of combining the loudnesses of the individual frequency components is heavily influenced by the loudest of the components of the sound.

Therefore in order to simulate this process it is necessary first of all to develop a method for estimating a partial loudness for each frequency band (octave or third octave) and then a method for combining these into an overall estimate of total loudness.

There are in fact two approaches to doing this, one originally developed by Stevens in the USA using octave bands, and another developed by Zwicker in Europe based on third octaves. Not surprisingly, in order to distinguish between them, the quantities that they lead to are sometimes called Stevens Phons and Zwicker Phons respectively.

The system developed by Stevens is much easier to use and is the more empirically based method than that of Zwicker which is more complicated and attempts to more closely follow what is known about the signal processing that occurs in the cochlea with regard to the masking of one component sound by another.

It is the Stevens-based method which is described below. It is based on the set of empirically developed curves shown in Figure 3.9 which give the partial loudness index in Sones for sound pressure levels in various octave bands. The method for determining the loudness is as follows (see Figure 3.9):

Step 1: Use the curves to read off the Sone value (S_i) for each octave band, and identify the highest of these values ($S_{i,\max}$).

Step 2: Combine the individual Sone values into the total Sone value (S_{tot}) using the formula:

$$S_{\text{tot}} = S_{i,\max} + 0.3((\textstyle\sum S_i) - S_{i,\max})$$

(i.e. add up all the Sone values except the maximum, multiply this total by 0.3 and then add the maximum value).

Step 3: Turn the total Sone value into Phons, either by using the vertical nomogram to the right of the set of curves, or by using the formula:

$$\text{Phons} = 40 + 33\log(\text{Sones})$$

An example of the method is given in Example 3.1.

To make it clear that this calculation method has been used, these are referred to as SonesOD and PhonsOD: the OD refers to octave band data in a diffuse sound field.

The above method can be incorporated into a digital sound level meter to give a direct read-out of Phons (and this is equally true of the more complicated Zwicker method).

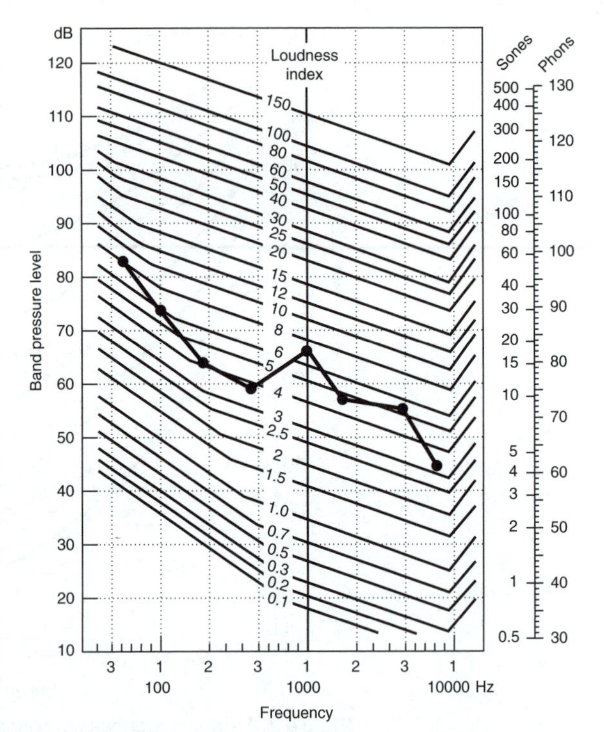

Figure 3.9 Chart for determination of loudness index

Example 3.1

Here we determine the loudness of the following sound in Sones and Phons.

Frequency	63	125	250	500	1000	2000	4000	8000
L_p dB	82	75	64	60	69	58	56	45
Partial loudness index, S_i	8.0	6.5	4.2	4.0	6.5	5.0	5.5	3.4

$S_{i,\max} = 8$ and the total of all other Sone values is 35.1, hence:

$$S_{\text{tot}} = S_{i,\max} + 0.3((\textstyle\sum S_i) - S_{i,\max})$$
$$= 8 + 0.3(35.1) = 18.5 \text{ Sones}$$

From the nomogram at the right of Figure 3.9, the calculated loudness level value corresponding to 18.5 Sones is 82 Phons.

You may wish to calculate the overall A-weighted sound pressure level determined for this same spectrum and compare (70 dBA).

Exercise

Following the example above, determine the loudness of a sound in Sones and Phons.

Frequency	63	125	250	500	1000	2000	4000	8000
L_p dB	45	56	58	69	60	64	75	82
Partial loudness index, S_i								

$S_{i,\max} = \ldots.$ hence $S_{tot} = S_{i,\max} + 0.3((\sum S_i) - S_{i,\max})$

$= \ldots\ldots\ldots = \ldots\ldots$ Sones

From the nomogram the loudness level is 109 Phons.

You may wish to calculate the overall A-weighted sound pressure level determined for this same spectrum and compare (82 dBA).

3.8 Noisiness, the Noy and PNdB

The Noy

When dealing with broadband sounds, that is to say the sounds of everyday life, the rating of sounds in terms of their respective noisiness instead of their loudness has a lot to commend it. This was certainly found to be the case with the early studies of aircraft noise. It seems that if people are asked to rate the noisiness of a sound a more consistent analysis results. In the public perception, noisiness is understood to carry connotations not only of loudness but also of intrusiveness or disturbance.

Following the way ideas about loudness were developed, it seems logical therefore to introduce a standard of noisiness and since this is about noise rather than pure tones, it makes sense to take a band of noise centred on 1 kHz, 1 octave wide and of measured sound pressure level 40 dB. Such a sound would be assigned a value as one unit of noisiness. The idea being that any sound judged twice as noisy is assigned a value of 2 units. The unit is called the Noy.

As was discussed for loudness, it may be possible but not practicable, to develop a procedure for the subjective assessment of noisiness. In this case a variable output generator producing noise of 1 octave centred on 1 kHz could be adjusted to a state of equal noisiness with the sound in question. A sound level meter would then measure the level and the Noy value read from the graph.

Noisiness from spectral data

We can follow the parallel with loudness even further by recalling the procedure for calculating loudness and loudness level from the spectrum of a sound. A similar empirical approach is applied by taking the spectrum of a sound and determining its overall noisiness by referring the spectrum to another set of empirical curves. Figure 3.10 shows a set of partial noisiness (or partial noisiness index) curves. Because the same 10 dB rule is generally found to apply here, an octave band of noise at a sound pressure level of 50 dB has a noisiness of 2 Noys. The calculation procedure is very similar to that for the calculated Sones. The spectral data is plotted onto the chart shown in Figure 3.10 and the partial noisiness for each octave band determined. The total noisiness of the sound is obtained by identifying the band with the highest partial noisiness and giving that a weighting factor or multiplier of unity; all the others are given a weighting of 0.3 and all these components summed and the total noisiness determined.

The shape of the partial noisiness index curves is not too different from that of the general shape of the equal loudness contours, except that in the higher frequency region, around 2 to 6 kHz, the curves descend quite dramatically; the implication of this is that any signal in those bands contributes more to the sensation of noisiness than would be expected on the basis of their loudness alone.

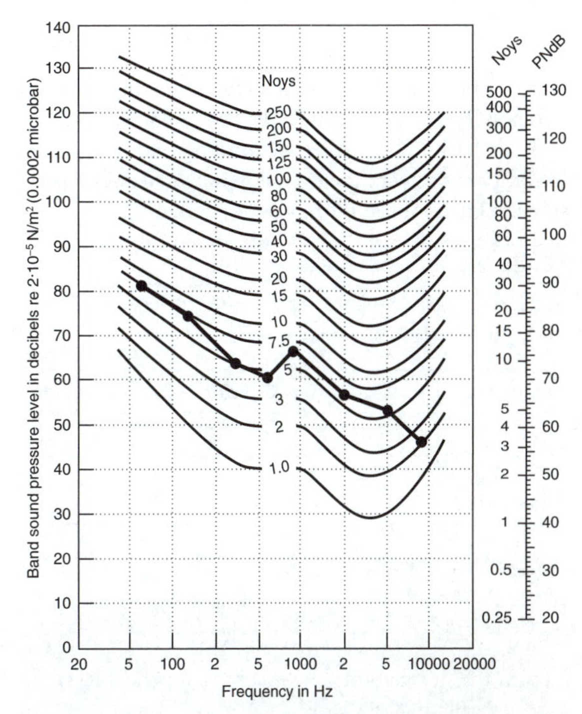

Figure 3.10 Example of the determination of noisiness of a sound in Noys and perceived noisiness level, PNdB

Perceived noisiness level, L_{PN}, expressed as a number of PNdB

If we further follow the structure of the section on loudness, it is logical to explore the possibility of a logarithmically-based equivalent to the noisiness expressed in Noys. The result is the introduction of the **perceived noisiness level**, L_{PN}, a quantity that will be familiar to those dealing with noise around an airport as it is commonly used in the process for the development of aircraft noise contours. This quantity is derived from the total Noy value by use of the nomogram on the graph or by using equations that look familiar from the discussion of loudness.

$$\text{Noys} = 2^{(L_{PN}-40)/10}$$

$$L_{PN} = 40 + 33\log(\text{Noys}_{tot})$$

Example 3.2

Here we determine the noisiness of a sound in Noys and the perceived noisiness level, PNdB.

Frequency	63	125	250	500	1000	2000	4000	8000
L_p dB	82	75	64	60	69	58	56	45
Partial noisiness N_i	7.5	7.5	4.5	4	7.5	6.5	6.5	2

$$N_{i,\max} = 7.5 \text{ hence } N_{tot} = N_{i,\max} + 0.3((\Sigma N_i) - N_{i,\max})$$
$$= 7.5 + 0.3(38.5) = 19 \text{ Noys}$$

From the nomogram at the right of Figure 3.9, the perceived noisiness level corresponding to 19 Noys is 83 PNdB.

Refer back to the earlier section on Sones and Phons to compare this with the overall A-weighted sound pressure level determined for this same spectrum.

Exercise

Determine the noisiness of a sound in Noys and the perceived noisiness level, PNdB.

Frequency	63	125	250	500	1000	2000	4000	8000
L_p dB	45	56	58	69	60	64	75	82
Partial noisiness N_i								

$$N_{i,\max} = \ldots. \text{ hence } N_{tot} = N_{i,\max} + 0.3((\Sigma N_i) - N_{i,\max})$$
$$= \ldots. + 0.3(\ldots\ldots)$$
$$= \ldots\ldots \text{Noys}$$

From the nomogram the perceived noisiness level is ...... PNdB. (95 PNdB)

Refer back to the earlier section on Sones and Phons to compare this with the overall A-weighted sound pressure level determined for this same spectrum.

Measurement of noisiness

Just as the shape of the A-weighting filter was developed from the shape of the equal loudness curves, so was a D-weighting filter developed from the shape of the Noy curves. Some sound level meters in the 1980s and 1990s had this frequency weighting filter and an estimate of the PNdB value could be obtained by adding a correction term to the D-weighted sound pressure level in accordance with the nature of the source. It was found, for example that:

L_{PN} = D-weighted sound pressure level + 7 for piston engine planes

L_{PN} = D-weighted sound pressure level + 13 for jet engine planes

L_{PN} = D-weighted sound pressure level + 13 (± 3) for aircraft generally.

Before the advent of the digital sound level meter in the 1990s the use of these approximate relationships gave a quick method for the direct measurement of the L_{PN}, (also referred to as PNdB), from a sound level meter. The use of the PNdB index for environmental noise assessment has since declined (in favour of the use of the A-weighted level) and the inclusion of the D-weighting in sound level meters has been discontinued. As is the case with loudness in Phons, modern sound level meters could in principle use digital signal processing and computational power to give a direct readout of an accurately determined PNdB value.

A refinement of the L_{PN} or PNdB is the effective perceived noise level which includes allowances for the duration of overflights and for the presence of tonal components in the aircraft noise. This index is still used in the aircraft noise certification process.

Noise and Number Index (NNI)

The perceived noise level was mainly developed in response to the increasing level of annoyance caused by the high frequency content of aircraft noise in the 1950s and 1960s. An index that was used in the past for the assessment of aircraft noise was the **Noise and Number Index (NNI)** which attempted to take into account the respective contributions of the noisiness of aircraft and the number of flights to the annoyance caused. For this, determinations of L_{PN} were made during an overflight so that the logarithmic average of the maximum L_{PN} values of various flights could be found. This is given the symbol $L_{PN,max}$. The NNI value is obtained from:

$$NNI = L_{PN,max} + 15\log N - 80$$

where N = number of aircraft heard during the daytime period.

Notice that the leverage implied in the multiplier of 15 is that the same increase in NNI (and hence of annoyance) can arise from either a twofold increase in the number of aircraft, N, or an increase of 4.5 dB in $L_{PN,max}$. The 80 value is introduced because annoyance is held to be zero when the sum of the first two terms does not exceed 80. NNI has now been replaced by $L_{Aeq,16h}$ as a measure of aircraft noise exposure, as explained in Chapter 4.

3.9 Acceptable noise levels inside buildings

A-weighted sound pressure levels

Within any building there are a range of spaces with different noise levels generated by the use of that space. For example the foyer area is only used for transit or general chatting when people are only a short distance apart. By contrast, in a board room it is essential that everyone present hears everything that is said even if they are metres apart at different ends of the room. A level of background noise in the building foyer or coffee area with A-weighted sound pressure level of around 50 dB would not unduly interfere with the use of that space, while such a level would be quite intrusive in the board room.

BS 8233:1999, provides guidance on acceptable noise levels for an extensive range of areas of occupancy within buildings, as shown in Table 3.1. These levels have been developed with a view to ensuring that the noise from outside the space does not interfere with the activity within the space. The level applies when the space is ready for occupancy but not actually occupied.

The main sources of noise from outside the space include building services such as air conditioning, lifts etc., plant and equipment in other rooms of the building, and intrusion of external noise coming through the façade.

Other sources of guidance for acceptable indoor ambient noise levels in spaces may be found in BS 8233, in the Chartered Institute of Building Services (CIBSE) Guide A, in Building Bulletin 93 (BB93) for spaces in schools, in the British Council for Offices (BCO) guides

Table 3.1 Indoor ambient noise levels in spaces when they are unoccupied, from BS 8233:1999

Criterion	Typical situations	Design range $L_{Aeq,T}$ (dB)	
		Good	Reasonable
Reasonable industrial working conditions	Heavy engineering	70	80
	Light engineering	65	75
	Garages, warehouses	65	75
Reasonable speech or telephone communications	Department store	50	55
	Cafeteria, canteen, kitchen	50	55
	Washroom, toilet	45	55
	Corridor	45	55
Reasonable conditions for study and work requiring concentration	Library, cellular office, museum	40	50
	Staff room	35	45
	Meeting room, executive office	35	40
Reasonable listening conditions	Classroom	35	40
	Church, lecture theatre, cinema	30	35
	Concert hall, theatre	25	30
	Recording studio	20	25
Reasonable resting/sleeping conditions	Living rooms	30	40
	Bedrooms[a]	30	35

[a] For a reasonable standard in bedrooms at night, individual noise events (measured with F time-weighting) should not normally exceed 45 dB L_{Amax}.

Best Practice in the Specification for Offices and *Office Fit-Out Guide*, and for hospitals in Health Technical Memorandum 08-01, *Acoustics*.

For some spaces where appropriate background noise levels are critical the acceptable levels are sometimes given in terms of the NR and NC descriptors discussed below.

NC and NR curves

The two sets of graphs presented in Figures 3.11 and 3.12 are referred to as the **noise criteria** and **noise rating** curves (and known as NC and NR curves) and are used for the rating of steady noise within buildings such as air conditioning noise. They are also essentially empirical, that is to say the result of research on people's response to noise rather than theoretical analysis. The NC curves originated in the USA and the NR curves (which are a smoother set of curves because they have been developed in such a way that an equation allows easy computation of their shape) originated in Europe based on the equal loudness contours in the superseded versions of ISO 226. A distinction between the two methods is that the NR curves are defined down to 31.5 Hz whereas the NC curves stop at 63 Hz.

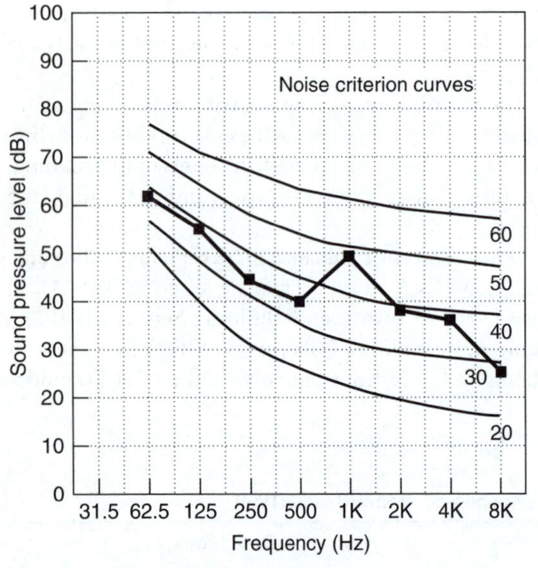

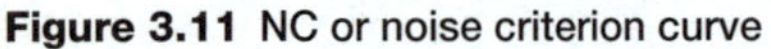

Figure 3.11 NC or noise criterion curve

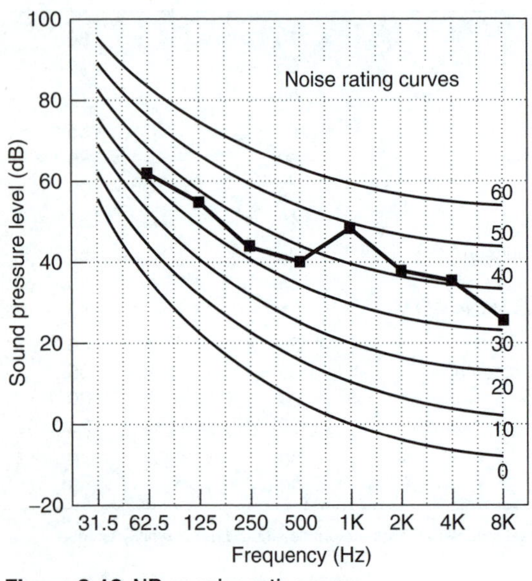

Figure 3.12 NR or noise rating curve

The method of use of the two sets of curves is similar. The spectrum of the noise in octave bands is plotted on the chart. The NC or the NR value for that spectrum is the curve with the lowest number (interpolating if necessary) that is just above the plotted spectrum. For example, the following spectrum has been plotted on both sets of curves shown below.

Frequency	63	125	250	500	1000	2000	4000	8000
L_p dB	62	55	44	40	49	38	36	25

From the NC curves, it can be seen that while the highest value for L_p is 62 dB at 63 Hz, when compared with the curves the level at 1 kHz is the highest with respect to the NC curves and this point is between NC45 and NC50. Interpolating between these two curves, the NC value for this spectrum is estimated at NC48. Similarly for the spectrum plotted on the NR curves, the NR value for the same spectrum is NR49. Specifications for acoustic comfort in buildings from noise from mechanical ventilation systems are sometimes expressed in terms of NR. Note that in some cases it is the sound pressure level in just one frequency band that can determine the value for NC or NR.

This approach has an advantage over just dealing with the A-weighted sound pressure level. The use of the NC or NR curve leads to the identification of the octave band that needs attenuating at first if a lower NC or NR value is to be achieved.

Example 3.3

NC40 has been set down as the limiting value in a contract for designing and installing an air conditioning system in a building, and the measured spectrum is that shown on Figure 3.11. This represents NC48, i.e. 8 dB above the specification. For what frequency should the maximum effort be applied for noise control?

Although the highest noise level is in the 63 Hz frequency band, it is the level of 49 dB at 1 kHz that defines the NC as 48. If the level in this frequency band was reduced by 8 dB to 41 dB, then NC40 would be achieved.

Exercise

Determine the NC and NR for the following spectrum

Frequency	63	125	250	500	1000	2000	4000	8000
L_p dB	65	62	58	49	40	38	42	40

NC = ... NR = ...

NR45 has been set down as the limiting value in a contract for designing and installing an air conditioning system in this building.

For what frequency, or frequencies, should the maximum effort be applied for noise control to meet this specification?

....

....

....

....

Inter-comparison of A-weighting, NC and NR values

Many acceptability criteria are specified in terms of A-weighted sound pressure levels but other references and specifications may be stated in terms of NR or NC value. As long as the comparison is only being made for air conditioning noise, it has been found that there is a sufficient commonality of the spectra and it is commonly assumed that the NC or NR value is some 6 or 7 units less than the A-weighted value. However, this should be applied with caution as much greater differences can be found for different frequency spectra.

A question arises about which of these approaches is the best for assessing a sound (PNdB can be put on one side because it is used primarily for aircraft noise purposes). The general answer is, it is a matter primarily of the spectrum involved. For general noise the A-weighted sound pressure level provides a good basis for assessment. For noise with substantial energy in the low frequencies, such as mechanical plant noise, NC and NR are generally preferred. They also have the benefit of indicating the frequency bands that need attenuation to achieve a lower NC or NR value.

Since the introduction of NC curves in 1957 and NR curves in 1962 many variants have been introduced, including: NCA, PNC, RC, NCB and LFNR. Some of these methods, and in particular the last one on the list (Low Frequency Noise Rating) are designed specifically to deal with improved assessment of noise containing a significant amount of low frequencies.

3.10 Other adverse effects of noise on health

The World Health Organization (WHO) has defined health as: 'a state of complete physical, mental and social well-being and not merely the absence of disease or infirmity'. This broad definition of health embraces the concept of well-being and, thereby, renders noise impacts such as population annoyance, interference with communication, and impaired task performance, as 'health' issues.

In its document *Guidelines for Community Noise* issued in 1999 the WHO lists the adverse effects of noise on health as:

- noise-induced hearing impairment
- interference with speech communication
- disturbance of rest and sleep
- psychophysiological, mental-health and performance effects
- effects on residential behaviour and annoyance
- interference with intended activities.

Table 3.2, taken from the Guidelines, lists the guideline values recommended by WHO for various environments.

Auditory and non-auditory effects of noise

Non-auditory effects of noise have been defined as 'all those effects on health and well-being which are caused by exposure to noise with the exclusion of effects on the hearing organ and effects which are due to the masking of auditory information (i.e. communication problems)'.[1] Such effects include performance effects, physiological responses and health outcomes, annoyance and sleep disturbance.

Auditory effects of noise are those related to hearing, i.e. NIHL, temporary threshold shift, damage to hair cells in the cochlea. This definition includes, for example, annoyance, sleep disturbance, task performance, social performance, cardiovascular effects, premature birth, psychiatric disorders, endocrine responses and chronic

[1] A. P. Smith and D. E. Broadbent, *Non-auditory Effects of Noise at Work: A review of the literature*, HSE Contract Research Report No. 30, 1992.

Table 3.2 WHO guideline values for community noise in specific environments

Specific environment	Critical health effect(s)	L_{Aeq} [dB(A)]	Time base [hours]	L_{Amax} fast [dB]
Outdoor living area	Serious annoyance, daytime and evening	55	16	–
	Moderate annoyance, daytime and evening	50	16	–
Dwelling, indoors	Speech intelligibility and moderate annoyance, daytime and evening	35	16	
Inside bedrooms	Sleep disturbance, night-time	30	8	45
Outside bedrooms	Sleep disturbance, window open (outdoor values)	45	8	60
School class rooms and pre-schools, indoors	Speech intelligibility, disturbance of information extraction, message communication	35	during class	–
Pre-school bedrooms, indoor	Sleep disturbance	30	sleeping-time	45
School, playground outdoor	Annoyance (external source)	55	during play	–
Hospital, ward rooms, indoors	Sleep disturbance, night-time Sleep disturbance, daytime and evenings	30 30	8 16	40 –
Hospitals, treatment rooms, indoors	Interference with rest and recovery	#1		
Industrial, commercial shopping and traffic areas, indoors and outdoors	Hearing impairment	70	24	110
Ceremonies, festivals and entertainment events	Hearing impairment (patrons: <5 times/year)	100	4	110
Public addresses, indoors and outdoors	Hearing impairment	85	1	110
Music and other sounds through headphones/ earphones	Hearing impairment (free-field value)	85 #4	1	110
Impulse sounds from toys, fireworks and firearms	Hearing impairment (adults) Hearing impairment (children)	– –	– –	140 #2 120 #2
Outdoors in parkland and conservations areas	Disruption of tranquillity	#3		

Notes
#1: As low as possible.
#2: Peak sound pressure (not LAF, max) measured 100 mm from the ear.
#3: Existing quiet outdoor areas should be preserved and the ratio of intruding noise to natural background sound should be kept low.
#4: Under headphones, adapted to free-field values.

health effects. The evidence for some of these effects, including reliable dose–response relationships, is much stronger than for others.

A useful further reference is the Medical Research Council/Institute of Environmental Health Report R10 (1997) *Non-auditory Effects of Noise.*

Questions

1 Describe the main components and function of the outer, middle and inner ear. What is meant by the term 'impedance matching' in connection with the middle ear.

2 The sketch shows the structure of the ear. Identify the parts labelled A to E. (adapted from IOA 2009)

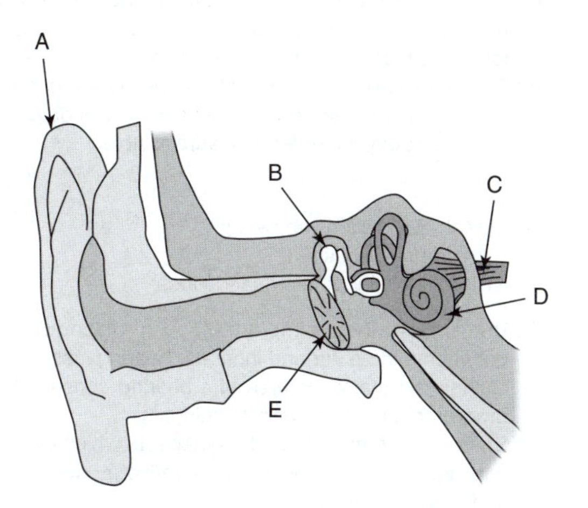

3 The sketch below shows the structure of the organ of Corti of the cochlea of the ear.

(a) Name the parts labelled A to E.
(b) Briefly describe how these structures are involved in the conversion of sound waves into nerve impulses.
(c) Distinguish between conductive and sensory hearing loss.
(d) Name two pathological conditions, other than noise-induced, associated with sensory hearing loss.
(e) Briefly describe the structural changes which occur in the inner ear as noise-induced hearing loss develops due to prolonged exposure to noise. (IOA)

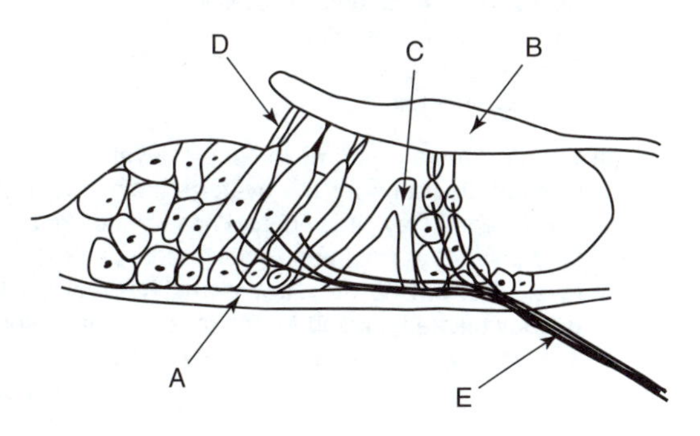

4 Describe the effect of noise on health distinguishing between auditory and non-auditory effects.

5 Discuss the criteria, based on different aspects of human response to noise, that may be used to set limits to maximum allowable levels of noise in various situations. Indicate approximately what levels of noise might be set: (i) to protect sleep disturbance in bedrooms at night, (ii) to avoid interference with speech, (iii) to minimize annoyance or disturbance to people in their homes in the daytime, and at night, (iv) to avoid disturbance to people working in offices, and (v) to prevent hearing loss.

6 Explain what is meant by an equal loudness contour and sketch the approximate shapes of three such contours for low, medium and high loudness levels respectively. Explain the relationship between these and the standard 'A' frequency weighting.

An octave analysis of the noise in a workshop yielded the following results:

OB	63	125	250	500	1000	2000	4000	8000
OB level (dB)	85	87	76	73	72	70	60	58

Calculate (a) the loudness in Stevens Sones, (b) the corresponding loudness in Phons, and (c) the sound level in dBA. ('A' weightings and loudness indices provided.) (IOA)

7 (a) Sketch audiograms to show clearly the conditions of:
(i) presbycusis
(ii) noise-induced hearing loss.

(b) Outline the damaging processes in the ear that cause these two conditions. How do the two conditions relate to one another?
(c) Discuss the practical and social implications on individuals of severe loss of hearing due to these conditions. (IOA)

8 (a) Discuss the structure of the ear with respect to the following dysfunctions:
(i) acoustic trauma
(ii) conductive hearing loss
(iii) tinnitus.

(b) Explain the principle behind the provision of ear defenders.
(c) Where protection is required for persons working underwater, what additional factors have to be considered?

9 (a) Describe the physical processes involved in the transformation of pressure waves in the inner ear into nerve impulses.
(b) Discuss the physiological characteristics associated with noise-induced hearing loss and indicate the frequency range at which it is most likely to be detected.
(c) Define conductive and sensory hearing loss.
(d) Discuss the advantages and disadvantages of the self-recording audiometer. (IOA)

10 (a) List and briefly discuss the effects of sound of both low and high intensity on people.

(b) Distinguish between conductive and sensorineural hearing loss.
(c) In what ways may noise cause (i) conductive, and (ii) sensorineural hearing loss?
(d) Briefly describe the following pathological conditions and in each case comment on their effect on the hearing mechanism: otosclerosis, Ménière's disease, wax plug, chronic catarrh.

11 (a) Explain the difference between conductive and sensorineural hearing loss.
(b) Give four examples of the cause of each type of loss, explaining the nature of the physiological changes associated with each cause. (IOA 2006)

12 (a) Outline the two main methods for establishing the daily personal noise exposure for individuals working in a noisy environment. Comment on the advantages and disadvantages of each method.
(b) Describe four factors that might be taken into consideration when choosing hearing protection for work in a vehicle body shop.
(c) Explain briefly what is meant by each of the following terms in the context of hearing disorders:
(i) conductive hearing loss
(ii) sensorineural hearing loss
(iii) presbyacusis
(iv) acoustic trauma
(v) tinnitus
(vi) temporary threshold shift
(vii) noise induced hearing loss
(viii) noise susceptibility.
(IOA 2008 (parts (a) and (b) relate to Chapter 9))

13 Calculate the loudness in Phons and perceived noise level in PNdB of noise which has the following octave band analysis:

Octave band centre frequency (Hz)	63	125	250	500	1000	2000	4000	8000
dB level	73	70	69	71	70	65	71	56

14 An octave band analysis of sound in a machine shop was made and the following results obtained:

Octave band centre frequency (Hz)	63	125	250	500	1000	2000	4000	8000
SPL (in dB)	68	72	90	87	86	88	90	84

Calculate the loudness in Phons and the perceived noise level in PNdB.

15 (a) Describe, with the aid of a labelled diagram, the components of the ear and describe their role in transforming sound in the external air to impulses in the auditory nerve.
(b) Name two pathological conditions, other than noise induced, associated with conductive hearing loss. In each case state the effect of the condition on the auditory mechanism.
(c) Briefly describe a method which may be used to distinguish between conductive and sensory hearing loss, and state the rationale of this method.
(d) Briefly describe the structural changes which occur in the inner ear as noise induced hearing loss develops as a result of prolonged exposure to noise.
(IOA 2004)

16 (a) Sketch audiograms showing:
(i) mild presbycusis
(ii) early noise induced hearing loss.

(b) Briefly describe a simple method of distinguishing between middle ear and inner ear hearing loss.
(c) Outline the major features of a hearing conservation programme in an industrial setting.
(d) State, and comment on, three aspects which may be relevant when considering the relative merits of earmuffs and earplugs.
(IOA 2003 (parts (c) and (d) relate to Chapter 9))

17 (a) Discuss with the aid of a suitable diagram, the way in which the detection of sound by humans varies with the amplitude and frequency of sound.
(b) Discuss what is meant by the equal loudness contours. Describe how they are derived and explain their relevance to the A-weighting scale. Use sketches to illustrate your answer. (IOA 2002 (part question))

18 (a) Describe the hearing process with particular reference to each of the following:
(i) the external ear
(ii) the middle ear and the ossicles
(iii) the transformer action
(iv) the aural reflex
(v) the inner ear.

(b) Sketch typical hearing threshold level versus frequency diagrams for the average male who is late middle aged (say circa 55) and has worked for a long time (say 40 years) in an A-weighted industrial noise level of about 80 dB. How is the HTL versus frequency diagram altered for the most and least sensitive (say 10%) part of a normal population?
(IOA 2000)

Chapter 4 Environmental noise

4.1 Introduction

Most noise whether occurring in the environment or in the workplace varies with time. Therefore this chapter starts with a discussion of how time varying noise may be described and measured using a variety of parameters, with particular emphasis on L_{Aeq}.

This chapter will explore features and characteristics of these different parameters and present an historical review of their development in order to explain why they are used. A brief history of how sound level meters have developed is also included.

The chapter concludes with a brief discussion of some of the wider aspects of environmental noise: its measurement, assessment, prediction and control. Since many of these topics are of great public interest much additional and more detailed information is available on websites, particularly, for example, regarding aircraft noise.

4.2 Time varying noise – continuous equivalent noise level, $L_{\text{Aeq},T}$

The level of many noises varies with time, for example noise from traffic (see Figure 4.1). It is not possible to find a single measure which can accurately quantify and correlate with what is heard, and a combination of measures relating to a time-averaged noise level and various percentile noise levels are used. These various measures are most usually made in dBA because this corresponds approximately to the frequency response of the ear, although in principle they could use the C, or any other weighting, or indeed octave or one third octave band values.

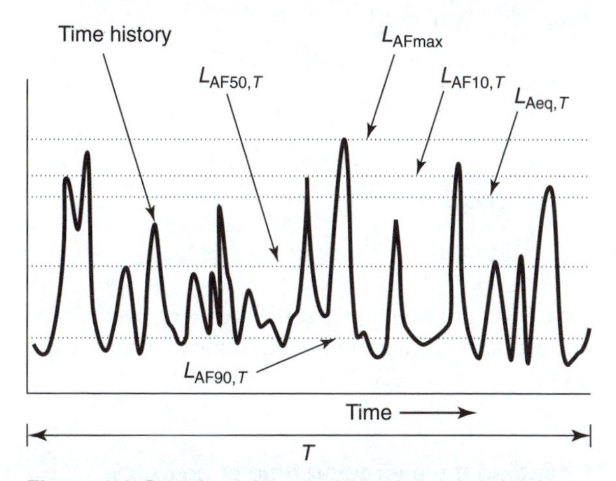

Figure 4.1 Sample of time varying noise, using the *A* frequency and *F* time weightings, over a sample duration of *T* seconds

The continuous equivalent, or time-averaged noise level, is discussed in more detail below. The statistical noise percentile level, $L_{\text{A}N,T}$ is the noise level, in dBA, which is exceeded for N% of the measurement time interval, T. Although in principle N may have any value between 0 and 100%, in practice L_{A10}, L_{A50}, and L_{A90} are the most commonly used values.

$L_{\text{AF10},T}$ is the sound level in dBA which is exceeded for 10% of the time interval T. It corresponds to the level of the higher peaks in the noise, and has been widely used in the UK for the measurement and assessment of traffic noise.

$L_{\text{A90F},T}$ is the sound level in dBA which is exceeded for 90% of the time interval T. It is widely used as a measure of background noise, i.e. the more or less constant lower level of noise which underlies bursts of higher levels of noise.

The indices $L_{\text{A99F},T}$ and $L_{\text{A1F},T}$ are also used as measures of the long-term background noise level and of the very highest noise levels which can occur, although the maximum noise level, $L_{\text{AFmax},T}$ is more commonly used for the latter purpose.

Continuous equivalent noise level $L_{\text{Aeq},T}$ (also known as the time-average noise level)

The continuous equivalent noise level, $L_{\text{Aeq},T}$ of a time varying noise is the sound pressure level, in dBA, of a steady sound that has, over the time period T, the same amount of A-weighted sound energy as the time varying noise.

If the time variation is fairly simple, consisting of different constant levels of noise for different periods of time (see Figure 4.2) the value of $L_{\text{Aeq},T}$ may be

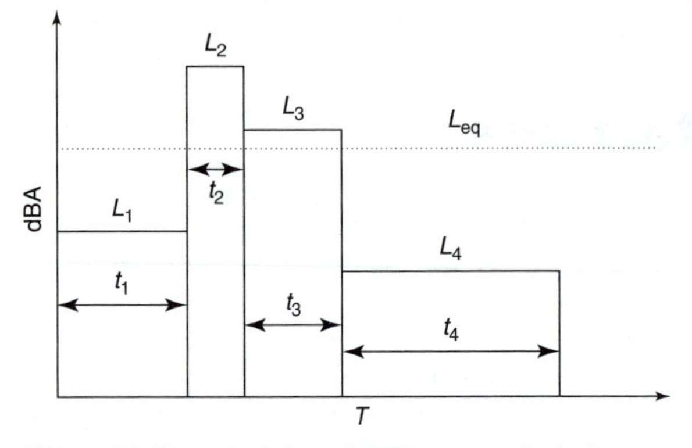

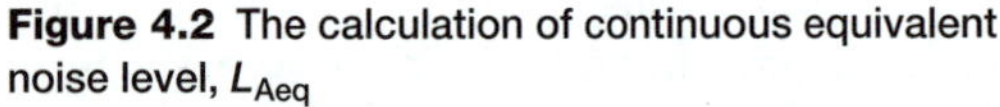

Figure 4.2 The calculation of continuous equivalent noise level, L_{Aeq}

calculated from the individual noise levels and time periods:

$$L_{Aeq,T} = 10\log[(t_1 \times 10^{L1/10} + t_2 \times 10^{L2/10} + t_3 \times 10^{L3/10} + \cdots + T_N \times 10^{LN/10})/T]$$

where t_1 is the time at noise level L_1 dBA
t_2 is the time at noise level L_2 dBA
t_3 is the time at noise level L_3 dBA, etc.

and T is the time over which the value is required.

The $L_{Aeq,T}$ is in effect a sort of average noise level over the measurement or assessment period. More specifically and precisely it is the noise level which corresponds to the time weighted average 'sound pressure-squared' value over the time T, which in turn is related to the average, time weighted sound intensity. Thus in the above equation it is the term inside the brackets $[(t_1 \times 10^{L1/10} + t_2 \times 10^{L2/10} + t_3 \times 10^{L3/10} + \cdots + T_N \times 10^{LN/10})/T]$ that represents the time weighted average 'sound pressure-squared', over the time T.

In more complicated patterns of time variation where the level of noise is continuously changing from moment to moment it is not possible to calculate the value of $L_{Aeq,T}$ but it may be measured using an integrating sound level meter capable of performing time integration of the instantaneous sound pressure squared according the defining equation:

$$L_{Aeq,T} = 10\log\left[\left(\int^{T} p^2(t)dt/T\right)\right]$$

The continuous equivalent sound level $L_{Aeq,T}$ is widely used for the measurement and assessment of both noise exposure levels in the workplace, and for environmental noise, as illustrated in the following examples.

Example 4.1

Calculate the continuous equivalent sound level, $L_{Aeq,8h}$, over an eight-hour working day, for an employee exposed to the following pattern of noise levels and exposure times:

94 dBA for 3 hours
89 dBA for 2 hours
98 dBA for 0.5 hours
83 dBA for 2.5 hours

$$\begin{aligned} L_{Aeq,T} &= 10\log[(t_1 \times 10^{L1/10} + t_2 \times 10^{L2/10} + t_3 \times 10^{L3/10} + \cdots + T_N \times 10^{LN/10})/T] \\ &= 10\log[(3 \times 10^{9.4} + 2 \times 10^{8.9} + 0.5 \times 10^{9.8} + 2.5 \times 10^{8.3})/8] \\ &= 92.034 = 92\,\text{dBA} \end{aligned}$$

Example 4.2

Noise from a building site is caused by five items of plant. The periods of operation of each item of plant during the working day and the noise level each produces at a noise sensitive property at the boundary of the site are shown below. Calculate the equivalent continuous noise level over a 12-hour working day.

Compressor	83 dBA operating for 5 h
Excavator	85 dBA operating for 2 h
Dumper truck	76 dBA operating for 6 h
Pump	75 dBA operating for 7 h
Pile-driver	88 dBA operating for 15 h

$$\begin{aligned} L_{Aeq} &= 10\log(t_1 \times 10^{L1/10} + t_2 \times 10^{L2/10} + \cdots + T_N \times 10^{LN/10})/T \\ &= 10\log(5 \times 10^{8.3} + 2 \times 10^{8.5} + 6 \times 10^{7.6} + 7 \times 10^{7.5} + 1.5 \times 10^{8.8})/12 \\ &= 84.032 \\ &= 84.0\,\text{dBA (to the nearest 0.5 dBA)} \end{aligned}$$

Note from this example that the averaging period, T, is not necessarily equal to the sum of the individual component periods (i.e. $t_1 + t_2 + t_3 + \cdots + t_N$).

Changing the averaging time of an $L_{Aeq,T}$ calculation

Sometimes it is required to average the same amount of noise energy over a different time period (see Figure 4.3).

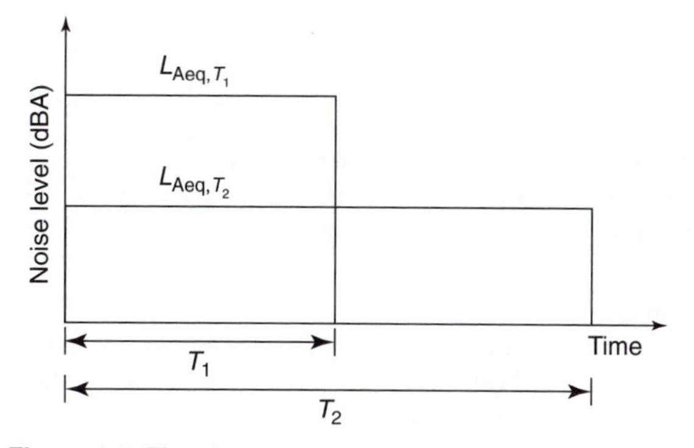

Figure 4.3 The change in L_{Aeq} values over different time periods

Example 4.3

Calculate the effect of spreading the building-site operations of the previous example over an 18-hour period, instead of a 12-hour period.

One could repeat the calculation, using the increased averaging time of 18 hours, but this is not necessary. It can be shown quite easily that the difference between two equivalent levels averaged over times T_1 and T_2 hours is given by (see Figure 4.3):

$$L_{Aeq,T_1} - L_{Aeq,T_2} = 10\log(T_2/T_1)$$

Applying this formula to the present case, where $T_2 = 18$ h and $T_1 = 12$ h:

$$L_{Aeq,T_1} - L_{Aeq,T_2} = 10\log(18/12) = 1.7609\ldots$$
$$= 2 \text{ dB (to the nearest 0.5 dB)}$$

so that the effect of extending the time period to 18 h in this case is to reduce the equivalent continuous noise level by 2 dB, to 82.0 dBA.

Example 4.4

What is the maximum time for which an employee may spend in a particular workshop where the noise level is 101 dBA without using hearing protection if his/her noise dose is not to exceed an L_{Aeq} of 85 dBA over the period of the eight-hour working shift?

(a) Assume that for the rest of the shift the employee works in a quiet environment; in practice this means less than 75 dBA.

(b) Assume that for the rest of the shift the employee is subjected to a constant level of 80 dBA.

Part (a)

A rough estimate of the allowed time may be obtained using the fact that the energy content of the noise exposure, and thus the L_{eq} value, will remain the same if the intensity of the sound is doubled, but for only half the duration, i.e. if the level is increased by 3 dB for half the time. This is called the **equal energy principle.**

Thus, 85 dBA for 8 h is equivalent to:

88 dBA for 4 h
91 dBA for 2 h
94 dBA for 1 h
97 dBA for 30 min
100 dBA for 15 min
103 dBA for 7.5 min
106 dBA for 3.75 min
and so on.

Therefore, in this case the exposure to 101 dBA may be allowed for somewhere between 7.5 and 15 minutes. It may be calculated more precisely as follows:

$$L_{Aeq,T_1} - L_{Aeq,T_2} = 10\log(T_2/T_1)$$

where T_1 is the required time (in hours)
$T_2 = 8$ h
$L_{Aeq,T_1} = 101$ dBA
$L_{Aeq,T_2} = 85$ dBA.

Therefore: $101 - 85 = 10\log(8/T_1)$
$16 = 10\log(8/T_1)$

From which (by dividing by 10 and taking antilogs):

$$8/T_1 = 10^{1.6} = 39.8$$

and $T_1 = 8/39.8 = 0.201\,\text{h} = 12.1$ minutes

(which agrees with the approximate estimate of between 7.5 and 15 minutes).

Part (b)

Let the required time in this case be t hours, so that the employee is exposed to 101 dBA for t hours and 80 dBA for $(8 - t)$ hours.

The eight-hour L_{Aeq} is given by:

$$L_{Aeq,8h} = 10\log[\{t \times 10^{10.1} + (8 - t) \times 10^{8.0}\}/8]$$

In this case this is 85 dBA, therefore:

$$85 = 10\log[\{t \times 10^{10.1} + (8 - t) \times 10^{8.0}\}/8]$$

This can be written in an alternative form (by dividing by 10 and taking antilogs):

$$t \times 10^{10.1} + (8 - t) \times 10^{8.0} = 8 \times 10^{8.5}$$

from which: $t = [8 \times (10^{8.5} - 10^{8.0})/(10^{10.1} - 10^{8.0})$
$= 0.1385$ h, or 8.3 minutes

This answer seems sensible; it is a few minutes shorter than the previous answer (12.1 minutes) because the employee already has a residual noise exposure before entering the noisy workshop.

A simpler method (less algebra) but more approximate method involves making the assumption that the exposure of 101 dBA for t hours is in addition to an exposure to 80 dBA for the full 8 h. If the 80 dBA is subtracted from 85 dBA (by decibel subtraction, it gives a value of 83.3 dBA, which is the L_{eq} value to be produced by the workshop exposure alone, the 101 dBA. The calculation then proceeds as in part (a) but using the value of 83.3 dBA instead of 90 dBA. This gives an allowed time of 0.1374 h or 8.2 minutes, which agrees with the more exact calculation to within a tenth of a minute.

Example 4.5

The noise level at a site on which it is proposed to build a housing estate arises mainly from trains on a nearby railway line. There are three types of train using the line – fast express trains, slower suburban trains and freight trains. It is proposed to predict the equivalent continuous noise level at the site over a 24-hour period from sample noise measurements of each of the three noise events. The results of these measurements are:

- for fast trains $L_{eq} = 85$ dBA over a period of 12 s
- for slow trains $L_{eq} = 78$ dBA over a period of 18 s
- for freight trains $L_{eq} = 76$ dBA over a period of 24 s.

During the 24-hour period there are 120 fast trains, 200 slow trains and 80 freight trains. Calculate the equivalent continuous noise level over a 24-hour period.

There are a number of alternative ways of doing this calculation.

Method 1

The 24-hour L_{eq} is separately calculated for each type of train, then these individual values are combined. The total duration for each type of event is obtained by multiplying the single event duration by the number of events.

For the fast trains the total duration is 1440 s (12 s per train × 120 trains) and:

$$L_{Aeq,24h} = 10\log[(120 \times 12 \times 10^{8.5})/(24 \times 60 \times 60)] = 67.2 \text{ dBA}$$

For the slow trains:

$$L_{Aeq,24h} = 10\log[(200 \times 18 \times 10^{7.8})/(24 \times 60 \times 60)] = 64.2 \text{ dBA}$$

For the freight trains:

$$L_{Aeq,24h} = 10\log[(80 \times 24 \times 10^{7.6})/(24 \times 60 \times 60)] = 59.5 \text{ dBA}$$

The total is obtained by combining these three levels:

$$L_{Aeq,24h} = 10\log(10^{6.72} + 10^{6.42} + 10^{5.94}) = 69.4 \text{ dBA}$$

Note that the 24-hour period has been converted to seconds because the individual event durations were in seconds.

Method 2

Method 2 simply involves combining the two stages of Method 1 and calculating the total $L_{Aeq,24h}$ for the three types of event, all in one go.

$$L_{Aeq,24h} = 10\log[(t_1 \times 10^{L1/10} + t_2 \times 10^{L2/10} + t_3 \times 10^{L3/10})/T]$$

$$L_{Aeq,24h} = 10\log[\{(120 \times 12 \times 10^{8.5}) + (200 \times 18 \times 10^{7.8}) + (80 \times 24 \times 10^{7.6})\}/(24 \times 60 \times 60)]$$

$$= 69.4 \text{ dBA as before}$$

Although the second method might seem quicker, the first method has the advantage that it also gives the individual contribution of each of the three types of event to the overall $L_{Aeq,24h}$ value.

Noise exposure from single discrete events (leading to Method 3)

In many situations, as in the last example, the total noise exposure over a period of time is made up from a number of different individual events such as the passing of an aircraft overhead or a train nearby, or a short burst of machinery noise, and L_{eq} measurements of the noise from different events will be made over different durations. It would be convenient for comparison of different types of event if the noise from all events could be averaged over the same duration. For convenience, a time of 1 s is chosen.

The **sound exposure level**, L_{AE}, of a single discrete noise event is the level which if maintained constant for a period of 1 s would contain as much A-weighted sound energy as is contained in the actual noise event. The term 'single event noise exposure level' and the symbols SEL and L_{AX} are also in use; they mean exactly the same. The idea of L_{AE} is illustrated in Figure 4.4.

The relationship between the $L_{Aeq,T}$ value produced by an event over a period of time, T, and the L_{AE} value for the event is given by:

$$L_{AE} = L_{Aeq,T} + 10\log T$$

where T must be in seconds.

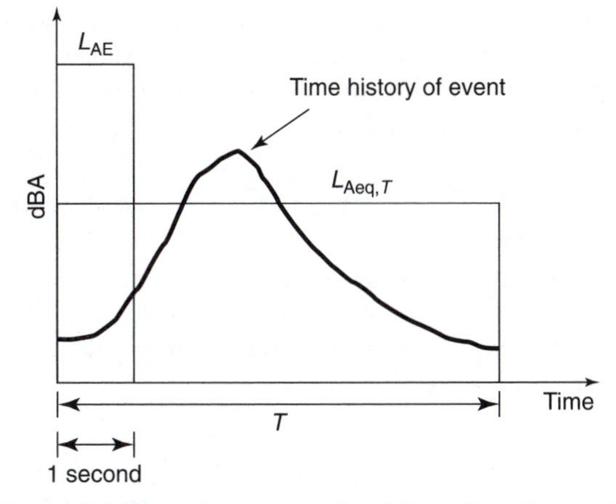

Figure 4.4 The noise exposure level, L_{AE}, of a noise event

Many sound level meters have the facility to measure and indicate L_{AE} values directly but, if not, the value may be calculated from the above formula.

Example 4.6

Calculate the L_{AE} values for the three types of train in the last example.

For the fast trains:

$L_{AE} = 85 + 10\log 12 = 95.8$ dBA

For the slow trains:

$L_{AE} = 78 + 10\log 18 = 90.6$ dBA

For the freight trains:

$L_{AE} = 76 + 10\log 24 = 89.8$ dBA

Note that if the L_{Aeq} of the event is measured over a time greater than one second, which is usually the case, then the L_{AE} value is higher than the $L_{Aeq,T}$ value.

We can use the L_{AE} value to calculate the value of $L_{Aeq,T}$ when N identical events all having the same L_{AE} value occur within time T:

$$\mathbf{L_{Aeq,T} = L_{AE} + 10\log N - 10\log T}$$

where, once again, T must be in seconds.

Example 4.7

Calculate the value of L_{Aeq} due to 120 fast trains passing over a 24-hour period, each with an L_{AE} value of 95.8 dBA, as in the previous example:

$$L_{Aeq,24h} = 95.8 + 10\log(120) - 10\log(24 \times 60 \times 60)$$
$$= 95.8 + 20.8 - 49.4$$
$$= 67.2 \text{ (as before)}$$

It is possible to calculate the L_{eq} over a given period of time from the number of different events occurring in that period, and their L_{AE} values.

$$L_{Aeq,T} = 10\log[\{(N_1 \times 10^{L_{AE1}/10}) + (N_2 \times 10^{L_{AE2}/10}) + (N_3 \times 10^{L_{AE3}/10}) + \cdots + (N_n \times 10^{L_{AEn}/10})\}/T]$$

where there are N_1 events with L_{AE1}, N_2 events with L_{AE2} etc. over time T seconds.

This gives a third method for doing Example 4.5 involving the three types of trains.

Method 3

In this case $N_1 = 120$, $L_{AE1} = 95.8$ dBA, $N_2 = 200$, $L_{AE2} = 90.6$ dBA, $N_3 = 80$, $L_{AE3} = 89.8$ dBA.
Therefore:

$$L_{Aeq,24h} = 10\log[\{(120 \times 10^{9.58}) + (200 \times 10^{9.06}) + (80 \times 10^{8.98})\}/(24 \times 60 \times 60)]$$
$$= 69.4 \text{ dBA (as before)}$$

Some noise indices based on $L_{Aeq,T}$

The useful property of the L_{Aeq} is that, as it is based on the energy average, individual values over specific time periods can be combined to produce an average over longer periods, or various descriptors, or metrics, which can be used in criteria as summarized below.

Occupational noise exposure $L_{Aeq,8h}$

This is in the form of a personal daily noise exposure level, $L_{Ep,d}$. It is used throughout Europe, and in the UK Control of Noise at Work Regulations 2005, as the basis for determining occupational noise exposure: the lower and upper exposure action values are 80 dBA and 85 dBA respectively.

Day–night level, L_{dn}

This is popular in the USA for environmental noise. It is a single descriptor to cover the 24-hour period but allows for the greater annoyance due to noise at night-time by applying a +10 dB for noise over the nine-hour period between 22.00 and 07.00 hours.

$$L_{dn} = 10\log[(15 \times 10^{(L_d/10)} + 9 \times 10^{(L_n+10)/10})/24]$$

Day–evening–night level, L_{den}

This is the main descriptor used in Europe following the introduction of the EU Directive 2002/49/EC of 25 June 2002. It is a single descriptor to cover the 24-hour period but allows for the varying annoyance for the day, evening and night with a weighting factor of 5 dB for evening and 10 dB for night. The day is 12 hours, the evening 4 hours and the night 8 hours. The default values are day from 07.00 to 19.00 hours, evening 19.00 to 23.00 hours and night 23.00 to 07.00 hours. The member states can adjust the clock times that apply for these periods to allow for the social environment in their country.

$$L_{den} = 10\log[(12 \times 10^{(L_{day}/10)} + 4 \times 10^{(L_{even}+5)/10} + 8 \times 10^{(L_{night}+10)/10}/24]$$

Day and night levels: L_d and L_n

In PPG24 (PAN566 in Scotland), used for planning assessments, $L_{Aeq,16hour(day)}$, from 07.00 hours to 23.00 hours, and $L_{Aeq,8hour(night)}$ from 23.00 to 07.00 hours are required.

The same $L_{Aeq,16hour(day)}$ descriptor is used for annual aircraft noise contours in the UK, although L_{den} contours are now also produced.

The value used in the UK for the assessment of noise from trains is 24-hour L_{Aeq} ($L_{Aeq,24hour}$).

Example 4.8

Twenty-four individual one-hour L_{Aeq} values are given below.

Calculate: $L_{Aeq,16hour(day)}$, $L_{Aeq,8hour(night)}$, $L_{Aeq,24hour}$, L_{dn}, L_{den}.[1]

Time	$L_{Aeq,1hour}$	Time	$L_{Aeq,1hour}$	Time	$L_{Aeq,1hour}$	Time	$L_{Aeq,1hour}$
00–01	45.4	06–07	48.4	12–13	52.3	18–19	52.8
01–02	42.2	07–08	51.8	13–14	51.3	19–20	52.7
02–03	42.0	08–09	51.0	14–15	51.2	20–21	51.6
03–04	43.3	09–10	52.6	15–16	51.6	21–22	51.5
04–05	41.6	10–11	53.7	16–17	52.6	22–23	49.9
05–06	43.6	11–12	51.9	17–18	51.7	23–24	49.2

4.3 A review of time varying noise descriptors

What is the best way of describing a period of time varying noise such as traffic noise? The most complete detailed description would involve a time history (see Figure 4.1 earlier), giving the time variation minute by minute over an hour, or every 15 minutes over a 24-hour period. Such a description is, however, rather detailed, and for convenience there is a need to be able to describe the period of time varying noise in terms of a single number or a series of numbers.

By using a range of descriptors it is possible to get a better feel for the characteristics of the noise climate over a period of time than would be possible from just one such value such as the L_{Aeq}.

Exercise

The table below shows the value of the various noise descriptors taken over a one-hour period for four different situations. What can you say about the noise climate in each case?

Case	L_{Aeq}	L_{A10}	L_{A50}	L_{A90}	L_{Amax}	L_{Amin}
A	56	57	55	54	58	53
B	65	55	54	53	82	51
C	61	64	59	54	79	53
D	47	50	44	38	78	36

Answer

In case A there is very little variation between the different values and this suggests that the noise climate is dominated by noise from one source emitting a more or less constant noise level, such as from an item of plant or machinery such as a pump, fan, motor or similar. The slight variations may arise from variations in noise from the machine itself or from other sources such as road traffic or aeroplanes.

In case B the fact that the L_{Aeq} value if is much higher than the L_{A10}, and that the three L_{AN} values are all similar suggests a short duration burst of high level noise on top of a more or less constant lower level of noise. This could be for example the noise climate close to a railway line late at night where maybe only one train passes during the hour.

Case C shows a wider variation between the three L_{AN} values than in A and B, indicating noise is from lots of different events or sources. The fact that the L_{A10} is 3 dB higher than the L_{Aeq} and that the L_{A50} is more or less midway between the L_{A10} and L_{A90} values indicates that this might be traffic noise.

Case D has the same features as case C, but at lower levels, and so may be traffic noise from a quieter road or from a quieter period in the day or night.

Some features and characteristics of the various descriptors

Constant level of noise

If the noise level is constant over a period of time the value of all the measures over the period will be the same constant values.

[1] Answers: $L_{Aeq,16hour(day)} = 52.0$ dB, $L_{Aeq,8hour(night)} = 45.4$ dB, $L_{Aeq,24hour} = 50.7$ dB, $L_{dn} = 54.1$ dB, $L_{den} = 54.3$ dB.

Why use L_{AN} values?

When the pattern of noise varies with time there is evidence that public annoyance is related to the occurrence of the higher level bursts of noise. The maximum level which occurred during the period is therefore obviously a useful descriptor, but there is no indication as to whether the noise event which produced this maximum value was uniquely high during the period, or whether it was one of several such events with similarly (but not nearly quite so) high noise levels. The second case is likely to be more annoying than the first, but the L_{Amax} value on its own will fail to distinguish between the two cases. This is where the use of a percentile value such as L_{A10} can be of value, and it has been shown to correlate reasonably well with public response, particularly for traffic noise.

Not also that the L_{A1} value can act as a useful alternative or supplement to the L_{Amax} value, and gives some indicate of the duration of the very highest noise levels in the noise climate.

When measuring high noise episodes in the environment you often have to measure the much lower background noise which prevails in between such noisy events. However, the background noise itself is rarely constant in level and so the use of a statistical descriptor such as the L_{A90} is useful way of ensuring a consistent approach to the measurement of background noise level. Over a long period of time of weeks and months the L_{A99} value is sometimes used as a long-term background noise level.

Limitations of L_{AN} values

To know, for example, that the L_{A10} level during a particular hour at a particular location is 62 dBA tells us that for six minutes in that hour the noise level will be above 62 dBA, but we will not know anything further about noise distribution during those noisiest six minutes, for example whether there is a constant noise level just above the L_{A10} value, or a series of even shorter duration higher level noise events (although knowing the L_{A1} and L_{Amax} values would give us some more information).

Similarly knowing that the L_{A90} value during an hour was 44 dBA tells us that for the quietest six minutes in the hour the noise level was below 44 dBA, but we do not know how low the noise levels are, or for how long, during these six minutes.

Sensitivity of parameters to changes in noise events

Any change in noise levels during a period will always cause a change in the L_{Aeq} value over the period, but may not necessarily change the L_{AN} values.

For example, if a more or less constant level of say 40 dBA over a period of one hour at a quiet location is disturbed by the onset of short bursts of higher levels noise, say at 70 dBA, then the L_{A10} value will not increase from 40 dBA unless the duration of the bursts exceeded 10% of the time period, i.e. 6 minutes per hour. Furthermore the L_{A10} value would also remain the same if the short burst of noise increased in level, say to 80 dBA, provided that its duration did not exceed 6 minutes. The value of the L_{Aeq} over the one-hour period would, however, be affected by the shorter durations of the higher level bursts of noise.

Exercise

What would be the value of $L_{Aeq,1hour}$ and $L_{A10,1hour}$ if the burst of higher level noise at 70 dBA noise lasted for three minutes in each hour, on top of the steady noise level of 40 dBA?

Answer

The $L_{Aeq,1hour}$ would be 57 dBA [$= 70 + 19\log(3/60)$], and the L_{A10} value would remain at 40 dBA. (Note the contribution of the steady level of 40 dBA to the L_{Aeq} value is negligible compared to that of the 70 dBA burst, and is ignored.)

A limitation of L_{Aeq}

How would the public respond to such a noise climate, i.e. one where the noise is at a fairly steady low level for almost all of the time but is interrupted by occasional short bursts of higher level noise, say three minutes every hour as in the above example?

The answer is that we cannot know precisely, but the indications are that the response is more likely to relate to the short bursts than it is to the average lower L_{Aeq} value, and that although L_{Aeq} has been shown to be a reasonably good indicator of human response in situations where the noise climate consists of a large number of events, it is not suited to situations such as in the above example, where the noise climate is dominated by a few occasional short bursts of high level noise.

Specification of noise limits

This in turn tells us that although L_{A10} may be a good indicator of annoyance from bursts of high level noise, an enforcement agency should be wary of using a L_{A10} value alone to set noise limits. If for example such an agency (a local authority planning or environmental health department, for example) were to seek to limit construction site noise simply by imposing a hourly limit on the L_{A10} at the site boundary this would allow the construction site to make very high levels of noise for up to six minutes each hour without exceeding any L_{A10} limit value. There would be similar pitfalls in using any other single value such as L_{Amax} or L_{Aeq} as a noise limiting condition, and for this reason a combination of values, most commonly involving separate limits on L_{Aeq} and on L_{Amax} are often used to specify limits on noise emissions.

Combining values for different periods

There is another way in which the L_{Aeq} value differs from the L_{AN} values. Because it is an average value, L_{Aeq} values can be aggregated and averaged in a way that period L_{AN} values cannot. If, for example, we monitor the parameters for 15-minute periods it is possible to combine the four $L_{Aeq,15min}$ values to obtain the $L_{Aeq,1hour}$ value, but it is not possible to do this for the L_{AN} values. Similarly, hourly values can be combined to give the L_{Aeq} over 24 hours, but not so the hourly L_{AN} values. The capability of being able to combine L_{Aeq} values in this way is one of the reasons for the popularity and widespread use of L_{Aeq} for measuring and assessing environmental noise. Notwithstanding the above limitation, which should generally apply, there are sometimes agreed standard procedures for aggregating L_{AN} values, such as that for obtaining the 18-hour L_{A10} value for noise insulation regulation purposes, which specifies the use of the arithmetic average of the 18 hourly L_{A10} values.

Generally if the pattern of noise is similar for the various component periods then averaging these values is likely to give a good approximation to the true value over the extended period, but it is important to understand that this may not necessarily always be the case, particularly if the noise pattern is significantly different during some of the component periods.

Relationships between different noise indices

In general there can be no fixed predictable relationship between L_{AN}, L_{Aeq} and L_{Amax} values, because differences will vary with the temporal pattern and distribution of the noise levels.

There are, however, some well known approximations, perhaps the best known of which is that for free flowing traffic the L_{A10} value is usually about 3 dB above the L_{Aeq} value.

Calculation of Road Traffic Noise (CRTN) gives a shortened measurement procedure for determining the $L_{A10,18hour}$ value from measurements made over any three consecutive hours between 10.00 and 17.00 hours:

$$L_{10(18hour)} = L_{10(3hour)} - 1 \text{ dBA}$$

PPG24 and BS 8233 give the following approximate relationship for converting $L_{A10,18hour}$ to $L_{Aeq,16hour}$ (for road traffic noise):

$$L_{Aeq,16h} = L_{A10,18h} - 2 \text{ dB}$$

This is generally correct with a 95% confidence interval of ±2 dB for moderate and heavy traffic flows.

The EC Environmental Noise Directive (END) 2006 requires EC countries to produce Noise Maps using the L_{den} noise index. Since Calculation of Road Traffic Noise (the UK traffic noise prediction method) predicts $L_{A10,18hour}$ there is a need for a method to convert this value to L_{den}.

The preferred approach[2] relies on determining hourly values of L_{A10} using the CRTN method and then converting these values to equivalent values of L_{Aeq} using the relationship:

$$L_{Aeq,1hour} = 0.94 \times L_{A10,1hour} + 0.77 \text{ dB}$$

except for non-motorway roads when hourly traffic flows are below 200 vehicles per hour during the period 24.00 to 06.00 hours, when the following relationship should be used:

$$L_{Aeq,1hour} = 0.57 \times L_{A10,1hour} + 24.46 \text{ dB}$$

The converted values obtained for the full 24 hours can then be used to derive the values of L_{den} and L_{night} as required by the END.

Alternative approaches are also given, including conversion from $L_{A10,18h}$ to L_{den}:

$$L_{den} = 0.92 \times L_{A10,18h} + 4.20 \quad \text{for non-motorway roads}$$

and

$$L_{den} = 0.90 \times L_{A10,18h} + 9.69 \quad \text{for motorways}$$

These approximate conversion procedures will be used until CRTN is modified or replaced (for example by a new Europe-wide prediction model, such as Harmonoise).

The use of fast (F) or slow (S) time weightings

All L_{AN} and L_{Amax} values are derived from sampling the instantaneous time weighted sound pressure levels, and therefore it is important always to state which weighting is being used, e.g. L_{AF10} or L_{AS10}, L_{AFmax} or L_{ASmax}, L_{AF90} or L_{AS90} etc. Although it is the maximum level that is the value most likely to be affected by the choice of time weighting the L_{AN} values will also be affected.

PPG24 requires use of the slow weighting for measuring the short bursts of noise at night-time, L_{ASmax}, and Slow weighting is also used for aircraft noise measurement in the environment. CRTN specifies the use of Fast

[2] P. G. Abbott and P. M. Nelson, *Converting the UK traffic noise index LA10,18h to EU noise indices for noise mapping* PR/SE/451/02 [EPG 1/2/37], TRL Report for Defra, 2002.

for measurement of L_{A10}, BS 4142 specifies the use of Fast weighting for measuring L_{A90}, and generally F is used rather than S.

For a constant steady noise the use of either Fast or Slow should give the same level, and in many cases where noise levels vary only slowly with time the difference between measurements made using F and S may be only 2 or 3 dB at most, and often less. However, larger differences can occur for situations where the noise level varies rapidly with time, and particularly for impulsive noise.

Note that L_{Aeq} measurements do not rely on the use of time weighted sound pressure level and so the use of F or S is not relevant.

Minimum sampling time for good statistical reliability

An important factor to keep in mind when dealing with percentile values is that a large sample size is necessary to ensure accurate data is obtained. To ensure the data set properly represents the noise, the usual minimum time period for such analysis for environmental noise is 10 to 15 minutes with a sampling rate of at least eight times per second. Some modern sound level meters will display a number for a percentile value even if it is over a time period of only a few seconds but this number may be quite meaningless if it lacks statistical validity. (The better quality sound level meters will not display a measurement parameter until sufficient data has been acquired to ensure statistical validity.)

4.4 The assessment of public response to environmental noise

Some possible methods are:

- analysis of complaints about the noise
- laboratory-based studies, in which a sample of people are exposed to various types and levels of noise under controlled conditions, and their response measured using interview or questionnaire techniques
- social surveys, in which a sample of people (usually a much larger sample than is possible for laboratory studies) are questioned about their response to the noise in their environment, with corresponding noise level surveys of that environment.

Noise indices

Social survey techniques will be briefly discussed in more detail (below) because many of the environmental noise indices currently in use have been developed and validated by social surveys (as described later) but first we need to define what we mean by a noise index, and what are the requirements of a 'good' noise index.

A noise index is a parameter which describes the noise and the value of which correlates with public response to the noise and which contains not only information about the measured physical attributes of the noise (level, frequency, content etc.) but also includes non-acoustic information which relates to the human response to the noise. Non-acoustic parameters could include, for example, the time period over which the noise is measured, and weighting factors that could be applied to it to reflect aspects of sensitivity of human response. Examples are $L_{Aeq,16hour}$, $L_{Aeq,8hour}$, $L_{Aeq,24hour}$, L_{den} and NNI (Noise and Number Index, although this is no longer in use).

What makes a 'good' noise index, i.e. what is a good way of measuring noise? Obviously the value of the index should correlate well with some measure of human response (annoyance for example) but, in addition it has to be practicable, i.e. reasonably easy to measure, and to predict, and also reasonable easy to understand and explain to members of the public.

Social surveys

The results of a social survey are usually resented as a graph (see Figure 4.5) of subjective response against an objective measure of the noise, i.e. the value of a noise index, often called the noise dose (and such a graph is sometimes known as the dose–response curve).

The measure of public response might take one of several different forms, as determined by the questionnaire survey. It might, for example, be a number (say 1 to 5), or it could be the percentage of respondents 'bothered by the noise', or the percentage of the respondents 'moderately annoyed' or 'highly annoyed'. The noise index could, for example be the $L_{Aeq,1hour}$, $L_{A10,18hour}$ or the L_{den} value.

The graph is derived from many, maybe several hundred, individual points, each representing the response

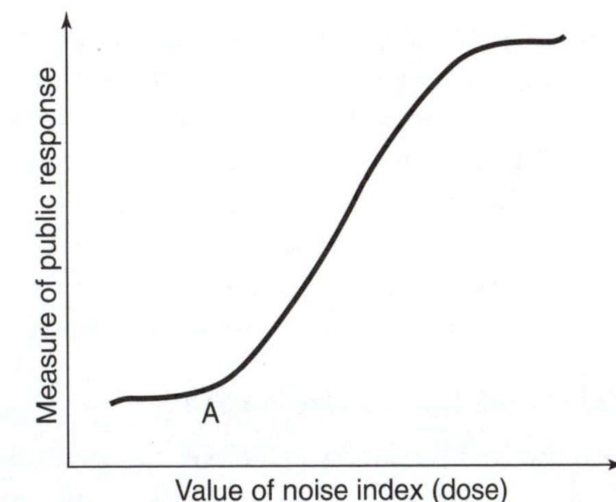

Figure 4.5 An idealized dose–response curve (with onset at A)

and noise exposure (dose) of each of the respondents to the survey, and the line shown is the line of best fit, derived using statistical techniques. The closeness of all the points to the line of best fit can be described by a correlation coefficient, a number which lies between zero and one. A coefficient of one represents perfect fit, i.e. each of the points fitting perfectly on the line of best fit, and a coefficient of zero means no correlation at all between dose and response, and no line of best fit. The value of the correlation coefficient indicates the reliability of the curve as an indicator of public response, and the suitability of the noise index for that purpose.

Once the dose–response curve has been established it may be used as a predictor of public response (e.g. that a value of $L_{Aeq,16hour}$ of X dB will result in Y% of the population exposed being highly annoyed).

It is important to understand, however, that this will be an indication of average public response, and it will not be possible to predict the response of any one individual, who may have greater or less than average sensitivity to the noise.

Onset of public response

Generally one would expect that public response to noise would increase as the dose increases. However, it might be the case that when the dose reaches a certain level the public response starts to increase at a faster rate. This is the case for the dose–response curve in Figure 4.5, where the public response suddenly starts to increase rapidly at point A, called the onset of the response. Such a characteristic in a dose–response curve (and it does not always occur) is useful for planners and those who set limits, as it identifies a natural value for such a limit.

4.5 A brief history of the development of noise indices and criteria

With dissatisfaction about increasing levels of environment noise in the late 1950s and 1960s attention turned towards developing the best methods for measuring and monitoring levels of noise in the environment, particularly from road traffic and from aircraft, and a number of large scale surveys were carried out. Some of these will be briefly described, together with other landmarks in the development of UK noise indices, laws, codes and standards.

The 1961–1962 London Noise Survey

In this survey, commissioned by the Wilson Committee, a sample of approximately 1400 people was questioned about noise and its importance relative to other factors, and noise levels were measured at 540 points equally spaced over 36 square miles of central London. The survey showed that traffic noise was judged to be by far the most important source of noise affecting people.

A survey was also carried out, again commissioned by the Wilson Committee, on people's reaction to aircraft noise around Heathrow. This led to the development of the NNI index, which remained in use until 1990.

Timeline

1963 The Wilson Report was published. This is described in more detail later.

1967 Greater London Area. 1400 residents at 14 sites adjacent to roads free from congestion and without intersection (Langdon and Scholes at BRE).

1968 Research into people's reaction to traffic noise suggested that at least half the occupants of dwellings affected by noise from free flowing traffic will normally be dissatisfied when L_{A10} outside their dwellings exceeds 70 dBA.

Attempts were made to develop noise indices which include terms to represent the degree of variation of noise levels throughout the day.

1968 Griffiths and Langdon developed the Traffic Noise Index (TNI) in an attempt to get best correlation with the 1967 survey.

1970 Robinson at NPL suggested the Noise Pollution Level (L_{NP}) as a method of estimating dissatisfaction with a wide variety of transportation noise sources, including traffic and aircraft.

Neither of these two indices proved to be significantly better correlators with annoyance than simpler indices such as L_{A10} or L_{Aeq}.

[The TNI was a combination of L_{A90} and L_{A10}:

$$\text{TNI} = 4 \times L_{A10} - 3 \times L_{A90} - 30]$$

the factor 30 being used simply to create a convenient numerical scale.

The noise pollution level, L_{NP} used a combination of the average (L_{Aeq}) value and the standard deviation, σ, of the distribution of sound levels in the time varying noise:

$$L_{NP} = L_{Aeq} + 2.56\sigma$$

For free flowing traffic the distribution of noise levels is often close to that of a random Gaussian distribution, for which the standard deviation can be shown to be related to L_{A10} and L_{A90} as follows: $\sigma = 2.56 \times (L_{A10} - L_{A90})$.

Therefore for free flowing traffic: $L_{NP} = L_{Aeq} + (L_{A10} - L_{A90})$.]

1972 The 1967 survey was repeated for a wider range of traffic conditions: 24 sites, 1459 respondents for free flowing traffic; 29 sites, 1574 respondents for non-free flowing traffic.

1972 Publication of Department of Employment Code of Practice for Reducing the Exposure of Employed Persons to Noise.

1974 Control of Pollution Act. Contained a section (section 4) on noise.

1974 Health and Safety at Work Act. Imposed duties on employers to protect employees against risk of hearing damage as a result of exposure to noise at work.

1978 The Noise Council (then a government supported advisory body, now no longer in existence) published 'A Guide to the Measurement and Prediction of Continuous Equivalent Noise L_{eq}' which promoted the use of L_{Aeq} for the assessment of environmental noise.

1984 Surveys by Fields and Walker in southern England established $L_{Aeq24hour}$ as the best index for predicting public response to railway noise.

1985 The UK Aircraft Noise Index Study (discussed in more detail below).

1988 Publication of latest revised version of Calculation of Road Traffic Noise (first published in 1975) providing the method for determining eligibility for compensation under the Noise Insulation Regulations 1967. May be replaced in due course by Europe-wide prediction method, such as Harmonoise.

1990 Report of the Noise Review Working Party (Department of the Environment). This was the report of a committee set up to review the framework of laws, regulations and standards in the UK and recommend any change necessary. Often regarded as a committee set up to review progress since the publication of the Wilson Report, nearly 30 years earlier. One recommendation was that there should be provision for financial support for sound insulation of dwellings close to new railway lines similar to that for road traffic. This led to the publication of the Calculation of Railway Noise procedure and of regulations in 1995.

1990 Environmental Protection Act. Incorporated revised and supplemented provisions of the Control of Pollution Act.

1991 Publication of *Railway Noise and the Insulation of Dwellings* (Mitchell Report) – a report of the committee formed to recommend to the Secretary of State for Transport a national noise insulation scheme for new railway lines.

1994 PPG24 *Planning and Noise* issued by Department of the Environment, replaced earlier Guidance Note 10(73).

1995 Noise Insulation (Railways and other Guided Transport Systems) Regulations 1995.

1995 Publication of Calculation of Railway Noise.

1997 Publication of latest revision of BS 4142.

1999 Publication of latest version of BS 8233.

2000 Publication of World Health Organization Guidelines for Community Noise, replacing an earlier document published in 1980.

2000 BRE/DEFRA Noise Incidence and Attitude surveys. A widespread survey was carried out across England and Wales into the noise climate in 1990 together with a corresponding survey of attitudes to noise. This was a follow-up to a similar survey 10 years earlier in 1990.

2002 The Environmental Noise Directive (EC Directive 2002/49/EC) in order to promote a common approach across the European Union towards the control of environmental noise has required that the following actions be implemented:

- the determination of exposure to environmental noise through noise mapping by methods of assessment common to member states
- ensuring that information on environmental noise and its effects is made available to the public
- adoption of action plans based upon the noise mapping results, particularly where exposure levels may induce harmful effects on human health.

2007 Attitudes to Noise from Aviation Sources in England (ANASE) study. A follow-up to the ANIS Study of 1985. Provides broad support for continued use of L_{Aeq} but suggests that significant disturbance occurs below 57 dB $L_{Aeq16\ hour}$. Some aspects of the report have been criticized and are controversial.

The Wilson Report

In April 1960 a committee was set up 'to examine the nature, sources and effects of the problem of noise and to advise what further measures can be taken to mitigate it'. The final report of the committee on the problem of noise, widely known as the Wilson Report, after the name of its chairman, was issued three years later in March 1963.

The report was 234 pages long, contained 14 chapters and 16 appendices, and covered all aspects of noise, and its recommendations shaped the approach to noise legislation and mitigation up to the present day.

Two major noise surveys were conducted during the period during which the committee sat and are reported: the 1961–62 London Noise Survey and the social survey in the vicinity of London (Heathrow) Airport, in 1961.

Many important initiatives and developments followed in the years following the Wilson Report, including:

- introduction of NNI for assessment of aircraft noise
- the Noise Insulation Regulations, requiring the measurement of $L_{A10,18hour}$ outside dwellings close to new roads, and leading to the development of Calculation of Road Traffic Noise (CRTN), the official UK method for predicting traffic noise levels
- BS 4142 first published in 1967 subsequently amended in 1975, 1980 and 1982, revised in 1990 to include L_{Aeq} and be in line with ISO 1996 and BS 7445, with latest revision in 1997
- Department of the Environment Circular 10/73 on Planning and Noise, subsequently replaced by PPG24 in 1994
- Department of Employment Code of Practice for Reducing the Exposure of Employed Persons to Noise, 1972, which subsequently, via EC Directives, became the 1989 Noise at Work Regulations and then the 2005 Control of Noise at Work Regulations.

1985 UK Aircraft Noise Index Study (ANIS)

By about 1980 the UK Department of Transport had become concerned that the NNI was out of line with various aircraft noise nuisance indices used in other countries, which tended to be based on the L_{Aeq} scale.

In the early 1980s the government commissioned a study to determine what noise index should be used to assess aircraft noise disturbance near major airports. This Aircraft Noise Index Study (ANIS), which was completed in 1984 and published shortly afterwards, included extensive social surveys and noise measurements around these airports plus detailed statistical analyses. The main result of the study was that the L_{Aeq} would be an appropriate index. Following this the decision to use the 16-hour L_{Aeq} for the UK aircraft noise index was announced in September 1990.

The survey reported that the number of people who found aircraft noise to be unacceptable increased from about 15% at 57 dBA to around about 75% at 69 dB, roughly in a straight line. Henceforth $L_{Aeq,16hour}$ aircraft noise contours were plotted in 3 dB steps starting at 57 dBA, which became an unofficial onset of annoyance, below which aircraft noise was deemed to be acceptable.[3]

The table below gives an approximate comparison between the NNI and $L_{Aeq,16hour}$ noise indices.

$L_{Aeq,16hour}$ (dB)	NNI	Annoyance
57	35	Low
60	40	Low
63	45	Moderate
66	50	Moderate
69	55	High
72	60	High

Some brief history: development of sound level meters

The development of the various noise indices described above has been interlinked with the development of sound level meters and their increasing noise measurement capabilities, and the change from analogue to digital signal processing.

The Wilson Report recommended in 1963 that people in their homes should not be exposed to noise levels that exceed 50 dBA for more than 10% of the time, or, putting it another way, that the L_{A10} should not exceed 50 dBA.

At this time sound levels meters were all analogue, i.e. with a display in which a moving needle moved over a scale. They measured and displayed instantaneous values of A-weighted sound level and in some cases dB and dBC as well, using either fast or slow time weighting. They were only really suitable for measuring noise levels which were more or less steady in level and were not impulsive in nature. More specialist meters were also available for measuring impulsive sound and were able to measure peak sound levels and the impulse time weighting (now no longer used), and with a 'peak hold' facility. Octave or third octave filters could be fitted to some of these meters.

The only way that the time history of a noise which varied in level over time could be tracked was to record samples of the instantaneous sound level either manually from the display, or by attaching the meter to a graphic recorder, or level recorder, and then subsequently examining the time variation from the graphical record. It was then possible to estimate maximum, minimum or percentile values from the graphical record, but L_{Aeq} had not yet appeared on the scene.

The term L_{Aeq} first came into prominence in the Health and Safety Code of Practice for Reducing the Exposure of Employed Persons to Noise in 1972 (which, via EC directives, eventually led to the 1989 Noise at Work Regulations, and then the Control of Noise at Work Regulations 2005). It was introduced as the appropriate method of estimating the noise exposure level of an employee whose exposure pattern varies widely with time over the working shift.

[3] Peter Brooker, 'The UK Aircraft Noise Index Study 20 years on', *IOA Acoustics Bulletin*, Vol. 29, No. 3, 2004, pp. 10–16.

Accurate measurement of L_{Aeq} required an instrument which automatically integrated the sound energy received by the microphone, at first developed in the form of small portable dosemeters, although approximate estimates could also be made from sampling the output of sound level meters. Soon, integrating sound level meters became available (i.e. capable of measuring L_{Aeq} accurately) and L_{Aeq} began to be used for measuring environmental noise as well as occupational noise exposure.

By the 1980s sound level meters with digital displays were available. These were devices in which most of the signal processing remained analogue in nature but with the instantaneous sound level (usually A-weighted) digitized, sampled and stored to enable calculation of parameters such as L_{Amax} and L_{AN} at regular intervals, e.g. every 5 minutes, every hour etc.

The first all-digital sound level meter appeared in the 1990s. Because all the signal processing, including frequency weighting and frequency spectrum analysis, was performed digitally a much wider range of parameters became available – octave band L_{eq} values, or octave band L_N values. It also became possible to compute and have direct display of indices such as Phons, PNdB, NR and NC values.

Noise measurement terminology

Units, scales, standards, indices, parameters, criteria, limits and metrics. What is the difference between some of these terms? Is there a difference, or do some of them mean the same thing? And can they be interchanged?

A **unit** is the physical quantity in which something is measured. Thus sound pressure is measured in pascals, sound power in watts and sound intensity in watts per square metre. Some quantities do not have any units, such as absorption coefficient or transmission coefficient.

Is the decibel a unit ? This is debatable – it is probably more correct to think of sound pressure level, for example, as being a sound pressure measured on a decibel scale, i.e. relative to a reference of 20 micropascals.

A **scale** is a method, or system, for measuring and comparing similar quantities, as for example the Celsius scale for measuring and comparing temperatures. There are several different decibel scales, for example for comparing sound pressures or sound powers, or sound intensities. There are also several different scales for environmental noise measurements – the L_{A10} scale and the L_{Aeq} scale for example.

What about criteria indices and limits?

A **noise index** is a way of measuring a noise in a way which allows us to make some sort of value judgement about the noise, i.e. gives us a way of assessing the noise, because for example the value of the index relates to some aspect of human response to the noise. Most commonly this may simply be the incorporation of a specific time interval into the measurement, so that $L_{A10,18hour}$ and $L_{Aeq,16hour}$ are examples of noise indices. It is also possible to incorporate weighting factors to take into account variation of human response to frequency (speech interference level, dBA or NR for example) or to allow for variation of sensitivity to noise at different times of day, such as L_{dn} or L_{den}. Another example is the rating level used in BS 4142, which may include a 5 dB factor to take into account 'noticeable' features of the noise.

A **limit** is a maximum or minimum permitted value; for example a local authority may impose a planning condition that the noise level from a particular development shall not exceed 35 dBA L_{Aeq} outside a certain residence at night-time, or that in order to be eligible for a grant towards sound insulation under the Noise Insulation Regulations, $L_{A10,18hour}$ outside a dwelling shall exceed 68 dB.

A **criterion** is a basis for forming a selection, judgement or assessment, so that for example a noise level in a particular situation may be assessed against a criterion of hearing damage, or of speech interference, or of annoyance. However, more specifically, limits may also be used as criteria, e.g. 'the criterion for achieving planning permission is that the noise from the development shall not exceed 35 dBA at night-time', or 'the acoustic criterion for eligibility for a sound insulation grant is that the dwelling $L_{A10,18hour}$ outside the dwelling shall exceed 68 dB'.

This leaves parameters, metrics and descriptors. A dictionary definition of a **parameter** is a 'measurable or quantifiable characteristic or feature', and of a **metric** is 'a system or standard of measurement'. These three terms would seem to be interchangeable in the context of noise measurement, e.g. the L_{A90} value may be chosen as the appropriate noise metric, noise measurement parameter or descriptor of the noise in a particular situation. However, the terms may have different distinct uses in a wider context, for example 'displacement, velocity and acceleration may be used as alternative descriptors of the magnitude of vibration'; or 'speed, traffic flow and % heavy vehicles' are important parameters in determining traffic noise.

There are many different dictionary definitions of 'standard', including: an object or quality or measure serving as a basis or example; or principle to which others conform or should conform, or by which the accuracy or quality of others is judged; the degree of excellence required for a particular purpose; a document specifying nationally or internationally agreed properties or procedures. In the context of noise measurement the word **standard** usually refers to an agreed method or procedure, as in a British or International Standard, but sometimes it is used in the same way as a limit or criterion.

4.6 Prediction assessment and mitigation of environmental noise

The main categories of environmental noise include transportation noise (road traffic, aircraft and train noise), industrial and commercial noise, leisure and entertainment noise, community noise and domestic noise.

Example 4.9

The noise from a construction site at a nearby property arises from three items of plant operating intermittently throughout a 12-hour day. The sound power levels, hours of operation and distance to the receiver position are given below. There is an unrestricted view of each item of plant from the reception position. Assuming free field sound propagation over hard ground conditions, calculate the $L_{Aeq,12hour}$ from the site.

Plant	L_W (dBA)	On-time (hours)	Distance (m)
A	106	10	40
B	110	6	60
C	116	2	80

Step 1

Calculate the sound pressure level at reception point for each item of plant using the formula based on the inverse square law (from Chapter 2):

$$L_p = L_W - 20\log r - 8$$

For A: $L_{pA} = 106 - 20\log(40) - 8 = 66.0$ dBA

Similarly for B and C: $L_{pB} = 66.4$ dBA and $L_{pC} = 69.9$ dBA

Step 2

Calculate $L_{Aeq,12hour}$:

$$L_{Aeq,12hour} = 10\log[(10 \times 10^{6.60} + 6 \times 10^{6.64} + 2 \times 10^{6.99})/12] = 68.5 \text{ dBA}$$

Prediction

This example serves as a basic model for the prediction of construction and other types of industrial noise from stationary sound sources, taking into account the effects of distance only. As shown in Chapter 2 the model may be adapted to include source directionality and other sound propagation effects such as ground attenuation, atmospheric refraction and the effects of barriers.

BS 5228-1:2009 suggests a variety of methods of predicting average noise level ($L_{Aeq,T}$) from construction sites based either on the sound power levels (as in the above example) or on sample L_{Aeq} values (at various distances) for the various construction site activities. The standard also suggests ways of adapting these methods for moving sources.

In general, prediction methods take into account factors relating to the source of noise, the factors that relate to the propagation of sound from source to receiver, and then factors relating to the reception point. A prediction model is usually based on a mix of empirical data (noise measurements) and theoretical considerations, and is a compromise between accuracy and complexity, with only factors that make a significant difference to the predicted outcome taken into account.

Road traffic noise

In the UK, the usual method for predicting noise from road traffic is to use the Calculation of Road Traffic Noise (CRTN) method which was originally devised for the purpose of determining eligibility for compensation under the Noise Insulation Regulations, but is now also used for planning investigations, noise impact assessments and for noise mapping. The method calculates the value of L_{A10} over 18 hours (06.00 to 24.00 hours).

The starting point is the calculation of the basic noise level from a road at a standard distance of 10 m from the edge of the nearside carriageway using the input parameters relating to the source of noise: traffic flow rate, average speed, percentage of heavy vehicles in the traffic flow, type of road surface and gradient. The effects of sound propagation (distance to receiver, ground attenuation, and screening by noise barriers) are then taken into account to calculate the noise levels at the reception point.

The next stage is to correct for any reception site layout features such as façade correction and reflections from nearby surfaces. The calculated basic noise level assumes that the road is straight and infinitely long and that the reception point has a clear unobstructed view of the entire road, i.e. a 180° angle of view. If this is not the case then the road must be split up into segments so that source and propagation parameters are constant for each segment. The angle of view (θ degrees) should be calculated and an angle of view correction of $10\log(\theta/180)$ applied to the noise level predicted for that segment. In the final stage of the prediction the noise level contribution from each road segment should be combined.

Train noise

Train noise at a site close to a railway line consists of a series of train noise events arising from each individual train passing by. A typical time history profile is shown in Figure 4.6 for a diesel engine train indicating that part of the noise is from the engine power unit (mainly the engine) and the remainder from the rolling noise (wheel–rail interaction). The peak at the start of the event may be missing for electric trains. The average level of train noise over a period of time ($L_{\mathrm{Aeq},T}$) is obtained by aggregating and averaging the noise from these events using the SEL values of train noise events at the reception point for each type of train and the numbers of events within the selected time period, T (as in Example 4.7 earlier). The SEL values must either be measured directly or predicted.

In the UK, the usual method for predicting noise from trains is to use the Calculation of Railway Noise (CRN) method. This follows broadly the same approach as for CRTN inasmuch as the first step is to predict a basic noise level (SEL value) at a fixed distance from the track (25 m) and then to make adjustments for sound propagation, reception site layout features and angle of view corrections for each segment of track. The source parameters relating to the prediction of basic SEL values are type of train, train length, average speed, and rail and track type. The SEL values for each segment are then used to calculate the average noise level for each type of train. The contributions from the various segments are combined to give the total noise from that track and the process repeated for all the different tracks, and these are combined to give the overall noise level from trains from all tracks at the reception point. The method also covers the prediction of L_{Amax} values, which is slightly different from that for $L_{\mathrm{Aeq},T}$ because the L_{Amax} value is usually determined by the engine noise which is modelled as a point source whereas the noise from the wheel–rail interaction is modelled as a line source.

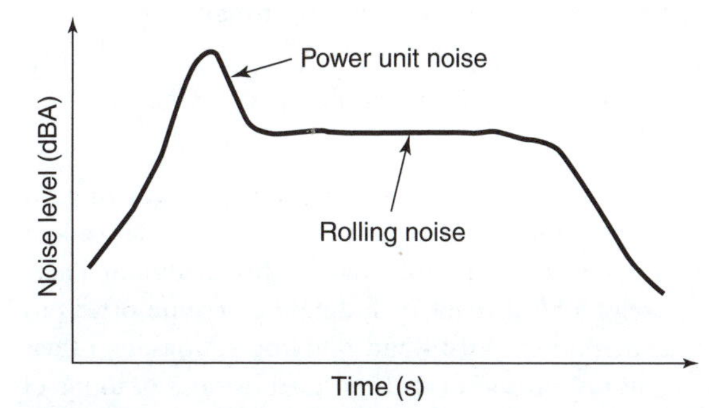

Figure 4.6 Typical time history of a diesel engine train pass-by

Integrated prediction methods

Eventually, for the purpose of future stages of noise mapping, the various national prediction schemes for road traffic and train noise (such as CRTN and CRN) used by individual counties within Europe will be replaced by Europe-wide Harmonoise noise prediction models. The EC research programme called IMAGINE (Improved Methods for the Assessment of the Generic Impact of Noise in the Environment) has extended this range with prediction noise models for aircraft and industrial noise sources. In addition to achieving improved prediction accuracy a feature of all of these models is the more detailed description of the noise source mechanisms and the increased separation of the source and propagation components of the prediction. This has the big advantage when, in the second stage of the European Noise Directive, it comes to the production of noise action plans, because it will allow much better prediction of the effectiveness of various noise mitigation measures. For road noise this has resulted in a separate description of the rolling noise and propulsion noise and because of these improved descriptions it will, for example, be possible to assess the effectiveness of using quieter types of tyres for road vehicles or using quieter road surfaces. In railway noise, rolling noise, traction noise and aerodynamic noise are contributing to the overall noise creation. Rolling noise is generated by the roughness of rail and wheels. Controlling the track roughness may have a noticeable effect for rail vehicles with smooth wheels, whereas for trains with rough wheels the effect is negligible. The source descriptions in the Harmonoise methods allow quick and easy assessment of the efficiency of such a measure, not only on a single vehicle level but also for a mixed traffic with both smooth wheel trains and rough wheel trains. This then allows cost–benefit studies, e.g. in terms of invested money against reduced annoyance, a work that would form the basis of noise action plans.

It was also considered a main priority in the Harmonoise project to improve the description of weather conditions and their influence on sound propagation. A description of sound propagation through a turbulent or layered atmosphere has led to short-term noise levels for 25 meteorological classes.

The prediction of aircraft noise

In the case of aircraft noise prediction, extra complexity arises because source related parameters have to be taken into account for each different aircraft type, including engine type, airframe type, weight including fuel load, sound power versus speed and thrust characteristics.

Sound propagation parameters will include different flight paths, atmospheric conditions (wind speed and direction, temperature, relative humidity, barometric pressure) and ground topography.

By treating each aircraft as a point source the noise at various positions around the airport may be predicted for each take-off or landing and these results combined for all events from all aircraft over time. It is usual to plot noise contours around the airport, typically in 3 or 5 dB steps centred on the runways.

Many different prediction packages are available, some of them used exclusively by various national government agencies to publish noise prediction contours around the world's major airports. In the UK such contours have been published annually for airports such as Heathrow and Gatwick first of all in terms of NNI and then $L_{\mathrm{Aeq,16hour}}$ and more recently in terms of L_{den}. One prediction package which is much more widely available is the Integrated Noise Model (INM) developed and made available by the US Federal Aviation Agency for general use in planning and impact assessment investigations, so that arguments between parties with opposing views about a proposed development may at least be based on predictions derived using the same method.

Limitations of noise prediction models

All noise level predictions are based on simplified models of sound generation and propagation. For example, diesel engines and aircraft are modelled as idealized simple sources and trains and motorway traffic as line sources. The sound propagation medium is usually modelled as uniform whereas in reality there may be sound velocity gradients and bursts of turbulence in the air. The equations and algorithms describing these models are based on assumptions and usually have stated ranges of validity. When these algorithms are translated into computer software to perform the calculations there is sometimes scope for interpretation in complex situations such as complicated road traffic noise predictions involving noise contributions from several different roads. Different software packages based on the same models and algorithms (such as CRTN for example) should give the same predictions within reasonable limits and 'round robin' tests are conducted to check that this is the case. Finally, of course, the validity of predictions will be dependent on the quality of the input data (traffic flow, speed, atmospheric and ground conditions etc.). For all of these reasons the results of sound level predictions should be considered to be subject to a range of uncertainties, similar to those associated with noise measurements. Measurement uncertainties are discussed further in Chapter 8.

Measurement v prediction of environmental noise

Despite the limitations outlined above, noise level prediction methods have some advantages over the alternative of noise measurement. The most obvious is of course that only prediction is possible for situations proposed but not yet in existence, but even for existing situations prediction has certain advantages:

- It is less expensive, particularly where sound levels are required at multiple reception positions and for contour maps.
- It enables the possibility of investigating changes (what-ifs), such as effect of changes to traffic speed, flow rate and composition, road surface type, construction of barriers etc.
- It is not subject to some of the practical difficulties associated with noise measurements such as weather conditions, background noise levels and access to suitable noise level positions.
- Results of predictions are more repeatable than measurements taken on different occasions because of fluctuations in flow rate, traffic composition, weather conditions etc. (which remain constant for repeated predictions).
- It is possible to investigate contributions from individual sources (or groups of sources) in multiple source situations, e.g. the contribution from one road in a complex road traffic noise situation, or from one particular source on a construction site.

There are situations for which it is very difficult to carry out accurate predictions and where noise measurement will be preferable, and noise measurement is the ultimate method of verifying the validity of prediction methods.

Assessment of environmental noise

Many developments require a noise impact assessment as part of a wider environmental impact assessment. This usually contains:

- a review of the proposed development, gathering together information about the development and about the geography and topology of the surrounding area likely to be affected, including the location of all potential noise sources and noisy operations, and their times of operation and noise emissions, locations of all noise sensitive properties (e.g. residences, schools, hospitals etc.)

- measurement of existing ambient and background noise levels in the area generally and at noise sensitive properties
- prediction of noise exposure levels in the area generally (e.g. by production of noise level contours) and specifically at noise sensitive properties
- a review of possible effects of noise and of methods of noise assessment and reasoned selection of most appropriate method
- assessment of the likely effects of the impact of the development on the area using the selected assessment method and based upon a comparison of predicted noise levels with existing measured baseline noise levels
- a review of possible noise mitigation measures and their likely effectiveness for reducing noise impact.

Sometimes, for transportation developments such as the creation of a new bypass road scheme for example, the assessment is expressed in terms of the numbers of people or households in the surrounding area predicted to be affected by various increments (or decrements) in noise level exposure bands (e.g. from 0 to 3 dBA increase, 3 to 5 dBA, 5 to 10 dBA etc.). These measures of impact can then be compared for different options for the route for the new road, including for the 'do nothing' option, taking into account the future effect of predicted increase in noise due to increased future traffic levels. In such cases it is usual to attach descriptors to the various noise level increments such as for example:

0 to 3 dBA increase – no significant impact
3 to 5 dBA increase – minor impact
5 to 10 dBA increase – significant impact
10 to 15 dBA increase – major impact.

The Department of Environment *Design Manual for Roads and Bridges* gives details of a method for assessing the impact of new road schemes along the lines outlined above.

The draft consultation document of the joint Institute of Acoustics (IOA) and Institute of Environmental Management and Assessment (IEMA), *Guidelines on Noise Impact Assessment*, gives guidance as to how basic noise changes may be categorized.

This document recognizes that each situation encountered is likely to be different and that there is no single set of subjective assessment criteria that would apply to all situations. The categorization shown in the table below (Table 2863/TX3 in the draft document) therefore takes this as its basis, with changes made as appropriate to reflect the site's own circumstances, as defined in the guidelines.

Table 2863/TX3 – Significance of Noise Level above Impact Assessment Criterion Noise Change (dBA) Category

0	No impact
0.1–2.9	Slight impact
3.0–4.9	Moderate impact
5.0–9.9	Substantial impact
10.0 and above	Severe impact

The mitigation or control of environmental noise

The most effective means of noise control is to reduce noise emission at source whether this be by producing quieter aircraft, road vehicles or trains, or by persuading people to leave entertainment premises quietly late at night. Much has been achieved since the 1970s to reduce noise emissions from individual aircraft and from road vehicles by engineering means driven by noise emission regulations.

Further noise control options may involve operational measures, selection of routes, timetabling and planning to separate noise sensitive receivers from sources of environmental noise. As a last resort noise mitigation measures may also be implemented at the point of reception such as barriers, improved layout to buildings and improved sound insulation.

Road traffic noise

In the 1970s the UK government initiated a programme of research to reduce noise from road traffic at source, with similar initiatives in other countries. In Europe this was driven in part by the impetus of EC directives on noise emissions, which specified target maximum noise levels for different classes of vehicle to be reduced in successive steps over the years.

Research had shown that noise from commercial vehicles (as opposed to cars) was a dominant factor and the UK government initiated a programme of research called the Quiet Heavy Vehicle Project to reduce noise emission from diesel engined heavy goods vehicles.

Vehicle noise

The main components of noise from vehicles are the noise from the power unit (mainly the engine) and rolling noise, i.e. noise associated with the movement of the vehicle along the road, which consists mainly of noise from the tyre–road interaction, and aerodynamic noise.

The power unit noise is mainly dependent upon engine speed, i.e. RPM, whereas rolling noise is road speed related, i.e. km/h, the two speeds being related by the gear in which the vehicle is being driven. The rate of increase of noise with speed is different for the two sources, with rolling noise increasing at a faster rate with speed increase than power unit noise. Therefore power unit noise is relatively more important at low speeds with rolling noise becoming relatively more important at high speeds, although the balance depends on the class of vehicle, with engine noise being relatively more important for diesel engined heavy vehicles than for petrol driven passenger cars. Back in the 1970s and 1980s power unit noise was the dominant source for most vehicles at most speeds but advances in the control of engine noise in particular have led to tyre–road noise being a much more important contributor for all speeds except for heavy vehicles at low speeds, and when vehicles are accelerating.

The main components of power unit noise are:

- engine noise
- noise from air intake and exhaust
- transmission noise, including gearbox
- noise from ancillary sources driven by the engine such as pumps, electric motors cooling fans etc.

Engine noise arises from two sources: combustion noise and mechanical noise. Combustion noise arises from the transmission of the transient bursts of high pressure in the engine cylinders caused by combustion of the air–fuel mix in the engine cylinder being transmitted via the engine structure to the outer noise radiating surfaces of the engine; and mechanical noise, arises from vibration of the moving parts of the engine (pistons, connecting rods, valves, timing belts or chains etc.).

Measures which may be used to reduce engine noise include:

- modification of the combustion process
- modification of engine structure
- treatments to noise radiating surfaces
- engine enclosure.

Modifications to the combustion process which achieve better mixing of the injected fuel with the air in the engine cylinder prior to combustion to achieve a smoother burning may result in noise reduction but may also have adverse consequences for the fuel efficiency of the vehicle and for air pollution emissions as well as for noise. Improvements to the combustion chamber design, the use of turbo charged engines and changes to injection timing can reduce combustion noise. Reducing the transmission of noise from the engine cylinders to the outer, noise radiating surfaces of the engine will also reduce noise but would require modifications to the structure of the engine cylinder block (e.g. by stiffening the engine structure). An alternative is to concentrate noise reduction effort on the treatments of those parts of the outer surfaces of the engine which radiate most noise (e.g. the engine sump, rocker cover, crankcase panels) using techniques of stiffening, damping, isolating or shielding the surfaces. These techniques are explained more fully in Chapter 9. A simpler alternative, in principle, is to surround the engine with a noise insulating enclosure. Such enclosures can achieve significant noise reductions but they may be bulky, add weight to the engine, can restrict the flow of air around the engine and so can cause heating problems, and can impair access to the engine for maintenance.

Tyre–road noise

When the vehicle is moving along the road the tyre treads in immediate contact with the road surface become distorted and air becomes trapped between these treads and is compressed. Moments later when this section of the tyre is released from the grip of the road surface these treads vibrate and radiate noise and the compressed, trapped air is released and is also set into vibration and radiates noise. It is thought that this 'air pumping' mechanism is a significant component of the tyre–road noise, but there will also be some noise radiated by the vibration of the entire tyre casing, and from the road surface and the vehicle as a result of vibration (or structure-borne noise) transmitted to these surfaces from the interaction of the tyre with the road.

Therefore a large number of parameters may affect the amount of noise radiated, including vehicle speed, tyre design (tread pattern and tyre casing structure), tyre wear, inflation pressure, type of road surface, vehicle suspension system. However, the main factors are vehicle speed, road surface and tyre design. The original 'quiet road surface' was of pervious macadam, which was developed because it had improved skid resistance and grip in wet weather conditions because water could drain away more quickly through the pores in the surface than it could from impervious surfaces. This type of porous surface was also found to have the benefit of reducing traffic noise because it provided more sound absorption than impervious surfaces. It was however claimed to be more expensive than the traditional macadam surface and it was less durable. Several other 'quiet road surfaces' have since been developed. Although these surfaces produce a reduction of only about 2 or 3 dBA, which subjectively is a relatively small change, it is worth remembering that, objectively, a 3 dB change would correspond to a halving of traffic flow. Research continues into the development of tyres which produce less noise.

Vehicle noise emission test

In order to check that new types of vehicles comply with noise emission regulations it is necessary to have a standard noise emissions test procedure.

Exercise

In such a procedure (called a 'drive-by test') the vehicle is driven along a road and the sound level is measured, in dBA. List all the other factors that would need to be specified in order to ensure standardization of the test procedure.

Answer

The various factors to be considered may be grouped according to source, sound propagation, and measurement and instrumentation:

- factors relating to the source: how the vehicle was driven (speed, gear, accelerating/cruising), road surface, type of tyres, road gradient
- factors relating to sound propagation from source to microphone: distance of microphone from road, type of ground (hard/soft), openness of site (absence of sound reflecting surfaces), weather conditions (wet/dry, wind speed and direction)
- factors relating to the instrumentation and measurement procedure: type of sound level meter (class 1, class 2) and microphone, position of microphone (distance from road, height above ground, orientation), calibration procedures, background noise level, noise measurement parameter (in this case maximum dBA level during the 'drive-by').

Test procedure

The test procedure requires that the maximum noise level is measured while the vehicle is driven along a designated stretch of straight flat road 20 m long, at a distance of 7.5 m from the roadside. The surrounding area should be flat hard ground with no sound reflecting surface within a distance of 50 m. The road surface should be dry, and background noise and wind speed low enough to avoid affecting the noise measurements. Noise measurements should be taken on both sides of the road (i.e. both sides of the vehicle). The vehicle should be driven in a manner which is repeatable and consistent with the emission of the highest noise level likely to be encountered in normal urban driving conditions. The exact specification varies with the class of vehicle but for passenger cars this requires that the driver arrives at the start of the designated stretch of road at a certain steady speed in top gear and then accelerates at full throttle over the test section of road.

The details of the noise measurement procedure may be found in BS ISO 362-1:2007.

Noise from tyres

In 2005 an EC directive (2001/43/EC) was introduced that established a test method for the type approval of tyres with respect to noise emissions and introduced limit values for different types and widths of tyre. The tyre noise emission limits will be reduced in future years and will apply to replacement tyres as well as those fitted to new vehicles from 2011. The tyre noise emission test (ISO 13325:2003) is broadly similar to the vehicle emission test except that the road test surface is controlled and vehicle noise is measured at specified speeds with the vehicle coasting with the engine switched off.

Measures to reduce traffic noise by operational measures

These apply the principle that noise increases with traffic flow, traffic speed and percentage of heavy vehicles, reduces with distance from the road and that smooth traffic flow is quieter than interrupted flow (when vehicles have to accelerate away from traffic lights and from other congestion points), and so may include:

- speed restrictions
- improved traffic management: traffic routing to avoid noise sensitive areas (one-way systems, traffic zoning)
- restrictions on heavy vehicles (lorry bans)
- traffic calming measures
- noise reducing road surfaces
- congestion charging
- measures to encourage increased use of public transport.

Traffic noise in future will be affected by the move away from purely internal combustion engine-based vehicles to hybrids, electric and other types of vehicles.

Aircraft noise

Control of noise from civil aviation requires international cooperation. An international meeting in Chicago in 1945, known as the Chicago Convention, led to the formation of the International Civil Aviation Organization (ICAO), as a specialist agency of the United Nations. The Chicago Convention is supported by 18 annexes which define standards and recommended practices (SARPs) that govern the operation of civil aviation. Annex 16 relates to aircraft noise and aircraft engine emissions. In the UK all matters relating to the control of civil aviation are dealt with by the Civil Aviation Authority (CAA).

The ICAO has recommended a 'balanced approach' to aircraft noise management consisting of four principal elements: reduction at source (quieter aircraft), land use planning and management, noise abatement operational procedures, and operating restrictions.

Noise control at source

Noise from aircraft in flight arises from the power unit and from the airframe. The power unit consists of a fan inducting air into the unit, a compressor driven by a turbine which drives compressed air into a combustion chamber where it is mixed and burnt with the aircraft fuel, and the exhaust gases ejected through a jet to provide the propulsion. Noise from the power unit is dominant during take-off and during level flight but airframe noise, i.e. aerodynamic noise arising from the flow of air around the aircraft, becomes important on approach to the airport and landing. Communities around the airport will also be subject to noise from aircraft starting up and taxiing on the ground, to ground running, engine testing, noise from auxiliary power units used while aircraft are on stand prior to take-off, and from vehicle noise on and around the airport.

Noise emission from the turbo jet engines of the 1970s has been reduced by the technique of using some of the air from the fan to bypass the turbine and combustion chamber and to mix with the jet exhaust efflux prior to emerging into the atmosphere. The first engines developed using this technique were called low bypass ratio (LBPR) engines and further noise reduction was achieved by increasing the amount of bypass air with the development of high bypass ratio (HBPR) engines, also known as 'turbofan' engines. Total noise reductions of up to 20 decibels have been achieved using the bypass technology and other improvements to engine design, as compared with the original turbo jet engines of the 1970s. The use of sound absorbing materials lining the inner casing of the power unit and of active noise control techniques also contributes significantly to noise reduction.

Silent Aircraft research project

A long-term joint research project led by Cambridge University and MIT Institute in the USA called the Silent Aircraft Initiative was launched in November 2003 with a bold aim: to discover ways to reduce aircraft noise dramatically, to the point where it would be virtually unnoticeable to people outside the airport perimeter.

Aircraft capable of controlled, quiet descent would greatly improve the quality of life near airports while allowing expansion of airport usage and flight paths, resulting in a clear economic and environmental benefit. The Silent Aircraft Initiative is working towards this goal by researching novel airframe configurations and techniques for controlling airflow, drag and descent, and including the possibility of screening of engine noise from listeners on the ground by the airframe (e.g. locating engines above the wings).

Operational techniques for reducing the impact of aircraft noise on communities

The general principle is to maximize the distance between the aircraft and noise sensitive areas on the ground below. Although procedures vary at different airports this involves take-off at full thrust, gaining height as quickly a possible, and then to cut back on thrust and climb more slowly, thereafter maintaining a minimum height. Some airports specify maximum noise limits during take-off which are monitored by permanent noise monitoring terminals close to the end of the runway. Many airports adopt minimum noise routes (MNRs) for departing aircraft, devised as far as possible to avoid flying over densely populated areas. Deviations from these routes are monitored by radar tracks linked to a noise and track keeping system.

Aircraft approaching the airport may be held in a queue waiting to land and are then required to circle in 'stacking areas' before beginning descent to the airport runway. These stacking areas are designated by NATS (National Air Traffic Control System, a part of the Civil Aviation Authority) and selected mainly on considerations of operational safety but with community noise also a consideration. During the final stages of descent the aircraft joins the airport's instrument landing system (ILS) which guides the aircraft along a constant glide slope (usually 3 degrees) towards the runway. In order to save fuel and reduce noise at ground level a continuous descent approach (CDA) to joining the ILS is increasingly being adopted by many airports. Prior to CDA, airport approach flight paths often involved periods of level flight alternating with periods of descent, eventually joining the ILS 'from below'. The CDA approach method involves descending along a path of constant slope, minimizing as far as possible any periods of level flight and eventually joining the ILS 'from above': the general principle being to 'remain higher for longer' (see Figure 4.7). The adoption of a CDA approach also facilitates the possibility of a 'low drag' approach with further noise and fuel saving benefits.

When the aircraft has landed, short bursts of 'reverse thrust' are sometimes used by the pilot to slow down the aircraft more quickly, for example to enable quicker arrival at the landing bay. Many airports have a policy of minimum use of reverse thrust, which produces extra noise, particularly at night-time.

The introduction of improved aircraft navigational techniques, based on satellite navigation and on-board electronic systems (as opposed to reliance on land-based radar beacons) will improve track keeping of individual aircraft and the future efficiency of airspace management giving better opportunities for directing more aircraft movements away from noise sensitive areas.

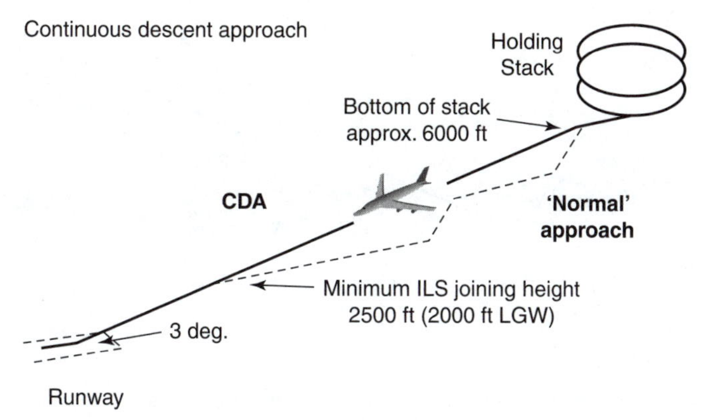

Figure 4.7 Adopting a CDA approach

Copyright © British Airways

All noise reducing operational procedures are subject to overriding safety considerations, including their suspension during inclement weather.

Other airport noise mitigation measures

Many airports have restrictions on the number of night-time flights which are allowed and some operate to a quota system which takes into account the amount of noise generated by different types of aircraft. According to such a system each aircraft is allocated a quota count (QC) which may range from 0.25 for the quietest aircraft, through values of 0.5, 1, 2, 4, 8 and 16, although very few of the noisiest aircraft with QC of 8 or 16 are operating in the UK. A doubling of the QC value corresponds to an increase in the noise exposure level per movement, so that for example one movement of an aircraft with a QC of 2 is equivalent in terms of noise exposure to two movements of another aircraft type with a QC of 1. Aircraft may have different QC values for arrival and departure movements.

Some airports use noise monitoring and NTK (noise and track keeping) systems to monitor compliance with airport flight procedures such as noise limits for take-off, and adherence to MNRs and CDA. Infringements of procedures can lead to discussions with airlines to effect improved compliance, and ultimately to fines imposed on airlines or on individual pilots. Some airports operate a policy of imposing increased landing charges on noisier types of aircraft.

Aircraft noise contours

These are useful for land use planning around airports to ensure that, as far as possible, land close to an airport is not used for noise sensitive development. Some airports operate schemes whereby householders living close to the airport may be eligible for improved sound insulation to their homes depending upon their location within a certain noise contour. The contours also provide local authorities and members of the public with information on the degree of noise exposure to aircraft noise in different areas around an airport. Note that these contours are predictions based on an average number of aircraft movements, and do not include noise from aircraft on the ground (e.g. starting up or taxiing at the terminal) or from other airport operations.

Community liaison

The control of aircraft and airport noise necessarily involves a balance between the benefits to the community arising from the operation of the airport and the environmental impact on those who live nearby. Many airports operate community liaison policies designed to keep the local community informed about environmental issues, including noise. These include regular meetings with local community groups, procedures to receive and respond to complaints and queries about aircraft noise, and information about noise published in leaflets, reports and on websites. Information available to the public can include data about noise infringements to operating procedures and fines, degree of compliance with NPRs and CDA, noise contours and results of noise monitoring. Complaints from members of the public are sometimes about individual flights being 'off track' or flying too low, and records of radar tracks of individual movements may be used to help in dealing with such complaints. It is possible on some airport websites to track individual flights (including aircraft heights) a day or so after they occurred (real time tracking could endanger aviation security).

Aircraft noise certification

International cooperation involving bodies such as the International Civil Aviation Organization (ICAO) have

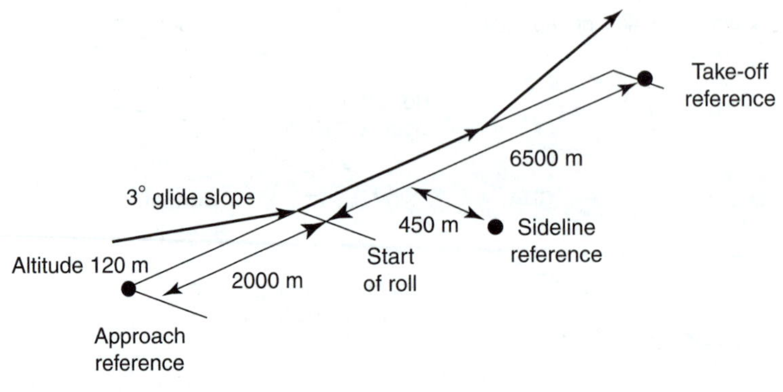

Figure 4.8 Aircraft noise certification measurement positions

led to a system of aircraft classification based on noise emission, leading to the gradual phasing out of older noisier types of aircraft at many airports. This has required the establishment of a standardized test procedure for measuring the noise emission of aircraft in flight, known as the noise certification test.

The certification test is a highly specialized procedure defined in ICAO Annex 16 and is administered in the UK by the Civil Aviation Authority (CAA).

Maximum noise levels are measured during both take-off and landing events in terms of effective perceived noise level (EPNdB), which is a rather complicated development of the perceived noise level described in Chapter 3, with the addition of corrections for tonality and for the variation in noise level which occurs over the duration of the event. Noise measurements are taken at three positions (see Figure 4.8) defined relative to the positions on the runway known as 'start of roll', i.e. the position where the aircraft's brakes are released and the aircraft starts its movement down the runway at full power prior to take-off; and the runway threshold (i.e. start of runway). These three positions are:

1. the take off reference position, at a point under the departure flight path at a distance of 6500 m from the point of start of roll
2. the sideline reference positions at which the peak noise is received at a point along a line parallel to and 450 m to the side of the extended runway centre-line during take-off
3. the approach reference position at a point under the approach flight path (which follows a 3-degree glide slope) at a distance of 2000 m from the start of the runway.

ICAO Annex 16 specifies maximum levels of EPNdB at these positions, depending on aircraft weight and type, as a certification requirement. These levels, which have been reduced over time, apply to large jet engined and propeller driven civil aircraft and to helicopters, but not to military aircraft. Modified and simplified arrangements based on dBA measurements apply to smaller civil aircraft and to microlights.

Aircraft classification according to noise emission is described in various chapters of ICAO Annex 16, Volume II, and are used to encourage the gradual phasing out of the use of older noisier aircraft types:

- 'Chapter 2' set limits for aircraft types certificated before 1977
- 'Chapter 3' set limits for aircraft types certificated after 1977
- 'Chapter 4' set limits for aircraft types certificated after 2006.

Sometimes noise mitigation measures, known a 'hush kits' may be applied to noisier aircraft to enable them to meet noise emission standards of later quieter aircraft types.

Noise from general aviation

It has been found that public response to noise from general aviation is less tolerant than it is to that from civil aviation at major civil airports, by about 5 dB. Why do you think this is? Possible reasons include: the perception that unlike civil airports the flights are for the benefit of a few individuals only rather than a benefit to the community as a whole, fears about safety, and loss of privacy (the possibility of being overlooked by flights).

Noise from railways

Noise from trains arises from the power unit, from rolling noise and from aerodynamic noise. For diesel powered trains the power unit is the dominant source at low speeds (and particularly at low frequencies) but rolling noise increases more rapidly with increase in speed and becomes the most important source over most of the speed range, except sometimes at the very highest speeds where aerodynamic noise may become the most important source. Occasionally bursts of high pitched tonal noise arise from 'wheel squeal' on bends for example. Although this can give rise to a great deal of annoyance it can be controlled by lubrication techniques at the point of contact between wheel and rail

In a diesel train the power derives from either one single diesel engine locomotive, or from several smaller engines in diesel multiple unit trains where each carriage has its own power unit. The engine drives a dynamo or alternator to produce electrical power which in turn drives the electric traction motor which drives the train. The engine also supplies power to other auxiliary units such as compressors and fans. All of these components produce some noise, together with the diesel engine exhaust, which is probably the most important noise source at low frequencies.

Electric trains derive their power either from an electrical third rail or from an overhead catenary cable and the power unit is much quieter than for diesels.

Rolling noise is caused by the interaction of surface irregularities in both the surfaces of the wheel and the rail. In track with jointed rails, as opposed to continuously welded rail, there will also be additional noise generated as the wheels pass over the joints in the rail. The degree of surface roughness of both rails and wheels is an important factor in determining the amount of rolling noise and both wheels and rails are regularly ground to minimize roughness. The type of braking system is also an important factor, with disc brakes being quieter than tread brakes. This is because tread brakes are in direct contact with the rolling surface and this causes a degree of surface roughness of the wheels which is greater than is the case for disc brakes, although some noise reduction of tread brakes may be achieved by replacing the traditional cast iron brake blocks with blocks made of a composite material.

The impact forces generated at the point of contact between wheel and rail are transmitted as vibration to the rest of the wheels and to the rails and the track supporting the rails, and all these surfaces radiate noise. Noise radiated by the track is most important at low frequencies and that from the wheels at high frequencies and this means that measures to reduce both components may be needed to achieve significant overall reduction in noise levels.

Research over several years has produced reductions of about 10 dBA, using a combinations of maintaining smooth wheels and rails, low noise designs for wheels and rails and track, and noise barriers attached to either the track or the train.

Modal analysis has been carried out into the vibration of train wheels, resulting in design changes to wheel thickness profile and wheel size (slightly smaller wheels) in order to minimize excitation of efficient noise radiating modes by the rail–wheel interaction. Other noise reduction measures include the introduction of tuned dynamic absorbers both to the wheels and to sections of rail, damping of wheels and rails, the use of concrete slabs to replace the traditional system of ballast as track foundation support, and the fitting of noise shields to the wheel web.

Research into public response to train noise has shown that it produces less annoyance than either road traffic or aircraft noise (see Figure 4.9), that diesel trains are more annoying than electric ones, and that freight trains are more annoying than passenger trains. In the 1970 Fields and Walker established that the noise index

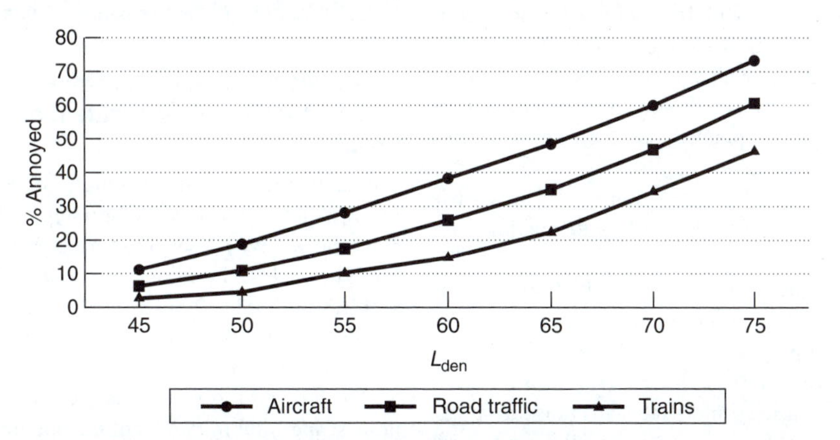

Figure 4.9 Comparison of public response to aircraft noise, road traffic noise and train noise

Data from EU's future policy, WG2 – Dose/effect position paper on dose response relationships between transportation noise and annoyance, 2002

which gave the best correlation to public annoyance was $L_{Aeq,24hour}$. They also found that the degree of annoyance increases with increasing noise levels but there was no threshold at which the train noise started to become annoying or suddenly increased in annoyance.

Noise Insulation Regulations similar to those for road traffic were introduced in the UK in 1996 following the introduction of the Calculation of Railway Noise prediction method in 1995. The EC has introduced noise emission tests and noise limits for high speed trains in the form of interoperability directives, to ensure common technical specification for trains that cross national borders. A technical specification for interoperability (TSI) for high speed trains was introduced in 2002 and a TSI for conventional trains has been produced which specifies future limits for starting noise, pass-by noise and stationary noise.

Other sources of railway noise

Environmental noise also arises from warning signals (horns and whistles) on board trains and at level crossings, from track repairs and maintenance, from railway depots and goods yards, and from station car parks. Annoyance from PA system used at stations can be reduced by commonsense measures such as careful location of loudspeakers, reducing volume to the minimum necessary for good speech intelligibility and reducing the number of announcements particularly at sensitive times such as early in the morning and late at night.

4.7 Legislation, regulations, standards and codes of practice

The framework of legislation and regulations provides not only the motivation and driving force for noise control but also often sets the targets required for noise reduction. These targets will in turn be based upon information about what are acceptable levels of noise from standards and codes of practice based on knowledge of human response to noise.

The targets required for occupational noise are clearly set out in the Control of Noise at Work Regulations and are discussed in Chapter 9.

In the case of environmental noise, the framework is more complex but two important aspects are control of nuisance, and planning. (Also see Chapter 10.)

Noise nuisance is dealt with by local authority environmental health officers and deals with noise from existing noise sources. Therefore any remedial noise control measures will be reactive and retrospective and thus constrained by what is economically and technically feasible in terms of modifying existing noisy activities.

New noisy developments require planning permission from the local planning authority. The planning process provides the ideal opportunity for minimizing the effects of noise, and for preventing the need for future noise nuisance actions. The local planning authority can set conditions aimed at minimizing noise disturbance and preventing nuisance. Planning conditions can be more restrictive than those which may be imposed in case of nuisance. These conditions can include noise limits, and restrictions on the types of plant and machinery to be used and on hours of operation.

If noise issues are not dealt with effectively at the planning stage an opportunity to minimize noise has been lost and it may be more difficult for the local authority to deal with noise nuisance problems which may occur in the future.

Therefore it is important that there is close liaison on noise issues between the local authority environmental health officers and planning officers. The control of noise by planning depends on the prediction of noise levels and on being able to relate the predicted noise levels to human response.

Noise nuisance is governed by the Environmental Protection Act 1990 supplemented by provisions of the Control of Pollution Act 1974 for construction sites. Department of Environment (now DEFRA) Planning Policy Guidance Note 24 (PPG24) gives guidance on noise and planning, and BS 5228 on noise from construction sites. BS 4142 is widely used for the assessment of commercial and industrial noise. Noise from certain industrial processes are controlled by the Environment Agency under the provisions of the Integrated Pollution Prevention and Control (IPPC) Regulations. Guidance on the control of noise inside buildings is given in BS 8233:1999, and will be discussed in Chapters 5 and 6.

BS 4142:1997 Method for rating industrial noise affecting mixed residential and industrial areas

This standard was first introduced in 1967 as one of the recommendations of the Wilson Report in 1963, and has since been revised several times. A revision in 1990 brought BS 4142 into line (via the introduction of L_{Aeq}) with BS 7445/ISO 1996 Parts 1, 2 and 3, which cover all aspects of environmental noise rather than just industrial and commercial noise.

The foreword to the standard clearly sets out the purpose, scope and limitations of the document:

> Response to noise is subjective and affected by many factors (acoustic and non-acoustic). In general, the likelihood of complaint in response to a noise depends on factors including

the margin by which it exceeds the background noise level, its absolute level, time of day, change in the noise environment etc., as well as local attitudes to the premises and the nature of the neighbourhood. This standard is only concerned with the rating of a noise of an industrial nature, based on the margin by which it exceeds a background noise level with an appropriate allowance for the acoustic features present in the noise. As this margin increases, so does the likelihood of complaint.

The standard is intended to be used for assessing the measured or calculated noise levels from both existing premises and new or modified premises. The standard may be helpful in certain aspects of environmental planning and may be used in conjunction with recommendations on noise levels and methods of measurement published elsewhere.

The standard is necessarily general in character and may not cover all situations. The likelihood that an individual will complain depends on individual attitudes and perceptions in addition to the noise levels and acoustic features present. This standard makes no recommendations in respect of the extent to which individual attitudes and perceptions should be taken into account in any particular case.

Although, in general, there will be a relationship between the incidence of complaints and the level of general community annoyance, quantitative assessment of the latter is beyond the scope of this standard, as is the assessment of nuisance.

It should be noted that noise assessment is a skilled operation and should be undertaken only by persons who are competent in the procedures.

It is also stated in the foreword that 'The user is reminded that this standard is not based on substantive research but rather on accumulated experience.'

It is instructive to read the scope of the standard set out at the beginning (paragraph 1):

> This British Standard describes methods for determining, at the outside of a building: (a) noise levels from factories, or industrial premises, or fixed installations, or sources of an industrial nature in commercial premises; and (b) background noise level.
>
> The standard also describes a method for assessing whether the noise referred to in (a) is likely to give rise to complaints from people residing in the building. The method is not suitable for assessing the noise measured inside buildings or when the background and rating noise levels are both very low.
>
> *Note.* For the purposes of this standard, background noise levels below about 30 dB and rating levels below about 35 dB are considered to be very low.

Assessment method

The assessment method described in the standard is based on a comparison between the level of noise from the source under investigation, called the specific noise, and the background noise level.

The specific noise may be determined either by prediction or by measurement, in which case it should be corrected for the influence of ambient noise on the measurement, if necessary. It may then be subject to a character correction of 5 dB, to determine the rating level, to be used in the assessment.

Rating level

Certain acoustic features can increase the likelihood of complaint over that expected from a simple comparison between the specific noise level and the background noise level. Where present at the assessment location, such features are taken into account by adding 5 dB to the specific noise level to obtain the rating level.

Apply a 5 dB correction if one or more of the following features occur, or are expected to be present for new or modified noise sources:

- the noise contains a distinguishable, discrete, continuous note (whine, hiss, screech, hum etc.)
- the noise contains distinct impulses (bangs, clicks, clatters or thumps)
- the noise is irregular enough to attract attention.

Note: The rating level is equal to the specific noise level if there are no such features present or expected to be present.

Likelihood of complaints

The likelihood of complaints is assessed by subtracting the measured background noise level from the rating level.

Note: More than one assessment may be appropriate.

The greater this difference the greater the likelihood of complaint:

- A difference of around +10 dB or more indicates that complaints are likely.
- A difference of around +5 dB is of marginal significance.
- If the rating level is more than 10 dB below the measured background noise level then this is a positive indication that complaints are unlikely.

Noise measurement positions

Choose measurement positions that are outside buildings and that will give results that are representative of the specific noise level and background noise level at the buildings where people are likely to be affected.

To minimize the influence of reflections, make the measurements at least 3.5 m from any reflecting surface other than the ground.

Note 1: The preferred measurement height is 1.2 m to 1.5 m above the ground.

Note 2: Where it is necessary to make measurements above ground floor level, choose a position which is 1 m from the façade on the relevant floor of the building.

Note 3: Report the measurement position, height and the distance from any reflecting structure other than the ground.

Other information

The standard is 20 pages long and contains much useful information about how the specific. residual and the background noise should be measured, in many difficult situations. It also provides advice about what information shold be recorded and reported.

Definitions

The standard provides definitions of the following terms (see Glossary, Appendix 1 to this book): specific noise source, reference time interval, specific noise level, measurement time interval, rating level, ambient noise, residual noise, residual noise level, background noise level.

Creeping background noise levels

Sometimes the BS 4142 rating level is used as a target for noise control from an industrial noise. For example an existing rating noise level of, say, 15 dB (i.e. 15 dB above background, and likely to give rise to complaints) may be required to be reduced to a rating level of 5 dB (of marginal significance) as a compromise between what would be ideal and what is practicable. In many cases such a reduction may be sufficient to eliminate the nuisance. It is sometimes claimed that this will nevertheless lead to a gradual increase in background noise levels.

Case studies involving BS 4142

Case study 1

A new factory has applied to move into a vacant site on the edge of an industrial estate, near to some residential properties, and will produce continuous noise throughout the daytime. There may be occasions in the future when night-time operations will become necessary. Background noise measurements made outside residential properties close to the new site give values of L_{A90} of 45 dBA during the daytime and 35 dBA late at night. Noise levels from the factory are made at its existing site, at positions that are a similar distance away to the houses at the new site, under similar prevailing wind conditions. The value of L_{Aeq} measured over one hour was 47 dBA. The noise contains a clearly audible hum. The noise level at the existing site, where there are no houses nearby, was obtained during a period when the plant was switched off for a short while, and was 40 dBA (L_{Aeq}) and 36 dBA (L_{A90}). Assess the likelihood of complaints if the planned move goes ahead, for both daytime and night-time operation.

Case study 2

A factory is to introduce a new noise source which operates for eight minutes each hour throughout the 24-hour period, into a neighbourhood where the existing noise levels are:

Daytime: 49 dB L_{Aeq} and 43 dB L_{A90}

Night-time: 40 dB L_{Aeq} and 33 dB L_{A90},.

During a test operation of the new source the noise level is measured outside the nearest residence over the eight-minute 'on' period on three occasions during the daytime and the L_{Aeq} is found to be 56 dBA. The noise is reasonably steady during its 'on time', with no tonal, impulsive or other noticeable characteristics. Assess the likely effects of the proposed change using BS 4142.

Solutions to case studies

Case study 1

Correction to measured specific noise level for background noise:

$$L = 10\log(10^{4.7} - 10^{4.0}) = 46 \text{ dBA}$$

Correction of +5 dBA for character of noise.
Therefore rating level = 46 + 5 = 51 dBA.

Daytime rating = rating level − background noise level
= 51 − 45 = 6 dBA (marginal significance for likelihood of complaints)

Night-time rating = 51 − 35
= 16 dBA (complaints are very likely)

Case study 2

Correction to measured specific noise level for background noise:

$$L = 10\log(10^{5.6} - 10^{4.9}) = 55 \text{ dBA}$$

Correction for duration (daytime only):

$$L = 55 - 10\log(60/8) = 46 \text{ dBA}$$

No correction for character of noise.

Daytime

Rating level = 46 dBA
Rating = rating level − background noise level
= 46 − 43 = 3 dBA (marginal significance for likelihood of complaints)

Night-time

Rating level = 55 dBA (no duration correction)
Rating = 51 − 33 = 22 dBA (complaints are very likely)

Planning Policy Guidance Note (PPG) 24

Department of Environment (now DEFRA) Planning Policy Guidance Note (PPG) 24 *Planning and Noise* (issued in 1994) gives guidance to local authorities in England on the use of their planning powers to minimize the adverse impact of noise without placing unreasonable restrictions on development or adding unduly to the costs and administrative burdens of business.

An underlying principle is the separation as far as possible of new noise sensitive developments from existing noise sources and conversely of new developments involving noisy activities from existing noise sensitive areas.

Where it is not possible to achieve such a separation of land uses, local planning authorities should consider whether it is practicable to control or reduce noise levels, or to mitigate the impact of noise, through the use of conditions or planning obligations.

Two different situations have to be considered: introducing a new potentially noisy development to an area where there are noise sensitive properties (e.g. residential dwellings), and introducing a new housing development in an area where there are existing noise sources.

In the first case a noise impact assessment should be carried out and in the second case the noise exposure category procedure should be used, as explained below.

Measures which are suggested to mitigate the impact of noise include using quieter machines or processes, use of barriers, improved sound insulation of buildings, and use of site layout to maximize screening of noise using non-noise-sensitive buildings.

The PPG recommends the use of BS 4142 for assessing noise from industrial and commercial premises, with the rating level defined in this standard as being an appropriate method of specifying permissible noise limits.

The guidance note also reviews the possibilities for noise mitigation by using the insulation of buildings against external noise, gives examples of planning conditions and reviews other legal and administrative measures (i.e. other than planning control) available for noise control.

Noise exposure categories

PPG24 suggests that sites for which planning applications for residential development are being considered should be characterized by the use of four different noise exposure categories (NECs) – A, B, C and D – depending upon the existing noise levels prevailing at the site. Sites in category A have the lowest noise levels and site D the highest. It is recommended that:

- For NEC A noise need not be considered as a determining factor in granting planning permission, although the noise level at the high end of the category should not be regarded as a desirable level.
- For NEC B noise should be taken into account when determining planning applications and, where appropriate, conditions imposed to ensure an adequate level of protection against noise.
- For NEC C planning permission should not normally be granted. Where it is considered that permission should be given, for example because there are no alternative quieter sites available, conditions should be imposed to ensure a commensurate level of protection against noise.
- For NEC D planning permission should normally be refused.

The PPG gives a recommended range of noise levels, expressed as equivalent L_{eq} levels, for each of the NECs for dwellings exposed to noise from road, rail, air or 'mixed sources'. These are shown in Table 4.1. Planning authorities may wish to choose category limits within the recommended bands or take local circumstances into account to define their own bands.

BS 5228-1:2009 Code of practice for noise and vibration control on construction and open sites – Part 1: Noise

This standard deals with both the occupational noise exposure of employees and with disturbance caused by construction noise at nearby noise sensitive properties. It aims to provide guidance to architects, contractors and site operatives, designers, developers, engineers, local authority environmental health officers and planners. It provides guidance concerning methods of controlling, predicting and measuring noise from construction sites and assessing its impact on those exposed to it. The importance of good community relations is emphasized and recommendations are given regarding procedures for the establishment of effective liaison between developers, site operators and local authorities.

Control of noise at source is suggested using a variety of approaches, including general (good housekeeping) measures, specification and substitution, modification of existing plant and equipment, use of enclosures, use and siting of equipment, and good maintenance; and by controlling the spread of noise using the effects of distance and screening. The recommended environmental noise descriptors are L_{A01}, L_{Amax} and $L_{Aeq,T}$. Prediction methods for $L_{Aeq,T}$ are described based upon information about sound power levels, distance from noise sensitive positions and 'on time' durations for each machine or process. A database of noise from plant and equipment is included and methods for assessing the significance of

Table 4.1 Recommended noise exposure categories for new dwellings near existing noise sources, from PPG24

Noise levels corresponding to the noise exposure categories for new dwellings $L_{Aeq,T}$ dB				
Noise source	Noise exposure category			
	A	B	C	D
Road traffic				
07.00–23.00	< 55	55–63	63–72	> 72
23.00–07.00	< 45	45–57	57–66	> 66
Rail traffic				
07.00–23.00	< 55	55–66	66–74	> 74
23.00–07.00	< 45	45–59	59–66	> 66
Air traffic				
07.00–23.00	< 57	57–66	66–72	> 72
23.00–07.00	< 48	48–57	57–66	> 66
Mixed sources				
07.00–23.00	< 55	55–63	63–72	> 72
23.00–07.00	< 45	45–57	57–66	> 66

Notes

1. Values in the table refer to noise levels measured on an open site at the position of the proposed dwellings, well away from any existing buildings, and 1.2 m to 1.5 m above the ground. The arithmetic average of recorded readings should be rounded up.
2. As a footnote to the above table PPG24 states that if short duration bursts of noise at night result in noise levels which regularly exceed $L_{Amax,Slow}$ 82 dB several times during a one-hour period at night then the site should be classified as NEC C regardless of the measured $L_{Aeq,8hour}$ dB (provided it does not already qualify for Category D).

the impact of construction noise is based both on fixed noise limits and on change in noise levels given. Advice about conducting noise surveys and a review of the legislative background to control of construction noise are also included.

Part 2 of the standard deals with vibration and will be discussed in Chapter 7.

BS 7445/ISO 1996 Description and measurement of environmental noise

This standard aims to provide a method for describing, measuring and rating environmental noise from all different sources (i.e. industry, transport, recreational and so on) which contribute to the total noise at a site, using L_{Aeq} as the basic measure. The standard lays down guidelines for specifying noise limits and describes procedures to be used for checking compliance with such limits, and for gathering environmental noise data which may be useful for making planning decisions about land use. The standard defines a rating level determined over a specified reference time interval, which is obtained by adding adjustments to the measured L_{Aeq} value based on both tonal and impulsive characteristics of the noise. Methods of determining a long-term average rating level and for representing the results in terms of noise zones on a map of the area are described.

BS 4142 can be regarded as a standard that describes methods and procedures that comply with BS 7445, but restricted to commercial and industrial noise.

Questions

1 Define and explain the difference between the following noise measures and explain how they are used in the assessment of various types of environmental noise: $L_{Aeq,T}$, $L_{A10,18hour}$, L_{A90}, L_{Amax}, L_{den}

2 Describe the main sources of noise from road vehicles and from trains and indicate how the relative importance of these sources may vary with speed. What are the main sources of noise from aircraft? How have the designers

and manufacturers of road vehicles, trains and aircraft attempted to reduce noise emissions?

3 Give an example of: (i) a physical measure of sound; (ii) a noise index and (iii) a noise rating procedure. (IOA 1999 (part question))

4 (a) Describe two accepted methods of assessing community noise exposure.
(b) Discuss the factors that affect subjective response to noise, and explain how these factors are taken into account in the two methods you have described in part (a). (IOA 1998)

5 Briefly describe (a) the noise certification procedure for aircraft, and (b) the noise emission test procedure (the 'drive-by' test) for road vehicles.

6 The data below are the results of a 15-minute traffic noise survey at a busy roadside in which 100 noise level samples were taken manually, at a sampling rate of approximately every three seconds. Assuming that the samples are representative of the noise climate at the site calculate $L_{A10,15min}$, $L_{A50,15min}$ and $L_{A90,15min}$.

75, 80, 67, 69, 81, 73, 84, 82, 79, 74, 74, 68, 71, 70, 84, 80, 74, 72, 71, 68, 66, 63, 81, 80, 75, 74, 70, 68, 69, 70, 83, 79, 76, 74, 75, 74, 83, 75, 71, 72, 79, 77, 77, 76, 82, 80, 82, 78, 69, 67, 69, 76, 74, 80, 81, 76, 76, 74, 75, 79, 78, 80, 74, 77, 83, 80, 79, 73, 75, 75, 69, 76, 73, 85, 80, 76, 73, 66, 69, 75, 70, 74, 81, 69, 81, 79, 70, 80, 83, 77, 74, 69, 75, 75, 69, 75, 74, 74, 71, 71

7 (a) $L_{Aeq,T}$ is widely used as a measure of environmental noise. Discuss the advantages and disadvantages of assessing noise using this parameter, illustrating your answer with examples of situations where it will be more, or less, appropriate.
(b) What other noise indices are commonly used to measure environmental noise? Explain how they help to provide a better, more complete, overall description of the noise climate than using $L_{Aeq,T}$ alone.

8 Noise levels are measured at a monitoring site at a house near to a railway line. The site is also subject to occasional bursts of noise from a nearby factory and background noise from traffic from a busy main road.

The railway traffic consists of three passenger trains and two freight trains each hour during the daytime. The average sound exposure levels for the trains, measured at the monitoring site, are 98 dBA for freight trains and 91 dBA for passenger trains. The factory emits short bursts of noise of 78 dBA for two minutes, three times each hour. The constant background noise from the traffic is 65 dBA.

Calculate the L_{Aeq} over each hour due to the trains and due to the factory, and also the total hourly L_{Aeq} due to all noise sources.

9 Data from a monitoring station near an airport suggests that three types of aircraft pass over a housing estate and that the number of flights per day and typical sound exposure levels L_{AE} are as follows:

Type	L_{AE}	Number/16 hours
A	102 dB	2
B	97 dB	8
C	95 dB	17

There is a proposal to double the number of flights by type B aircraft. Calculate the present and possible future values of the16-hour L_{Aeq}. (IOA 2001 (part question))

10 (a) What is meant by the terms noise, annoyance and statutory nuisance.
(b) Describe six factors that might be considered when assessing the likelihood of annoyance by noise.
(c) Give brief descriptions of three indices or rating procedures for the assessment of environmental noise. Your descriptions should identify which of the factors in your answer to (b) are taken into account in these indices or procedures and how they are incorporated.
(d) A survey was carried out at the site of a proposed housing development. Eight hourly $L_{eq,1h}$ levels are given in the table below.

Time period	$L_{eq,1h}$	Time period	$L_{eq,1h}$
2300–2400	46.5	0300–0400	40.7
2400–0100	46.0	0400–0500	41.6
0100–0200	45.2	0500–0600	62.3
0200–0300	43.0	0600–0700	49.4

(i) Calculate the $L_{eq,8h}$.
(ii) During the period 0500–0600, a taxi stopped close to the measurement position with its engine running for five minutes and it was observed that the sound pressure level then remained steady at 72.9 dB(A). Estimate what the $L_{eq,8h}$ would have been if this five-minute period was excluded from the measurement. (IOA 2008)

11 (a) Describe briefly the rating principles contained in BS 4142.
(b) What advice or guidance does the British Standard give about the limits to the applicability of the rating method?
(c) A firm wishes to introduce a new item of equipment at an existing site. It is intended that the equipment will only operate during the normal working day, and apart from shutdowns for maintenance will be in continuous operation. There is a sensitive residential façade on a line of sight with, and approximately 30 metres from, the proposed location of the new equipment. What data are required in order to carry out an assessment in accordance with BS 4142? How would you obtain the necessary data?

(d) What information should be reported regarding subjective impressions? (IOA 2009)

12 Annoyance has been defined as 'a feeling of displeasure evoked by a noise' and as 'any feeling of resentment, displeasure, discomfort and irritation occurring when a noise intrudes into someone's thoughts and moods or interferes with activity'.

Many factors must be considered when assessing the likelihood of annoyance by noise. Discuss these factors, including also those not directly associated with the character of the noise itself, and illustrate your answer with examples of how they are reflected in the measurement parameters, indices and rating procedures used for the assessment of environmental noise.

(IOA 2003 (part question))

13 A factory has begun to operate during the daytime only in a rural residential area. As a result of complaints about noise from the factory, the investigating local authority officer measures the L_{Aeq} and L_{A90} for several minutes on a calm, cloudy day just before the factory begins to operate. The levels recorded were 37 dB and 35 dB respectively. The officer measures the L_{Aeq} again at the nearest noise-affected location, for a representative period while the factory is operating. The level recorded was 51 dB. The officer notes that the noise during operation is fairly steady and that there is a distinctive hum present.

(a) Define specific noise level, rating level, ambient noise level, residual noise level and background noise level according to BS 4142 and comment on whether there is any difference between the residual noise level and background noise level in this case.

(b) What would be the BS 4142 assessment of the noise level in this case? State any assumptions made. (IOA 2000)

14 There have been many attempts to develop indices which relate noise exposure to human response.

(a) Discuss in detail the factors that need to be considered when developing such an index to quantify the exposure, and the standard expressed in that index for pollution control. Include in your discussion particular reference to the difficulties in establishing the dose/response relationship and the factors which need to be considered when determining an index. Include in your discussion the role that economics plays in the Standard.

(b) Outline, and discuss with reference to the issues and factors mentioned in (a) above, any index currently in use in the UK. (IOA 1993)

15 (a) Road traffic noise may be considered to be composed of a component arising from the engine, intake, exhaust and transmission of vehicles and a component arising from the tyre and road surface interaction.

(i) Describe qualitatively how the magnitudes of these two components vary with vehicle speed, acceleration and the proportion of heavy vehicles in the traffic flow.

(ii) Describe briefly the main mechanisms involved in the production of tyre/surface noise.

(iii) How do 'quiet' road surfaces function and how are they made? Your answer should include reference, in particular, to stone mastic asphalt (SMA) and porous asphalt road surfaces.

(b) A local authority plans to introduce 300 new buses on a certain road that carries 3000 light vehicles and 300 heavy vehicles every day over the period 0700 to 2300 hours. The local authority proposes to lay a noise-reducing surface to ensure that there is no net increase in noise resulting from their proposal. At an observation point, the single event level of each light vehicle is 70 dB SEL, and of each bus and heavy vehicle it is 80 dB SEL. Assuming that it produces the same noise reduction for all vehicles, how much noise reduction must the new surface provide? Comment on whether this would be feasible.

(IOA 2008)

16 (a) (i) Compare and contrast the temporal characteristics of road and railway noise, considering both short-term and long-term periods of time. Why are there separate daytime and night-time noise insulation criteria for railways, whereas there is only a single criterion for roads?

(ii) Describe the difficulties that might arise if a statistical noise index is used for railway noise.

(iii) Explain how in the case of road traffic noise it is possible to convert a statistical noise index into an equivalent continuous noise level. Why cannot a similar conversion be derived for railway noise?

(iv) The effects of transport noise on people could be categorized under the broad headings of disturbance and annoyance. Explain what is meant by these terms.

(v) In what ways do road traffic noise and railway noise differ in terms of the disturbance and annoyance that they generate?

(b) A railway operator intends to introduce new rolling stock on an electric-only line passing close to a residential area. There will be 25 of these trains per day with an SEL of 80 dB(A), whereas there are currently 50 trains per day using a rolling stock with an SEL of 86 dB(A). How many of the old trains would need to be withdrawn if the total noise exposure of the residents is not to be increased?

(IOA 2008)

17 A small general aviation aerodrome is used mainly by a flying school and private pilots. The owner wishes to apply for permission to extend the runway to allow for use by small business aircraft. One end of the runway is in line with a small village.

(a) What information would you need to make a noise impact assessment of the proposed runway extension?
(b) How would you use this information to carry out the noise impact assessment?
(c) What criteria would you apply?
(d) The owner of the airfield has told an objector that the proposal will have negligible effect on him because there is currently a daily average of 100 movements of light aircraft audible from his property and that there will only be an additional three business movements per day. You determine that the SEL of individual light and business aircraft movements at this location is 86 dB and 103 dB respectively.

Is the owner correct in his claim? Justify your answer by means of a calculation and by reference to established criteria. (IOA 2008)

18 The noise immission from a factory at a nearby dwelling during daytime follows a regular pattern. Once every hour, starting on the hour (e.g. at 9.00, 10.00 hours etc.) there is a burst of noise lasting one minute of 62.0 dBA. Once every hour, starting on the half hour (e.g. at 9.30, 10.30 hours etc.) there is a burst of noise of 56 dBA lasting five minutes, and twice every hour, on the quarter hour (e.g. starting at 9.15, 9.45 hours etc.) there is a burst of noise of 53 dBA, lasting 12 minutes.

At other times during each hour the noise level from the factory is well below the background noise from a motorway, which is a constant and continuous 50 dBA.

(a) Calculate the value of L_{Aeq} of the total noise each hour received at the dwelling, and the contribution due the factory alone. Also calculate the hourly L_{A10}, L_{A50} and L_{A90} values.

(You may assume for the purpose of this part of the question that the factory noise data is from the factory noise alone, i.e. is unaffected by background noise.)

(b) Explain how in situations such as that described in part (a) above, an accurate indication of the factory noise at a nearby dwelling might be obtained, in the presence of comparable levels of background noise.

19 A site near to the proposed route of a new railway is subject to noise from over-flying aircraft and from a busy main road. The level of aircraft noise at the site, measured between 07.00 and 23.00, is 62 dBA $L_{Aeq,16h}$; the road traffic noise, measured between 06.00 and 24.00, is 66 dBA $L_{A10,18h}$.

The railway traffic will consist of freight and passenger trains. It is expected that 36 passenger trains and 12 freight trains will run during the day (07.00–23.00) and eight freight trains will run at night (23.00–07.00). The average sound exposure levels for the trains are 98 dBA for freight trains and 91 dBA for passenger trains.

Calculate the daytime (07.00–23.00) and night-time (23.00–07.00) L_{Aeq} values due to the trains; the total 24 hour L_{Aeq} due to all noises at the site when the railway is built, assuming there are no over-flying aircraft during the night, and ignoring any road traffic noise at night.

20 The table below shows the results (1 to 5) of five environmental noise surveys. The surveys refer to five different situations (A to E) also described below. Match each of the survey results (1 to 5) to one of the situations (A to E), giving, in each case, the reasons for your choice, and as far as possible an explanation of the differences in the various values.

	L_{A1}	L_{A10}	L_{A50}	L_{A90}	L_{Amax}	L_{Aeq}
1	63.2	54.9	54.6	54.2	63.6	54.7
2	72.3	37.4	36.7	35.4	74.2	52.4
3	79.2	70	60	50	79	67
4	72.8	72.4	55.0	48.6	79.1	62.4
5	63	48	43	39	65	46

The five situations are:

A. Monitoring on the roof top of commercial premises during the night-time to establish existing noise levels prior to making a planning application to change plant. Unfortunately a fan close to the microphone which was thought to be switched off at night was later found to have remained on during the monitoring period.

B. The survey as in A above was repeated on another night when the noise climate could be assumed to be similar, but not identical, but with the fan switched off.

C. Measurements taken close to a railway line during the night-time when two trains passed by during the one hour monitoring period with no other significant noise sources. The trains on this line take approximately 20 seconds to pass by during which the sound level is fairly constant. It is known that the SEL value the train events is approximately 85 dBA.

D. At the same location as for C but for a one-hour daytime period during which 20 trains (of the same type as in the night-time) passed by. Although the noise climate was clearly dominated by the trains there was also an increased level of noise from nearby road traffic, and other sources, compared with the night-time situation.

E. Daytime free flowing traffic noise measured outside a house close to a busy main road, in order to establish whether the house was eligible for a grant towards double glazing under the Noise Insulation Regulations, for which a criterion is that $L_{A10,18h} > 68$ dBA.

Chapter 5 Room acoustics

5.1 Introduction

Having learnt about the basic physics of sound and how it is described, measured and assessed in Chapters 1 to 4, Chapter 5 is about room acoustics, i.e. how sound behaves in confined spaces of all sorts, which includes rooms in dwellings, offices, theatres, classrooms etc., and is predominantly about the effects of the absorption of sound at the surfaces of the room – floors, walls and ceilings.

Reflection, absorption and transmission of sound

Sound absorption is the property of a material (e.g. a wall) to absorb sound, i.e. turn sound energy into (very very small amounts of) heat energy, with the aim of reducing the amount of sound energy that is reflected back into the room. Sound insulation is the property of a material (e.g. a wall) which determines how much sound energy will be transmitted through the wall into the adjacent space (e.g. the next door room). Sound insulation is dealt with in Chapter 6.

Consider a beam of sound incident upon a wall, i.e. striking the wall (see Figure 5.1). The sound energy incident upon the wall can be either reflected back into the room, transmitted through the wall or absorbed within the wall (or by an acoustic lining applied to the wall). If we represent the sound intensity of the incident beam by I_i and that of the reflected, transmitted and absorbed sound by I_r, I_t, I_a respectively then we can write simply from the law of conservation of energy that:

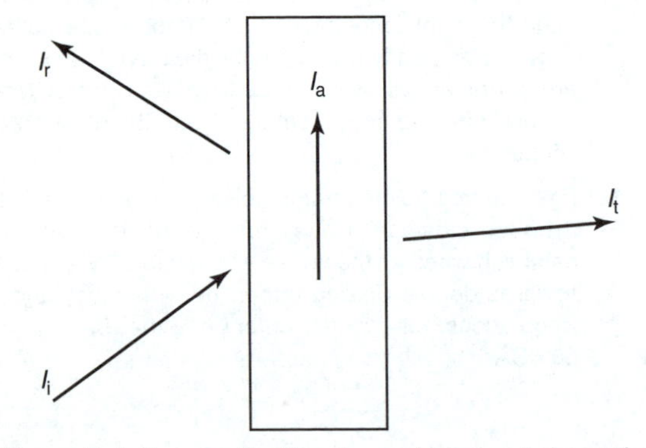

Figure 5.1 Sound reflection, transmission and absorption at a wall

$$I_i = I_r + I_t + I_a$$

The fractions of the sound energy which are reflected and transmitted by the surface (e.g. the wall) are called the reflection coefficient (r) and transmission coefficient (t) respectively:

$$r = I_r/I_i \quad \text{and} \quad t = I_t/I_i$$

It is important to note that since these coefficients are defined by the sound intensities they are related to the ratios of pressure squared (p^2) and *not* by ratios of sound pressures, i.e.

$$r = \left(p_r^2/p_i^2\right) = (p_r/p_i)^2 \quad \text{and} \quad t = \left(p_t^2/p_i^2\right) = (p_t/p_i)^2$$

Absorption coefficient (α)

The absorption coefficient is defined differently than the reflection and transmissions coefficient. The absorption coefficient is the fraction of incident sound intensity which is either absorbed or transmitted, i.e. the fraction of incident sound intensity which is not reflected:

$$\alpha = (I_a + I_t)/I_i \quad \text{and} \quad \alpha = 1 - r$$

The reason for this definition is that, as far as room acoustics is concerned, when incident sound strikes an area of wall, all that matters is how much (i.e. what proportion) of the incident sound is reflected back into the room, i.e. it does not matter whether the sound which is not reflected is transmitted or absorbed. This is nicely illustrated by Sabine's description of the amount of absorption of a surface as the equivalent area of 'open window'. In other words Sabine likened an area of perfect absorption to an open window, in which there is no reflection because all the sound is transmitted to the outside and no part of it is reflected back into the room.

The absorption coefficient, then, is a number which lies between 0 and 1; 1 being a perfect absorber and 0 being a perfect reflector (i.e. no absorption at all).

Variation of properties with frequency and with angle of incidence

The absorption coefficient α varies with frequency and values are required and usually given in octave or one third octave bands. It also depend on the angle incidence of the sound striking the partition. Values are usually measured and quoted either for normal or for random incidence.

Weighted absorption coefficient, α_W

This is a single figure value derived from one third octave band values, by a procedure defined in ISO 11654, in which the spectrum of α values is compared to a series of reference curves according to a specified criterion. The value of α_W is the value of the matching reference curve at 500 Hz.

Terminology – absorption, insulation and isolation

Sound absorbing materials are used to minimize reflection of airborne sound by converting sound energy to heat through some frictional process in the absorber. The most commonly used sound absorbing materials are porous materials such as open-celled polymer foams and fibrous materials such as slabs of mineral fibre. They are commonly applied to walls and ceilings to improve their sound absorbing properties and minimize sound reflection and thus minimize the amount of reverberation that occurs in the room. They are also used inside ventilating and air conditioning ducts to minimize sound reflection at the duct walls and so reduce sound levels inside the duct, and in the cavities of lightweight double leaf partitions to minimize the build-up of reverberant sound caused by sound reflection inside the cavity.

Sound insulating materials are used to minimize the transmission of airborne sound, e.g. from one room to another via a sound insulating wall. Good sound insulating materials are dense heavy materials such as masonry (brickwork, concrete) and sheet metal, because these materials have a high mass per unit area. Sound insulating materials are used as noise screens or barriers, e.g. alongside roads, as partitions, walls, floors and ceilings, to minimize sound transmission between rooms, and as enclosures around noisy machines.

Although slabs of mineral fibre are very good sound absorbers they are lightweight materials and therefore tend to be poor at preventing transmission of sound, i.e. they are poor sound insulators. Conversely, although masonry walls and sheet metal panels are good sound insulators because they are dense and heavy, they usually have hard sound reflecting surfaces and so tend to be poor absorbers of sound (unless of course they are lined with a layer of sound absorbing material).

Isolation

Vibration isolation is the reduction of structure-borne sound and vibration by the use of resilient material (mats or pads of cork rubber or mineral fibre or metal springs) interposed between the vibrating surface and the surface to be protected (isolated). Vibration isolation is discussed in Chapter 7, but briefly, the isolators (springs) and the surface to be isolated form a mass–spring vibrating system with a its own natural frequency, and if the frequency to be isolated is well above this natural frequency then isolation will occur.

Correct use of terminology

The correct use of the terms absorption, insulation and isolation is important. The situation is not helped by the fact that mineral fibre slabs, used as acoustic absorbers, and also sometimes for vibration isolation, are also widely used for thermal insulation (but not for sound insulation).

But much more important than matters of terminology is the underlying understanding of the difference between the requirements for reducing reflection of sound, the transmission of airborne and structure-borne sound, and the different types of materials needed to achieve these reductions.

5.2 Sound absorption

Amount of absorption, *A*

An area of surface (e.g. a carpet lying on a floor) with a surface area S m^2 and having an absorption coefficient α provides an amount of absorption A m^2, such that:

$$A = S\alpha$$

For example, if the carpet has an area of 10 m^2 and an absorption coefficient of 0.6, then the absorption provided by the carpet is 6 m^2, i.e. it is equivalent in terms of sound absorption, to 6 m^2 of perfect absorber, or open window units.

Alternatively if we know that a sample of material with surface area S m^2 produces an amount of absorption of A m^2 then we can calculate its absorption coefficient:

$$\alpha = A/S$$

In particular, if a room has lots of different surfaces, of areas S_1, S_2, S_3 etc. and absorption coefficients α_1, α_2, α_3

and absorptions A_1, A_2, A_3 then the total amount of absorption in the room A_{TOTAL} is given by:

$$A_{TOTAL} = A_1 + A_2 + A_3 + \cdots$$
$$= S_1\alpha_1 + S_2\alpha_2 + S_3\alpha_3 + \cdots$$

And if the total surface area of the room is S_{total} then the average absorption coefficient of all the surfaces in the room, α_{AVGE}, is given by:

$$\alpha_{AVGE} = A_{TOTAL}/S_{TOTAL}$$

Usually the total surface area of the room S_{total} is taken as the area of the six surfaces of the room, i.e. four walls ceiling and floor, so that if the room has dimensions (length, breadth and height) of L, B and H:

$$S_{TOTAL} = 2(LB + LH + BH)$$

Types of absorber

There are three main types of sound absorbers: porous, panel/membrane and Helmholtz/cavity absorbers.

Porous absorbers are broadband absorbers, i.e. they work over a wide range of frequencies (although they are much more effective at high frequencies than at low frequencies), whereas the other two types are effective over a more limited range of frequency centred around the resonance frequency of the absorber (Figure 5.2).

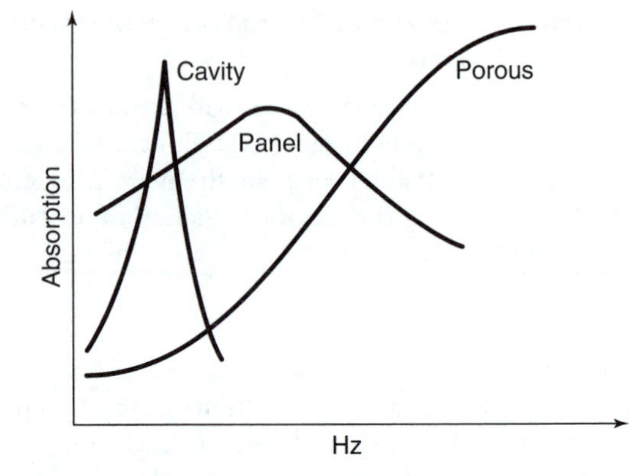

Figure 5.2 Variation of absorption with frequency for porous, panel and cavity absorbers

Porous absorbers

Sound waves cause air particles to vibrate. Vibrating air particles rub against fibres or foam, and friction causes sound energy to be converted to heat, with a reduction of amplitude of sound waves and of sound energy.

The sound absorption depends on the porosity of the foam or fibre material, which in turn depends on the fibre density and average spacing between fibres, or in the case of open-celled polymer foams on the thickness and interconnectivity of the pores in the foam. There is an optimum porosity which allows sound to be transmitted into the porous layer (rather than being reflected at the surface) but to be efficiently absorbed once it has entered the foam/fibre layer. Less space between fibres promotes friction and sound absorption, but if fibre density is too high the increased acoustic impedance of the porous layer causes reflection of sound at the surface of the material, and therefore less absorption inside the material.

Any surface films or layers applied to the surface of a porous layer can reduce the amount of absorption which occurs. Examples are thin layers of polymer film used over mineral fibre layers to protect the mineral fibre and to prevent the shedding of fibres, or layers of gloss paint applied over plaster-based acoustic tiles.

Typical values of absorption coefficients, in octave bands, are shown in Table 5.1.

Porous absorbers are broadband absorbers, i.e. they absorb sound over a wide range of frequencies (as compared with resonant absorbers) but the absorption coefficient increases with frequency, from very low values at low frequencies to close to one at high frequencies. The

Table 5.1 Some typical values of absorption coefficients

	Typical absorption coefficients in octave bands					
Material	125	250	500	1000	2000	4000
Concrete, unpainted	0.01	0.01	0.02	0.02	0.02	0.05
Wood floor on joists	0.15	0.11	0.10	0.07	0.06	0.07
Brickwork	0.02	0.03	0.03	0.04	0.05	0.07
Plastered wall	0.02	0.02	0.03	0.04	0.05	0.05
Glazing (4 mm)	0.30	0.20	0.10	0.07	0.05	0.02
Cord carpet	0.05	0.05	0.10	0.20	0.45	0.65
Axminster carpet	0.08	0.8	0.30	0.60	0.75	0.80
75 mm mineral wool, 60 kg/m^3	0.34	0.95	1	0.82	0.87	0.86
25 mm acoustic plaster on solid base	0.03	0.15	0.50	0.80	0.85	0.80
13 mm ceiling tile on slab	0.2	0.25	0.70	0.85	0.70	0.60
13 mm ceiling tile over 500 mm air space	0.75	0.70	0.65	0.85	0.85	0.80

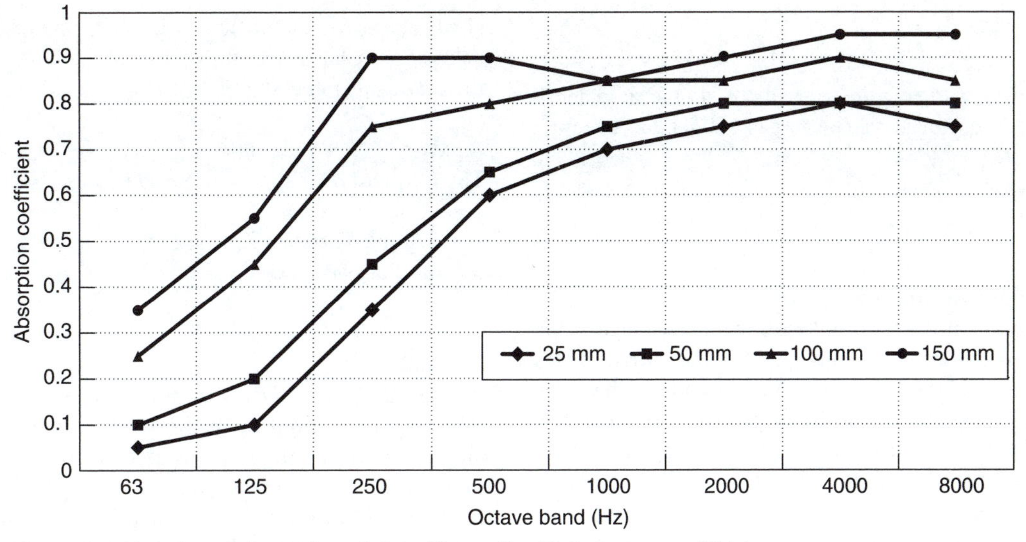

Figure 5.3 Variation of absorption of glass fibre quilt with frequency and thickness

low frequency performance improves with the thickness of the absorbing layer (see Figure 5.3). Porous absorbers are therefore mainly used for reducing reflected sound at medium and high frequencies. The optimum position for a slab of sound absorbing material is a quarter of a wavelength from a wall – this is where acoustic particle velocity is highest. More movement means more loss of energy due to friction, i.e. more absorption.

Resonant absorbers, i.e. panel absorbers and Helmholtz resonators

Note that panel absorbers are also sometimes called membrane absorbers, and that Helmholtz absorbers (or Helmholtz resonators) are also sometimes called cavity resonators (absorbers). Also note that these are both types of resonant absorbers in which the absorption is selective, being highest at or close to a resonance frequency.

Both types of absorbers can be modelled as mass–spring systems, which is the standard model for a simple vibrating system. These are discussed in Chapter 7, but they consist of three essential elements, mass (or inertia), stiffness (or resilience, provided by the spring) and some damping or frictional mechanism, which converts vibrational energy into heat. The system has a natural frequency which depends mainly on the mass and the stiffness, and if the system is forced to vibrate the greatest amplitude of vibration occurs at resonance, when the driving or forcing frequency equals the natural frequency (and hence is also called the resonance frequency). Since the absorbing mechanisms rely on movement (of the air particles in the neck of the Helmholtz resonator, or in the material of the panel in the panel resonator) the maximum absorption occurs at the resonance frequency.

Panel absorbers

In the case of the panel (or membrane) absorber (Figure 5.4) the mass is the mass of the panel (which will depend on the panel thickness and density of the panel material) and the stiffness is the stiffness of the air gap between the panel and the wall (and depends on the thickness of the air gap – the thicker the gap the lower the stiffness). Sometimes the cavity contains sound absorption (as shown).

The natural frequency, f, of a panel absorber is given by: $f = 60/\sqrt{(m \times d)}$, where m is the surface density of the panel in kg/m^2, and d is the thickness of the air gap in m.

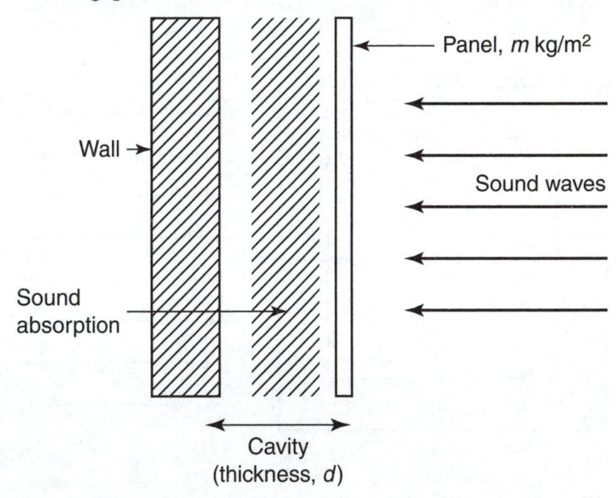

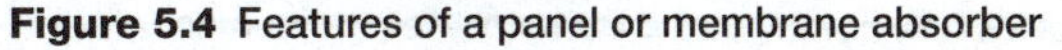
Figure 5.4 Features of a panel or membrane absorber

Exercise

Using this formula, estimate the natural frequency of a panel absorber made from plywood of surface density 2.5 kg/m^2 laid over an air gap of thickness 50 mm.

Answer
170 Hz

Helmholtz resonators

In the case of the Helmholtz resonator (see Figure 5.5) the mass of the mass–spring system is the mass of the air in the neck of the cavity and hence depends on the diameter and length of the neck, and the stiffness is the stiffness of the air in the cavity and hence depends on the volume of the cavity (the greater the volume the lower the stiffness).

The natural frequency, f, of a Helmholtz resonator is given, approximately, by:

$$f = (c/2\pi)\sqrt{(S/(L'V))}$$

where c is the velocity of sound in air, V is the volume of the cavity (m^3), S is the cross-sectional area of the neck (m^2), L' is the 'effective' length of the neck:

$$L' = L + 1.5 \times r \text{ (approximately)}$$

where L is the length of the neck (m) and r is the radius of the neck (m).

Exercise

Using this equation estimate the natural frequency of a milk bottle (or any other sort of bottle with a well defined neck), and if you have access to a sound level meter with third octave filters, or a musical ear, compare your estimate with the note obtained by blowing gently across the top of the bottle, or by pressing the palm of your hand over the mouth of the bottle and then suddenly removing it.

Answer
The frequency for the milk bottle was estimated to be 278 Hz, based on $L = 20$ mm, $r = 14$ mm and $V = 570$ ml. The sound produced was in the 250 Hz one third octave band.

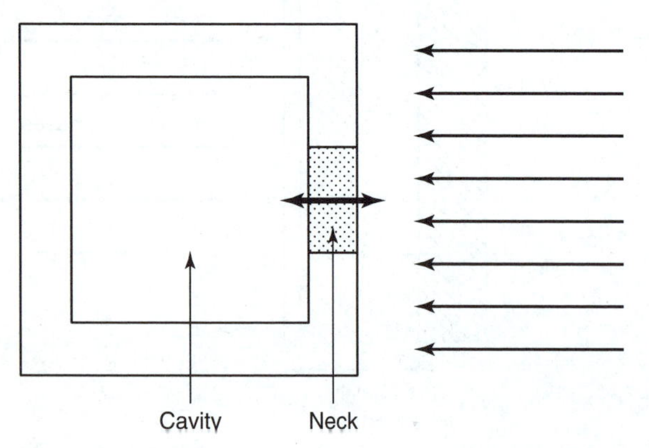

Figure 5.5 Features of a cavity or Helmholtz resonator

Historical

Although the name derives from the use of a series of flasks used by Helmholtz in the 1860s to detect pure tones, it is said that the Romans used cavity resonators formed from clay urns to tune the acoustic characteristics of amphitheatres. Cavity resonance is important in the acoustics of violins and guitars, and in some ancient musical instruments such as the ocarina and the djembe. A more modern application occurs in the design of some loudspeaker cabinets to enhance low frequency response.

Selective low frequency absorption

Both Helmholtz resonators and panel resonators are used mainly for absorbing sound at medium and low frequencies and it is possible to design both types of resonators so that the resonance frequency (and hence maximum sound absorption) occurs at the chosen, required frequency.

Damping and width of resonance

An important difference, in practice, between the two types is that panel resonators usually have a lot of damping (within the panel material) so that the resonance is usually rather broad, i.e. the absorption is spread over a wide range of frequencies above and below the resonant frequency, whereas in the case of Helmholtz resonators there is often very little damping and so the resonance is very sharp and absorption occurs over a very narrow range of frequencies. It is possible to broaden the absorption range of the Helmholtz resonator by placing some sound absorbing material (e.g. polymer foam or mineral fibre) in the cavity. This produces less absorption at the resonance frequency, but spreads the absorption over a wider range of frequencies.

Windows

In the purpose built panel absorbers described above the air gap provides the stiffness and ideally the panel should be limp, i.e. have very low stiffness of its own. However, there are many surfaces within rooms, such as windows, wardrobes and cupboards, that can also provide low frequency resonant absorption. In the case of windows, for example, the glass provides both mass and stiffness and its own natural frequencies and resulting absorption.

Multiple Helmholtz resonators

Figure 5.6 illustrates an important type of acoustic treatment applied to walls, comprising a perforated panel (i.e. with holes in it) separated from the wall by an air gap, or cavity, with a mineral fibre sound absorbing layer in the cavity. The perforated panel may for example be of 'pegboard' (perforated hardboard) or perforated sheet metal.

The parameters defining the construction are: cavity thickness *d*, panel thickness *t*, mass per unit area of panel *m*, hole diameter *h*, the percentage (%) of open area of perforations, and the amount of sound absorption in the cavity.

First let us consider the percentage of open area of the panel perforations. There are three situations to consider. If the percentage of perforations is greater than about 25% the perforated panel in effect becomes 'acoustically invisible', i.e. it has no effect, and the construction is in effect a wall with a sound absorbing layer. It is for this reason that industrial sound absorbing panels, used to line the inside of acoustic enclosures, and in-duct silencers, are usually provided with a very thin perforated sheet steel layer, which holds the mineral fibre in place and protects it against knocks and abrasion, and, provided the % open area is greater that about 25% will not significantly reduce the sound absorption produced by the mineral fibre layer. The second case is when there are no perforations, i.e. zero % open area. In this case we have a panel or membrane absorber, described above.

The third case is the case when the % open area is below 25%, when the construction behaves as a multiple Helmholtz resonator. It is as though the cavity in Figure 5.6 was divided into cells, each with an area of panel containing just one hole, shown by the horizontal lines in the sketch. Each cell would the be an individual Helmholtz resonator with the air in the hole acting as the mass and the volume of the cavity cell acting as the spring. Therefore the parameters determining the resonance frequency will be: percentage open area, hole diameter, panel thickness (i.e. hole length) and cavity thickness.

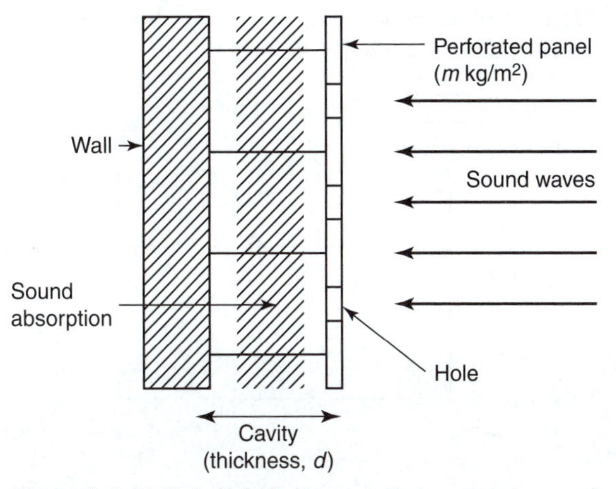

Figure 5.6 Multiple Helmholtz resonator

It is instructive to consider how the performance of the perforated panel absorber will change with variation of the other parameters listed above.

5.3 Room acoustics

The problem of room acoustics may be stated simply enough: we have a room and we can accurately define all the surfaces in the room and their absorption coefficients, we have one (or more) sound source whose output (e.g. sound power and directivity) is also known. We would like to know (i.e. be able to predict), the sound pressure level everywhere in the room, either when the source has been switched on long enough for the sound pressure level to have become steady – the 'steady state' case – or in the transient situation just after the source(s) have been switched on or off.

The problem is almost impossible to solve accurately and precisely in all but the very simplest cases (e.g. a rectangular room with no other surfaces, i.e. one that is completely empty). This is simply because it is too complicated; there are too many reflections, diffraction and scattering of sound waves to consider.

There are three approaches to room acoustics: the geometric (ray tracing), wave theory (room modes) and statistical (e.g. Sabine's) approaches.

The geometric or ray (or beam) tracing approach

This is an approximate approach to predicting room acoustics which works for medium and high frequencies. It is based on the assumption that sound travels in straight lines as rays or beams, and it is possible to follow or track each ray after each reflection and calculate the sound pressure over any small area by calculating the contribution of all the rays landing on that area taking into account the absorption coefficient at each reflection (i.e. at each surface) and the distances travelled. An alternative, but essentially equivalent, approach to tracing rays is to represent each reflection by an 'image' source. The limitation to this approach to room acoustics is that this behaviour only happens, approximately, when the size of the surface is large compared to the wavelength, i.e. for large surfaces or high frequencies.

Architects and designers have always used ray tracing to investigate the effects of a few important reflections (e.g. from stage via ceiling to rear of hall, or from balconies, curved surfaces, and domes or 'barrel' roofs), which may causes focusing effects which are considered to be undesirable.

With the advent of powerful computing techniques, several commercial ray-tracing room acoustics software packages have become available in recent years. It is necessary to construct a three-dimensional virtual model of the room, assigning an absorption coefficient (in octave bands) to every surface, and defining the position, sound power and directionality of every noise source. The software then sends out rays from each noise source and calculates the value of various acoustic parameters at any specified points (or series of points) defining a receiving surface (e.g. audience area) with output in graphic as well as numeric form. The effects of any changes to the model (e.g. the effects of adding extra absorption) may also be demonstrated to a listener using auralization techniques.

The model should be simplified to include only the most significant acoustically reflecting surfaces.

A limitation of the technique is that it cannot properly take into account diffraction by surface edges, although it is possible to assign diffusion or scattering coefficients to surfaces. Some packages attempt to include the effects of sound transmission through partitions, although using some fairly basic assumptions.

The wave theory approach

This is the only one of the three approaches which is capable of bring theoretically correct. It is based on solving the wave equation which describes the behaviour of sound in the room. This shows that the sound travels in the room as a series of three-dimensional standing waves or room modes, each at its own frequency, and each with its own 'shape', showing how the sound pressure in the room varies between regions of pressure maxima and pressure minima (sound pressure nodes and antinodes).

At any point in the room, and at any frequency, the sound pressure level will be the sum of the contributions from of all the modes of the room (in theory an infinite number of them) at that point.

This approach works well for small rooms of simple shape at low frequencies, but not for large rooms or rooms with a complicated shape at high frequencies.

The first problem is that although it is possible to calculate the room mode frequencies for simple 'rectangular box' shaped rooms it is much more difficult to calculate frequencies for rooms with a more complicated shape, although computer software packages based on 'finite element' techniques are available that can do this.

The second difficulty is that as frequency increases the number of modes increases rapidly so that the modes become spaced more and more closely together in frequency, and eventually they merge together.

Room modes are discussed in more detail later in this chapter.

The statistical approach

In principle the ray and wave approaches give the values of acoustic parameters at different positions within a room. They can get very complicated for complex situations and require a lot of computation.

The statistical approach abandons all attempts at a detailed, position-related prediction, and limits itself to predicting statistical average values of acoustic data (e.g. sound pressure level or reverberation time) which are assumed to be constant throughout the entire room. Essentially they are based on an acoustical energy balance between the amount of energy entering the room from the sound source and the total of the energy contained within the sound in the room plus that absorbed during reflections at room surfaces.

The most common statistical method is that developed by Sabine, and so often known as Sabine acoustics.

The water tank analogy

The analogy shown in Figure 5.7, may be helpful in understanding Sabine's statistical theory. The rate of flow of water is analogous to the sound power of the source feeding sound energy into the room and the leakage of water from the tank is analogous to the effect of sound absorption at the room surfaces. The level of water in the tank is analogous to the sound pressure level in the room.

Note that the analogy works for both the steady state situation and the transient situation, when the tank is filling up or is emptying (Figure 5.8), representing the growth and decay of sound in a room when the sound source is either switched on or switched off.

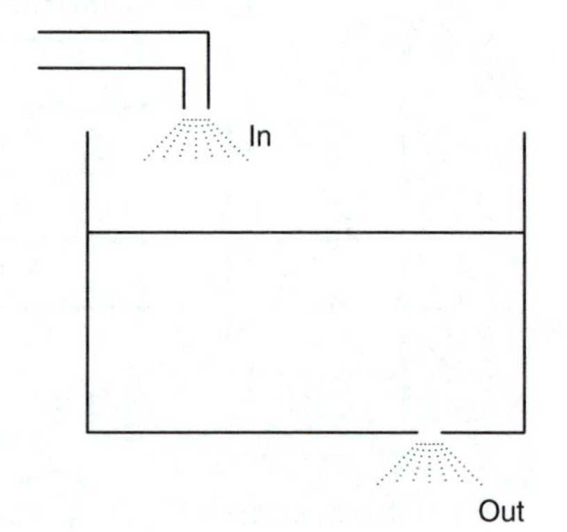

Figure 5.7 The water tank analogy

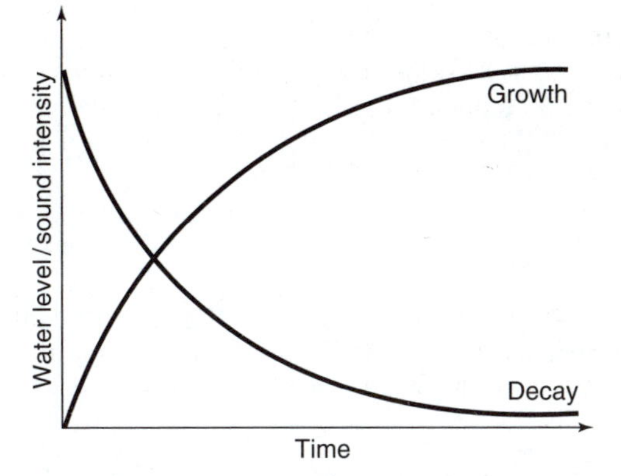

Figure 5.8 Variation of water (and sound intensity) with time

Sabine acoustics: assumptions and limitations

1. Sound in the room is diffuse, i.e. sound energy density is constant throughout the room, and at any point the distribution of sound intensity is random in direction, i.e. sound waves are travelling equally in all directions.
2. Sound absorption is spread evenly throughout room surfaces.
3. Room dimensions are proportionate, i.e. no one room dimension is very much larger (or smaller) than the others, so that corridors, stairwells and large open plan offices are non-Sabine spaces.

5.4 The transient case of Sabine acoustics

In a perfectly diffuse sound field the intensity of sound in a room decays exponentially when the sound source is switched off, but if the intensity is expressed in decibels the decay of sound level versus time becomes linear, i.e. a straight line graph of sound level versus time (see Figure 5.9). It is the slope of this line which is measured to determine the reverberation time, defined below.

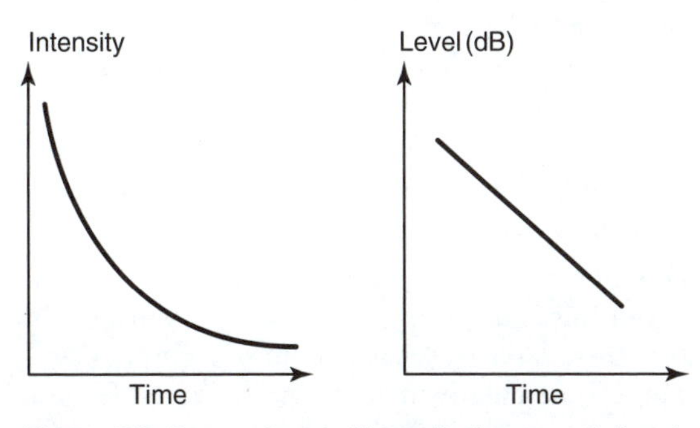

Figure 5.9 Decay of sound intensity (exponential) and level (linear) with time

W. C. Sabine was the first to conduct a systematic investigation into room acoustics starting in 1895. He defined the reverberation time as the time taken for the sound intensity in a room to drop to one millionth of its original value (i.e. by 60 dB) measured from the time that the sound source is switched off.

He showed that the reverberation time of a room, T (in seconds) depends on its volume, V (m^3) and the amount of sound absorption it contains, S (m^2), according to the formula:

$$T = 0.16 \times V/A$$

Sabine's formula is the simplest formula for predicting reverberation time. Eyring's formula is more accurate but more complicated:

$$T = 0.16 \times V/[-S\alpha_{\mathrm{AVGE}} \times \ln(1 - \alpha_{\mathrm{AVGE}})]$$

The derivation of Eyring's formula, from which Sabine's formula may also be deduced, is given in Appendix 5.2.

There are even more accurate (and complicated) formulae (e.g. by Millington and Sette), which allow non-uniform distribution of absorption (up to a certain extent) to be taken into account.

The direct derivation of Sabine's formula is given in Appendix 5.3. It involves some mathematics to set up an equation balancing the rate at which sound energy is being absorbed by the room surfaces with the rate at which the sound intensity in the room (assumed to be the same everywhere in the room) is decreasing with time. This equation leads to the conclusion that the sound intensity in the room is decreasing exponentially with time, and the reverberation time is calculated on this basis. The most significant aspect of this is that if the sound intensity decays exponentially with time it is easily shown that the sound intensity level, in decibels, decreases linearly with time, i.e. the graph of sound intensity level versus time is a straight line, as shown in Figure 5.9 (until the level drops to being close to the background level, when a 'levelling off' occurs).

The practical application of this is that if a decay curve is available for inspection, when measuring the reverberation time it is possible to see how well, or not, the decay curves approach the theoretical ideal straight line decay, and reverberation time values from very poor decay curves can be rejected.

Example 5.1

For a room 8 m long × 6 m wide × 4 m high determine the reverberation time using the Sabine equation when the absorption coefficient for the floor is 0.02, for the walls 0.04, and 0.1 for the ceiling.

Volume, $V = 8 \times 6 \times 4 = 192\ \text{m}^3$
Area of walls $= 2 \times [(8 \times 4) + (6 \times 4)] = 112\ \text{m}^2$
Area of floor and ceiling $= 48\ \text{m}^2$
Total area $= 208\ \text{m}^2$

	α	S	$A = S\alpha$
Walls	0.04	112	4.48
Floor	0.02	48	0.96
Ceiling	0.1	48	4.8
Total		208	10.24

$$\text{Average } \alpha\ (\alpha_{\text{AVGE}}) = \text{Total } A/\text{Total } S$$
$$= 10.24/208 = 0.049$$

Using Sabine's formula:

$$T = 0.16 \times V/A$$

$$T = 0.16 \times 192/10.24 = 3.0 \text{ seconds}$$

Using Eyring's formula:

$$T = 0.16 \times V/[-S \times \ln(1 - \alpha_{\text{AVGE}})]$$

$$T = 0.16 \times 192/[208 \times \ln(1 - 0.049)]$$

$$= 2.9 \text{ seconds}$$

Exercise

In order to reduce the reverberation time in the room in the above example the floor is covered with carpet having an α value of 0.8 in a particular frequency band, the walls are lined with sound absorbing material with an α value of 0.2, and the ceiling is covered with acoustic tiles with an α value of 0.3, in the same frequency band. Calculate the reverberation time of the room using both Sabine's and Eyring's formulae.

Answers

$T = 0.41$ seconds using Sabine's formula and $T = 0.33$ seconds using Eyring's formula. Average α value $= 0.36$.

Exercise

Calculate the reverberation time using Sabine's and Eyring's formulae for a room which has double the dimensions of the above room, i.e. 16 m long × 12 m wide × 8 m high, and the sound absorption coefficients remain the same as originally, i.e. 0.02 for the floor, 0.04 for the walls, and 0.1 for the ceiling.

Answers

$T = 6.0$ seconds using Sabine's formula and $T = 5.9$ seconds using Eyring's formula. Average α value $= 0.049$.

Exercise

Repeat the calculations for the larger room when treated as before, i.e. with α values of 0.8 for the floor, 0.2 for the walls, and 0.1 for the ceiling.

Answers

$T = 0.82$ seconds using Sabine's formula and $T = 0.66$ seconds using Eyring's formula. Average α value $= 0.36$.

Exercise

What general lessons can you learn from this sequence of examples?

Answer

- When the room is 'live' (average α value $= 0.049$) Sabine and Eyring give very similar results.
- When the room is 'dead' (average α value $= 0.36$) Sabine and Eyring give very different results.
- The reverberation time predicted by Eyring's formula is always smaller than that given by Sabine's formula.
- Doubling the size of the room dimensions with the same α values doubles the reverberation time (because although volume increases by eight times, surface area and absorption also increase, by four times). The average α value remains the same because although the amount of absorption doubles, so does the total surface area of the room.

Note that this example does not have the sound absorption material well distributed over the surfaces and hence caution should be applied with the use of the Sabine equation.

Sabine v Eyring – when is it necessary to use Eyring's formula?

Sabine's formula is very widely used because it is so simple, but it is really only accurate for fairly 'live' rooms where there is only a small amount of absorption and a lot of reflection of sound. It does not work well for 'dead' rooms, i.e. rooms containing a lot of absorption. To demonstrate this, re-write Sabine's formula using

$A = S\alpha_{AVGE}$ where α_{AVGE} is the average value of absorption coefficient, so that Sabine's formula becomes:

$$T = 0.16V/(S\alpha_{AVGE})$$

Common sense tells us that if α_{AVGE} equals 1.0 we have free field conditions and T must be zero, but Sabine's formula predicts a finite, i.e. non-zero, value for T, which must be wrong.

A rule of thumb is that Sabine's formula should not be used, and Eyring's formula should be used instead, if α_{AVGE} is greater than 0.2.

Why are the two formulae different?

The key difference is that in the derivation of Eyring's formula sound energy is only absorbed in discrete steps, every time a sound wave meets a reflecting surface (e.g. a wall), whereas in the derivation of Sabine's formula it is assumed that the absorption is taking place continuously. Apart from this difference all the basic assumptions that have to be true for Sabine's formula (uniform distribution of absorption etc.) must also be true for Eyring's formula to work.

Note that it has been assumed that the only sound absorption that takes place is at the surfaces of the room, i.e. that air absorption is negligible. For the very largest spaces (e.g. cathedrals and very large concert halls) air absorption might become significant at high frequencies. This is easily incorporated into both Sabine's and Eyring's formulae.

Reverberation time with air absorption:

$$T = 0.16V/\{(S \times \alpha_{AVGE}) + kV\}$$

where k is humidity and frequency dependent.

For example $k = 20 \times 10^{-3}$ at 4 kHz and 60% humidity.

Example 5.2

A cathedral has a volume of 60,000 m^3 and internal surface area of 42,000 m^2 with an average α of 0.02. Estimate the reverberation time (a) assuming surface absorption alone, and (b) including air absorption using the value of k given above.

Solution

A due to surface absorption $= 42{,}000 \times 0.02 = 840$ m^2

$T = 0.16 \times 60{,}000/840 = 11.4$ seconds

Including air absorption $A = 840 + (60{,}000 \times 20 \times 10^{-3}) = 840 + 1200 = 2040$ m^2

$T = 0.16 \times 60{,}000/2040 = 4.7$ seconds

Optimum reverberation time values of spaces

For every space there is an optimum reverberation time value, depending on the function (i.e. for performance of speech or for music, and, if music, then the type of music), and to a lesser extent, the volume of the space.

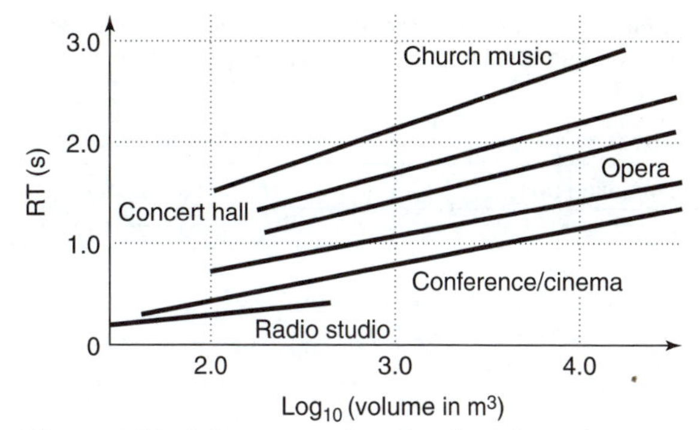

Figure 5.10 Optimum reverberation times for various types of space

If the reverberation time of a room is too low then there will not be enough reflected sound to adequately support the direct sound and the sound level will not be high enough at the back of the room, and music will lack the fullness that arises when reverberation causes notes to blend together: the room is said to be too dry for music. Whereas if the RT is too high there is too much reverberation, speech becomes unintelligible and music can lack clarity.

Figure 5.10 shows optimum values for various types of music and for various room volumes. Guidance is also given in BS 8233 for various types of space, including offices (see Table 5.2), and in BB 93 for rooms (e.g. classrooms) in schools (Table 5.3).

A suitable value for speech is about one second (varying from about 0.6 seconds for small rooms to 1.2 seconds or more for large spaces), with values for music being up to double these values.

Optimum volume

In order that the optimum listening conditions are obtained, it is essential that a hall has the correct order of

Table 5.2 Guide to reverberation time at 500 Hz in unoccupied rooms for speech and music (from BS 8233)

	Reverberation time in seconds	
Room volume (m^3)	Speech	Music
50	0.4	1.0
100	0.5	1.1
200	0.6	1.2
500	0.7	1.3
1000	0.9	1.5
2000	1.0	1.6

Table 5.3 Some recommended maximum mid-frequency reverberation times for rooms in schools (from BB93)

Type of room	Reverberation time in seconds
Swimming pool	Less than 2.0
Indoor sports halls, gymnasia	Less than 1.2
Large lecture rooms (more than 50 people), libraries, drama studios, music classroom, dining rooms, offices, staff rooms	Less than 1.0
Small lecture rooms (less than 50 people), classrooms (secondary school), individual study rooms, small music practice rooms	Less than 0.8
Primary school classrooms, nursery school playrooms	Less than 0.6
Classrooms for hearing impaired students and for speech therapy	Less than 0.4

Table 5.4 Optimum volume/person (m^3) for various types of hall

	Minimum	Optimum	Maximum
Concert halls	6.5	7.1	9.9
Italian-type opera houses	4.0	4.2–5.1	5.7
Churches	5.7	7.1–9.9	11.9
Cinemas	–	3.1	4.2
Rooms for speech	–	2.8	4.9

volume for its use. These values are given in Table 5.4. It can be seen that the volume per person is dependent upon the purpose for which the building is to be used. Music played in a hall with too small a volume is likely to lack fullness, whereas speech in a hall with a very large volume for its seating capacity can be expected to lack clarity. Tables 5.5 and 5.6 give a list of the vital acoustic statistics for a few of the well known concert halls and opera houses. Nearly all of the best of these fit into the general pattern of volumes recommended in Table 5.4.

Volumes of churches vary enormously, with many famous cathedrals having a capacity up to four times that recommended. St Paul's Cathedral, for example, has a volume of roughly 150,000 m^3. Many other English cathedrals have a volume of about 30,000 m^3 or more.

Uses of Sabine's formula

- To predict RT of spaces from room surface materials (α values and areas).
- To estimate room absorption A from measured values of reverberation time, using $A = 0.16V/T$.
- To estimate amount of absorption to be added or removed from a room in order to modify the RT to an optimum value.
- To measure absorption coefficient, α, in a reverberant test room (using the method described later) using the formula:

$$\alpha = (0.16V/S)[(1/T_1) - (1/T_2)]$$

- To correct measured sound insulation values (level differences) to take into account the amount of absorption in the receiving groom (this will be dealt with in Chapter 6).

Example 5.3

A workshop of dimensions 8 m long × 6 m wide × 4 m high with hard sound reflecting surfaces is to be refurbished as a room for meetings. The reverberation time is measured and found to be 3.0 seconds in a particular octave band. Calculate how much extra sound absorption must be introduced into the room in order to reduce the reverberation time to 0.8 seconds, suitable for speech.

If this is to be achieved using carpet on the floor with an α value of 0.2 in the same band and by covering areas of

Table 5.5 Acoustical data for some well known concert halls

Name	Volume (m^3)	Audience capacity	Volume per audience seat (m^3)	Mid-frequency RT in seconds (full hall)
St Andrew's Hall, Glasgow (built 1877)	16,100	2133	7.6	1.9
Carnegie Hall, New York (1891)	24,250	2760	8.8	1.7
Symphony Hall, Boston (1900)	18,740	2631	7.1	1.8
Tanglewood Music Shed, Lennox, Mass. (1938)	42,450	6000	7.1	2.05
Royal Festival Hall (1951)	22,000	3000	7.3	1.47
Liederhalle, Grosser Saal, Stuttgart (1956)	16,000	2000	8.0	1.62
F. R. Mann Concert Hall, Tel Aviv (1957)	21,200	2715	7.8	1.55
Beethovenhalle, Bonn (1959)	15,700	1407	11.2	1.7
Philharmonic Hall, New York (1962)	24,430	2644	9.3	2.0
Philharmonic Hall, Berlin (1963)	20,000	2200	11.8	2.0

Table 5.6 Acoustical data for some well known opera houses

Name	Volume (m^3)	Audience capacity	Volume per audience seat (m^3)	Mid-frequency RT in seconds (full hall)
Teatro alla Scala, Milan (1778)	11,245	2289	4.91	1.2
Academy of Music, Philadelphia (1857)	15,090	2836	5.32	1.35
Royal Opera House (1858)	12,240	2180	5.6	1.1
Theatre National de L'Opera, Paris (1875)	9,960	2131	4.67	1.1
Metropolitan Opera House, New York (1883)	19,520	3639	5.36	1.2

the wall and ceiling with sound absorbing panels having an α value of 0.4, calculate the area of these panels which must be used. (Use Sabine's formula rather than Eyring.)

Solution

Since $T = 0.16 \times V/A$ it follows that: $A = 0.16 \times V/T$.

Existing amount of absorption, $A = 0.16 \times 192/3.0 = 10.2\ m^2$.

Amount of absorption needed for required T value of 0.8 seconds:

$$A = 0.16 \times 192/0.8 = 38.4\ m^2$$

Therefore additional absorption required $= 38.4 - 10.2 = 28.2\ m^2$.

Additional absorption due to carpet $= S\alpha = 48 \times 0.2 = 9.6\ m^2$.

Therefore, additional absorption, A, required from sound absorbing panels $= 28.2 - 9.6 = 18.6\ m^2$.

Since $A = S\alpha$, $S = A/\alpha = 18.6/0.4 = 46.5\ m^2$. Therefore an area of $46.5\ m^2$ of acoustic panels are needed.

A refinement

Strictly speaking we are covering up some of the absorption of the existing surfaces. Although we do not know the α value of the original surfaces we can calculate the average value:

$$\text{Surface area of room} = 2 \times [(8 \times 4) + (6 \times 4) + (8 \times 6)] = 208\ m^2$$

$$\text{Average } \alpha = \text{Total } A/\text{Total } S = 10.2/208 = 0.05$$

Area of original surfaces covered by carpet and acoustic panels $= 48 + 46.5 = 94.5\ m^2$. Therefore original absorption which has been covered up $= 94.5 \times 0.05 = 4.7\ m^2$. So to replace this an additional $4.7/0.4 = 11.8\ m^2$ of acoustic panels are needed.

Example 5.4

A board room of volume $150\ m^3$ has a reverberation time of 0.8 seconds when empty. Estimate the reverberation time during a board meeting when 12 people are present. Assume an absorption of $0.45\ m^2$ per person in the required octave band.

Solution

A when empty $= 0.16 \times 150/0.8 = 30\ m^2$.

A when occupied $= 30 + (12 \times 0.45) = 35.4\ m^2$.

T when occupied $= 0.16 \times 150/35.4 = 0.68$ seconds.

Example 5.5

In a large hall or theatre the total amount of sound absorption is often determined by the amount of sound absorption provided by the seats and audience which may far exceed the absorption produced by the remaining fabric of the building.

If the amount of absorption per seat is $0.5\ m^2$ in a certain octave band, estimate how many seats may be accommodated in a hall of volume $2000\ m^3$ if the reverberation time is not to fall below the optimum required value of 1.0 seconds.

Solution

$$T = 0.16 \times V/A \text{ so that } A = 0.16 \times V/T$$

Therefore the absorption required to produce optimum $T = 0.16 \times 2000/1 = 320\ m^2$.

So the maximum number of seats $= 320/0.5 = 640$.

Volume per seat $= 2000/640 = 3.1\ m^3$ per seat.

5.5 Sabine acoustics for steady state situations

Recalling the water tank analogy, if sound energy, from a loudspeaker for example, is entering a room continuously then after a certain time a steady state situation will be reached where the rate of sound energy entering the room is balanced by the rate at which sound energy is absorbed from the room surfaces (with, usually, a negligibly small additional contribution from the air in the room). The steady sound pressure level at which this balance occurs is called the reverberant sound pressure level in the room.

Reverberant sound

The Sabine statistical theory of room acoustics shows that the reverberant sound pressure level, L_{REV} is given by:

$$L_{REV} = L_W + 10\log(4/R_C)$$

where L_W is the sound power level entering the room. A derivation of this formula is given in Appendix 5.4.

Room constant

R_C is called the room constant and is defined by:

$R_C = S\alpha_{AVGE}/(1 - \alpha_{AVGE})$, measured in m^2 units

where S is the total area of the room surfaces (usually taken as the area of the walls, floor and ceiling) and α_{AVGE} is the average absorption coefficient of those surfaces.

Note that the reverberant sound pressure level depends only on the source and the room surfaces, it does not depend on distance, i.e. on position within the room. Thus according to Sabine's theory, the result of the very many sound reflections, and diffraction and scattering that occurs within the room, results in the reverberant sound intensity and sound pressure level being the same everywhere throughout the room, i.e. diffuse sound.

Diffuse sound

Another important feature of the reverberant sound field, as it is called, is that it is also diffuse. This means that at any point in the room the sound energy is arriving with equal intensity from all directions.

Direct sound

In addition to the reverberant sound resulting from reflections with room surfaces a receiver in the room will also receive sound directly from the sound source in the room.

The direct sound pressure level is that obtained for the free field situation which has already been discussed earlier in this chapter and in Chapter 2:

$$L_{DIRECT} = L_W + 10\log(Q/4\pi r^2)$$
$$= L_W - 20\log r - 11$$

In contrast to the reverberant sound pressure, the direct sound pressure depends on the distance from the source but not at all on the room surfaces. It is also completely directional, i.e. it arrives from the direction of the sound source.

Total sound pressure level

The total sound pressure level at any point in the room is the combination of these two, i.e. reverberant and direct, sound pressure levels, and may be obtained either by calculating each separately and then combining them using logarithmic or decibel addition, or by using the single formula:

$$L_{TOTAL} = L_W + 10\log[(Q/4\pi r^2) + (4/R_C)]$$

The above equation indicates how the total sound pressure level at any point depends on the distance from the source. Close to the source the total sound pressure level is mainly due to the direct level, and reduces with distance (at a rate of 6 dB per doubling of distance). As the distance from the source increases the sound pressure level falls at a slower rate until it eventually becomes constant. At these distances from the source the total sound pressure level becomes mainly due to the reverberant sound. The reverberant level will depend on the room constant.

The variation of total sound pressure level with distance, for rooms with different room constants, is illustrated in Figure 5.11.

The above ideas have some important practical consequences:

- The same sound source may produce significantly different sound pressure levels at the same distance from the source in different rooms, therefore care must be taken when comparing noise emissions from sound sources.
- Noise levels in a room may be reduced by increasing the amount of sound absorption (and therefore room constant) but only for the reverberant component of the total sound field, and will be ineffective close to the sound source.

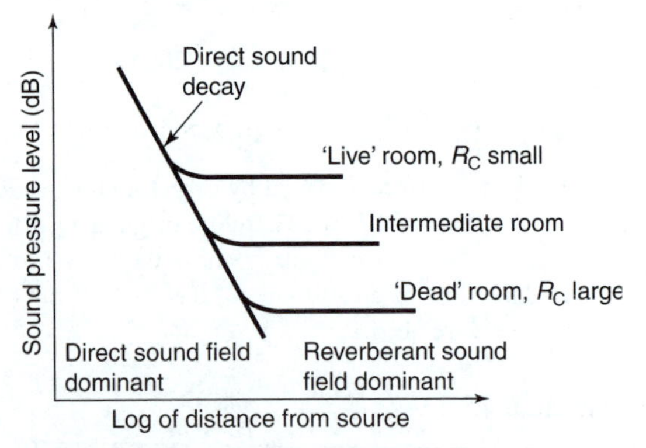

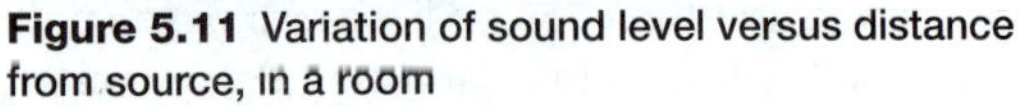
Figure 5.11 Variation of sound level versus distance from source, in a room

Approximation of R_C to A for live spaces

Since $R_C = S\alpha_{AVGE}/(1 - \alpha_{AVGE})$ when the $\alpha_{AVGE} \ll 1$, $1-\alpha_{AVGE} \approx 1$ and $R_C \approx S\alpha_{AVGE}$, i.e. $R_C \approx A$, so that the room constant has approximately the same value as the amount of absorption in the room, A.

A rule of thumb is that the approximation should only be used if $\alpha_{AVGE} < 0.2$. This is the same rule as that for deciding whether to use Sabine's or Eyring's formula for estimating reverberation times.

Room radius

The distance from the source at which the direct and reverberant sound pressure levels are equal is called the room radius, R, given (from the above equation) by:

$$Q/4\pi R^2 = 4/R_C$$

$$\text{i.e. } R = \sqrt{[QR_C/16\pi]}$$

Example 5.6

A room 8m long × 6 m wide × 4 m high has surfaces with the following absorption coefficients in a particular octave band: floor 0.02, the walls 0.04 and ceiling 0.1.

An omni-directional sound source with a sound power of 0.05 W is located in the centre of the room. Calculate (i) the room constant, (ii) the reverberant sound pressure level, (iii) the direct and total sound pressure levels at distances of 1 m and 2 m from the sound source and (iv) the room radius.

Solution

Sound power level of source = $10\log(0.05/10^{-12})$ = 107 dB re. 1×10^{-12} W.

From the previous examples for the same room: $A = 10.2\ \text{m}^2$ and $\alpha_{AVGE} = 0.05$, $S = 208\ \text{m}^2$.

Room constant = $208 \times 0.05/(1 - 0.05) = 10.8\ \text{m}^2$.

Reverberant sound pressure level, $L_{REVERB} = 107 + 10\log(4/10.9) = 102.7$ dB re. 2.0×10^{-5} Pa.

At a distance of 1 m direct sound pressure level $L_{DIRECT} = 107 - 20\log 1 - 11 = 96$ dB, and at a distance of 2 m, $L_{DIRECT} = 90$ dB.

At 1 m, total sound pressure level = $107 + 10\log[(1/4\pi \times 1^2) + (4/10.9)] = 103.5$ dB re. 2.0×10^{-5} Pa.

At 2 m, total sound pressure level = $107 + 10\log[(1/4\pi \times 2^2) + (4/10.9)] = 102.9$ dB.

Room radius = $\sqrt{[QR_C/16\pi]} = \sqrt{[1 \times 10.9/16\pi]}$ = 0.46 m.

So, for this very 'live' room the values of A and R_C are within a few per cent of each other and the sound is almost entirely reverberant except for very close to the sound source (less than 0.5 m away).

Exercise

Repeat the question for the room in the above example when the floor is covered with carpet having an α value of 0.8 in a particular frequency band, and the walls are lined with sound absorbing material with an α value of 0.2 and the ceiling is covered with acoustic tiles with an α value of 0.3, in the same frequency band.

Answers

$A = 75.2\ \text{m}^2$ and $\alpha_{AVGE} = 0.36$, $R_C = 117.8$, L_{REVERB} = 92.3 dB, room radius = 1.5 m.

The direct sound pressure levels will remain the same (96 and 90 dB), and the total sound pressure levels are 97.5 dB at 1 m and 94.3 dB at 2 m.

So, for this fairly 'dead' room the room constant is much larger than the room absorption, the reverberant sound pressure level has been reduced from 102.7 dB to 92.3 dB and the room radius has increased from 0.5 m to 1.5 m.

Effect of absorption on reverberant sound level

In spaces such as factories, workshops and plant rooms the main purpose of adding sound absorption to a room will be to reduce the level of reverberant sound rather than to reduce the reverberation time (although this will happen as well).

In order to predict the reduction which may be achieved, consider a room in which the reverberant sound level is L_1 and the existing room constant and sound absorption are R_{C1} and A_1. When the amount of absorption is increased to A_2, and the room constant to R_{C2}, the reverberant level in the room reduces to L_2.

Therefore:

$$L_1 = L_W + 10\log(4/R_{C1})$$

and

$$L_2 = L_W + 10\log(4/R_{C2})$$

so that

$$L_1 - L_2 = 10\log(4/R_{C1}) - 10\log(4/R_{C2})$$

Therefore reduction in reverberant noise level is:

$$L_1 - L_2 = 10\log(R_{C2}/R_{C1})$$

Since factories, workshops and plant rooms are very often highly reverberant spaces with very little intrinsic absorption, the approximation that $R_C \approx A$ often applies, so that reduction in reverberant noise level is:

$$L_1 - L_2 = 10\log(A_2/A_1)$$

Example 5.7

A workshop of dimensions 20 m × 8 m × 4 m has a reverberation time of 3.8 seconds in the 1000 Hz octave band. The reverberant sound pressure level in the workshop is 86 dB in this same band. It is proposed to reduce the noise level in the workshop by hanging slabs of sound absorbing material in the room.

Estimate the reduction in noise if 50 m^2 of sound absorption is introduced into the workshop.

Solution

The volume of the workshop is 640 m^3. Using Sabine's formula we can estimate the amount of absorption in the workshop:

$$A_1 = 0.16 \times V/A = 0.16 \times 640/3.8 = 26.9 \text{ m}^2$$

After the addition of the absorption the slabs of material:

$$A_2 = 26.9 + 50 = 76.9 \text{m}^2$$

Therefore reduction in reverberant noise level is:

$$L_1 - L_2 = 10\log(A_2/A_1) = 10\log(76.9/26.9)$$

$$= 4.6 \text{ dB}$$

Therefore the level of reverberant noise in the workshop will be reduced from 86 dB to 81.4 dB.

Special acoustic test rooms – anechoic and reverberant rooms

An ordinary room such as an office, classroom or a room in a residential dwelling is often called semi-reverberant, because there will usually be a mix of both direct and reverberant sound in such rooms. Acoustic testing often requires rooms of an extreme design in which either the direct field only, or the reverberant field only, is present; these are called anechoic or reverberant rooms.

An **anechoic room** is one in which the walls, floor and ceiling have been lined with sound absorbing material so that there is no reflection of sound and so measurements on sound sources may be carried out under free field conditions. Anechoic rooms may be used to measure the sound power levels and directivities of machines and other sound sources, and for testing microphones and loudspeakers.

A **semi-anechoic room** is one with anechoic walls and ceiling but with a hard, sound reflecting floor.

A **reverberant room** (or reverberation chamber) is one with hard sound reflecting surfaces, designed to produce as diffuse a distribution of sound as possible within the room. Such rooms usually have fairly long reverberation times, of several seconds. In order to promote diffusion and minimize the influence of standing waves, reverberant rooms are sometimes built with opposing surfaces (opposite walls, or floor and ceiling, slightly non-parallel). An alternative approach is to use large sheets of sound reflecting material within the room to act as sound diffusers. Reverberant rooms are used to measure the absorption coefficients of materials and the sound power levels of sound sources, but not their directivity.

A pair of two adjacent reverberant rooms (called a transmission suite) is used for laboratory testing of the sound insulation of panels and partitions, which are built into an aperture in the common wall between the two rooms. This is discussed in Chapter 6.

Both anechoic and reverberant test rooms are designed to have a very high degree of sound insulation, to eliminate interference from noise from external sources, such as traffic, and their performance is governed by British and International Standards.

How might you test the acoustic performance of an anechoic room?

A good test would be to plot sound level versus distance from a sound source (loudspeaker) in the room, and compare this with the perfect free field case of 6 dB reduction per doubling of distance. Any departures from perfect free field behaviour would show up close to the surfaces of the room, as a result of residual sound reflections from imperfectly sound absorbing surfaces. Since the absorption coefficient of the material used to line the room surfaces will increase with frequency, the limitations of the room will become more apparent in the lower octave bands, and the room will have a 'cut-off' frequency below which its 'free field' performance will be considered to be unacceptable.

A perfect anechoic room would have a reverberation time of zero, but this would not be so easy to verify by measurement as the level versus distance method.

How might you test the acoustic performance of a reverberant room?

A reverberation chamber is likely to have a high reverberation time but its essential feature is that a sound source in such a room should cause a uniform sound to be set up, i.e. in an ideal room the sound pressure level should not vary with position in the room (except very close to the sound source). Any standing waves which did occur would of course produce the opposite effect of regularly spaced sound pressure maxima and minima as a result of constructive and destructive patterns of interference.

Therefore the appropriate test would be to measure the sound distribution produced by a loudspeaker radiating octave bands of noise in the room and measure the variation of sound pressure level at different positions, in terms of standard deviations from the mean level.

Standing waves are more likely to cause problems at low frequencies and therefore, as with the anechoic rooms, there will be a 'cut-off' frequency below which its 'diffuse field' performance will be considered to be unacceptable.

British and International Standards specify the performance of reverberant rooms in terms of variability of sound field (in terms of standard deviations of sound pressure levels), permitted range of reverberation times, and also in terms of room volumes – the bigger the volume the lower the 'cut-off' frequency.

Note: There are websites which will allow you to listen to sound recordings made in anechoic rooms and reverberation chambers.

5.6 Room acoustics measurements

Measurement of reverberation time

Many years ago reverberation times were estimated by hand from the slope of a trace on a paper chart, as shown in Figure 5.12. This was time consuming and involved some judgement in deciding on the best straight line, and which part of the decay to use, in the case of a non-ideal decay. Modern instrumentation calculates automatically from the decay curve and presents the reverberation time value directly, but some instruments will also allow the decay curve to be seen (Figure 5.13), and it is good measurement practice to sample the decay curves occasionally, particularly at low frequencies.

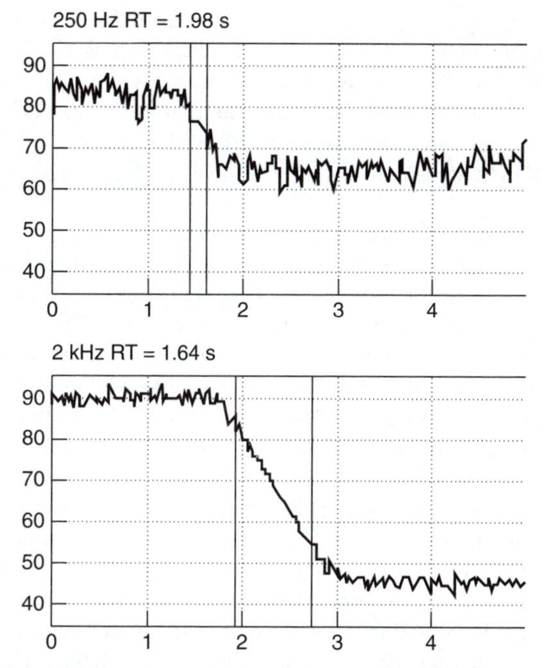

Figure 5.13 A better quality of decay curve is achieved at high frequency (2 kHz) than at low frequency (250 Hz)

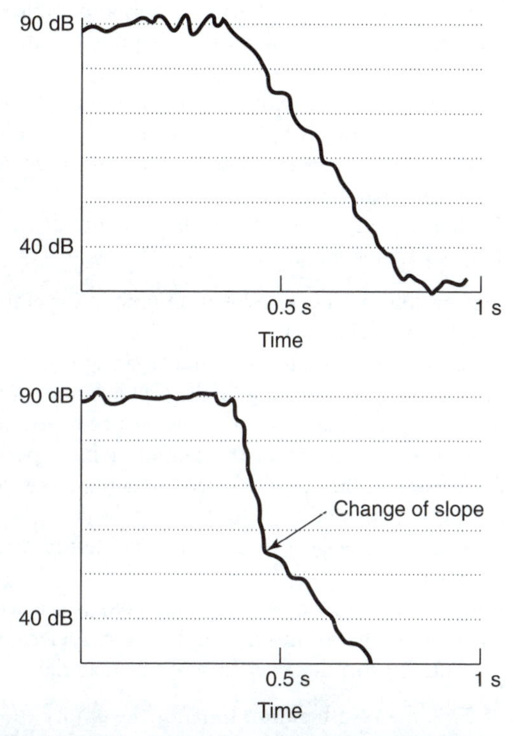

Figure 5.12 Examples of 'paper chart' reverberation decay curves. The lower decay curve shows a change of slope caused by a non-diffuse sound field and may indicate coupled room modes or a flutter echo

It is often the case that the decay curves for a room will be good (i.e. straight line decays) at high frequencies, say at 1000 Hz and above, but become poor at low frequencies, say at 100 Hz to 250 Hz. Why do you think that this is?

Faults in decay curves can arise because of coupled spaces, e.g. a room with an alcove, or because of prominent room modes or simply because the original sound level was not high enough above the background noise level.

T_{15}, T_{20} and T_{30} etc.

Many instruments give values such as these. In all cases they are 'proper' estimates of the reverberation times (i.e. the time taken for the sound level to decay by 60 dB) but measured over differing parts of the decay curve, i.e. the first 15 dB, the first 20 dB etc. For example, if the decay curve shows that the sound level drops by 20 dB in 0.3 seconds, then, assuming straight line decay, it will take 0.9 seconds to decay by 60 dB, i.e. the T, measured as the T_{20} in this case, will be 0.9 seconds. Similarly if, on the same decay curve, the sound level drops by 30 dB in 0.45 seconds, then the T_{30} will be also 0.9 seconds.

The practical application of all of this is that if an instrument gives several different values such as T_{15}, T_{20} and T_{30} then a simple comparison gives a check on the quality of the decay curve, without needing to look at it – i.e. if the three values are very similar this indicates good straight line decay.

Standard measurement procedures

Measurements of the reverberation time in a room can be used to quantify the acoustic characteristics of the room. There are various guides to the acceptable reverberation times for rooms with particular purposes, as discussed earlier. For example a room to be used for speech is required to be more absorbent, i.e. have a lower reverberation time, than a room to be used for music.

The procedure for measurement of reverberation time is given in ISO 3382-2:2008 *Acoustics – Measurement of room acoustics – Part 2: Reverberation time in ordinary rooms.* The reverberation time is defined as the time sound takes to decay by 60 dB, so the basis of the measurement is to produce a loud sound in the room then stop the source and track the slope of the decay with time. The sound can be produced with a noise generator and loudspeaker or, for field measurements, it is sometimes more convenient to use an impulse sound such as from a bursting balloon or a starting pistol. The sound source is usually located in the room at a position that will lead to as diffuse a sound field as possible, e.g. facing into a corner of the room. Alternatively, on some occasions it may be more appropriate to place the sound source in a location applicable to the use of the room, such as on the stage.

Measurement should be made at a number of microphone locations distributed around the space. In order to avoid the effects from the near field these should be more than $\lambda/4$ from the room surfaces and outside the direct field of the sound source. In each location a record of the decaying sound field is made and the reverberation time determined from the slope of the decay between the points 5 dB below the sound source and 5 dB above the background level in the room. It is rarely possible to achieve a full 60 dB drop between these points so the usual method is to measure over 30 dB and double this for the quoted value of reverberation time.

Ideally the decay of sound would be an almost straight line but in practice this is rarely the case as the sound fields in rooms are not evenly diffuse. In some cases this can arise from an uneven shape or distribution of the sound absorbing. In other cases, a clear split slope in the decay curve (see Figure 5.12) can indicate the presence of parts of the space with very different sound fields and described as coupled spaces. An example could be a theatre where some sound from the stage goes directly into the audience area while some sound goes into the fly tower and then feeds back at a later time into the audience area.

In the low frequencies it is often difficult to achieve a sufficiently loud sound level to be even 30 dB above the background level. Also the decays usually show great variability related to the large wavelengths for the low frequency sound.

The record of the decay can be made on a chart for later analysis. It is more common to use digital storage of the decay of the sound signal for immediate or later determination of the reverberation time. Various software packages use techniques to apply a straight line fit to the decaying sound level. The limitations and assumptions involved in this determination should be carefully assessed to ensure that the output data is applicable for the purpose of the measurements.

Measurement of the random incidence sound absorption coefficient in a reverberation room

To estimate the reverberation time in spaces it is important to have values of the absorption coefficient of the range of materials and objects in the room. The impedance tube method, described below, is only applicable for certain types of materials and can only provide data on the normal incidence absorption.

BS EN ISO 354:2003 defines the method for determination of the sound absorption coefficients for building materials based on measurements in a specially designed reverberation chamber.

The basic principle of the method is, for each one third octave frequency band:

- Measure the average reverberation time in the empty reverberation chamber.
- Install the material to be tested in an appropriate manner, i.e. curtains hung on the wall or floor covering placed on the floor. Ceiling tiles can be tested in a suspension system from the ceiling or in a special framework on the floor designed to maintain the correct airspace behind the tiles.
- Measure the average T in the room with test material installed.
- From the difference between the values for T (i.e. with and without the test material in the room), determine the sound absorption coefficient of the material.

So, if T_1 is the reverberation time for the empty room and T_2 is the reverberation time with the sample installed and the area of the sample is S:

$$T_1 = 0.161V/A_1 \quad \text{and} \quad T_2 = 0.161V/(A_1 + A_2)$$

where A_1 is the absorption of the empty room and A_2 is the absorption of the test material.

Rearranging these equations:

$A_1 + A_2 = 0.16 \times V/T_2$ and
$A_2 = 0.16 \times (V/T_2) - 0.16 \times (V/T_1)$, which becomes
$A_2 = 0.16 \times V[(1/T_2) - (1/T_1)]$

from which the absorption coefficient α of the test material can be determined:

$$\alpha = A_2/S = (0.16 \times V/S)[(1/T_2) - (1/T_1)]$$

Example 5.8

A reverberant room has dimensions of 10 m × 5 m × 4 m and has a reverberation time of 8.2 seconds when empty. When a sample of 10 m^2 of sound absorbing material is laid on the floor the reverberation time is reduced to 3.4 seconds. Calculate the α of the sample.

$V = 10 \times 5 \times 4 = 200\ \text{m}^3$

$A_2 = 0.16 \times 200 \times [(1/3.4) - (1/8.2)] = 5.5\ \text{m}^2$

$\alpha = 5.5/10 = 0.55$

Measurement of the normal incidence sound absorption coefficient

These measurements are made in a *standing wave tube* (more properly it should be 'partial standing wave tube') or an *impedance tube,* shown in Figure 5.14. BS EN ISO 10534-1:2001 defines the measurements procedure. The loudspeaker generates a pure tone at the frequency of interest and this is partially reflected from the surface of the sample. The incident and the reflected wave will both be sine waves of the same frequency and their superposition will occur. There will be points along the tube where the two components will be in phase, i.e. maximum positive

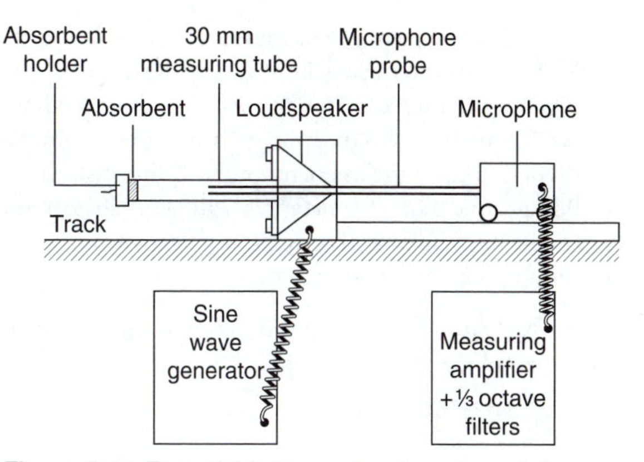

Figure 5.14 Essential features of an impedance tube

sound pressure, and other points where they will be out of phase, i.e. minimum sound pressure, as shown in Figure 5.15. The absorption coefficient at the frequency under test can be determined from the ratio of the maxima to the minima sound pressures along the tube, as measured with a probe microphone.

Some limitations and precautions with the use of the impedance tube include:

- Plane wave incidence – measurements using such a method only provide information on the sound absorption for plane wave incidence.
- Sample – the impedance tube method depends on the reflection of sound from what is a small sample of the material. Thus it is valid only for determination of the sound absorption of materials with effectively a uniform surface. For materials with surface variation it is necessary to repeat the measurements with a number of samples randomly selected from the material to ensure a representative value for the absorption.
- Frequency limits – the diameter of the measurement tube should not exceed 0.59 × the wavelength of

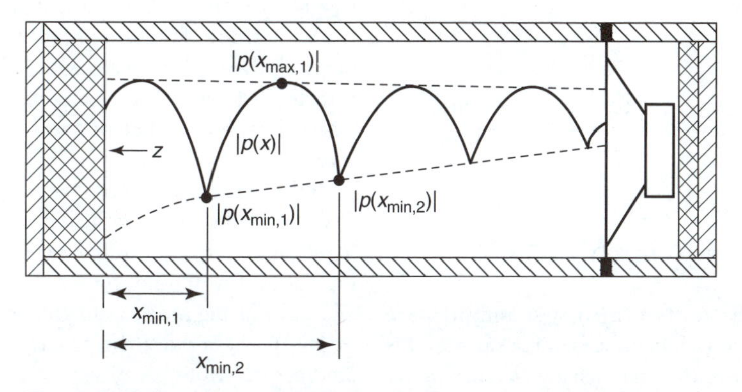

Figure 5.15 Standing wave pattern set up in an impedance tube
From BS EN ISO 10534-1

the highest frequency to be measured. There is also effectively a low frequency limit in that the length of the tube should exceed 0.75 × the wavelength of the lowest frequency. This means several tubes could be needed to cover the frequency range of interest.
- Background noise – materials with low absorption coefficients involve the determination of very low sound pressure variations along the tube.

If $n = p_{max}/p_{min}$ = standing wave ratio, it can be shown (see Appendix 5.1 for proof) that:

$$\alpha = 4n/(n^2 + 2n + 1)$$

Example 5.9

The maximum sound pressure level in the tube was 76 dB and the minimum was 50 dB at a certain frequency. What is the value of α for the test sample?

First of all it is necessary to find the standing wave ratio, n. Two ways of doing this are shown below. The value of α is then determined using the above formula.

Method 1

$L_{max} = 20\log(p_{max}/p_0)$, from which (see Chapter 1):

$$p_{max} = p_0 \times 10^{L_{max}/20} = 2.0 \times 10^{-5} \times 10^{76/20}$$
$$= 2.0 \times 10^{-5} \times 10^{3.8} = 0.126 \text{ Pa}$$

Similarly:

$$p_{min} = p_0 \times 10^{L_{min}/20} = 2.0 \times 10^{-5} \times 10^{50/20}$$
$$= 2.0 \times 10^{-5} \times 10^{2.5} = 0.0063 \text{ Pa}$$

Therefore:

$$n = p_{max}/p_{min} = 0.126/0.0063 = 20$$

Method 2

$$n = p_{max}/p_{min} = 10^{(L_{max} - L_{min})/20}$$
$$= 10^{(76-50)/20} = 10^{26/20} = 10^{1.3} = 20, \text{ as before}$$

Then:

$$\alpha = (4 \times 20)/[(20^2 + (2 \times 20) + 1)]$$
$$= 80/[400 + 40 + 1] = 80/441 = 0.18$$

By measuring the distance of the first sound pressure minimum from the end of the tube (sample surface) it is also possible to deduce the phase change occurring on reflection, and the acoustic impedance, Z, of the sample. A more detailed analysis of the theory is given in BS EN ISO 1054-1. In a modified version of the impedance tube the moving probe microphone is replaced by two microphones set flush with the inner walls of the tube, a known distance apart, and the sine wave signal is replaced by a broadband noise. By measuring the transfer function between the two microphone signals, i.e. the ratio of the two sound pressures, it is possible to deduce the normal incidence absorption coefficient and acoustic impedance of the sample over a wide range of frequencies. The method is explained in BS EN ISO 1054-2.

5.7 Room modes

Large versus small spaces

The discussion thus far has related to sound in large spaces. The term 'large' is relative and generally in acoustic considerations implies that the longest dimension is several times the longest wavelength being considered. If, for example, the lowest frequency for concern is 200 Hz (wavelength about 1.7 m) then the foregoing analysis relates to rooms of dimension greater than about 4 m. With 'small' enclosures consideration now turns to the other end of the scale, where individual room modes are important.

Standing waves in a room

This section extends the discussion of one- and two-dimensional standing waves in Chapter 1 to the three-dimensional case of standing waves in a room.

A stretched string, of length L, held between rigid ends (the imposed boundary conditions) is capable of vibrating preferentially in one or more modes. The simplest mode involved a vibration in the form of one half of a sine wave as this movement pattern satisfies the limits imposed by the fixed ends. This mode has a particular frequency and if the string is disturbed at that frequency it will exhibit that pattern of vibration. There is also a whole family of higher modes described by order N; these are patterns with N half sine waves fitted into the length L. Each has a frequency of oscillation which is N times that of the lowest or fundamental mode.

If the string is disturbed in a random manner, then any frequency component of the induced vibration that is not identical in frequency to one of these modes will quickly die out as it would not have zero amplitude at the fixed ends. The string proceeds to vibrate happily as a mixture of the special modes (the precise distribution of energy among the various permitted modes is determined by the manner of initiation) until air resistance etc. damps it out.

A mental picture of the behaviour of any of these modes can be obtained by likening it to the standing wave pattern produced when two identical waves of the mode frequency pass along the string in opposite directions, as though originating from outside the fixed limits.

Now consider the rectangular room with hard reflecting walls; this is necessarily rather more complex than the string case because there are three dimensions to consider. The boundary conditions imposed by the walls require that adjacent to a wall, there shall be no movement of those air molecules in a direction perpendicular to that wall. This implies that at a hard wall the component of the particle velocity of the sound field perpendicular to the wall must permanently be zero. This condition can only be achieved at certain acoustic frequencies; each being associated with a particular mode, that is, with a particular spatial pattern of sound levels.

Consider the longest dimension of a rectangular room to be its length, L. The simplest mode and, consequently the one with the lowest frequency, corresponds in pattern along the length of the room to the stretched string vibrating with one half cycle of a sine wave. This correspondence implies the particle velocity over the end walls of the room will be permanently zero and at every point on the mid-plane, a maximum.

Two points can be made about this simple case.

1. As $L = \lambda/2$, the sound frequency, f_1, needed to excite this mode must be given by $f_1 = c/2L$. Thus for a room 5 m long, for this resonance the λ is 10 m and frequency is given by $344/10 = 34$ Hz.
2. Where the particle velocity pattern exhibits a node, the acoustic pressure exhibits an antinode. This implies that this pattern will be very loud over the entire end walls, but very quiet over the entire mid-plane. At all other vertical planes between the two end walls the loudness will be constant over any one plane, with that loudness being somewhere between the extremes.

This argument can be applied to all three axes. Thus if the width W (= 4 m say) is the next largest dimension then another mode of simple structure is identified, and since for $\lambda = 8$ m $f_2 = 344/8 = 43$ Hz, it follows that if the loudspeaker were driven at 43 Hz a corresponding loudness pattern would be heard, with the side walls loud and the median plane quiet. Thirdly, if the height was 3 m and the loudspeaker excited at 57 Hz, the floor and ceiling planes would be loud, any horizontal plane would show a constant loudness and the half-height plane would be very quiet.

Those are examples of the simplest (in the sense that the geometrical variation of levels is very simple) normal modes of the room. Consider now the next frequency that will comply with the boundary conditions that will have a complete wavelength between the surfaces. The respective frequencies would now be $2 \times 34 = 68$ Hz, $2 \times 43 = 86$ Hz and $2 \times 57 = 114$ Hz respectively. So if the source produced 86 Hz then it would vary in loudness moving across the width, being loud at the wall and in the middle but quiet at the 1/4 and 3/4 marks.

The six modes explored so far could be described as the (1 0 0), (0 1 0), (0 0 1), (2 0 0), (0 2 0), (0 0 2) modes. It follows, therefore, that even for a simple rectangular room the whole picture is quite complicated but it is useful to consider it as a series of standing wave patterns within each plane. But the sound is travelling in all directions from the source so it is possible to imagine that standing wave patterns can also be set up for sound rays striking the walls at various angles. In the most general case one could envisage a sound wave being reflected off all pairs of walls; at certain frequencies and propagation angles, the boundary conditions on all walls could be satisfied. The interaction of all these reflections produces very complex patterns of loud and quiet zones.

Calculating the frequencies of the normal modes of a room

The analysis is done for a simple rectangular room by ascribing to each mode a unique set of integers N_L, N_w and N_H. The frequency required to excite a mode is given by:

$$f = (c/2)\sqrt{[(N_L/L)^2 + (N_W/W)^2 + (N_H/H)^2]}$$

Table 5.7 lists some of the normal mode frequencies for the room we have been discussing above, with dimensions 5 m × 4 m × 3 m. The first two columns show these modes in 'mode number' order, and the second two columns show these same 29 modes rearranged in order of increasing frequency. For comparison the final two columns show the frequencies of the same modes for a cubic room of dimensions 4 m × 4 m × 4 m having very similar size (volume) to the first room, but a different shape.

Study Table 5.7. What general observations can you make?

- The six, simple, one-dimensional' modes identified above (34, 43, 57, and 68, 86 and 114 Hz can all be identified in Table 5.7, and the three modes 3 0 0, 3 0 3 and 0 0 3 also follow the harmonic sequence.
- As the mode numbers generally rise the frequencies rise but progressively crowd together.
- For the cube shaped room several of the modes have the same frequency. This means that frequencies at which modes occur are fewer and more widely spaced

Table 5.7 Mode frequencies for two rooms

Modes of room 3 m × 4 m × 5 m		Modes rearranged in ascending order		Modes of room 4 m × 4 m × 4 m	
Mode	Frequency	Mode	Frequency	Mode	Frequency
1 0 0	56.7	0 0 1	34.0	1 0 0	42.5
0 1 0	42.5	0 1 0	42.5	0 1 0	42.5
0 0 1	34.0	1 1 0	54.4	0 0 1	42.5
1 1 0	70.8	1 0 0	56.7	1 1 0	60.1
1 0 1	66.1	1 0 1	66.1	1 0 1	60.1
1 1 0	54.4	0 0 2	68.0	1 1 0	60.1
1 1 1	78.6	1 1 0	70.8	1 1 1	73.6
2 0 0	113.3	1 1 1	78.6	2 0 0	85.0
0 2 0	85.0	0 1 2	80.2	0 2 0	85.0
0 0 2	68.0	0 2 0	85.0	0 0 2	85.0
0 1 2	80.2	1 0 2	88.5	0 1 2	95.0
0 2 1	91.5	0 2 1	91.5	0 2 1	95.0
1 2 0	102.2	1 1 2	98.2	1 2 0	95.0
1 0 2	88.5	0 0 3	102.0	1 0 2	95.0
2 1 0	121.0	1 2 0	102.2	2 1 0	95.0
2 0 1	118.3	1 2 1	107.7	2 0 1	95.0
2 1 1	125.7	0 2 2	108.9	2 1 1	104.1
1 2 1	107.7	2 0 0	113.3	1 2 1	104.1
1 1 2	98.2	2 0 1	118.3	1 1 2	104.1
0 2 2	108.9	2 1 0	121.0	0 2 2	120.2
2 0 2	132.2	1 2 2	122.7	2 0 2	120.2
2 2 0	141.7	2 1 1	125.7	2 2 0	120.2
1 2 2	122.7	0 3 0	127.5	1 2 2	127.5
2 1 2	138.8	2 0 2	132.2	2 1 2	127.5
2 2 1	145.7	2 1 2	138.8	2 2 1	127.5
2 2 2	157.1	2 2 0	141.7	3 0 0	127.5
3 0 0	170.0	2 2 1	145.7	0 3 0	127.5
0 3 0	127.5	2 2 2	157.1	0 0 3	127.5
0 0 3	102.0	3 0 0	170.0	2 2 2	147.2

than for the non-cubic shaped room. The shape of the room has made a big difference to the modal distribution of frequencies for the two rooms, although they are similar in volume.

Note: When a precise value is required, the correct speed of sound for the temperature of the space should be used in the calculation (see Chapter 1).

Exercise

How will the mode frequencies be affected if the room dimensions are (a) doubled, (b) halved?

Answer

If the room dimensions were halved then the mode frequencies would all be shifted upwards by a factor of two, and conversely if the room dimensions were doubled the modes would all be shifted downwards in frequency by a factor of two.

The important consequence here is that for small rooms the lowest modes have relatively high frequencies and modes remain well spaced and exist as separate modes into the low to mid audio frequency range for small rooms whereas for large rooms the lowest modes are much lower in frequency and so become very close and merge together at relatively low audio frequencies.

Axial, tangential and oblique modes

Modes with two of the mode numbers, N_L, N_W or N_H, equal to zero can be thought of as arising from constituent waves propagating in both directions along the axis of the mode number that is not zero. These modes are called axial modes.

Modes with one of the mode numbers N_L, N_W, N_H zero can be thought of as arising from constituent waves propagating parallel to a pair of walls. These modes are called tangential modes.

Modes with none of the mode numbers N_L, N_W, N_H zero are called oblique modes.

Mode spacing

Frequencies in the source sound that correspond with room modes will show considerable spatial fluctuations in level. Also the sound at those frequencies will continue to 'ring' for considerably longer than others. The effects are most noticeable when the room modes are widely spaced, usually at the lower frequencies.

Although the cube shaped room investigated above was an extreme example, it can be shown more generally that if the room dimensions are related to each other by simple whole number ratios, e.g. 1:1, 2:1, 3:2 etc., then several different modes will have the same frequency. This means that the modes will not be evenly spaced out in frequency but bunched together.

Golden ratios

It is better for achieving good room acoustics if room modes are distributed fairly evenly along the frequency scale, and it has been shown that in order to achieve this it is best if the ratios between room dimensions are not related by simple integer numbers. Various 'golden ratios' for room dimensions have been suggested which will lead to the most even frequency spacing of modes in a room.

The effects of room modes in small rooms may be minimized by:

- avoiding rectangular rooms with dimensions related by simple ratios
- adding either sound absorption or sound diffusion to one of each pair of parallel surfaces
- making opposite surfaces (e.g. walls) slightly non-parallel.

5.8 Design of rooms for a good acoustic environment

Some of the more basic requirements for good room acoustics are descried below.

Optimum background noise level

This is the starting point in achieving a good acoustic environment. If the background noise level is too high, annoyance, poor working efficiency and interference with communication (poor speech intelligibility) can occur. If the background noise level is too low there may be problems with lack of acoustic privacy. Background noise may arise from intrusion of external noise (e.g. road traffic noise, aircraft noise etc.) or from within the building (e.g. building services noise) or from within the room itself (occupancy noise). Some guidelines for background noise were discussed in Chapter 3.

Optimum reverberation time

This depends mainly on the purpose for which the room has been designed (e.g. mainly for listening to speech, or for music) and its volume, and this has been discussed in detail earlier in this chapter.

Adequate supply of sound to all parts of the room

This will be particularly important for spaces which rely on natural acoustics, i.e. on the sound power from the human voice (or voices) or from musical instruments without electronic amplification. The requirement is, for example, that the human voice from the front of the room can be clearly heard and intelligible to the furthest listener at the back. This may set a limit to the size of the space, but will also depend on control of background noise and on good design to produce as much direct sound as possible to all areas, supplemented by reverberant sound.

Uniform distribution of sound throughout the space

The acoustic quality should be good throughout all areas, i.e. for all seats in a hall or theatre, classroom or church. This means that the level of wanted sound from the speaker or performer should be sufficient in all areas and, ideally, should not vary too much with position, and this should also apply to reverberation times and background noise levels and to other acoustic parameters described later. It also means that there should not be any shadow zones where sound is prevented from reaching the listener directly from the source, by pillars, alcoves or balconies, for example.

The absence of acoustic faults

The presence of room modes, focusing effects of sound by concave curved surfaces, and flutter echoes from parallel sound reflecting walls can all produce aural effects which are undesirable for the listener. These effects are also likely to be very position dependent so that the acoustic quality in some seats may be very different from that in other seats.

Speech intelligibility and privacy

If speech in a room is audible but unintelligible (i.e. muffled and unclear) then this indicates that although in principle there may be a sufficient amount of speech sound reaching the listener, the speech signal is being masked, either by reverberation or by high levels of background noise. Therefore obtaining good speech intelligibility in a room depends on achieving a combination of a good level of speech signal at the listener together with sufficiently low levels of background noise and of reverberation. High levels of speech intelligibility can, in different circumstances, give rise to poor levels of acoustic privacy. Therefore the requirements for good privacy are the opposite to those for good speech intelligibility, so that acoustic privacy can be a problem in 'dead' rooms with very low levels of background noise.

Line of sight, raised performers and raked seating

In rooms where the audience is arranged in rows of seats which are all at the same level then the level of speech sound reaching the audience at the back of the room is attenuated not only as a result of 'distance' (at a rate of 6 dB per doubling of distance) but also because of absorption by the intervening rows of audience. The absorption by the audience can be significant at certain frequencies. The use of tiered or raked seating not only gives much better line of sight, visually, but also gives much better provision of direct sound to all rows of seats. For the same reason, it is helpful if the speaker at the front of the audience is raised on a stage.

Other requirements

There are some other requirements for rooms designed for audiences to listen to speech or music which depend on an understanding of human perception. These are explained below.

Subjective aspects

Both speech and music consist of series of individual transient sounds (syllables of speech or musical notes). For each individual sound the ear will receive the sound directly (the first arrival) followed by a series of successive reflections. The first reflections will be those from nearby reflecting surfaces close to the source. They will have the greatest sound intensity (because they have travelled the shortest distance) and may be separated out as distinct sound events. Later reflections will be lower in intensity (travelled further) and are likely to be crowded together, until eventually they become merged into the 'reverberation'.

This so far is an objective description. Psychoacoustics research tells us how humans respond subjectively to such a sequence of transient sounds. Some important features of human response are briefly described below but it must be understood that this will be a simplification of a complex subject.

Reflections which arrive at the ear fairly soon after the direct sound (called early reflections) are perceived by the ear–brain system as being part of the direct sound and produce an agreeable sensation, and they also add to the intensity and loudness of the sound. Conversely, sounds which arrive at the ear a long time after the direct sound (and called 'late' sound or late reflections) often produce an unpleasant and discordant sensation. In some cases (depending on the combination of level and delay time) late reflections are perceived as separate sound events, i.e. as echoes.

The clear message to the acoustic designer is that design should promote and encourage early reflections whereas late reflections should be avoided. Therefore surfaces close to the source which produce early reflections for the listener should be hard and sound reflecting, whereas surfaces which may cause long delayed reflections (distant rear walls for example) should either be covered with sound absorption or with sound diffusers.

How early is early and how late is late? There are many factors involved in determining human response to any particular set of reflections, but generally early sound is considered to be that which arrives within 50 ms (i.e. a twentieth of a second) of the direct sound for speech, and 80 ms for music; and sound arriving any later is considered to be 'late sound'.

Another factor of subjective importance is the direction from which a sound appears to be coming. It appears that audiences listening to music in concert halls prefer the feeling of envelopment which occurs if the sound is arriving from 'all around' rather than from the direction of the performers at the front of the hall. Therefore the acoustic designer will attempt to encourage 'lateral reflections', i.e. from the side walls inwards towards the centre of the hall.

On the other hand, audiences listening to speech prefer that sound appears to be coming from the human voice at the front, even though in a large hall most of the sound energy may be reaching them from loudspeakers on the side walls, or from behind them. Here another psycho-acoustic effect is important, the so-called 'Haas effect' which has to do with (among other things) the attribution of direction for a series of reflections. According to the Haas effect the ear may attribute the direction from which a sound appears to be coming to the direction of the 'first arrival' it receives, even though the intensity of the first arrival may be less than that of subsequent arrivals. This means that if the sound signal from a microphone close to the human voice at the front of the hall to the loudspeaker at the back is electronically time-delayed by more than the transit time of the direct speech it may appear to the listener at the rear that the sound is mainly coming directly from the voice at the front of the hall.

Acoustic quality parameters

A number of parameters have been developed to relate to some of the subjective effects described above. They will be position dependent and so allow estimates to be made of how the acoustic performance of the space varies in different areas. Some of these are related to the relative amount of early and late sounds received at a particular position in the room. An example is the 'clarity' parameter used mainly for assessing quality for music performance. Another parameter is the 'lateral energy fraction' which relates to the relative amount of sound energy reaching a listener from the front as compared to from 'all around'. Early decay time is the reverberation time based only on the very first part of the decay (the first 10 dB) and therefore is determined by only the first few reflections reaching the listener, and is another indicator of the subjective acoustic quality of a space. Unlike the reverberation time discussed earlier in this chapter, which is largely position independent, the early decay time will vary with seating position in the space.

These parameters may be measured in the real space, or may be predicted by building acoustics models of the space using ray or beam tracing software packages.

Acoustic quality descriptors

Various terms such as live, dead, richness, warmth, fullness, dryness are used to describe the acoustic behaviour of performance spaces such as theatres and concert halls and these may, to some extent, be related to some of the parameters discussed above.

Design parameters

The parameters which are available to the acoustic designer to achieve good acoustic conditions in a room are: the volume of the room, the shape, and the acoustic treatments that may be applied to the room surfaces.

Volume

We have already seen that the size of a room for unaided speech is limited by the level of speech at the furthest distance from the speaker. In large halls and theatres the acoustic absorption is mainly provided by the audience and their seats. Since the reverberation time depends on both the volume of the space and the amount of absorption, achieving the optimum reverberation time for a space with a particular volume will lead to an optimum number of seats for that volume, i.e. to an optimum volume per seat. More seats will result in a reverberation time which is too low, and, conversely, fewer seats in a time which is too high.

Shape

We have already seen that shapes which contain concave surfaces can lead to undesirable focusing effects and that in small rooms certain rectangular shapes can cause problems with room modes. Two common shapes for theatres are (in plan): the (rectangular) 'shoe box' shape (with the audience arranged in rows, each with the same number of seats, determined by the width of the theatre, and the number of rows by its length), and the 'fan' shape where the number of audience seats increases further from the front.

The fan shape has the advantage of gathering the audience closer, on average, to the stage than for the shoe box, and so increasing the level of direct sound received, whereas the shoe box shape is better at promoting lateral reflections from the side walls towards the centre of the rows.

Disposition of surface treatments

Each room surface (wall, floor or ceiling) may be used to either reflect, absorb or diffuse sound according to its surface treatment and, for diffusion, its shape or profile.

Diffusion is a useful option where it is required to limit specular reflection from a surface, but the use of absorptive treatment might lead to a reverberation time which is too low. Examples could include the prevention of undesirable flutter echoes, focusing effects or long delayed reflections, or reflections from a balcony being sent back to actors on a stage.

Acoustic issues that can arise in the design of spaces to achieve a good acoustic environment

These include:

- optimum background noise level
- intrusion of external noise
- internal noise levels
- reverberation time (optimum)
- sound insulation between internal spaces
- façade sound insulation
- absorption
- flanking transmission/structure-borne sound transmission
- standing waves (room modes)
- direct/reverberant sound balance
- speech intelligibility
- acoustic privacy
- internally generated noise (building services)
- occupancy noise.

Non-acoustic factors and constraints on acoustic design

These include considerations relating to: risk of fire, other health and safety issues (e.g. resulting from shedding of fibres), cleaning and hygiene, ability to withstand water and other adverse environments, physical robustness, cost, appearance and aesthetics, lighting, line of sight, evacuation procedures, cost, thermal insulation.

Exercise

Pick out the main acoustic issues and constraints most relevant to each of the following types of building:

Theatre	Doctors' waiting room
Cinema	Music practice room
Library	Airport lounge
Lecture room	Railway/underground station
Conference/meeting room	Office
Dwellings	Schools
Hotels	Law courts
Council; chamber	Churches (various types)
Sports grounds and buildings	Restaurant
Shopping mall/precinct	TV/radio studio
Audiometric testing booth	

BS 8233:1999 *Code of practice for sound insulation and noise reduction for buildings* is a useful reference.

5.9 Sound intensity and energy density in a diffuse sound field

Sound intensity is a measure of the rate of flow of sound energy in a given direction. It is the sound power per unit area passing through an area at right angles to the direction in which the sound wave is travelling (and measured in W/m^2).

For a plane wave in a free field the sound intensity (I) and sound pressure (p) at any point are related via the specific acoustic impedance of the medium (ρc) by:

$$I = p^2/\rho c$$

The sound energy in a sound wave is the energy of the vibrating elements of the medium through which the wave is passing. Each small element of the medium has kinetic energy because it is moving (vibrating) and potential energy because it is being successively expanded and compressed (in the same way as a stretched spring possesses potential energy). The total amount of energy per unit volume in a wave is called the energy density, ε, in J/m^3. It is determined by calculating the total of the potential and kinetic energies of all the elements of the medium within a known volume. This will vary from moment to moment throughout the cycle of the sound wave, just as the sound pressure and acoustic particle velocities also vary, and ε, like I, is usually taken as the average over a cycle.

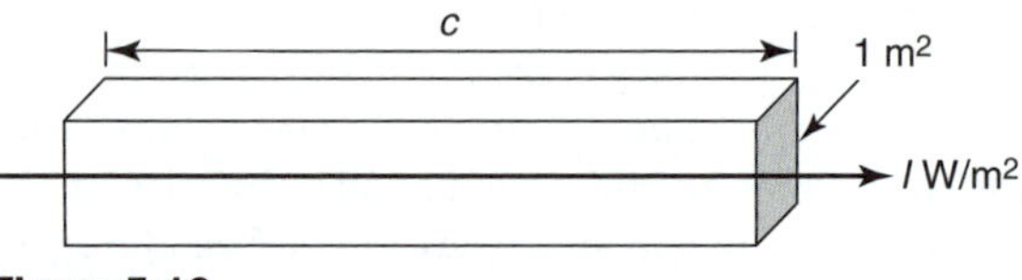

Figure 5.16

It can be shown that the energy density, ε, in a plane wave is given by:

$$\varepsilon = p^2/\rho c^2 = I/c$$

Although a rigorous proof involves some mathematics (not given here but available in many acoustics texts)[1] a simple and plausible explanation can be achieved by considering the amount of energy in a box of length c and cross-sectional area 1 m^2 through which a plane wave of intensity I m^2 is passing (see Figure 5.16).

Since the wave travels a distance c in one second the box contains 'one seconds' worth' of sound energy. The amount of energy in the box is obtained first of all by multiplying the sound intensity by the cross-sectional area to obtain the sound power, i.e. the rate at which sound energy is passing through the box, and then multiplying the power by the time to obtain the energy.

Therefore sound energy contained within the box = $I\,W/m^2 \times 1\,m^2 \times 1\,s = I\,J$

Volume of box = $1\,m^2 \times c\,m = c\,m^3$

Therefore energy density = energy/volume = I/c, as stated above.

Diffuse sound field conditions

In a perfectly diffuse sound field sound waves are arriving with equal intensity from all directions (see Figure 5.17). Therefore at any point within such a sound field the acoustic intensity must be zero because the flow of sound energy in any particular direction is balanced

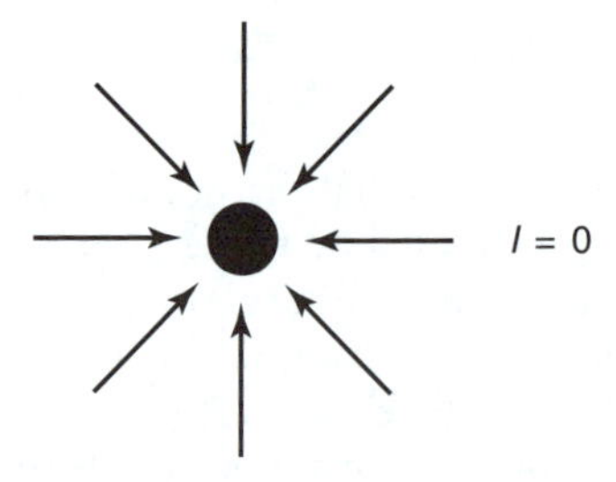

Figure 5.17

[1] See for example Kinsler, L. E., Frey, A. R., Coppens, A. B. and Sanders, J. V. (1999) *Fundamentals of Acoustics*, Wiley; Bies, D. A. and Hansen, C. H. (2009) *Engineering Noise Control*, Spon; Fahy, F. (2001) *Foundations of Engineering Acoustics*, Academic Press; Hall, D. E. (1987) *Basic Acoustics*, Wiley.

by an identical flow of sound energy in exactly the opposite direction.

It is, however, possible to consider an 'effective' sound intensity in a particular direction, at a point where the sound pressure, p, is perpendicular to an imaginary plane surface passing though the point. It can be shown that:

$$I_{EFF} = p^2/4\rho c = \varepsilon c/4$$

A similar situation will arise at the boundary of the diffuse sound field, i.e. at the sound reflecting wall of the room within which the diffuse sound field is contained (see Figure 5.18).

Once again a proof of this formula involves some mathematics (available again in many acoustics texts).

Converting to decibels:

$$L_I = L_p - 6$$

where L_I is the sound intensity level at the boundary of the reverberant room and L_p is the sound pressure level in the room.

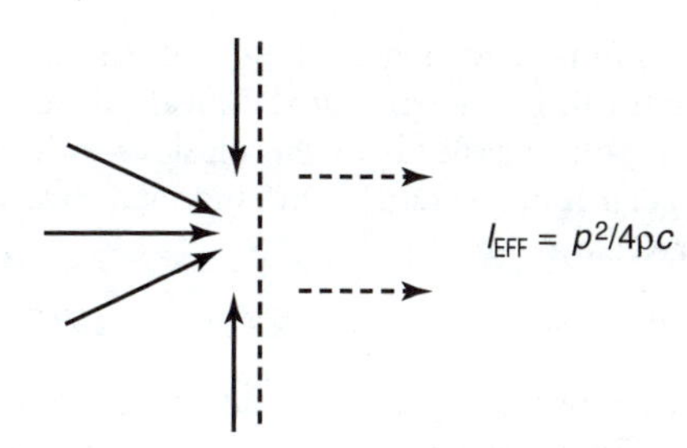

Figure 5.18

Considerations of energy density and of sound intensity in and at the boundary of a diffuse sound field feature in the derivation of Sabine's formula, and in that of the formula for predicting sound pressure level in a diffuse sound field ($L_p = L_W + 10\log(4/R_C)$), which are of central importance in this chapter, and in the transmission of sound across a boundary (e.g. a wall) which is considered in Chapter 6.

Appendix 5.1 Derivation of standing wave tube formula

A = Amplitude of incident wave

B = Amplitude of reflected wave

Maximum sound pressure, $p_{max} = A + B$

Minimum sound pressure, $p_{min} = A - B$

Standing wave ratio $= p_{max}/p_{min} = n$

$$= (A + B)/(A - B)$$

Algebra:

$$n = (A + B)/(A - B) = [(A/B) + 1]/[(A/B) - 1]$$

$$n\,[(A/B) - 1] = (A/B) + 1$$

$$n(A/B) - n = (A/B) + 1$$

$$(A/B)(n - 1) = n + 1$$

$$(A/B) = (n + 1)/(n - 1)$$

$$(B/A) = (n - 1)/(n + 1)$$

Reflection coefficient, $r = (B/A)^2$

Absorption coefficient, $\alpha = 1 - r = 1 - (B/A)^2$

$$\alpha = 1 - [(n - 1)/(n + 1)]^2$$

$$= [(n + 1)^2 - (n - 1)^2]/(n + 1)^2$$

$$= [(n^2 + 2n + 1) - (n^2 - 2n + 1)]/(n^2 + 2n + 1)$$

$$\alpha = 4n/(n^2 + 2n + 1)$$

Appendix 5.2 Derivation of Eyring's formula

Eyring's formula is derived below, and then Sabine's formula is deduced from it. A direct derivation of Sabine's formula is given in Appendix 5.3.

Let I_0 be the sound intensity at time $t = 0$ when the source is disconnected and the average absorption coefficient of the surfaces of the enclosure is α_{ave}.

Then:

The intensity after 1 reflection $= I_1 = (1 - \alpha_{ave})I_0$
The intensity after 2 reflections $= I_2 = (1 - \alpha_{ave})^2 I_0$
The intensity after n reflections $= I_n = (1 - \alpha_{ave})^n I_0$

The reverberation time is the time for the sound intensity to fall to 1/1,000,000 of the original, so:

$$I_n/I_0 = (1 - \alpha_{ave})^n = 10^{-6}$$

Taking $\log_{10}$ of each side:

$$\log_{10}(1 - \alpha_{ave})^n = -6$$

$$n \times \log_{10}(1 - \alpha_{ave}) = -6$$

$$n = -6/\{\log_{10}(1 - \alpha_{ave})\}$$

The question is, how long do these n journeys take, on average? It can be shown that the mean distance (the mean free path) travelled between collisions with walls in a rectangular room, volume V m^3 and total surface area S m^2 is given by:

Mean path in metres between collisions $= 4V/S$ m

So n reflections take $4Vn/cS$ seconds, so the reverberation time, T, is given by:

$$T = 4Vn/cS = (4V/cS)n$$

$$= (4V/cS)[-6/\{\log_{10}(1 - \alpha_{ave})\}]$$

From which:

$$T = 0.16\ V/\{-S \log_e(1 - \alpha_{ave})\} \qquad \text{(Eyring's formula)}$$

In the final stages, if the derivation $\log_{10}$ has been converted to $\log_e$ using $\log_e X = 2.303\log_{10} X$, then when $\alpha_{ave} \ll 1$, i.e. for 'live' rooms, $[-\log_e(1 - \alpha_{ave})]$ can be approximated by α_{ave}, in which case Eyring's formula simplifies to:

$$T = 0.16V/\{S\alpha_{ave}\}$$

This is Sabine's equation.

Appendix 5.3 Derivation of Sabine's formula

Assumptions:

- sound energy density is uniform throughout the room
- sound energy is transmitted equally in all directions
- sound source maintains constant supply of sound energy
- sound energy loss occurs only at room walls (no air absorption)
- sound absorption distributed evenly over room surfaces.

Decay of sound in a room (after sound source switched off)

Rate at which sound is being absorbed by room surfaces $= \alpha IS$

Total amount of sound energy in room $= \varepsilon V$

Rate of change of sound energy

$$= -\frac{d}{dt}(\varepsilon V) = -V\frac{d\varepsilon}{dt}$$

(*Note:* there is a $-$ sign because ε is decreasing with time.)

But $\quad \varepsilon = \dfrac{4I}{c}$

Therefore rate of decay of energy

$$= -V\frac{d}{dt}\left(\frac{4I}{c}\right) = -\frac{4V}{c}\frac{dI}{dt}$$

Rate of decay $=$ rate of absorption

$$-(4V/c)\frac{dI}{dt} = \alpha SI$$

therefore $\quad \dfrac{dI}{dt} = -(\alpha Sc/4V)I$

$$= -(Ac/4V)I \text{ (since } \alpha S = A)$$

therefore $\quad \dfrac{dI}{dt} = -(Ac/4V)I$

This is a standard differential equation, known to represent the exponential decay with time of a quantity (in this case I), with a standard solution:

$$I = I_0 e^{-(Ac/4V)t}$$

$$I/I_0 = e^{-(Ac/4V)t}$$

$$\ln(I/I_0) = e^{-(Ac/4V)t}$$

Using the relationship between ln ($\log_e$) and $\log_{10}$, which is $\log N = \ln N/(2.3)$:

$$10\log(I/I_0) = -(Ac/4V)t(10/2.3)$$

When $t = T$ (reverberation time), $10\log(I/I_0) = 60$ dB

Therefore: $\quad 60 = (10 \times c/2.3 \times 4)(AT/V)$

Take $c = 340$ m/s:

$$T = 0.16V/A$$

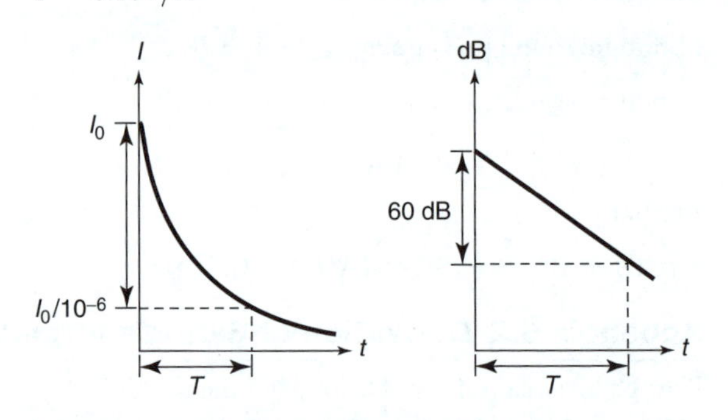

Growth of sound in a room (from the moment sound source is switched on)

For the sound energy in the room:

Rate of supply by source $=$ rate of increase in room $+$ rate of absorption by walls

$$\text{i.e. } W = \frac{d}{dt}(V\varepsilon) + SI\alpha_{AVGE}$$

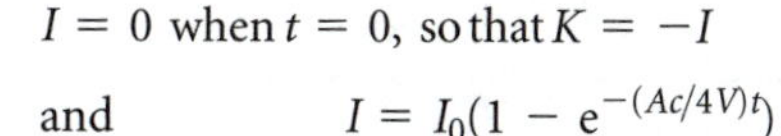

When the steady state intensity I_0 is reached:

$W = SI_0\alpha_{AVGE}$

and therefore $SI_0\alpha_{AVGE} = Vd\varepsilon/dt + SI\alpha_{AVGE}$

and since $\varepsilon = 4I/c$: $SI_0\alpha_{AVGE} = (4V/c)dI/dt + SI\alpha_{AVGE}$

and since $A = S\alpha_{AVGE}$ and rearranging:

$$(4V/c)\frac{dI}{dt} + AI = AI_0$$

or: $$\frac{dI}{dt} = -(Ac/4V)I + (Ac/4V)I_0$$

This is a standard form of differential equation, with known solution:

$I = Ke^{-(Ac/4V)t} + I_0$

where K is a constant which depends upon the initial conditions:

$I = 0$ when $t = 0$, so that $K = -I$

and $I = I_0(1 - e^{-(Ac/4V)t})$

which is sketched in graphical form below.

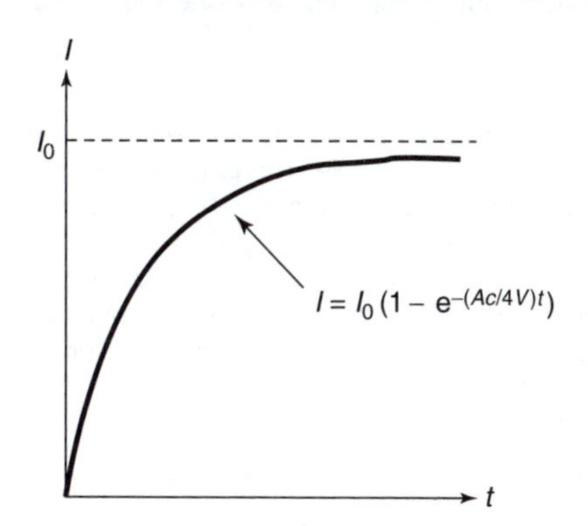

Appendix 5.4 Derivation of formula for reverberant sound level in a room

Sound power contributing to the reverberant sound field, W_R:

$W_R = W(1 - \alpha_{AVGE})$

Rate of sound energy supply = rate of sound absorption: $W_R = I_R S\alpha_{AVGE}$

(I_R is the reverberant sound intensity field and S is the total area of the room surfaces)

Therefore

$I_R S\alpha_{AVGE} = W(1 - \alpha_{AVGE})$

and

$I_R = W(1 - \alpha_{AVGE})/(S\alpha_{AVGE})$

If we define room constant

$R_C = (S\alpha_{AVGE})/(1 - \alpha_{AVGE})$

then

$I_R = W/R_C$

Remembering (from section 5.9) that the sound intensity, I_R, at the boundary of the room in which the reverberant sound pressure level is p_R, is given by:

$I_R = p_R^2/4\rho c$

then

$p_R^2/4\rho c = W/R_C$

and

$p_R^2 = W\rho c4/R_C$

Converting into decibels, reverberant sound pressure level L_R:

$L_R = 10\log\left(p_R^2/p_0^2\right) = 10\log(W\rho c4/R_C p_0^2)$

Using $p_0 = 2 \times 10^{-5}$ Pa, $\rho c = 420$ kgm^{-2}s and $W_0 = 1 \times 10^{-12}$ W/m^2:

$L_R = L_W + 10\log(4/R_C)$

Direct sound pressure p_D and direct sound intensity I_D at distance r from source (of directivity factor Q) are given by:

$I_D = p_D^2/\rho c = QW/(4\pi r^2)$

so:

$p_D^2 = W\rho cQ/(4\pi r^2)$

and total sound pressure p_{TOTAL} is given by:

$p_{TOTAL}^2 = p_D^2 + p_R^2 = W\rho cQ/(4\pi r^2) + W\rho c4/R_C$

$p_{TOTAL}^2 = W\rho c[(Q/4\pi r^2) + (4/R_C)]$

from which total sound pressure level, L_p, in the room is given by:

$L_p = L_W + 10\log[(Q/4\pi r^2) + (4/R_C)]$

Questions

1 A lecture room 16 m long, 12.5 m wide and 5 m high has a reverberation time of 0.7 s. Calculate the average absorption coefficient of the surfaces using the Eyring formula.

2 A room 16 m long, 10 m wide and 5 m high which was previously used as a laboratory is to be converted to use as a lecture room for 200 people. The original wall and floor surfaces are hard plaster and concrete whose average absorption coefficient is 0.05. Acoustical tiles of absorption coefficient 0.75 are available for wall or ceiling finishes. The optimum reverberation time for the new use of the room is 0.87 s. The absorption of seated audience (per person) is 0.4 m^2 units. Calculate the area of the tiles to be applied to achieve this.

3 The absorption coefficient of a certain material was measured in a reverberation chamber of volume 1300 m^3 and the following average reverberation times were obtained:

Frequency (Hz)	RT room empty (s)	RT with 30 m^2 of sample (s)
125	16.8	10.2
250	20.1	10.4
500	18.5	9.4
1000	14.5	8.0
2000	9.1	6.1

The average absorption coefficient of all the surfaces is less than 0.2. Find the absorption coefficients of the material at the frequencies given.

4 A canteen of dimensions 8 m × 4 m × 3.5 m high contains an extractor fan situated at ceiling height half way along one of the long walls. The sound power level of the fan is 48 dB in the 125 Hz octave band. Calculate the total sound pressure level in this band at a point 3 m from the fan. Also calculate the direct sound pressure level at this point. The reverberation time of the canteen has been measured and is 0.8 seconds in this octave band.

5 A room of 4000 m^3 has a reverberation time of two seconds. How much sound-absorbing material would need to be added to reduce the reverberant noise level by 6 dB? Of what importance is the noise spectrum and the absorption frequency characteristic?

6 A factory workshop with hard uniform surfaces have dimensions 30 m by 15 m by 6 m high and a dB(A) weighted reverberation time of three seconds. The noise level in the workshop is 98 dB(A). Predict the approximate new noise level if the ceiling is covered with acoustic absorbent, having a dB(A) weighted absorption coefficient of 0.8.

7 A workshop of volume 1000 m^3 has a reverberation time of 3 seconds. The sound pressure level with all the machines in use is 105 dB at a certain frequency. If the reverberation time is reduced to 0.75 seconds what would the sound pressure level be?

8 Compare the reverberant room method and the standing wave tube methods for measuring the absorption coefficient of a material.

During a standing wave tube measurement of absorption coefficient at a particular frequency, a difference of 16 dB was observed between the first maximum and minimum from the sample surface. Calculate the standing wave ratio and the absorption coefficient. (IOA 1997)

9 (a) Explain the difference between porous and resonant absorption. Give an example of the application of each type of absorption.
(b) Describe the physical mechanism by which sound energy is lost in a cavity resonator.
(c) Why do some practical forms of cavity resonator incorporate porous absorbents?
(d) Explain with the help of labelled sketches the difference between a perforated plate resonant absorber and a membrane absorber. (IOA 2006)

10 An exhibition hall has been built with a length of 60 m, a width of 30 m, and a height of 9 m. It has a sound absorbent finish to the ceiling (with an absorption coefficient of 1.0 at mid-frequencies), but no other sound absorbent surfaces. It is required to have an RT of no more than 1.4 s at mid frequencies.
(a) Is it likely to meet the requirement? Use two methods of calculation and comment on their likely accuracy in this case. What other approach could be used?
(b) After a while the operator installs (empty) metal shelving along two adjacent walls of the hall and the measured mid-frequency RT is significantly reduced. Why might this happen? (IOA 2006 (part question))

11 (a) What are the different assumptions behind Eyring's and Sabine's formulae relating to the theoretical prediction of reverberation time? Where and why might one be used in preference to the other?
(b) Sketch and annotate graphs of sound absorption against frequency showing the important characteristics of (i) porous, (ii) panel and (iii) Helmholtz absorbers.
(c) A studio has a carpeted floor and discrete absorbent panels on the ceiling and one wall. When measured, the reverberation time was found not to be in close agreement with the predictions undertaken using both Eyring's and Sabine's formulae. Give two possible reasons why this might be. (IOA 2005)

12 The RT of an unoccupied conference room of volume 500 m^3 and the desired optimum value are given below for three octave bands, together with the absorption per person. Calculate how many people are required in the audience to achieve the optimum value.

Frequency (Hz)	125	500	2000
Optimum RT	1.6	1.2	1.1
Actual RT	1.9	1.7	1.6
Absorption per person (m^2)	0.17	0.43	0.47

(IOA 2001 (part question))

13 The dimensions of a conference room are 8 m × 5 m × 3 m high. The floor is carpeted and the walls are made of concrete blockwork with a plaster finish. The room has three windows, each with dimensions 1.5 m × 1.0 m. The ceiling is made of plasterboard, and the door, of dimensions 2 m × 1.0 m, is of solid hardwood. Using Sabine's equation, calculate the reverberation time at 1000 Hz of the empty room, and also when occupied by 10 directors and the company secretary during board meetings.

	Absorption coefficient at 1000 Hz
Absorption area per person	0.45
Plasterboard	0.08
Hardwood	0.07
Glazing	0.05
Plaster on concrete blockwork	0.05
Carpeting	0.64

(IOA (part question))

14 A reverberation chamber has dimensions 5.8 m × 4.5 m × 9.5 m. The reverberation time was measured at 500 Hz as 12.5 s with the room empty and 3.7 s when a specimen of material of area 10 m^2 was included. Calculate the absorption coefficient of the specimen (you may neglect the absorption of the room surfaces).

15 An omni-directional source of sound power 5 W is placed on the platform of a concert hall. The hall has dimensions 20 × 16 × 12.5 m^3 and a reverberation time of 4 s for the frequency bands being considered.

What is the sound pressure level at distances of 4 m and 20 m from the source?

If the absorption of the hall were doubled what would be the new levels at the two locations?

(IOA 1990 (part question))

16 The clean room in a new production facility has the dimensions: length 15.4 m, width 6.2 m, height 2.9 m, windows 2, each 1.3 m × 3.1 m, doors 2, each 1.5 m × 2.0 m and surfaces given in the table below. It contains an air handling outlet on the ceiling which has a rated sound power level at 500 Hz of 49.2 dB.

(a) Calculate the total absorption of the room surfaces at 500 Hz.
(b) Estimate the reverberation time and room constant at this frequency.
(c) Calculate the reverberant sound pressure level resulting from the air handling outlet and the sound pressure level 1 m above the floor directly beneath the air handling unit, stating any assumptions made in these calculations.

Surface	Material	500 Hz absorption coefficient
Walls	Plaster	0.11
Floor	Plastic material	0.20
Ceiling	Suspended	0.68
Windows	Sealed unit	0.02
Doors	Metal clad	0.05

(IOA 2008)

17 (i) Starting with a description of the factors affecting speech intelligibility, discuss the important parameters to be considered when designing a theatre for speech, including background noise, reverberation time and loudness.
(ii) You are designing a theatre where the auditorium has a rectangular floor plan, and the ceiling and floor are both flat and level. The audience of 500 people occupies the entire floor area ($\alpha = 0.8$ at 500 Hz). The proscenium (opening between the auditorium and stage) can be considered to be a perfectly absorbing surface of area equal to 20% of the floor plan area. Use the Sabine formula to calculate a suitable ceiling height (ignore air absorption and any other absorption that might occur).
(iii) In what ways does this situation depart from the assumptions upon which the Sabine formula is based?
(iv) More money is found to provide a more sophisticated design than that described above. Describe briefly how you would improve the listening conditions for the audience. (IOA 1994)

18 You are advising on the design of music practice rooms which range in plan area from 10 m^2 to 20 m^2 and are 2.8 m high.

(a) Describe the principal issues relating to room acoustics and say how you would resolve these by considering each of the following:
 (i) room dimensions
 (ii) sound absorbing treatment
 (iii) diffusion
 (iv) room shape.
(b) An orchestral rehearsal room of 50 m^2 plan area, also 2.8 m high, for 40 musicians is also to be Included in the scheme. Make comments on this proposal related to room acoustics. (IOA 1996)

19 (a) Distinguish between axial, tangential and oblique room modes and calculate the frequency of the lowest mode of each type for a rectangular space measuring 4 m × 4 m × 2 m.

(b) Interpret and give reasons for the form of the decay curve shown in the figure below.

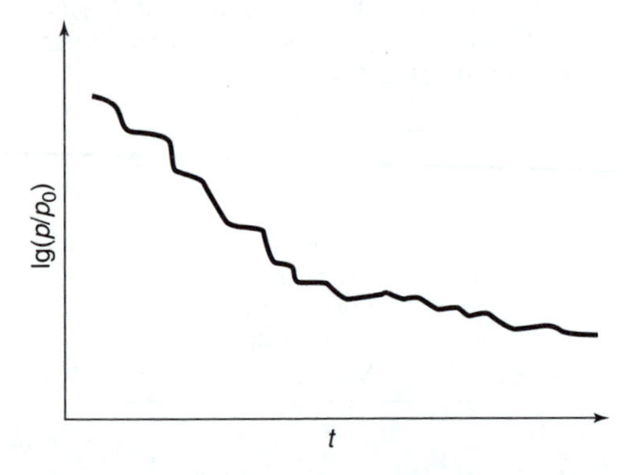

(c) Discuss the conditions to be fulfilled for the sound field in a room to be described as diffuse.

(d) A sound source in a reverberation room of volume 200 m^3, with a reverberation time at 1 kHz of 2.0 s, produces a reverberant sound pressure level, at 1 kHz, of 78 dB. What is the sound power level of the source at 1 kHz? (IOA 2007)

20 A box-shaped classroom has a carpet and a sound absorbent ceiling, and no other sound absorbent surfaces. It is to be made 40% longer (with the same carpet and ceiling finishes throughout) although the teacher's position shall remain unchanged.

(a) Ignoring specular reflections, what will be the effect on speech intelligibility for a child who sits in the same place before and after the extension? Explain your answer.

(b) BB93 provides guidance on restricting low frequency RTs in relation to mid-frequency RTs. If the classroom ceiling treatment consists of sound absorbent 'acoustic plaster' sprayed onto the underside of a concrete slab to a depth of 30 mm, what other type of acoustic element would you consider might be necessary in order to comply with the BB93 guidance? Provide a brief sketch of such an element and explain how it works.

(IOA 2006 (part question))

21 (a) You have recently been appointed by a client to design a new office building which is to contain a 200 seat lecture theatre. Describe the key acoustic issues that will need to be considered and, where appropriate, include relevant parameter values.

(b) In the completed lecture theatre there will be a ventilation system with a supply diffuser mounted at the base of each seat. Considering both the direct and reverberant sound, calculate the maximum allowable sound power level of each diffuser such that the sound pressure level at a distance of 1.5 m above each seat is no greater than 25 dB re. 20 μPa. For the direct sound you need only consider one diffuser.

Use the following data:

Directivity factor (Q) of each diffuser = 2
Total equivalent area of lecture theatre = 150 m^2
Total surface area of lecture theatre surfaces = 710 m^2.

(IOA 2005)

22 (a) A woodworking router machine has a sound power level of 100 dB(A). It is installed in a workshop 20 m × 8 m × 5 m high, where the average reverberation time is 1.5 seconds. Calculate the sound pressure level at the operator position, 1 m from the machine, and at a coffee break area, 8 m from the machine.

(b) An absorbent treatment is applied to areas of the internal surface of the workshop roof, reducing the reverberation time to 1 second. What are the revised SPLs at the operator location and coffee area?

(c) Suggest three alternative methods of noise control for this machine and workshop, indicating the likely effectiveness of each suggestion. (IOA 2007)

Chapter 6 Sound insulation

6.1 Introduction

This chapter deals with transmission of sound between two spaces through a dividing partition such as a wall, floor or ceiling. Much of the chapter is concerned with airborne sound insulation, although impact and structure-borne sound transmission are also briefly discussed. In connection with airborne sound insulation it is useful to think of the incident sound wave generated by a loudspeaker on the source side of a partition as exciting the partition into vibration. The vibrating partition then, in effect, acts as a large loudspeaker radiating sound into this space.

The chapter starts with a reminder of the definitions of sound transmission coefficient and sound reduction index, the basic parameters underlying much of this chapter, and the calculation of the sound reduction index of composite partitions, which then leads to a discussion of the effect of holes and gaps on sound insulation.

Simple calculation models are described for calculating sound transmission between adjacent spaces and also via the façade from inside to outside and vice versa. The assumptions on which these models are based and their consequent limitations are discussed.

The procedures for the measurement of sound insulation according to ISO 140 are described together with the ISO 717 method for calculating single figure sound insulation rating values.

Factors affecting sound reduction index of partitions for both single leaf and double leaf partitions are examined, including effects of the mass law, panel resonance, and coincidence.

The chapter concludes with a summary of Building Regulations Approved Document E, and structure-borne sound and flanking paths are briefly considered.

6.2 Sound reduction index of composite partitions

The direct transmission of sound between two rooms is determined by the sound reduction index, R, of the partition (also called the transmission loss), measured in decibels. This is closely related to the transmission coefficient, t, of the partition which gives the fraction of incident sound energy transmitted by the partition.

Sound reduction index $R = 10\log(1/t)$ dB

As examples: if $t = 0.001 = 10^{-3}$ then $R = 30$ dB, and if $t = 0.00001 = 10^{-5}$, then $R = 50$ dB.

Composite sound reduction index values

A wall, partition or façade may consist of more than one material or building element; for example there may be a wall, a window and a door. The effective or average sound reduction index of the entire area may be calculated by first finding the individual transmission coefficients of each element, then calculating the area weighted average transmission coefficient, and then converting this into a sound reduction index. In other words there are three steps:

Step 1: calculate the individual t values from each of the R values: $t_1 = 1/10^{R/10} = 10^{-R/10}$, e.g.:

if $t = 3.5 \times 10^{-5}$ then $R = 44.6$ dB

Step 2: calculate the area weighted average t value, t_{AVGE}:

$t_{AVGE} = (t_1S_1 + t_1S_1 + t_1S_1 + \cdots + t_NS_N)/S_{TOTAL}$

where $S_1, S_2, S_3, \ldots, S_N$ are the individual component areas, and S_{TOTAL} is the total area.

Step 3: calculate the effective or average sound reduction index, R_{AVGE}:

$R_{AVGE} = 10\log(1/t_{AVGE})$

Example 6.1

A partition between two rooms has dimensions 6 m × 3 m and contains a door of dimensions 2 m × 1 m and two windows, each of dimensions 1.5 m × 1 m. In a certain octave band the wall has an R value of 45 dB, the door of 35 dB, and the window of 20 dB. Calculate the effective or average R value of the partition.

Step 1: for wall, $t_W = 10^{-4.5} = 3.16 \times 10^{-5}$; for door, $t_D = 10^{-3.5} = 3.16 \times 10^{-4}$; for glass (window), $t_G = 10^{-2.0} = 0.01$

Step 2: $t_{AVGE} = [(13 \times 3.16 \times 10^{-5}) + (2 \times 3.16 \times 10^{-4}) + (3 \times 0.01)]/18 = 0.00173$

Step 3: $R_{AVGE} = 10\log(1/0.00173) = 27.6$ dB

Alternatively the calculation may be presented in the form of a table:

Element	*R* value (dB)	$t = 10^{-R/10}$	Area S(m²)	*St*
Wall	45	3.16×10^{-5}	13	0.000411
Door	35	3.16×10^{-4}	2	0.000632
Glass	20	1×10^{-2}	3	0.03
Total			18	0.0310
			t_{AVGE}	0.00173
			R_{AVGE}	27.6 dB

The effect of holes and gaps

A value of $t = 1$ is assigned to the hole/gap, which is then included as one of the components in the calculation of the average *t* value (i.e. composite partition consisting of wall, door, window and hole/gap).

Note that in assuming that all the energy striking the partition passes through it (i.e. that $t = 1$ and $R = 0$ dB for the hole) we are ignoring the effects of diffraction at the edges of the hole, but this will in most cases be a reasonable assumption, and allows an approximate estimate of the significance of air gaps to the effectiveness of sound insulation.

Example 6.2

Repeat the first example above but allowing for a 2 mm air gap around the door.

Solution

Area of air gap = perimeter of door × thickness of gap = $6 \text{ m} \times 0.002 \text{ m} = 0.012 \text{ m}^2$.

Assuming all other areas remain unchanged:

Element	*R* value (dB)	$t = 10^{-R/10}$	Area S(m²)	*St*
Wall	45	3.16×10^{-5}	13	0.000411
Door	35	3.16×10^{-4}	2	0.000632
Glass	20	1×10^{-2}	3	0.03
Air gap	0	1	0.012	0.012
Total			18	0.043044
			t_{AVGE}	0.002391
			R_{AVGE}	26.2 dB

Answer is 26.2 dB.

Strictly speaking we should account for the area of the gap either in the reduced area of wall (or of the door) or increased total area but, if you have the patience to check it out, you will find it makes a negligible difference as long as the gaps are small.

Exercise

What would be the effect of a 5 mm gap?

Answer

24.7 dB.

Holes in walls

Example 6.3

What is the average or effective sound reduction index of a wall with sound reduction index of 40 dB which contains a hole of 0.1% of the wall area?

Solution

We proceed as before except that we can set the total area to 100 (as in 100%) and the component areas accordingly:

Element	*R* value (dB)	$t = 10^{-R/10}$	Area S(m²)	*St*
Wall	40	1.0×10^{-4}	99.99	0.00999
Air gap	0	1	0.01	0.01
Total			100	0.01999
			t_{AVGE}	0.0002
			R_{AVGE}	37.0 dB

Answer is 37.0 dB.

Exercise

Put the above calculation into a spreadsheet, and then investigate the effect of various size of hole (e.g. 10%, 1%, 0.1% etc.) for various values of sound reduction index of wall (e.g. 20 dB, 30 dB, 40 dB, 50 dB etc. (calculated to the nearest 0.1 dB).

Answer

See the following table.

% area of hole	Effective or average sound reduction index of wall (dB)					
	10 dB	20 dB	30 dB	40 dB	50 dB	60 dB
0.01%	10	20	29.6	37	39.6	40
0.1%	10	19.6	27.0	29.6	30	30
1%	9.6	17.0	19.6	20	20	20
10%	7.2	9.6	10	10	10	10
20%	5.5	6.8	7.0	7	7	7
50%	2.6	3.0	3	3	3	3

What general conclusions can you draw about the effects of holes on sound reduction index from this table?

The general conclusions to be drawn are that:

- The greater the sound reduction index of the wall the greater the effect of the hole, so that for very high R values, e.g. of 50 and 60 dB, even very small holes have a severely limiting effect on the sound insulation which can be achieved. The effect of holes and gaps is less severe for lower values of R.
- The relative size of the hole compared to the size of the wall sets an upper limit to the effective sound reduction index which can be achieved, e.g.:
  - if the hole/gap occupies 10% of the total area, the R value is limited to 10 dB or less
  - if the hole/gap occupies 1% of the total area, the R value is limited to 20 dB or less
  - if the hole/gap occupies 0.1% of the total area, the R value is limited to 30 dB or less
  - if the hole/gap occupies 0.01% of the total area, the R value is limited to 40 dB or less.

Effect of open windows on façade sound insulation

PPG24 (Planning Policy Guidance Note 24, see Chapters 4 and 8) suggests that as a rough rule of thumb the sound insulation provided by the façade of a residential property with windows open is typically between10 dB and 15 dB.

Exercise

Comment on this rule of thumb in the light of the above exercise.

Answer

Assuming an open window to have an R value = 0 dB (and $t = 1$ as in the above examples), the effective sound reduction index of the façade will depend on the relative glazing area in the façade. The above rule of thumb would appear to be based on a % of open window area of between 10 and 20%.

Exercise

A partition between two rooms has dimensions 8 m × 3 m and contains a door of dimensions 2 m × 1 m and two windows, each of dimensions 2 m × 1 m. In a certain octave band the wall has an R value of 50 dB, the door of 38 dB, the window of 25 dB. Calculate the effective or average R value of the partition (a) assuming no gap around the door and then (b) assuming a 2 mm air gap and (c) a 5 mm gap.

Answers

(a) 32.6 dB, (b) 29.8 dB, (c) 27.5 dB.

The effect of glazing area on the sound insulation of a façade

To illustrate the effect of different proportions of glazing consider a façade composed of eight panels of equal area (i.e. 12.5% each) which are either wall with R = 50 dB or glazing of R = 20 dB in a certain octave band. Using the calculation methods described above, estimate the average sound reduction index if the following numbers of panels are glazed: 0, 1, 2, 3, 4 and 8. The solution is shown in the following table.

No. of glazed panels	% area of glazing	Composite R value
0	0	50
1	12.5	29
2	25	25
3	37.5	23
4	50	22
8	100	20

Note that the introduction of the first small area of glazing has a dramatic effect on R_w compared with the relatively smaller changes when the area of the lower performance glazing component is doubled, tripled, etc.

Calculation of maximum allowable area of glazing

Example 6.4

The façade of a building has an area of 25 m^2 and is made up of brick wall having a sound reduction index of 50 dB in a particular octave band, and a window of sound reduction index of 20 dB in the same band. Estimate the maximum area of window which is allowable if the average sound reduction index is to be at least 30 dB in this same band, in order to comply with a planning condition relating to the ingress of external noise.

Solution

This is a variation on the above calculations which requires a little algebraic manipulation.

We start by assuming the required area of glazing to be S_G so that the area of wall is $(25 - S_G)$. Since we know the required average value of sound reduction index to be 30 dB this also allows us to determine the required average value of t, t_{AVGE}, which is $10^{-30/10} = 1.0 \times 10^{-3}$. The values are tabulated in the usual way as shown below:

Element	R value (dB)	$t = 10^{-R/10}$	Area S (m^2)	St
Wall	50	1.0×10^{-5}	$25 - S_G$	$(25-S_G)\times1.0\times10^{-5}$
Glazing	20	1.0×10^{-2}	S_G	$S_G \times 1.0 \times 10^{-2}$
Total			25	$(25 \times 1.0 \times 10^{-3})$
			t_{AVGE}	1.0×10^{-3}
			R_{AVGE}	30.0 dB

The final column of the table is expressed by the following equation:

$$S_W \times t_W + S_G \times t_G = S_{TOTAL} \times t_{AVGE}$$

$$[(25 - S_G) \times 1.0 \times 10^{-5}] + [S_G \times 1.0 \times 10^{-2}] = 25 \times 1.0 \times 10^{-3}$$

Rearranging to find S_G:

$$[(1.0 \times 10^{-2}) - (1.0 \times 10^{-5})] \times S = 25 \times [1.0 \times 10^{-3} - 1.0 \times 10^{-5}]$$

$$9.99 \times 10^{-3} = 25 \times 9.9 \times 10^{-4}$$

$$S = (25 \times 9.9 \times 10^{-4})/(9.99 \times 10^{-3}) = 2.48 \text{ m}^2$$

or more generally: $S_W \times t_W + S_G \times t_G = S_{TOTAL} \times t_{AVGE}$

and $S_G = S_{TOTAL} \times [t_{AVGE} - t_W]/[t_G - t_W]$

where S_{TOTAL} is the total area of the façade, and t_{AVGE}, t_W and t_G are, respectively the average transmission coefficient ('t' value) and those for wall and glass.

Exercise

The façade of a building has an area of 45 m^2 and is made up of brick wall having a sound reduction index of 55 dB in a particular octave band, and a window of sound reduction index of 30 dB in the same band. Estimate the maximum area of window which is allowable if the average sound reduction index is to be at least 40 dB in this same band, in order to comply with a planning condition relating to the ingress of external noise.

Answer
4.4 m^2

6.3 Sound transmission calculations

Direct and indirect sound transmission

The airborne sound in a room strikes all the six room surfaces, not just the partition with the adjacent wall. Sound striking a party wall sets this into vibration and this causes sound radiation into the adjacent room. This is known as the direct sound transmission path or simply the direct path. But the sound striking the other room surfaces causes other transmission paths into the adjacent room as shown in Figure 6.1. These are called indirect or flanking paths.

Two direct paths are shown (one into each adjacent room); all other paths are indirect. Note that Figure 6.1 could equally be either a plan view representing four rooms on the same level, or a section, representing two rooms below and two rooms above.

The calculation of direct airborne sound transmission between rooms

The level difference between two adjacent rooms arising from direct sound transmission via the partition between them depends not only on R, but also on the surface area of the partition S, and on the amount of absorption, A, in the receiving room. It can be shown that for direct path sound transmission only:

$$L_1 - L_2 = R - 10\log S + 10\log A$$

This equation can be rearranged as:

$$L_2 = L_1 - R + 10\log S - 10\log A$$

or as:

$$R = L_1 - L_2 + 10\log S - 10\log A$$

where L_1 and L_2 are the reverberant sound levels in the source and receiving rooms respectively.

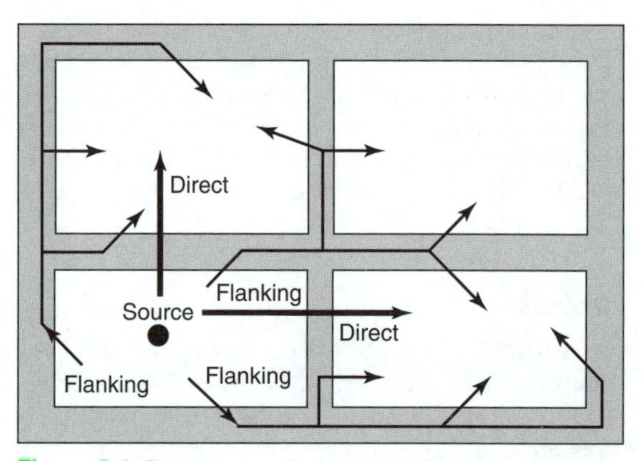

Figure 6.1 Sound transmission paths from one room to two adjacent rooms

It can be seen from this equation that if the partition area is doubled, L_2 increases by 3 dB; if the absorption area in the receiving room is doubled L_2 falls by 3 dB.

Assumptions and limitations

This equation:

- applies to direct sound transmission only and therefore does not take into account transmission via indirect and flanking paths, and
- assumes reverberant sound in both rooms.

A derivation of the equation is given in Appendix 6.1.

Applications of the sound transmission equation

- To predict the sound level in the receiving room (given all other variables).
- To predict the relative contribution of sound transmission through the different elements in a composite partition.
- To predict the sound reduction index required in order to achieve a certain level (L_2) in the receiving room (given all other variables).
- To measure the sound reduction index of the partition (in a special test facility called a reverberant suite).

Example 6.5

Reverberant noise from a plant room is transmitted to an adjacent office through a party wall. The dimensions of the office are 10 × 7 × 3 m high, and the partition separating it from the plant room is of dimensions 7 × 3 m^2. The reverberation time in a particular octave band is 0.7 s for the office. In the same octave band the reverberant sound pressure level in the plant room is 85 dB and the transmission loss of the party wall is 50 dB. Calculate the sound pressure level in the office.

Solution

The first step is to calculate the absorption in the receiving room using the reverberation time and Sabine's formula:

$$A = 0.16 \times V/T = 0.16 \times 210/0.7 = 48 \text{ m}^2$$

$$L_2 = L_1 - R + 10\log S - 10\log A$$

$$L_2 = 85 - 50 + 10\log(21) - 10\log(48) = 31.4 \text{ dB}$$

Example 6.6

A large workshop is to be divided using a partition to create an office space. In a particular octave band the reverberant sound pressure level in the workshop is 85 dB and this is not expected to change when the partition is in place. The dimensions of the office will be 5 m × 6 m × 3 m and the partition will have dimensions 5 m × 3 m. From measurements in a similar office it is estimated that the average absorption coefficient of the office surfaces will be 0.1. For comfortable working conditions it is necessary that the reverberant sound level in the office, arising from sound transmission through the partition, shall be limited to 35 dB in this particular octave band. Estimate the minimum sound reduction index of the partition required to meet this criterion.

Solution

The first step is to calculate the sound absorption in the receiving room using the average absorption coefficient and the room surface area.

$$S = 2 \times [(5 \times 6) + (5 \times 3) + (6 \times 3)] = 126 \text{ m}^2$$

$$A = S \times \alpha_{AVGE} = 126 \times 0.1 = 12.6 \text{ m}^2$$

$$R = L_1 - L_2 + 10\log S - 10\log A$$

$$R = 85 - 36 + 10\log(15) - 10\log(12.6) = 49.8 \text{ dB}$$

Example 6.7

In a sound insulation test of a sample of office partition carried out in a transmission suite the average sound levels in the source and receiving rooms are 96 dB and 57 dB in a particular frequency band. The area of the test sample is 10 m^2, and the reverberation time in the receiving room in this band is 2.5 seconds. Calculate the sound reduction index of this partition in this frequency band.

Solution

The first step is to calculate the absorption in the receiving room using the reverberation time and Sabine's formula:

$$A = 0.16 \times V/T = 0.16 \times 200/2.5 = 12.8 \text{ m}^2$$

$$R = L_1 - L_2 + 10\log S - 10\log A$$

$$R = 96 - 57 + 10\log(10) - 10\log(12.8) = 37.9 \text{ dB}$$

Example 6.8

A foreman's office of dimensions 4 m × 3 m × 2.5 m high is to be built, as an acoustic refuge in a paper mill, where the reverberant sound level is 85 dB in the 1000 Hz octave band. The office is to be situated in the corner of the workshop so that only two sides and the flat ceiling have to be built onto the existing walls. The office is to be made of chipboard with an area of glazing 1.5 m^2 and a door of area 2 m^2.

The sound reduction index values in the 1000 Hz bands are: office wall (and ceiling) 40 dB, door 30 dB and window 20 dB. The average absorption coefficient of the inside surfaces of the office is 0.1.

Calculate the reverberant sound pressure level inside the office in this octave band resulting from sound transmission from the workshop noise outside.

Solution

The first step is to calculate the amount of sound absorption, A, inside the office:

$$S = 2 \times [(4 \times 2.5) + (3 \times 2.5) + (4 \times 3)] = 59\ \text{m}^2$$

$$A = S\alpha_{AVGE} = 59 \times 0.1 = 5.9\ \text{m}^2$$

The appropriate equation is: $L_2 = L_1 - R + 10\log S - 10\log A$.

There are two alternative approaches to the next step: either to calculate the composite sound reduction index of the office (i.e. two walls and ceiling, door and window) and calculate the transmission via the total office area, or, to calculate the sound transmission through each component of the office separately.

Method 1

The first step is to calculate the average sound reduction index of the office using the method described earlier. This gives an average value for R of 31.8 dB.

Then: $L_2 = 85 - 31.8 + 10\log(29.5) - 10\log(5.9)$

$L_2 = 60.2$ dB

Method 2

Applying the same equation for each element of the office:

via walls and ceiling:

$$L_2 = 85 - 40 + 10\log(26.0) - 10\log(5.9) = 51.4\ \text{dB}$$

via office door:

$$L_2 = 85 - 30 + 10\log(2.0) - 10\log(5.9) = 50.3\ \text{dB}$$

via office window:

$$L_2 = 85 - 20 + 10\log(1.5) - 10\log(5.9) = 59.1\ \text{dB}$$

The total reverberant level inside the office is obtained by combining these three levels:

$$L_{TOTAL} = 10 \times \log[10^{5.14} + 10^{5.03} + 10^{5.92}]$$

$= 60.2$ dB, as for method 1

The second method has the advantage of giving information about the relative importance of the three component transmission paths, i.e. that most of the sound is transmitted via the window and least via the door.

Sound transmission via building façade; from inside to outside

There are three stages to this calculation, which may relate either to sound transmission via the entire façade, or via an element of the façade such as a window.

Stage 1

Calculate the sound level, L_{OUT}, just outside the façade arising from sound transmission through the façade (see Chapter 5, section 5.9):

$$L_{OUT} = L_{IN} - R - 6$$

where L_{IN} is the reverberant sound level inside the room and R is the sound reduction index of the façade or façade element.

Stage 2

Calculate the sound power level, L_W, of the façade element acting as a sound source radiating sound towards the reception point:

$$L_W = L_{OUT} + 10\log S$$

where S is the area of the façade or façade element.

Stage 3

For reception point at a distance, r, much greater than the dimensions of the façade, it may be assumed that the façade (or façade element) behaves as a point source radiating into a free field, with a directivity index D:

$$L_R = L_W - 20\log r - 11 + D$$

The first two stages are exactly as for the room to room transmission case, and are explained in Appendix 6.1.

Example 6.9

A small single storey factory building contains a window of sound reduction index 20 dB and area 4 m^2. The reverberant sound level inside the factory is 85 dB in a particular octave band. Calculate the sound level radiated through the façade via the window to a noise sensitive reception point 40 m away opposite the window.

Solution

Stage 1: $L_{OUT} = L_{IN} - R - 6 = 85 - 20 - 6 = 59$ dB

Stage 2: $L_W = L_{OUT} + 10\log S = 59 + 10\log 4 = 65$ dB

Stage 3: $L_R = L_W - 20\log r - 11 + D$
$= 65 - 20\log 40 + 3 = 42$ dB

(assuming, in this case, a value of $D = +3$ dB)

A little algebra will show that all three stages may be combined into one equation (again assuming $D = +3$ dB):

$$L_R = L_{IN} + 10\log S - 20\log r - 14$$

Sound transmission via building façade: from outside to inside

The reverberant sound level inside a room, L_{IN}, arising from the transmission of external noise through the façade of a building depends on: the external sound level, the sound reduction index, R, and area, S, of the façade or façade element, and the amount of sound absorption, A, in the room.

The external sound level to be used in the prediction of L_{IN} may be a free field value, L_{FF}, measured or calculated in the absence of the façade (e.g. if prediction is necessary before the façade is built), or it may be measured with the façade in place, in which case the level 2 m from the façade is used, L_{2m}.

These two possible external noise levels are related by:

$$L_{2m} = L_{FF} + 3$$

The internal sound level may then be related to the external level by:

$$L_{IN} = L_{2m} - R + 10\log S - 10\log A$$

or by: $$L_{IN} = L_{FF} - R + 10\log S - 10\log A + 3$$

This equation assumes diffuse sound incident on the partition which may be approximately true when the external noise is from road traffic or from passing trains.

If the external noise arises from a point source such as a compressor or diesel generator an adjustment of up to 6 dB may need to be made to the above equation, depending on the angle of incidence of the sound onto the façade.

If the external noise is measured very close to the façade the level, L_{CLOSE}, will be 6 dB higher than the free field value, and 3 dB higher than the level at 2 m, L_{2m}.

$$L_{CLOSE} = L_{FF} + 6 = L_{2m} + 3$$

The effect of façade reflections on sound levels measured at 3 m or more from the façade may be considered to be insignificant, and such measurements may be regarded as free field measurements.

Example 6.10

The traffic noise at the site of a proposed dwelling is measured in free field conditions, and is 75 dB, in a particular octave band. Calculate the reverberant sound level to be expected in a room facing the road when the dwelling is built. The sound reduction index of the façade, of area 24 m^2, is 30 dB in the same octave band, and the volume of the room is 120 m^3 and its reverberation time is 0.5 seconds.

Solution

The first step is to calculate the amount of sound absorption in the room using Sabine's formula:

$$A = 0.16 \times V/T = 0.16 \times 120/0.5 = 38.4 \text{ m}^2$$

and then:

$$L_{IN} = L_{FF} - R + 10\log S - 10\log A + 3$$
$$= 75 - 30 + 10\log(24) - 10\log(38.4) + 3$$
$$L_{IN} = 46 \text{ dB}$$

The above approach is a simplified version of one given in BS EN 12354-3:2000 *Building acoustics – Estimation of acoustic performance of buildings from the performance of elements – Part 3: Airborne sound insulation against outdoor sound.* A more complicated example, also involving sound transmission via ventilators (briefly discussed later in this chapter) is given in BS 8233, and this same equation is the basis of the calculation of ingress of external noise into school buildings, in BB93.

6.4 The measurement of sound insulation

Laboratory and field measurements

We have already seen that sound transmission between adjacent rooms may occur not only as a result of direct transmission through the common shared partition (the party wall or floor) but also via many indirect transmission paths involving flanking walls and floors and ceilings.

Special sound transmission test suites consisting of a pair of adjacent reverberant rooms (either horizontally or vertically adjacent) are designed and built so that the only significant transmission between the two rooms is via the direct transmission path and that the contribution of all indirect paths is negligible. These test suites are built with very thick concrete walls and floors designed to have much better sound insulation than the partitions to be tested, which are then built into the partition (wall or floor) between two adjacent rooms.

A sound insulation test carried out under such conditions is called a **laboratory test** and the results of such a test relate only to the test partition itself and may therefore be generally applied to any other situation involving the same type of test partition.

All other tests, for example between two adjacent dwellings, are called **field tests** because the sound transmission may involve a significant and unknown contribution from indirect sound paths (even though in many cases the direct path may be the dominant one). This means that field test results are valid only for the particular test situation, and cannot, for example, be generalized to other situations involving the same type of party wall, because the flanking walls, floor and ceilings may be different and therefore the contribution of indirect transmission to the overall sound insulation between the two spaces may be different.

The field test measurement of airborne sound insulation between adjacent rooms

The basic idea is to create a loud sound in one of the rooms (called the source room) using a loudspeaker and measure the sound pressure level in both the source room (L_1) and in the adjacent (receiving) room (L_2). The basic parameter to be measured in this test is the difference between these two, the level difference, $D = L_1 - L_2$.

In order to minimize the influence of standing waves in the rooms the sound from the loudspeaker should be random noise (pink noise) and because sound insulation depends on frequency, sound pressure level measurements should be carried out in one third octave bands.

Ideally the sound fields in both rooms should be as diffuse as possible. This means, in effect, as reverberant as possible and that the loudspeaker should be positioned in the source room so as to promote reverberant sound and to minimize the influence of direct sound on the party wall and on indirect transmission paths. A good position for the loudspeaker is therefore pointing towards a non-party corner (as this excites the greatest number of room modes) and positioning the loudspeaker close to the party wall should be avoided. The requirement for reverberant sound also means that both rooms should be unfurnished.

Because the sound in both rooms can never be perfectly reverberant the sound pressure level will vary with position in the rooms. Therefore it will be necessary to make sound pressure level measurements at several positions in each room (for each third octave band) to produce a 'space–time' average value, either by using several discrete microphone positions or by moving the microphone (the sweep method) and measuring third octave L_{eq} values over the duration of the microphone sweep.

The loudspeaker in the source room must be capable of generating sufficient sound power to produce a sound level in the receiving room which is, ideally, at least 10 dB above the background in the receiving room. Background noise levels should be measured (in each one third octave band) and if necessary the level L_2 in the receiving room adjusted for the influence of background noise.

The level L_2 in the receiving room, and therefore the level difference, D, will depend on the amount of sound absorption in this room, as well as on the amount of sound energy transmitted from the source room. To control this influence the reverberation time in the receiving room, which is related to the receiving room absorption, is also measured (in each one third octave band).

This allows three field measurement parameters, based on the level difference D, to be calculated: the standardized level difference, D_{nT}, the normalized level difference, D_n, and the apparent sound reduction index, R'; and a fourth parameter, sound reduction index, R, can be measured under laboratory test conditions. These parameters will be defined and discussed later, after we have considered what needs to be specified in order to set up a standard test procedure.

A standardized test procedure

The above has been a general outline of the test method.

What test parameters do you think need to be specified and controlled in order to produce a standard test procedure which will ensure reliable measurements and valid comparability of test results for different operators and for different situations? (More specific terms such as repeatability and reproducibility will be introduced and explained later in Chapter 8, on measurement.)

The Airborne sound insulation field test method is defined and described in detail in BS EN ISO 140-4:1998 *Acoustics – Measurement of sound insulation in buildings and of building elements – Part 4: Field measurements of airborne sound insulation between rooms.* The standard sets out the specification for: the selection of source and receiving rooms; type, number and positioning of loudspeakers; selection of microphone positions; measurement of reverberation time; and correction for background noise level.

Source and receiving rooms

If the rooms are of different volumes the larger one should be chosen as source room, unless a contradictory procedure is agreed upon.

Sound source

If a single loudspeaker is used it shall be operated in at least two positions. If more than one sound source is used simultaneously the sources shall be of the same type and driven at the same level by similar but uncorrelated signals.

Qualification procedures are given for the loudspeakers and the loudspeaker positions in Annex A of the standard, which includes: loudspeaker positions with regard to microphone positions, test procedure for loudspeaker radiation directivity, and guidance on the selection of optimum loudspeaker positions.

Microphone positions

The average sound pressure level in each room may be determined either by using a number of fixed microphone positions or by using a moving microphone method.

The following minimum separation distances are specified:

- 0.7 m between microphone positions
- 0.5 m between any microphone position and room boundary or diffusers
- 1.0 m between any microphone position and the sound source.

If using fixed microphone positions a minimum of five positions shall be used, for each loudspeaker position, i.e. 10 in all.

If using a moving microphone the sweep radius shall be at least 0.7 m. The plane of the traverse shall be inclined in order to cover a large proportion of the space permitted for measurement. The plane of traverse shall not lie within 10 degrees of any plane of the room (wall, floor, ceiling).

The duration of the traverse period shall not be less than 15 s.

A minimum of two sweeps shall be used, i.e. at least one for each source position, unless multiple sound sources operating simultaneously are used, in which case a minimum of one sweep may be used.

Measurement averaging times

For fixed microphone positions the averaging time shall be at least 6 s at each frequency band below 400 Hz and at least 4 s for higher frequency bands.

Using a moving microphone the averaging time shall cover the whole number of traverses (i.e. minimum of two) and shall not be less than 30 s.

Frequency range of measurements

The usual range is from 100 Hz to 3150 Hz (16 bands) but the standard recommends that this range may be extended upwards to include the 4000 and 5000 Hz bands, and may be extended downwards to include the 50 Hz, 63 Hz and 80 Hz.

Exercise

What measurement problems might occur in extending the measurements down to these lower one third octave bands? And how may they be dealt with?

Answer

It will be more difficult to produce a reverberant (and diffuse) sound at low frequencies because the effect of room modes will become more significant. In order to minimize these effects more measurement positions should be used (in order to improve the standard deviation of the average level) together, maybe, with use of diffusers to improve the uniformity of the sound field.

Guidance is given in Annex D of the standard on carrying out measurements at lower frequencies in the 80, 63 and 50 Hz bands.

Measurement of reverberation time

The reverberation time measurements should be carried out in accordance with ISO 354 using a loudspeaker as sound source. This requires the evaluation of the reverberation time from the decay curve from about 0.1 s after the sound source has been switched off, or from a sound pressure level a few decibels lower than at the beginning of the decay. A decay range should be used which is neither less than 20 dB, nor so large that the observed decay cannot be approximated by a straight line. The bottom of this range should be at least 10 dB above the background noise level.

A minimum of six decay measurements are required for each frequency band. At least one loudspeaker position and three microphone positions with two readings in each case should be used.

Note: The above list of measurement details is not complete. The above extracts have been included to illustrate the level of detail that can be included in an acoustic measurement standard.

Note: If you are ever responsible for carrying out a test according to this, or any other standard, you should always obtain and read through the latest version of the standard itself, rather than relying on reports or summaries in textbooks, notes or other sources.

The airborne sound insulation performance parameters D_{nT}, D_n and R'

The definitions of these terms are:

$$D_n = D - 10\log(A/A_0)$$

$$D_{nT} = D + 10\log(T/0.5) = L_1 - L_2 + 10\log(T/0.5)$$

and $R' = D + 10\log(S/A) = D + 10\log S - 10\log A$

Note that the value of $T = 0.5$ seconds, in effect a 'reference' value of reverberation time, is chosen because it is, approximately, the average reverberation time in living rooms in UK dwellings. A_0 is the reference absorption value, of $10\ m^2$.

Exercise

How will the D_{nT} value differ from the D value if (a) $T > 0.5$ and (b) $T < 0.5$?

Answer

If $T > 0.5$ seconds, which means, in relative terms, a 'lively' receiving room with less absorption and a higher reverberant sound level than for a 'standard' ($T = 0.5$) room the correction will be positive, i.e. $D_{nT} > D$. If $T < 0.5$ seconds the reverse will be true and $D_{nT} < D$.

The production of D_{nT} (or D_n or R') values is the purpose of this test, with an end product of D_{nT} (or D_n or R') values in each one third octave band, in both tabular and graphical form.

The measurement of impact sound insulation

This test is designed to simulate the effects of impact noise such a footsteps or the moving of furniture on a floor above heard in the room below. It involves the use of a standard impact sound source, commonly also known as a footsteps or tapping machine, shown in Figure 6.2. This machine has five metal cylinders (or hammers) which are lifted and dropped onto the floor in sequence by the action of cams attached to a rotating shaft. The resulting noise is measured, in one third octave bands, in the room below and the average sound pressure level, L_i, is called the impact sound pressure level.

As with the airborne sound insulation test the reverberation time is measured in the room and used to generate two additional parameters: the normalized impact sound pressure level, L'_n, and the standardized impact sound pressure level, L'_{nT}. These are defined and discussed later.

The standardized test procedure is defined in BS EN ISO 140-7:1998 *Acoustics – Measurement of sound insulation in buildings and of building elements – Part 7: Field measurements of impact sound insulation of floors.*

The tapping machine should be used in at least four positions on the floor, randomly distributed, at least 0.5 m from the walls.

Many of the requirements of the measurement procedure relating to the microphone positions, one third octave band frequency range and measurement of reverberation time are the same as for the airborne sound insulation test described in Part 4 of the Standard:

> Using moving microphones: The minimum number of measurements is 4, i.e. one for each tapping machine position.

Annex A of the standard gives advice regarding the mounting of the tapping machine on soft floor coverings.

The impact sound insulation performance parameters L_n and L'_n

Normalized impact sound pressure level $L_n = L_i + 10\log(A/A_0)$
where $A_0 = 10\ m^2$ of sound absorption

Standardized impact sound pressure level $L'_n = L_i - 10\log(T/T_0)$
where $T_0 = 0.5$ seconds

Note that the $'$ denotes that these are field test parameters. For laboratory impact sound insulation of floors, measured according to ISO 140 Part 6, the normalized

Figure 6.2 Tapping machine used to determine the impact sound insulation of floors (note the five 'hammers' that strike the floor sequentially).

Courtesy of Brüel and Kjær Ltd

Table 6.1 Summary of parameters and notation

Parameter	Symbol	Lab or field	Airborne or impact	Defining equation
Sound reduction index	R	Lab	Airborne	$R = L_1 - L_2 + 10\log S - 10\log A$
Apparent sound reduction index	R'	Field	Airborne	$R' = L_1 - L_2 + 10\log S - 10\log A$
Level difference	D	Either	Airborne	$D = L_1 - L_2$
Normalized level difference	D_n	Field	Airborne	$D_n = L_1 - L_2 - 10\log(A/A_0)$
Standardized level difference	D_{nT}	Field	Airborne	$D_{nT} = L_1 - L_2 + 10\log(T/0.5)$
Normalized impact sound pressure level	L_n	Lab	Impact	$L_n = L_1 - L_2 + 10\log(A/A_0)$
Normalized impact sound pressure level	L'_n	Field	Impact	$L'_n = L_1 - L_2 + 10\log(A/A_0)$
Standardized impact sound pressure level	L'_{nT}	Field	Impact	$L'_{nT} = L_1 - L_2 - 10\log(T/0.5)$

impact sound pressure defined in exactly the same way as for the field test above, is given the symbol L_n. There is no laboratory test equivalent to L'_n.

A summary of parameters and notation relating to sound insulation testing is given in Table 6.1.

ISO 140 Part 5: Field measurements of airborne sound insulation of façade elements and façades

This part of the standard requires simultaneous sound pressure level measurements (L_{eq} values) inside and outside a building, using either an external loudspeaker or road traffic noise as the sound source. The standard distinguishes between element methods and global methods of measurement.

Element methods are for determining sound reduction index values, or apparent sound reduction index values. The loudspeaker is the preferred source, with road traffic as an alternative. The external sound pressure level is measured at the façade using a flush-mounted microphone.

Global methods are for determining outdoor to indoor sound level difference. Road traffic noise is the preferred source but with a loudspeaker as an alternative. The external sound pressure level is measured with a microphone placed 2 m in front of the façade. Noise from rail or air traffic may be used as an alternative to road traffic.

There are a variety of measurement parameters to identify the various options. Those relating to global methods are: $D_{tr,2m,nT}$, $D_{rt,2m,nT}$ and $D_{at,2m,nT}$, and $D_{ls,2m,nT}$ and $D_{ls,2m,n}$. Those relating to element methods are: $R'_{45°}$, $R'_{tr,s}$, $R'_{rt,s}$ and $R'_{a,s}$. Here, 'ls' relates to loudspeaker, 'tr' relates to road traffic, 'rt' to rail traffic and 'at' to air traffic. The subscript '2 m' refers to the external measurement position, 2 m from the façade. The subscript 45° relate to the loudspeaker source being positioned at an angle of 45° incident to the façade element.

ISO 140 Part 10: Laboratory measurements of airborne sound insulation of small building elements

This part of the standard deals with the sound insulation of small building elements of area less than 1 m^2 such as grilles and ventilators.

As with a laboratory test for sound reduction index the small element is placed in the partition between the two rooms in a transmission suite, so that the only way that sound can be transmitted from source to receiving room is via the element (indirect paths are negligible). The difference between the two rooms is measured. The small element level difference, D_{NE} value is defined by the equation:

$$D_{NE} = L_1 - L_2 + 10\log(A_0/A)$$

where A is the amount of sound absorption in the receiving room and A_0 is the reference amount of sound absorption (10 m^2). In other words the D_{NE} is the level difference which would be obtained if the amount of absorption in the receiving room during the test was 10 m^2.

If the element is then used in another situation between two rooms where the absorption in the receiving room is A m^2 the level difference may be predicted from:

$$L_1 - L_2 = D_{NE} + 10\log(A/A_0)$$

Example 6.11

A ventilator with a D_{NE} value of 24 dB in a particular octave band is inserted in the wall between two rooms.

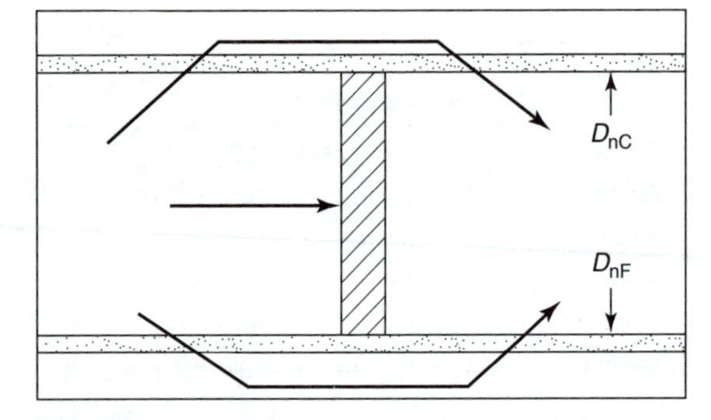

Figure 6.3 Sound transmission via floor (as in ISO 140 Part 9) and via ceiling (as in ISO 140 Part 12)

The sound absorption in the receiving room in the band is 5 m^2. A machine operating in one of the rooms creates a reverberant sound pressure level of 70 dB. Calculate the sound pressure level in the other room arising from sound transmission via the ventilator.

$$L_1 - L_2 = D_{NE} - 10\log(A_0/A)$$

$$= 24 - 10\log(10/5) = 21 \text{ dB}$$

$$L_2 = 70 - 21 = 49 \text{ dB}$$

ISO 140 Parts 9 and 12

These two parts of the standard deal with airborne sound transmission via suspended ceilings (Part 9) and via access floors (Part 12), as illustrated in Figure 6.3. They specify laboratory test procedures and define the two quantities D_{nC} and D_{nF}, defined in a similar way to D_{ne} for small building elements:

$$D_{nC} = L_1 - L_2 + 10\log(A_0/A)$$

and $D_{nF} = L_1 - L_2 + 10\log(A_0/A)$

6.5 Rating of sound insulation

Single figure sound insulation rating values

In the same way as it is sometimes more convenient to describe a sound in terms of a single figure weighted value, such as dBA, rather than by a full octave or one third octave band spectrum, single figure weighted values have been developed to describe the sound insulation performance of building elements such as walls and floors. This allows a rank ordering of partitions by acoustic performance which is easier than comparing data over the entire frequency range.

Therefore for each of the airborne sound insulation parameters listed in Table 6.1 the single figure ratings D_{nw} $D_{nT,w}$ and R_w and R'_w have been developed. The procedure for deriving a single figure weighted sound reduction index, R_w, from 16 one third octave band values of R is described in ISO 717-1 and exactly the same procedure applies to the determination of D_{nw} $D_{nT,w}$ and R'_w from their respective one third octave band values.

$D_{nT,w}$ has been used since 1992 to demonstrate compliance with Building Regulations for airborne sound insulation between dwellings. It ignores the area of the separating partition but includes flanking sound.

R_w is based on laboratory measurements and is used to compare the airborne sound insulation performance of different types of construction. Flanking sound is minimized by including isolation between the test sample and the structure of the sound transmission suite.

R'_w is based on field measurements which includes flanking and is used to check that the required amount of sound insulation has been provided, generally in commercial buildings. (Note that specifying R'_w as the performance target eliminates any argument that the 'failure' is due to flanking sound rather than the poor performance of the partition.)

The ISO 717-1 method

ISO 717-1 specifies a single number rating system based on the Weighted Sound Reduction Index, R_w. The R_w is determined by comparing the 16 individual R values from 100 to 3150 Hz with values for a standard shaped curve. The shape of the standard curve is based on typical noise spectra encountered in buildings and on human hearing sensitivity, and is listed in Table 6.2 and shown in

Table 6.2 Reference values for the determination of R_w (from ISO 717-1:2004 *Acoustics – Rating of sound insulation in buildings and of building elements – Part 1: Airborne sound insulation*)

	Reference value	
Band centre frequency	1/3 octave	Octave
100	33	
125	36	36
160	39	
200	42	
250	45	45
315	48	
400	51	
500	52	52
630	53	
800	54	
1000	55	55
1250	56	
1600	56	
2000	56	56
2500	56	
3150	56	

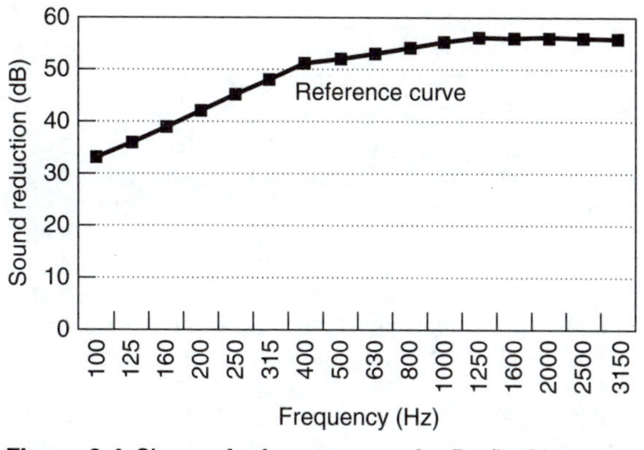

Figure 6.4 Shape of reference curve for R_W (in this case equal to 52)

Figure 6.4. Table 6.2 can be used to represent a whole family of different reference curves, all of exactly the same spectrum shape, by adding (or subtracting) a constant number of decibels to each row in the table. The particular reference curve is denoted by the value of the 500 Hz band, so that the figures shown in Table 6.2 describe the 52 dB R_w reference curve. Adding 5 dB to each of the values in Table 6.2 would, for example, produce the 57 dB R_w reference curve.

The measured values of R for the partition are compared with a selected reference curve by calculating the difference in dB between the measured data and the reference curve values in each band. Bearing in mind that the higher the number for R the better the sound insulation provided by the partition, the difference in those bands where the measured value is less than that of the reference curve (i.e. where the measured sound insulation performance has failed to meet the standard set by the reference curve), is called an adverse (or unfavourable) deviation. The unfavourable deviations for all one third octave bands is summed to produce the aggregate adverse deviation (AAD), or averaged to provide the mean unfavourable deviation (MUD). The criterion given in ISO 717-1 for matching the measured values of the data to a particular reference curve is that the AAD shall be as close as possible to, but shall not exceed, 32 dB or that the MUD shall be as close as possible to, but shall not exceed, 2.0 dB. The measured test data is compared to various reference curves, i.e. the selected reference curve is altered until this requirement is met. The R_w is then the value for the reference curve at 500 Hz. Note that any favourable deviations, i.e. when the measured value is above the value for the reference curve, are not taken into consideration in this determination.

The method for manually determining the R_w value is illustrated in Table 6.3 and Figure 6.5. In Table 6.3 the test data is presented in column 3, and is first compared with

Table 6.3 Illustrating the calculation of single figure rating values for airborne sound insulation

			First try		Second try		Third try	
1/3 OB (Hz)	52 dB Ref values	Test values	Ref curve 32	AD	Ref curve 30	AD	Ref curve	AD
100	33	20	13	Pass	11	Pass		
125	36	16	16	Pass	13	Pass		
160	39	18	19	1	17	Pass		
200	42	23	22	Pass	20	Pass		
250	45	22	25	3	23	1		
315	48	23	28	5	26	3		
400	51	25	31	6	29	4		
500	52	27	32	5	30	3		
630	53	28	33	5	31	3		
800	54	31	34	3	32	1		
1000	55	32	35	3	33	1		
1250	56	32	36	4	34	2		
1600	56	33	36	3	34	1		
2000	56	33	36	3	34	1		
2500	56	31	36	5	34	3		
3150	56	26	36	10	34	8		
				56 dB		31 dB		

Result: $D_{nT,w} = 30$ dB

Aggregate adverse deviation (AAD) = 31 dB, AD of 10 dB at 3150 Hz

Mean unfavourable deviation (MUD) = 31/16 = 1.9 dB.

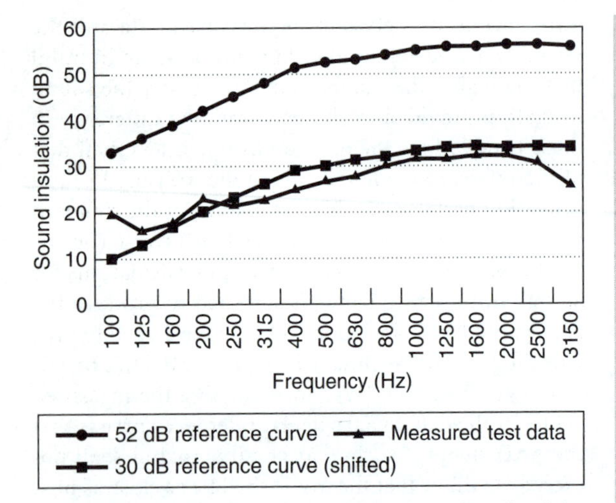

Figure 6.5 Method for determining R_w based on the data in Table 6.3

a reference curve of 30 dB, resulting in an AAD of 56 dB which is too high (i.e. greater than 32 dB). The next attempt is a selection of the 30 dB reference curve which gives an AAD of 31 dB. Increasing the value of the reference curve to 31 dB will increase the AAD to more than 32 dB, so that the R_w of the building element under test is 30 dB, and a third attempt is not needed.

Octave band test data

When only the data at octave bands is available the R_w can be determined in a similar manner from the position of the octave band reference curve so that the total of the unfavourable deviations does not exceed 10 dB. The reference curve only applies for very general types of noises and does not necessarily provide a good rank ordering of the performance of partitions for other types of spectra.

Exercise

Practice Example Questions on $D_{nT,w}/R_w$ values

Determine the single figure weighted value for the following sets of test results.

For examples 1 to 5 the single figure values are given. For example 6 the D_{nT} values have first to be calculated from the one third octave band values of L_1, L_2 and T and then the $D_{nT,w}$ value calculated.

	D_{nT}/R_w values					L_1	L_2	T
	1	2	3	4	5		6	
100	57	41	28	17	31	88	63	0.45
125	52	40	27	20	30	87	61	0.53
160	51	37	23	21	32	85	60	0.59
200	54	34	28	24	32	86	62	0.53
250	57	37	33	25	27	84	59	0.49
315	56	41	35	27	28	82	57	0.45
400	57	45	38	30	28	84	54	0.40
500	58	48	43	34	34	85	51	0.39
630	54	47	45	39	32	87	49	0.38
800	57	47	50	40	32	88	48	0.40
1000	59	51	52	43	36	89	46	0.42
1250	64	52	59	49	39	85	43	0.45
1600	52	52	60	52	38	86	42	0.40
2000	63	52	62	56	37	94	39	0.36
2500	63	55	63	59	38	87	38	0.35
3150	65	58	65	62	37	95	36	0.32

Answers

1. 59 dB, −26 dB
2. 49 dB, −29 dB
3. 44 dB, −30 dB
4. 37 dB, −28 dB
5. 35 dB, −28 dB
6. 37 dB, −29.3 dB (also see below)

1/3 OB (Hz)	D_{nT}	1/3 OB (Hz)	D_{nT}
100	24.5	630	36.8
125	26.3	800	39.0
160	25.7	1000	39.2
200	24.3	1250	41.4
250	24.9	1600	43.0
315	24.5	2000	53.6
400	29.0	2500	47.5
500	32.9	3150	47.1

Single figure rating values for impact sound insulation

A form of single figure index L_{nw}, the normalized impact sound level, is derived using the standard shaped curve as defined in ISO 717-2:2004 *Rating of sound insulation in buildings and of building elements – Part 2: Impact sound insulation*, as given in the Table 6.4 and Figure 6.6.

The single number is found by comparing the difference when the measured value for the one third octave band data is above the value for the standard curve and

Table 6.4 Reference values for the determination of L_{nw} (from ISO 717-2:2004)

Band centre frequency	Reference value 1/3 octave	Reference value Octave
100	62	
125	62	67
160	62	
200	62	
250	62	67
315	62	
400	61	
500	60	65
630	59	
800	58	
1000	57	62
1250	54	
1600	51	
2000	48	53
2500	45	
3150	42	

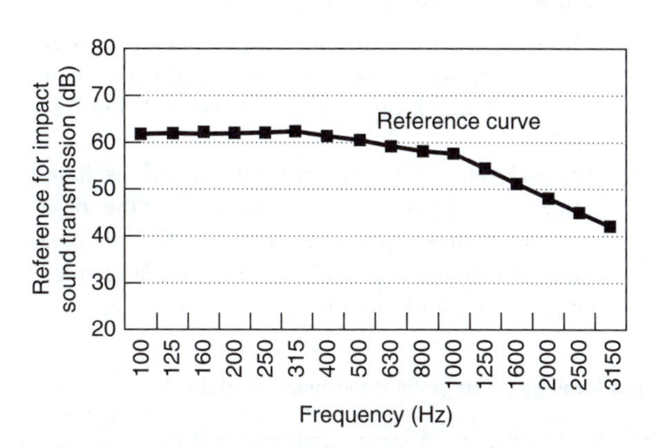

Figure 6.6 Shape of the reference curve for L_{nw} (= 60 for one third octave band data in this case)

moving the curve until the sum of these deficiencies does not exceed 32 dB. The L_{nw} is found from the value for 500 Hz once this position has been found (see Figure 6.7).

Impact noise transmission can be minimized by increasing the vibration isolation in a system. For floors this can be achieved by use of resilient layers on the upper surface (like carpet) or within the floor/ceiling construction. International Standard ISO 140 Part 11 describes laboratory based methods for measuring the reduction of transmitted impact noise by floor coverings on lightweight reference floors.

Dissatisfaction with sound insulation standards

It became apparent during the 1980s and 1990s that the airborne sound insulation in new dwellings – mainly based on 'deemed to satisfy' constructions – failed to meet the minimum performance standards set out in the Building Regulations. Over the same period, sound output from hi-fi systems (particularly at lower frequencies) increased, as did the expectations of quiet by the occupants of dwellings. The poor compliance and low satisfaction with existing sound insulation standards resulted in a review of the Regulations and, after a period of consultation, they were amended. These amended Building Regulations 2000 came into force on 1 July 2003 with changes to the method of rating the airborne sound insulation of separating walls and floors. A brief review of these Regulations is given later in this chapter.

Spectrum adaptation terms *C* and C_{tr}

These were developed, for airborne sound insulation only, in order to provide a link between the single frequency weighted values, such as $D_{nT,w}$ and R_w and the level difference, D_A, in dBA, between source and receiving room.

The aim was to link the single figure weighted sound insulation value (e.g. R_w) with the dBA level difference (D_A) using a correction factor, *C*:

$$D_A = R_w + C$$

It is possible to do this, i.e. to find a correction factor (in dB) but the factor will be different for different shaped source noise spectra.

The method for determining the spectral adaptation terms is specified in Annex C of ISO 717-1.

There are two terms. The first, *C*, is intended for living activity noise, children playing, railway traffic at medium and high speed, highway (> 80 km/h) road traffic, and jet aircraft at short distances (A-weighted pink noise).

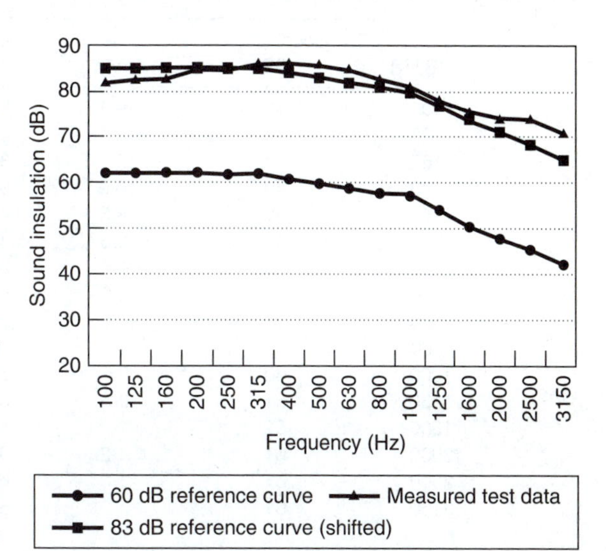

Figure 6.7 The method for determining L_{nw}

The second, C_{tr}, is intended for lower frequency noise such as urban road traffic, low speed railway traffic, aircraft at large distances, pop music, and factories which emit low to medium frequency noise (A-weighted urban road traffic noise).

The airborne sound insulation rating of a partition is given by the sum of the values of the single figure index and the appropriate spectrum adaptation term. The correct presentation of the result should be, for example:

$$R_w(C; C_{tr}) = 42(-6; -4)\text{ dB} \quad \text{or}$$

$$D_{nT,w}(C; C_{tr}) = 42(-6; -4)\text{ dB}$$

The Building Regulations 2000 – Approved Document E *Resistance to the Passage of Sound* (2003 edition) specifies the use of the spectrum adaptation term C_{tr}. This gives more emphasis to sound insulation at low frequencies.

The performance of separating walls and floors for airborne sound insulation is specified by the single figure rating $D_{nT,w} + C_{tr}$.

The method for determining the spectral adaptation terms, *C* and C_{tr}

Two different source spectra have been defined in ISO 717, spectrum 1 and spectrum 2, and these lead to two different corrections: C (for spectrum 1) and C_{tr} (for spectrum 2).

The two spectra used in the ISO standard have negative values (see Table 6.5, column 6). This is because they have been normalized, i.e. adjusted so that their total overall A-weighted value (obtained by combining all bands), is 0 dBA. This normalizing process is illustrated in Table 6.5 for spectrum 1, which corresponds to A-weighted pink noise. For pink noise the level in each one third octave band is constant. In column 2 in Table 6.5 the values correspond to A-weighted levels of pink noise for which the unweighted level in each band is 60 dB. Columns 3, 4 and 5 show the process of combining these band levels to produce the total level, which in this case is 70 dBA. In column 6 the normalized band levels are shown, obtained by subtracting 70 dB from the original (column 2) spectral values. The result of combining the (negative) values in column 6 will be 0 dBA

A similar approach leads to the normalized values for spectrum 2 which are based on a typical A-weighted spectrum for traffic noise.

Calculation method for *C* and C_{tr}

1. The single figure rating R_w, R'_w, D_{nw} or $D_{nT,w}$ is determined from the measured one third octave band values (R, R', D_n or D_{nT}) as described below (see Table 6.6).
2. For each one third octave band the sound insulation value (R, R', D_n or D_{nT}) is subtracted from the (negative) normalized spectrum value. For the '*i*th' band this will produce a level of $(L_{i1} - R_i)$ for spectrum 1 or $(L_{i2} - R_i)$ for spectrum 2.

Table 6.5 Showing how the normalized spectrum 1 values (used in the determination of the spectrum adaptation term, *C*) are obtained

1	2	3	4	5	6
OB(Hz)	Spectrum 1 OB	A-wtg	A-wtd OB	$10^{L_A/10}$	Normalized spectrum 1
100	41	−19.1	21.9	154.9	−29
125	44	−16.1	27.9	616.6	−26
160	47	−13.4	33.6	2290.9	−23
200	49	−10.9	38.1	6456.5	−21
250	51	−8.6	42.4	17378.0	−19
315	53	−6.6	46.4	43651.6	−17
400	55	−4.8	50.2	104712.9	−15
500	57	−3.2	53.8	239883.3	−13
630	58	−1.9	56.1	407380.3	−12
800	59	−0.8	58.2	660693.4	−11
1000	60	0	60	1000000.0	−10
1250	61	0.6	61.6	1445439.8	−9
1600	61	1	62	1584893.2	−9
2000	61	1.2	62.2	1659586.9	−9
2500	61	1.3	62.3	1698243.7	−9
3150	61	1.2	62.2	1659586.9	−9
			Sum =	10530968.8	
				L_A = 70 dB	L_A = 0 dB

Table 6.6 The calculation of both spectrum adaptation terms

1	2	3	4	5	6	7	8	9	10
1/3 OB Hz	R_i dB	52 dB Ref values shifted by −22 dB	Unfavourable deviation	Spectrum No.1 (L_{i1}) dB	$X_{i1} = (L_{i1} - R_i)$ (5 − 2)	$10^{X_{i1}/10}$ $10^{(6)/10}$ dB × 10^{-5}	Spectrum No. 2 (L_{i2}) dB	$X_{i2} = (L_{i2} - R_i)$ (8 − 2)	$10^{X_{i2}/10}$ $10^{(9)/10}$ dB × 10^{-5}
100	20.4	11		−29	−49.4	1.15	−20	−40.4	9.12
125	16.3	14		−26	−42.3	5.89	−20	−36.3	23.44
160	17.7	17		−23	−40.7	8.51	−18	−35.7	26.92
200	22.6	20		−21	−43.6	4.37	−16	−38.6	13.80
250	22.4	23	0.6	−19	−41.4	7.24	−15	−37.4	18.20
315	22.7	26	3.3	−17	−39.5	10.72	−14	−36.7	21.38
400	24.8	29	4.2	−15	−39.8	10.47	−13	−37.8	16.60
500	26.6	30	3.4	−13	−39.6	10.96	−12	−38.6	13.80
630	28.0	31	3.0	−12	−40.0	10.0	−11	−39.0	12.59
800	30.5	32	1.5	−11	−41.5	7.08	−9	−39.5	11.22
1000	31.8	33	1.2	−10	−41.8	6.61	−8	−39.8	10.47
1250	32.5	34	1.5	−9	−41.5	7.08	−9	−41.5	7.08
1600	33.4	34	0.6	−9	−42.4	5.75	−10	−43.4	4.57
2000	33.0	34	1.0	−9	−42.0	6.31	−11	−44.0	3.98
2500	31.0	34	3.0	−9	−40.0	10.00	−13	−44.0	3.98
3150	25.5	34	8.5	−9	−34.5	35.48	−15	−40.5	8.91
	Sum = 31.8 < 32.0 R_w = 52 − 22 dB = 30 dB			Sum = 147.62 × 10^{-5} −10log(147.62 × 10^{-5}) = 28.3 C = 28 − 30 dB = −2 dB			Sum = 206.06 × 10^{-5} −10log(206.06 × 10^{-5}) = 26.9 C_{tr} = 27 − 30 dB = −3 dB		

Source: Taken from Table C.1 from Annex C of EN 717-1:1990, for measurements in the specified frequency range 100 Hz to 3150 Hz

3. These differences are then summed logarithmically, to obtain either X_{A1} or X_{A2}.

$$X_{A1} = 10\log\Sigma(10^{(L_{i1}-R_i)/10})$$

$$X_{A2} = 10\log\Sigma(10^{(L_{i2}-R_i)/10})$$

4. The correction C or C_{tr} is found by subtracting the weighted sound insulation value (R_w, R'_w, D_{nw} or $D_{nT,w}$) from either X_{A1} or X_{A2}.

$$C = X_{A1} - D_{nT,w} \quad \text{and} \quad C_{tr} = X_{A2} - D_{nT,w}$$

This process is illustrated in Table 6.6, taken from ISO 717-1, for sound reduction index values R (but could equally well apply to R', D_n and D_{nT} values).

The first stage is to calculate the weighted value as shown in columns 2, 3 and 4, resulting in a value of R_w = 30 dB.

The second stage is shown in columns 5 and 6 for spectrum 1 and in columns 8 and 9 for spectrum 2.

The third and fourth stages are shown in columns 7 and at the foot of columns 5, 6 and 7 for spectrum 1 (and in column 10 at the foot of columns 8, 9 and 10 for spectrum 2).

The result should be stated as:

$$R_w(C; C_{tr}) = 30(-2; -3) \text{ dB}$$

Exercise

Repeat the previous exercise (Practice Example Questions on $D_{nT,w}/R_w$ values) but this time calculate the C and C_{tr} values.

Answers

1. 59(−1, −2) dB
2. 49(−1, −4) dB
3. 44(−2, −7) dB
4. 37(−1, −6) dB
5. 35(−1, −2) dB
6. 37(−2, −5) dB

Uses and limitations of single figure rating values

Single figure R_w values are sometimes used to give a quick and very approximate estimate of sound transmission in terms of overall dBA levels. For example, if the noise level outside a building is 75 dBA and the R_w value of the façade is 25 dB then, at a rough estimate, the level inside the building arising from the ingress of external noise will be about 50 dBA (i.e. 75 − 25). Conversely, if the external noise level and the maximum allowable level of internal noise level are known then a rough estimate of the required R_w value of the façade may be obtained. BS 8233 states that 'the use of R_w values in this way is likely to underestimate the level inside by up to 5 dBA. Where the estimate is within 5 dBA of the limit, a more rigorous calculation should be carried out using octave bands.'

As discussed earlier, the use of an '$R + C_{tr}$' value might reduce the error, but for the most accurate sound insulation design, octave band values of *R* should always be used in calculations.

The difference between lab and field test data and their use

Walls and floors are always likely to give a better sound insulation performance when tested under laboratory conditions than when tested in the field, because of the possibility of flanking transmission and of imperfect workmanship and detailing during installation.

For this reason, acoustic designers will usually de-rate laboratory test values of a sound reduction index by *X* dB when performing design calculations. The value of *X* will depend upon the particular situation and on the designer's experience, but could be up to 8 dB. BB 93 suggests a value of 5 dB.

6.6 Factors affecting sound reduction index of partitions

The sound transmission process

When sound waves strike a partition such as a wall or floor the pressure fluctuations of the wave cause the partition to vibrate (unless the partition is perfectly rigid, when, by definition there will be no movement, i.e. no vibration). On the other side of the partition the vibration induced by the sound wave will cause the partition to radiate sound waves. This is the sound transmission process.

It follows that factors and situations which lead to an increase in the level of vibration of the panel caused by the sound waves on the source side will cause it to radiate more sound waves into the receiving space, and so lead to a reduction in sound reduction index or transmission loss of the partition. Thus panels which are efficient sound radiators are poor sound insulators.

The three physical properties which are relevant to the vibration of any structure are the mass and stiffness of the structure and the amount of damping in the structure, i.e. the loss of vibrational energy which occurs as a result of frictional processes occurring within the structure.

It follows that the sound reduction index of a single leaf partition depends in the mass, stiffness and damping of the partition, as well as on the frequency of the sound waves causing the panel to vibrate.

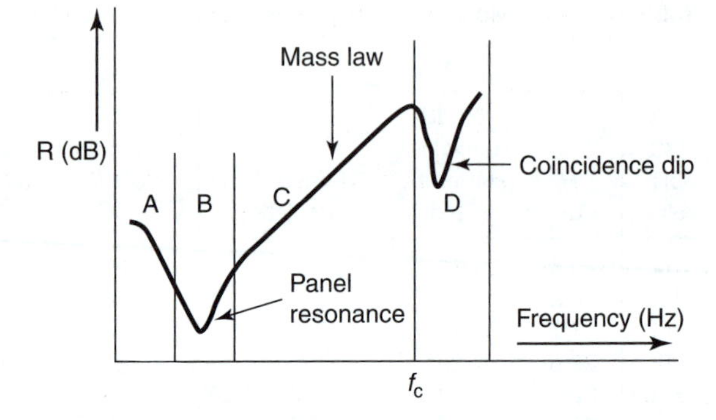

Figure 6.8 Typical variation of sound reduction index with frequency for a single leaf panel

In the case of multiple leaf partitions, i.e. those with two or more partitions separated by a cavity or air gap, such as cavity walls or double glazing, another factor is the degree of physical separation, or isolation, between the leaves.

There is one other factor which is relevant to the radiation of sound by a vibrating panel, and that is its radiation efficiency. This will be discussed later in the chapters on vibration and on noise control.

Single leaf partitions

A sketch illustrating how the sound reduction index (R in dB) might vary with frequency for a typical partition is shown in Figure 6.8. It can be divided into different frequency regions where the different factors, mass, stiffness and damping, each have the main influence.

In region A, at low frequencies before the main 'panel resonances' of the partition occur, stiffness plays a major part in determining the value of *R* so that in this region making the partition stiffer would increase the value of *R*.

In region B, where the panel resonances occur, the damping has a major influence. In this region, increasing the damping would reduce the amplitude of vibration at resonance and thus increase the value of *R*.

At frequencies above the panel resonances, region C, the mass of the panel is the most important factor, and increasing the mass of the partition would increase the sound reduction index. At even higher frequencies, in region D, another type of resonance occurs, arising from flexural or bending waves travelling in the partition. The effect of these waves on the sound reduction index is called the **coincidence effect;** the drop in the value of *R* which occurs in this region (the coincidence region) is called the **coincidence dip**, which starts at frequencies above a **critical frequency**, f_c. The effects of damping will be important in this region, in limiting resonant vibration amplitudes and minimizing the depth of the coincidence dip.

There are a number of schemes for the prediction of the sound reduction index based on the partition behaviour in these different regions, the simplest of which is based on the mass law.

The mass law approximation

The mass law assumes that the sound reduction index, R, depends only on the mass of the partition. It therefore ignores the influence of stiffness and damping. According to the mass law, if we were to construct partitions of the same area from different materials such as glass, lead sheet, wood, concrete or steel and adjust their thicknesses so that all had the same surface density (kg/m^2), then they would all have the same sound insulation properties.

This is obviously not completely true, but nevertheless many partitions do approximately obey such a law over a limited frequency range, as indicated in Figure 6.8.

According to the mass law the sound reduction index can be predicted simply from a knowledge of the surface density of a partition and the sound frequency:

$$R = 20\log(fm) - 48$$

This equation is in part based on theoretical considerations (Newton's second law of motion, force = mass × acceleration). Inspection of the equation indicates that R increases by 6 dB when either the superficial mass or the frequency is doubled. For example, taking a single sheet of plasterboard of 7 kg/m^2 superficial mass, at 500 Hz R = 23 dB.

Superficial mass or surface density

This is the mass per unit area, in kg/m^2, of a partition. It is related to the density, ρ, which is measured in kg/m^3 and partition thickness, t (in m) by:

$$m = \rho t$$

Exercise

If a certain type of glass has a density of 2500 kg/m^3 what is the surface density of a sheet of glass 6 mm thick?

Answer

$m = 2500 \times 0.006 = 15$ kg/m^2

It has been found in practice that the rate of increase of R with mass is often closer to 5 than 6 dB per doubling of mass. Such evidence arises from analyses of empirically based graphs plotting the variation of R against surface density. A good example is the graph of R_w against m given in BB93 (in Appendix Figure 3.5).

Exercise

Inspect Figure 3.5 of BB93. Note that the horizontal axis is plotted on a log scale. Estimate the rate of increase of R with mass in terms of X dB per doubling of mass, and attempt to derive an equation to describe the 'line of best fit' in Figure 3.5. (In case you do not have ready access to Figure 3.5 here are three points read off from the line of best fit: when m = 10, R = 27 dB; when m = 20, R = 32 dB; when m = 100, R = 42 dB (approximate values only).)

Answer

We are looking for a relationship such as: $R = A + B10\log(m)$. Where A and B are constants to be determined from the graph. B represents the slope of the graph.

Taking the two points a decade apart we have:

$$27 = A + B\log 10 \quad \text{and} \quad 42 = A + B\log 100$$

Eliminating A from these two equations leads to B = 15 dB per decade of mass, and A can then be found to be 12. Therefore the equation is: $R = 12 + 15\log(m)$.

The rate of increase of 15 dB per tenfold increase of mass corresponds to a rate of 4.5 dB per doubling of mass (from $4.5 = 15 \times (\log 2/\log 10)$), as compared to the rate of 6 dB per doubling or 20 dB per tenfold increase predicted by the earlier theoretically based equation.

The preferred use of test data for *R* values

It must be emphasized that mass law, and similar formulae, are approximations, and to be used as a last resort to obtain an approximate value for R when data from tests carried out in accordance with BS EN ISO 140 Part 3 is not available.

Stiffness and panel resonance effects

Although the mass law may be approximately obeyed by many materials over a certain frequency range the neglect of the influence of damping and stiffness leads to discrepancies at higher and lower frequencies as shown in Figure 6.8. At low frequencies there are panel resonances and at high frequencies above the mass law region flexural waves in the panel also cause resonances, an effect called the coincidence effect which occurs above a certain critical frequency.

Two-dimensional standing waves in panels (and membranes) leading to panel resonances were discussed in Chapter 4. For these frequencies the partition will radiate very efficiently on the receiver room side and the effect is a degradation in the sound reduction, i.e. dips occur in the sound reduction index curve. The physical properties of the materials and the dimensions of most panels in buildings mean that these panel resonances usually occur in the low frequencies.

At frequencies below the panel resonances the sound transmission is governed by the stiffness of the panel. This is the case in very thin partitions such as the metalwork of an air conditioning duct which is known to transmit low frequency sound readily.

Exercise

If stiffness is the main consideration in low frequency sound transmission through ducts, what difference do you think that the shape of the duct (e.g. circular or rectangular) might have?

Answer

Circular ducts are stiffer than rectangular ones of the same thickness and so their sound reduction index at low frequencies is higher, even though at high frequencies, where mass becomes the main factor, there is no difference.

Coincidence or bending wave effects

These have a very important effect on the sound transmission of many common building materials. A non-rigid, i.e. flexible, partition will support bending or flexural waves. These will be excited by sound waves moving along the panel in the source room (Figure 6.9).

The speed of sound in air is independent of frequency but the speed of bending waves in the panel increases with frequency. The two wave speeds become equal at a frequency called the coincidence frequency which depends on the angle of incidence at which the airborne sound waves meet the partition. Below the coincidence frequency the panel wave velocity is lower than the sound speed (and these panel waves are sometimes called 'acoustically slow panel waves') and in the circumstances the sound waves are not efficient at exciting the panel waves, but become very efficient at doing so at and above the coincidence frequency.

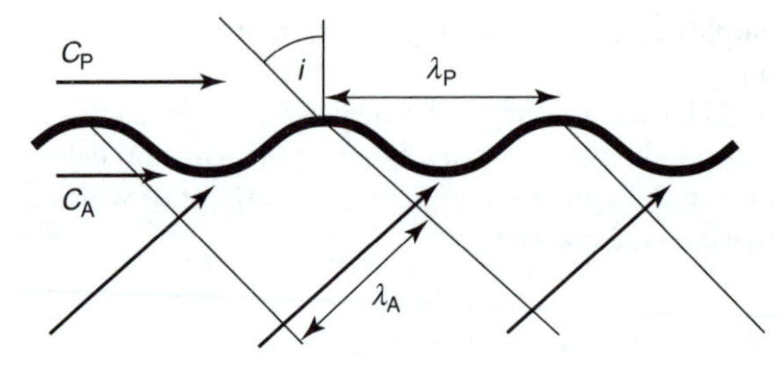

Figure 6.9 The coincidence effect

A sound wave which strikes a partition at grazing incidence (i.e. 90° angle of incidence) will travel along the surface of the partition at a speed c (i.e. typically 340 m/s), but sound at an angle of incidence θ will be $c/\cos\theta$).

The **critical frequency** is the lowest frequency for which coincidence occurs for sound at grazing incidence. Note also that at coincidence the wavelength of the incident sound wave equals the bending wavelength of the waves in the panel and resonant excitation of the panel occurs. This results in large radiation of sound into the receiving room.

Coincidence between sound waves in air (A), and in panel (P): $\lambda_P = \lambda_A/\sin i$.

At other angles of incidence coincidence, efficient coupling between the sound waves and the panel occurs at higher frequencies. The overall result is a significant degradation in performance from mass law expectations over a range of about 1 octave above the critical frequency. This drop in the sound reduction is referred to as the **coincidence dip**.

For heavy partitions such as masonry walls the critical frequency is quite low, probably in the region where elasticity and inertial effects cancel. For plywood, plasterboard, window glazing etc. the critical frequency can be in the 1 to 4 kHz region and thus quite important for the design of partitions with high noise reduction.

By way of example, for concrete and steel partitions of identical superficial masses of 55 kg/m^2, the critical frequencies are 600 Hz and 1.8 kHz respectively.

The relationship between critical frequency and panel stiffness

The speed of bending waves in panels depends on their bending stiffness as well as on frequency, so that at any given frequency the wave speed is higher for a stiffer panel than for one with lower stiffness. This in turn means that for a stiffer panel the frequency at which panel waves reach the speed of sound in air is lower than for a less stiff panel, i.e. the critical frequency reduces as panel stiffness increases.

Table 6.7 Critical frequencies for some panels of different materials and thickness

Material	$f_C \times t$	Density (kg/m^3)
Plywood	20	600
Glass	12.7	2300
Plasterboard	45	750
Concrete	19	2600
Steel	12.4	7700
Lead	53.8	11,300
Brick	22.1	1900
Chipboard	24.1	650

The relationship between critical frequency, f_C, and panel thickness, t, for some common materials is shown in Table 6.7 (approximate values only).

Exercise

What is the critical frequency for 6 mm thick glass sheet?

Answer

$f_C = 12.7/0.006 = 2116.7$ Hz

And so we would expect 6 mm glass sheet to have a coincidence dip in the 2000 Hz octave band.

Multiple leaf partitions

Suppose you want to increase the sound insulation performance of a single leaf partition of modest sound insulation performance, say $R = 20$ dB. According to the mass law, doubling the thickness will achieve a 5 or 6 dB increase, i.e. the R value will increase to 25 or 26 dB. Another doubling of mass, i.e. to four times the original, may achieve up to 30 or 32 dB. It can be seen that this incremental approach leads to a situation of diminishing returns.

An alternative approach would be to use two leaves separated by an air gap or cavity, as in the case of double glazing or a cavity wall construction. A simple minded view might expect a doubling of the sound insulation, i.e. 20 dB from each leaf, so we might expect a sound reduction index of 40 dB from the double leaf construction. This is unlikely to be achieved. Can you suggest reasons why it is unlikely to be achieved? And remedies to improve the sound insulation?

One reason is because there will be a structural connection between the two leaves via the building structure (e.g. the window frame) allowing structure-borne sound to create an alternative flanking transmission path which bypasses direct transmission through the cavity.

Another reason is that reflection of sound inside the cavity between the two leaves will lead to a build-up of reverberant sound inside the cavity, which will result in an increase in sound transmission through the second leaf into the receiving space.

A third reason is that acoustic coupling between the two leaves will occur via the stiffness of the air in the cavity. In effect the structure behaves like a mass–spring–mass vibrating system with each of the leaves acting as a mass connected to the other leaf by the air in the cavity acting as the spring. Such a system will have a natural frequency at which resonance will occur, when sound will be transmitted efficiently between the leaves resulting in a poor sound insulation performance and a dip in the sound reduction index value.

The natural frequency, f_0, is given by the equation:

$$f_0 = 60\sqrt{\{(m_1 + m_2)/m_1 m_2 d)\}}$$

where m_1 and m_2 are the superficial masses (kg/m^2) of the leaves and d the separation (in m).

Exercise

Estimate the natural frequency for double glazing in which each leaf of 6 mm glass is separated by an 8 mm air gap (known as 6-8-6 glazing).

Answer

From an earlier example, assuming a density of 2500 kg/m^3 for glass gives $m_1 = m_2 = 15$ kg/m^2 from which:

$$f_0 = 60\sqrt{\{(15 + 15)/(15 \times 15 \times 0.008)\}} = 245 \text{ Hz}$$

which, unfortunately is close to the peak in the typical spectrum of road traffic noise.

The remedies are (as illustrated in Figure 6.10):

- to attempt to isolate the two leaves from the building structure and hence from each other, using resilient materials
- to put sound absorbing material into the cavity to reduce reflection and minimize the build-up of reverberant sound, and
- to widen the cavity gap so that the natural frequency is no longer at a subjectively sensitive frequency range.

For such apparently simple systems such as cavity walls and double windows, the full acoustic analysis is surprisingly complex. It is true however that the wider the air gap, the weaker is the coupling. Masonry cavity walls show a significant decrement in performance, for example, when the cavity is reduced from 75 mm to 50 mm.

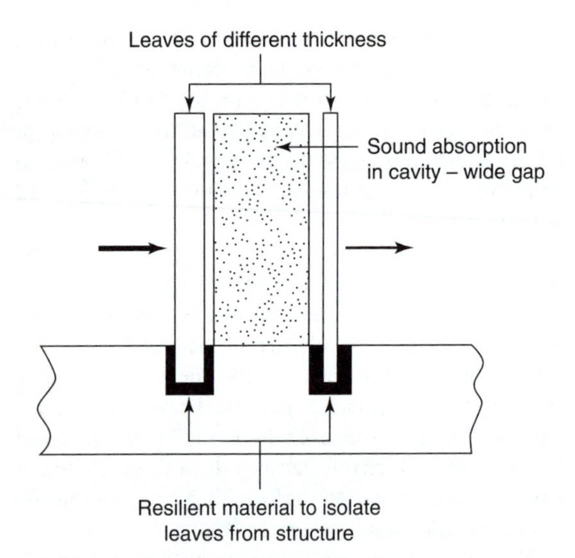

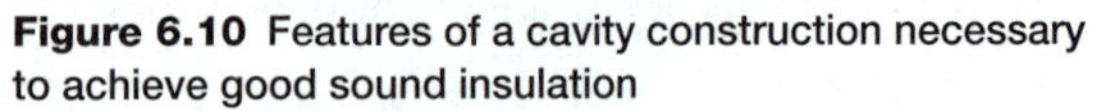

Figure 6.10 Features of a cavity construction necessary to achieve good sound insulation

There are, in addition, cavity resonances in each of the three dimensions appearing at a variety of frequency ranges. Each effect tends to degrade the overall performance. For double glazing, lining the reveals of the cavity with sound absorption pieces of absorbing ceiling tile is a popular way of limiting the effects of resonances in the height and width directions. If possible it is always best to arrange for a laboratory test of the sound reduction index to be carried out rather than to rely on predicted values.

In practice it is advisable to have the leaves of the partition made of different thickness and even mount them in a non-parallel manner. A popular glazing configuration for modestly high external noise levels would be 10-12-6 which means 10 mm and 6 mm glasses with a 12 mm air space. A 10.4-12-6 configuration would have the 10 mm glass laminated with a thin sheet of transparent plastic in the middle. This is allegedly better because the plastic sheet helps to damp out resonances in the glass itself.

Ideally the two leaves should be perfectly isolated from each other, both structurally and acoustically. In practice this is impossible, but well designed, detailed and constructed double leaf partitions should achieve a considerable improvement on single leaf constructions for the same total mass per unit area. In the context of the earlier example, using a double leaf construction where each leaf has a sound reduction index of 20 dB is unlikely to achieve an R value of 40 dB, but it should be possible to significantly improve on the 25 or 26 dB achievable using a single leaf construction and the mass law solution.

6.7 Summary of Building Regulations 2000 Part E (2003 edition)

The Approved Document came into force on 1 July 2003 and deals with the requirements of Part E of Schedule 1 to the Building Regulations (as amended by S.I.2002/2871).

The Document is 76 pages long with many pages of information about different types of constructions for walls and floors. Hence these notes provide just a very brief overview.

There are *four* requirements, E1, E2, E3 and E4.

Protection against sound from other parts of the building and adjoining buildings

E1. Dwelling-houses, flats and rooms for residential purposes shall be designed and constructed in such a way that they provide reasonable resistance to sound from other parts of the same building and from adjoining buildings.

Protection against sound within a dwelling-house etc.

E2. Dwelling-houses, flats and rooms for residential purposes shall be designed and constructed in such a way that: (a) internal walls between a bedroom or a room containing a water closet, and other rooms; and (b) internal floors, provide reasonable resistance to sound.

Note: Requirement E2 does not apply to: (a) an internal wall which contains a door; (b) an internal wall which separates an en suite toilet from the associated bedroom; (c) existing walls and floors in a building which is subject to a material change of use.

Reverberation in the common internal parts of buildings containing flats or rooms for residential purposes

E3. The common internal parts of buildings which contain flats or rooms for residential purposes shall be designed and constructed in such a way as to prevent more reverberation around the common parts than is reasonable.

Note: Requirement E3 only applies to corridors, stairwells, hallways and entrance halls which give access to the flat or room for residential purposes.

Acoustic conditions in schools

E4. (1) Each room or other space in a school building shall be designed and constructed in such a way that it has the acoustic conditions and the insulation against disturbance by noise appropriate to its intended use.

(2) For the purposes of this Part, 'school' has the same meaning as in section 4 of the Education Act 1996; and 'school building' means any building forming a school or part of a school.

Section 0 Performance: Performance standards

These are shown in the following tables (Tables 1a, 1b and 2 of Approved Document E).

Note that performance is specified in terms of a minimum value $D_{nT,w} + C_{tr}$ for airborne sound insulation and in terms of a maximum value of $L'_{nT,w}$ for impact sound insulation.

Note the slight relaxation (by 2 dB) in the standard of sound insulation required for 'material change of use' constructions (i.e. conversions) as compared with those for 'new-build'. This is because in the case of conversions existing flanking walls and floors may necessarily remain unchanged and so flanking transmission may degrade performance as compared with new-build.

Section 1 Pre-completion testing

Separating walls and floors between dwellings (houses, flats and rooms for residential purposes) in **new constructions** have to be shown by sound insulations tests to comply with the performance standards of Table 1a and 1b of Part 1 of the Regulations (below), **unless** they have been constructed in accordance with **robust standard details**, and registered as such.

Robust standard details constructions cannot be used for conversions (or 'material change of use constructions')

Table 1a Dwelling-houses and flats – performance standards for separating walls, separating floors, and stairs that have a separating function

	Airborne sound insulation sound insulation $D_{nT,w} + C_{tr}$ dB (Minimum values)	Impact sound insulation $L'_{nT,w}$ dB (Maximum values)
Purpose built dwelling-houses and flats		
Walls	45	–
Floors and stairs	45	62
Dwelling-houses and flats formed by material change of use		
Walls	43	–
Floors and stairs	43	64

Source: Building Regulations Approved Document E, 2003

Table 1b Rooms for residential purposes – performance standards for separating walls, separating floors, and stairs that have a separating function

	Airborne sound insulation sound insulation $D_{nT,w} + C_{tr}$ dB (Minimum values)	Impact sound insulation $L'_{nT,w}$ dB (Maximum values)
Purpose built rooms for residential purposes		
Walls	43	–
Floors and stairs	45	62
Rooms for residential purposes formed by material change of use		
Walls	43	–
Floors and stairs	43	64

Source: Building Regulations Approved Document E, 2003

Table 2 Laboratory values for new internal walls and floors within: dwelling-houses, flats and rooms for residential purposes, whether purpose built or formed by material change of use

	Airborne sound insulation R_w dB (Minimum values)
Walls	40
Floors	40

Source: Building Regulations Approved Document E, 2003

because flanking walls and floors may have to remain in place (i.e. cannot be constructed according to robust standard detail requirements) hence sound insulation performance cannot be assured. Thus pre-completion testing must be carried out in all cases of separating walls and floors between dwellings (houses, flats and rooms for residential purposes) created by material change of use construction.

Section 1 includes details of the pre-completion testing requirements, i.e. how many samples of each type of interface must be tested.

Pre-completion testing is not required for demonstrating compliance with parts E2 and E3 of the Regulations (sound insulation of internal walls and floors, and reverberation of common parts of building). In these cases compliance must be demonstrated by appropriate design.

Section 2 Separating walls and associated flanking constructions for new buildings

2.1 This section gives examples of wall types which, if built correctly, should achieve the performance standards set out in Section 0: Performance – Table 1a.

2.2 The guidance in this section is not exhaustive and other designs, materials or products may be used to achieve the performance standards set out in Section 0: Performance – Table 1a.

Four main types of wall construction are described with appropriate junction requirements in each case (see Figure 6.11).

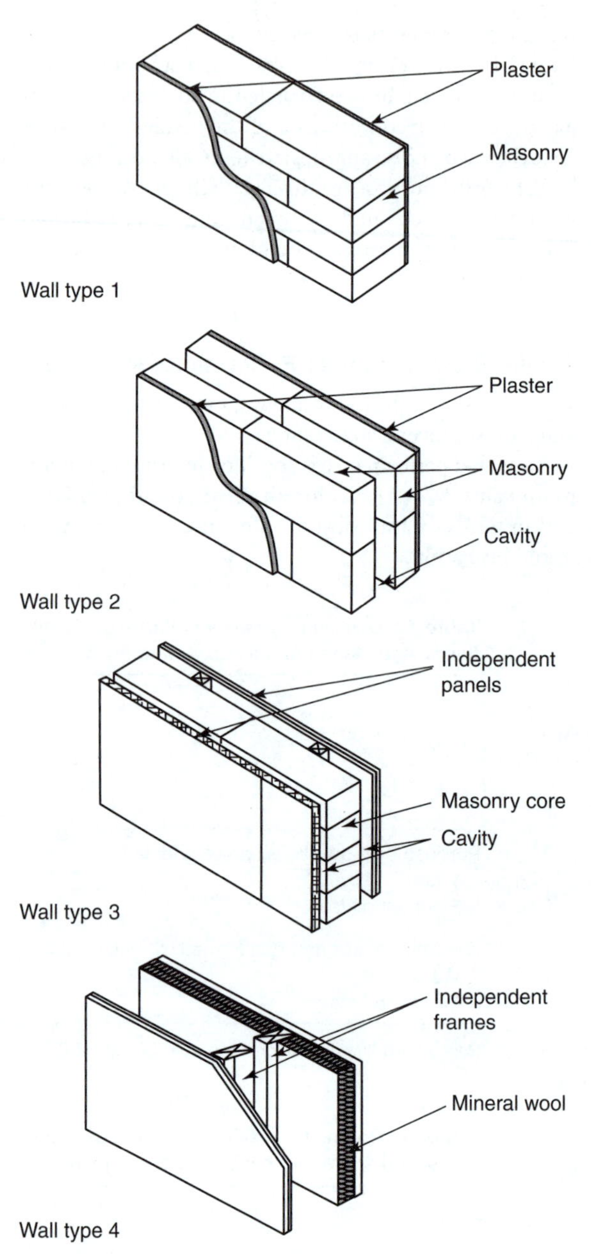

Figure 6.11 Types of separating wall

Wall type 1: Solid masonry

The resistance to the passage of airborne sound depends mainly on the mass per unit area of the wall.

Wall type 2: Cavity masonry

The resistance to the passage of airborne sound depends on the mass per unit area of the leaves and on the degree of isolation achieved. The isolation is affected by connections (such as wall ties and foundations) between the wall leaves and by the cavity width.

Wall type 3: Masonry between independent panels

The resistance to the passage of airborne sound depends partly on the type and mass per unit area of the core, and partly on the isolation and mass per unit area of the independent panels.

Wall type 4: Framed walls with absorbent material

The resistance to the passage of airborne sound depends on the mass per unit area of the leaves, the isolation of the frames, and the absorption in the cavity between the frames.

Section 3 Separating floors and associated flanking constructions for new buildings

3.1 This section gives examples of floor types which, if built correctly, should achieve the performance standards set out in Section 0: Performance – Table 1a.

3.2 The guidance in this section is not exhaustive and other designs, materials or products may be used to achieve the performance standards set out in Section 0: Performance – Table 1a.

Three main types of floor are described with appropriate junction requirements in each case (see Figure 6.12).

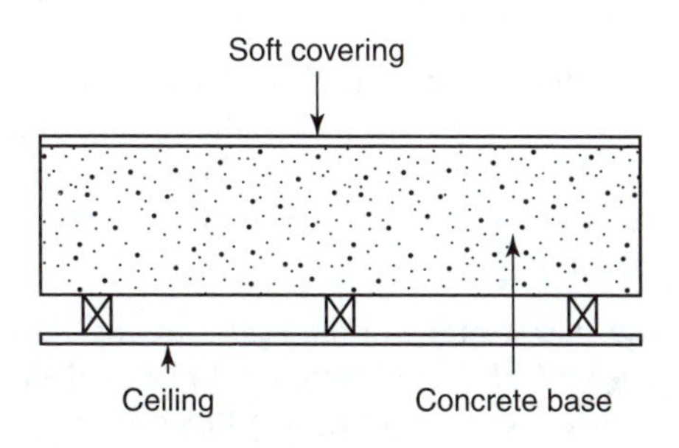

Floor type 1

Floating floor Floating layer Resilient layer

Ceiling

Concrete base

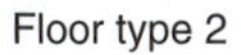

Floor type 2

Platform floor Floating layer Resilient layer

Timber frame base

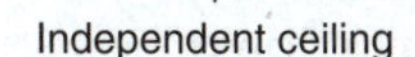

Independent ceiling

Floor type 3

Figure 6.12 Types of separating floor

Floor type 1: Concrete base with ceiling and soft floor covering

The resistance to the passage of airborne sound depends mainly on the mass per unit area of the concrete base and partly on the mass per unit area of the ceiling. The soft floor covering reduces impact sound at source.

Floor type 2: Concrete base with ceiling and floating floor

The resistance to the passage of airborne and impact depends on the mass per unit area of the concrete base, as well as the mass per unit area and isolation of the floating layer ceiling. The floating floor reduces impact at source.

Floor type 2 requires one of the floating floors described in this section. The description of floor type 2 contains a suffix (a), (b) or (c) which refers to the floating floor used.

Floor type 3: Timber frame base with ceiling and platform floor

The resistance to the passage of airborne and impact sound depends on the structural floor base and the isolation of the platform floor and the ceiling. The platform floor reduces impact sound at source.

Each floor type requires one of three types of ceiling treatments, A, B or C (A gives best performance, then B, then C) (Figure 6.13):

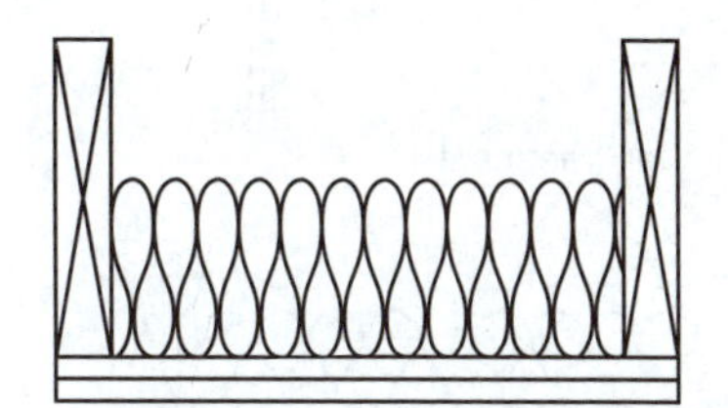

Ceiling treatment A (independent joists)

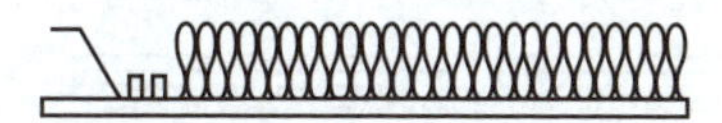

Ceiling treatment B (resilient bars)

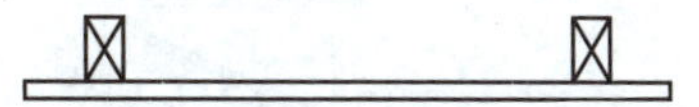

Ceiling treatment C (timber battens)

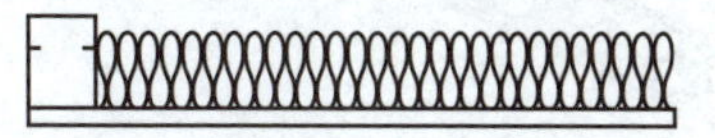

Ceiling treatment C (resilient channels)

Figure 6.13 Ceiling treatments A, B and C

- ceiling treatment A: independent ceiling with absorbent material
- ceiling treatment B: plasterboard on resilient bars with absorbent material
- ceiling treatment C: plasterboard on timber battens or proprietary resilient channels with absorbent material.

Details and specifications are also given for: soft floor coverings, floating and platform floors.

Section 4 Dwelling-houses and flats formed by material change of use

4.4 For situations where it is uncertain whether the existing construction achieves the performance standards set out in Section 0: Performance – Table 1a. This section describes one wall treatment, two floor treatments and one stair treatment as shown in Figure 6.14. These constructions can be used to increase the sound insulation.

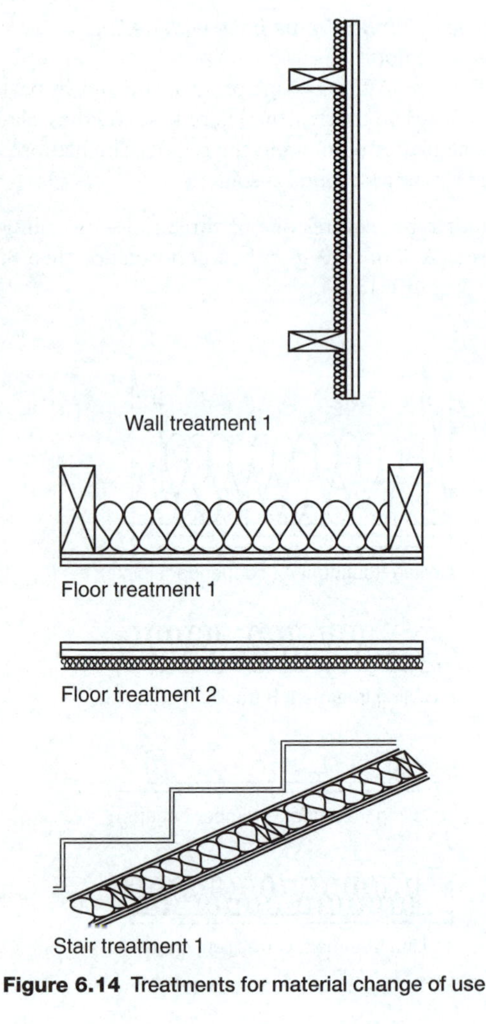

Figure 6.14 Treatments for material change of use

4.5 The guidance in this section is not exhaustive and other designs, materials or products may be used to achieve the performance standards set out in Section 0: Performance – Table 1a.

4.6 Wall treatment 1 ***Independent panel(s) with absorbent material*** The resistance to the passage of airborne sound depends on the form of existing construction, the mass of the independent panel(s), the isolation of the panel(s) and the absorbent material.

4.7 Floor treatment 1 ***Independent ceiling with absorbent material*** The resistance to the passage of airborne and impact sound depends on the combined mass of the existing floor and the independent ceiling, the absorbent material, the isolation of the independent ceiling and the airtightness of the whole construction.

4.8 Floor treatment 2 ***Platform floor with absorbent material*** The resistance to the passage of airborne and impact sound depends on the total mass of the floor, the effectiveness of the resilient layer and the absorbent material.

4.9 Stair treatment 1 ***Stair covering and independent ceiling with absorbent material*** To be used where a timber stair performs a separating function. The resistance to airborne sound depends mainly on the mass of the stair, the mass and isolation of any independent ceiling and the airtightness of any cupboard or enclosure under the stairs. The stair covering reduces impact sound at source.

The other sections of the Regulations (pages 60 to 76) are: Section 5: Internal walls and floors for new buildings; Section 6: Rooms for residential purposes; Section 7: Reverberation in the common internal parts of buildings containing flats and rooms for residential purposes; Section 8: Acoustic conditions in schools; Annex A: Method for calculating mass per unit area; Annex B: Procedures for sound insulation testing; Annex C: Glossary; and Annex D: References.

6.8 Flanking paths and structure-borne sound

Much of this chapter has been concerned with the direct sound transmission path through partitions acting as a separating wall or floor between two rooms. This final section briefly considers some aspects of indirect transmission paths, involving both airborne and structure-borne sound transmission.

At the beginning of this chapter Figure 6.1 illustrated multiple sound transmission paths via the flanking walls, floor and ceiling for transmission between two adjacent

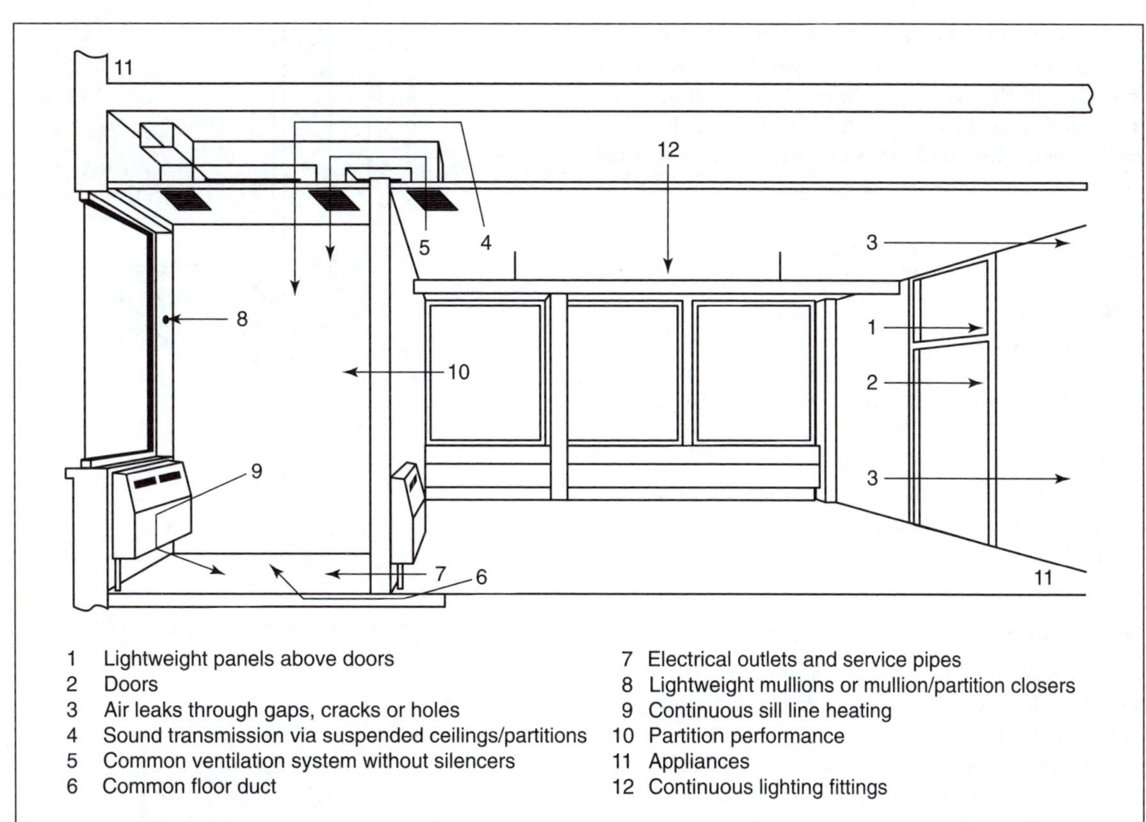

NOTE: This figure is derived from the British Gypsum White Book and is reproduced by kind permission of British Gypsum Ltd.

Figure 6.15 Sound transmission paths between rooms

From BS 8223

rooms (in addition to the direct path via the separating wall). Many additional paths can occur, between offices for example, via corridors, ventilators and ventilation ducts, suspended ceilings and sub-floors for example. Figure 6.15 illustrates some of the possible sound transmission paths between two offices.

Measurement of flanking transmission

Annex C of BS EN ISO 140-4:1998 describes methods for measuring the contribution of flanking paths to the transmission of sound between two rooms. The two techniques (a and b below) involve either blocking off the direct transmission by constructing an additional independent wall between the two rooms, or by measurement of vibration levels in the flanking walls, floor and ceiling, and then converting the vibration levels into sound pressure levels. Another technique which could be used involves the use of a sound intensity meter to estimate the flow of sound energy through each surface.

Technique a

By covering the separating element on both sides with additional flexible layers, for example 13 mm gypsum board on a separate frame at a distance which gives a resonance frequency of the system of layer and airspace well below the frequency range of interest. The airspace should contain sound-absorbing material.

Technique b

By measuring the average surface velocity levels of the specimen and the flanking surfaces in the receiving room.

Specification of flanking structure and details in the Building Regulations

Building Regulations Approved Document E (ADE) emphasizes the importance of controlling flanking transmission in order to achieve acceptable levels of performance.

Examples are measures specified to reduce sound transmission via a cavity wall flanking the separating wall between two spaces. This is achieved in two ways: by structurally connecting the separating wall to the inner leaf of the flanking wall, and by the use of a cavity block across the cavity at the point where the separating wall and flanking wall meet. The specified measures require the separating and cavity walls to be connected together either by the use of builders' wall ties or by structural bonding of the two walls, see Figures 6.16 and 6.17. The cavity stop is made of dense mineral fibre.

Paragraph 2.65 of ADE on the construction for wall type 2 (cavity masonry) states: 'Do not build cavity walls off a solid continuous slab floor.' Why is this? And what is the solution? Basically to prevent a flanking path bypassing the separating wall via the floor. The solution is to design a break in the floor at the junction of wall and floor, as shown in Diagram 2.24 and paragraph 2.85 of ADE.

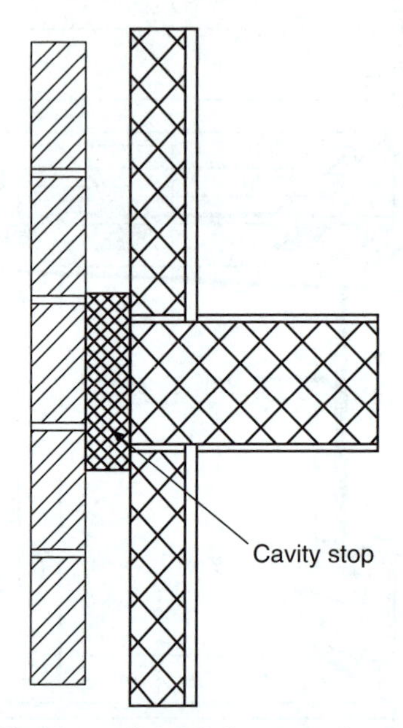

Figure 6.16 Flanking wall details for separating wall type 1

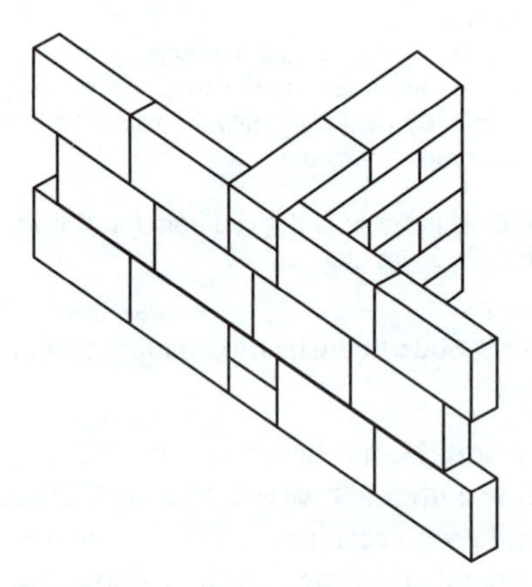

Figure 6.17 Flanking wall details for separating wall type 1

Prediction of flanking transmission

We have shown in this chapter that it is possible to predict the effects of airborne sound transmission via a partition in some simple cases (room to room, inside to outside and outside to inside). The prediction of transmission via flanking walls, floors and ceilings, which involves structure-borne transmission as well, is much more difficult, and generally beyond the scope of this text. However, such prediction techniques are available, and are likely to become increasingly used in the future as predictive software packages improve and become more widely available.

Predicting flanking transmission involves understanding how the vibration set up in the separating wall by the incident sound is transmitted to the flanking wall (or floor or ceiling) and then radiated as sound. The vibration in each partition will be the sum of the vibration of all of its modes of vibration, in much the same way as the sound pressure in a room is the sum of the sound pressure in all of the modes of the room (as discussed in Chapter 5). Each flanking partition will have its own distribution of modes and the vibration will vary with position on the partition.

This is obviously very complicated. The statistical energy analysis (SEA) technique attempts to simplify matters in much the same way as Sabine's approach (also statistical) to sound in rooms simplifies room acoustics theory. Rather than attempting to predict the vibration of individual modes, SEA deals only with average vibration levels in each partition (in the same way as Sabine's theory deals with a single reverberant sound levels in each room). Also like Sabine acoustics, SEA deals with the flow and balance of energy, i.e. the flow of vibrational energy from one partition to its neighbour, and the balance between the rate at which vibrational energy is supplied to the partition and the rate at which it is used up in overcoming damping in the partition and in radiating sound. Therefore important input data to any SEA calculation are the damping of each panel or partition and the energy transfer function between each pair of panels. As in the acoustics case these will be frequency dependent and therefore will need to be specified in octaves or third octave bands. The input variable to the calculation will be the force applied to the panel (which causes it to vibrate), which will be related to the incident sound pressure, and the output variable will be the average value of the square of vibration velocity v^2_{AVGE} which can then be converted to a radiated sound pressure level (see later in Chapters 7 and 9).

Again like Sabine acoustics, the SEA method works best at higher frequencies when the modal density is high, i.e. at frequencies where there are very many modes closely spaced together in frequency so that the averaging approach is valid and individual modes do not have to be considered.

BS EN ISO 12354-1 adopts a simplified version of the SEA approach to the prediction of flanking transmission, particularly in Part 1, the room to room case.

Appendix 6.1 Derivation of room to room transmission equation

This derivation is based on four sets of relationships which we have already met:

1 The sound intensity at the surface of a reverberant room, L_I

$L_I = L_p - 6$ (A)

where L_I is the sound intensity level a the boundary of the reverberant room and L_p is the sound pressure level in the room (discussed in Chapter 5, section 5.9).

2 Transmission of sound energy across a partition

$t = I_2/I_1$ $\quad R = 10\log(1/t)$

$L_{I1} = 10\log(I_1/I_0)$ and $L_{I2} = 10\log(I_2/I_0)$

So that $L_{I1} - L_{I2} = 10\log(I_1/I_2) = 10\log(1/t) = R$

Therefore: $L_{I2} = L_{I1} - R$ (B)

3 Sound power level radiated by a surface at which sound intensity is I W/m^2

$W = I \times S$

Converting to decibels:

$L_W = L_I + 10\log S$ (C)

4 Sound pressure level in a reverberant room

$L_p = L_W + 10\log(4/R_C)$

Assuming that for a 'live' room $R_C \approx A$:

$L_p = L_W + 10\log(4/A)$

$L_p = L_W - 10\log A + 6$ (D)

Symbols to be used in the derivation (Figure 6.18):

L_1 = sound pressure level in the source room

L_2 = sound pressure level in the receiving room

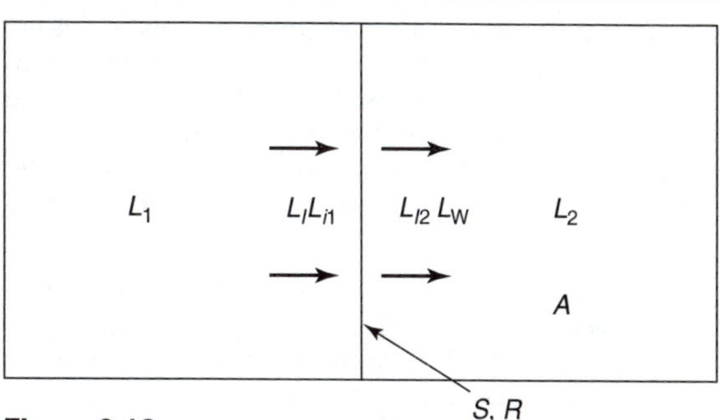

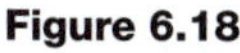

Figure 6.18

L_{I1} = sound intensity level at partition wall on source room side

L_{I2} = sound intensity level at partition wall on receiving room side

L_w = sound power level radiated by the partition into the receiving room

S = area of partition (m^2), A = sound absorption in receiving room (m^2)

Derivation

Sound intensity level at partition on source room side:

$L_{I1} = L_1 - 6$ from (A)

Sound intensity level at partition on receiving room side:

$L_{I2} = L_{I1} - R$ from (B)

Therefore: $L_{I2} = L_1 - 6 - R$

Sound power level being radiated into receiving room:

$L_W = L_{I2} + 10\log S$ from (C)

Therefore: $L_W = L_1 - 6 - R + 10\log S$

Reverberant sound pressure level in receiving room:

$L_2 = L_W - 10\log A + 6$ from (D)

Therefore: $L_2 = L_1 - 6 - R + 10\log S - 10\log A + 6$

from which: $L_2 = L_1 - R + 10\log S - 10\log A$

or: $L_1 - L_2 = R - 10\log S + 10\log A$

Questions

1 An external wall, of area 4 m by 2.5 m, in a house facing a motorway is required to have a sound reduction of 50 dB. The construction consists of a 280 mm cavity wall containing a double-glazed (and sealed) window. The sound reduction indices are 55 dB for the 280 mm cavity wall and 44 dB for the sealed double-glazed window with a 150 mm cavity. Calculate to the nearest 0.1 m^2 the maximum size of window to achieve the required insulation.

2 A partition across a room 6 m by 5 m includes a timber doorway of area 2 m^2. Determine by how much the partition will be a better insulator of sound if constructed in 115 mm brickwork instead of 100 mm building blocks, each plastered both sides, at the particular frequency where the sound reduction indices of brick, building block and timber are respectively 48 dB, 34 dB and 26 dB.

3 Calculate the average sound reduction index of a partition made of 24 m^2 of 115 mm brickwork and 6 m^2 of plate glass where the sound reduction indices of brickwork and glass at a certain frequency are 45 dB and 30 dB respectively.

4 Calculate the sound reduction index of a partition of total area 20 m^2 consisting of 15 m^2 of brickwork, 3 m^2 of windows and 2 m^2 of door. The sound reduction indices are 50 dB, 20 dB and 26 dB for the brickwork, windows and door respectively.

5 If an average sound reduction of 20 dB is required by a barrier, what is the approximate minimum mass per unit area required?

6 In a field test carried out to assess the transmission loss through a partition separating two adjoining dwellings, the following results were obtained. At a frequency of 250 Hz the sound pressure levels at six positions in the source room were respectively 99, 100, 95, 95, 97 and 98 dB. In the receiving room the levels were 60, 58, 59, 63, 59 and 61 dB. The measured reverberation time in the receiving room was 1.53 seconds. Apply the appropriate correction factor and calculate the normalized level difference between the two rooms.

7 The sound pressure level of noise from a stationary item of plant on a construction site was 81 dBA at a distance of 10 m. The window of a flat is 40 m from this item. It is of single-glazed construction and of area 1 m^2 with a sound reduction index of 19 dB. If the total area of exposed external brick wall of a room is 5 m^2 find the level in the room. (Sound reduction index for the brickwork may be taken as 51 dB and sound absorption in the flat is 5 m^2.)

8 A door of 20 dB sound reduction index in the 2000 Hz octave band is 2 m high by 1 m wide. It is set in a wall 6 m long by 2.5 m high, which has a sound reduction index of 50 dB in the same frequency band. Estimate the sound reduction index of the composite partition if:

(i) the door fits perfectly

(ii) there is a gap of 3 mm around the edge of the door.

9 An office is to be partitioned off from the rest of a busy print room, in which the noise levels in the 1000 Hz band can be as high as 76 dB. Calculate the sound reduction index required of the partition if the intrusive noise from the typing pool is to be kept below 40 dB in this band. The dimensions of the partition are to be 10 m × 4 m and it is estimated that the amount of absorption in the newly created office will be 25 m^2 in the 1000 Hz band. Explain the assumptions underlying your calculation.

10 A diesel engine is situated in a plant room which contains one window of dimensions 3 m × 2 m. The walls and door of the plant room are massively constructed so that the window as 'weak link' is the major transmission path for noise to the outside. Calculate the sound pressure level in the 500 Hz octave band at a point 20 m from the window on a line perpendicular to the surface of the window, given the following information. Measurements of the SPL from the machine, in the 500 Hz octave band, at 1 m from the engine surface, at four points around the engine, under anechoic conditions, gave the following readings: 110 dB, 96 dB, 108 dB, 101 dB.

The dimensions of the plant room are 10 m × 6 m × 3 m high and the absorption coefficients of the inside surfaces, at 500 Hz are: walls $\alpha = 0.02$, floor $\alpha = 0.01$, ceiling $\alpha = 0.04$. The sound reduction index of the window at 500 Hz is 30 dB.

(*Hint:* Use the four given measurements of the engine noise as the basis for calculating the sound power level of the engine (approximately), and assume that the sound inside the plant room will be completely reverberant.)

11 (i) Discuss the factors which affect the sound reduction index of a single-leaf partition, including how it varies with frequency. Explain the principles underlying the performance of double-skin partitions and the design measures necessary to achieve the maximum possible sound insulation.

(ii) Sound transmission through building façades can be predicted by:

$$L_2 = L_1 - R - 6$$

where L_2 = sound pressure level just outside the building

L_1 = level inside

R = sound reduction index of the building façade.

Explain the basis of this formula and state any assumptions involved.

(iii) The sound pressure level inside a building is 98 dB in a certain octave band. Calculate the noise level in that octave band radiated from the façade of the building at a point which is 100 m from the centre of the façade in a direction at right angles to the façade.

The dimensions of the building façade are 5 m × 2.5 m and its sound reduction index in the required octave band is 25 dB. The ground between the façade and the reception point is soft and sound absorbing. Explain the steps in your calculation.

(IOA)

12 (a) Define, for a panel, (i) the sound transmission coefficient, τ and (ii) sound reduction index, R, and (iii) field sound reduction index, Rf.

(b) When building elements, for example modular partitions, are installed in a practical situations it is often found that the effective sound reduction index is lower than that measured under controlled laboratory conditions. Give reasons why this might be.

(c) Two cellular offices each have a volume of 40 m^3 and a reverberation time of 0.9 s in the 500 Hz octave band. They have a common wall, of area 12 m^2, with a sound reduction index in the 500 Hz octave band of 38 dB. When commissioning tests are performed the sound level difference between the offices is found to be 33 dB in the 500 Hz octave band.

(i) What level difference would you have expected if the only sound transmission path were through the common wall?

(ii) What fraction of the total sound power getting from one office to the other is via the common wall? (IOA 2006)

13 (a) State two factors that influence (i) airborne sound insulation of a party wall, and (ii) impact sound insulation of a ceiling/floor combination.

(b) Describe the main features of the procedures for measuring room-to-room airborne and impact sound insulation.

(c) A computer room and an office, both having dimensions of 6 m × 6 m × 3.5 m high, are separated by a common block wall with a sound reduction index of 31 dB at 250 Hz. The reverberant sound level in the computer room, at 250 Hz, is 87 dB. What is the reverberant sound level in the office if the average absorption coefficient, at 250 Hz, is 0.2? Ignore all other transmission paths. (IOA 2007)

14 Noise from a compressor in a plant room is transmitted to an adjacent office through a party wall. The dimensions of the plant room are 10 × 8 × 3 m high, and those of the office are 5 × 5 × 3 m high, so that the partition separating them is of dimensions 5 × 3 m. The reverberation times in a particular octave band are 2.1 s for the plant room and 0.5 s for the office. In the same octave band the sound power radiated by the compressor is 0.01 W, and the transmission loss of the party wall is 50 dB.

(a) Calculate the sound pressure level in the office, explaining the stages in your calculation and any assumptions on which they are based.

(b) Calculate the effect of increasing the absorption in the plant room by 18 m^2.

(c) The sound from the compressor is also transmitted to the outside through a window of dimensions 2 × 1 m^2 and 20 dB transmission loss. Estimate the sound pressure level 20 m from the window.

15 The façade of a building measures 38 m × 16 m and is constructed from cavity brickwork with doors of area 8 m^2 and 25 6 m^2 single glazed windows.

Calculate:

(i) the average sound reduction index of the façade at 500 Hz

(ii) the maximum area of single glazing that could be installed if the average sound reduction index is not to fall below 30 dB.

Sound reduction index at 500 Hz: cavity brickwork 49 dB, door 21 dB, single glazed window 27 dB.

(IOA 1995)

16 A partition between an office and a workshop contains a window of area 2 m^2 and sound reduction index 20 dB. Calculate the sound power level radiated into the office through the workshop window if there is a reverberant sound pressure level of 97 dB in the workshop.

Estimate the reverberant sound level in the office which is 8 m × 5 m × 3.5 m high and has a reverberation time of 1.2 seconds. (IOA 1996)

17 A cellular office has a partition with a single door to an adjacent open plan area. Calculate the sound reduction index (SRI) required from the door if the overall partition and door are to provide an SRI of 33 dB.

You may assume the following:

Partition (including door)	3.0 m wide × 2.5 m high
Door	1.2 m wide × 2.1 m high
Partition SRI	38 dB

The building services consultant wishes to include a 10 mm gap along the full width of the door to allow return air from the cellular office into open plan area. Calculate the reduction in sound insulation that would result from this. (IOA 2001)

18 The results of an impact sound insulation measurement (at 500 Hz) for a party floor are as follows:

Position	Receiving room noise level (dB)	Receiving room reverberation time (s)		Receiving room background noise level (dB)
		Measurement 1	Measurement 2	Measurement 1
1	45.4	0.60	0.61	35.8
2	44.8	0.62	0.61	
3	44.5	0.59	0.58	
4	45.6	n/a	n/a	

Calculate the standardized impact sound pressure level (L'_{nT}) at 500 Hz, showing clearly how this is obtained from the above results. (IOA 2005 (part question))

19 The results of an airborne sound insulation measurement (at 500 Hz) for a party floor are as follows:

Source room noise level (two positions) — 94.9 dB, 93.8 dB

Receiving room noise level (two positions) — 43.1 dB, 42.4 dB

Receiving room background noise level — 34.8 dB

Receiving room reverberation time (six positions) — 0.52, 0.54, 0.54, 0.56, 0.53, 0.55 seconds

Calculate the standardized level difference (D_{nT}) at 500 Hz, showing clearly how this is obtained from the above results. (IOA 2004 (part question))

20 A building has a façade which is 20 m long × 5 m high. Openable windows with a sound reduction index of 32 dB comprise 20% of the cladding area and the remainder comprises a lightweight cladding system with a sound reduction index of 40 dB. Calculate the change in overall sound reduction index if 5% of the window area were to be opened. (IOA 2004 (part question))

21 The sound pressure level inside a factory workshop is 108 dB at the 500 Hz octave band. The noise is emanating from a long lightweight façade of the factory. Calculate the noise level in that octave band at reception points 20 m, 50 m and 200 m away from the façade in a direction perpendicular to the façade.

The dimensions of the factory façade are 100 m × 25 m and its sound reduction index in the required octave band is 20 dB. The ground between the façade and the reception points is soft. Explain the steps in your calculation. (IOA 1996 (part question))

22 A bedroom of a dwelling is to incorporate a double glazed window within a masonry cavity wall. The window is to have a small slot to provide ventilation to the bedroom when it is closed. Calculate the sound insulation required from the ventilation slot if a composite performance of 30 dB is to be achieved.

You should assume the following:

Size of wall (including window and slot) — 4.5 m wide × 2.5 m high

Ratio of window (including slot) to wall — 1.3

Size of slot — 1.5 m wide × 0.02 m high

Wall SRI — 50 dB

Window SRI — 38 dB

(IOA 2002 (part question))

23 (a) Sketch and describe the sound insulation characteristics of a single glazed window.

(b) A 4 m wide × 2.8 m high wall between a studio and a control room contains a 2 m wide × 1.2 m high single glazed window. If the overall sound insulation of the wall including the window is 45 dB and the performance of the wall alone is 55 dB, estimate the sound insulation of the window.

(c) It has been decided that the sound insulation of the window needs to be improved by installing an additional pane of glass separated from it. Outline the principles of how the sound insulation of this window could be maximized. (IOA 1999)

24 A new building is to contain a plant room and a computer room, both of which are to be located adjacent to offices. In each case the two spaces are to be separated by a full height partition (5.0 m wide × 3.0 m high with a sound insulation of 50 dB) containing a door (1.2 m wide × 2.1 m high). The overall sound insulation required is as follows:

- plant room to office 44 dB
- computer room to office 40 dB.

The contractor has informed you that a manufacturer can supply doors with three sound insulation ratings (28, 34 and 38 dB) but wants to know which would be the most relevant to use since there is a significant cost differential. For both cases, calculate the sound insulation requirements for the door and identity the most appropriate door to use. (IOA 2003)

25 (a) Explain the following terms used in connection with the sound reduction index of single and double leaf constructions: mass law, critical frequency, mass spring mass resonances.

(b) The table below gives sound reduction index data for various partitions made from sheet steel: single leaf panels of 3 mm and 6 mm thickness, and a double leaf panel consisting of two 3 mm thick panels separated by a cavity of 100 mm thickness, without and with sound absorbing material in the cavity.

Making use of the formulae and data given below explain and compare the data for the different panel types.

Formulae and data:

Mass law: $R = 20\log(mf) - 47$ dB

Mass–spring–mass frequency:

$f = 1900 \times [(m_1 + m_2)/m_1 m_2 d]^{0.5}$ Hz

where m_1 and m_2 are the surface densities of the two leaves, in kg/m^2 and d is the cavity thickness, in mm.

Product of critical frequency and panel surface density (Hzkg/m^2), for steel = 97,700.

Density of steel = 7700 kg/m^3

Panel type	63 Hz	125 Hz	250 Hz	500 Hz	1 kHz	2 kHz	4 kHz	8 kHz
3 mm	16	22	28	34	40	45	32	41
6 mm	22	28	34	40	45	37	42	51
3-100-3 without absorption	29	36	48	60	60	60	60	66
3-100-3 with absorption	21	27	47	62	75	75	76	78

Chapter 7 Vibration

7.1 Introduction

The chapter starts with a description of the nature of vibration and how vibration magnitude can be described in terms of either a displacement, velocity or acceleration, and the relationships between these parameters, and how they may be expressed in terms of decibels.

The behaviour of a simple mass–spring system is then discussed in some detail because it introduces many of the ideas which can be transferred to the study of more complex systems. A qualitative description is given in the main text and a more mathematical treatment in Appendix 7.1.

The discussion of forced vibration leads on to the theory of vibration isolation and to a simple design procedure of isolated systems, and its practical limitations. Human response to vibration is then discussed, leading to a review of the various ways in which vibration levels may be assessed: hand-arm and whole body vibration, vibration dose value and building damage criteria.

The chapter closes with a discussion between the link between the vibration of a surface and the resulting radiated sound pressure. The Appendix gives a more detailed mathematical treatment of the vibration of the mass–spring–damper system.

The measurement of vibration is discussed in Chapter 8.

You may be able to find some interesting demonstrations of vibration – on standing waves, modes of vibration etc. using the internet.

7.2 The nature of vibration

A vibration is a type of motion in which a particle or body moves to and fro, or oscillates, about some fixed position. The distance of the particle or body from its fixed, or reference position is called the displacement, which may be positive or negative. The average displacement of a vibrating object is usually zero, so that the fixed position is also called the mean position. Any motion which results in a change in the mean position is called a translation. Any vibration may be described by a graph showing how the displacement varies with time (see Figure 7.1).

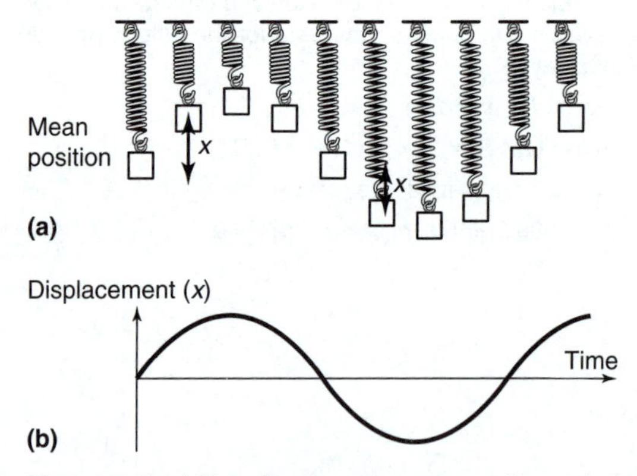

Figure 7.1 (a) The vibration of a mass on the end of a spring at 10 equally spaced instants in time; (b) the graphical representation of the vibration

Types of vibration – periodic, random and transient

Vibrations may be classified as either periodic, random or transient (see Figure 7.2). In a periodic vibration the motion repeats itself exactly, after a time interval called the period. For the simplest type of periodic vibration, called simple harmonic motion, the displacement–time graph is a sine wave. A motion of this type can be described in terms of a single frequency: examples are the motion of the prongs of a tuning fork, or of the bob of a simple pendulum. More complex periodic motions can be made up from a combination of different sine waves, and so these vibrations contain a number of different frequencies and have a frequency spectrum consisting of a number of lines representing a fundamental frequency and its harmonics. Examples are the motion of a piston in an internal combustion engine, or vibrations produced by regularly repeating forces in rotating machinery, such as motors, generators and fans.

In a random vibration the oscillations never repeat exactly. Examples are vibrations produced in structures by wind, or wave forces, or vibrations produced in a motor car as a result of its ride along a bumpy road. Random

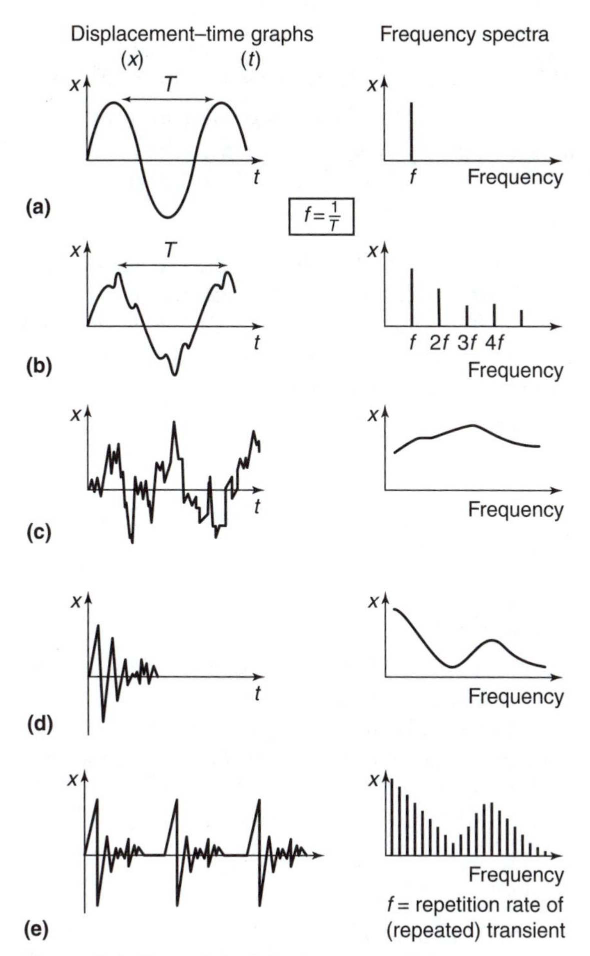

Figure 7.2 Examples of displacement–time graphs for various vibrations and their frequency spectra

vibrations contain a little of every frequency, and so the frequency spectrum is a continuous curve, called a broadband spectrum.

A periodic vibration is said to be deterministic because its displacement at any time can be predicted. In the case of a random vibration it is not possible to do this, and the vibration can only be described in statistical terms. Random vibrations may be further subdivided into stationary and non-stationary types. In a stationary random process the statistical characteristics of the vibration, such as the RMS displacement, for example, remain constant in time. For a non-stationary process this is not so.

Transient vibrations die away to zero after a period of time. Examples are the vibrations in a building caused by the passage of a heavy vehicle, or the vibrations of a plucked violin string. The simplest type would be represented by a decaying sine wave. The frequency content of transient vibrations is complicated, but obviously in the case of the violin string it would contain the fundamental frequency and its harmonics. In the case of a repeated transient such as the impacts between teeth in a gear mechanism, the repetition rate of the impacts, and its harmonics will also be important.

7.3 Displacement, velocity and acceleration

The vibration of an object may be described in terms of either a displacement, or a velocity or an acceleration (see Figure 7.3). The interrelationship between these three descriptors may best be explained by considering the simplest sort of vibration – a single frequency, i.e. simple harmonic motion. Lengthy consideration of such a simple example is fully justified because the French

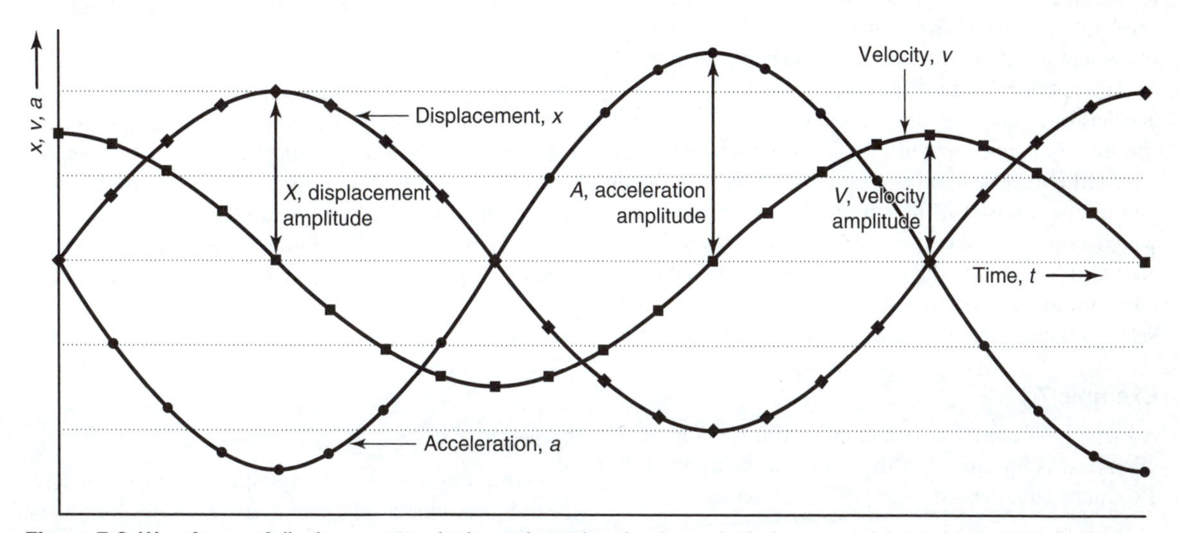

Figure 7.3 Waveforms of displacement, velocity and acceleration for a single frequency (simple harmonic) vibration

mathematician Fourier showed that any periodic motion, however complex, can be made up from a series of simple harmonic motions. Displacement is usually measured in metres (m) but this unit is too large to be conveniently used for vibrations.

The variation of displacement, velocity and acceleration may be described mathematically as shown below, and graphically as shown in Figure 7.3.

$$\text{Displacement, } x = X\sin(2\pi ft) = X\sin(\omega t)$$

$$\text{Velocity, } v = V\cos(2\pi ft) = V\cos(\omega t)$$

$$\text{Acceleration, } a = -A\sin(2\pi ft) = -A\sin(\omega t)$$

where t is time (seconds), f is frequency (Hz), $\omega = 2\pi f =$ angular frequency, in radians per second, and X, V and A are the amplitudes, or peak values, of displacement velocity and acceleration respectively, while x, v and a represent the instantaneous values of these time varying quantities.

Note that:

- The velocity waveform is ¼ cycle (90°) out of phase with the displacement waveform.
- The acceleration waveform is ¼ cycle (90°) out of phase with the velocity waveform, and therefore ½ cycle (180°) out of phase with the displacement waveform.

In summary, the relationships between displacement, velocity and acceleration amplitudes are:

$$V = 2\pi fX$$
$$A = 2\pi fV$$
$$A = 4\pi^2 f^2 X$$

These relationships, which only apply to single frequency (sinusoidal) vibrations, also apply to the RMS values (amplitude/$\sqrt{2}$). For broadband vibration the measured acceleration signal may be converted to velocity and displacement signals using an electronic integrating circuit.

The relationship between displacement, velocity and acceleration may be illustrated by considering the motion of a simple pendulum. At the central, lowest position, the displacement is zero but the velocity of the pendulum bob is at its maximum. However, the rate at which velocity is changing, i.e. the acceleration, is also zero at this point. At the end positions, displacement is a maximum, and velocity is zero, but the rate at which velocity is changing, i.e. the acceleration, is a maximum.

Example 7.1

A panel is vibrating a frequency of 100 Hz. The RMS vibration is measured at the centre and found to be 1.0 mm/s. Find the RMS and peak displacement.

$V = 2\pi afX$, from which $X = V/2\pi f$

Solution

$$\text{RMS } X = [1.0/(2 \times 3.142 \times 100)] = 1.6 \times 10^{-3}\text{ mm}$$
$$= 1.6 \times 10^{-6}\text{m} = 1.6\,\mu\text{m (1.6 microns)}$$
$$\text{Peak } X = \text{RMS } X/0.707 = 1.6/0.707 = 2.25\,\mu\text{m}$$

Example 7.2

One of the walls of a building is subject to vibrations of 100 Hz frequency. The peak displacement is measured and found to be 0.01 mm. Calculate the peak and RMS velocities of the wall at the measurement point, assuming the vibrations to be sinusoidal.

$V = 2\pi fX$

Solution

$$\text{Peak } V = 2 \times 3.142 \times 100 \times 0.01 = 6.3\text{ mm/s}$$
$$\text{RMS } V = 0.707 \times 6.3 = 4.4\text{ mm/s}$$

Example 7.3

An annoying whine from a machine is identified as being produced by a sheet-metal cover which is performing resonant vibrations at a frequency of 1000 Hz. The vibrations are measured using an accelerometer, and the RMS acceleration is found to be 5.0 m/s^2. Calculate the RMS velocity and displacement of the cover.

Solution

$$A = 2\pi fV\text{, from which } V = A/2\pi f$$
$$= 5/(2 \times 3.142 \times 1000) = 0.8 \times 10^{-3}\text{ m/s}$$
$$V = 2\pi fX\text{, from which } X = V/2\pi f$$
$$= 0.8/[(2 \times 3.142 \times 1000)] = 1.3 \times 10^{-4}\text{ mm}$$
$$= 0.13\,\mu\text{m}$$

This example illustrates the effect of frequency on the relationship between X, V and A. The acceleration is fairly high and the vibration produces considerable noise, but because of the frequency factor it corresponds to a displacement which is only miniscule.

Alternatively, acceleration and displacement may be directly related by combining equations $V = 2afX$ and $A = 2afV$ to eliminate V: $A = 4\pi f^2 X$ from which $X = A/4\pi f^2$, and this could have been used to calculate X directly from A.

Example 7.4

The vibration level near to a suspected worn bearing is measured using a displacement meter and is found to be 0.5 mm peak. The vibration is predominantly of one frequency, 20 Hz. The measurement is checked using an accelerometer and the peak acceleration found to be 9.0 m/s^2. Do the two measurements agree?

The equation $A = 4\pi f^2 X$ can be used to find the value of peak acceleration which corresponds to a displacement amplitude of 0.5 mm at 20 Hz. This turns out to be 8.0 m/s, so the two measurements are in reasonable but not complete agreement.

This example also illustrates the fact that any vibration can in principle be measured in terms of either x or v or a; the three measurements would be related and not be independent of each other.

Warning: Relationships between RMS (or peak) values of x, v and a can only be calculated using the previous equations if the vibration is of one single known frequency. It may be extended, approximately, to vibration measurements made in frequency bands, e.g. third octaves, but could not be used for overall vibration levels containing a wide range of frequencies. However, electrical signals which represent overall acceleration, incorporating a wide range of frequencies, may be converted electrically into a signal which represents velocity or displacement using an integrating circuit or integrator in the measuring instrument.

Vibrations and waves

A sound wave in air causes the air particles to vibrate, as described in Chapter 1.

Exercise

How are the vibration of neighbouring particles related in (a) a single frequency plane wave, and (b) a standing wave caused by the interference of plane waves travelling in opposite directions?

Answer

In a plane wave all the particles will vibrate with exactly the same amplitude (assuming that there is no absorption of the sound) and there will be a constant phase difference between the vibration of adjacent particles such that particles one half wavelength apart will be half a cycle (or 180°) out of phase, and particles one complete wavelength apart will be in phase. In a standing wave the amplitude of the particles will vary depending on their position with respect to the 'mode shape', from a maximum at the antinodes to a minimum at the nodes. All particles between adjacent nodes will be vibrating in phase and will be half a cycle out of phase with those between the next pair of nodes.

Example 7.5

Calculate the vibration RMS velocity displacement and acceleration of an air particle in a 100 Hz pure tone plane wave in which the sound pressure level is 94 dB? (Take the characteristic acoustic impedance of the air, ρc, to be 415 rayls.)

Solution

The RMS sound pressure corresponding to 94 dB is 1 Pa.

Therefore acoustic particle velocity $= 1/415$

$= 0.0024$ m/s (or 2.4 mm/s)

Acoustic particle displacement, $X = V/2\pi f$

$= 0.0024/(2 \times 3.142 \times 100)$

$= 3.8 \times 10^{-6}$ m (or 3.8 microns)

Particle acceleration, $A = 2\pi f V$

$= [(2 \times 3.142 \times 100 \times 0.0024)] = 1.5$ m/s^2

Reference levels and use of decibels

The logarithmic scale is often used to compare and measure vibration levels. The decibel system may be used in two ways: firstly, simply to compare two levels. For example, when comparing two levels, 1 and 2, we might state that level 2 is, say, 16 dB above level 1. This description, however, does not allow us to know the level 2 absolutely; to do this we must always compare our level to a standard or reference level. The following reference levels for vibration measurement have been recommended:

- velocity 10^{-9} m/s, 10^{-6} mm/s
- acceleration 10^{-6} m/s^2.

It can be shown that the total energy of a vibrating object is proportional to the square of the amplitude (cf. intensity of a sound wave is proportional to square of pressure). This applies, of course, whether amplitude is expressed in terms of displacement, velocity or acceleration.

Since the decibel system is used to compare the energy, intensity or power of two quantities, the difference, N in decibels between two vibration levels is given by:

$$N\,(\text{dB}) = 10\log(A/A_0)^2 = 20\log(A/A_0)$$

where A is the level being described and A_0 is the reference level.

Alternatively, if N and A_0 are known, A may be found using:

$$A = A_0 10^{(N/20)}$$

Similar formulae may be used for displacement and velocity.

Example 7.6

The displacement amplitude of a vibration is measured and found to be 250 μm. Convert this to dB re. 10^{-11} m

$$N = 20\log(X/X_0) = 20\log(250 \times 10^{-6}/10^{-11})$$
$$= 20\log(250 \times 10^5) = 20 \times 7.4$$
$$= 148 \text{ dB re. } 10^{-11} \text{ m}$$

Example 7.7

The vibrational velocity amplitude of a machine is measured and quoted as 96 dB re. 10^{-6} mm/s. Calculate the velocity amplitude in absolute terms.

$$V = V_0 \times 10^{(N/20)}$$

In this case $N = 96$ dB and $V_0 = 10^{-6}$ mm/s

$$\text{Therefore: } V = 1 \times 10^{-6} \times 10^{(96/20)}$$
$$= 1 \times 10^{-6} \times 10^{4.8} = 0.063 \text{ mm/s}$$

Example 7.8

The RMS level of vibrational acceleration on the handle of a power tool is measured as 15 dB above that of a standard vibration source whose RMS level is known to be 6.94 m/s^2. Calculate the acceleration on the handle in m/s^2.

$$A = A_0 \times 10^{(N/20)}$$

In this case, $A_0 = 6.94$ m/s^2 and $N = 15$ dB

$$\text{Therefore: } A = 6.94 \times 10^{(15/20)} = 6.94 \times 10^{0.75}$$
$$= 39.0 \text{ m/s}^2$$

Example 7.9

The vibration amplitude of a resonating sheet panel is measured and found to be 15.0 m/s^2. After treating the panel with a spray-on proprietary damping compound, the level was reduced to 0.85 m/s^2. Express the reduction in decibels.

$$N = 20\log(A/A_0)$$
$$= 20\log(15/0.85) = 25.0 \text{ dB}$$

The vibration level has been reduced by 25 dB.

7.4 The behaviour of a mass–spring–damper system

All vibrating systems possess three essential features: **mass, stiffness** and **damping**. In this section we will consider the behaviour of the simplest possible vibrating system, consisting of a single mass, spring and damping element, called a dashpot (Figure 7.4). Consider first of all the free or natural vibrations of an undamped system,

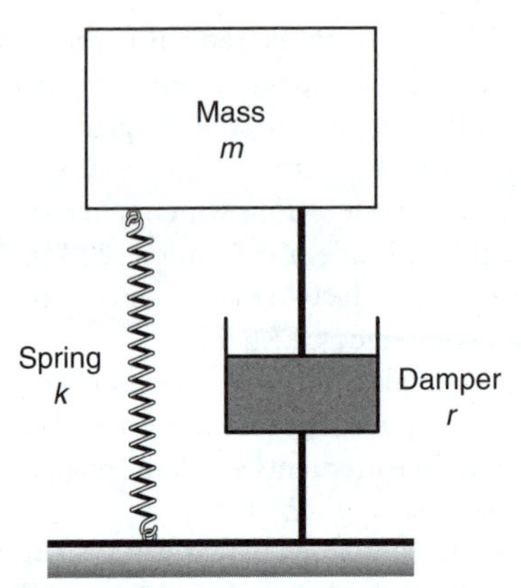

Figure 7.4 Mass–spring–damper model of a simple vibrating system

and then we will investigate the effects of damping. Finally we will consider forced vibration of the system. An understanding of these basic principles is relevant to many areas of acoustics, including the frequency response of microphones and accelerometers, the mass law of sound insulation and the theory of vibration isolation.

The characteristics of mass, stiffness and damping

The system is defined by the mass, m in kg, the stiffness of the spring, k, in N/m, and the damping constant, r is in Ns/m.

When a force F is applied to the mass (m) an acceleration, a, results, according to Newton's second law of motion:

$$F = ma$$

where F is in newtons, a in m/s/s or m/s^2 and m is in kg.

The spring stiffness, k, is defined by Hooke's law for springs, according to which the extension, x, of the spring length caused by a force F is proportional to that force, i.e.

$$F = kx$$

where F is in newtons, x in metres and k in N/m.

In the case of viscous damper the damping force F is related to the velocity of the motion via the damping constant, r:

$$F = rv$$

where F is in newtons, v in m/s and r is in Ns/m.

Free or natural undamped vibrations

First let us consider the behaviour of the system in qualitative terms. Free or natural vibrations occur when the system is disturbed from its rest, or equilibrium, position in some way, but this disturbance is only transient. In this case, because there is no damping the mass will perform simple harmonic motion (i.e. with its displacement from the rest position varying sinusoidally with time) with a constant amplitude. This is the sort of motion shown in Figure 7.4 above. The frequency of the vibration is the system's own natural frequency, depending only on the mass, and on the stiffness of the spring.

In order to understand why this occurs consider what happens if the spring is extended by pulling the mass away from its rest position, and then letting go. The spring exerts a restoring force on the mass, which causes the mass to move back towards its rest position. The restoring force depends on the amount the spring is extended, and as the mass approaches the rest position this force reduces. However, the mass has now picked up speed, and momentum, and so even though the restoring force will actually have reduced to zero when the mass reaches its former rest position, because of the inertia of the mass it overshoots and moves through the rest position. The spring is now being compressed (it was formerly being stretched) and the restoring force has now switched direction, but is still opposing the motion of the mass, and trying to return it to its rest position. Eventually as the extension, and with it the magnitude of the restoring force, increases, the momentum of the mass is overcome, and the mass comes, momentarily, to rest. The restoring force of the spring now causes the mass to pick up speed, moving back towards the rest position, which it reaches, and overshoots. And so the cycle continues, as a carefully balanced interplay between the restoring force in the spring, and the inertia, i.e. resistance to change of motion of the mass.

The motion may also be considered from an energy viewpoint. The initial disturbance, which caused the motion will have given a certain amount of energy to the system. If there is no damping (i.e. no loss of energy through frictional processes) then the total amount of energy in the system remains constant, and is continually changing, throughout the cycle from potential energy in the stretched or compressed spring, to kinetic energy of the moving mass.

The amplitude of the natural vibrations depends only upon the magnitude of the initial disturbance, and not upon the mass and stiffness, and conversely the natural frequency depends only upon the mass and stiffness, and not upon the initial disturbance.

The natural frequency of an undamped mass–spring system

The natural frequency f_0 of a system without damping, i.e. consisting only of a mass, m, and spring of stiffness k, is given by:

$$f_0 = (1/2\pi)\sqrt{(k/m)}$$

where k is in N/m (Nm^{-1}), m is in kg and f_0 in Hz.

The theory which leads to this equation is discussed in Appendix 7.1.

Note: The above formula shows that the natural frequency increases as stiffness increases and that it decreases as mass increases. Although the above formula for natural frequency relates to a simple mass–spring system, this general relationship between mass, stiffness and natural frequency applies to more complicated vibrating systems.

It follows that in order to achieve a very high natural frequency a vibrating system would have to be at the same time both very stiff and very light. A good example is the diaphragm of a condenser microphone which consists of a very thin (and therefore very light) metal diaphragm just a few microns thick but stretched very tightly so that it is also very stiff.

Example 7.10

Calculate the natural frequency and period of a system consisting of a mass of 200 kg attached to a spring of 800 kNm^{-1} stiffness.

$$f_0 = (1/2\pi)\sqrt{(k/m)} = (1/2\pi)\sqrt{(800{,}000/200)}$$
$$= 10.1 \text{ Hz}$$

Combinations of springs

In the above example the mass might be some sort of machine such as a motor or a fan, supported off the ground by springs. As we shall discover later, the action of the mass–spring system can be designed to reduce transmission of the vibration from the machine to the floor or vice versa, i.e. as a vibration isolation system. In the example, a number of different combinations might have been used to support the mass of the machine, for example one at each corner (i.e. four springs), or eight springs, or 10 springs etc.

The total stiffness, k_T, of a combination of springs used in this way is simply the sum of their individual stiffnesses, i.e.

$$k_T = k_1 + k_2 + k_3 + \cdots + k_N$$

If the weight of the machine is uniformly distributed it is likely that all the springs will have the same stiffness and the total stiffness will simply be the stiffness of one spring multiplied by the total number. In the above example there might have been four springs each of stiffness 200 kNm^{-1} or eight springs of stiffness 100 kNm^{-1}. If there is an uneven weight distribution, for example if the front part of the machine is heavier than the rear part, then a combination of springs of different stiffness might have to be used to achieve stability.

Example 7.11

A mass of 20 kg attached to a spring performs a natural vibration with a frequency of 2.5 Hz. Calculate the stiffness of the spring.

$2.5 = (1/2\pi)\sqrt{(k/20)}$
Therefore $\sqrt{(k/20)} = (2.5) \times 2\pi = 15.7$
Squaring: $k/20 = (15.7)^2 = 246.7$
and so $k = 20 \times 246.7 = 4935\ \text{Nm}^{-1}$,
or $4.94\ \text{kNm}^{-1}$, or $4.94\ \text{Nmm}^{-1}$.

Example 7.12

The natural frequency of a simple mass–spring system is 10 Hz. What is the effect on the natural frequency of: (a) doubling the spring stiffness, (b) doubling the mass?

Solution
Because of the square root in the formula for f_0, doubling the stiffness increases f_0 by a factor of $\sqrt{2}$, to14.1 Hz, and doubling the mass reduces f_0 by the same factor, i.e. to 7.1 Hz. Note that a fourfold increase in stiffness will double the natural frequency, and a fourfold increase in mass will halve it.

Example 7.13

For a mass–spring system for which $m = 10$ kg, $k = 200\ \text{Nm}^{-1}$ and $f_0 = 0.71$ Hz, calculate the effect of: (a) reducing k to $120\ \text{Nm}^{-1}$, (b) reducing m to 7.5 kg?

(a) New value of $f_0 = (0.71) \times \sqrt{(120/200)} = 0.55$ Hz

(b) New value of $m = (0.71) \times \sqrt{(10/7.5)} = 0.82$ Hz

Static deflection

When a mass is attached to a vertical spring the extension caused by the weight of the spring is called the static deflection X_S – see Figure 7.5.

The value of X_S is related, through Hooke's law, to the weight of the mass (mg newtons, where $g = 9.81\ \text{ms}^{-2}$, the acceleration due to gravity) and the stiffness of the spring:

$$mg = kX_S$$

from which: $X_S = mg/k$

The natural frequency f_0 and the static deflection X_S are related, as shown below.

Rearranging: $k/m = g/X_S$

Substituting for k/m in the formula for f_0:

$$f_0 = (1/2\pi)\sqrt{(g/X_S)}$$

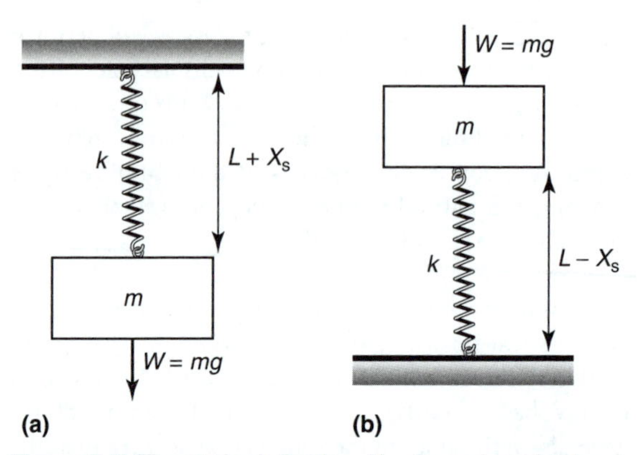

Figure 7.5 The static deflection, X_S of a spring of natural length L and stiffness k when subjected to the weight W of an attached mass m: (a) hung from the ceiling, with spring in tension and (b) supported from the floor, with spring in compression.

So far X_S is in metres but if it is converted to millimetres:

$$f_0 = (1/2\pi)\sqrt{(9.81 \times 1000)/X_S)} = 15.8\sqrt{(1/X_S)}$$

where X_S is now in mm.

Example 7.14

A mass of 5 kg attached to a spring produces a static deflection of 2 mm. What is the stiffness of the spring, and the natural frequency of the system?

$f_0 = 15.8\sqrt{(1/X_S)} = 15.8\sqrt{(1/2)} = 11.2$ Hz
$mg = kX_S$, from which $k = mg/X_S = 5 \times 9.81/0.002$
$= 24500\ \text{Nm}^{-1}$, or $24.5\ \text{kNm}^{-1}$

Note that X_S reverts back to metres in the Hooke's law formula.

Example 7.15

A mass–spring system has a natural frequency of 20 Hz. What is the static deflection produced in the spring by the weight of the mass?

Using the formula: $20 = 15.8\sqrt{(1/X_S)}$
Squaring both sides: $400 = (249.6)(1/X_S)$
from which: $X_S = 249.6/400 = 0.6$ mm

Figure 7.6 illustrates the relationship between natural frequency and static deflection.

Static and dynamic stiffness

Some materials used for making springs, such as rubber, may behave differently under static and dynamic (i.e. vibration) load conditions. In such cases the stiffness obtained from static deflection tests may differ from that

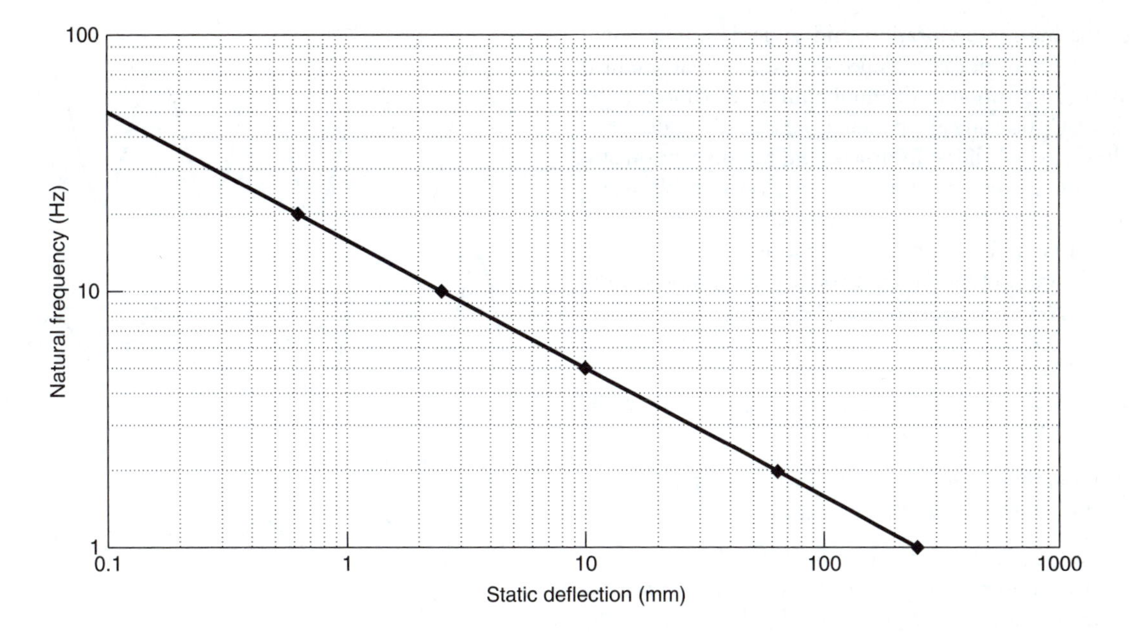

Figure 7.6 The relationship between static deflection and natural frequency for a mass–spring system

obtained from the natural frequency formula, by measuring f_0 and mass. Usually the dynamic stiffness is larger than the static stiffness, often by a factor in the region of 1.2 to 1.5. The stiffness of some materials may also be non-linear, i.e. the stiffness can vary with the magnitude of the loading force.

The effect of damping

Damping is the term given to any mechanism, occurring either within or between the components in a vibrating system, which leads to the conversion of vibrational energy into heat energy, and to a reduction in vibration amplitude. There will always be some damping in any system – otherwise perpetual motion would occur.

Hysteretic damping occurs within materials to an extent depending on their atomic, molecular and crystal structure. It is the result of phase changes between stress and strain which occur during a vibration cycle – a process called hysteresis – and is rather similar to the classical thermal absorption which occurs for sound waves in air, because heat flow cycles fall out of step with the compressions and rarefactions of the sound pressure cycle. **Friction damping**, as the name implies, occurs because of mechanical contact between components. **Viscous damping**, beloved of textbook writers, is the type of damping which would occur if a mass–spring system was fitted with a dashpot (see Figure 7.4 earlier) where a plunger moves through a viscous fluid. This type of damping occurs between lubricated surfaces. It is also the type of damping for which the equations are easiest to solve.

The three different cases of **overdamping**, **underdamping** and **critical damping** can be best explained by considering what happens if the mass in a mass–spring system is pulled aside from its rest position, and let go. The three cases are illustrated in Figure 7.7.

In the case of overdamping (also known as dead-beat damping) the damping force is so large that vibration does not occur, and the mass gradually moves back towards the rest position, without overshoot. When the amount of damping is small, underdamping occurs and the mass oscillates about the rest position, but with an amplitude which decreases with time until it eventually comes to rest. Critical damping is the situation which is between the other two, i.e. any more damping would cause overdamping, and any less would cause overshoot

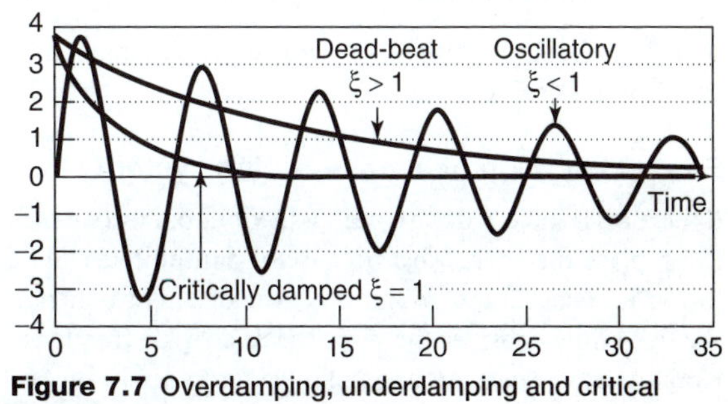

Figure 7.7 Overdamping, underdamping and critical damping

of the mass, i.e. vibration, and underdamping would occur. In the critically damped situation the mass returns to its rest position in the minimum time, without overshoot. Critical and overdamping are used in the design of 'door closers', shock absorbers, and the suspension systems of vehicles and of moving coil electrical voltmeters. It is the case of underdamping which is usually of most concern in noise and vibration control.

The value of the damping coefficient r_C, which causes critical damping is given by:

$$r_C = 2\sqrt{(k/m)}$$

The amount of damping in a vibrating system may be described by its damping ratio, ξ:

$\xi =$ damping coefficient of the system/damping coefficient when critically damped

The value of $\xi = 1$ for a critically damped system, $\xi > 1$ for overdamping, and $\xi < 1$ for underdamping. Values of ξ can be as low as 0.01 (or 1% of critical) for very lightly damped materials, such as mild steel, and as high as 0.2 (or 20% of critical) for highly damped materials, such as some plastics. For the case of underdamping the motion is rather like a sinusoidal vibration, but with an amplitude which decays with time at a rate which depends on the value of the damping ratio, as shown in Figure 7.8.

The damped natural frequency, f_d is related to its undamped counterpart, f_0, by:

$$f_d = f_0\sqrt{(1 - \xi^2)}$$

Therefore, if for example $\xi = 0.1$, then $f_d = 0.99f_0$, i.e. the natural frequency has been reduced by 1%.

The amount of damping may be measured from the rate of decay of the damped oscillations. The natural logarithm (ln) of the ratio of successive amplitudes is a constant, called the logarithmic decrement, δ:

$$\delta = (1/n)\ln(x_1/x_n)$$

where x_1 and x_n are two amplitudes n cycles apart.

It can be shown that the log decrement, δ, and the damping ratio, ξ, are related:

$\delta = 2\pi\xi/\sqrt{(1 - \xi^2)} = \delta = 2\pi\xi$, approximately, for small values of ξ.

Forced vibration of a mass–spring system

If the mass is subjected to a continuous vibratory force which can replace the energy lost because of damping it will be forced into continuous vibration at the driving frequency. We need only consider the response to a single frequency vibratory force, since more complicated forces can be broken down into a combination of sinusoidal forces.

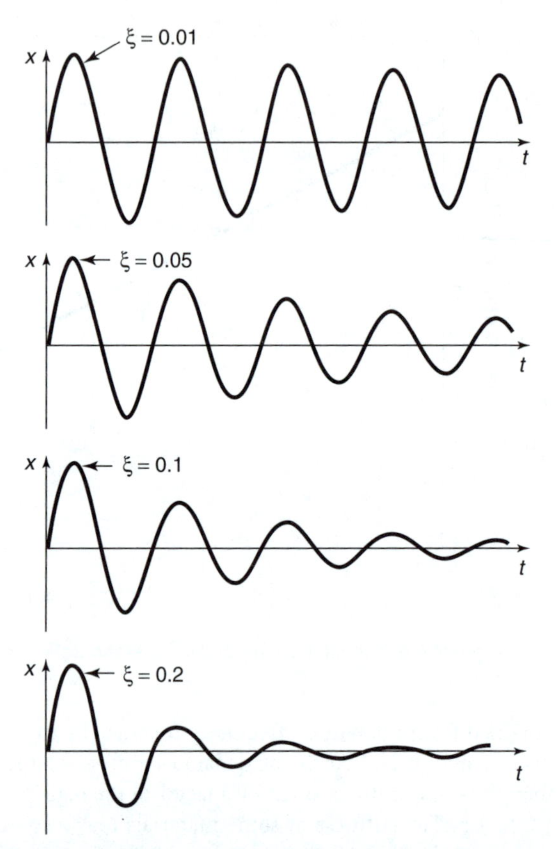

Figure 7.8 Graphs showing displacement (x) against time (t) of damped vibration for various values of damping ratio, ξ

Figure 7.9 shows how the amplitude of forced vibration varies with the forcing frequency for various degrees of damping, with the force amplitude remaining constant.

The graph shows that the vibration amplitude increases sharply when the driving frequency, f, approaches the natural frequency of the system, f_0. This is the phenomenon of **resonance**, and the forced vibration amplitude is a maximum at the resonance frequency, which, for a lightly damped system is almost the same as the natural frequency.

The amplitude of vibration at resonance depends on the amount of damping in the system, i.e. on the damping ratio, ξ, as shown in Figure 7.9. The **Q factor** of the system, also related to the amount of damping, may be obtained from the 'resonance' curve of Figure 7.9 in two ways: from the maximum amplitude, at resonance, and from the width of the resonance peak

$$Q = \frac{\text{displacement amplitude at resonance}}{\text{static displacement produced by the same force}}$$

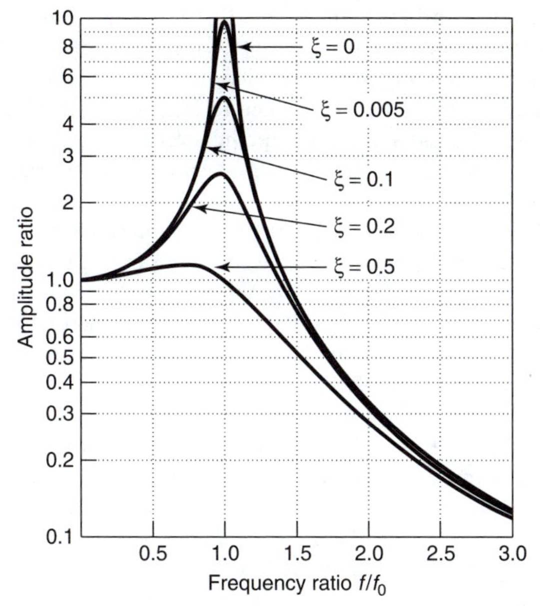

Figure 7.9 Graph of forced vibration amplitude response *v* forcing/driving frequency (the resonance curve), for various degrees of damping ratio, ξ

Therefore a Q factor of 10 means that the forced vibration amplitude at resonance is 10 times the static deflection. Alternatively Q may be found from the width of the resonance curve, using the formula:

$$Q = f_0/(f_2 - f_1)$$

where f_1, and f_2 are frequencies at which the forced vibration amplitude has dropped to $\sqrt{1/2}$ (i.e. 0.707) times its resonant value – the so-called half-power points. In practice this is a better method for measuring Q than the resonant magnification method. The Q factor, and the damping ratio ξ are simply related:

$$Q = 1/(2 \times \xi)$$

Looking at Figure 7.9, note the following:

- The amplitude of the vibratory force is held constant, only its frequency is being changed.
- Horizontal axis: f/f_0 is the ratio of forcing or driving frequency and the natural frequency f_0.
- Vertical axis: is the ratio of the forced vibration amplitude produced by vibratory force of frequency f to the static deflection produced by the same magnitude of steady force.

Stiffness, damping and mass controlled frequency regions

It is convenient to think of the curve of Figure 7.9 as being divided into three different frequency regions.

The stiffness controlled region

Well below resonance (i.e. $f \ll f_0$ or $f/f_0 \ll 1$), the vibration of the system is controlled mainly by its stiffness. In this frequency region the response of the system is fairly flat, i.e. it does not change very much with frequency. Therefore the frequency response of a microphone, or of an accelerometer, both of which behave like mass–spring systems, will be flat only well below its natural frequency. This means that these devices must have high natural frequencies, well above the audio range, if they are to have a flat frequency response up to 20 kHz.

The damping controlled region

This is at or close to resonance (i.e. $f = f_0$ or $f/f_0 = 1$). This is the resonance condition where large amplitudes of vibration occur when the forcing/driving frequency coincides with the natural frequency of the system. The amplitude at and near to resonance is controlled mainly by damping. Low damping produces a very sharp resonant peak and very large amplitude of vibration at resonance. More damping reduces the amplitude at resonance and produces a broader response curve.

The mass controlled region

Above resonance (i.e. $f_0 \gg f$ or $f/f_0 \gg 1$) the vibration amplitude reduces with increasing frequency, and well above resonance is proportional to $1/f^2$, i.e. reduces at a rate of 6 dB per octave, eventually falling below the static deflection value. In this frequency range the vibration amplitude is mainly determined by the mass of the system. Sound insulating panels and partitions often obey the 'mass law'[1] because they are being forced to vibrate at frequencies well above resonance by the airborne sound pressure incident upon them. It is also in the mass controlled region that vibration isolation occurs, but this can best be explained by describing the forced vibration response curve in terms of transmissibility, explained further below.

Degrees of freedom of vibrating systems

The simple mass–spring–dashpot system that we have been studying is sometimes called a **single degree of freedom system**, because it contains one mass, one spring, one damper, has one natural frequency and one mode of

[1] The mass law is an approximate law used in predicting the sound insulation of single leaf partitions, according to which, over a range of frequencies it is the mass of a partition which is the dominant influence in determining the sound reduction index (as opposed to its stiffness and damping). Refer to Chapter 6 for more details.

vibration. A two degree of freedom system will have two masses, two springs and two dampers and it will have two natural frequencies and two modes of vibration (one in which the two masses are both moving in the same direction and one in which they are moving in opposite directions). Similarly an N degree of freedom system will have N natural frequencies and N modes of vibration.

All such vibrating systems are called **lumped parameter systems** because the mass stiffness and damping are presented as separate, discrete entities. In a vibrating rod, beam or plate each small portion of material will contain all three elements (mass, stiffness and damping) and such a system is called a **distributed parameter system**, because the mass, stiffness and damping are distributed continuously throughout the material of the rod, beam or plate. Such systems will have an infinite number of degrees of freedom and an infinite number of natural frequencies.

Lumped parameter systems may be used as models of the vibration of distributed parameter systems such as rods, beams and plates. The higher the number of degrees of freedom in the model the closer it approximates to the vibration of the distributed freedom system.

7.5 Vibration isolation

The principle underlying vibration isolation is that the amplitude of forced vibration of a vibrating system reduces to low values (see Figure 7.9) when the system is operating well above resonance, i.e. when the driving frequency, f, is well above the natural frequency f_0. In practice, as a rule of thumb, the value of f should be at least three times the value of f_0 (i.e. $f/f_0 > 3$) in order to achieve a significant amount of isolation.

The application of this principle may be illustrated by the following situation, shown in Figure 7.10.

In part (a) of Figure 7.10 a motor is connected directly to the floor. Because the rotating parts of the motor can never in practice be perfectly balanced the rotation results in a vertical component of an out of balance force shown as F_i in the sketch, and this vibratory force is efficiently transmitted to the floor, where it can cause disturbance. In sketch (b) this same motor has been placed on springs (called vibration isolators or anti-vibration mounts). This has the effect of creating a mass–spring system, with a natural frequency f_0 in which the motor acts as the mass and the isolators as springs.

Let us suppose that the motor is rotating at a speed of 3000 RPM (revolutions per minute). This corresponds to $3000/60 = 50$ revolutions per second and so results in a vibratory driving force with a driving frequency, f, of 50 Hz. In order that the mass–spring system is driven well above resonance, using the rule of thumb, the natural frequency f_0 of the system should be arranged to be no more than $50/3 = 16.7$ Hz.

This is arranged by selecting the stiffness of the springs in conjunction with the mass of the motor (using the formula $f_0 = (1/2\pi)\sqrt{(k/m)}$. When this is done the transmitted force F_t will be much reduced i.e. the vibration produced by the motor will have been isolated, as shown in part (b) of Figure 7.10.

Transmissibility

The forced vibration amplitude response curves of Figure 7.9 (the resonance curves) have been useful in demonstrating that vibration isolation may be achieved when $f/f_0 > 1$. However, vibration isolation design calculations are easier if the forced vibration isolation response is expressed in terms of the transmissibility parameter T.

Definition of transmissibility

The definition is illustrated in Figure 7.11, which shows the two different vibration isolation situations for which the transmissibility may be defined: (a) the case of the vibrating machine for which the vibration is to be isolated from the vibration sensitive floor below (this is the case illustrated in the previous example), and (b) the case of the vibrating floor for which the vibration is to be isolated from the vibration sensitive device above.

In both cases the transmissibility is defined as the amplitude ratio across the isolators (or springs), but in

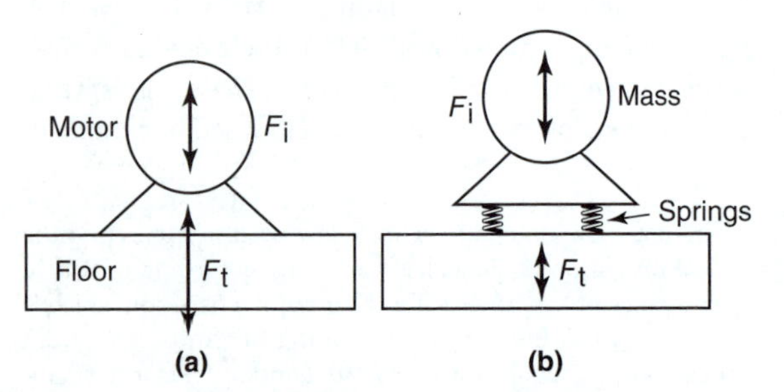

Figure 7.10 The application of vibration isolation

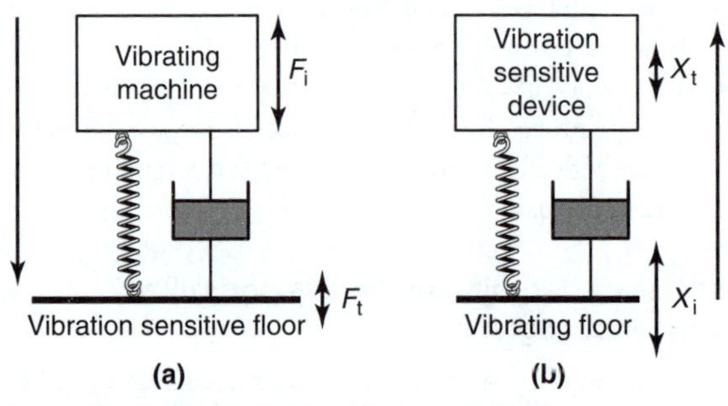

Figure 7.11 The definition of transmissibility

case (a) it is the amplitude of the vibrating force which is used, to define a force transmissibility, T_F, whereas in case (b) it is the amplitude of the vibration displacement which is used, to define a displacement transmissibility, T_X.

$T_F = F_t/F_i$ = force transmitted to the floor/force at the vibrating machine

$T_X = X_t/X_i$ = displacement transmitted to the sensitive device/displacement at the floor

Fortunately for the simple mass–spring–damper model of a vibrating system that we are going to discuss the two are identical and we can use a single transmissibility, T, i.e.

$$T_F = T_X = T$$

$T > 1$ means amplification of vibration across the isolators

$T < 1$ means reduction of vibration i.e. isolation.

Isolation efficiency, η

The isolation efficiency, η (in %), is defined as:

$$\eta = (1 - T) \times 100\%$$

For example, if $T = 0.1$ then $\eta = (1 - 0.1) \times 100\% = 90\%$ isolation efficiency, and, similarly, if $T = 0.01$ then $\eta = 99\%$ isolation efficiency.

Vibration reduction expressed in decibels

The transmissibility T is a vibration amplitude ratio. We learnt in Chapter 1 that decibels are based on the square of amplitude ratios. Therefore the transmissibility is related to the decibel reduction, N, in dB by:

$$N = 10\log(1/T)^2 = 20\log(1/T)$$

Example 7.16

The transmissibility of a vibration isolation system is 0.1. What is the vibration reduction achieved, in decibels?

$$N = 20\log(1/0.1) = 20\log(10) = 20 \text{ dB}$$

The variation of transmissibility with driving frequency, and with damping

The variation of transmissibility with driving frequency and with damping ratio is illustrated as a family of curves, in Figure 7.12. There are some similarities with the forced vibration amplitude response curves shown earlier (see Figure 7.9), notably the presence of the peak at resonance, whose height and width is dependent on the damping ratio, but above resonance there is an important difference.

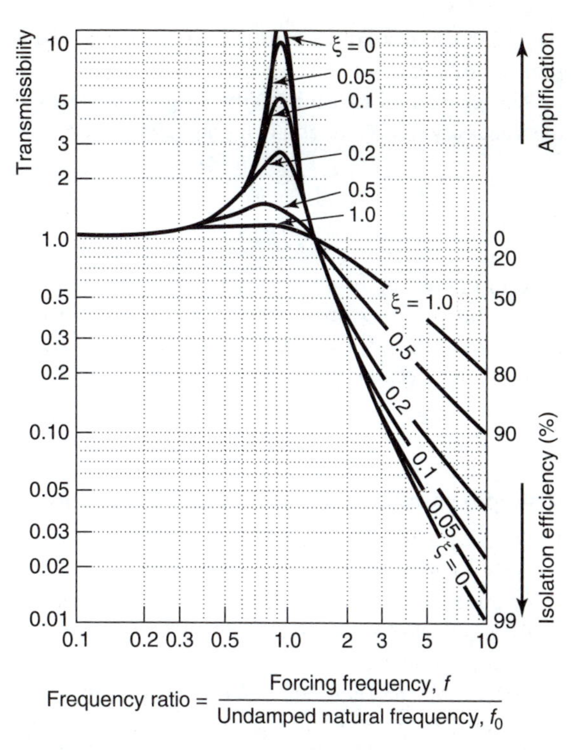

Figure 7.12 Graph of transmissibility, T, versus frequency ratio, f/f_0, for various degrees of damping ratio, ξ

The important difference is that above resonance, in the vibration isolation region, lower values of damping ratio result in lower, i.e. better, isolation. This is in contrast to the situation at or close to resonance, where higher damping is helpful in reducing the vibration amplitude at resonance. Therefore designers face a conflict: lower damping gives better isolation well above resonance but also gives higher vibration amplitude at resonance.

The significance of this is that the driving frequency is related to machine speed for many vibration producing machines. Therefore as machines speed up when they are switched on they move through the 'resonance speed' and are subjected to high levels of vibration before they attain operating speed well above resonance, and become isolated, and similarly when they are switched off. In practice a compromise may have to be adopted in selecting springs which have sufficient damping to limit vibration at resonant speeds during switch on and switch off, while still achieving the required amount of isolation.

An alternative way of limiting vibration amplitudes is to use some physical stop mechanism (called a snubber) to limit the resonant vibration amplitude. Another solution is to ensure that the machine is accelerated rapidly through resonance on starting up, and braked sharply on switching off, so that resonant vibrations do not have time to build up.

The formula for transmissibility T is given in terms of the frequency ratio (f/f_0) and damping ratio ξ by:

$$T = \sqrt{\frac{1 + 4\xi^2(f/f_0)^2}{(1 - (f/f_0)^2)^2 + 4\xi^2(f/f_0)^2}}$$

It is sometimes useful to do a quick and approximate calculation assuming zero damping, i.e. that $\xi = 0$, when, for the case of $f > f_0$ (i.e. above resonance, where vibration isolation occurs):

$$T = 1/[(f/f_0)^2 - 1]$$

A simple vibration isolation design procedure

1. Determine the lowest possible driving frequency, f. For a rotating machine, where vibration is produced by out of balance forces, this may be RPM/60, where RPM is the lowest possible rotational speed in revolutions per minute.
2. Determine the required isolator performance, i.e. the amount of vibration reduction required. This amounts to determining the required transmissibility T (or, if performance is given in some other way, e.g. as an isolation efficiency or decibel reduction, then convert this to a value of T).
3. From formulae relating T to f/f_0 (or from graphs of T against f/f_0) determine the ratio f/f_0. Knowing f (Step 1) and f/f_0 (Step 3) calculate the required natural frequency f_0 of the machine/isolator system.
4. Knowing f_0 and m (mass of machine), and number of isolators, determine the required stiffness of the isolators.
5. Select springs from those available with suitable stiffness taking into account factors such as: load, frequency (f_0) range, damping, cost and environmental conditions.

Vibration isolation calculations: summary of formulae

The following formulae are those required for simple vibration isolation design calculations:

Springs (k_1, k_2, k_3 etc.) in parallel: combined or total stiffness, $k_T = k_1 + k_2 + k_3 + \cdots$

Natural frequency of mass–spring system: $f_0 = (1/2\pi)\sqrt{(k/m)}$

Transmissibility of mass–spring system:

$$T = \sqrt{\frac{1 + 4\xi^2(f/f_0)^2}{(1 - (f/f_0)^2)^2 + 4\xi^2(f/f_0)^2}}$$

where ξ = damping ratio. For zero damping ($\xi = 0$) this simplifies to: $T = 1/[(f/f_0)^2 - 1]$

Isolation efficiency $\eta = (1 - T) \times 100\%$

Isolation expressed in decibels (N dB): $N = 20\log(1/T)$, from which $T = 1/10^{N/20}$

Relationship between natural frequency f_0 and static deflection X_s:

$f_0 = 15.8/\sqrt{(X_s)}$ or $X_s = (15.8/f_0)^2$ (where X_s is in mm).

Example 7.17

(a) A fan is bolted to a base-plate which in turn is mounted evenly on steel springs in conjunction with a viscous damper. Explain why it is necessary to select the characteristics of both springs and viscous damper appropriately in order to minimize the transmission of vibrations to the floor. Illustrate your answer by reference to the concept of *transmissibility* and sketch typical transmissibility curves for such a system.

(b) If the total mass of the fan and its base-plate is 1000 kg, and the fan rotates at 1500 RPM, calculate the total stiffness required by the springs in order to reduce the vibratory force transmitted to the floor by a factor of 20. Assume that the damping ratio is zero.

Also calculate the natural frequency of the system, and the static deflection of the springs.

(c) Springs are available with a combined stiffness of 987 kN/m and damping ratios of 0.1, 0.05 and 0.01. Perform calculations, using formulae given below, to decide on the most suitable choice of spring for the above application, bearing in mind that the fan will need to be turned on and off periodically. What would be the consequences of choosing the alternative springs?

Solution

(a) Definition of transmissibility and sketch of graph of T against f/f_0 – see earlier.

The explanation should include effect of damping at/near to resonance, and well above resonance.

Selection of spring stiffness is determined by the need to achieve a high enough value of f/f_0 in order that T shall be low enough to provide required isolation.

Selection of spring damping is a compromise between the desire to achieve maximum possible isolation at fan normal running speed, and the need to limit the amplitude of resonant vibration when the fan moves through the resonant speed during switch on and switch off.

(b) $f = 1500$ RPM/60 = 25 Hz
Use formula $T = 1/[(f/f_0)^2 - 1]$:
$T = 1/20 = 0.05 = 1/[(f/f_0)^2 - 1]$ when $\xi = 0$ (for $f/f_0 < 1$)
From which $f/f_0 = 4.58$, and $f_0 = 25/4.58 = 5.5$ Hz
$f_0 = 15.8/(X_s)^{0.5}$, from which $X_s = (15.8/f_0)^2 = (15.8/5.5)^2 = 8.3$ mm static deflection

$$f_0 = \frac{1}{2\pi}\sqrt{\frac{k}{m}}$$

therefore $k = 4\pi^2 f_0^2 m$ (where $m = 1000$ kg)
from which: $k = 1180$ kN/m

(c) We have to recalculate f_0 and f/f_0 because of the slightly different stiffness of the springs available:
If $k = 987$ kN/m then:

$f_0 = (1/2\pi)\sqrt{(987{,}000/1000)} = 5.0$ Hz, and for normal running speed $f/f_0 = 25/5 = 5$

Substitution into the formula for T, for values of ξ and f/f_0 gives:

	$f/f_0 = 5$	$f/f_0 = 1$
$\xi = 0.01$	0.04	50.0
$\xi = 0.05$	0.05	10.0
$\xi = 0.1$	0.06	5.1

Therefore the springs with $\xi = 0.05$ will give the required isolation provided that the resonant amplification value of 10.0 can be tolerated. The lower value of spring damping ($\xi = 0.01$) gives better isolation but much larger resonant amplification. The higher value of damping ($\xi = 0.1$) gives much lower amplification at resonance, but fails to give required degree of isolation.

Therefore select springs with $\xi = 0.05$.

A critique of the simple vibration isolation design procedure

1. Simple theory assumes a rigid floor or base – a non-rigid base gives less isolation, and problems can arise at the resonance frequency of the floor.
2. Simple theory assumes a point mass vibrating in the vertical (z) direction only. A rigid body has six degrees of freedom – three translational modes of vibration, i.e. in x and y directions (as well as z), and three rotational modes roll, pitch and yaw. The simple design procedure only provides isolation against vibration in the z direction. Vibration in the horizontal (x and y) direction may become a problem if there are vibration producing forces in these directions.
3. Vibration in the rotational modes may become a problem if the machine has a non-uniform weight distribution, or a high centre of gravity. (An inertia base will lower the centre of gravity.)
4. The amount of isolation will depend on the damping in the isolators. Although choosing isolators with low damping will maximize the amount of isolation this will also cause high levels of vibration, when the machine speed passes through resonance during switch on and switch off.
5. The selected isolators will only have the predicted stiffness if they are used within the manufacturer's specified load range, but not if they are overloaded or underloaded. Some springs may also be non-linear. In the case of rubber isolators there may be a difference between static and dynamic stiffness.
6. In order to avoid bridging of isolation (i.e. flanking paths) all connections to the machine (pipes, cables, ducts, etc.) must be via flexible connectors.

Types of isolators

A wide variety of resilient materials may be used as isolators of vibration. At the 'high frequency' end of the range, at 25 Hz and above, where static deflections are small, cork, cork composites, felt, foamed plastic and foamed rubber may be used in the form of pads or mats. All these materials derive part of their springiness from the air they contain and they should be used in compression.

In the intermediate frequency range, 5–35 Hz, rubber and elastomer materials are used, in a wide variety of shapes. They are used either in compression or in shear. Natural rubber can only operate over a limited temperature range and is attacked by oil. These and other disadvantages may be overcome by using one of a wide variety of synthetic rubbers. An important property of rubber, relevant to its use as an isolator, is that its great compressibility arises from its change of shape, hence the frequent use of rubber in shear isolators. A solid rubber cube squeezed between two opposite faces will only be compressible if the other four faces are allowed to bulge outwards. Therefore consideration of the shape of rubber isolators is very important.

Metal springs are used in the lowest frequency range, 2–15 Hz, where static deflections are greatest. Advantages of metal springs are that they can be designed and fabricated into a variety of shapes and configurations to give any required stiffness, and they are able to withstand high loads. They have very good resistance to adverse environments, being unaffected by oil and high temperatures.

A disadvantage is that although they are effective in isolating low frequency vibration, high frequencies travel along the coils of metal springs and are not isolated. For this reason metal springs are often fitted with a pad of neoprene or similar material to reduce the transmission of the high frequencies. Another disadvantage of metal springs is that they have very low damping compared with the other materials mentioned above. This can cause problems when a machine passes through its resonant speed during starting and stopping. To reduce these problems a damping mechanism using some form of friction, or a viscous dashpot can be built into the metal spring isolator.

Pneumatic isolators, in which an air cushion provides the springiness, may also be used to isolate vibrations of the very lowest frequencies.

Inertia bases

Sometimes it is useful to mount a machine such as a motor or fan, or a motor–fan assembly, on a heavy concrete base, called an inertia base, which increases the mass, and then to isolate the assembly from the floor using springs. Some of the advantages of using inertia bases are that:

- they limit the amplitude of the vibration motion
- they give more stability to the system
- they lower the centre of mass of the system
- they provide rigidity between equipment parts
- they minimize reaction torque effects
- they give a more even weight distribution and so provide a more foolproof installation
- they reduce the coupling between the six rigid body modes
- they minimize errors in centre of mass location
- they minimize height variations from variable loading or reaction forces
- they act as a local acoustic barrier, i.e. provide additional sound insulation and so reduce airborne sound transmission through the floor below.

The isolation of a transient vibration

The theory we have considered so far relates to the isolation of continuous vibration, on the basis that isolation occurs if the driving frequency is much higher than the natural frequency. A different approach is required for the isolation of transient vibration such as produced by an industrial punch press, used for example to stamp out shapes from sheet metal. Every time such a punching occurs a transient pulse of force from the press is transmitted to the floor and causes vibration. The key issue determining the effectiveness of isolation is the duration of the transient pulse (T) compared to the period of natural vibrations ($T_0 = 1//f_0$) of the mass spring isolation system.

If $T \ll T_0$, i.e. the exciting pulse only lasts for a small part of the natural cycle of vibration, and does not last long enough to fully excite the natural resonance of the system, then the vibration energy absorbed during the short period of shock excitation is released slowly over a longer period of time, but at much lower vibration amplitudes. This is the condition under which the isolation is effective This situation is illustrated in Figure 7.13 (bottom).

If $T \gg T_0$ then the excitation continues over several cycles of natural vibration resulting in a large amplitude of vibration transmitted to the floor, and the isolation is ineffective (see Figure 7.13 (top right)).

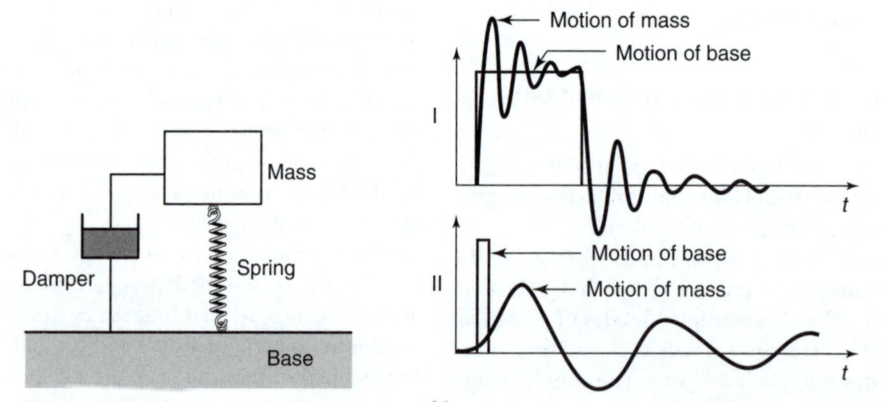

Figure 7.13 Response of a single degree of freedom system to transient vibration excitation

7.6 The assessment of vibration

The effects of vibration on people and on buildings

Depending upon the level, and a variety of other factors, vibration may affect people's comfort and well-being, impair their efficiency at performing a variety of tasks, or even at very high levels become a hazard to their health and safety. A well known example of the harmful effects of vibration is the white finger syndrome (also known as Raynaud's disease) in which prolonged use of hand-held equipment producing vibration, such as certain types of power tools and chain-saws, produces loss of sensation in the fingers. In the first stages of the condition the vibration exposure produces restriction of flow of blood to the tips of the fingers. With further prolonged exposure the damage spreads to the connective tissues of the hands and arms, and in extreme cases damage to the bones of the fingers can occur. The symptoms are exacerbated when the vibration exposure occurs in cold conditions.

Whole body vibration exposure at work, which can arise from riding in vehicles over uneven ground, may affect drivers, particularly of off-road vehicles and of construction equipment, and can cause back and spinal problems.

An EC directive requires member states to take action to control vibration at work from both hand-arm and whole body vibration, and this is enacted in the Control of Vibration at Work Regulations 2005.

The vibration produced by the various forms of transportation (e.g. road traffic, trains, aircraft, helicopters, ships and boats) is of great interest for a variety of reasons. In addition to problems of whole body vibration exposure of drivers, mentioned above, there is the effect of vibration levels on the comfort of passengers; and thirdly, there is often great concern among members of the public about vibration produced in buildings, including domestic dwellings adjacent to roads or railway lines, or near to air routes. A great variety of industrial machinery produces vibration which is experienced by people at work. Particular sources which can cause vibration to be experienced by the occupants of nearby buildings, and thus often give rise to concern among members of the public, include heavy-duty air compressors, forge hammers, pile-driving and quarry-blasting operations.

Recommendations concerning the measurement and assessment of vibration exposure levels are discussed in general terms in the British and International Standards BS 6841 and ISO 2631. The usual preferred method of measurement is in frequency weighted RMS acceleration, in m/s^2, and these standards define a range of different frequency weightings for different assessment situations. When the vibration exposure includes transient or impulsive vibration and shocks, the use of the vibration dose value (VDV) is recommended and is the basis of the assessment method given in BS 6472 for the assessment of human response to vibration in buildings. A different measurement parameter, peak particle velocity, in mm/s, is recommended for the assessment of the potential of damage to buildings caused by vibration, as described in BS 7354 Parts 1 and 2.

Some general information on vibration magnitudes and human response

Both BS 6841 and ISO 2631 give the following indications relating to human perception and comfort when exposed to vibration magnitudes expressed in RMS acceleration, in m/s^2:

Less than 0.315 m/s^2	not uncomfortable
0.315 m/s^2 to 0.63 m/s^2	a little uncomfortable
0. 5 m/s^2 to 1.0 m/s^2	fairly uncomfortable
0. 8 m/s^2 to 1.6 m/s^2	uncomfortable
1.25 m/s^2 to 2.5 m/s^2	very uncomfortable
Greater than 2.0 m/s^2	extremely uncomfortable.

Experience in many countries has shown that occupants of residential buildings are likely to complain if the magnitudes are only slightly above the perception threshold.

Perception thresholds

Fifty per cent of alert fit persons can just detect a W_k weighted vibration with a peak magnitude of 0.015 m/s^2. (Various types of frequency weightings, including W_k, are discussed later in this chapter.)

There is a large variation between individuals in their ability to perceive vibration. Although the median perception threshold is approximately 0.015 m/s^2, the interquartile range of responses may extend from about 0.01 m/s^2 to 0.02 m/s^2 peak.

Broadly similar information about human response to vibration is given in BS 5228-2:2009, but expressed in terms of peak particle velocity (ppv) in mm/s:

> Human beings are known to be very sensitive to vibration, the threshold of perception being typically in the PPV range of 0.14 mms^{-1} to 0.3 mms^{-1}. Vibrations above these values can disturb, startle, cause annoyance or interfere with work activities. at higher levels they can be described as unpleasant or even painful. In residential accommodation, vibrations can promote anxiety lest some structural mishap might occur. Guidance on the effects on physical health of vibration at sustained high levels is given in BS 6841, although such levels are unlikely to be encountered as a result of construction and demolition activities.

Guidance on effects of vibration levels (Table B.2 of BS 5228-2:2009)

Vibration level	Effect
0.14 mms^{-1}	Vibration might be just perceptible in the most sensitive situations for most vibration frequencies associated with construction. At lower frequencies people are less sensitive to vibration
0.3 mms^{-1}	Vibration might be just perceptible in residential environments
1.0 mms^{-1}	It is likely that vibration of this level in residential environments will cause complaint, but can be tolerated if prior warning and explanation has been given to residents
10 mms^{-1}	Vibration is likely to be intolerable for any more than a very brief exposure to this level

Estimation of vibration exposure: RMS, RMQ and VDV

A vibration exposure level is a quantity which is related to the amount of vibration received by an individual over a period of time. It involves a combination of the level of vibration, usually incorporating some form of frequency weighting, and the way in which the level varies with time during the exposure, and the overall duration of the exposure.

The way in which the vibration level varies with time is important; it may include, for example, short bursts of high levels of vibration, such as shocks and impulses, as well as longer periods of lower levels of vibration. The question then arises as to how much emphasis should be given to each of these components?

The usual way of measuring a time varying quantity is to estimate the root mean square (RMS) value over the measurement period T, i.e. in the case of the time varying and frequency weighted acceleration, $a_W(t)$:

$$\text{RMS value of } a_W(t) = \left[\frac{1}{T}\int_0^T a_W^2(t)\,dt\right]^{0.5}$$

The frequency weighting (w) will be one of the possible weightings given in either ISO 2631, BS 6841 or BS 6472.

The RMS value is also known as the 'equivalent' acceleration, $a_{Weq,T}$ because it is the constant value of acceleration which would contain the same amount of vibrational energy as (i.e. be equivalent to) the time varying acceleration, over the measurement time interval. The corresponding quantity for sound pressure, expressed in decibels, would be the equivalent sound pressure level $L_{Aeq,T}$.

Although root mean square (RMS) time-averaging seems to work well for the assessment of human response to steady, continuous vibration, it has been shown to underestimate the subjective effects produced by high peak levels of short duration contained in impulsive types of vibration, and it has been found that using 'fourth power time averaging' gives better correlation, giving rise to the quantities RMQ and VDV.

The root mean quad (RMQ) value of a vibration event is the fourth root of the mean fourth power value, analogous to RMS, which is the square root of the mean square value:

$$\text{RMQ value of } a_W(t) = \left[\frac{1}{T}\int_0^T a_W^4(t)\,dt\right]^{0.25}$$

Note that changing the index from 4 to 2, or from 0.25 to 0.5, in the above expression, will give the definition of RMS value.

Like the RMS, the RMQ is an average value, so it will have the units of acceleration, m/s^2. For a sine wave of amplitude A, the RMS value is $0.7071A$ and the RMQ value is slightly higher: $0.7825A$.

Vibration dose value, VDV

The vibration dose value (VDV) is a measure of vibration exposure used in ISO 2631, BS 6841 or BS 6472, which gives better correlation with human response than RMS based measures when the vibration includes short bursts of high amplitudes such as from impulses and shocks. It is the fourth root of the integral of the fourth power of vibration value with respect to time. The mathematical definition of VDV is given below:

$$\text{VDV} = \left[\int_0^T a_W^4(t)\,dt\right]^{0.25}$$

The units of VDV are $m/s^{1.75}$ because, dimensionally, it is composed as follows:

$$[(ms^{-2})^4 s]^{0.25} = [m^4 s^{-7}]^{0.25} = ms^{-7/4} \text{ or } m/s^{7/4}$$

The following example, although rather simplistic (because it only involves 10 samples), illustrates the process of calculating VDV from a waveform sample.

Example 7.18

Suppose there were 10 acceleration samples in the waveform of a short event: $0, -1, -2, -1, 0, +1, +4, 0, -1, 0$ (all in m/s^2).

The arithmetic mean of these 10 samples is zero. The table below shows the deviations from the mean (d), and the square (d^2) and fourth power (d^4) of these deviations and their means.

	d	d^2	d^4
a_1	0	0	0
a_2	−1	1	1
a_3	−2	4	16
a_4	−1	1	1
a_5	0	0	0
a_6	+1	1	1
a_7	+4	16	256
a_8	0	0	0
a_9	−1	1	1
a_{10}	0	0	0
Totals	**0**	**24**	**276**

Mean deviation $= 0/10 = 0$

Mean $d^2 = 24/10 = 2.4$, and root mean square (RMS) $= \sqrt{2.4} = 1.55\ \text{ms}^{-2}$

Mean $d^4 = 276/10 = 27.6$ and root mean quad (RMQ) $= (27.6)^{0.25} = \sqrt{\sqrt{27.6}} = 2.29\ \text{ms}^{-2}$

Vibration dose value $= (276)^{0.25} = 4.08\ \text{ms}^{-1.75}$

Note that the VDV is very similar to that of the highest sample (a_7) The example therefore illustrates that the fourth power time averaging process emphasizes the peaks in the signal more than the usual RMS time averaging.

Combining dose values

In both BS 6841 and BS 6472 the VDV concept is used to evaluate the cumulative effects of bursts of intermittent vibration and of impulsive vibration. The cumulative effects can be estimated by combining the VDVs of individual events according to the fourth-power law:

$$V_T = [V_1{}^4 + V_2{}^4 + \cdots + V_N{}^4]^{0.25}$$

where V_T is the total VDV and $V_1, V_2, \ldots, V_N$, are the VDVs of the individual events.

If there are N identical events, then:

$$V_T = [NV^4]^{0.25} = N^{0.25}V$$

Estimated vibration dose value, eVDV

In situations where the magnitude of the vibration signal is essentially constant (a) with time over the duration of the vibration event (T) it is possible to simplify matters because the mathematical expression simplifies to:

$$\text{VDV} = aT^{0.25}$$

where 'a' is the average value measured using fourth power time averaging, i.e. the RMQ value. A further simplification can be made to allow the vibration magnitude to be measured as an RMS value, with an adjustment factor of 1.4, to yield the estimated vibration dose value, eVDV:

$$\text{eVDV} = 1.4a_{\text{rms}}T^{0.25}$$

It is considered that eVDV is a reasonably good approximation to the true VDV for vibration signals with a low crest factor (less than 6).

Note: This formula illustrates how the fourth power time averaging process leads to a much greater importance for vibration magnitude than for event duration: one train with double the magnitude (i.e. $2a_{\text{rms}}$) will be equivalent in terms of eVDV to 16 trains with magnitude a_{rms} (the event duration in both cases the same).

Example 7.19

During the passage of a train the vibration measured using an accelerometer attached the ground at a plot of land near to the railway line is $0.0032\ \text{ms}^{-2}$ (RMS), for a period of 15 seconds. Calculate the eVDV of the event.

$$\text{eVDV} = 1.4aT^{0.25} = 1.4 \times 0.0032 \times (15)^{0.25} = 0.0088\ \text{ms}^{-1.75}$$

Combining vibration dose values

If, during a certain period (e.g. a 16-hour daytime period), there are N different vibration events which occur and are measured at a certain location, with VDVs $V_1, V_2, V_3, \ldots, V_N$, then the total VDV is given by:

$$\text{VDV}_{\text{total}} = [(V_1)^4 + (V_2)^4 + (V_3)^4 + \cdots + (V_N)^4]^{0.25}$$

Example 7.20

Three bursts of construction activity produce VDVs of 0.1, 0.2 and 0.4 $\text{ms}^{-1.75}$ during a morning period. Calculate the total VDV during the period.

$$\text{VDV}_{\text{total}} = [(0.1)^4 + (0.2)^4 + (0.4)^4]^{0.25} = 0.42\ \text{ms}^{-1.75}$$

Note that the fourth power averaging process means that the total is dominated by the highest individual value of 0.4.

It follows that if there are N identical events (e.g. trains of the same type) in a certain period, each with a VDV of V_e, then the total VDV during the period is given by:

$$\text{VDV}_{\text{total}} = [N(V_e)^4]^{0.25} = N^{0.25}V_e$$

Example 7.21

Fifty trains of the type similar to that in the previous example (i.e. with an eVDV of $0.0088\ \text{ms}^{-1.75}$) pass by that site during the daytime. Calculate the total eVDV due to the trains during the daytime period.

$$\text{eVDV}_{\text{total}} = (50)^{0.25} \times 0.0088 = 0.023\ \text{ms}^{-1.75}$$

Changing time period over which VDV (or eVDV) is to be assessed

Suppose the VDV is measured during a typical and representative period of vibration activity, for a sample time t_s, and it is required to estimate the VDV over the entire period T, for which this activity takes place:

$$VDV_T = VDV_{ts} \times (T/t_s)^{0.25}$$

Example 7.22

The VDV arising from a construction site at a nearby house is measured for a typical 30-minute period, and is 0.22 $ms^{-1.75}$. Assuming that similar vibration producing activity occurs for the rest of the 10-hour working day, calculate the total VDV from construction during the daytime.

$$VDV_{total} = 0.22 \times (10/0.5)^{0.25} = 0.47\ ms^{-1.75}$$

A brief summary of some regulations and standards relating to vibration exposure

The brief notes below are only intended to give an indication of the main features of each document. As always it is recommended that the full text is studied prior to application.

The Control of Vibration at Work Regulations 2005

(S.I. 2005/1093, came into force on 6 July 2005)

Exposure limit values and action values

These are expressed in terms of an employee's daily vibration exposure level A(8), which is the frequency weighted RMS vibration acceleration in m/s^2 averaged over the working day but normalized to an 8-hour period (irrespective of the actual duration of the working day).

Different frequency weightings are used for hand-arm and whole body vibration as specified in appropriate standards.

For hand-arm vibration:

(a) the daily exposure limit value is 5 m/s^2 A(8)
(b) the daily exposure action value is 2.5 m/s^2 A(8).

For whole body vibration:

(a) the daily exposure limit value is 1.15 m/s^2 A(8)
(b) the daily exposure action value is 0.5 m/s^2 A(8).

Exposure limit values should not be exceeded. Where exposure action values are exceeded, duties are imposed upon employers.

Duties of employers

Where exposure levels are above the action values, employers must:

- carry out assessments of the risk to health created by vibration at the workplace
- eliminate or control exposure to vibration at the workplace
- carry out health surveillance of employees
- provide information, instruction and training to employees.

BS EN ISO 5349-2:2002 – Measurement and evaluation of human exposure to hand-transmitted vibration – Part 2: Practical guidance for measurement at the workplace

Measurement

The basic quantity to be measured is the RMS single axis acceleration value of the frequency weighted hand transmitted vibration, a_{hw}, measured in m/s^2.

This is measured in three directions (x, y, z), preferably simultaneously using a tri-axial accelerometer, to give the three values a_{hwx}, a_{hwy} and a_{hwz}, and the vibration total value (also known as the vector sum) a_{hv} is calculated, being the square root of the sum of the squares of the component values:

$$a_{hv} = \sqrt{(a^2_{hwx} + a^2_{hwy} + a^2_{hwx})}$$

(Where tri-axial measurements cannot be made the Standard (Part 1) suggests how single axis measurements may be adapted to estimate a_{hv}.)

Measurement directions

The three mutually perpendicular axes, x, y, z, are defined in section 4.2.3 and Figure 1 of Part 1 of the Standard, with respect to the clenched fist: x through the hand from back to palm; y along the knuckles (from 4th to 1st); z along the back of the hand from wrist to knuckles. A more precise but more complicated definition is given in the note to Figure 1 in the Standard.

Post-processing of measurement results

The value of a_{hv} is determined for each significant component of the subject's exposure pattern (e.g. each vibration producing process or machine) a_{hv1}, a_{hv2}, . . . , a_{hvi}, a_{hvj} . . . , a_{hvN}, and using the exposure times (t_1, t_2, . . . , t_i, . . . , t_N). For each component the energy equivalent vibration total value, $a_{hv(eq,8h)}$, also known as the A(8) value, is computed over an eight-hour reference period (in a similar way to the calculation of L_{Aeq} for noise, but without the decibels):

$$a_{hv(eq,8h)} = A(8)$$
$$= \sqrt{\{[(a^2_{hv1} \times t_1) + (a^2_{hv2} \times t_2) + \cdots + (a^2_{hvN} \times t_N)]/8\}}$$

which is best illustrated by an example where there are three components to the daily exposure: 2.0 m/s^2 for 1 hour, 3.5 m/s^2 for 3 hours and 10.0 m/s^2 for 0.5 hour, so that:

$$A(8) =$$
$$\sqrt{\{[(2.0^2 \times 1) + (3.5^2 \times 3) + (10.0^2 \times 0.5)]/8\}}$$
$$= 3.4 \text{ m/s}$$

(The A(8) is in effect the vibration exposure equivalent to $L_{EP,d}$ for noise exposure.)

Evaluation/assessment

The estimated values of A(8) should be compared with the action and exposure levels in the Control of Vibration at Work Regulations 2005.

Note that although the need to use measured data in making an assessment of A(8) is acknowledged in Guidance to the Regulations, much information is given on typical acceleration levels produced by various types of machines, in order, as far as possible, to minimize the need for employers to carry out or commission vibration measurements. One reason for this is that although vibration measurements on items of machinery may be accurately carried out, it is difficult to relate them exactly to the levels of vibration entering the hand-arm system, bearing in mind factors such as the way in which the tool may be held, strength of grip etc.

ISO 2631-1:1997 Mechanical vibration and shock – Evaluation of human exposure to whole body vibration – Part 1: General requirements

The scope of the standard

Except for vibration with substantial peaks, with crest factors greater than 9, the method of evaluation adopted in this standard is based on frequency weighted RMS values of acceleration in m/s^2.

The standard defines methods for the measurement of periodic, random and transient vibration and it indicates the principal factors that combine to determine the degree to which a vibration exposure will be acceptable. Guidance is provided on the possible effects of vibration on health, comfort perception and motion sickness. The potential effects of vibration on human performance are not specifically covered although most of the guidance given does apply to this area as well. The principle of preferred methods of mounting transducers for determining human exposure are also defined.

The frequency range considered is:

- 0.5 Hz to 80 Hz for health, comfort and perception, and
- 0.1 Hz to 0.5 Hz for motion sickness.

The standard applies to vibration from sources such as vehicles, machinery and buildings transmitted to the human body as a whole (whether standing, sitting or recumbent) via feet, back or buttocks. It does not apply to extreme magnitude single shocks such as occur in vehicle accidents.

Coordinate system and frequency weightings

A basi-centric coordinate system relative to the human body is used:

- *z* from toe to head
- *x* from back to chest
- *y* from right side to left side.

Two principal frequency weightings are used for evaluation of health comfort and perception: W_k for the *z* direction and W_d for the *x* and *y* directions, with additional frequency weightings used for special cases: W_c (seat back measurements), W_e (rotational vibration) and W_j (under head of recumbent person).

The W_f weighting is used for motion sickness.

The vibration measurement shall be made at the point where the vibration enters the human body, if necessary using a rigid surface interposed between the human body and any resilient material, e.g. a seat cushion or head rest, on which to mount the vibration transducer.

The duration of the vibration measurement shall be long enough for reasonable statistical precision and to be representative of the activity being measured. For a measurement error of less than 3 dB (with a 90% confidence limit) for a lower limiting frequency of 1 Hz the minimum measurement time is 108 seconds.

Additional vibration assessment measures

The basic method of vibration evaluation is to determine the weighted RMS acceleration level, a_w. Where the vibration contains bursts of shock, transient vibration or high crest factors (greater than 9), the basic method may underestimate the severity of vibration with respect to discomfort, and one of two alternative measures, the maximum transient vibration value (MTTV) of the running RMS value, or the vibration dose value (VDV) should also be determined.

The running RMS method involves the use of an integration time, and the standard recommends the use of an integration time of one second for the evaluation of MTTV.

Where either of the alternative measures MTTV or VDV are determined, the value of the basic assessment measure, a_w shall also be measured and recorded. The ratio of the alternative to the basic measure may be used as an indication of the importance of the alternative measure in evaluating human response, particularly if either:

$$MTTV/a_w > 1.5 \quad \text{or} \quad VDV/a_w > 1.75$$

Motion sickness and low frequency vibration exposure

This is covered in Part 9 and Annex D of ISO 2631-1. The use of weighting network W_f is recommended. Guidance on assessment is given in Annex D, a motion sickness vibration value, MSDV, is defined:

$$MSDV = a_W T^{0.5}$$

where a_W is the weighted (W_f) acceleration in m/s^2 and T is the duration of the vibration event in seconds.

Note that because of the low frequency range the vibration measurement periods should be at least four minutes.

The percentage of people who may be affected by motion sickness (i.e. who may vomit) is proportional to k_m MSDV in the z direction, where k_m is typically 0.33. The likelihood of vomiting increases if $a_W > 0.5$ m/s^2.

BS 6472-1:2008 Guide to evaluation of human exposure to vibration in buildings – Part 1: Vibration sources other than blasting

The appropriate measurement parameter is the vibration dose value, VDV, using the appropriate frequency weighting. Where the vibration is continuous and does not vary in magnitude with time and has a crest factor of between 3 and 6 the estimated vibration dose value, eVDV, may be used as an approximation of the VDV. The use of eVDV is not recommended for time varying vibration or for shocks.

Frequency weightings

Use frequency weightings W_d for horizontal directions (x and y axes) and W_b for vertical direction (z axis). Weightings are as defined in BS 6841. Measure vibration in each direction unless it can be shown that vibration in one direction is dominant. Assess vibration exposure in each direction separately.

Measurement position. Measure, as far as possible, at point of entry of the vibration into the human body, e.g. for a standing person measure vibration on the floor. Where the exposed person is moving about measure at 'worst case' position, i.e. usually at the centre of the floor. Where it is not possible to measure at the point of entry to the human body a transfer function between the measurement point and the point of entry to the human body must be used.

Post-processing of measurement results. The measurement results will be samples of different vibration events or activities which occur through the day or night. They should be aggregated or combined to produce the VDV over the 16-hour daytime or eight-hour night-time period to produce the period values, with subscripts used to indicate the frequency weightings and periods, e.g. $VDV_{d,night}$ or $VDV_{b,day}$.

Evaluation or assessment. The estimated day or night dose values should be compared with Table 1 of the standard (below) to indicate the likelihood of adverse comment.

Report. As with many measurement procedure standards there is section at the end usually called 'Information to be recorded and reported'. In this case it is contained in Annex A and called 'Suggested format and content of an assessment report'. This section should be used as a checklist when reporting your measurements and assessment, to ensure that all necessary information has been included.

Table 1 BS 6472 vibration assessment criteria

Vibration dose ranges which might result in various probabilities of adverse comment within residential buildings

Place	Low probability of adverse comment*	Adverse comment possible	Adverse comment probable**
Residential buildings 16-hour day	0.2 to 0.4	0.4 to 0.8	0.8 to 1.6
Residential buildings 8-hour night	0.1 to 0.2	0.2 to 0.4	0.4 to 0.8

* Below these ranges adverse comment is not expected.

** Above these ranges adverse comment is very likely.

BS 7385-1:1990 (ISO 4866:1990) Evaluation and measurement for vibration in buildings – Part 1: Guide for measurement of vibrations and evaluation of their effects on buildings; Part 2: 1993 Guide to damage levels from groundborne vibration

Where to measure

For building damage assessment the preferred position is at the foundation, a typical location being at a point low on the main load-bearing external wall at ground floor level when measurements on the foundations proper are not possible. Measurements should be taken on the side of the building facing the source of vibration. Where it is not feasible to measure at the foundation of the building, the measurement should be obtained on the ground, outside of the building. One of the horizontal vibration components should be in the radial direction between the source and the building in the case of ground measurements or oriented parallel with a major axis of the building.

Where a detailed engineering analysis of the building vibration is required, vibration measurements at locations other than the base of the building should be taken. Where a building is higher than four floors (12 m), subsequent measuring points should be added every four floors and at the highest floor of the building. Where a building is more than 10 m long, measuring positions should be installed at horizontal intervals of approximately 10 m.

What to measure

Peak particle velocity (ppv) has been found to be the best single descriptor for correlating with case history data on the occurrence of vibration-induced damage.

The preferred method of measuring ppv is to record simultaneously unfiltered time histories of the three orthogonal components of particle velocity, which allows any desired value to be extracted at a later stage. The maximum of the three orthogonal components should be used for the assessment.

Assessment

The case history data suggests that the probability of damage tends towards zero at levels below 12.5 mm/s peak component particle velocity. The limit for cosmetic damage varies from 15 mm/s at 4 Hz to 50 mm/s at >40 Hz for measurements taken at the base of the building. Different low frequency limits (<40 Hz) are given for two different types of buildings. The limits for cosmetic damage should be doubled for minor damage, and doubled again for major damage. The limits for cosmetic damage are shown in Figure 7.14 and the table below.

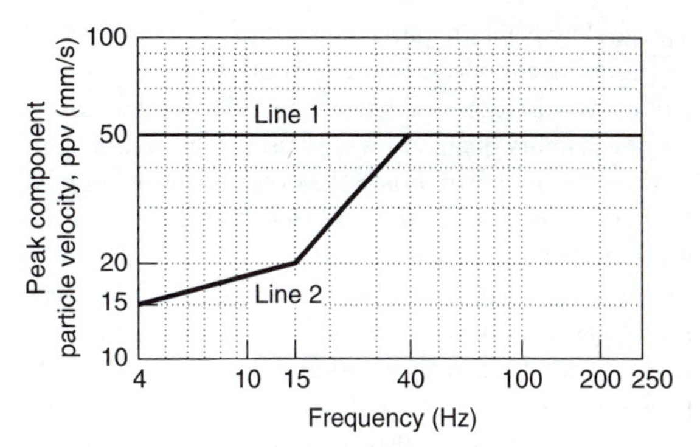

Figure 7.14 Transient vibration guide values for cosmetic damage

Line	Type of building	Peak component particle velocity (mm/s) in frequency range of predominant pulse	
1	Reinforced or framed structures Industrial and heavy commercial buildings	50 at 4 Hz and above	
2	Unreinforced or light framed structures	4 Hz to 15 Hz	15 Hz and above
	Residential or light commercial type buildings	15 at 4 Hz increasing to 20 at 15 Hz	20 at 15 Hz increasing to 50 at 40 Hz and above

BS 6841:1987 British Standard Guide to the measurement and evaluation of human exposure to whole-body mechanical vibration and repeated shock

The British Standard Guide gives methods for quantifying vibration and repeated shocks in relation to human health, interference with activities, discomfort, the probability of vibration perception and the incidence of motion sickness. The guide is applicable to motions transmitted to the body as a whole through the supporting surfaces: the feet of a standing person, the buttocks, back and feet of a seated person and the supporting area of a recumbent person. The four principal effects of vibration considered by the guide are:

- degraded health
- impaired activities (e.g. hand manipulation, effects on vision)
- impaired comfort (and perception)
- motion sickness.

The guide specifies requirements for the measurement of vibration magnitude, measurement of the frequency content of the vibration and the direction of measurement.

The primary quantity for expressing vibration magnitude is the root mean square (RMS) acceleration, in m/s^2 for translational vibration and rad/s^2 for rotational vibration. RMS values give good correlation with human response for steady, continuous vibration, but it has been found that the severity of vibrations which are intermittent, or impulsive, with occasional short duration, high peak values will often be underestimated by RMS values. Therefore the standard also gives alternative methods for measurement and evaluation in these cases, leading to the establishment of a vibration dose value.

The guide specifies measurements in a direction relative to the axes of the human body. The basi-centric coordinate system is used similar to that in ISO 2631. Measurements should always be taken as close as possible to the interface between the human body and the source of the vibration. Human response to vibration depends on the frequency of the vibration, and the guide defines six different frequency weighting networks (rather like the A-weighting used for sound). Designated W_b, W_C, W_d, W_e, W_f and W_g, they cover the frequency range from 0. 1 Hz to 100 Hz, although frequencies below 0.5 Hz relate only to travel sickness and to the W_f weighting. The choice of the appropriate weighting depends on the different effects of vibration which are being assessed (e.g. health, activity, comfort) and on the direction of vibration measurement.

Note: This standard is under revision.

ISO 8569:1996 Mechanical vibration and shock – Measurement and evaluation of shock and vibration effects on sensitive equipment in buildings

The standard defines methods of measurement and of reporting data. The frequency range of interest is between 0.5 Hz and 250 Hz, and vibration amplitudes are typically in the range 10^{-4} to 2×10^{-2} m/s.

As far as possible, vibration measurement should be made with the vibration sensitive equipment in place and both with the equipment running and with it not running. If necessary a dummy with the same mass and dynamic behaviour as the vibration sensitive equipment shall be used in place of the equipment.

Vibration should be measured on the surface supporting the equipment (i.e. floor or wall etc.) as close as possible to the equipment (less than 0.2 m away), or, if necessary, on the equipment itself.

The minimum measurement requirement for a field survey is that either the peak particle velocity or the peak acceleration level shall be recorded. Equipment shall be calibrated. For a full engineering analysis, time histories (in three orthogonal directions) and frequency analysis are also required. If the vibration source is repetitive the measurement shall include several cycles.

A proforma for recording data is provided (Annex A of the standard).

7.7 Radiation of sound from a vibrating surface

The plane wave radiator

It is possible to predict the sound pressure level fairly easily for the case of vibrating surface which radiates plane waves, in effect from an infinite rigid vibrating plane surface.

Assume that the vibration velocity of the vibrating surface is V m/s. The task is to predict the acoustic sound pressure, p, the acoustic particle velocity, U, the acoustic intensity, I, and the sound intensity level, L_I and sound pressure level, L_p at a point in front of the vibrating surface. See Figure 7.15.

Step 1

In a plane wave there is no divergence of the wavefront and so, assuming also that there is no absorption, the particle velocity U does not vary with the distance from the source.

Since there must be physical continuity of particle velocity at the rigid vibrating plane, this means that $U = V$. This is a key step: $U = V$.

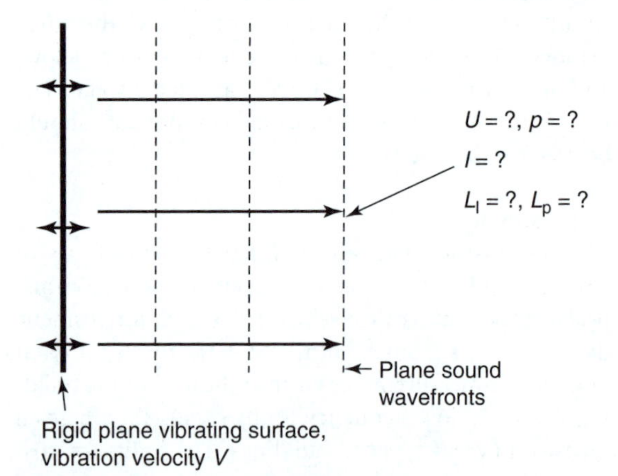

Figure 7.15 The radiation of plane waves by a vibrating surface

Step 2

Plane wave relationships, between p, U and I are:

$$p = Uz$$
$$I = pU$$
$$I = p^2/z$$
$$I = zU^2$$

where $z = \rho c$.

Step 3

The vibration velocity V of the source may be related to the acoustic particle velocity in the medium, and via the plane wave relationships it is possible to relate the acoustic particle velocity, U, to the acoustic pressure and intensity and sound pressure level.

There are various alternative ways to proceed:

(a) using $p = \rho c\, U$ and $L_p = 10\log(p/p_0)$
(b) or using $I = \rho c z^2$ and $L_I = 10\log(I/I_0)$, and in a plane wave $L_p = L_I$.

Example 7.23

The vibration of a machine panel is measured and found to be 1 g rms ($= 9.81$ m/s^2) with the dominant vibration frequency being 1000 Hz. Estimate the sound pressure level radiated by panel assuming that the panel behaves like a plane wave radiator.

Solution

$$A = 2\pi f V$$
$$V = A/2\pi f$$
$$V = (9.8)/2\pi(1000) = 1.6 \times 10^{-3} \text{ m/s}$$

Therefore: $U = V = 1.6 \times 10^{-3}$ m/s $= 1.6$ mm/s
Using option (a), using $\rho c = 415$ kgm^{-2}s:

$$p = \rho c V = 415 \times 1.6 \times 10^{-3} = 0.65 \text{ Pa}$$

from which:

$$L_p = 20\log(p/p_0) = 20\log(0.65/0.00002) = 90 \text{ dB}$$

Or, using option (b):

$$I = zV^2 = 415 \times (1.6 \times 10^{-3})^2$$
$$= 1.1 \times 10^{-3} \text{ W/m}^2$$

Sound intensity level:

$$L_I = 10\log(I/I_0) = 10\log(1.1 \times 10^{-3}/1.0 \times 10^{-12})$$
$$= 90 \text{ dB}$$

and, assume $L_p = L_I = 90$ dB.

Radiation from real vibrating surfaces

In reality, vibrating surfaces are not infinite in extent and may not be rigid, and the sound waves they radiate may not be ideal plane waves.

The above calculation procedure may be adapted by assigning a radiation efficiency factor, σ, to the vibrating surface. The radiation efficiency is the ratio of the sound energy radiated by a vibrating surface divided by the sound energy that would be radiated by the same surface assuming it radiated plane waves.

Therefore by definition $\sigma = 1$ for an ideal plane wave radiator, and for real sources in practice, σ is a number which lies between 0 and 1.

Note: Theoretically σ is related to the real part of radiation impedance of the surface, and it is possible, theoretically, for the radiation efficiency sometimes to be (very slightly) greater than 1, but in practice the sound pressure level predicted using a value of $\sigma = 1$ should be regarded as an upper limit, and possibly an overestimate in most cases, where σ will be less than 1.

Since σ is defined in terms of sound energy which is more directly related to sound intensity than sound pressure, the incorporation of σ into calculations is best suited to option (b) above, as follows:

$$I = \sigma z V^2$$

Sound power radiated by a vibrating surface

Real surfaces are finite in extent, and for a vibrating surface of area S m^2 the sound power radiated may be calculated using the relationship that $I = W/S$ and $W = IS$:

$$W = \sigma S z V^2$$

This is a key formula often used as the starting point for calculations.

The sound power level may be calculated using $L_W = 10\log(W/W_0)$, where $W_0 = 1.0 \times 10^{-12}$ W/m^2.

The effect of σ in decibels

If we use option (a) above we need to adjust the answer obtained assuming plane wave radiation to take into account radiation efficiencies of less than one by adding a factor $10\log(\sigma)$. For example, in the above example if the radiation efficiency is known to be $\sigma = 0.5$ then:

$$\text{Corrected } L_p = \text{Ideal } L_p + 10\log(\sigma)$$
$$= 90 + 10\log(0.5) = 90 - 3 = 87 \text{ dB}$$

Prediction of sound pressure levels at distances from real radiating surfaces

In the above example we were told to assume that a vibrating machine panel behaved like a plane wave radiator. We were not given the dimensions of the panel. The answer we obtained, 90 dB, would correspond to sound pressure level measured close the panel surface, but in reality the sound pressure would drop off with distance because the panel was not in fact infinite in extent.

If we know that the dimensions of the (rectangular) panel are 0.1 m × 0.2 m and that the radiation efficiency is 0.5 we can proceed as follows (recalling that the vibration velocity of the panel was 1.6×10^{-3} m/s).

Sound power radiated by the panel:

$$\begin{aligned} W &= \sigma S z V^2 \\ &= 0.5 \times 415 \times (0.1 \times 0.2) \times (1.6 \times 10^{-3})^2 \\ &= 1.1 \times 10^{-5}\ \text{W} \end{aligned}$$

and sound power level, L_W is given by:

$$L_W = 10\log(1.1 \times 10^{-5}/1.0 \times 10^{-12}) = 70.3\ \text{dB}$$

If we want to calculate the sound pressure level at a distance of 10 m from the panel we can assume that at this distance the panel will behave like a point source (because distance 10 m is very much larger than the largest source dimension, 0.2 m), and so we can use:

$$\begin{aligned} L_p = L_W - 20\log r - 11 &= 70.3 - 20\log 10 - 11 \\ &= 39.3\ \text{dB} \end{aligned}$$

Radiation from rigid finite surfaces

Initially we had to assume that in order to radiate plane waves the vibrating surface was both rigid and infinite.

In the case of rigid finite surfaces the radiation efficiency depends on the size of the radiating panel compared to the wavelength of the sound being radiated. If the panel size is large compared with the wavelength then σ will be large (close to 1) and the panel will be an efficient radiator, whereas if the wavelength is small compared to panel dimensions then σ will be small and it will be an inefficient radiator.

Figure 7.16 indicates how the radiation efficiency varies with frequency for a rigid circular piston in an infinite baffle (see more advance acoustics textbooks, such as Kinsler and Frey, *Fundamentals of Acoustics*, for accurate graph).

Radiation from a non-rigid (i.e. flexible) surfaces

The first effect of having a non-rigid surface is that the vibration velocity, V, will not be the same everywhere on the surface. If the panel is clamped at its edges the velocity will be greatest in the centre of the panel and zero at the edges.

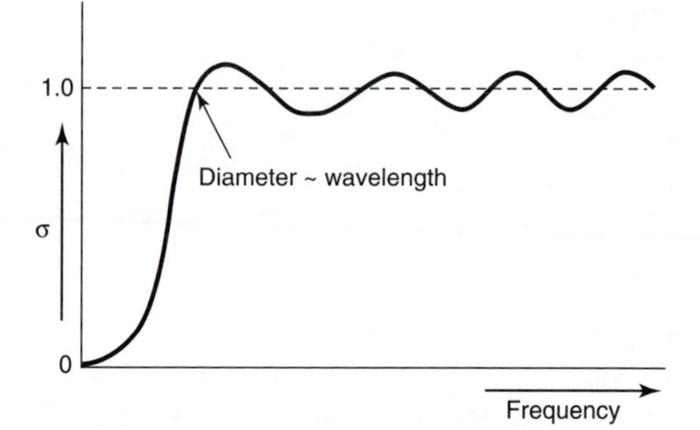

Figure 7.16 Variation of radiation efficiency of a vibrating surface with frequency

The approach used earlier can be adapted by taking several measurements of the panel velocity at different positions and using an average value. However, it is the average V^2 value that must be used in the formula for W (and not the square of the average value of V):

$$W = \sigma S z V^2_{\text{avge}}$$

This equation can be expressed in decibel form to give the sound power level L_W:

$$L_W = 10\log(V^2_{\text{avge}}) + 10\log S + 10\log\sigma + 146$$

or as $L_W = L_V + 10\log S + 10\log\sigma - 34$

where L_V is the vibration level in decibels relative to a reference of 10^{-9} m/s^2.

Sound cancellation for higher flexural panel modes

The above approach will be suitable for low frequency flexural modes such as the (1, 1) mode where all parts of the vibrating surfaces will be vibrating in phase, albeit with different amplitudes. However, at higher frequencies vibration modes occur (e.g. the (2, 2) mode – see Chapter 4) where the vibration of certain parts of the surface are out of phase with that from other parts. In these circumstances the sound from parts of the surface are cancelled (because of destructive interference) by radiation from other parts. The estimation of radiation efficiency in such cases is complicated but in general the effect of such cancellation is likely to reduce the radiation efficiency.

Relationship between radiation efficiency and critical frequency

We learnt in Chapter 6 that above the critical frequency, f_C, sound waves on one side of a partition are effective in exciting panel vibration, which then results in sound being radiated into the space on the other side, and a resulting reduction in the sound insulation produced by the partition. In other words, above the critical frequency the partition is an efficient radiator of sound, but a poor sound insulator, with the situation being reversed below the critical frequency. Therefore, in general, and in the absence of more specific predictions, radiation efficiencies of panels and partitions are usually regarded as being high (close to 1) at frequencies above the critical frequency, and low at frequencies below f_C.

Appendix 7.1 The mathematical theory of the mass–spring–damper system

The purpose of this appendix is to introduce the theoretical basis underlying the discussion of the motion of a mass–spring–damper system introduced in this chapter. It is not intended to be a full detailed mathematical treatment, but rather an introduction to the mathematical ideas for those unfamiliar with them and to act as a stepping stone to those wishing to move on to more advanced texts.

Relationship between displacement, velocity and acceleration – the terminology of calculus

Velocity is the rate of change of displacement with respect to time, and acceleration is the rate of change of velocity with respect to time. Differential calculus is the branch of mathematics concerned with determining the instantaneous rates of change of quantities and it may therefore be used to determine the relationship between these three quantities.

In the terminology of calculus the instantaneous rate of change of variable x with respect to time, t, is written as either dx/dt (dee x by dee t) or as $\dot{x}$ (x dot).

In this case x represents displacement and therefore dx/dt or $\dot{x}$ represents velocity, v.

The process of finding dx/dt or $\dot{x}$ for any mathematical function x is called differentiation, and there are straightforward rules for doing this for simple functions.

Acceleration, a, is the rate of change of velocity with respect to time, i.e. $a = dv/dt$ or $\dot{v}$.

It is also the rate of change of the rate of change of displacement, so that acceleration, a, may also be written as d^2x/dt^2 or $\ddot{x}$ or dv/dt or $\dot{v}$.

Relationship between amplitudes of displacement, velocity and acceleration

For a single frequency sinusoidal vibration the instantaneous vibration displacement, x, varies with time, t, according to:

$$x = X\sin 2\pi ft = X\sin\omega t$$

where X is the amplitude of displacement, and $\omega = 2\pi f$.

Velocity, v, is the rate of change of displacement with respect to time.

Using the rules for differentiation gives in this case:

$$v = \dot{x} = X\omega\cos\omega t = V\cos\omega t$$

where V (velocity amplitude) $= \omega X = 2\pi fX$

The instantaneous acceleration, a, is obtained by differentiating the expression for the instantaneous velocity with respect to time. Using the rules for differentiation gives in this case:

$$a = \dot{v} = -V\omega\sin\omega t = -A\sin\omega t$$

and also, since $a = -\omega^2 x$, $a = -4\pi^2\omega^2 f^2 x$

$$= -4\pi^2\omega^2 f^2 X\sin\omega t$$

where A (acceleration amplitude) $= \omega V = 2\pi fV$.

In summary: $V = 2\pi fX$

$$A = 2\pi fV$$

$$A = 4\pi^2 X$$

The theory of the mass–spring system

Figure 7.17 shows the forces and direction of motion of the mass–spring system.

The mathematical theory of the mass–spring system is based on the principle that the vibrating mass must obey Newton's second law, i.e. at any point in its cycle the total force acting on the mass (kx) and its acceleration (x), are related by:

Force $=$ *mass* $\times$ *acceleration*

i.e. $-kx = m\ddot{x}$

The minus sign indicates that the force is always opposing the motion, i.e. always in the opposite direction to that of the acceleration and always directed towards the equilibrium ($x = 0$) point, so that when the displacement is increasing the mass is slowing down, and vice versa.

The above equation is more often written in the form:

$$m\ddot{x} + kx = 0$$

This equation which is known as the equation of motion for the mass is a very well known second order differential equation, and it has a solution of the form:

$$x = A\sin\omega t$$

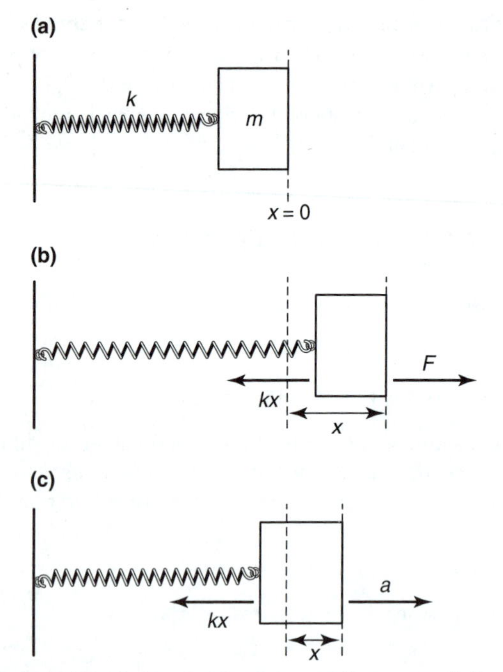

Figure 7.17 (a) showing the mass–spring system in its undisturbed state, i.e. when displacement $x = 0$; (b) showing the system at some arbitrary position in its vibration cycle after being disturbed by the force F; (c) at another point in the cycle after the force F has been removed. The direction of the acceleration a is shown.

where A is an unknown constant. It is easy to verify, using calculus to differentiate the expression for x (twice, in order to obtain $\ddot{x}$), that the solution does in fact satisfy the differential equation, provided that the following condition is met:

$$\omega^2 = k/m$$

$$\text{i.e. that } \omega = \sqrt{(k/m)}$$

$$\text{i.e. that } 2\pi f_0 = \sqrt{(k/m)}$$

$$\text{i.e. that } f_0 = (1/2\pi)\sqrt{(k/m)}$$

Note: The expression for f_0 is independent of the constant A, which will be determined by the initial conditions, i.e. the initial disturbance causing the vibration. It can also be shown that $x = B\cos\omega t$ is also equally valid as a solution to the differential equation, and so is any combination of these two, such as $x = A\sin\omega t + B\cos\omega t$, where B is another unknown constant.

Exercise

Can you explain the physical significance of the different (sine and cosine) solutions?

Answer

The $A\sin\omega t$ solution represents a motion which started from rest at time $t = 0$ (because $\sin 0 = 0$), with $x = 0$ but with $v = V$, i.e. the disturbance was in the form of a 'push' to the mass, giving it a velocity V at the zero displacement position. The $B\cos\omega t$ solution represents a motion which starts from one of its two 'extreme' positions, i.e. the system was disturbed by pulling the mass away from its rest position, and letting go, at time $t = 0$ (because $\cos 0 = 1$), so that at $t = 0$: $x = X_0$ and $v = 0$. All possible combinations of these two situations can also occur.

Damped natural vibration

The presence of damping means that there is an additional force which acts on the mass in the mass–spring system, in addition to the restoring force, kx, of the spring. For viscous damping, this damping force will be proportional to the velocity of the mass ($\dot{x}$) and may be written as $r\dot{x}$, where r is called the damping coefficient. This force acts in the same direction as the spring force, i.e. opposing the motion of the mass, and can be included in the equation of motion of the mass:

$$-kx - r\dot{x} = m\ddot{x}$$

$$\text{or } m\ddot{x} + r\dot{x} + kx = 0$$

Note: The damping coefficient, r, is measured in units of Nsm^{-1}, or since the newton has units of kgms^{-2}, in kgs^{-1}.

We shall not give the detailed mathematical solutions to this equation here, but only describe them. There are three different solutions, depending upon the amount of damping, i.e. depending upon the value of the damping coefficient, r, and they correspond to the three physical situations of underdamping, overdamping and critical damping.

The value of damping coefficient r_c, which causes critical damping is given by:

$$r_c = 2\sqrt{(km)}$$

The amount of damping in a vibrating system is described by its damping ratio, ξ:

$$\xi = \text{(damping coefficient of the system)}/\text{(damping coefficient when critically damped)}$$

The value of $\xi = 1$ for a critically damped system, $\xi > 1$ for overdamping, and $\xi < 1$ for underdamping. Values of ξ can be as low as 0.01 (or 1% of critical) for very lightly damped materials, such as mild steel and as high as 0.2 (or 20% of critical) for highly damped materials.

The solution to the differential equation of motion ($m\ddot{x} + r\dot{x} + kx = 0$) for the case of underdamping has the following form:

$$x = A \sin \omega_d t e^{-(r/2m)t}$$

or, more generally:

$$x = (A \sin \omega_d t + B \cos \omega_d t) A e^{-(r/2m)t}$$

This solution describes damped sinusoidal vibration. It differs from the undamped case in two ways:

1. The angular frequency of the damped free vibrations, ω_d is slightly less than the ω_0 of the undamped case, ($\omega_d = \omega_0 \sqrt{(1 - \xi^2)}$).
2. The amplitude of the vibration is no longer constant, but decays exponentially with time, at a rate which depends on the damping coefficient, r (via the decay constant $r/2m$).

Forced vibration of a mass–spring system

If the mass is subjected to a continuous vibratory force which can replace the energy lost because of damping it will be forced into continuous vibration at the driving frequency. We need only consider the response to a single frequency vibratory force, since more complicated forces can be broken down into a combination of sinusoidal forces. The inclusion of the driving force term, $F \sin\omega t$, into the equation of motion of the mass gives:

$$-r\dot{x} - kx + F \sin \omega t = m\ddot{x}$$

or

$$m\ddot{x} + r\dot{x} + kx = F \sin \omega t$$

The solution of this equation is of the form $x = A\sin(\omega t - \alpha)$ where A is the amplitude of the forced vibration, which is no longer a constant as in the case of free undamped vibrations, but which varies with frequency, and α is an angle which represents the phase difference between the driving force and the response.

The frequency dependent amplitude, A, is given by:

$$A = \frac{F/k}{\sqrt{[1 - (f/f_0)^2]^2 + [2\xi(f/f_0)]^2}}$$

Where F is the amplitude of the force, and (F/k), the numerator in the above expression, represents the static deflection which would be produced by this force (i.e. at zero Hertz). Therefore the ratio $A/(F/k)$, which is represented by the denominator in the above expression, is the ratio of the forced vibration amplitude to the static deflection, called the dynamic magnification, and is plotted, as a function of the frequency ratio, f/f_0, in Figure 7.9. The graph shows that the vibration amplitude increases sharply when the driving frequency, f approaches the natural frequency of the system, f_0. This is the phenomenon of **resonance,** and the forced vibration amplitude is a maximum at the resonance frequency, which, for a lightly damped system is almost the same as the natural frequency.

The amplitude of vibration at resonance depends on the amount of damping in the system, i.e. on the damping ratio, ξ, as shown in Figure 7.9. The **Q factor** of the system, also related to the amount of damping, may be obtained from the 'resonance' curve of Figure 7.9 in two ways: from the maximum amplitude, at resonance, and from the width of the resonance peak.

$$Q = \frac{\text{displacement amplitude at resonance}}{\text{static displacement produced by the same force}}$$

Therefore a Q factor of 10 means that the forced vibration amplitude at resonance is 10 times the static deflection. Alternatively Q may be found from the width of the resonance curve, using the formula:

$$Q = f_0/(f_2 - f_1)$$

where f_1 and f_2 are the so-called half power points, i.e. the frequencies either side of resonance at which the forced vibration amplitude has dropped to $1/\sqrt{2}$ of its resonance value.

The Q factor and the damping ratio are related by:

$$Q = 1/(2\xi)$$

Example 7.24

The natural frequency of a mass spring system is 150 Hz, and its half power points determined from its resonance curve are 155 Hz and 145 Hz. Calculate the Q factor and damping ratio.

$Q = 150/(155 - 145) = 150/10 = 15$

Damping ratio $\xi = 1/(2 \times 15) = 0.033$

Transmissibility

The transmissibility is the ratio of the force transmitted via the spring to the base to the force applied to the mass, F_t/F.

The transmitted force is the combination of the restoring force of the spring and the damping force:

$$F_t = kx + r\dot{x}$$
$$T = F_t/F$$

From earlier we have:

$$x = A\sin(\omega t - \alpha)$$

where $A = (F/k)/D$

and $D = \sqrt{\{[1 - (f/f_0)^2]^2 + [2\xi(f/f_0)]^2\}}$

By differentiating the expression for x to obtain $\dot{x}$ and then with some algebra it is possible to obtain the formula for transmissibility used in the main text:

$$T = \sqrt{\frac{1 + 4\xi^2 (f/f_0)^2}{(1 - (f/f_0)^2)^2 + 4\xi^2 (f/f_0)^2}}$$

Use of complex exponential notation

The algebra is simplified considerably by the use of the complex exponential notation.

This allows the solution of the equation of motion:

$$m\ddot{x} + kx = 0$$

to be expressed in the form:

$$x = Ae^{j\omega t} = A[\cos\omega t + j\sin\omega t]$$

The combination of a sine function (e.g. $A \sin \omega t$) and a cosine function (e.g. $B \cos \omega t$) is here written as a complex number $[A \sin \omega t + jB \cos \omega t]$, where j is the complex operator $\sqrt{(-1)}$. The significance of the 'j' operator is to indicate that the sine and cosine components are 90° out of phase with each other.

Using this complex number notation considerably simplifies the differentiation of the solution to obtain $\dot{x}$ and $\ddot{x}$ from x, as compared to using the sine and cosine version of the solution:

$$\dot{x} = Aj\omega e^{j\omega t} = j\omega x$$

$$\text{and } \ddot{x} = Aj^2\omega^2 e^{j\omega} = j^2\omega^2 x = -\omega^2 x$$

The reader may wish to confirm that the solution expressed in this form does indeed satisfy the equation of motion for free undamped motion, and leads to the well known formula for the resonance frequency.

Questions

1 The vibration acceleration amplitude of the floor of an office situated next to a workshop is 0.3 m/s^2 at a frequency of 30 Hz. Find the velocity and displacement amplitudes.

2 The vibration displacement amplitude measured on the ground floor of a house near to a building site is found to be 0.002 mm at a frequency of 50 Hz. What are the corresponding amplitudes of velocity and acceleration?

3 The maximum vibration level allowed in a certain working area is 10 mm/s. Find the corresponding maximum permitted displacement and acceleration levels for vibrations of 10 Hz frequency.

4 Two vibration levels are measured and the results quoted in terms of decibels relative to 1 *g* (9.81 m/s^2). The first level is +7 dB and the second is −5 dB. Convert both levels into absolute values (m/s^2) and into decibels relative to 10^{-6} m/s^2.

5 The acceleration vibration level is measured on the casing of a power tool, using a sound level meter adapted for vibration measurements. The level is found to be 84 dB, in arbitrary decibel units. The meter is then calibrated using a vibrating calibration table which produces a level of 7.07 m/s^2. This produces a reading of 97 dB on the meter. What is the level on the casing in m/s^2?

6 The vibrational velocity of a surface is given as 103 dB re. 10^{-6} mm/s. Express this in absolute terms (i.e. in mm/s).

7 The resonance frequency of a thin machine panel has been estimated as 1200 Hz, and its effective mass as 120 g. The vibration levels at the centre of the panel are to be measured using an accelerometer of mass 80 g. Calculate the resonance frequency of the panel with the accelerometer attached.

8 A large machine is vibrating at a frequency of 850 Hz. Measurement of the vibration level indicates an RMS acceleration of 12.3 ms^{-2}. Estimate the sound pressure level radiated from the panel at a position close to the panel. Explain the assumptions made in your calculation. (Take the specific acoustic impedance of air to be 410 rayls.) (IOA 1990)

9 (a) Describe and explain the relationship between the vibration of a surface and the amplitude of the radiated sound.

(b) Explain why the directivity of the radiation from a vibrating rigid body depends on the ratio of the dimensions of the body to the wavelength of the radiated sound.

(c) A steel panel (6 m × 3 m) situated in the side of a ship (beneath the water line) is found to vibrate in the 125 Hz one third octave band due to structural excitation from a nearby pump. The spatially averaged one third octave acceleration level is measured as 100 dB re. 10^{-5} ms^{-2}, and the radiation efficiency is 0.004. Estimate the sound power radiation to the water (in W), and hence the expected underwater sound pressure level at a distance of 60 m from the panel and express the result in terms of dB re. 1 μPa. (Use (ρc)water = 1.58 × 106 kgm^{-4}s^{-1}) (IOA 2007)

10 (a) Describe **four** undesirable effects which can result from the exposure of the human body to whole-body vibration.

(b) What is the relevant British Standard for assessing the acceptability of intermittent whole body vibration exposure in buildings? Give the suggested assessment quantity and its units.

(c) The operator of a civil engineering machine is exposed to undesirable levels of vibration at 06 Hz, oriented along the vertical axis. It is suggested that the operating platform should be mounted on springs to reduce this exposure. What should be the natural frequency for the system if the amplitude of the vibration is to be reduced by a factor of 4? Damping may be ignored. (IOA 2007)

11 A water company has installed a diesel engine and pump on a steel frame with a total mass of 19,000 kg which has to be isolated from the pump house floor. The machine operates at a rated speed of 3000 rpm.

(a) If an isolation efficiency of 90% is required, express the required vibration reduction in decibels.

(b) A simple spring–mass–damper model is used to select the anti-vibration mounts. According to the simple model, what spring constant would be required if damping is neglected and there are 10 mounts distributed uniformly around the frame?

(c) What would be the static deflection of the vibration isolators under the weight of the engine and pump?

(d) What isolation efficiency would be achieved if the pump and engine were operated at half the rated speed? (IOA 2008)

12 A machine with an operating speed of 300 rpm and mass of 50 kg is mounted on four parallel springs. Calculate the required stiffness of each spring to give a reduction in the transmitted force of 20 dB. (IOA 2004)

13 A large diesel engine generator set of mass 18,000 kg running at 1500 rpm is to be isolated from the floor using eight identical springs. Calculate the required spring stiffness in order to achieve an isolation efficiency of 95%. Assume zero spring damping.

(IOA 2003 (adapted))

14 (a) Describe using sketch graphs how the amplitude of oscillation of a damped oscillatory system varies with frequency. How does the response of the system change with increasing damping? Indicate with some explanation the frequency range over which the system is: stiffness controlled, damping controlled, mass controlled.

(b) Distinguish between force and displacement transmissibility of anti-vibration systems. What are the advantages and disadvantages of incorporating damping into the design of practical anti-vibration systems.

(c) An anti-vibration system has a static deflection of 4 mm. Over what frequency range will the isolation efficiency exceed 95%? (Ignore damping.) (IOA 1990)

15 The RMS vibration in the concrete floor of a studio above an underground railway is 3×10^{-6} ms^{-1}. The floor area is 30 m^2 and the sound power level it radiates due to the vibration is 50 dB re. 10^{-12} W. Calculate the radiation efficiency of the floor.

Use the formula $W = \sigma S \rho c V^2$. Take the specific acoustic impedance of air as 415 kgm^{2}s^{-1}. (IOA)

16 An employee of a civil engineering company uses hand tools for extended periods. In doing so he is exposed to hand-arm vibration at 15 Hz with an RMS particle velocity of 3×10^{-2} ms^{-1}.

Calculate the RMS acceleration and hence the time for which the employee can be so exposed during a working day before reaching a normalized exposure A(8), of 2.8 ms^{-2}. Use a weighting at 15 Hz of 0.95. (IOA 2006)

17 An employee typically uses a hand-held grinder for three hours during the course of a shift. Frequency weighted vibration measurements at the tool's single handle are as follows:

x-axis 2.3 ms^{-2}

y-axis 3.4 ms^{-2}

z-axis 1.3 ms^{-2}

Calculate (i) the combined acceleration, a_{hv}, at the tool handle, and (ii) the equivalent eight-hour acceleration A(8). (IOA 2003)

Chapter 8 Measurement and instrumentation

8.1 Introduction

An introduction to sound measurement, including peak and RMS values of sound waveforms, frequency analysis, the decibel scale and frequency weightings was given in Chapter 1. This chapter describes the operation and use of sound level meters and the measurement of sound levels. The operation and performance of microphones, accelerometers and associated instrumentation, and of vibration is discussed, including an introduction to the use of digital signal processing in sound and vibration measurement.

The measurement of sound intensity and sound power level is described and the chapter concludes with a discussion of calibration and of measurement uncertainty.

8.2 The measurement of sound levels

A sound level meter consists of a microphone, which produces an electrical signal, various stages of electronic signal conditioning and signal processing, and an output and display. These may all be contained within one hand held unit, or as several different linked components, with a microphone and signal conditioning system connected to a variety of different data capture and storage devices including computers and printers. The main functional blocks are briefly described below.

The microphone converts the sound pressure waveform into an analogous electrical waveform signal. The faithfulness with which it does this (expressed technically in terms of its frequency response, linearity and dynamic range) determines the accuracy of the entire instrument.

Microphone preamplifier and amplification

This stage is sometimes called signal conditioning, and its function is to deliver the signal to the subsequent stage in an appropriate condition for further processing. The preamplifier acts as an electrical impedance matching device, between the very high electrical impedance of the microphone and the much lower electrical impedance of the following stages of processing, and the following amplifier provides any necessary increase in signal amplitude.

Signal processing delivers the appropriate values of the required acoustic measurement parameter to the output and display stage. It is convenient to consider the signal processing requirements in two stages, frequency signal processing and time series signal processing.

Frequency content signal processing

Two forms of frequency signal processing are commonly used: frequency weighting and frequency analysis.

A- or C-frequency weighting produces a single figure measurement of a broadband noise which takes the frequency spectrum into account by weighting the signal according to standardized frequency weighting curves. The Z or zero weighting gives the overall noise level when the frequency weighting is zero in all frequency bands from 5 Hz to 20,000 Hz, and the Flat weighting performs the same function when the response is zero over a more limited range from 31.5 Hz to 8000 Hz.

Frequency analysis provides the capability to measure the sound pressure level in octave or third octave bands.

Time series signal processing

Three forms of signal processing are commonly used:

- time weighted sound pressure level (instantaneous, maximum and percentile values)
- peak sound pressure level
- time average sound pressure level.

Time weighted sound pressure level

This is the mode of measurement one would select in order to investigate a moment to moment variation in sound level, for example when trying to locate the most important sources of noise coming from a machine, or transmission paths from one area to another. The variation in noise level indicated by the meter should correlate with what is heard by the ear.

Time weighted sound pressure level is the instantaneous value of a running average in which the RMS signal has been multiplied by a time dependent weighting factor, which decreases exponentially with elapsed time.

The time constant of the time weighting is 1.0 second for the slow time weighting (S) and 0.125 s for the fast (F) time weighting.

The F and S time weightings have their origins in an era when sound level meters had an analogue display consisting of the needle or pointer of a galvanometer which moved over a scale, rather like the speedometer or fuel gauge of a car dashboard. If the sound level fluctuated rapidly the movement to the needle could be slowed down by selecting S, making the sound level easier to read off the scale. Fast would be selected to measure the highest level during a short duration event such as a bang or clatter from an impact, a handclap or a fast motorcycle passing nearby. If the noise was steady then the same reading would be obtained using either Fast or Slow time weightings.

The effect of F and S time weightings may perhaps best be understood by considering the time weighted response to the sudden introduction of a tone burst several seconds long such as that which occurs when a calibrator signal is applied to the microphone of a sound level meter (see Figure 8.1). At the start of the burst the indicated sound pressure level will not immediately jump to the level corresponding to the RMS level of the tone burst, but will increase gradually towards this level, more quickly in the case of fast than for slow. Similarly at the end of the burst the indicated level will not suddenly drop to zero but will reduce (or decay) gradually, more rapidly for F than for S.

BS EN 61672-1 requires that, for a 4000 Hz tone burst, the sound level indicated using F should be within 2.6 dB of the continuous level of the tone after 0.1 second from the start of the burst, within 1.0 dB after 0.2 seconds and within 0.1 dB within 0.5 seconds; the corresponding figures for the slow time weighting being 10.2 dB, 7.4 dB and 4.1 dB. At the end of the tone burst the same standard requires an initial rate of decay of 25 dB per second for fast and between 3.4 and 5.3 dB per second for slow.

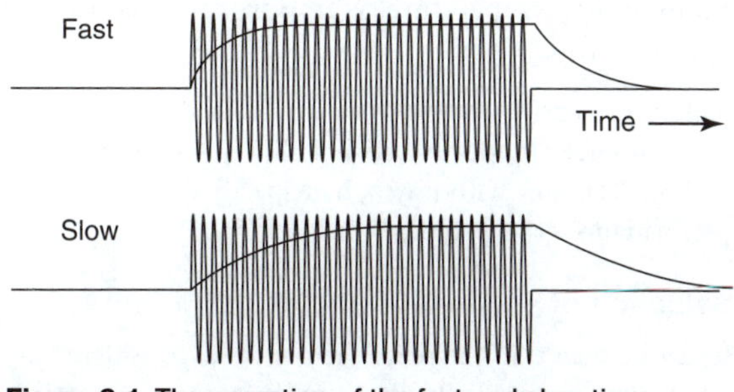

Figure 8.1 The operation of the fast and slow time weightings

Percentile and maximum values of time weighted sound pressure level

The highest value of the time weighted sound pressure level during a measurement time period is called the maximum noise level. This will be different depending upon whether the fast or slow time weighting has been used during the measurement, and this should always be indicated by use of the appropriate symbols: $L_{AFmax,T}$, $L_{ASmax,T}$, $L_{CFmax,T}$, $L_{CSmax,T}$ where T is the measurement period.

The difference between the fast and slow L_{max} values will depend on the 'crest factor' of the signal waveform, i.e. the ratio of its peak to RMS value. For very short duration impulsive signals, of high crest factor, where the waveform increases very rapidly, such as for impact noise, or noise from gunfire, the difference will be much larger than for more steady continuous noise having lower crest factors.

The values of percentile noise levels will also depend upon whether the fast or slow time weighting has been used and so this should also be always specified when stating a measurement result, e.g. as: $L_{AF10,T}$, $L_{AS10,T}$, $L_{AF90,T}$, $L_{AS90,T}$, $L_{AF10,T}$ etc. It is more usual for the F weighting to be specified for percentile measurements but the S weighting is sometimes used for L_{max} measurement.

Peak sound pressure level

The peak sound pressure level is the sound pressure level corresponding to the highest value of the waveform of the sound. This will be very different from, and much higher than, the L_{AFmax} or L_{ASmax} value, which is based on the time weighted RMS sound pressure. The waveform signal may be unweighted or Z-weighted (L_{peak} or L_{zpeak}), or C frequency weighted (L_{peak}), but not A-weighted.

L_{peak} or L_{Cpeak} is used for assessment of hearing damage risk from high levels of impulsive noise, and its use for this purpose is specified in the UK 2005 Control of Noise at Work Regulations, described in Chapter 9.

Time averaged sound pressure level, L_{AT}

This is also known as the continuous equivalent noise level, $L_{Aeq,T}$.

The measurement of L_{AT} requires the use of an integrating sound level meter, which integrates the instantaneous value of the square of the sound pressure with respect to time. The time weightings F and S are not involved in the determination of L_{AT}.

Although the time averaged level is most commonly an A-weighted value, there is no reason on principle why any other frequency weighted sound pressure level, or indeed any octave or one third octave band sound pressure level, should not be determined as a time averaged value.

The sound level meter and calibrator

There is a bewildering range of hand-held sound level meters on the market costing from a few hundred to several thousand pounds. They may be classified by the range of functions they possess, i.e. what they can measure, and also by their type or class, according to British and International Standards which set down, among other things their limits of accuracy.

The simplest, cheapest and most basic instrument is one which will only be able to measure instantaneous sound pressure level in dBA. This will certainly be useful for checking on levels in the workplace to determine if they ever exceed 85 or 90 dBA. They will only be able, however, to measure noise levels which are reasonably constant or steady. If noise of this type occurs in bursts it will be possible to estimate L_{Aeq} values by a combination of measurement, and calculation. If, however, the noise level fluctuates rapidly or is impulsive then this type of sound level meter cannot be used to give an accurate measurement. If the meter measures dBC as well as dBA this is useful because the difference between these two levels will give an indication of whether the noise has a predominantly high or low frequency spectrum. A more detailed measurement of the frequency spectrum requires a meter with a set of octave or one third octave filters. Sometimes meters of this basic type can be fitted with octave bands.

For measurements of fluctuating noise an integrating sound level meter is needed, which will allow $L_{A,T}$ (or $L_{Aeq,T}$) values to be determined.

If it is required to measure impulsive noise, and, in particular if it is necessary to check compliance with the peak action level of the 2005 Control of Noise at Work Regulations, then it is necessary to have a sound level meter with the facility to measure the true peak value of the impulsive noise. This value should either be unweighted (or linear), or Z-weighted or C-weighted, but not A-weighted.

The capability of being able to measure L_{Aeq} over a period of a few minutes (perhaps up to one hour) will probably suffice for sampling workplace activities, but environmental noise monitoring often requires measurements over several hours, whole days and even several days. Instruments differ in the way the measurement period may be controlled and initiated, from simple manual control of both start and cessation of measurement to completely automated measurements over preset intervals starting at pre-selected times. Some meters will also have the facility to start measuring and/or to capture audio recordings when pre-set sound levels are exceeded. Additionally some meters will have the facility to measure statistical percentiles (L_{AN} – the level exceeded for N% of the time, e.g. L_{10} or L_{90}). L_{90} measurements are required for BS 4142 assessments and L_{10} for measuring traffic noise.

Some sound level meters have output signals suitable for audio recorders, headphones, chart recorders and for downloading data to computers for further signal processing.

Calibrators

An essential accessory to every sound level meter is a calibrator which when fitted over the microphone of the sound level meter produces an accurately known sound level. The calibrator produces a check that the meter is working properly and the meter should be calibrated before and after every measurement, and sometimes in between if a long programme of measurements is being undertaken. Minor fluctuations from the calibration level due to temperature changes can be corrected by adjustment to the meter, but consistent discrepancies could indicate malfunction and the meter should be sent back to the manufacturer for repair.

In addition to day to day checks both sound level meter and calibrator should be periodically checked and calibrated by an accredited laboratory at least once every two years.

Performance of sound level meters

International Standard IEC 61672-1:2003 *Electroacoustics – Sound level meters – Part 1: Specifications* specifies the performance of sound level meters and defines two different types, Class 1 and Class 2.

The standard deals with: the directionality of the sound level meter, i.e. the extent to which it has different sensitivity to sound approaching the microphone from different directions; the accuracy of the A- and C-weighting networks within the meter; the time response characteristics; and the sensitivity of the meter to influences such as humidity, temperature, vibration, and electric and magnetic fields.

Note: the earlier corresponding sound level meter performance standards (BS EN ISO 60651 and BS EN ISO 60804), now withdrawn, had specified four different performance grades: types 0, 1, 2 and 3.

Selection of sound level meter

It can be seen that the main factors to be considered are what the meter will do (e.g. instantaneous (time weighted) sound pressure level only, L_{eq}, peak, octave bands, etc.) and how accurate it is, and the cost. Other factors are portability, how easy it is to use, ruggedness, battery life (this is important if the meter is to be used for several hours at a time), service arrangements, and flexibility in terms of possible upgrade (e.g. to the microphone or signal processing software) at a later date.

8.3 Microphones

Microphones used for accurate sound measurement need to be sensitive, stable, have a good frequency response and be able to operate over a very wide range of sound levels. The two types of microphones which best satisfy these requirements are the condenser microphone and the electret microphone, sometimes called the prepolarized microphone, which is an adapted form of condenser microphone.

Other types of microphone are the electrodynamic, the piezoelectric and the MEMS microphone. The electrodynamic (or simply the dynamic) microphone takes its output signal from the electrical voltage induced in an electric coil attached to the vibrating microphone diaphragm in a magnetic field supplied by a permanent magnet. This is the microphone type that is most widely used for audio recordings; it is cheap, robust and sensitive, but because it often has an uneven frequency response it is not suitable for accurate sound level measurement. Piezoelectric microphones utilize the piezoelectric effect discussed in more detail later in this chapter in connection with accelerometers for vibration measurement. They are sometimes used in less expensive and less accurate equipment. MEMS microphones (microelectronic mechanical systems) have a pressure sensitive diaphragm etched directly onto a silicon chip. They have the advantage that electrical signal conditioning and signal processing may be built into the same chip. They are used in hearing aids and in some mobile phones, which with the addition of applications software may be used to carry out sound level measurements.

The condenser microphone is the most delicate and easily damaged part of the sound level meter and is expensive to replace, and so needs to be handled with great care. It is the most important part of the instrument, determining overall accuracy of the entire measuring system. In some cases a sound level meter may be upgraded, say from Class 2 to Class 1 accuracy, by changing the microphone. Condenser microphones need to be supplied with a polarizing voltage (typically 200 volts DC) whereas electret microphones do not, and so different preamplifier stages are needed for the two types. Nevertheless, many sound level meters will accommodate both types provided the correct polarizing voltage is selected. Condenser microphones are sensitive to damage when the sound level meter is operated in wet and humid environments because of the possibility that the polarization voltage across the very small gap (of a few microns) between the metal diaphragm and backing plates of the condenser will cause arcing and sparking if condensation occurs. Electret microphones do not suffer from this problem because a prepolarized polymer film between the plates means that a polarizing voltage is unnecessary.

Overall, the condenser and electret types are comparable in cost and performance. A good quality microphone will have a flat frequency response and will have the same sensitivity for different frequencies of sound over a range from about 20 Hz to an upper limit in excess of 10 kHz. The high frequency response will depend on the size of the microphone and on the type of sound field for which it is calibrated. Half inch (12.5 mm) diameter microphones have a better high frequency performance than the one inch types which are gradually falling into disuse in modern instruments, despite their increased sensitivity. Quarter inch (6 mm) diameter microphones will allow measurements to even higher, ultrasonic frequencies. Many microphones are calibrated for free field use and designed to be pointed at the source of sound so that the sound waves strike the diaphragm at 0°, or normal incidence. Pressure response microphones are designed to be used at 90° or grazing incidence. When the sound field is diffuse, e.g. in reverberant situations, the ideal microphone is one which has been calibrated for random sound incidence. The manufacturer's handbook will give guidance as to which type has been fitted and how it should be used.

Windshield

A suitable windshield should always be fitted over the microphone when in use to prevent winds (externally) and draughts of air from cooling fans fitted to machinery (internally) from producing spurious wind generated noise in the microphone.

Correct use of the sound level meter

This involves:

- preliminary inspection and checks to ensure that the meter is working properly
- selection of meter functions and settings appropriate to the measurements to be taken, having decided what, when and where to measure
- precautions to avoid unwanted influence on the meter readings
- post-measurement checks and recording of all relevant measurement details.

Before use each time the meter and the microphone should both be examined for any obvious sign of damage. Most meters have a facility for checking the state of the battery and after this the meter should be switched to measure sound pressure level and a simple test carried out to check that it seems to be responding to ambient noises such as fingers clicking, whistling, talking into the microphone, etc. Instantaneous sound pressure level mode is best for this rather than L_{eq}. The meter should then be calibrated, the windshield fitted, and it is ready for use.

Each different type of sound level meter will have different types of controls and the manufacturer's handbook should be consulted, kept safely and its instructions always followed carefully. However, in most cases the following selections have to be made:

- appropriate time function: SPL, F, S or I, L_{eq}, peak
- appropriate frequency weighting function: A, C, linear (unweighted), octave bands
- appropriate range: some meters will be fitted with overload and underload indicators which assist in range selection.

Care has to be taken to ensure that the meter is not being affected by wind, unwanted reflections, background noise and extremes of temperature and relative humidity. The use of the windshield will protect against the influence of draughts indoors or mild breezes outdoors but care must be taken to ensure that wind generated noise, even using a windshield, is at least 10 dB below the measured value. Sometimes this can be checked by taking a measurement with some temporary wind barrier in place or by noting how the indicated sound level fluctuates with gusts of wind. Generally, valid measurements cannot be taken if the wind speed exceeds about 5 m/s, but different measurement standards specify various values. Sound reflected by nearby objects, including the body of the operator, can be detected by the microphone and influence the measured value of sound pressure level. For this reason the sound level meter should always be held with outstretched arms away from the body of the operator towards the sound source, or mounted on a tripod. Unless there is a good reason the microphone should be positioned well away, at least one metre, from hard reflecting surfaces. If this is not possible then the presence of such a surface(s) should be noted in the measurement report. Unless there is a good reason for doing so, noise measurements are not usually taken outdoors when it is raining. Apart from the danger of malfunction and damage due to ingress of water into the meter and microphone the sounds produced by the rainfall may influence the reading, and if noise from a distant noise source is being measured, sound propagation conditions will be very different from those in dry weather conditions. If noise from a specific source, such as a machine, is being measured then it will be necessary to check that background noise levels are at least 10 dB below the measured value. Some meters have a 'pause' button which is useful for cutting out the effects of passing vehicles or aeroplanes etc. on the measurement. The manufacturer's handbook should be consulted with regard to using the sound level meter in extremely hot or cold weather.

When all the measurements have been completed the battery condition and the calibration should be checked again before leaving the measurement site and all the relevant measurement details should be recorded. It is useful to devise a measurement report form which can also act as a prompt and a checklist to ensure that all checks have been made and necessary details recorded.

Measurement aims

The starting point of any noise measurement is to decide on the main purpose or aim of the measurement, since this will influence further decisions. As an example, consider the possible reasons that there could be for needing to measure noise levels produced by a piece of machinery, such as a compressor. These could include:

- to compare with noise levels from other, alternative equipment
- to compare with manufacturer's noise specification
- to compare with company's own (i.e. customer's) noise specification
- to assess the need for, or effect of, machine maintenance
- to assess the need for, and to quantify noise reduction requirements
- to diagnose noise sources and mechanisms and specify control methods
- to assess the effectiveness of noise control measures
- to measure the sound power level of the machine
- to assess the noise exposure level of the machine operator
- to use as a basis for predicting noise levels at other distances
- to use as part of a noise complaint investigation
- to use as part of a planning application investigation
- to use as part of an environmental impact assessment
- to assess whether a machine is a major contributor to the noise level at the receiver.

Measurement reports

Although, depending on the circumstances, many different report formats may be appropriate, all reports should contain some or all of the following information:

- summary of main points, including conclusions and recommendations
- aims and/or objectives stating the purpose of the measurements
- information about the noise source, e.g. information about machinery or processes being measured – machine type, load, speed, type of material being processed etc. These details should be sufficient to enable the noise measurements to be repeated under similar conditions at some later date
- information about the acoustic environment in which the measurements are made, e.g. a map or plan showing layout of workplace, measurement positions, hard reflecting or sound absorbing surfaces etc.
- information about the measurement equipment: type and serial number of sound level meter and calibrator
- details of the noise measurement programme, i.e. what was measured, how, when and where
- results of the measurements (tables, graphs etc.)
- analysis and discussion of results
- conclusions and recommendations.

8.4 Linearity, frequency response and dynamic range

Before describing the different types of microphone, their main characteristics will be discussed. Frequency response and dynamic range, for example, are important properties not only of microphones but also of many other kinds of instrumentation. These are illustrated in Figure 8.2.

Sensitivity

Sensitivity is the ratio of the output voltage produced by the signal to the sound pressure striking the microphone. It is measured in volts per pascal (V/Pa) or in dB relative to a sensitivity of 1 V/Pa. Ideally the microphone sensitivity should be as high as possible. It is also usual to express microphone sensitivities in decibels, either in relative terms to compare the sensitivities of different microphones, or in absolute terms, using a reference sensitivity.

Note: Since the output voltage produced by a microphone will depend on the electrical impedance of the device to which it is connected, sometimes called the electrical load (e.g. to the preamplifier of a sound level meter), it is usual to quote the open circuit sensitivity of the microphone, which is the maximum voltage signal it would deliver when connected to a load with infinite impedance, i.e. one which draws zero electrical current from the microphone.

Example 8.1

Microphone A has a sensitivity of 50 mV/Pa and microphone B has a sensitivity of 12.5 mV/Pa. Compare these two sensitivities on a decibel scale, relative to each other and relative to a reference sensitivity of 1 V/Pa.

Two different sensitivities, S_1 and S_2 may be compared on a decibel scale by treating them as any other signal (pressure, acceleration, voltage etc.) using the following formula to calculate N, the number of decibels:

$$\begin{aligned} N &= 20\log(S_2/S_1) \\ &= 20\log(50/12.5) = 12 \text{ dB} \end{aligned}$$

Thus microphone A is 12 dB more sensitive than B, that is to say, if both microphones were subjected to exactly the same sound pressure then the electrical signal from A would be 12 dB higher than from B.

The sensitivity of microphone A, relative to 1 V/Pa (i.e. relative to 1000 mV/Pa) is:

$$N = 20\log(50/1000) = -26 \text{ dB}$$

The sensitivity of microphone A is therefore −26 dB relative to 1 V/Pa. A similar calculation for microphone B gives −38 dB. Thus the sensitivity of microphone A is 26 dB below the reference value, and microphone B is a further 12 dB lower.

Example 8.2

Microphone C has a sensitivity of −48 dB relative to 1 V/Pa. Express this in mV/Pa and calculate the magnitude of the microphone signal, in mV, produced by a sound pressure level of 74 dB.

In this case $N = 20\log(S/S_0)$ where S is the unknown sensitivity of microphone C and S_0 is the reference sensitivity.

It follows that:

$$(S/S_0) = 10^{(N/20)} = 10^{-48/20} = 0.004$$

S_0 is the reference sensitivity of 1000 mV/Pa. Therefore the sensitivity of microphone C is:

$$S = 0.004 \times 1000 = 4 \text{ mV/Pa}$$

A sound pressure level of 74 dB relative to 20 Pa corresponds to a sound pressure of: $20 \times 10^{-6} \times 10^{74/20} = 0.1$ Pa. Output voltage from microphone = 4 mV/Pa × 0. 1 Pa = 0.4 mV.

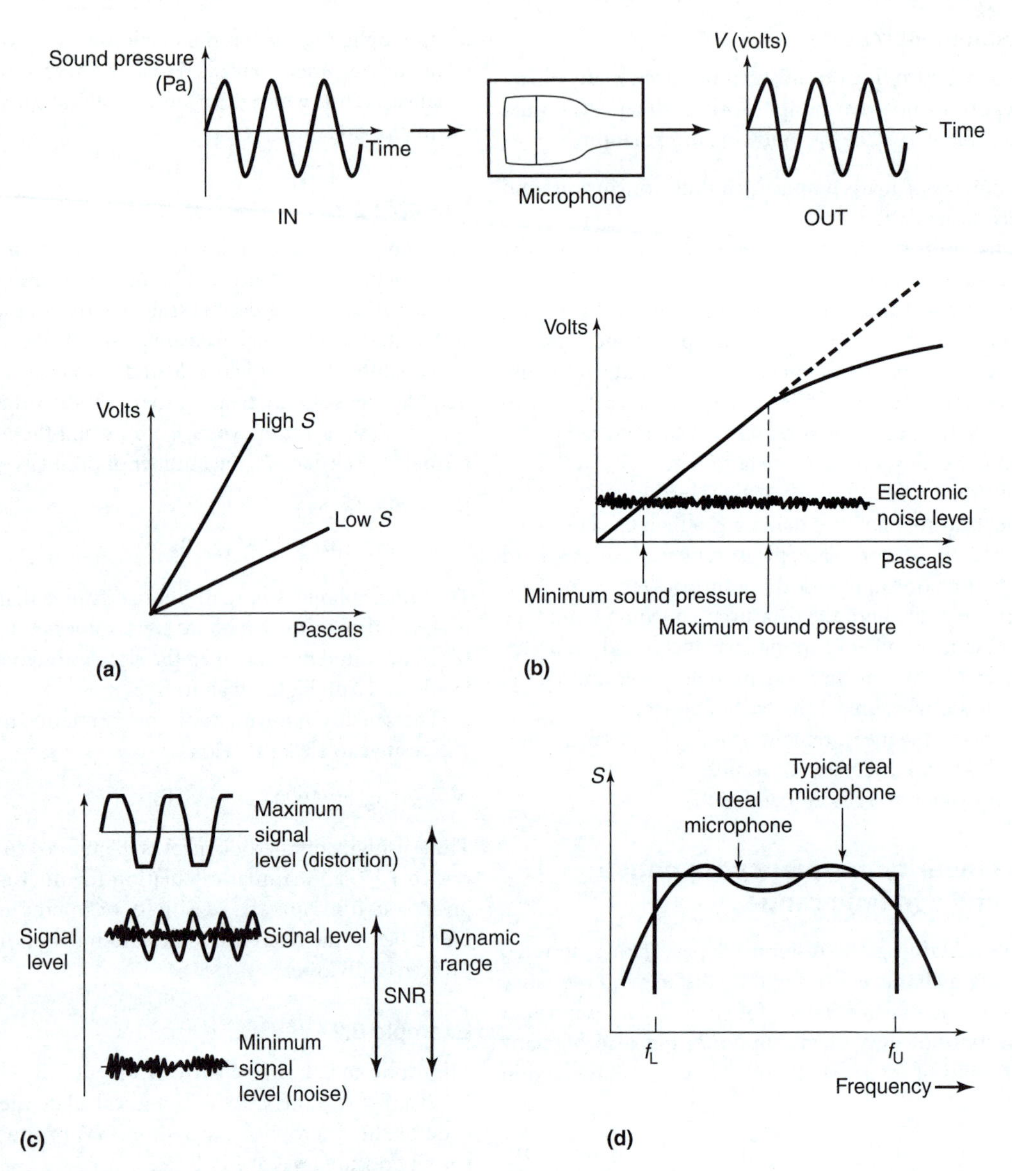

Figure 8.2 Illustrating important properties of a microphone: (a) sensitivity S = output/input = volts/pascals (represented by the gradient of the two lines on the graph); (b) linearity; (c) signal-to-noise ratio (SNR) and dynamic range; (d) frequency response, f_U = upper cut-off frequency and f_L = lower cut-off frequency

Output impedance

Condenser microphones have a very high electrical output impedance. This means that they are only capable of delivering very small electrical currents (and very small amounts of electrical energy) to the circuits that follow them. Ideally they need to be connected only to devices with a very high input impedance, which means a device which will only take a very small amount of electrical current (and energy) from the microphone. Such a device is a specially designed microphone preamplifier, which matches the very high output impedance of the microphone to the lower input impedance of the following sound level meter electronics. Without such a preamplifier to buffer it from the rest of the sound level meter, the loading effect of the following electronics would severely reduce the microphone output voltage signal. In practice, condenser microphones for sound measurement are always supplied with a preamplifier to which they are directly connected. See also Appendix 4 and the Glossary.

Linearity

The microphone output signal must be linear. This means that the sensitivity must remain constant throughout the measured range of sound pressures, or in other words, the output voltage signal of the microphone must be directly proportional to the sound pressure. This is necessary if the microphone is to give a faithful reproduction of the sound pressure waveform (see Figure 8.2).

Dynamic range

Dynamic range is the measurement range of the microphone – the range of sound pressures over which the microphone operates and faithfully reproduces the sound pressure waveform (i.e. over which it remains linear). Measurement is limited at low levels by the electronic noise of the microphone and its preamplifier.

Electronic noise produces a random output even when there is no sound pressure acting on the microphone. Measurement is limited at high levels because the microphone eventually becomes non-linear and produces a distorted waveform. At even higher levels the microphone may become damaged. The dynamic range is the range between these two extremes; it is often expressed in decibels. A more precise definition requires specification of the signal-to-noise ratio at one extreme, and the permitted percentage of distortion at the other extreme. The dynamic range of the microphone and preamplifier assembly usually determines the measurement range of the entire sound measurement system.

The dynamic range is an important property of all instruments, e.g. tape recorders, amplifiers, rectifier circuits and accelerometers. In all cases there are upper and lower limits for the signal, and the device will function according to its specification only if the signal lies within these limits.

Signal-to-noise ratio

Dynamic range is a property of the instrument, e.g. the microphone. A closely related term is the signal-to-noise ratio, which is a property of the signal. This again may be expressed in decibels. The two terms are closely related because the maximum signal-to-noise ratio is determined by the dynamic range of the instrument, as shown in Figure 8.2.

Frequency response

Ideally the microphone should respond equally to all frequencies, i.e. it should have the same sensitivity to all frequencies. Together with linearity, this requirement is essential if the microphone is to give a faithful reproduction of complex sound waveforms, which will contain a range of frequencies. The frequency response of a microphone is represented by a graph of sensitivity versus frequency, and ideally this should be horizontal, i.e. a 'flat' frequency response, over the range of frequencies it is required to measure. In practice the frequency response of a good microphone will be approximately flat between upper and lower frequency limits, but will drop off at upper and lower frequency limits (see Figure 8.2).

Like dynamic range, the frequency response is an important property of amplifiers, loudspeakers, accelerometers, tape recorders and many other items of instrumentation. In all cases it is usual to represent the frequency response on a graph, for example a graph of amplifier gain versus frequency, or a plot of audio recorder response (i.e. the ratio of replayed signal level/recorded signal level) versus frequency.

Some other important characteristics of microphones

Ideally the microphone should have good stability, i.e. its sensitivity should remain absolutely constant with the passage of time, unaffected by all other changes in the environment (e.g. temperature, humidity, pressure, electric and magnetic fields, vibration) except sound pressure. And as far as possible, the microphone should be able to withstand, within limits, the effects of such changes in the environment, without suffering damage. The microphone should be as rugged as possible, so it can withstand day-to-day handling without damage. Although some microphones are more rugged than others, all microphones are extremely delicate and require the most careful handling and storage in order to avoid damage. Another important property of a microphone is its cost.

The condenser microphone

A condenser or capacitor is an electrical component which prevents the passage of electric current in one direction (direct current) but allows the transmission of electrical current which alternates in direction (alternating current). In its most common form called the parallel plate condenser it consists of two parallel metal (i.e. electrically conducting) plates separated by a gap between them which is filled with a non-conducting material (called a dielectric). The device acts as a mechanism for storing electrical charge. When it is connected to a battery which provides a constant electrical voltage, DC current flows to each plate (but not between the plates) until the condenser is fully charged to the supply voltage, when further flow of current ceases. The amount of charge stored by a capacitor is proportional to the voltage across its plates, and is indicated by the capacitance of the condenser, measured in farads (F), as the amount of charge per volt. The capacitance of a

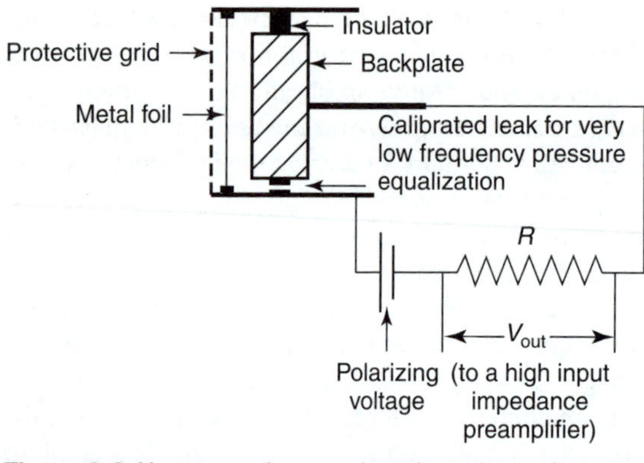

Figure 8.3 How a condenser microphone is used as part of an electrical circuit to produce a signal V_{out} in response to the sound pressure at the microphone diaphragm

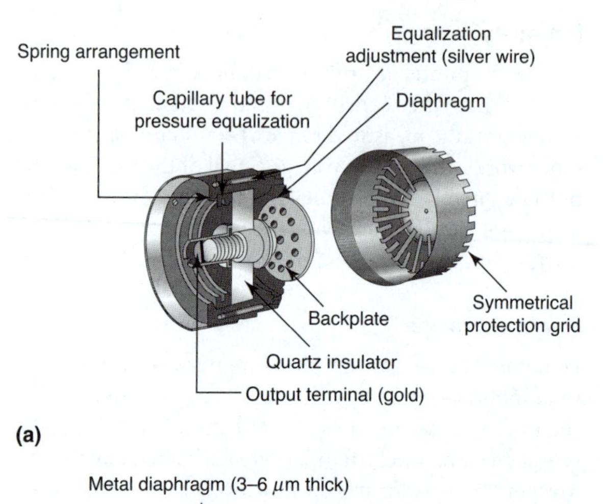

(a)

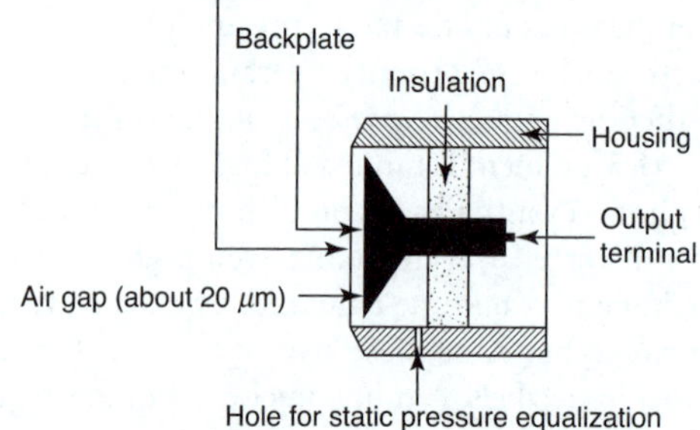

(b)

Figure 8.4 The condenser microphone: (a) construction details and (b) schematic diagram, not including protective grid

condenser depends on the area of the plates, and the distance between them. Therefore if for a given amount of charge on the plates the capacitance were to vary, this would produce a variation in the voltage across the plates. The capacitance will vary if the distance between the plates varies. This is the principle underlying the operation of the condenser microphone.

Condenser microphones are very widely used in sound level meters and are essential if the highest possible degree of accuracy is required.

A diagram of an electrical circuit showing how a condenser microphone is used is shown in Figure 8.3, and Figure 8.4 shows typical constructional details. Sound waves strike the flexible diaphragm, which forms one of the plates of a parallel plate condenser. A rigid metal backing-plate forms the other plate, and there is a very small air gap between the two. A small 'pressure equalization' hole behind the diaphragm ensures that the static atmospheric pressure is the same on either side of the diaphragm. The sound pressure fluctuations vibrate the diaphragm, causing variations in the thickness of the air gap between the plates. These in turn cause changes in the capacitance of the condenser. The condenser is incorporated into an electric circuit with a steady DC voltage, called the polarizing voltage, across its plates. The changes of capacitance produced by the sound are converted into changes in the voltage across the condenser plates. These electrical voltage changes form the microphone output signal, which is fed into the preamplifier, then to the rest of the sound level meter circuits.

The vibrating diaphragm of the microphone behaves rather like a mass–spring system, with its own natural or resonance frequency. Such a system only has a flat frequency response in the range of frequencies well below the natural frequency. This means that, in order to have a flat region extending across the whole audio frequency range (20–20,000 Hz), it is necessary for the microphone diaphragm to have a very high natural frequency, requiring a diaphragm which is simultaneously very stiff and very lightweight. This is achieved by making the diaphragm of a very thin sheet of a specially developed metal alloy, and stretching it like a drum skin to achieve the required stiffness.

The design of the microphone involves a compromise between its various requirements. The sensitivity depends on the stiffness of the diaphragm, the capacitance of the condenser and the magnitude of the polarizing voltage.

Increasing the diameter of the diaphragm would have the advantage of increasing the sensitivity, increasing the capacitance and decreasing the output impedance, but it would also have the disadvantage of decreasing the high frequency performance and of making the microphone more directional in its response. Increasing the stiffness of the diaphragm will increase the natural frequency but will decrease the sensitivity.

The condenser microphone is generally considered to have a superior performance (stability, dynamic range, frequency response, sensitivity etc.) compared to other types of microphone used for measurement purposes. But one disadvantage is its vulnerability to malfunction and damage in humid environments. Inside the microphone there is a fairly high potential difference (up to 200 V) across a very small air gap, and the problem arises from the possibility of condensation occurring within the air gap, leading to an electric discharge across the plates through the moist, conducting air. A discharge causes loss of signal, but more important, can lead to permanent damage to the microphone diaphragm. For this reason, unless special precautions are taken, measurements should never be taken in the rain or in other very humid situations. If malfunction is suspected when operating under such conditions, the sound level meter should be switched off immediately, to remove the polarizing voltage from the condenser.

It is possible to reduce these humidity problems by incorporating special modifications to the basic microphone design. They include a heating resistor in the preamplifier circuit, a thin quartz (insulating) coating to the diaphragm and a change to the pressure equalization arrangements so that all the air entering the microphone is dried by a silica gel dehumidifier. In addition it is possible to provide the microphone with a waterproof rain shield (Figure 8.5). Despite their basic vulnerability to high humidities, condenser microphones have been used successfully for permanent outdoor all-weather noise monitoring, near to airports, for many years. The vulnerability to humidity problems may be eliminated by the use of an electret microphone.

The electret microphone

The electret microphone is a modified form of the condenser microphone (see Figure 8.6). The diagram consists of a polymer film coated on one surface with a thin metal film; the other surface is in contact with a metal backing-plate. These two metallic surfaces form the plates of the condenser; the polymer film in the gap takes the place of the air in the condenser microphone. The polymer is a specially developed material that is permanently polarized. Electric charges permanently reside on the two plates of the condenser – positive charge on one plate, negative charge on the other. This has the advantage that there is no need for the polarizing voltage required in the condenser microphone. The electret microphone is not vulnerable to the humidity problems of the condenser microphone because there is no air gap and no polarizing voltage. They are more rugged than condenser microphones and they have a higher capacitance because

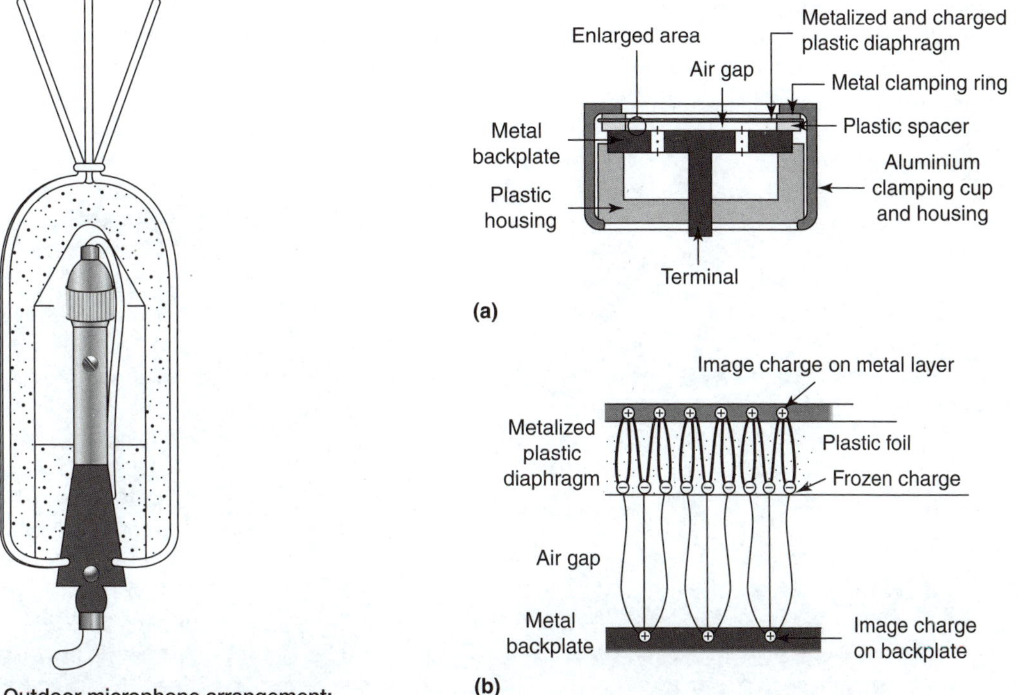

Figure 8.5 Outdoor microphone arrangement: half inch (12.5 mm) microphone and preamplifier fitted with rain cover and dehumidifier, and mounted inside a windscreen with bird spikes

Figure 8.6 Schematic diagram of an electret (prepolarized) condenser microphone: (a) construction and (b) enlargement to illustrate charge distribution

they use a polymer dielectric. Electret microphones were developed in the early 1960s and since then have been widely used in hearing aids, mobile phones and domestic audio recorders. Their technical performance has been improved considerably over the years and they are now used for accurate sound measurement purposes. Many sound level meters are designed to be used with either condenser microphones or electret (prepolarized) microphones.

Exercise

At which frequency will the sound wavelength be equal to the diameter of a half inch microphone?

Answer

Using $c = f\lambda$, so that $f = c/\lambda = 330\ \text{ms}^{-1}/12.7 \times 10^{-3}\text{m} = 26\ \text{kHz}$.

The behaviour and performance of microphones in sound fields at high frequencies

At very low frequencies, where the sound wavelength is very much greater than the size of the microphone, the microphone can be considered as having almost point size; the microphone does not noticeably interfere with the sound wave. This is no longer true at higher frequencies; diffraction, scattering and reflection from the surface of the microphone become significant, and the presence of the microphone actually disturbs and changes the sound field it is intended to measure. These changes become important at frequencies where the wavelength of the sound becomes similar in magnitude to the dimensions of the microphone. These changes have nothing to do with the microphone's type or principles of operation; they occur as a result of diffraction simply because the microphone acts as an obstacle in the path of the sound waves.

Directionality of microphones

At low frequencies, sound waves spread evenly over the microphone and the sound pressure will be almost the same all over its surface. Under these circumstances the microphone will respond equally to sound approaching from all directions. As the frequency increases and the wavelength decreases, the distribution of sound over the microphone becomes less uniform, so the microphone becomes more directional. Figure 8.7 shows the directionality pattern of a half inch microphone at different frequencies. At low frequencies the pattern is approximately circular, indicating equal sensitivity to sound from all directions, but as the frequency increases, the region of maximum sensitivity becomes more and more confined to sound striking the microphone at small angles of incidence. At these high frequencies the orientation of the microphone relative to the direction of the sound becomes an important measurement consideration.

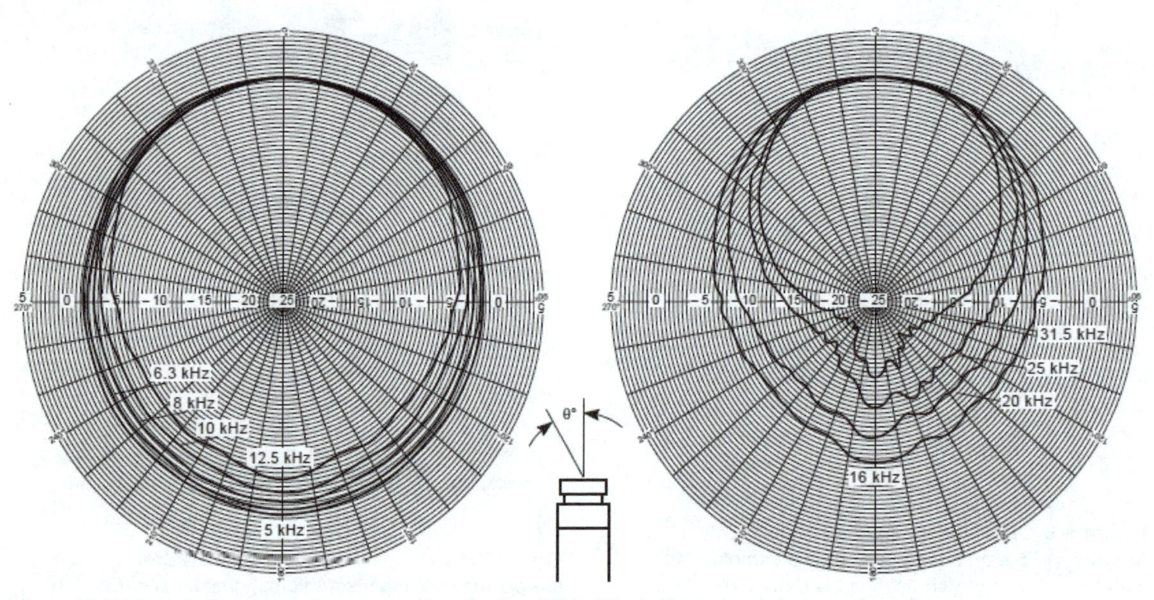

Figure 8.7 Typical directional response pattern for the GR half inch electret condenser microphone

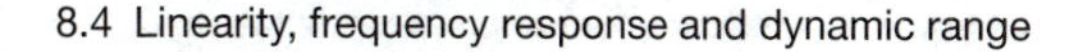

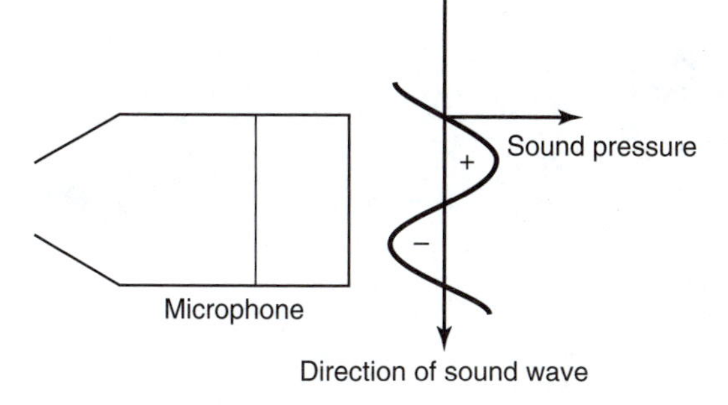

Figure 8.8 The effect of high frequency sound on microphone response

Frequency response

The frequency response of a microphone depends on details of its internal construction, which determine the natural frequency of the diaphragm. But besides construction, the high frequency behaviour of all microphones is determined by the size of the diaphragm compared to the wavelength. This is easily understood by considering a microphone placed in a plane-wave sound field where the sound wavelength equals the microphone diameter (see Figure 8.8). The microphone is positioned so that sound waves approach at a 90° angle of incidence, i.e. so that the sound waves graze over the microphone surface. Half of the diaphragm is subjected to a positive sound pressure (a compression) while the other half is being subjected to a negative sound pressure (a rarefaction); the sound pressure averaged over the microphone surface is equal to zero. The situation changes as the angle of incidence changes, until at 0° the sound pressure is in phase over the whole area of the diaphragm.

The general situation is that the sensitivity of the microphone drops off at high frequencies, where the sound wavelength becomes similar in magnitude to the microphone diameter. And at high frequencies the sensitivity depends on the angle of incidence of the sound. It follows that smaller microphones have a better high frequency response than larger microphones, although they tend to be less sensitive. At a given frequency, smaller microphones are also less directional than larger microphones.

Free field, pressure and random incidence responses

Because the microphone acts as an obstacle to the very sound field it is to measure, the sound pressure at the microphone diaphragm will be different from the sound pressure at that point when the microphone is absent. The difference is known as the free field correction factor and depends on the angle of incidence, the frequency of the sound and the microphone diameter. It may be calculated using diffraction theory. Manufacturers of microphones usually designate three different types of microphone, calibrated for use in different types of sound field: free field response, pressure response and random incidence response (see Figure 8.9).

In Figure 8.9(a) free field measurements indicate the sound pressure as though the microphone were not present; pressure response measurements indicate the sound pressure actually present at the microphone diaphragm. In part (b) free field microphones include the free field correction in their calibration and should be used at 0° incidence. Pressure response microphones should be used at 90° incidence and random incidence microphones should be used in reverberant sound fields.

Free field microphones are intended to give a reading which indicates the sound pressure level at the measurement position in the absence of the microphone. This is achieved by including the free field correction factor in the calibration of the microphone. Free field microphones are usually designed for use at 0° incidence and are the correct type to use for measuring direct sound which comes predominantly from a particular direction; the microphone should be pointed in that direction towards the source of sound.

Pressure response microphones do not include the free field correction in their calibration; they indicate the sound pressure actually present at the microphone diaphragm, including a component caused by scattering and diffraction due to the presence of the microphone itself. This type of microphone is the most suitable for measuring sound in confined spaces, e.g. when using acoustic couplers or when calibrating audiometers, and for measuring the sound pressures acting on surfaces, e.g. at the wall of a duct, when the microphone would be flush-mounted in the wall. When they are used for measuring free field sound, pressure response microphones are intended to be oriented so that the sound strikes the microphone at a 90° angle of incidence so that when measuring traffic noise at a roadside, for example, the microphone axis should be aligned vertically, i.e. with the diaphragm horizontal.

Random incidence response microphones are calibrated and intended for use in situations where the sound is diffuse and strikes the microphone randomly from all directions, e.g. in a reverberant sound field indoors. Free field microphones may also be adapted to measure diffuse sound by fitting a diffusing device called a random incidence corrector to the microphone.

Figure 8.10 shows the differences between the frequency responses of the three microphone types. They only become appreciable at high frequencies, being

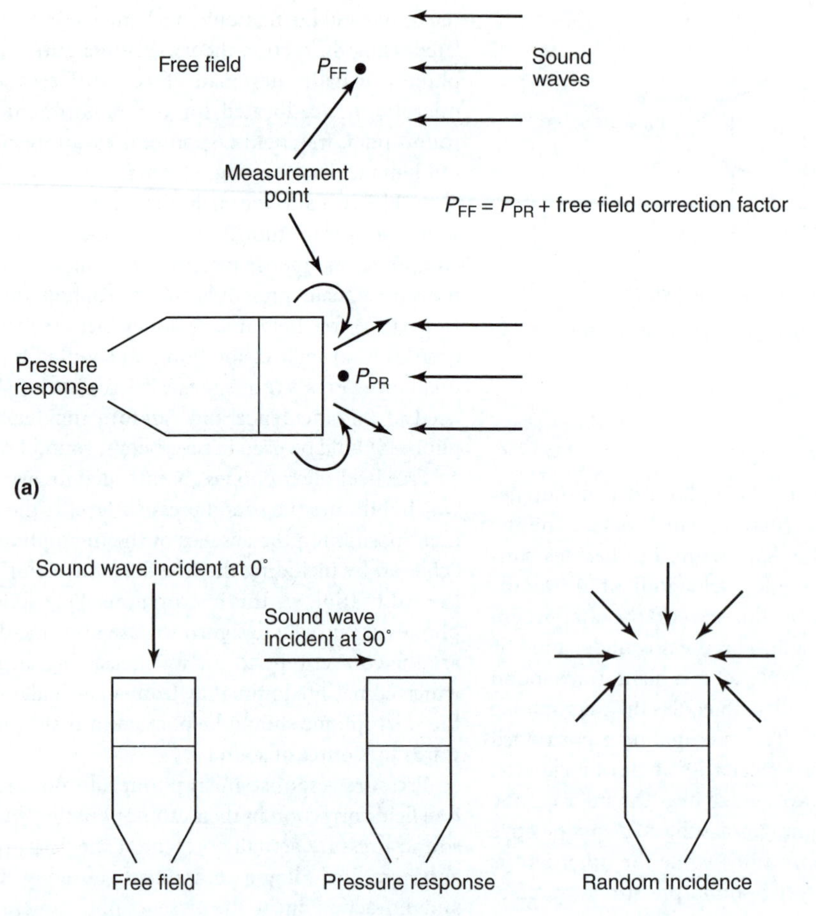

Figure 8.9 Free field, pressure response and random incidence microphones

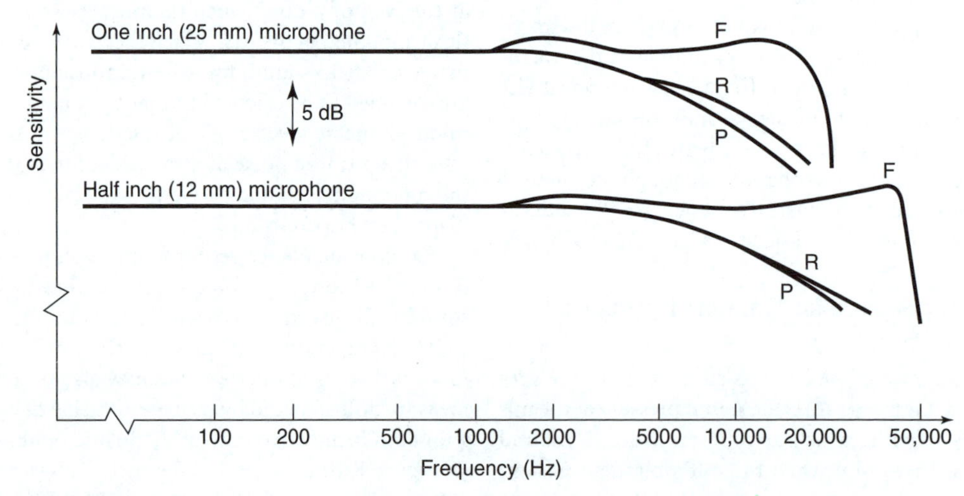

Figure 8.10 Typical frequency response characteristics of condenser microphones: (F) free field, usually 0° (R) random incidence (P) pressure response

approximately 1 dB at 2000 Hz, 3 dB at 5000 Hz, and 10 dB at 10,000 Hz for a one inch diameter microphone.

It is therefore important when measuring sounds with a significant high frequency content to choose the correct type of microphone for the type of sound field to be measured, and to orientate the microphone correctly with respect to the direction of the sound. The manufacturer of the sound level meter usually gives advice concerning the correct choice and use of microphone in the instrument handbook.

8.5 The use of analogue and digital signal processing in sound level meters

We learnt earlier in this chapter that in a sound level meter the signal produced by the microphone is subjected to signal conditioning and signal processing before being presented for output and display.

The signal conditioning usually means preamplifier and amplifier stages (and sometimes high or low pass frequency filtering), and the signal processing includes either waveform processing (usually frequency weighting or frequency filtering) and time series processing (usually RMS, time weighting or time averaging).

Originally all this signal processing was performed using analogue techniques, but modern sound level meters rely on digital signal processing. Signal conditioning is performed by analogue methods.

An introduction to analogue and digital signal processing

The voltage signal produced by a microphone is called an analogue signal because it is exactly similar or analogous in its waveform to the sound pressure waveform. An analogue signal varies continuously both in size and in time, whereas a digital quantity can only vary in discrete steps, rather like a small quantity of money which can only increase or decrease in steps of one penny. In an analogue instrument all the signal processing (amplifying, averaging, filtering etc.) is performed by circuits which leave the processed signal still in its analogue form. In a digital sound level meter some of the processing is performed on a digital version of the signal. At the heart of any digital instrument is the analogue-to-digital converter (ADC). The ADC samples the continuous analogue signal at regular time intervals and digitizes it, i.e. divides it up into a number of discrete intervals or steps (Figure 8.11). The digitized signal now consists of a series of numbers representing the value of the signal at successive time intervals. These numbers are stored in binary code instead of the decimal code which is in everyday use, i.e. to base 2 instead of to base 10.

Two very important properties of the ADC are the number of steps into which the signal is digitized and the rate at which the continuous signal is sampled. Both properties determine the degree to which the digitized signal faithfully resembles the original analogue form. The number of digitization steps is related to the number of bits in the ADC. A bit is one piece of digital information, i.e. one digit in a binary number. An eight-bit ADC, for example, divides the signal up into $2^8 = 256$ different levels. The ADC is designed to work with analogue signals within a certain voltage range, and the signal conditioning circuits prior to the ADC perform the important task of ensuring that the signal is always amplified by the correct amount to fill this range, thus ensuring the maximum possible resolution of the digitized signal.

The dynamic range of a digitized signal from an N-bit ADC is from 1 to 2^N, i.e. from the lowest to the highest possible value. Expressed in decibels this is 20 log 2^N, or approximately 6 dB per bit. The earliest ADCs were eight-bit devices with a dynamic range of 48 dB, which was a severe limitation for use with wide range signals.

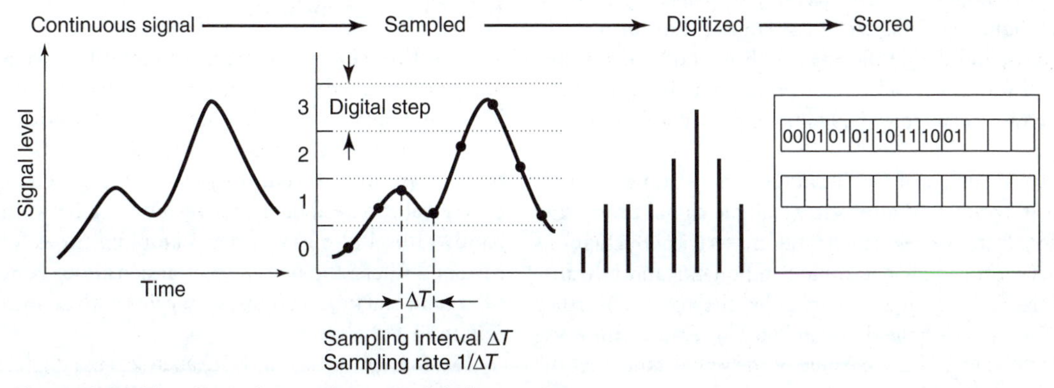

Figure 8.11 The effect of an analogue-to-digital converter

Modern devices operate with 16 bits or more, i.e. a dynamic range of at least 96 dB, and so are no longer a limit to the dynamic range of the instrument.

Shannon's sampling theorem and the Nyquist frequency

If the sampled signal contains frequency components, e.g. if it represents a sound waveform, then the rate at which the ADC samples the signal determines the upper frequency limit of signals that may be faithfully processed by the converter. An important result of signal processing theory, called Shannon's sampling theorem, requires that a minimum of two samples per hertz is needed adequately to represent the highest frequency component in the signal. A sample rate of at least 40,000 samples per second would therefore be needed adequately to represent a signal with frequency components of up to 20,000 Hz. It is important that this condition is met, otherwise false frequencies, called aliases, start to appear in the analysis. These arise at submultiples, e.g. one-half of those highest frequencies in the signal which are not adequately sampled because the sampling rate is too low. In order to reduce the possibility of aliasing, a low pass filter called an **anti-aliasing filter**, is used to limit the upper frequency content of the signal entering the ADC to less than one-half of the sampling rate. The upper frequency limit of the signal, half the sampling rate, is called the **Nyquist frequency**.

There are many mathematically based signal processes, such as time averaging of the signal and statistical analysis, which may be performed much more easily with digital information than with analogue signals. Therefore parameters such as L_{eq}, SEL and L_N (e.g. L_{10} and L_{90}) which describe variability of sound level may be determined by digital processing of the digitized RMS signal. In this case a sample rate of a few tens of samples per second will be adequate to represent time variations in the signal level.

Frequency analysis of signals may be accomplished by either analogue or digital techniques. There are two distinct digital methods called fast Fourier transform (FFT) analysis, and digital filtering. Both methods require the sound waveform to be sampled at a much higher sampling rate, i.e. up to 40,000 samples per second for the full audio range.

An advantage of digital circuitry in instrumentation is that it avoids limitations in analogue signal processing arising from the very small random electrical fluctuations which occur in all electronic components and in transducers. These fluctuations are called electronic noise; they are mixed in with the signal and set a lower limit, the noise level, to any meaningful value of the signal. Note that the term 'noise' in this context has nothing to do with acoustic noise, but simply means 'non-signal'. Each piece of analogue signal processing involves more electronic components, so it adds more noise to the signal. The advantage of digital processing is that, once the signal has been digitized, all the subsequent processes are effectively arithmetic or counting processes; they do not create extra noise, so there is no further loss or degradation of the information in the signal. However, any detail in the signal which is lost in the original digitizing process can never be replaced later, so the performance of the instrument depends very much on the quality of the ADC.

Signal processing in sound level meters

The earliest sound level meters were entirely analogue, i.e. all the signal processing and the output and display (a moving needle of a voltmeter) were carried out using analogue methods. In some fully digital sound level meters all the signal processing is carried out digitally after the analogue signal from the microphone has been conditioned by preamplification, further amplification and anti-aliasing filter stages. In other, earlier, types a combination of analogue and digital signal processing was used; the waveform signal processing (frequency weighting and frequency analysis) and the RMS and time weighting (fast and slow) were carried out using analogue methods and then the ADC digitized and stored the RMS signal, and digital signal processing was used to calculate time series analyses such as L_{10}, L_{50}, L_{90}, L_{max}, L_{min} etc. over various periods as required.

Information may be transferred from the digital sound level meter in digital form to other equipment (computers, graphical recorders, printers etc.) for further post-processing, analysis and display and storage. The analogue signal may be reconstructed from the digital information using a digital-to-analogue converter (DAC).

8.6 Frequency analysis

This section continues the introduction to frequency analysis contained in section 1.6 of Chapter 1, where the ideas of Fourier series for harmonic waveforms, frequency spectra and octave and third octave frequency bands were introduced. It discusses the difference between real and ideal filters, the relationship between filter bandwidths and measurement sampling times for narrowband filters, the power spectral density of band pass filters and real time analysis, but begins with an introduction to FFT analysis.

Fast fourier transform (FFT) analysers use digital signal processing techniques to produce very rapid narrowband

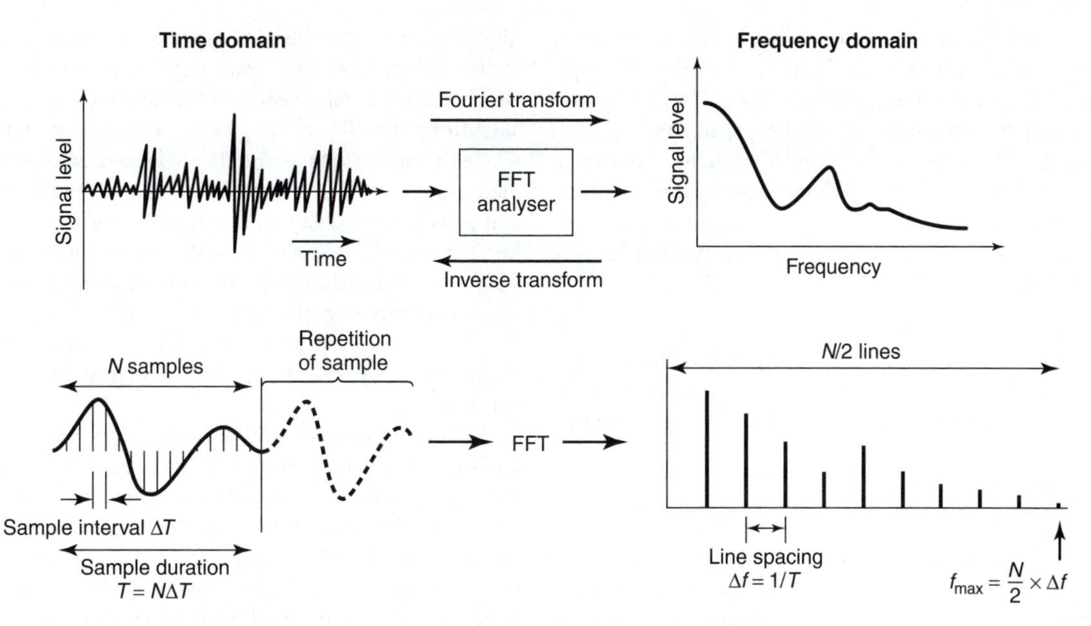

Figure 8.12 The Fourier transform: relationships between parameters in an FFT analyser

frequency analysis of acoustic signals. The method has its origin in the work of the French mathematician J. B. Fourier (1768–1830) who showed that any complex periodic signal, with repetition time T seconds, can be broken down or analysed into a series of component times having frequencies which are multiples (harmonics) of the repetition rate $1/T$, called the fundamental frequency (see Figure 8.12). Conversely, complex harmonic signals may be built up or synthesized from combinations of harmonically related single frequencies.

In theory the Fourier series for a harmonic waveform may be infinite, i.e. it may consist of an infinite number of harmonics, but usually it is strongly convergent, so the amplitude of the higher harmonics reduces rapidly with increasing harmonic number. The amplitude and phase (compared to the fundamental) of each frequency component may be obtained from a formula derived by Fourier, which involves integration over one complete cycle of the signal, i.e. over period T. The Fourier series for simple periodic signals is covered in many mathematical textbooks. The result for a square wave signal is illustrated in Figure 1.12; the signal is broken down into a series of odd numbered harmonics (1, 3, 5 etc.). Fourier's theorem allows periodic waveform signals to be represented by a line spectrum of frequencies. The frequency spacing between the lines is $1/T$; if the period T of the signal increases then $1/T$ decreases and the line spacing decreases, so the lines move closer together.

The extension of Fourier's theorem to non-repetitive signals involves the extension of the period T to infinity, and the evaluation of an integral over this time, called the Fourier transform, produces a continuous frequency spectrum, i.e. a spectrum in which an infinite number of lines are separated by infinitesimally small frequency increments. The Fourier transform may be thought of as a mathematical process which converts a description of the signal in the time domain, i.e. a waveform, into a description of the signal in the frequency domain, i.e. a frequency spectrum. Conversely, the inverse Fourier transform converts a frequency spectrum into its corresponding waveform (Figure 8.12).

Although the Fourier transform of some simple signals may be evaluated exactly, using techniques of mathematical analysis, most complex signals require the use of numerical methods of integration which involve the performance of very large numbers of calculations. In the 1960s Cooley and Tukey, in the United States, devised a method for vastly reducing the calculation time for evaluating the Fourier transform; fast fourier transform (FFT) analysis derives from their algorithm. In an FFT analyser a sample of the signal is digitized, and the FFT calculations performed on the digitized sample to produce a line spectrum. The fundamental frequency in this spectrum is $1/T$, where T is the total duration of the sample in seconds; the frequency spacing between the lines, $\Delta f = 1/T$. Thus, in effect, the FFT produces the spectrum which would result from an indefinite repetition of the sample (rather like analysing an endlessly playing loop of audio tape).

An FFT performed on a sample of N points produces a spectrum of $N/2$ frequency lines, because each line has to be defined by two pieces of information, amplitude and phase. The time interval between each digital sample is T/N, so the sampling rate is N/T. The maximum frequency component in the line spectrum, f_{max}, is $(N/2)(1/T)$, which is the Nyquist frequency relating to the requirements of Shannon's sampling theorem, described earlier.

In summary these variables and their relationships are (see Figure 8.12):

- number of samples $= N$
- interval between successive samples $= \Delta T$ seconds
- total duration of sample $= T = N\Delta T$ seconds
- sampling rate $= N/T = 1/\Delta T$ samples per second
- line spacing, $\Delta f = 1/T$ Hz
- number of frequency lines $= N/2$
- maximum frequency, $f_{max} = (N/2)(1/T)$ Hz.

The value of N is usually fixed for a particular type of FFT analyser. Any one of the three variables f_{max}, Δf and T may then be selected over a range of values; this selection then fixes the value of the other two variables. The sampling rate is automatically adjusted to be at least twice the value of f_{max}, and an anti-aliasing filter having an appropriate cut-off frequency is automatically selected.

An example may illustrate the interrelationship between these variables and their selection ranges. A particular FFT analyser operates on 1024 data sample points (note that N is usually an integral power of 2). In theory this should produce a 512-line spectrum, but in practice only 400 lines are used; this allows for the fact that the anti-aliasing filter cannot have a perfectly sharp cut-off characteristic. The frequency range, i.e. from zero to f_{max}, may be selected within the range 20–20,000 Hz in steps of 2, 5, 10 (i.e. 20 Hz, 50 Hz, 100 Hz, 200 Hz etc.). On the highest range the 400 lines are spread over a frequency range of 20,000 Hz, so that the line spacing, Δf, is $20{,}000/400 = 50$ Hz. This means that the total duration of the waveform sample, during which 1024 values are collected, is 1/50 second or 20 ms, and the sampling rate is therefore $1024 \times 50 = 51{,}200$ samples per second, which is, as required, more than twice the value of f_{max}. At the other extreme, if an f_{max} value of 20 Hz is selected, this gives a much higher frequency resolution with $\Delta f = 0.005$ Hz, but with a much higher total sample duration of 20 s. The sampling rate in this case is 51 samples per second which is again, as required, more than twice the value of f_{max}.

The fact that the Fourier transform is carried out on a finite number of data samples rather than on a continuous waveform has some inevitable consequences for the accuracy of the results of FFT analysis. The possibility of aliasing has already been discussed. Another difficulty arises because random signals usually begin and end on different non-zero values. The repetition of these samples in the FFT process will create steps or discontinuities, which will cause false values in the frequency spectrum. The difficulty will not be present in transient signals, where the capture of the sample can be arranged to completely include all of the transient, so the sample starts and ends with zero values. It may be partially overcome for non-transient signals by using a time window which weights or shapes the sample in a way which smooths out the discontinuities at the start and end of the sample, but which has minimum effect in between. (Some examples are the Hanning, Hamming, Rectangular and Gaussian windows.)

To improve the statistical reliability of the results, FFT analysers usually have the facility to compute and display the average spectrum derived from analysis of several samples of the signal. Some analysers also have the facility for zoom-FFT which enables a limited portion of the frequency range to be analysed with a greater degree of frequency resolution, i.e. with reduced line spacing.

Dual-channel FFT analysers are able to sample two signals simultaneously and compute various joint functions of the two input signals such as correlation, coherence and transfer functions. The two signals could be the vibration from a particular machine, or part of a machine, and the noise signal from a nearby microphone; or they could be vibration (or noise) from two different parts of a structure.

The main use of FFT analysis is for performing very rapid frequency analysis of signals which have prominent tonal and harmonic characteristics, such as the noise or vibration from turbines and gearboxes. It enables noise sources to be identified from a detailed examination of their frequency spectrum. All vibrating machines, including motors, pumps and fans, produce a vibration spectrum which contains components at the rotation speed and its harmonics. The level of these components gives a good indication of wear and onset of faulty operation, so FFT analysis is widely used in condition monitoring of machinery.

This has been a simplified and non-mathematical introduction to FFT analysis. More advanced texts should be consulted for a more detailed and rigorous treatment.

Digital filters

In a digital filter a sequence of digital operations involving addition, multiplication and time delay are performed on the digitized signal. The effect of these operations on the digital signal is intended to be equivalent to the effect of an analogue filter on the corresponding analogue signal. There are a number of important differences between digital filtering and FFT analysis.

Digital filtering is a continuous process, in which a continuous input of signal to the filter results in a continuous output of filtered signal, exactly as for an analogue filter. In contrast, FFT analysers operate on discrete blocks of data; each sample block is captured then analysed while the next block is being captured, and so on. FFT analysers produce a line spectrum, whereas the digital filtering process is usually used for octave and one third octave analysis. Digital filters may be designed to produce band pass filters of any desired bandwidth and shape, including those that meet the requirements of British and international standards, which have been based on the performance of analogue filters. The centre frequency of the band is determined by the signal sampling rate, and the filter shape by the details of the sequence of filtering operations. Changes in the configuration of digital filters may therefore be accomplished very rapidly by changes to software, whereas changes in physical electronic components would be required for the equivalent analogue filter.

Real and ideal band pass filters

Ideal filters would exhibit 100% transmission throughout the pass band and 0% transmission outside those limits. This is not achievable in practice and the sides of the filter characteristic exhibit a finite slope. The allowed departure from the ideal is set down in various IEC standards. Figure 8.13 illustrates these ideal and practical filter characteristics.

Measurement durations for narrowband analysis

Narrowband analysis may be useful in some cases, where for example a one third octave filter fails to identify a suspected pure tone or narrowband component in the sound. These are typically available as 1%, 3% and 10% filter bandwidths (i.e. having a bandwidth of 1% etc. of the band centre frequency) with a continuously variable centre frequency.

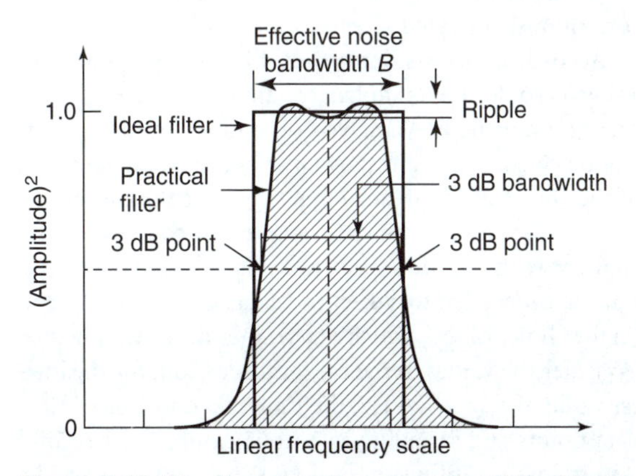

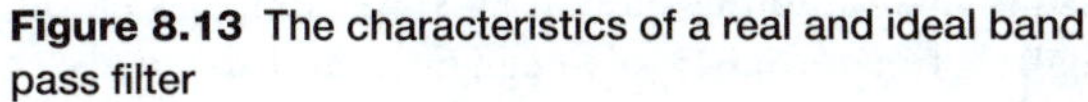
Figure 8.13 The characteristics of a real and ideal band pass filter

Simple observations with octave or third octave bands show that with the lower frequency narrower bands (e.g. the 63 Hz band) the output display tends to fluctuate more than for a higher frequency wider band (e.g. 1000 Hz). This means that a longer measurement time must be used for narrower than for wider bandwidths in order to gain the same statistical reliability, and therefore the same measurement accuracy, for the average value. This effect will be increased when using narrower bandwidths. This is encapsulated in the Time Bandwidth Product Theorem of signal processing which states that the standard deviation divided by the mean value of a set of individual readings of a sound level, ε, is given by:

$$\varepsilon = 1/(2\sqrt{(BT)})$$

where B = bandwidth in Hz and T is the averaging time in seconds.

Example 8.3

For a 3% filter with band centre frequency = 100 Hz then B = 3 Hz.

If the averaging time T = 10 s, then:

$$\varepsilon = 1/\left(2\sqrt{(3 \times 10)}\right) = 0.09.$$

If the mean value of a statistically significant number of independent observations of a sound level = 80 dB, then the standard deviation would be $0.09 \times 80 = 7.2$ dB. Assuming the normal or Gaussian statistical distribution for ransom fluctuations in the signal this means that about 56% of all the data points would lie between $(80 - 7.2)$ dB and $(80 + 7.2)$ dB, i.e. between 72.8 and 87.2 dB. If the averaging time had been 40 s the range would have halved.

The message is that adequate measurement time must be allowed when using narrowband filters in order to obtain accurate measurements and that longer averaging times are needed for lower frequency signals.

Power spectral density

Suppose we have three noise sources A, B and C and that the sound pressure levels measured arising from each under identical conditions has been measured as follows: A: 90 dB in the 100 Hz 1/3 octave band, B: 90 dB in the 100 Hz octave band, and C: 90 dB in the 1000 Hz 1/3 octave band, assuming that in each case noise levels from all other bands are insignificant. What can you say about the sound power emitted by each of these sources (i.e. what similarities and differences are there)?

The sound pressure level is related to RMS sound pressure and therefore also to the mean square pressure and therefore the sound intensity at the measurement point. Since these values are the same and since the measurement conditions are the same for all three sources this means that the sound power and sound power levels are the same for all three sources. What is different is that this sound power is spread over a different band of frequencies in each case. The 100 octave band of source B is three times as wide as that of source A, and that of source C is 10 times as wide as that of source A. Therefore although the total sound power is the same for all three sources the amount of sound power per unit of frequency bandwidth is different. This difference is expressed in terms of the **power spectral density** of each source.

Power spectral density (PSD), sometimes simply called spectral density, describes how the power of a signal varies with frequency. It is the amount of power per unit of frequency in a signal. It is measured in watts per Hz, or more generally for a voltage signal in V^2/Hz (since power is proportional to V^2). PSD is therefore itself a quantity which varies with frequency and the integration of PSD over a given frequency range will give the average power in the signal over that frequency band.

Since the power in a signal is proportional to its mean square value (e.g. either in V^2/Hz or, for an acoustic signal in Pa^2/Hz):

PSD = mean square value/bandwidth

Note: The amount of sound power might vary with frequency within the frequency range of an octave or third octave band and therefore in a more strict definition of PSD, using the ideas of differential calculus the PSD at any frequency is the ratio of mean square values to bandwidth measured over a very small bandwidth centred around the particular frequency.

Example 8.4

A band limited white noise signal has a spectral density of 0.04 V^2/Hz from 0 to 200 Hz and is zero at all other frequencies. What is the RMS value of the signal?

From the above:

$$\text{Mean square value} = \text{PSD} \times \text{bandwidth} = 0.04 \times 200$$

$$\text{RMS} = \text{Root mean square value} = \sqrt{(0.04 \times 200)} = 2.83 \text{ volts}$$

Power spectral densities are useful for comparing narrow band levels from analysers using different bandwidths.

Exercise

How does the PSD of white noise and pink noise vary with frequency? (If necessary refer back to Chapter 1 for explanation of white and pink noise.)

Real-time frequency analysis

The level and frequency content of the noise from a machine may change slowly, over days and weeks, and these changes may be adequately followed using a simple sound level meter fitted with a manually operated set of octave band filters. In contrast, it takes only a few seconds for changes to occur in noise level and frequency content as an aircraft flies over. Changes during human speech or during a handclap take place over even shorter time periods; they require the use of a real-time analyser, which gives a very quick, almost instantaneous analysis and display of frequency spectra. This enables rapid changes in the signal to be observed as they happen, i.e. in real time.

Analogue real-time analysers use a band of analogue filters, in parallel, so that the signal is divided and passes through all the filters at the same time. More modern devices use digital filters which can operate so quickly that a complete band spectrum may be obtained rapidly enough to follow any changes as they occur. The spectrum, which may be octave, one third octave or narrowband, is usually displayed on a screen. There are facilities to compute and display various noise indices, to store the analyses, and to produce hard copies of the spectra. And it is often possible to build up a three-dimensional picture (level, frequency and time) called a waterfall plot; this shows how the frequency content of the signal changes with time. One definition says that real-time frequency analysis includes all of the signals in all of the frequency bands at all times, displaying the results on a continuously updated screen.

According to this definition, FFT analysers are not always capable of operating in real time. The time taken for carrying out the FFT analysis on a block of data points will be constant for any particular analyser, independent of the choice of frequency range, f_{max} (see the earlier discussion). Let us assume for example that for a particular analyser with 400-line spectrum this is typically 0.045 s. But the time taken to collect the data sample will depend on the choice of f_{max} which determines the sampling rate. Consider the values in the earlier discussion: for the lowest value of f_{max} = 20 Hz, the time taken to collect 1024 data points at a sampling rate of 51 samples per second results in a sampling time of 20 s. By comparison, the analysis time of 0.045 s is negligible, and the analyser operates in real time. At the highest frequency range, when f_{max} is 20,000 Hz, the sampling rate is 51,200 samples per second, and the sample is collected in 0.02 s, which is shorter than the analysis time. Under these circumstances the analyser cannot possibly display analysis of all the signal, so it does not operate in real time. On this basis such an analyser would operate in real time only for frequencies up to about 5000 Hz.

8.7 The performance of sound level meters

The performance of sound level meters is specified by BS EN 61672 *Electroacoustics – Sound level meters – Part 1: Specifications*; *Part 2: Pattern evaluation tests*; *Part 3: Periodic testing*:

- Part 1 of the standard deals with the specification of the performance of the sound level meter.
- Part 2 deals with the method of testing that national metrology institutes and the manufacturers of the sound level meter should carry out to show that that particular type of meter meets the performance specification of Part 1.
- Part 3 describes a series of tests, less detailed than those of Part 2, that should be commissioned periodically (every one or two years) by the user of the sound level meter to ensure that it continues to meet the performance specification for that class of instrument.

Part 1: Specification

This standard gives electroacoustical specifications for three types of sound measuring instruments:

- a conventional sound level meter that measures exponential time-weighted sound level
- an integrating average sound level meter that measures time-average sound level
- an integrating sound level meter that measures sound exposure level.

This standard supersedes BS EN 60651:1994 (which covered ordinary sound level meters) and BS EN 60804: 2001 (which covered integrating sound level meters), which are withdrawn.

Two performance categories, Class 1 and Class 2, are specified in the standard, replacing the former types 0, 1, 2 and 3. The specification (i.e. the design goal) is the same for both classes but wider tolerances are permitted for Class 2 as compared with Class 1.

The standard is applicable not only to self-contained hand-held sound level meters but also to multi-component instrumentation, including computers, printers etc.

Terminology

Time-weighted sound level

It is suggested that when describing a time-weighted sound level the time constant (F or S) and frequency weighting (A, C or F) should both always be specified. Thus: L_{AF}, L_{AS}, L_{CF}, L_{CS}, L_{AFmax}, L_{ASmax}, L_{CFmax}, L_{CSmax}, L_{AF90}, L_{AS90}, L_{AF10} etc.

Time-average sound level

The standard defines time-average sound level, L_{AT}, as an alternative term, synonymous with equivalent continuous sound level L_{AeqT}.

Note 3 of section 3.9 of the standard states that: 'in principle, time weighting is not involved in the determination of time-average sound level'.

Minimum requirements

As a minimum a conventional sound level meter shall have A-frequency weighting and F-time weighting. A Class 1 sound level meter shall also have C-frequency weighting.

A zero-frequency weighting (Z-weighting) is also defined in the standard, but is optional. The Z-weighting (zero dB in all frequency bands) is defined for the range from 10 Hz to 20,000 Hz although a limited frequency range version, called FLAT, from 31.5 Hz, to 8000 Hz, is also defined.

Performance specifications

As with the previous standard, separate tolerance limits on design goals are specified for Class 1 and Class 2 sound level meters for: frequency weightings, directional response, and the F/S time weighting characteristics (based on response to onset and cessation of 4000 Hz tone bursts). Tables 8.1 and 8.2 show the limits for directionality and for the frequency weightings.

Other aspects of performance which are described include: level linearity, self-generated noise, overload indication, under-range indication, peak C sound level L_{Cpeak}, reset, thresholds, display, analogue or digital output, timing facilities, radiofrequency emissions and disturbances to a public power supply, crosstalk, power supply requirements, environmental criteria, auxiliary devices, handbook requirements.

Note that tolerance limits in this standard now include allowance for measurement uncertainty, not included in the previous standards.

Table 8.1 of BS EN 61672-1:2003 Directional response limits including maximum expanded uncertainty of measurement

Frequency (kHz)	Maximum absolute difference in displayed sound levels at any two sound-incidence angles within +/− θ degrees from the reference direction (dB)					
	$\theta = 30°$		$\theta = 90°$		$\theta = 150°$	
	Class					
	1	2	1	2	1	2
0.25 to 1	1.3	2.3	1.8	3.3	2.3	5.3
>1 to 2	1.5	2.5	2.5	4.5	4.5	7.5
>2 to 4	2.0	4.5	4.5	7.5	6.5	12.5
>4 to 8	3.5	7.0	8.0	13.0	11.0	17.0
>8 to 12.5	5.5		11.5		15.5	

Table 8.2 of BS EN 61672-1: 2003 Frequency weightings and tolerance limits including maximum expanded uncertainty of measurement

Nominal frequency (Hz)	Frequency weightings (dB)			Tolerance limits (dB)	
				Class	
	A	C	Z	1	2
10	−70.4	−14.3	0.0	+3.5; −∞	+5.5; −∞
12.5	−63.4	−11.2	0.0	+3.0; −∞	+5.5; −∞
16	−56.7	−8.5	0.0	+2.5; −4.5	+5.5; −∞
20	−50.5	−6.2	0.0	+/−2.5	+/−3.5
25	−44.7	−4.4	0.0	+2.5; −2.0	+/−3.5
31.5	−39.4	−3.0	0.0	+/−2.0	+/−3.5
40	−34.6	−2.0	0.0	+/−1.5	+/−2.5
50	−30.2	−1.3	0.0	+/−1.5	+/−2.5
63	−26.2	−0.8	0.0	+/−1.5	+/−2.5
80	−22.5	−0.5	0.0	+/−1.5	+/−2.5
100	−19.1	−0.3	0.0	+/−1.5	+/−2.0
125	−16.1	−0.2	0.0	+/−1.5	+/−2.0
160	−13.5	−0.1	0.0	+/−1.5	+/−2.0
200	−10.9	0.0	0.0	+/−1.5	+/−2.0
250	−8.6	0.0	0.0	+/−1.4	+/−1.9
315	−6.6	0.0	0.0	+/−1.4	+/−1.9
400	−4.8	0.0	0.0	+/−1.4	+/−1.9
500	−3.2	0.0	0.0	+/−1.4	+/−1.9
630	−1.9	0.0	0.0	+/−1.4	+/−1.9
800	−0.8	0.0	0.0	+/−1.4	+/−1.9
1000	0	0	0	+/−1.1	+/−1.4
1250	+0.6	0.0	0.0	+/−1.4	+/−1.9
1600	+1.0	−0.1	0.0	+/−1.6	+/−2.6
2000	+1.2	−0.2	0.0	+/−1.6	+/−2.6
2500	+1.3	−0.3	0.0	+/−1.6	+/−3.1
3150	+1.2	−0.5	0.0	+/−1.6	+/−3.1
4000	+1.0	−0.8	0.0	+/−1.6	+/−3.6
5000	+0.5	−1.3	0.0	+/−2.1	+/−4.1
6300	−0.1	−2.0	0.0	+2.1; −2.6	+/−5.1
8000	−1.1	−3.0	0.0	+2.1; −3.1	+/−5.6
10,000	−2.5	−4.4	0.0	+2.6; −3.6	+5.6; −∞
12,500	−4.3	−6.2	0.0	+3.0; −6.0	+6.0; −∞
16,000	−6.6	−8.5	0.0	+3.5; −17.0	+6.0; −∞
20,000	−9.3	−11.2	0.0	+4.0 ; −∞	+6.0; −∞

8.8 Introduction to the principles of the direct measurement of acoustic intensity

The intensity of an acoustic wave is defined as the rate at which acoustic energy passes through unit area perpendicular to the direction of propagation of the wave. It is measured in W/m^2.

The instantaneous intensity (I) at a point in a wave can be shown to be the product of the acoustic pressure (p) and the acoustic particle velocity (u) at that instant:

$$I = pu$$

To find the average intensity of the wave at that point it is necessary to find the 'time average' of this product because both p and u vary from moment to moment throughout the cycle of the sound wave, so:

$$I_{\mathrm{AVG}} = (pu)_{\mathrm{AVGE}}$$

The acoustic intensity is proportional to the square of the sound pressure ($I \propto p^2$) and for plane waves $I = p^2/pc$. Sound pressure may be easily measured using a microphone, and this relationship is the basis of indirect measurements of acoustic intensity. There is, however, a very important distinction between the intensity and the square of the pressure to which it is so closely related. Since intensity describes a rate of flow (of energy), it is a quantity which has a direction, like force, momentum or velocity. It is called a vector quantity. The quantity square of the pressure has no direction attached to it. It is a scalar quantity like mass or volume. Therefore indirect measurements of acoustic intensity based on sound pressure measurements may give the correct magnitude of the intensity but can never indicate the direction of flow of acoustic energy. Returning to the equation $I_{\mathrm{AVG}} = (pu)_{\mathrm{AVG}}$, the direct measurement of acoustic intensity at a point in a sound field requires two transducers at that point, one to measure p (i.e. a microphone) and another to measure u. The signals from these two transducers must then be multiplied together and time averaged to produce an analogue signal which directly represents the acoustic intensity.

Unfortunately it is not easy to measure acoustic particle velocity directly. There is, however, a relationship between p and u (not involving the unknown I) which can help. The correct statement of this relationship (called the equation of motion) involves the use of the notation and symbols of differential calculus, but the general idea may be obtained by considering a small box of air in the path of a plane wave striking one face of the box (of area S) at right angles (Figure 8.14).

The length of the box is Δx; the symbol Δ indicates a small increment in the direction x in which the wave is travelling. The pressure and particle velocity at the front face of the box (facing the approaching sound wave) are p and u, respectively; at the opposite face they have undergone very small changes and are $p + \Delta p$ and $u + \Delta u$, respectively.

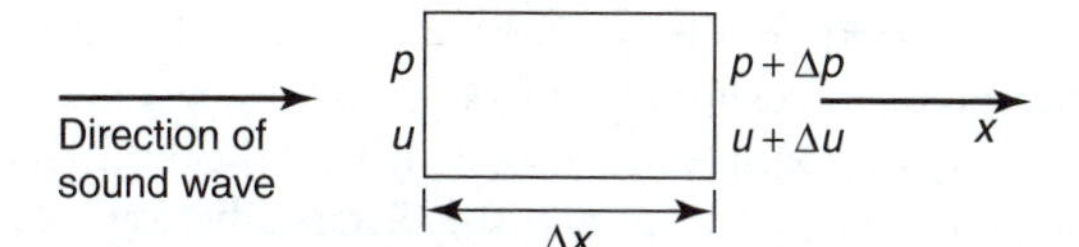

Figure 8.14 Incremental changes in acoustic pressure and particle velocity in a plane wave occurring over a small distance

The required relationship is obtained by applying Newton's second law to the motion of the air in the box caused by the passage of the sound wave: force = mass × acceleration.

Remember that:

$$\text{Force} = \text{pressure} \times \text{area}$$

So the net force on the box, to the right is:

$$pS - (p + \Delta p)S = -S\Delta p$$

The mass of air in the box is:

$$\text{density} \times \text{volume} = \rho S \Delta x$$

where ρ = density of the air.

Now recall the formula for acceleration:

$$\begin{aligned}\text{Acceleration} &= (\text{change in velocity})/(\text{time taken})\\ &= \Delta u/\Delta t\end{aligned}$$

where Δt = small interval of time in which the particle velocity has changed from u to $u + \Delta u$. Combining everything together into the force equation:

$$-S\Delta p = \rho S \Delta x(\Delta u/\Delta t)$$

S cancels leaving:

$$-\Delta p = \rho \Delta x(\Delta u/\Delta t)$$

and on rearranging:

$$\rho(\Delta u/\Delta t) = -(\Delta p/\Delta x)$$

This is the required relationship which shows that the rate of change of particle velocity (with time) is related via the density of the medium to the rate of change of acoustic pressure with distance, i.e. to the sound pressure gradient. Therefore, if the sound pressure gradient can be measured, the rate of change of particle velocity can be found, and by integrating this with respect to time, the particle velocity itself can be obtained.

The sound pressure gradient in a sound field may be measured, approximately at least, by having two microphones close together in the sound field. If the two sound pressures are p_1 and p_2 and the distance between the microphones is Δx then:

$$\text{Sound pressure gradient} = (\Delta p/\Delta x) = -(p_1 - p_2)/\Delta x$$

The two-microphone arrangement is the basis of the sound intensity meter. The sound pressure used for the calculation of $I_{AVG} = (pu)_{AVG}$ is the average of the two signals, i.e. it is given by:

$$p = (p_1 + p_2)/2$$

The sound intensity may therefore be measured by processing the two signals and combining them in the following way. The difference between the two signals $(p_1 - p_2)$ is obtained then integrated with respect to time to produce a signal which is proportional to u. The sum $(p_1 + p_2)$ of the two signals is obtained to produce a signal which is proportional to p. The signals proportional to p and u are then multiplied together and time averaged to produce a signal which may be calibrated to represent the average acoustic intensity.

A block diagram of a sound intensity meter is shown in Figure 8.15.

The vector nature of acoustic intensity means that the two-microphone intensity probe is directional in its response, giving the maximum signal when the axis of the probe (i.e. the line joining the two microphones) is lined up in the direction of the acoustic energy flow. It is this ability to identify the direction of flow of acoustic energy, as well as to measure its magnitude, which is the basis of many of the applications of the sound intensity meter.

One of these applications is in the measurement of the sound power emitted by noise sources. Using an acoustic intensity meter it is possible to measure the sound power level of a noisy machine in situ, in the presence of background noise and without the need for a specialist acoustic environment, i.e. an anechoic or reverberant room. Other applications include the location and

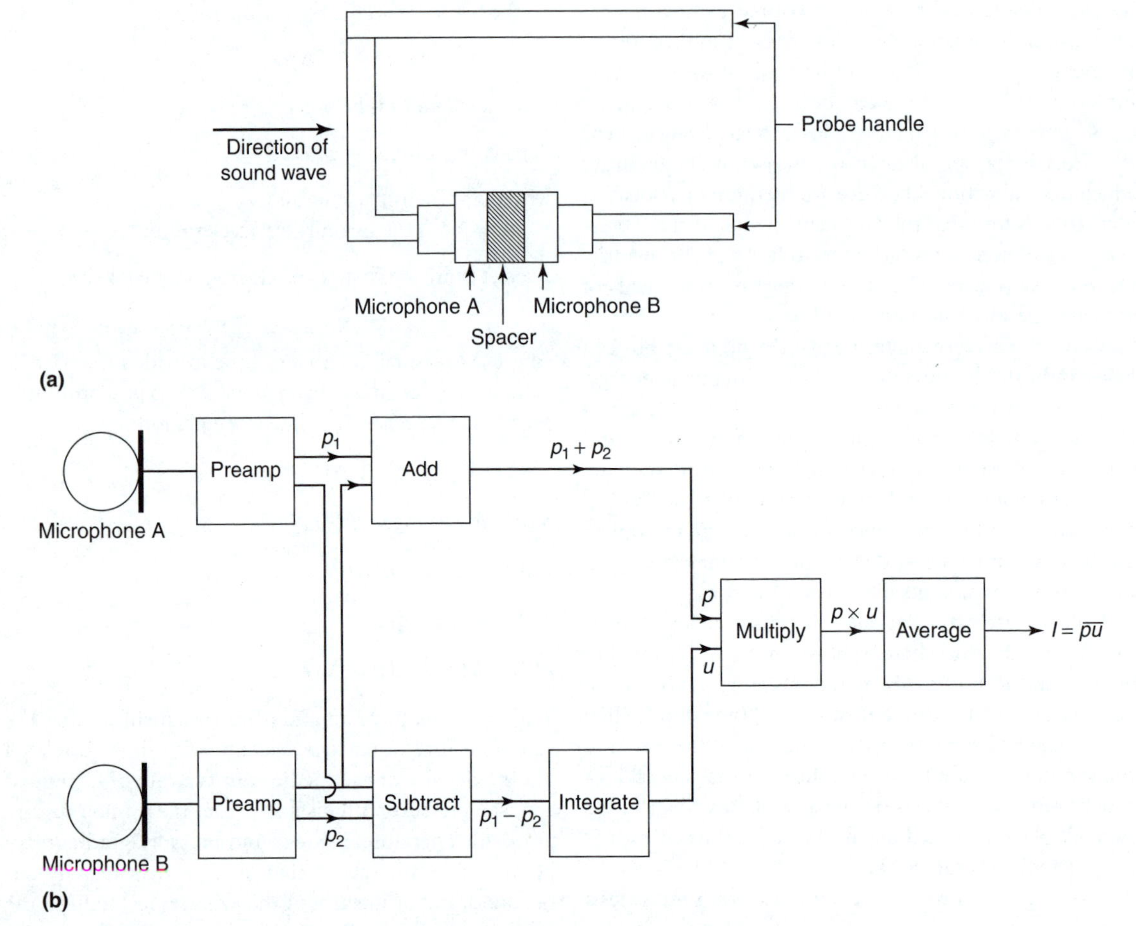

Figure 8.15 (a) A sound intensity meter and probe with (b) its block diagram

identification of noise sources and the detection of leaks and weak-link areas in the transmission of sound through partitions and panels.

8.9 The measurement of sound power levels

There are, in principle, four different test methods:

1. Free field methods require that the measurement of sound radiated from the source is not affected by reflected sound. Such an acoustic environment may sometimes be achieved by performing tests outdoors, or in an anechoic or semi-anechoic test room.
2. Sound intensity methods use a sound intensity meter, which is sensitive to the direction of flow of sound energy; they allow the measurement of sound power levels in non-anechoic environments using the same method as for the free field tests.
3. Reverberant methods require the specialist test environment of a reverberant room.
4. Substitution methods require a calibrated sound power source.

Free field methods

An anechoic room is a specially constructed chamber in which the walls, floor and ceiling are covered with sound absorbing material, usually in the form of wedges, which are effective in eliminating reflected sound. Thus measurements of sound pressure levels produced by a noise source in such a room are of the direct sound only, and can be used to compute the sound power level of the source. The directivity of the source can also be measured. The method is convenient for small portable machines but the size of the room limits the size of machines which can be tested since, ideally, measurements should be made in the far field of the source.

A semi-anechoic room (also called 'hemi-anechoic') has sound absorbent walls and ceiling but a hard sound-reflecting floor.

The test method involves taking measurements of sound pressure level at spatial intervals over a surface surrounding the machine within the far field of the source. This test surface may be a hemisphere for small machines or a parallelepiped for large machines, following the main contours of the machine shape. Ideally the surface would be checked to confirm that, in the vicinity of the measurement positions, the level is falling by approximately 5 or 6 dB per doubling of distance, i.e. that the measurements positions are in the far field of the machine, but this will often be difficult to achieve in practice.

The sound power level, L_W, is related to the area of the measurement surface (S in m^2) and the average sound pressure level over the surface ($L_{p,AVGE}$) by the equation.

$$L_W = L_{p,AVGE} + 10\log S$$

If the measurement surface is a hemisphere of radius r then $S = 2\pi r^2$ and $10\log S$ becomes $20\log r + 8$, giving the 'hard-ground' version of the direct field equation:

$$L_W = L_{p,AVGE} + 20\log r + 8$$

Approximate free field method for large machines

For large machines that cannot easily be accommodated in an anechoic room, or which can only be tested in situ, an approximation to the free field method may be carried out.

Example 8.5

The manufacturer's data for the noise emission for a chiller (dimensions 5 m long × 2.5 m wide × 2.5 m high) has been supplied to you in the form of sound pressure levels measured outdoors at a distance of 1 m from the unit over a sound reflecting plane, as follows:

End 1	87 dB
End 2	86 dB
Side 1	83 dB
Side 2	84 dB
Top	92 dB

Estimate the sound power level of the chiller.

We have to calculate the area, S, of a five sided box surrounding the chiller at a distance of 1 m from it surfaces:

$$S = (2 \times 7 \times 3.5) + (2 \times 4.5 \times 3.5) + (7.0 \times 4.5) = 112\ m^2$$

Log average sound pressure level, $L_{p,AVGE}$, measured over the five sides:

$$L_{p,AVGE} = 10\log[(10^{8.7} + 10^{8.6} + 10^{8.3} + 10^{8.4} + 10^{9.2})/5] = 87.7\ dB$$

Sound power level, $L_W = L_{p,AVGE} + 10\log S = 88 + 10\log(112) = 108$ dB

Sound intensity methods

The measurement of sound power level from a noise source using a sound intensity meter is similar in principle to the free field method using an anechoic or semi-anechoic room. The sound intensity is measured over a test surface, usually a hemisphere or parallelepiped surrounding the noise source. The sound power level (L_W) is obtained from the average sound intensity over the test surface (L_I) and its area (S):

$$L_W = L_I + 10\log S$$

A necessary requirement of the free field method, in which a sound level meter is used, is that sound energy only travels outwards from the source through the test surface. The great advantage of the sound intensity meter is that it allows measurements to be carried out in an ordinary room, where the presence of reflecting surfaces means that some reflected sound energy will pass through the test surface in the opposite, inwards direction. However, provided that there is no acoustic absorption inside the test surface, all of this reflected sound energy will eventually pass out through the test surface again, either directly or maybe after reflection or scattering by the surfaces of the noise source. The sound intensity meter indicates the net flow of acoustic energy through the test surface, so the value of the average sound intensity measured over the entire surface will correspond only to that provided by the outward flow of energy from the source. The reflected intensity sound passing through the surface cancels out, making a net zero contribution to the flow of acoustic energy through the surface. For similar reasons, measurements can also be taken in the presence of background noise from another source, provided that it produces a constant level and is located outside the test surface, so that it also produces a net zero flow of sound energy through the test surface. The appropriate ISO standard gives details of the measurement method, including recommendations for the size and shape of the test surface, number of measurement points to be used, and checks on the adequacy of the equipment and measurement procedures.

Reverberant room methods

A truly reverberant room is one in which the energy from a noise source in the room is diffused completely throughout the room, so that ideally the sound pressure level would be the same everywhere. In practice, measurements are taken at several points in the room and the average sound pressure level is obtained. This is related to the sound power level of the source L_W and to the room constant, as described in Chapter 5. Expressing the room constant in terms of the reverberation time T (which can be measured directly), by using Sabine's formula the following relationship is obtained:

$$L_W = L_p + 10\log10 V - 10\log10 T - 14$$

where V is the room volume in m^3. A modified and more accurate version of the above formula is given in the ISO standard. Only sound power levels can be obtained from reverberant-room methods – the directivity information can only be obtained from free field (anechoic and semi-anechoic) measurements.

The appropriate standard gives qualification procedures in respect of the size, dimension ratios and frequency dependent reverberation times of a chamber to be used for this purpose. It also details the number of source positions and microphone positions needed to get a reliable value for the average reverberant level. L_{eq} is often used by way of temporal averaging.

Measurements are made in one or one third octave bands over a frequency range that relates to the qualification conditions for the chamber. Sources suspected of having pure tone or narrowband components are given special treatment. The possibility of modal behaviour is dealt with by increasing the number of measurement points to be averaged. If a calibrated sound source is used, as in the substitution method described below, the need to measure the reverberation time and exact volume of the reverberant chamber is avoided.

Substitution method

In the substitution method of sound power level measurement, a reference noise source of known sound power level is used. Comparisons are made of the noise levels produced in the room by the machine under test and the standard source. This type of test may be carried out in a special reverberant room, or with lesser accuracy in the actual location of the test machine. The reference source should be capable of being calibrated and should be omni-directional, as far as possible, over the frequency range of interest.

The principle of the method is that the difference in the two sound pressure levels is the same as the difference in their two sound power levels. The method is illustrated by Example 8.6.

Example 8.6

Measurements were made in a reverberant room of the noise produced by an electric drill. The average sound pressure level in the 1000 Hz octave band was 85 dB. When the drill was replaced by a calibrated noise source which has a sound power level of 100 dB, in this octave band, the average level produced was 89 dB. Calculate the sound power level of the drill in the 1000 Hz band.

$$L_{W\ \text{Test source}} - L_{W\ \text{Standard source}} = L_{p\ \text{Test source}} - L_{p\ \text{Standard source}}$$

Sound power level of drill $- 100 = 85 - 89 = -4$ dB

Therefore sound power level of drill $= 100 - 4 = 96$ dB.

8.10 The measurement of vibration

Transducers

A transducer is a device which produces an electrical signal proportional to the physical quantity of interest. A variety of transducers are available for the measurement of mechanical vibrations. These include displacement gauges which work on a capacitative principle, with the vibrating surface forming one plate in a variable air gap condenser, and electrodynamic (coil and magnet type) gauges which give a direct measurement of vibration velocity. However, the most commonly used device for vibration measurement in the audio-frequency range is the piezoelectric accelerometer, which gives an electrical signal which is proportional to the vibration acceleration.

The piezoelectric effect

When a slab of piezoelectric material is subjected to a stress, the material becomes polarized, with electric charge collecting at the stressed faces of the slab. These materials are ceramics, and therefore non-conducting, and in order to utilize the effect the charged slab faces are silvered to create electrodes. The piezoelectric effect then results in an electrical potential difference (i.e. a voltage difference) being set up between the electrodes which is proportional to the stress. If the stress changes from a compression to a tension, then the polarity of the voltage difference also changes, and so it follows that an alternating stress across the slab results in the production of an alternating signal. It is also possible to arrange for piezoelectric materials to respond to shear stresses in a similar way.

The piezoelectric effect was discovered by Pierre and Jacques Curie round about 1880 and the first important application was as underwater sound transducers in the development of sonar in the First World War (1914–1918). Although some naturally occurring crystalline materials such as quartz, tourmaline and Rochelle salt are piezoelectric, modern transducers are usually made in ceramic form from a combination of lead zirconate and lead titanate, known as PZT. All piezoelectric materials have a Curie point, which is a temperature above which all piezoelectric properties are permanently lost. In practice, operating temperatures must be limited to well below the Curie point to prevent significant loss of sensitivity. Most accelerometers can be used at up to 250°C, and certain special types to much higher temperatures.

The piezoelectric accelerometer

The compression type of accelerometer (Figure 8.16) consists of two piezoelectric discs sandwiched between a mass and the base of the device. The base is thick and stiff so that bending of the base (base strain) is minimized. The piezoelectric discs form part of a mass–spring system whose resonance frequency is designed to be well above the measurement frequency range. The motion of the vibrating surface is transmitted via the base and the discs to the mass.

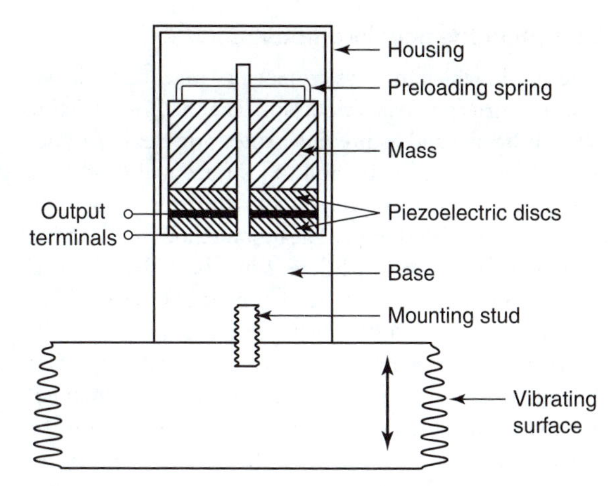

Figure 8.16 The construction of a compression type of piezoelectric accelerometer

In reaction to this the mass exerts a force on the piezoelectric discs which is proportional to the mass and its acceleration (force = mass × acceleration). The mass, the base and the vibrating surface all move together with the same acceleration, at frequencies well below resonance. Therefore the force on the piezoelectric discs, and the voltage signal it produces, are proportional to the acceleration of the vibrating surface:

Force on piezoelectric discs $\propto$ acceleration of vibrating surface

Voltage signal across piezoelectric discs $\propto$ force on piezoelectric discs

Therefore:

Voltage signal across piezoelectric discs $\propto$ acceleration of vibrating surface

The preloading spring ensures that the discs remain under compression during the entire cycle.

Accelerometer frequency response

This is reasonably flat (to within 12%) for frequencies up to about one third of the resonant frequency, and this sets the upper frequency limit of the device. The lower frequency limit is set by the electronics of the measurement system (preamplifiers and cables).

Mounting the accelerometer

The way in which the accelerometer is attached to the vibrating surface is very important. The method of fixing should be as rigid as possible, since any flexibility between the accelerometer and the surface will produce a 'mounting resonance' which will reduce the upper frequency limit of the device. The best method of mounting is to use a steel stud which screws into both the vibrating surface and the base of the accelerometer. Other methods include the use of studs which screw into the base of the accelerometer and are attached to the vibrating surface by cement, by a thin layer of beeswax or by a permanent magnet. Thin double-sided adhesive tape may also be used. The cement should be as stiff as possible: certain epoxy resins and cyanoacrylates are recommended, but soft setting glues or gums should be avoided. The beeswax method is good provided that the surfaces are clean and smooth, and the wax layer thin and even, in which case measurements may be made up to about 8 kHz. However, a disadvantage is that the wax softens at higher temperatures, and cannot be used above about 40°C. The magnet method can be used for frequencies up to about 3 kHz. The accelerometer may be attached to a hand-held probe, but although this method is very convenient for carrying out quick vibration surveys it should only be used for frequencies below about 1 kHz because of the low mounting resonance frequency. Figure 8.17 shows a typical frequency response curve of an accelerometer, including the effect of the mounting.

Accelerometer size

The sensitivity of an accelerometer depends upon the type of piezoelectric material used, and also on the mass which produces the stress in the discs. Large accelerometers, with a greater mass, are more sensitive than smaller ones. Increased mass, however, reduces the resonance frequency of the device and therefore its upper working frequency limit. Accelerometers are available in a wide range of sizes, with total masses (i.e. the whole device) ranging from 2 g to 100 g giving a corresponding range of resonant frequencies from 80 kHz to 10 kHz. The size of the accelerometer should also be considered in relation to the surface to which it is to be attached. If it is too large it can significantly alter the vibration level and the resonance frequency of the surface. To avoid this, a good rule is that the mass of the accelerometer should be less than one tenth of the mass of the vibrating surface to which it is attached. This means that only the smallest devices are suitable for measuring the vibration of small, thin sheet-metal panels.

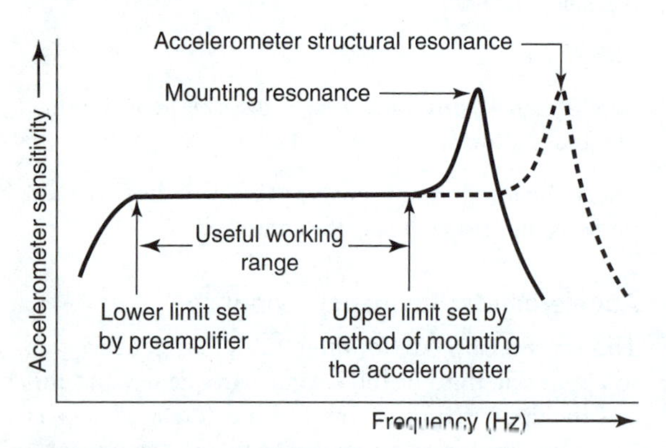

Figure 8.17 A typical frequency response curve for a piezoelectric accelerometer

The effect of adverse environments

Piezoelectric accelerometers may have to operate at high temperatures or in the presence of corrosive substances, high humidity, nuclear radiation or high acoustic noise levels. The accelerometer has been designed to withstand and be insensitive to these adverse environmental influences, but under extreme conditions special types may have to be used. The manufacturer of the accelerometer should be consulted in these cases.

Sensitivity to base strain, temperature transients, transverse vibrations

When very low-level vibration signals are being measured it is important that spurious signals, i.e. those which are unrelated to the acceleration, are kept to a minimum. Such signals can be generated in the accelerometer as a result of transient changes in ambient temperatures, particularly for very low frequencies. Ideally accelerometers should be sensitive only to motion in the axial direction (perpendicular to the base) and should have zero sensitivity in other directions. In practice there is a small cross-sensitivity (a few per cent of the axial value) to vibration in transverse directions. The third way in which the accelerometer can generate spurious signals (i.e. noise) is as a result of flexure of the base of the device.

Accelerometers which are designed to subject the piezoelectric element to shear strains are usually less sensitive to all of these influences than compression type accelerometers.

Cable noise

Spurious signals may be generated in the accelerometer cables as a result of cable vibration. These result from the generation of electrical charges caused by relative motion between various internal parts of the cable. They can be reduced by the use of specially designed (graphited)

cables but it is always good practice to avoid cable whip by clamping down the cable, with tape or adhesive, as close to the accelerometer as possible, and at other points along its length as well. Another way in which cable noise can be generated is as a result of ground loops between the separate ground levels of the accelerometer and its associated measuring equipment. This can be minimized by using a mounting method which electrically isolates the accelerometer from the vibrating surface, e.g. by using an insulating washer beneath the accelerometer base. Sometimes induced currents are generated in the cable as the result of the operation of electrical machinery nearby. The use of properly shielded cables reduces this pick-up which can in extreme cases be further reduced by using special accelerometers and preamplifiers which allow both terminals of the accelerometer to be isolated from the case of the accelerometer, and from instrument ground level. In order to minimize the effect of pick-up the signal should be amplified as early in the measurement chain as possible, preferably before the signal is transmitted down long cables. The ideal position for the preamplifier is, therefore, as close to the accelerometer as possible.

Dynamic range of acceleration measurements

The upper amplitude measurement limit is dependent ultimately upon the accelerometer's structural strength, and also, of course, on the way it is fixed to the vibrating surface. A typical general-purpose accelerometer can respond linearly to levels of up to about 50,000 m/s^2, assuming the fixing method is satisfactory at these levels. Using the beeswax method the upper limit will be about 100 m/s^2. The lower limit of measurement is governed by some of the factors discussed above, i.e. noise produced in accelerometer and cables, and also by the preamplifier circuitry. A typical lower limit value is about 0.01 m/s^2.

Electronic instrumentation for use with accelerometers

The essential parts of a vibration measuring system, in addition to the accelerometer, are a preamplifier, further amplifier stages and filters for signal conditioning and analysis, and some form of read-out, display or recording device, e.g. a meter. A range of specialist vibration meters are available incorporating these and additional features but in addition some sound level meters can be used for vibration measurement. To do this the microphone of the sound level meter is removed and replaced by the accelerometer and its cable, using a suitable adaptor. The meter scale of the instrument will then read in arbitrary decibel units, but can be calibrated using a known vibration level. Alternatively, some instruments have a variety of removable meter scales and allow vibration levels to be read off directly in appropriate units (e.g. in m/s^2).

Preamplifiers

The accelerometer has a very high electrical output impedance, which in effect means that it delivers a signal of very low power, and so the size of the output signal can be reduced by the electrical loading effect of the cables and amplifiers which follow it. The purpose of the preamplifier is not only to amplify the weak signals from the accelerometer, but more importantly to present the signals at a lower, more convenient output impedance to further amplifier stages.

There are two types of preamplifier which are suitable: the **voltage preamplifier** and the **charge preamplifier**. One essential difference between the two is the loading effect of the accelerometer cable. With a voltage preamplifier, the sensitivity of the system depends on the length and type of cable being used (in fact, upon the electrical capacitance of the cable). This means that if the cable is changed the sensitivity of the measuring system changes. This may be inconvenient but may be overcome provided a standard vibration source is always available for recalibration. With a charge preamplifier the sensitivity of the system is independent of cable length. The low frequency limit of a charge preamplifier is lower, although with most voltage preamplifiers measurements can be made accurately down to about 2 Hz, which is low enough for most vibration work. The preamplifier in the sound level meter is of the voltage type whereas in specialist vibration meters, charge amplifiers are often used. Accelerometer manufacturers specify the sensitivity of their devices in two ways to suit either preamplifier, i.e. in volts (or millivolts) per m/s^2 (voltage sensitivity for a given cable length) or in picocoulombs, pC, per m/s^2 (charge sensitivity). The two sensitivities may be related if the total capacitance of the system (accelerometer, cable and preamp) measured in picofarads, pF, is known, using the relationship:

Charge sensitivity = voltage sensitivity × capacitance
[picocoulombs per m/s^2 = volts per m/s^2 × picofarads]

For example, an accelerometer with a charge sensitivity of 1.5 pC/ms^{-2} and a capacitance (including its cable) of 1000 pF connected to a voltage preamplifier will have a voltage sensitivity of $1.5\ pC/ms^{-2}/1000\ pF = 1.5/1000 = 1.5 \times 10^{-3} V/ms^{-2} = 1.5\ mV/ms^{-2}$. If the cable, and hence the total capacitance, was to change then the voltage sensitivity would change, but the charge sensitivity (if a charge amplifier were to be used) would not change.

Some accelerometers contain built-in integrated circuit preamplifiers and maybe some amplification as well,

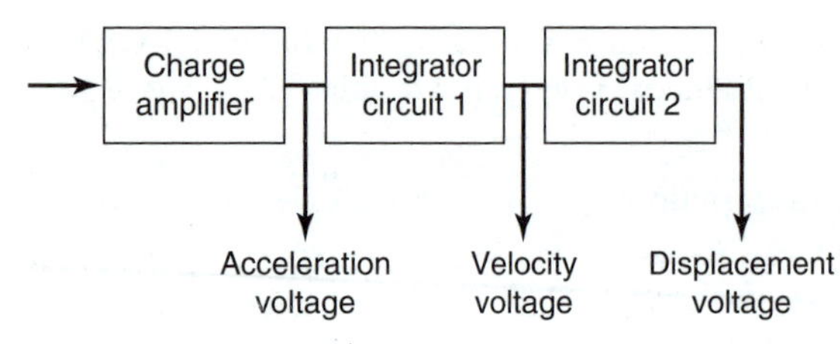

Figure 8.18 The operation of an electronic integrator for acceleration signals

in which case they will have a lower output impedance and increased sensitivity, but will require electrical power to be supplied via the signal cable, and this may affect maximum operation time if battery operated.

Integrating devices

These are circuits placed after the preamplifier to electrically integrate the accelerometer signal; once to convert it into the analogue of vibration velocity, and twice to convert it into vibration displacement (as shown in Figure 8.18). The perfect integrator would in effect have to have a frequency response characteristics with a slope of −6 dB/octave to produce velocity, and −12 dB/octave for displacement.

Practical integrating circuits approximate to this above a certain frequency. They are suitable for use with continuous random signals but not for shock signals and transients.

High pass and low pass filters for signal conditioning

The purpose of these is to limit measurement errors caused by spurious signals or by signals outside the working range of the accelerometer and its instrumentation. The low pass filter, for example, may be used to cut out high frequency signals from the mounting resonance of the accelerometer. The high pass filter may be used to cut out frequencies below the range of interest which may contain spurious temperature-transient signals or electrical interference, or to reduce measurement error from frequencies below that of the integrator cut-off.

Mechanical overloading

At very high vibration levels, the accelerometer response becomes non-linear and measurements will then be in error, maybe by large amounts. High amplitude shocks and impacts which may excite the accelerometer resonance can thus give rise to measurement problems, even though the frequencies of interest may be well below those frequencies in the excitation which causes the overload. This is a problem which cannot be overcome by electrical filtering.

One possible solution is to use a mechanical low pass filter between the accelerometer base and the vibrating surface, to remove the high frequencies before they reach the accelerometer. Such a filter is in effect a layer of suitable resilience. An overload indicator in the input preamplifier can also be used to warn of possible measurement inaccuracies.

Band pass filters

The incorporation of band pass filters into the system allows the frequency spectrum of the vibration to be measured either in octave, third octave or narrow bands. Narrowband analysis can enable the frequency of pure tones in the signal to be measured, and knowledge of the frequency can help to identify the source of the vibration. Special frequency weighting networks can be incorporated into the meter to indicate human response to vibration as discussed in Chapter 7.

Detector and indicator circuits

Indications of RMS and either peak or peak-to-peak values are usually provided, with peak-hold facilities for impulsive signals. Vibration signals are often of much lower frequency than many noise signals, so much longer averaging times such as 1 s or 10 s are used in order to provide a reasonably steady output display, instead of the fast and slow responses used in sound level meters. Some meters will have time integration for measuring time varying vibration signals, either using RMS time averaging to give an energy equivalent value, similar to L_{eq} for noise, or using fourth-power law, root mean quad time averaging to give vibration dose values as discussed in Chapter 7.

The display may be analogue (i.e. a meter with a pointer) or digital, and will be calibrated directly in vibration units, m/s^2, or m/s. Vibration meters usually also have a variety of output facilities to allow the vibration signal to be connected to an oscilloscope, tape recorder, printer or computer.

Background vibration levels

As with sound measurements it is always good practice to measure background levels in the absence of the vibration source under investigation. As well as any ambient levels of vibration the vibration meter will also be affected by electronic noise, or other unwanted signals in the instrumentation system, arising either from the electronics (e.g. random thermal movement of electrons) or from the accelerometer (e.g. from transient temperature fluctuations) or from the cable (e.g. electromagnetic pick-up, ground loops or triboelectricity).

A good practice when measuring vibration levels, especially very low levels, is to check the background levels produced by the complete measurement system – by measuring the signal produced – i.e. the apparent vibration level, when the accelerometer is mounted on a non-vibrating surface. This should be done as far as is possible with all other environmental conditions the same as for the actual measurement. For reasonable accuracy the background noise level of the instrument system should be less than one third of (or 10 dB less than) the measured vibration signal.

8.11 Calibration of noise and vibration measurements

The principle behind the use of the acoustic calibrator is that the whole measurement system from the transducer right through to the display device should be calibrated and not just individual parts of the system. It is always important to note the settings of all relevant instrument controls, e.g. the positions of gain control switches; adjustments need to be made to the calibration if the settings are subsequently changed.

Many acoustic calibrators operate at only one level and at one frequency – common types generate either 94 or 114 dB at 1000 Hz at the microphone. Confidence that the sound level meter also reads correctly at other levels, say 84 dB and 74 dB, is based on the accuracy and reliability of the attenuators; the above case produces attenuations of 10 dB and 20 dB respectively. The reliability of measurements at other frequencies is based on the frequency response of the whole instrument, especially the microphone. Some calibrators allow calibration over a range of different levels and frequencies.

Automated calibration

Fully automated noise measuring stations operating continuously over long time periods have to be calibrated in a different way. It is obviously impossible to automate the process of fitting an acoustic calibrator over the microphone. A fine grid, called an electrostatic actuator, is permanently fitted over the diaphragm of the microphone. Calibration is achieved by remotely applying a known high voltage to the actuator. The resulting electrostatic force between diaphragm and actuator deflects the diaphragm in a similar way to the pressure from a sound wave. The signal level produced by the actuator voltage is displayed and may be compared to that expected from the known microphone sensitivity.

The principle of total calibration of the entire measurement system applies equally to vibration measuring equipment; a calibrated vibrating table is the equivalent of an acoustic calibrator. The vibration transducer is mounted on the vibrating table, which then produces an accurately known vibration level.

Internal calibration

Sound level meters and vibration meters contain their own internal calibration systems. They consist of voltage signals of accurately known amplitude, generated within the instrument, which can be applied to the input amplifiers instead of the transducer signal. Alternatively a calibrated signal from an external signal generator may be used. In either case this form of calibration can serve only to check the signal processing and display systems; it cannot detect changes in sensitivity of the transducer.

Traceability and verification

In addition to the day-to-day checking of sound level meters using acoustic calibrators, it is good measurement practice and a requirement of some measurement standards, including BS 4142, to ensure that both sound level meters and calibrators are periodically calibrated by a laboratory belong to the United Kingdom Accreditation Service (UKAS). In the laboratory the equipment is calibrated against more accurate equipment, which itself has been calibrated against even more accurate equipment, and so on, thus ensuring traceability of calibration to a national or international standard.

Laboratory calibration

An acoustic calibrator is a transfer standard which has itself been calibrated using a microphone of accurately known sensitivity. The calibrator is then used to transfer the calibration of other uncalibrated microphones, but with a slightly lesser degree of accuracy. Somewhere along the line a microphone must be calibrated by other means, without using a transfer standard. There are three methods available:

- electrostatic actuator method
- comparison method (i.e. direct comparison with other microphones)
- reciprocity method.

The electrostatic actuator method has already been described. It is also useful for measuring the frequency response of microphones and sound level meters in the laboratory as well as for remote calibration.

The comparison method is simple in principle. The sound pressure level is measured at a certain point using a calibrated microphone. This is then replaced by the uncalibrated microphone, which is thus experiencing the same sound pressure. The indicated levels are compared. Although simple in principle, an anechoic room is usually needed in practice to achieve sufficient control of the acoustic environment and to ensure a high degree of accuracy.

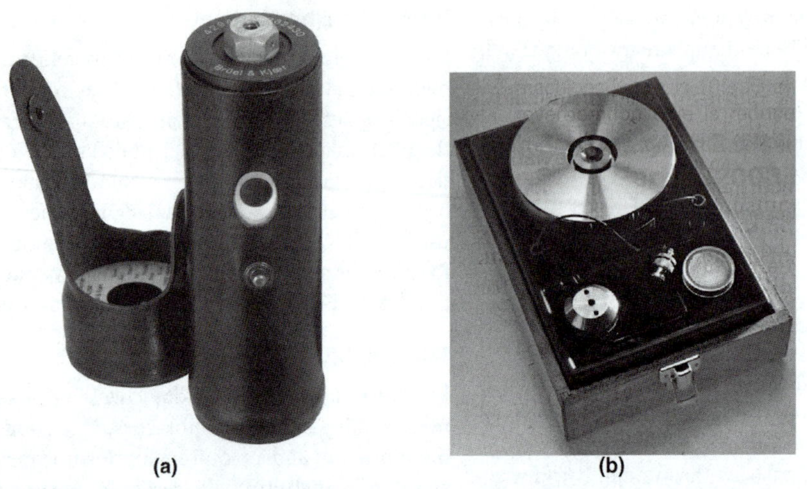

(a) (b)

Figure 8.19 (a) Portable calibrator for checking performance of accelerometer and the measurement chain. This unit provides an acceleration of 10 ms^{-2}. (b) Calibration exciter with built-in calibration accelerometer for laboratory checking of the performance of accelerometer and the measurement chain

Courtesy of Brüel and Kjær Ltd

The reciprocity technique is an absolute method, i.e. it does not rely on the availability of an already calibrated microphone. The method, which will not be described here, relies on the relationships between the performance of the device when operated as a microphone and when operated in reverse as a loudspeaker.

Calibration of vibration measurement systems

The best way is to calibrate the entire measurement system by using a vibrational source of known level. Various vibrating tables onto which the accelerometer is mounted are available for this purpose (see Figure 8.19). Some of these operate at one vibration level and frequency, others allow calibration over a range of frequencies and levels. If a permanent record of the vibration signal is to made (e.g. a tape recording) for subsequent analysis, then the recording should also include a calibration signal, and a note should be made of all instrumentation settings during the recording of the calibration and of the subsequent vibration signals.

Calibration of the electronic instrumentation may be performed by feeding an accurately known voltage into the system. Many instruments supply a highly stabilized reference voltage for this purpose. This type of calibration can allow for correction of any changes in the sensitivity of the instrumentation (the electronics) but not for changes in sensitivity of the accelerometer or its cable.

8.12 Measurement uncertainty

Suppose we take a measurement of the noise level at a certain distance from a machine in a certain room. We obtain a reading of, say, 76.0 dBA. If we repeat the measurement on two more occasions we may get slightly different readings, say, 75.6 and 76.5 dBA. If we were to repeat the measurement, say up to 10 times, we may get a scatter of results, say from 75.4 to 76.6 dBA. This variation in the measurement results is an indication of the **uncertainty** in the result of the measurement.

Can you think of some possible reasons for this uncertainty, i.e. to explain the fact that we obtain slightly different readings when we repeat the measurement?

Stop reading – and make a list of some of the possible causes of this uncertainty.

Some possible causes are:

- slight variations in the position of the microphone
- slight variations in the amount of noise produced by the source on the different measurement occasions
- slight variations in the arrangement of the room, i.e. in the position of possible reflecting surfaces
- slight variations in the performance of the sound level meter on the different occasions.

Uncertainty or error?

Sometimes the term measurement error, or experimental error, is used instead of uncertainty, to describe these variations in the measurement result. The use of the word error, does, however, imply that the variation is, in some way, a mistake, which can be eliminated if only we had been more careful, and this is not so. It is important to understand that there will always be a certain degree of uncertainty associated with any measurement, no matter how accurate the measuring instrument, and no matter how careful the measurement procedure – although, of course, the degree of uncertainty can be reduced (but never eliminated completely) by using more accurate instruments and improved measurement procedures.

Reducing measurement uncertainties

How could the measurement uncertainty, i.e. the scatter of the results, be reduced in the simple measurement described above? Stop reading and make a list.

We could:

- take more care over the microphone positioning
- make sure that the orientation of the microphone was the same each time
- make sure that the orientation of the machine was the same each time
- make sure that the room surfaces (tables, chairs etc.) were in the same position each time
- make sure that the sound level meter had been calibrated accurately each time, that the battery had been checked, and that the sound level meter had been switched on for the same time, say five minutes, before taking the measurements
- make sure that the machine was operated in exactly the same way each time, and had been running for the same time, say five minutes, before taking the measurements
- make sure that the room was at the same temperature each time (because this could affect both the machine and the sound level meter).

If we did all of these things we would expect that the scatter of our results would be reduced, but not eliminated – there would still be some uncertainty.

Random and systematic uncertainties

The causes of measurement uncertainties can be either random or systematic. Random uncertainties are caused by slight changes in measurement conditions each time we repeat a measurement, e.g. in the positioning of the microphone. Random uncertainties can never be eliminated completely, although they can be reduced by paying careful attention to measurement procedures, and their magnitudes may estimated, so that their consequences can be allowed for, by examining the scatter of repeated results. Systematic causes of uncertainty are those which are always the same, and so cannot be detected by repeating the measurement. An example would be if, because of faulty calibration, a sound level meter always read, consistently, 0.2 dB too high. A systematic uncertainty may only be detected when there is some change in the measurement procedure, e.g. using either a different sound level meter or different calibrator. Once it has been detected, a systematic uncertainty can be eliminated from all future measurements, and all previous measurement readings may be adjusted accordingly.

Repeatability and reproducibility

A distinction can be made between the different ways in which a measurement may be repeated.

Suppose in our simple example the noise from the machine was to be repeated several times by the same observer using the same sound level meter. In this case the variation in results would give an indication of the **repeatability** of the measurement.

If, however, the measurement were to be repeated by different observers, each using their own sound level meters and calibrators (but all having been given the same instructions about measurement position and distance, and machine operation), then in this case the variation in results would give an indication of the **reproducibility** of the measurement.

Not surprisingly the variation in reproducibility tests is usually greater than that from repeatability measurements. British and International Standards for measurements carried out under laboratory conditions, sound reduction index of partitions, absorption coefficients of sound absorbing materials, and sound power level measurements often contain requirements for a statement of measurement uncertainty based on either repeatability or reproducibility tests. Sometimes 'round robin' tests are carried out where the same test sample is sent in turn to different laboratories for testing.

Quantifying measurement uncertainties

Random errors are best quantified by repeating measurements. The mean value is taken as the best estimate of the quantity being measured and the scatter of the readings from this mean value can be used to provide an estimate of the uncertainty. If there are sufficient measurements for a statistical approach to be valid then the standard deviation from the mean value may be used as the basis of a statement of uncertainty. Otherwise the extreme values may be used. Sometimes if it is not possible to repeat measurements it may be possible to arrive at an estimate in other ways.

Combining uncertainties

Suppose a certain noise measurement contains two main causes of uncertainty, perhaps due to the positioning of the microphone, and due to day to day variation in the noise output from the source, and these have been estimated as giving rise to variations of, say ±8% and ±6% from the mean value. It can be seen that the maximum possible variation from the mean, which could occur in any particular measurement, would be ±14% (because 8 + 6 = 14). However it is also possible that the uncertainties might, at least in part, cancel each other out, giving a minimum possible variation from the mean of ±2% (because 8 − 6 = 2). If the uncertainties are random in nature it can be shown, using statistics, that the most likely outcome will be a combined uncertainty of $\sqrt{(8^2 + 6^2)} = \sqrt{100} = \pm 10\%$.

Uncertainty budgets

In the above example there were only two major causes of uncertainty, but often there are several different causes. In such situations a list of the various causes of uncertainty, and their estimated magnitudes, should be drawn up. Such a list is called an uncertainty budget, and may be used to estimate the total combined uncertainty in the measurement.

Accuracy and precision

The accuracy of a reading is the degree to which it might be different from some notional 'true' value. The precision of a measurement is an indication of the smallest change in the measured quantity. Thus sound level meter A may have an indicator scale which allows the sound level to be measured to the nearest dB, whereas sound level meter B may indicate to the nearest 0.1 dB. Thus sound level meter B has greater precision, because it allows smaller changes in sound level to be detected. This does not necessarily mean, however, that sound level meter B is more accurate.

The significance of measurement uncertainty

If the purpose of a measurement is to compare the measured quantity with some standard or limit, then it becomes clear that consideration of measurement uncertainty should be taken into account. As an example, suppose that it is required that the sound level from a certain machine, measured under specified conditions, shall be less than, say, 80 dBA, and it is estimated that the measurement uncertainty is ±2 dB. In this situation a measurement of 86 dBA would mean that the noise from the machine would clearly be unacceptable, and a measurement of 84 dBA would also be unacceptable, but a measurement value of 79 dBA or 81 dBA would be more difficult to assess since the measured value is within the uncertainty of the limit value.

Until now estimates of uncertainty in acoustical measurements have only been required of laboratory measurements, but there can also be situations where an idea of the degree of uncertainty in 'field' measurements can be important as well. Sometimes estimates of uncertainty in field measurement of sound level (often taken to be ±2 dB) have been, implicitly, 'built into' noise limits.

Some of the British and International Standards for the measurement of sound power level give estimates of measurement uncertainties. It is noticeable that the uncertainty in the overall dBA values is much lower than the individual octave (or third octave) band values. Also the uncertainties in the low (say 63, 125 and 250 Hz bands) and very high octave bands (400 and 800 Hz bands) are higher than those in the middle of the frequency range.

These variations in uncertainty are generally true for sound level measurements, i.e. overall dBA measurements are subject to less measurement uncertainty than the individual frequency bands.

NPL/Salford University Guide to Measurement Uncertainty (November 2001)

This guide explains, with the aid of examples, how to estimate uncertainties in acoustic measurements and how to draw up uncertainty budgets.

Possible causes of uncertainties in acoustic measurements

Chapter 5 of the guide, the longest chapter, discusses the possible cause of uncertainty which can arise in acoustic measurements. These different causes are discussed under the headings of: those which are associated with the source of the noise, those associated with the sound transmission path, and those associated with the measurement receiver position, and its immediate surrounding environment.

The noise source

A major source of uncertainty may be the noise source and its operation.

It is important to consider the following:

- the spectral content of the noise emission
- the nature of the noise source: point/line/area
- running condition/method of operation/machine load and speed etc.
- state of repair and maintenance
- source height
- whether sources are stationary or moving
- enclosure and barriers close to the source
- environmental conditions (e.g. weather)
- number of sources in operation and their positions relative to the measuring positions
- interaction between each source
- location and state of any doors and louvres in any source enclosures

- variations of noise emission with time, both in the short term (e.g. machine operating cycles) and in the long term.

Short-term variations in the noise emission will influence the duration of the measurement required to obtain a satisfactory sample. In general, the duration of the measurement should be representative of a single or several complete cycles of operation.

Longer-term changes can usually be accounted for by suitable sampling strategies, and should be considered in detail when comparing two measurements, or considering a single measurement to be representative of a period longer than that actually measured.

Spectral content

Low frequency measurements may be subject to uncertainties due to standing waves, high background noise levels, wind noise, structure/ground-borne vibration. Measurement of high frequencies may be subject to uncertainties due to directionality of sources, changes in relative humidity, scattering and reflections from even fairly small objects.

The type of source (point, line, area etc.) may determine the extent of uncertainty due to variation in measurement position. Uncertainties may be caused by any changes in source operating conditions and state of repair and maintenance, and any movement of the source. Changes to the height of the source, and to any surfaces close to the source may affect sound propagation to the receiver. The weather causes variations in the noise output from the source, e.g. causing fan loading to vary with wind or temperature.

The noise transmission path

For many environmental noise measurements the prevailing weather conditions constitute a major source of uncertainty, especially over medium to long distances. Changes in the weather may occur suddenly, within the duration of a normal measurement.

It is important to consider the following:

- weather
- ground effects
- barriers.

Changes to the ground surface or to barriers are unlikely to occur suddenly; however, these should always be considered when comparing medium/long-term noise levels.

Changes in weather conditions can cause changes in refraction, absorption and scattering of sound in the atmosphere, and to the acoustic impedance of the ground surface.

Receiver (and immediate surrounding environment)

All measuring processes have an associated degree of uncertainty, determined by the accuracy of the instrumentation and the competence of the operator.

It is important to consider the following:

- microphone position
- instrumentation
- choice of measurement position
- background noise level
- environment immediately surrounding receiver (e.g. presence of reflecting/absorbing surfaces).

Measurement uncertainty arises from the possibility in variations in any of these.

The selection of an appropriate measurement position is important. This will determine what the measurement actually represents and how relevant the result is to the purpose of the survey. Microphone positions should always be well defined to ensure that repeated measurements are truly comparable.

8.13 A measurement and instrumentation exercise

A number of different measurement situations are listed below. In each case consider:

- the measurement procedure(s) you would use
- specific requirements of instrumentation and other facilities and equipment you would need (including factors such as signal capture and storage arrangements, signal conditioning and processing requirements, ease of post-processing analysis, and various day to day practical factors)
- the noise parameters or indices that you would measure, and any relevant standards or codes of practice
- factors which would determine the measurement limits (e.g. maximum or minimum values that could be measured), and any constraints
- factors which will affect the uncertainty of the measured result(s):
  - (i) the measurement of sound power level of a machine
  - (ii) the measurement of traffic noise at a roadside, for planning purposes
  - (iii) the measurement of the frequency spectrum of random noise using an FFT analyser
  - (iv) the measurement of airborne sound insulation between two dwellings
  - (v) the measurement of noise levels in a busy sheet metal works, in order to carry out noise exposure assessments

(vi) measurements taken outside a dwelling in order to investigate a complaint of noise from a nearby industrial source

(vii) measurements of noise from a distant industrial complex (e.g. oil refinery or power station) taken over a long period of time in order to establish ambient noise levels.

(viii) to continuously monitor noise outdoors near to an airport over a six-month period

(ix) to measure short bursts of impulsive environmental noise (e.g. from a clay pigeon shoot, or quarry blasting or pile driving) giving rise to complaints by nearby householders

(x) to measure and identify the source of an annoying whine or hum from a piece of industrial machinery

(xi) to monitor noise inside the bedroom of a dwelling from a nearby industrial estate, in order to investigate allegations of occasional bursts of noise which occur in the middle of the night.

Questions

1 Draw a block diagram of a simple sound level meter and describe the function and characteristics of the various parts. Describe the special features of (a) impulsive sound level meters; (b) integrating sound level meters; and (c) digital sound level meters.

2 Describe the principle of operation of the following types of microphones and compare and contrast their properties: (a) condenser microphone; (b) electret microphone; and (c) piezoelectric microphone.

3 Explain the difference between analogue and digital processing of signals. Discuss the impact of digital techniques on sound measuring instrumentation.

4 Explain the meaning of the following terms used in connection with the performance of microphones: sensitivity, linearity, stability, output impedance, frequency response and dynamic range.

5 How may acoustic equipment be calibrated so that confidence may be placed in the reliability of measurements?

6 Explain the principles of acoustic intensity measurement. What are the advantages of the direct measurement of sound intensity over the indirect method, based on sound pressure measurements? Give some applications of direct intensity measurement.

7 Explain why it is sometimes necessary to make audio recordings of sound and vibration signals. Discuss the performance requirements of the audio recording method.

8 State, with reasons, the equipment you would choose for the measurement and analysis of noise in the following situations, giving any special requirements:

(a) To assess the noise received by a factory worker who is continually moving from place to place; the noise level in the factory varies at different positions and throughout the day.

(b) To monitor the noise inside the bedroom of a dwelling from a nearby factory, over a 24-hour period; you are particularly interested in measuring and identifying bursts of noise from the factory which occasionally occur in the middle of the night.

(c) To continuously monitor noise outdoors near to an airport over a six-month period.

(d) To measure and identify the source of an annoying whine from a piece of factory machinery.

(e) To measure the noise from a clay pigeon shoot; L_{AE}, peak sound pressure and maximum octave band levels are required.

9 Explain why the sensitivity of a microphone at high frequencies depends on the size of the microphone and the angle of incidence of the sound. Explain free field response, pressure response and random incidence response.

10 Explain the meaning and significance of the following terms used in connection with sound measurement: frequency weighting network, crest factor, averaging time, real-time analysis and filter bandwidth.

11 (a) Describe the construction and operating principles of an instrumentation condenser microphone. State two advantages and two disadvantages of this type of microphone compared with a piezo-resistive pressure sensor for environmental noise measurements.

(b) Identify two factors which are important in controlling the frequency range of a microphone and describe how these factors affect the design of a microphone for ultrasound measurements.

(c) Sketch the frequency response of a piezoelectric accelerometer. Describe **one** factor that affects the frequency range and **one** factor that affects the sensitivity of this type of sensor.

(d) An accelerometer produces an output on a level display of 35.6 dB when mounted on a calibrator producing 9.81 ms^{-2} rms. at 80 Hz. The accelerometer is then placed on a table vibrating harmonically at 50 Hz with peak velocity 0.024 ms^{-1}. What level would you expect to see indicated on the display?

(IOA 2008)

12 (a) Explain the terms sensitivity, linearity and frequency response used in connection with instrumentation microphone performance.
(b) Explain what is meant by the term 'free field correction' used in conjunction with instrumentation microphones, and indicate those parameters on which it depends. Explain the difference between instrumentation microphones designated as (i) free field and (ii) pressure response.
(c) An instrumentation microphone has a sensitivity of −40 dB re. 1 V/Pa. What output voltage will be produced by such a microphone when exposed to a sound pressure level of 114 dB re. 20 μPa?
(d) When used to measure A-weighted sound levels a certain sound level meter is said to have a dynamic range from 20 dB to 110 dB. Explain what this means and describe two factors that will have a strong influence on the dynamic range of a sound level meter? (IOA 2005)

13 (a) Discuss 10 factors or practical problems that should be considered when selecting and using a piezoelectric accelerometer to measure mechanical vibrations.
(b) Explain why a preamplifier is introduced into the measurement circuit when using a piezoelectric accelerometer. Give the advantages of a charge preamplifier over those of a voltage preamplifier.
(c) A piezoelectric accelerometer has a capacitance of 1000 pF and a charge sensitivity of 12 pC/g, where g is the acceleration due to gravity. What is the open-circuit voltage output of the accelerometer for an input r.m.s. acceleration of 1 ms^{-2}? (IOA 2005)

14 When a piezoelectric accelerometer is calibrated, the output voltage is 96 mV rms for a sinusoidal input of peak amplitude of 9.81 ms^{-2} at 80 Hz. The accelerometer is then placed on an object vibrating sinusoidally at 50 Hz. The output voltage is now 54 mV rms. Calculate the peak velocity. (IOA 2004, GPA (part question))

15 Discuss the main principles of operation and constructional details of the capacitor microphone. Account for the differences between the free field and pressure responses for this type of microphone and indicate how these responses can be used to advantage.

Explain what design modifications are necessary for capacitor microphone systems used for external, unattended noise monitoring purposes.

If open circuit output voltages of 7.94 μV and 25.2 V were recorded corresponding to the internal noise level and the 3% distortion limit respectively for a microphone (sensitivity −38 dB re. 1 V per N/m^2) estimate the dynamic range for the microphone. (IOA 1986)

16 (a) A microphone amplifier is calibrated to read dB re. 20 μPa when used with a microphone of sensitivity 50 mv/Pa.
(i) Calculate the rms acceleration if the meter reads 78 dB when the microphone is replaced by an accelerometer of sensitivity 2 mV/ms^{-2}.
(ii) If the vibration frequency is known to be 50 Hz calculate the corresponding rms velocity and displacement.
(b) A one third octave filter has a centre frequency of 1 kHz. What are its upper and lower frequency limits?

Describe briefly with the aid of sketches, how real band pass filter amplitude transfer functions (attenuation curves) differ from ideal ones. (IOA 1985)

17 (i) Explain why the frequency response of a microphone is directional at higher frequencies and distinguish between pressure response and free field response.
(ii) Describe the characteristics of octave band filters suitable for the analysis of noise.
(iii) A sound level meter correctly registers 124 dB SPL when undergoing a calibration check. If the microphone sensitivity is −26 dB re 1 V/Pa, what is the input voltage to the meter? (reference pressure = −20 μPa). The microphone is then replaced by an accelerometer, with a sensitivity of 2 mV/ms^{-2}, attached to a vibrating surface. If the reading is 85 dB, what is the rms acceleration of the surface? (IOA 1981)

18 (a) Explain the terms frequency response and dynamic range when applied to a sound-level meter. Discuss the factors which are likely to affect the frequency response of the meter.
(b) What determines the lowest sound level that can be measured accurately by a sound-level meter?
(c) By means of a block diagram, explain the functioning of a digital sound-level meter. Why does such a meter have a mode in which the maximum sound level in a measuring period is stored in the display?
(d) Comment upon the statement: 'It is not the inherent performance of the sound-level meter but the method of application that is the dominant factor in determining the quality of measured and reported data'. (IOA 1979)

19 A manufacturer quotes the dynamic range of a sound level meter as 38 to 140 dB when used with a microphone of sensitivity 50 mVPa^{-1}. Estimate the internal electrical noise generated in the instrument as an equivalent input voltage. What would be the limits of the dynamic range of the instrument when fitted with a $\frac{1}{4}$ inch microphone of sensitivity 4 mVPa^{-1}?
(IOA 1989 (part question))

20 Briefly describe the basic characteristic of a digital frequency analyser based on the fast Fourier transform (FFT), and mention one application.

Assuming that the sampling frequency is $2f_u$ samples per second, where f_u is the upper limit of the frequency range of interest, and the FFT execution time for a 1024 sample FFT is 0.2 s, determine whether the analyser can be used for real time analysis for the frequency range from 0 to 4000 Hz. (IOA 1992 (part question))

Chapter 9 Noise control

9.1 Introduction

This chapter reviews and explains the range of methods which are available for reducing noise, starting with commonsense 'good housekeeping' methods. The control of noise at source starts with an understanding of how noise is produced and so the noise generating mechanisms involved in many sources of noise in industry are explained, and the principles of noise reduction described.

The reduction of noise during its transmission path between source and receiver, using the techniques of sound insulation, sound absorption, vibration isolation and damping are then discussed. Methods for diagnosing which are the most important noise sources and paths, and for the specification of noise from machinery are described.

There then follows a discussion of hearing protection policies and the use of personal ear protectors. The chapter concludes with a review of the control of noise using legal and administrative procedures, although this will be dealt with more comprehensively in Chapter 10.

The measurement of sound power level has been described in Chapter 8 on measurement and instrumentation.

Much useful material relating to this chapter is contained in the Health and Safety Executive Guidance to the Regulations, *Controlling Noise at Work*, issued in 2005 (HSE Books), particularly Part 3 on practical noise control and Part 5 on hearing protection. The HSE website also gives links to many illustrative case studies on noise control in industry.

9.2 Control of noise by good planning and management

Although there are many technical solutions to noise problems the best approach is to avoid or minimize problems wherever possible. Much can be done to avoid noise problems by good noise control management, practised on a day to day basis, and by giving careful consideration, in advance, to situations which might possibly give rise to noise problems. Some of the possibilities are as follows:

- Avoid making unnecessary noise, e.g. leaving machines running when not required, leaving doors open when not needed.
- Select the quietest procedure for an operation or process, where alternatives exist.
- Include noise level requirements in the specification for new machines and equipment.
- Carefully consider the layout and siting of noisy processes and equipment, in relation to noise sensitive areas and personnel.
- Aim to arrange that such processes and equipment do not operate at the most noise sensitive times (e.g. evenings and weekends).
- Arrange for regular maintenance of machines and of noise control equipment so that unnecessary increases in noise levels are avoided. This also applies to the fabric of buildings so that doors and windows, for example, are always a good fit and do not transmit noise through holes and gaps.
- Arrange for regular noise surveys to be carried out by properly trained personnel.
- Educate and inform so that people are aware of the purpose of noise control equipment, and how it should be used. This might simply be a matter of ensuring that noise shielding hoods or enclosures are always replaced after removal for inspection or maintenance, and that vibration isolation is not bridged inadvertently.

The source–transmission path–receive model of noise control

This simple model (Figure 9.1) immediately suggests that there are three opportunities to reduce noise: either at the source or during the transmission path or at the receiver.

Noise control at source may be reduced either by modification to the source, replacement by a quieter source performing the same process, or by using an alternative process. Once the noise has left the source it is transmitted to the receiver as airborne or structure-borne sound, and the standard techniques of using sound absorption or sound insulation for the reduction of airborne sound, or isolation or damping to reduce structure-borne sound,

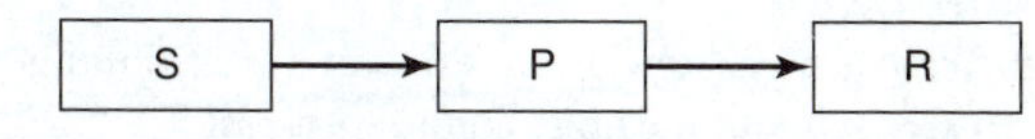

Figure 9.1 Source–transmission path–receiver model

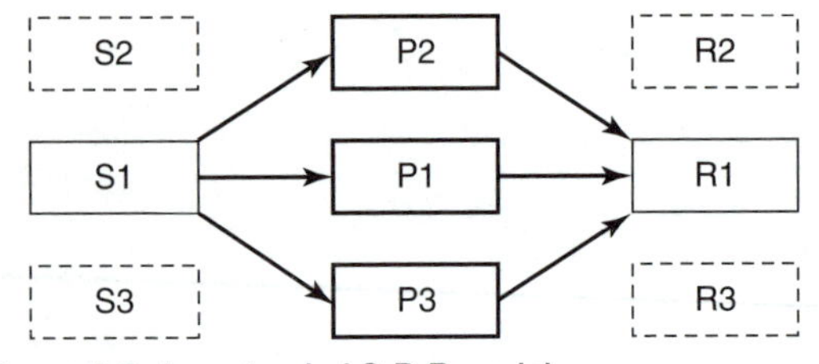

Figure 9.2 An extended S-P-R model

are employed. Finally noise control at the receiver might include measures such as provision of noise shelters or refuges, improvement of sound insulation (e.g. fitting of double glazing) or use of hearing protection.

The model becomes more useful when it is extended to include multiple sources or paths or receivers, as shown in Figure 9.2.

A simple example would be where a source such as a machine produces noise which reaches a receiver in the same room by three different paths: direct airborne transmission, airborne transmission after reflection from the room surfaces, and as a result of structure-borne sound via transmission of vibration from the machine through the floor or walls.

The solutions in this case are a sound insulating screen to reduce direct sound between source and receiver, sound absorption applied to the walls and ceiling of the room to reduce reflected and reverberant sound, and vibration isolation of the machine from the floor and/or walls of the room to reduce structure-borne sound.

The use of this more extended version of the model highlights the need for diagnosis techniques to evaluate the relative importance of the different transmission paths.

Exercise

Diagnosis has shown that in a particular case sound from the source reaches the receiver via three paths equally, and the level at the receiver is 70 dB. Three separate noise control treatments are available to reduce sound transmission via each of three paths by 20 dB.

(a) By how much will the noise at the receiver be reduced if first just one, then two and finally all three treatments are used?
(b) If instead of treating the three paths there is an alternative treatment which will reduce the source level by 20 dB what would be the level at the receiver?
(c) What general conclusions can you draw from this exercise about the noise control process?

Answer

(a) If the total noise arriving at the receiver from three equal contributions is 70 dB, then the noise via each pathway must be 65 dB (decibel addition). Therefore the level arriving via just one path is 65 dB, and via two paths is 68 dB.

Each noise control treatment will reduce the sound level transmitted by that path by 20 dB to 45 dB, which is insignificant compared with the level of 65 dB via an untreated path. Therefore if only one noise control treatment is used the level at the receiver will be 68 dB, i.e. a noise reduction of 2 dB. If two paths are treated then the level will be 65 dB i.e. a reduction of 5 dB. If all three paths are treated the level via each path will be 45 dB and the total level will be 50 dB, i.e. a reduction of 20 dB.

(b) If control at source is used instead of three separate pathway treatments a reduction of 20 dB at source also achieves a reduction of 20 dB at the receiver.

(c) Therefore the example illustrates the advantage of noise control at source wherever possible because only one treatment is needed rather than three, and that noise control may be a multi-stage process, i.e. more than one treatment may be needed for effective noise control if there are multiple sources or multiple paths.

The situation becomes more complex, requiring further diagnosis, if multiple sources and multiple receivers are involved as well as multiple receivers.

9.3 Noise control strategy for multiple sources

Exercise

Ten identical noise sources contribute to the total noise in a workshop. At a particular point in the workshop each noise source produces a level of 80 dBA when it alone is operating. Calculate the cumulative noise at the reception point when each source is switched on in turn, i.e. from one on, two on, etc. until all 10 are switched on.

Answer

This is a simple decibel addition problem (see Chapter 1). We start by combining 80 and 80 to make 83 dB, then combine the total, 83 dB with 80 dB, to give 85 dB, and so on. Or we can simply use the equation that for N sources 'on' the combined level will be $80 + 10\log N$. The way in which the total level increases with the number of sources is shown below (to nearest half of a dB).

1	2	3	4	5	6	7	8	9	10
80	83	85	86	87	88	88.5	89	89.5	90

Exercise

Suppose that there are two alternative options for noise control of these machines. The less expensive but less effective option will reduce the noise of each machine, when treated, by 3 dB, and the more expensive option will provide a 10 dB reduction. Funds are available either to treat three of the machines using the more expensive option, or to treat all 10 machines using the less expensive option. What would be the best noise reduction strategy?

Answer

From the above table it can be seen that even if three of the machines were turned off altogether the total reduction in noise level would only be 1.5 dB. Therefore it would be better to choose the cheaper treatment for all of the machines – this would reduce the total level by 3 dB. The best strategy is to apply noise control equally to all sources.

What if an eleventh source were now to be introduced which produces a noise level of 90 dB on its own. How should the noise control strategy change?

The total noise with all 11 machines on would now be 93 dB, but the noise control strategy should change completely – all noise control effort should now be directed towards reducing the noise from the 11th source until it is reduced by 10 dB to the level of all the others, when the original strategy should again be adopted.

The above are two simple and extreme examples, but in more complicated cases the contributions to the total from each source will be different and it will be important to rank the different sources by their contribution to the total noise and hence their priority for noise control treatment. The noise control may be a multi-stage process and at each stage it will be necessary to reassess the strategy and prioritize for noise control. It may be necessary to carry out the assessments in octave frequency bands as well as in overall noise levels.

9.4 Mechanisms of noise generation

Sound is generated when air particles are caused to vibrate and pass their vibration on to neighbouring particles of air. It is useful for practical purposes to identify three main types of noise generation:

- that produced by a vibrating surface such as a machine panel
- by turbulent fluid flow as generated by fans, pumps, jets, valves and whistles, and known as either aerodynamic or hydrodynamic noise
- or as a result of the vibration of impacting solid surfaces – both the noise from the impact itself and the ringing of components which subsequently occurs.

Sound radiation from vibrating surfaces

This is often the case with noise produced by machinery when some of the forces generated within the machine are transmitted to the outer surfaces of the machine, which radiate noise, but is also what happens when sound is radiated by a loudspeaker or by the façade of a building. The prediction of the radiated sound pressure and sound power levels was described in Chapter 7 on vibration. The control of machinery noise is described in more detail below.

Aerodynamic and hydrodynamic noise

If a fluid such as air or water is caused to flow slowly along a smooth pipe or duct the flow will be laminar or streamlined. Very little noise is produced by laminar flow.

If the flow velocity is increased, viscous forces between adjacent layers of fluid eventually cause the laminar flow to break down into turbulent vortices or eddies (Figure 9.3). Such turbulent flow is very noisy. The onset of turbulence will occur at lower flow velocities if the surface of the pipe or duct is rough, or if there are any obstacles in the path of the flow, or if the flow has to change direction suddenly such as happens at bends, or as a result of changes of cross-section of the duct or pipe (Figure 9.4). Therefore important principles in minimizing flow generated noise are to ensure that flow is slow and smooth.

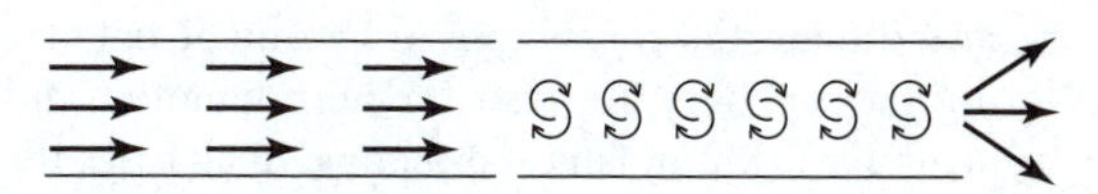

Figure 9.3 Flow generated turbulence in a pipe or duct. At low flow velocity the flow is laminar (left); high flow velocity results in turbulence (right)

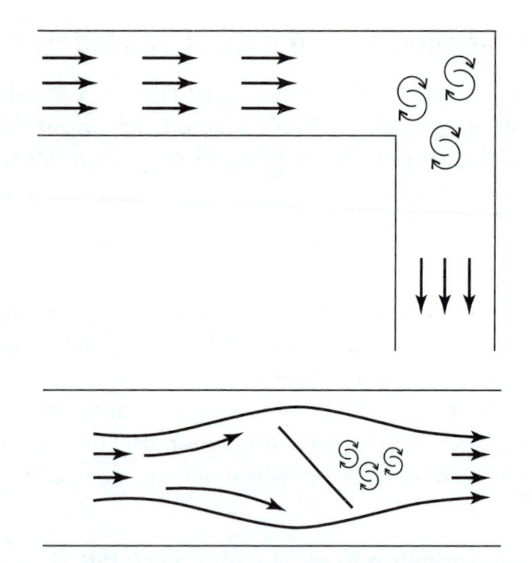

Figure 9.4 Turbulence generated at bends and by obstacles in the flow

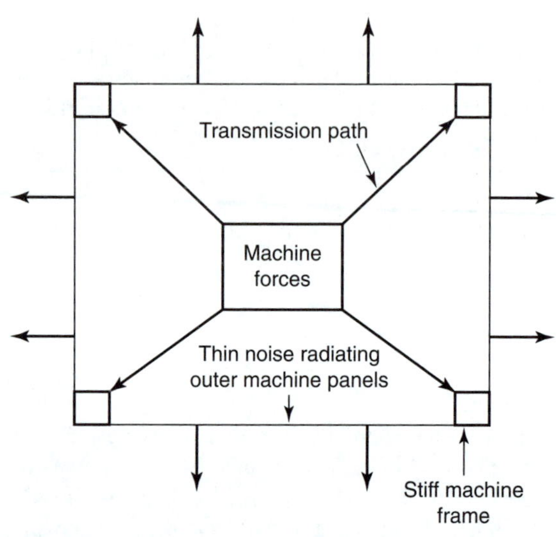

Figure 9.5 Schematic model of noise generation by machines

Noise from impacts

Impact noise consists of (minimum) noise from the impact plus additional noise from 'ringing' of components. Noise from impact depends on the momentum of impacting parts, i.e. on mass and impact velocity. Therefore noise may be reduced by the reduction either of the mass of impacting parts and/or of their impact velocity (e.g. through cushioning of impacts, reduction of 'drop heights' etc.). The 'ringing' of impacting parts may be reduced by improved damping of the 'ringing' parts.

9.5 Noise from machinery – simple model

Somewhere within the machine, forces are generated which enable the machine to perform its function. The machine forces are contained by a stiff framework, and some small proportion of these forces are transmitted to the thin outer surfaces of the machine, which act like loudspeakers and radiate noise. A simple schematic model of how a machine produces noise is shown in Figure 9.5.

What sort of forces are we talking about? All rotating machines are sources of noise producing forces such as the cyclic mechanical forces produced at bearings and gears, the electromagnetic forces in motors and alternators, and the inertial forces caused by out-of-balance rotating parts. There are also the aerodynamic and hydrodynamic forces in fans and pumps. In all cases the noise depends strongly on the rotational speed, and one way of achieving noise reduction is to reduce speed where possible. Other examples include combustion forces in engines, magnetic forces in motors, impact forces in punching and hammering operations, turning, lifting, pressing, cutting and grinding forces in many material handling and shaping processes.

What opportunities for noise reduction does this model suggest? The possibilities are for:

- reducing or modifying the forces (while still allowing the machine to function properly)
- minimizing the transmission of these forces to the outer noise radiating surfaces of the machine
- treating the outer surfaces so that they radiate less noise.

Modification of forces

Modification of the noise producing forces involves fundamental changes to the machine design. Examples include helical cutter blades for woodworking machines, specially designed punch and press tools and highly damped circular saw blades. Quieter gears, pumps and bearings are also available. In many cases the quieter design involves better quality engineering (e.g. to produce closer tolerances between moving parts or better balancing of rotating parts) and is therefore, of course, more expensive.

It is not only the magnitude of the forces but also their rate of change of force with time which is important in noise production. Impacts between hard surfaces produce very short bursts of impact force which result in high frequency noise whereas for softer, cushioned surfaces the impact forces are spread over a longer period of time and produce lower frequency sounds (the difference between a 'sharp crack' and a 'dull thud'). An example is the difference between the combustion forces which occur in the cylinders of petrol and diesel engines. In a petrol engine the rate of rise of pressure inside the cylinder is more gradual than in the diesel engine when at the point of 'firing' there is a sudden sharp increase in pressure which is responsible for more high frequency content in diesel engine noise, which gives rise to the characteristic 'diesel knock' which is particularly noticeable at low speeds and at idling.

The ultimate example is where short duration impact forces can be replaced entirely by steady or constant pressing or cutting forces, as when a nail or staple is pressed home quietly rather than being hammered or when a guillotine is used to cut sheets of metal using a 'scissor action'. There are quiet concrete breakers which employ very large but steady twisting and bending forces (rather like breaking a bar of chocolate) rather than using much noisier impact methods, and similarly quieter options are also available for driving piles into the ground on construction sites.

Attenuation of the force transmission path

If it is not possible to modify the noise producing forces produced by a machine it might be possible to modify the machine structure in order to attenuate their transmission, in the form of vibration, to the outer noise radiating surfaces. Such modification could involve changes in either mass, stiffness or damping of the machine structure, including the possibility of vibration isolation.

9.6 The reduction of noise from sound radiating surfaces

Given that certain forces are acting on a panel, the aim should be to minimize the amplitude of panel vibration, and its radiating surface area and radiation efficiency. The possible noise reduction treatments to the panel are stiffening, damping, isolating or shielding (i.e. covering or acoustic lagging). Increasing the stiffness can increase the radiation efficiency, so the most effective treatment for a thin, floppy panel might be to increase its damping rather than to stiffen it. For panels which are already fairly stiff, damping treatments may not be very effective because the mechanism of damping requires a large vibration amplitude; the damping forces can then convert vibration energy into heat. A last resort is to close-shield or cover the panel with an acoustic lagging material (see Figure 9.8 later).

Isolation of machine panel

If the panel is already fairly stiff, it may already be an efficient sound radiator; further stiffening will reduce vibration amplitudes and radiated noise levels. The best noise control treatment for very stiff panels is to try to isolate them from the surrounding framework in order to prevent the vibration producing forces from the framework reaching the panel (see Figures 9.6 and 9.7).

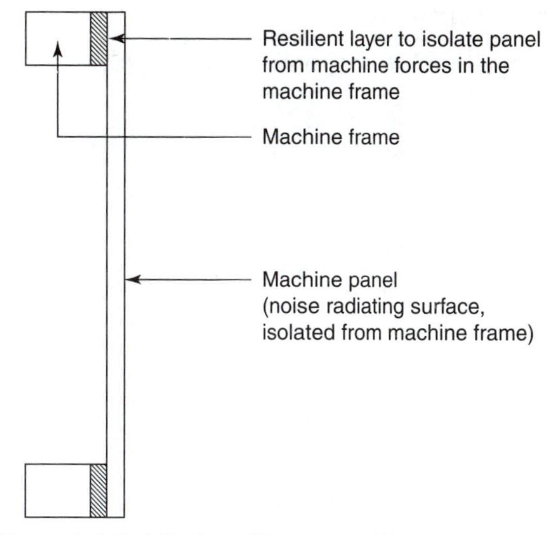

Figure 9.6 Detail of panel from a machine frame

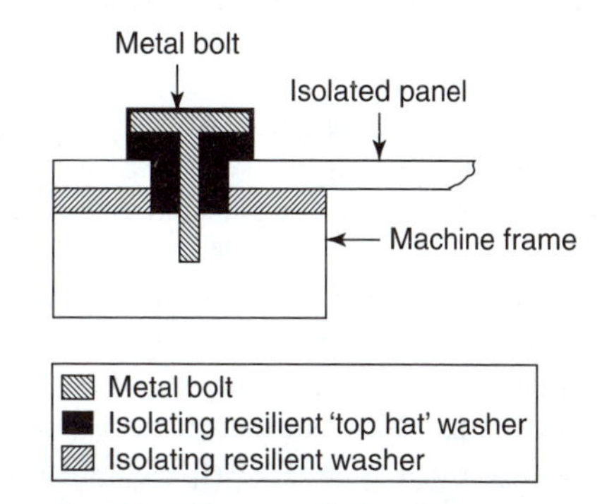

Figure 9.7 How an isolated panel might be secured to a machine frame without bridging the isolation

Acoustic lagging

A last resort is to close-shield or cover the panel with an acoustic lagging material positioned very close to the noise radiating surface (see Figure 9.8). The shield consists of a layer of light resilient material attached to a second layer of heavier material. The resilient layer may be foam or a fibrous material such as mineral wool. The outer layer should be as massive as possible and ideally be limp and well damped. Thin lead sheet is ideal, but other materials can be used, including damped sheet steel. Commercial 'sound barrier mats' are available which consist of a heavy but flexible layer of fabric or PVC which is loaded with particles of lead and lined with foam rubber.

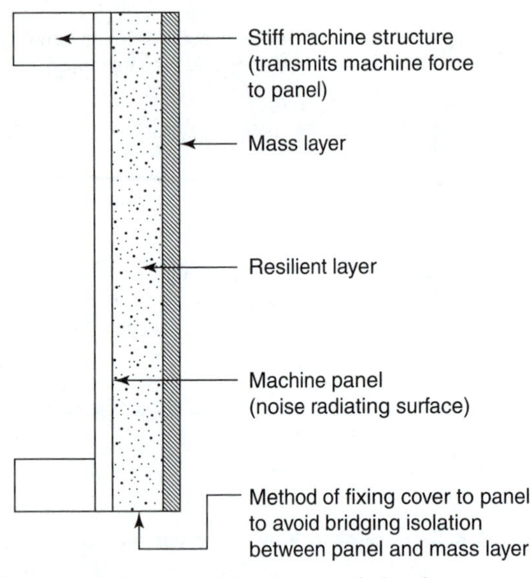

Figure 9.8 Close shielding or acoustic lagging

The shield acts in a complex way, incorporating in various degrees, at different frequencies, the mechanisms of sound insulation and absorption, and vibration isolation and damping. The outer layer serves to reduce the radiation of sound by insulation and so it needs to be as heavy as possible. The foam or fibrous layer absorbs sound trapped between the outer layer and the radiating surface. In addition, the air in the fibrous or foam layer acts as a spring which at certain frequencies isolates the outer layer from the radiating surface. The presence of the shield may cause some damping of the vibration which is causing the noise radiation.

It is important that the shield is used with the resilient absorbing layer adjacent to the vibrating surface (and not the other way around) and that the method of fixing and supporting the shield does not allow contact between the shielded surface and the outer shielding layer. Any such bridging of the isolation between these two layers will reduce the effectiveness of the shield.

This type of shielding can be used for reducing noise from certain parts of a machine such as thin metal panels, and also for 'cladding', or 'lagging' pipes and ducts which are significant noise radiators. The advantage of a flexible shield is that it can easily be cut and shaped to curved surfaces. This type of noise reduction treatment can be effective at high frequencies but gives only small reductions at low frequencies. In fact, increases in noise levels can arise at the frequency where there is a resonance of the mass of the heavy outer layer and the springiness of the air filled foam or fibrous layer. To achieve bigger noise reductions at low frequencies, carefully designed cladding is required with a thick fibrous layer and very heavy outer skin. Alternatively, a double-skin system may be used with a second resilient layer and a second massive skin built onto the first layer of cladding.

9.7 Diagnosis – sources, mechanisms, paths and radiating areas

Multiple noise sources

If there are many different noise sources, e.g. several different machines in a workshop, then it is necessary to rank them in priority order for noise reduction treatment according to their contribution to the overall noise level with all sources on. The best method is, if possible, to turn all the sources off, and then to turn each one on in turn, and measure the noise level produced individually by each source. The background noise level, with all sources off, should also be measured.

For operational reasons it may not always be possible to do this and so a second-best approach is to try turning sources off in turn, and assessing the effect on the overall noise level. However, as we have seen from the earlier '10 equal sources' example, this may not be effective. Another approach would be to compare noise levels measured close to each machine.

Diagnosis of noise producing mechanisms and of noise radiating surfaces

One machine may itself be a multiple noise source. A road vehicle for example produces noise from the engine and exhaust, from the tyres, and aerodynamic noise from the movement of the vehicle through the air. Sometimes it is important to know how the noise is being generated and where it is coming from.

A knowledge of how a machine operates is important in being able to identify noise producing mechanisms. Using such knowledge it is possible to gain insight into noise producing mechanisms by investigating the effect on the noise of variation of machine working parameters such as speed, load, type of material being processed etc. Specialist acoustic analysis techniques may also be used to aid diagnosis, such as narrowband analysis, correlation and acoustic intensity techniques. Sometimes, for example, the knowledge of the exact frequency of a pure tone in a band of noise can be used to link the cause of the tone to the blade passing frequency of a particular fan, or to the gear meshing frequency of a particular set of gears.

One technique to identify the most important noise radiating areas of a machine involves covering each of the possible areas with acoustic lagging, as already described, and then removing the covers from each area in turn and comparing the effects on the measured noise levels. Another method would be to measure the level of vibration of each surface and, taking into account each area and an assumed radiation efficiency, to estimate the sound power radiated for each, or to use a sound intensity meter to measure the sound powers directly.

Identification of sound transmission paths

Sometimes the source of the noise is obvious and it is deciding which is the most important sound transmission path which is the diagnosis problem.

Exercise

The top floor of a two storey commercial building contains a large refrigeration plant with many connections to service pipes, ducts and cables serving the entire building. The noise from the plant is causing disturbance to employees working on the ground floor below and is also clearly audible in the basement floor. Measures to reduce structure-borne transmission from the floor below would involve a comprehensive programme of work to isolate not only the plant but also all of the service pipe connections and would be difficult and expensive. Reduction of airborne sound transmission from the plant to the space below would involve improving the sound insulation of the concrete floor/ceiling slab separating the first and second floors by installing a suspended ceiling. This would also be difficult and expensive.

What methods could you suggest to estimate the relative magnitudes of the airborne and structure-borne sound transmission?

Answer

With the plant in operation, measure (in octave bands) the sound level in both the source and receiving rooms and determine the level difference $L_1 - L_2$. This value may then be compared with:

- the calculated value of $L_1 - L_2$ using 'room to room' transmission formula and estimated values of the sound reduction index of the floor slab and of the absorption in the receiving room
- measured values of $L_1 - L_2$ using a loudspeaker source instead of the chiller as a source, on the basis that (provided that the loudspeaker is isolated from the floor) this will measure only the airborne component or sound transmission.

Two alternative methods involve taking either vibration or sound intensity measurements in the receiving room in order to compare the sound radiated from the underside of the floor slab with that from the walls in the receiving room.

9.8 Control of sound transmission: absorption, insulation, isolation and damping

Once the sound has left the source the next set of opportunities for noise control involve attenuating the noise during its transmission to the receiver. The four standard techniques for doing this involve the use of sound absorbing and/or sound insulating materials for the reduction of airborne sound, and the use of vibration isolation and damping for the use of structure-borne sound.

The use of sound absorption was described in Chapter 5, sound insulation in Chapter 6 and vibration isolation and damping in Chapter 7. A summary of the uses of the four techniques in noise control is given below.

Sound absorption

Sound absorbing materials are used to reduce airborne sound. They convert airborne sound energy into heat via some frictional mechanisms, e.g. in the pores of a porous sound absorber. Generally they are used to reduce the reflection of sound from hard, sound reflecting surfaces and to reduce the build-up of reverberant sound in rooms or in other enclosed spaces such as air conditioning ducts.

Noise control uses include:

- control of reverberation time in built spaces
- reduction of reverberant sound levels
- when used with diffusers of sound, to reduce the undesirable effects of standing wave resonances in spaces
- for lining the inside of ductwork, and in splitter and other types of resistive or absorptive (or dissipative) silencers
- for lining the inside of enclosures, and barriers (on the noise source side)
- for use in acoustic louvres
- for plugging small gaps and holes in sound insulation
- for filling the cavity of double-skin partitions and in the reveals of double glazing, to reduce the build-up of sound in the cavity between skins
- in conjunction with a massive layer as a cladding or close shield for pipes, ducts and panels.

Sound insulation

Sound insulating materials are used to reduce transmission of airborne sound between spaces. They are used for noise reduction by screens, barriers, partitions and enclosures, as well as for acoustic lagging of ducts and pipes and machine panels.

Vibration isolation

Isolation employs resilient materials to reduce the transmission of structure-borne sound or vibration. In Chapter 7 the focus was on the isolation of vibration from machines from the supporting structure, but the same principles may be used for many other isolation situations, including:

- the isolation of sound radiating panels from vibrating machine frames
- the isolation of motors, pumps and fans from diesel engine generators and compressors
- the separation of walls in double leaf constructions and of floors and ceilings in floating floors and suspended ceilings
- the isolation of entire rooms (e.g. box in box constructions)
- the isolation of entire buildings (e.g. from ground vibration from trains).

Damping

Damping is the process whereby because of some frictional process vibrational or structure-borne sound energy is converted to heat, thereby reducing the level of vibration and sound. It is the counterpart to sound absorption, which converts airborne sound energy into heat via frictional processes thus reducing the level of airborne sound.

Damping plays an important part in vibration isolation as discussed in Chapter 7, controlling the amplitude of vibration at resonance, but also affecting the transmissibility achieved well above the resonance frequency of the mass–spring resonance.

In many cases adequate damping may be achieved by selection of materials with good intrinsic damping, but sometimes a damping element has to be supplied separately, as in the case of some types of vehicle shock absorbers, in door closers, and in some cases for metal springs used for vibration isolation.

Damping can be important in reducing the level of vibration and therefore of radiated noise from thin sheet panels. There are two cases for which the amount of damping of the panel material will affect the level of vibration and noise:

- when the panel is excited into transient vibration by an impact, as in the case of panels used to form chutes, hoppers, bins, bottle banks, conveyor belts etc.
- when panels such as those used for the outer surfaces of machines, such as the outer casings of fans and motors, are forced into resonant vibration when the machine speed produces forced vibration at the natural frequency of the panel.

The damping of thin sheet panels may be increased:

- by changing the material to one with higher intrinsic damping, e.g. when a bin or hopper is made from a well damped plastic material instead of thin mild steel, which has very low intrinsic damping
- by adding a damping layer of some visco-elastic material
- by the method of fabrication used to join sheets together to form structures.

It is very easy to reduce the vibration of thin relatively undamped structures. Simple experiments will show that the ringing of a wine glass or piece of china may be noticeably reduced by touching lightly with the fingers or applying a patch of duct tape. The attachment of a damping layer to a thin mild steel sheet panel causes the vibration of the panel to be transmitted to the damping layer, where the damping takes place. A wide variety of forms of damping treatments are available. The damping layer may be applied in sheet form and attached using adhesive, or it may be supplied attached to an adhesive or magnetic backing, or may be sprayed on in wet form as a paste or liquid. In constrained layer damping the damping layer is constrained by a thin layer of the undamped panel material which improves the effectiveness of the damping, as compared to without the constraining layer (unconstrained layer damping) – see Figure 9.9.

Mild steel sheets which have been joined together by using rivets, or where the sheets have been bolted or crimped together, are much more highly damped than the individual sheets, or where the sheets have been welded together. This is because vibration causes microscopic movements and therefore friction at the points

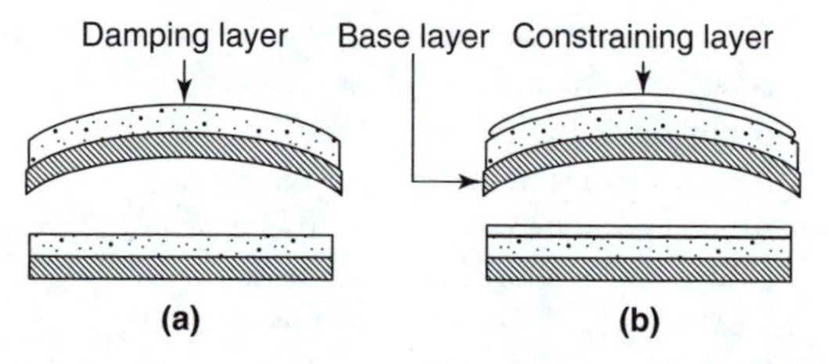

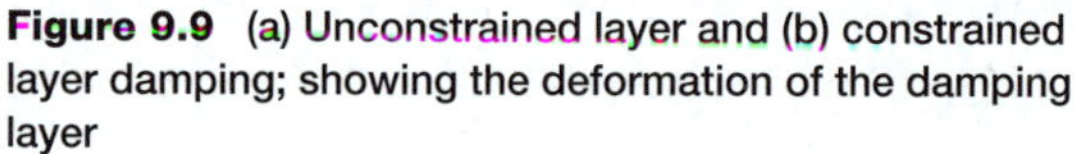

Figure 9.9 (a) Unconstrained layer and (b) constrained layer damping; showing the deformation of the damping layer

where the sheets are joined, whereas a good welded joint produces very little increase in damping.

Note: These techniques (absorption, insulation, isolation and damping) are often used in conjunction, particularly the use of sound insulating and absorbing materials. A good example is in the design of acoustic enclosures.

9.9 Acoustic enclosures

The use of an enclosure made of a good sound insulating material to reduce the noise from a machine is an obvious and attractive solution to a noise problem. With very careful design and use of materials with a high sound reduction index, large noise reductions outside the enclosure can be achieved. However, there are disadvantages, since enclosures can create problems of access to the machine (e.g. for maintenance and setting up), inconvenience to the operator and create cooling problems; they take up floor space and can be expensive.

All of these factors have to be considered by the acoustics engineer, as well as the acoustic parameters such as required noise reductions, and the frequency content of the noise source. In many straightforward situations modest reductions of, say, 10 to 20 dB may be achieved in the medium to high frequency ranges by careful choice of material and design. However, if much larger reductions are required, and particularly at low frequencies, considerable engineering skill and attention to detail are required.

Principles

- The material of the enclosure must have a high enough sound reduction index at all frequencies to give the required noise reductions.
- The inside of the enclosure should be lined with sound absorbing material to reduce the build-up of reverberant sound inside the enclosure, which would cause an increase in the sound transmitted through the enclosure walls to the outside.
- Any 'weak links' in the enclosure (as far as sound insulation is concerned) can seriously reduce the amount of noise reduction achieved. These could include gaps, cracks and holes, poor sealing around windows and doors, and openings for access, ventilation and services.
- Flanking paths for the transmission of sound must be avoided or minimized. This means that if there is any possibility of structure-borne sound being transmitted then the machine must be effectively isolated from the floor and from the enclosure.

Practice

In practice this means that all pipes and ducts (for water, air, fuel etc.) should be mechanically isolated from the machine, using flexible connectors. Where these pass through the enclosure, care should be taken so that access holes are as small as possible, and air-sealed with a flexible, mastic type of material. The aim should be to minimize the leakage of airborne sound and yet to prevent transmission of vibration. Similarly it is important that any gaps around doors and windows in the enclosure are well sealed.

If the air cooling requirements of the machine are such that either natural or forced ventilation is needed, then the air can be led through absorbent lined ducts, preferably with one or two 90° bends to help minimize the escape of airborne sound from the enclosure. Sometimes access for materials entering and leaving the machine can be through absorbent lined tunnels. The ends of the tunnels can be covered with push-aside flaps of heavy sound barrier material, which close automatically. The number of practical problems which may be encountered, and the ways of solving them, are endless, but it is attention to these details, as outlined above, which is essential if the simple concept of a box around the machine is to be turned into a practical working solution, and if the noise reduction performance of the enclosure is to approach a value which is theoretically achievable for the material being used for the enclosure. Figure 9.10 illustrates some of these features.

Although an enclosure may be successful in reducing noise levels outside, noise energy is 'bottled up' within the enclosure and noise levels within may be much higher than they would be in the absence of the enclosure. To reduce this high level of reverberant sound it is common practice to line the inside walls of enclosures with sound absorbing materials. The benefits of reducing noise levels inside the enclosure in this way are obvious for large enclosures which may contain working personnel, such as the machine operator.

However, even when this is not the case it is still useful to use absorption in this way since a reduction of noise within the enclosure leads to a corresponding reduction in levels outside. Although these reductions are modest compared with the total reduction, the use of sound absorbing material in this way does provide a useful supplement to the noise reduction produced by the insulation of the enclosure walls. The use of sound absorbing materials also helps to reduce the effect of small gaps and leaks in the enclosure.

Prediction of enclosure performance: noise reduction, insertion loss

If the enclosure is large compared with the machine it is enclosing and if it is contained within a large space such as a workshop, then the theories of room acoustics may apply, and in particular if both spaces are reverberant the

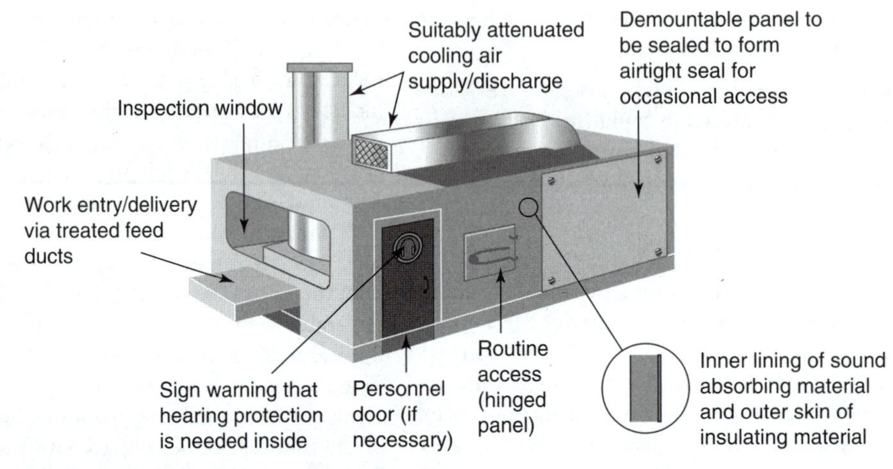

Figure 9.10 Some features of a good enclosure

'room to room' sound transmission formula discussed in Chapter 6 may be used to predict the difference in reverberant sound levels between the inside of the enclosure L_{IN} and the level outside the enclosure, L_{OUT} in the workshop. This difference, $L_{IN} - L_{OUT}$, is sometimes referred to as the noise reduction of the enclosure, NR (not to be confused with NR standing for noise reduction curves met in Chapter 3). As shown in Chapter 6:

$$NR = L_{IN} - L_{OUT} = R_E - 10\log S_E + 10\log A_R$$

where R_E is the sound reduction index of the enclosure walls and roof (this may be a composite value made up from those of the walls, windows, door etc. of the enclosure), S_E is the area of the enclosure (four walls and roof) and A_R is the acoustic absorption in the receiving room, i.e. in the workshop.

Another way of assessing the acoustic performance of the enclosure, called the insertion loss (IL) is the difference between the sound level from the noisy machine measured at a point outside the enclosure, i.e. in the workshop before (L_{BEFORE}) and after (L_{AFTER}) the machine has been enclosed. The insertion loss can also be predicted using the room acoustic theory:

$$IL = L_{BEFORE} - L_{AFTER} = R_E - 10\log S_E + 10\log A_E$$

These two formulae, for *NR* and for *IL*, appear to be similar but note that in the second case it is the absorption inside the enclosure which is important, whereas in the first case it is the absorption within the outer room, i.e. the workshop, which is relevant.

Note that in both cases the performance will depend on frequency and the values of NR and IL should be estimated or measured in octave bands. The derivation of the above equation for predicting IL is given at the end of this chapter.

Example 9.1

An enclosure around a machine in a workshop has dimensions 4 × 3 × 2.5 m and contains a door of area 2 m^2 and a window of area 2 m^2. The walls and ceiling of the enclosure are lined with sound absorbing material except for the door and window. The workshop has dimensions of 20 × 15 × 6 m. In a certain octave band the average absorption coefficient of the sound absorbing material used inside the enclosure is 0.65 and that of the workshop surfaces and of the door and window of the enclosure are 0.05. In the same octave band the composite sound reduction index of the enclosure walls, including the door and window, is 35 dB. Estimate both the noise reduction (NR) and the insertion loss (IL) of the enclosure.

Area of enclosure walls and ceiling (including door and window) $= S_E = 47$ m^2

Area of sound absorbing panels inside enclosure = 43 m^2

Sound absorption inside enclosure:

$$A_E = (43 \times 0.65) + (4 \times 0.05) = 28 \text{ m}^2$$

Sound reduction index of enclosure walls, $R_E = 35$ dB

$$\text{Insertion loss of enclosure} = R_E - 10\log S_E + 10\log A_E = 35 - 10\log(35) + 10\log(28) = 33 \text{ dB}$$

Area of workshop surfaces = 1020 m^2

$$\text{Absorption of workshop surfaces} = A_R = 1020 \times 0.05 = 51 \text{ m}^2$$

$$\text{Noise reduction of enclosure} = R_E - 10\log S_E + 10\log A_R = 35 - 10\log(35) + 10\log(51) = 35 \text{ dB}$$

It can be seen that the two measures give slightly different performance values. The insertion loss may be closer to the real noise attenuation benefit enjoyed by the listener outside the enclosure but it may be more difficult to measure accurately because the L_{BEFORE} and L_{AFTER} levels would be measured on different occasions, maybe separated by a considerable period of time (to allow for the design and installation of the enclosure) and conditions may have changed between the two measurement occasions.

These predictions should be used with caution because the assumptions of Sabine room acoustics on which they are based may not always apply. For example the sound fields may not be reverberant and direct sound may become important. If the enclosure is close fitting around the machine there may be acoustic coupling between the walls of the machine and the walls of the enclosure.

Insertion loss is used widely to specify the performance of other items of noise control equipment such as noise barriers, silencers etc.

Acoustic refuges and specialist test facilities which require very low noise levels such as anechoic rooms and audiometric testing booths employ the same noise reducing principles as an acoustic enclosure except that in these cases the noisy environment is on the outside and the purpose is to protect the persons within the enclosure.

9.10 Fan and duct noise and attenuators

Aerodynamic noise overview

Laminar or streamline flow, which is relatively quiet, becomes turbulent (very noisy) when flow speed becomes too high or when the flow path is not smooth enough, i.e. contains too many bends, obstacles or changes of cross-section. Therefore the basic principle in order to minimize noise is keep airflow as *slow* and *smooth* as possible.

Variation with speed

Aerodynamic noise is very strongly dependent on air flow speed, V (proportional to V^N where N may be as high as 6 or 8). This means that there is a very big decibel increase or decrease when doubling or halving the speed of a fan, so speed reduction at off peak times gives good noise reduction, and larger slower fans will be quieter than smaller faster fans which move the same amount of airflow.

$$W \propto V^N$$

$$W_2/W_1 = (V_2/V_1)^N$$

So for example if $N = 8$ and $(V_2/V_1) = 2$, $W_2/W_1 = 2^8 = 256$ and this corresponds to a change in sound power level of $10\log 256 = 24$ dB.

Or, combining this into one calculation:

$$L_{W1} - L_{W2} = 10\log(V_2/V_1)^N = N \times 10\log[(V_2/V_1)]$$
$$= 8 \times 10\log[2] = 24 \text{ dB}$$

In other words a reduction of 24 dB occurs for a halving of speed (for $N = 8$).

Fans and blowers

Fans are part of a class of rotating machines (including blowers, propellers, turbines, pumps) which produce noise as a result of rotating blades passing through the surrounding fluid (air or water) and creating turbulent, noisy fluid flow at the blade tips. Fans of one sort or another are responsible for about 30% of complaints about noise nuisance from industry and commerce. They are used for moving air for ventilation, removing fumes and waste gases and for transporting particulate material in the form of either process material (powders, granules etc.) or waste particulate material. The axial and centrifugal types widely used for moving air and waste gases produce a broadband noise but tonal components can be produced by interaction with obstacles such as guide vanes placed too close to the fan blades. For other types, such as propeller fans, turbines and high pressure industrial blowers for transporting particulate material, the tonal component may be much more prominent.

Noise problems from fans may be minimized by:

- careful location of the fan, and of inlet and outlets to avoid noise sensitive areas
- planning and prediction of noise levels before installation, if possible, so that noise control can be incorporated at the design stage, rather than retrospectively
- careful selection of the most appropriate type and size of fan, operating at lowest possible speed
- reduction of fan speed, if possible, at periods of reduced load and at sensitive times (e.g. at night-time)
- correct installation to ensure that fan is operating at maximum efficiency and minimum noise emission
- design of a duct system to ensure a smooth airflow into the fan and a smooth airflow at the lowest possible speed through ductwork to minimize additional noise generation
- use of silencers, fan enclosure and acoustic lagging of ductwork, if necessary, to provide additional attenuation.

Blade passing frequency

When a fan emits a pure tone it is most likely to be at the blade passing frequency, i.e. the number of times a tip of

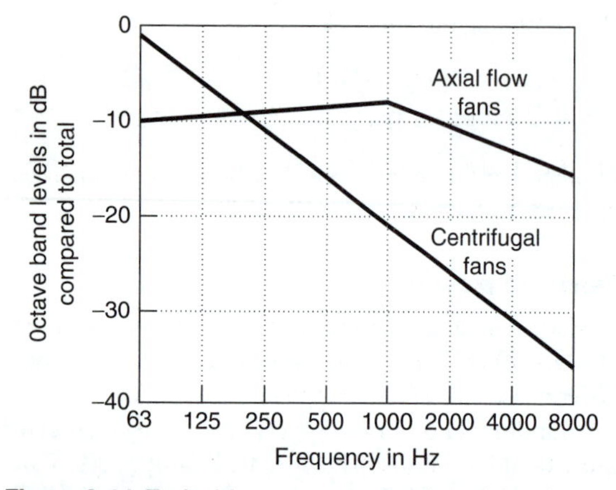

Figure 9.11 Typical frequency spectra for axial and centrifugal fans

a fan blade passes a nearby surface (e.g. the fan casing, or a support strut) when a pulse of turbulent air (vortices or eddies) is created. The frequency, f, is related to the speed of the fan, in revolutions per minute, RPM, and the number of fan blades, N:

$$f = \text{RPM} \times N/60 \text{ Hz}$$

It follows that any change that causes the minimum distance between the tips of the fan blades and nearby static objects to decrease will lead to an increase in the pure tone noise emitted by the fan.

Fan selection

The control of noise from fans starts with the selection of the fan. The main criterion for selection is the volume flow rate of air which has to be moved and the pressure drop against which it has to work. Within these parameters a fan with the lowest noise output, when working at its optimum efficiency, should be selected. This may need to take into account the shape of the noise spectrum (see Figure 9.11), because it may be better to accept a fan which has the lowest sound power level at lower frequencies which are difficult to reduce even though it may produce higher levels at the more easily controlled high frequencies. The size of the fan is also an important factor because in general larger fans operating at lower speeds move the same amount of air with less noise than smaller, higher speed fans. The noise output increases very rapidly with speed so that variable speed fans can be used to reduce speed and noise by reducing the speed when the load requirement is low.

Correct installation of the fan

Once the fan has been selected it should be installed so as to make the minimum noise possible and this requires a smooth flow of air into the fan since input turbulence can increase the noise level by several decibels. There should be no obstructions to flow, no bends or changes of cross-section within at least one duct diameter, either up or downstream of the fan. Noise from the inlet and supply side of the fan may be reduced by using in-line silencers, and airborne noise radiated from the casing of the fan itself may be reduced by an enclosure around the fan. The transmission of structure-borne sound should be reduced by fitting anti-vibration mounts between the fan and the building structure, usually the floor, and using flexible connectors between fan and ductwork.

The sound power produced by the fan may be augmented by noise generated by the flow of air through the duct system, and this should be minimized by designing the system to keep the airflow as smooth and as slow as possible. If the airflow carries particulate material then there is additional noise generated by impact of the particles with the duct walls. Noise generated within and emanating from the duct system may be reduced using a secondary silencer, and noise breaking out from the duct through the duct walls of the duct may be reduced by application of acoustic lagging treatment to the duct.

Sources of noise generation in a fan/duct system

The main, primary source of noise is the fan. A secondary source is airflow generated noise, sometimes called secondary noise.

Main components of a fan/duct system delivering air conditioning

A schematic sketch illustrating some features of an air distribution system is shown in Figure 9.12(a), and Figure 9.12(b) illustrates typical sound transmission paths.

A fan (or air handling unit) located in plant room draws in air via a hole in the external wall of the plant room (left edge of sketch), and delivers this air via a duct system with a branch, bends, and a lined plenum chamber and a diffuser or grille to a ventilate a room (bottom of Figure 9.12). The aperture in the external wall is fitted with louvres which may be lined with sound absorbing material. The main features illustrated in Figure 9.12 are:

- the fan
- primary silencers either side of the fan in the plant room – attenuation requirements in octave bands to be determined by calculation/design
- components of air delivery system: lengths of straight duct, bends, branches, grilles/diffusers which deliver air into ventilated rooms. Also plenum chambers, dampers/flow control valves).

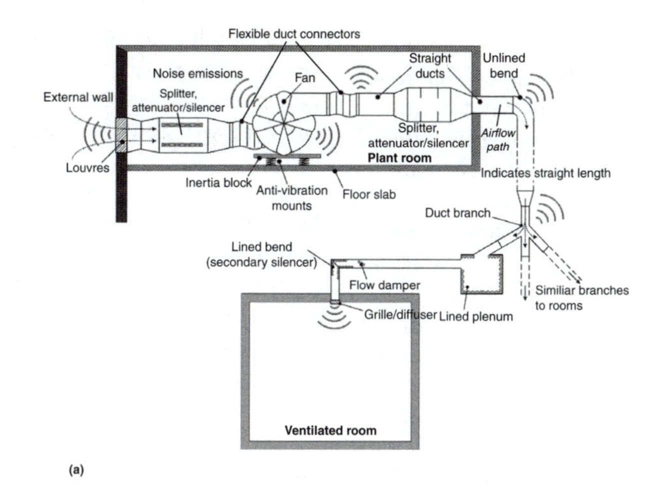

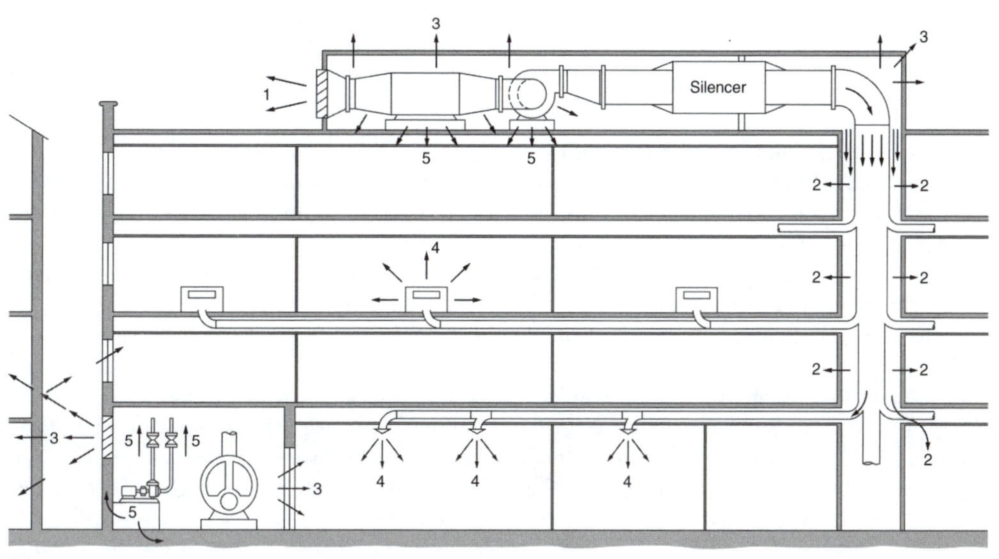

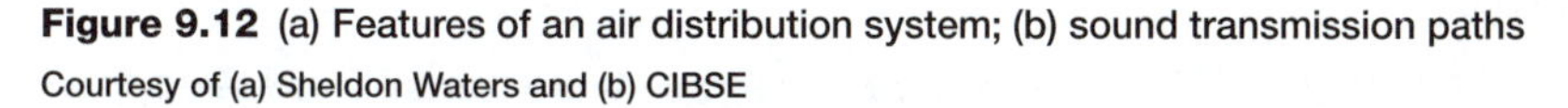
Figure 9.12 (a) Features of an air distribution system; (b) sound transmission paths

Courtesy of (a) Sheldon Waters and (b) CIBSE

Each of these components provides some attenuation of noise but also some generation of flow noise.

Straight duct runs

A smooth, rigid-walled duct would not provide any attenuation of sound striking it. However, sheet metal ducts are not rigid and the sound causes the duct walls to vibrate and in doing so to remove energy from the sound waves in the duct. Some of this energy is reradiated from the duct walls and some is dissipated by damping of the walls. The attenuation will depend on the mass per unit area of the duct walls, upon the frequency and the duct size. At low frequencies the stiffness of the duct walls will be more important than the mass and as a result low frequency attenuation will be different for round and circular cross-section ducts, but at high frequencies it is only the mass per unit area of the duct walls that matters, and not the shape.

The attenuation of unlined sheet metal ducts is very small, ranging from 0.6 dB/metre for small rectangular ductwork at low frequencies to less than 0.1 dB/metre at high frequencies. If the duct walls are lined with sound absorbing material (at extra cost) the attenuation will be considerably increased. It is, however, more usual in the UK to have unlined ducts but to insert lengths of absorptive silencers, or attenuators, at strategic points in the ductwork.

Bends

The attenuation produced by a right-angled bend results mainly from the reflection of sound back towards the fan, thus reducing the amount of sound energy moving forward towards the ventilated room. It varies with the size of the duct and the frequency of the sound. For a square duct the bend attenuation is a maximum of some 7 or 8 dB at the frequency (octave band) for which the wavelength of sound in air is twice the duct width. At higher frequencies (an octave above) the attenuation drops to 3 or 4 dB and at lower frequencies it can fall to less than 1 dB.

If the bend contains turning vanes to help smooth the airflow around the corner then the attenuation produced will be very much reduced (and so, of course, will the amount of noise regenerated by the bend). If the bend is lined with acoustically absorbing material then the attenuation will be greatly increased.

Branches

At a point where a duct splits into two branches it is assumed that the acoustic energy divides between the two branches in the same ratio as the division of airflow. Thus if the main duct divides into two equal branches the sound power level in each branch just below the junction is 3 dB less than in the main branch just above the junction. If the flow had divided into two parts down one branch and three parts down the other, then in the smaller branch the branch attenuation would be $10\log(5/2) = 4$ dB, i.e. the sound power level in the smaller branch would be 4 dB less than in the main branch. In the other branch the attenuation would be 2.2 dB ($= 10\log(5/3)$).

End reflection at grilles/diffusers

The sound near the end of the duct travelling towards the opening into the room 'notices' a change of acoustic impedance at the opening. This causes some of the sound energy to be reflected back down the duct rather than being transmitted into the room. This in effect provides an attenuation which depends on the wavelength of the sound and the dimensions of the opening.

The attenuation is largest when the wavelength is much greater than the dimensions of the opening, i.e. for low frequencies and small openings. The attenuation is very small when the opening dimensions are greater than the wavelength, i.e. for large openings and high frequencies. *Note:* More specialized texts (e.g. CIBSE or ASHRAE Guides) should be consulted for data on attenuation by bends, straight duct runs and end reflections.

As a result of these various forms of attenuation as the air and noise travels from the fan through the delivery system towards the ventilated rooms the noise from the fan decreases and the noise generated from airflow increases.

Secondary silencers (shown as a lined bend in Figure 9.12a) are sometimes needed at the end of the air distribution system, i.e. close to the ventilated rooms, to control noise from airflow generation. These are much smaller than the primary silencers because the airflow noise is often at a medium/high frequency as opposed to the fan noise which is at a low/medium frequency. Secondary silencers also help to control 'crosstalk' between rooms.

Other important features

The fan should be isolated from the floor in the plant room using anti-vibration mounts and from other parts of building structure by the use of flexible connectors, in order to minimize the transmission of structure-borne noise.

The main noise transmission paths are:

- airborne transmission from fan in plant room, and from plant room to adjacent rooms and to the exterior environment
- structure-borne noise from the fan to other rooms in the building
- transmission to ventilated rooms via the air distribution (i.e. duct) system
- duct break-out which can affect non-ventilated (as well as ventilated) spaces if ducts pass overhead in ceiling void spaces
- duct break-in (e.g. from high noise levels in the plant room)
- crosstalk between rooms via the duct system.

Prediction of noise levels transmitted via a duct system

In each octave band:

- Determine the fan sound power level entering the duct system in the plant room.
- Estimate the duct system attenuation (from straight duct runs, bends, branches, end reflection at grille/diffusers).
- Subtract the system attenuation from the fan sound power to find the sound power level entering the ventilated room.
- Calculate the sound pressure level in the room as for any other noise source, i.e. as a combination of direct and reverberant sound pressure levels.

- Repeat for the next octave band.
- Compare with the noise limit/target for acceptable noise levels in room (e.g. NR 40 or 45 dBA).
- Hence determine the additional attenuation to be provided by the primary silencer in the plant room.
- Combine octave band SPLS into dBA, NR or NC as required.

Note: A more detailed prediction might also include the prediction of flow generated, or secondary noise, and of duct break-out noise.

Example 9.2

Air enters a room via a single grille situated in one of the top corners of the room. A target level of NC 35 has to be achieved by the ventilation system which consists of a fan connected to a main duct of length 5 m. The airflow then splits into two equal parts, one of the branches of length 3 m serving the room in question via the single grille. There is one bend in the main duct prior to the division of the airflow, and one bend in each of the branches.

Calculate the extra attenuation required in the 250 Hz octave band to meet the NC 35 target at a point in the room 4 m from the grille, given the following information, all of which refers to the 250 Hz octave band:

- sound power level of fan = 88 dB
- attenuation of main duct = 0.3 dB/m
- attenuation of branch duct = 0.5 dB/m
- attenuation of bend in main duct = 6 dB
- attenuation of bend in branch duct = 3 dB
- room constant = 50 m^2
- end reflection at grille = 4 dB.

Step 1

Calculate the attenuation provided by the system:

Main duct, 0.3 × 5	1.5 dB
Bend, main duct	6 dB
Branch (50% of total airflow)	3 dB
Branch duct, 0.5 × 3	1.5 dB
Bend, branch duct	3 dB
End reflection	4 dB
Total attenuation of system	19 dB

Therefore sound power level entering room = 88 − 19 = 69 dB.

Step 2

Calculate the sound pressure level in the room, using the usual formula for the total of both the direct and reverberant sound level

$L_p = L_W + 10\log[(Q/4\pi r^2) + (4/R_C)]$

where $Q = 8$, $L_W = 69$ dB, $r = 4$ m and $R = 50$ m^2.

This gives $L_p = 59.8$ dB or 60 dB to the nearest decibel.

Step 3

Compare this with the target and estimate the extra attenuation required.

The 250 Hz octave band level for NC 35 should be below 55 dB (refer back to Chapter 3 for NC curves). Therefore the extra attenuation required = 60 − 55 = 5 dB.

A similar calculation should be carried out for each octave band in turn, based on appropriate attenuation values for each band. The above example is of a very simple situation and is intended to illustrate the calculation method. A more accurate procedure also calculates the extra flow noise generated at bends, grilles and other obstacles.

Design of in-duct silencers/attenuators

(These are sometimes called splitter silencers.)

A duct lined on the inside with sound absorbent material is, in effect, an attenuator or silencer, but rather than using continuous lengths of lined ducts the more common practice in the UK is to use inserted lengths of ductwork which, in addition to having two opposite walls lined with sound absorbing material, also have one or more additional slabs of sound absorbing material (called splitters, because they split the airway into two or more sections) inserted into the duct (see Figure 9.13). This is the conventional attenuator or 'splitter' silencer.

The attenuation produced depends on:

- the length of the attenuator (performance is often specified in dB per metre, for any particular octave band)
- the number of slabs of sound absorbent material (wall linings and splitters)
- the absorption coefficient of the sound absorbing material, which depends on the type of material, the thickness of the absorbent slabs and on the frequency of the sound
- the area of sound absorbent material exposed to noise (related to the duct dimensions).

The attenuation increases with frequency from low to medium frequencies as the value of the absorption coefficient increases, but then decrease at higher frequencies as sound becomes more directional and is 'beamed' down the centre of the duct, and has less contact with the sound absorbent material (see Figure 9.14).

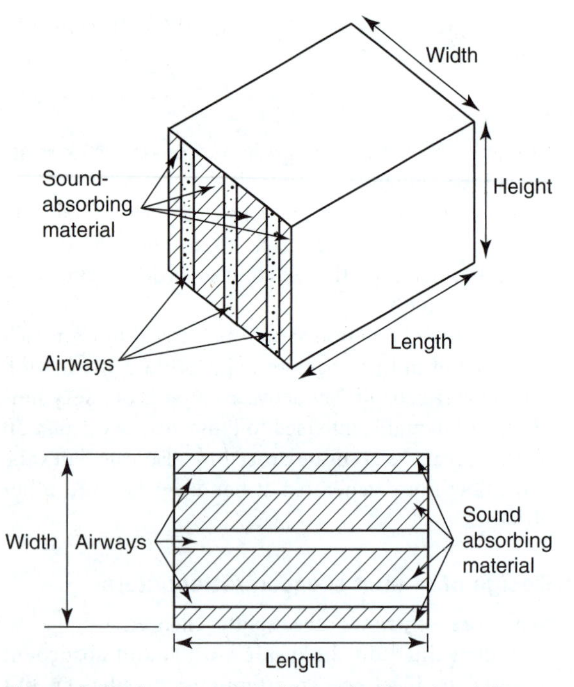

Figure 9.13 An attenuator with two 'splitters' and three airways

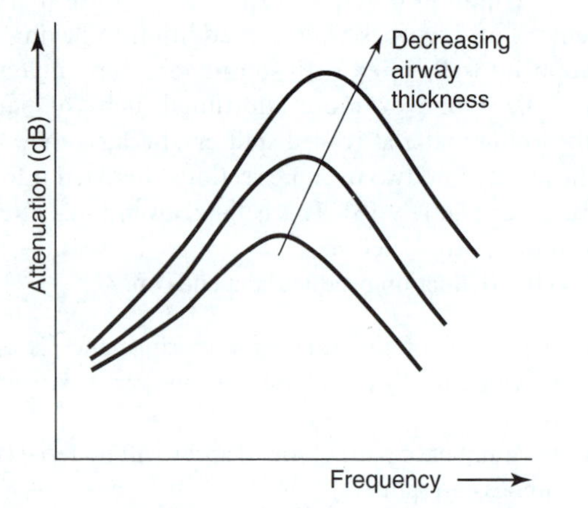

Figure 9.14 Showing how the attenuation of an attenuator varies with frequency and airway thickness

Types of sound attenuators: active/passive, absorptive/dissipative, reactive

The above types are absorptive or dissipative attenuators which rely on the use of sound absorbing material and produce attenuation over a wide range of frequencies (albeit more at some frequencies than at others), i.e. broadband absorbers. There are other types of attenuators called reactive attenuators which are tuned to provide attenuation only over a narrow range of frequencies. Although ultimately, as for the dissipative types, attenuation is achieved through the conversion of sound energy into heat via frictional processes, these work by using reflections, destructive interference (noise cancelling) and resonances (and anti-resonances) to achieve highly frequency selective attenuation. Note that both of these types of attenuators are 'passive' attenuators, as opposed to active noise control systems, which employ a system of microphones and loudspeakers to produce 'out of phase' cancelling noise.

There are also hybrid attenuators which use a combination of these methods.

Control of duct break-out noise

This may be partly achieved through use of in-duct attenuators, but also if necessary by acoustic lagging of the ductwork (resilient layer/mass layer covering of the ductwork), or by using stiffer and heavier ducts, i.e. having a higher sound reduction index.

9.11 Reactive silencers

A sudden increase in the cross-section of a pipe or duct produces a change in acoustic impedance which causes sound energy to be reflected back towards the noise source. The amount of attenuation produced depends on the ratio of the cross-sectional areas of the expanded and original sections of pipe. The expansion has the effect of a high pass acoustic filter, providing most attenuation at low frequencies when the wavelength of sound is much greater than the duct or pipe dimensions. A more gradual change of section, as in the conical connector (see Figure 9.15(a)) works on the same principle. The attenuation produced is slightly less than for abrupt changes of section, but the gradual transformation allows a smoother airflow to be maintained. This type of device can be used to introduce a splitter silencer into a duct of different cross-sectional area.

A simple expansion chamber, as shown in Figure 9.15(b), is the most basic form of reactive silencer. The attenuation it produces is a maximum for the frequency at which the length of the chamber is a quarter of the wavelength of sound in the gas in the chamber. At this frequency the chamber presents the greatest acoustic impedance mismatch to the sound in the inlet tube. The attenuation drops off away from this frequency and is a minimum when the chamber length equals half a wavelength. At this frequency there is resonant standing wave amplification of the sound waves in the chamber. Higher up the frequency scale there are other maxima (where chamber length is an odd number of quarter wavelengths) and minima (chamber length equal to an even number of quarter wavelengths) in the attenuation. The maximum attenuation at the quarter-wavelength condition depends on the expansion ratio.

The situation is complicated by the presence of the tailpipe. Resonances and anti-resonances associated with the length of the tailpipe, and its relationship to the wavelength of the sound, produce more peaks and troughs in the attenuation spectrum of the silencer. Further expansion chambers can be used to provide more attenuation (see Figure 9.15(c)). These may be of the same length, or of different lengths in an attempt to fill in gaps in the attenuation provided by the first chamber. A further complication is provided by the presence of interconnecting pipes between chambers. There are yet more peaks and troughs in the silencer attenuation spectrum associated with the presence of these interconnecting pipes. The complete spectrum of the attenuation produced by a device such as that illustrated in Figure 9.15(c) will therefore be very complicated, necessitating computer techniques for its prediction. It will depend on the lengths of the different expansion chambers and of the tail and interconnecting pipes and on the expansion ratio.

A reactive silencer of a different type can be designed to give high attenuation at a particular frequency. This type of silencer employs a resonant system which extracts sound energy from the duct at the resonance frequency in similar fashion to the Helmholtz or cavity resonator of Figure 5.5. The resonant system may be a Helmholtz resonator in which a series of holes in the pipe wall communicate with a cavity surrounding the pipe. The resonator can be considered as a mass–spring system, in which the air in the holes acts as the mass and the volume of air in the cavity acts as the spring. Alternatively, a tuned length of closed-ended pipe acts as the resonant

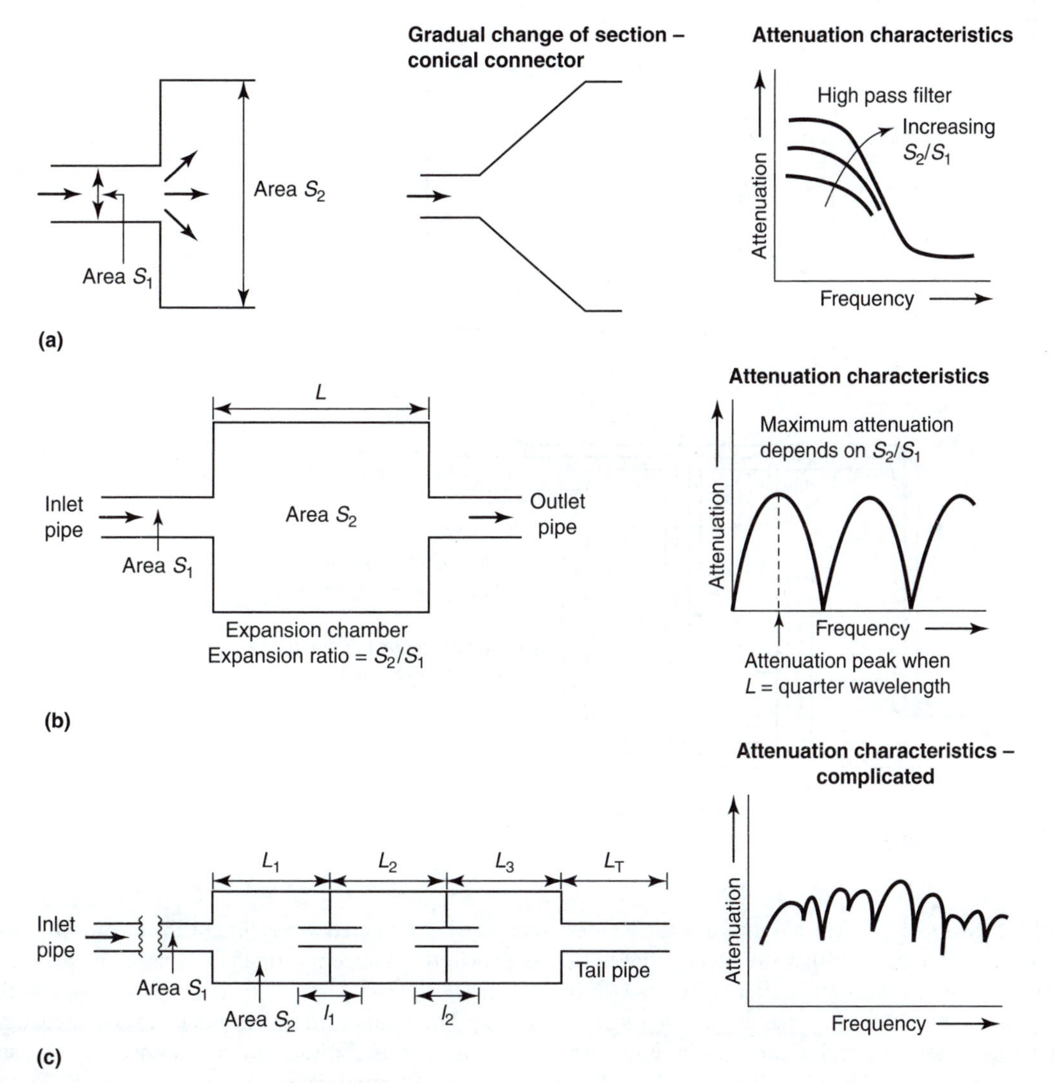

Figure 9.15 Various types of reactive attenuators

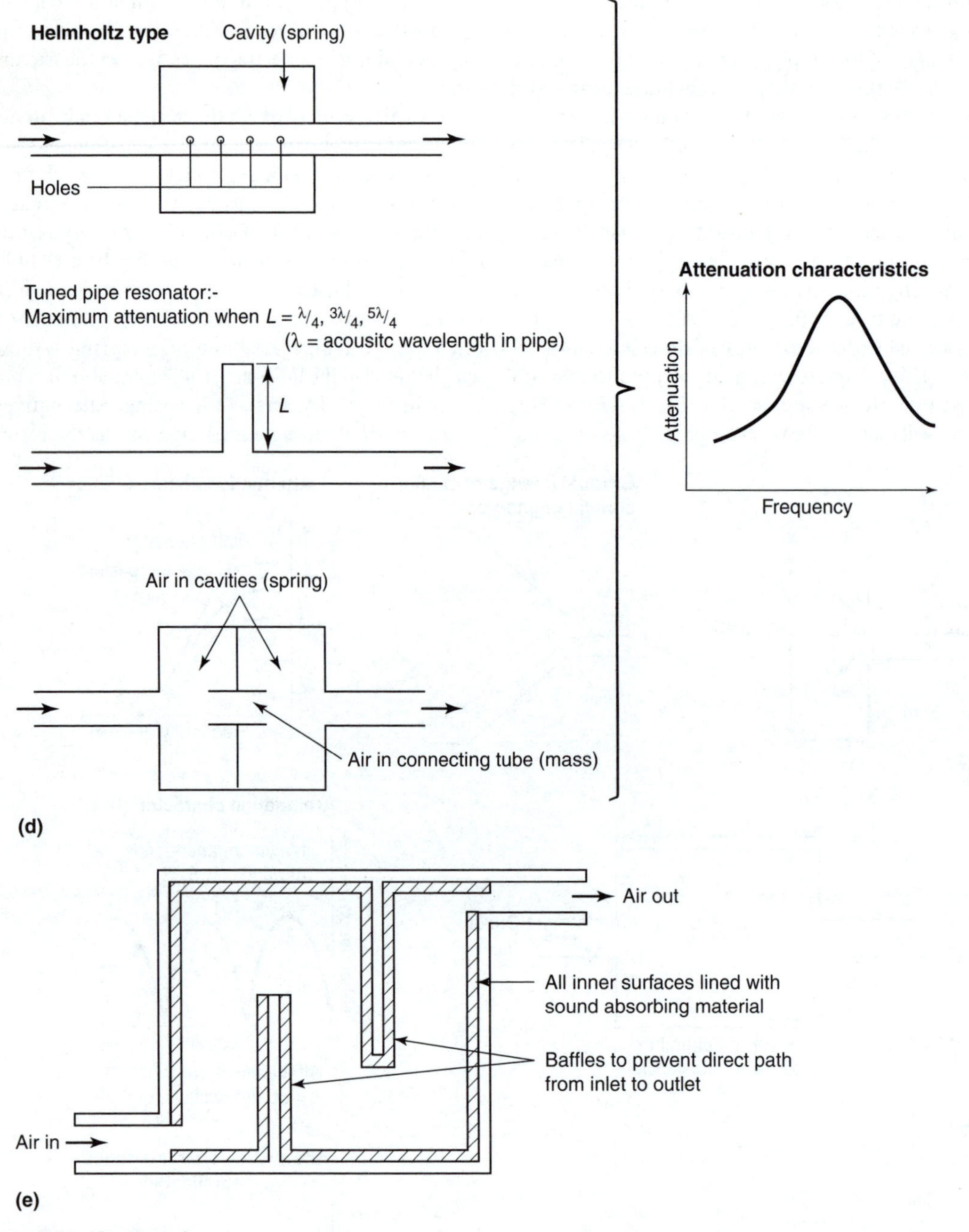

Figure 9.15 *(Continued)*

system (see Figure 9.15(d)). The advantage of these 'side-branch' types of resonator is that the silencer does not produce high flow resistance and back pressures which is sometimes the case with other types of reactive silencer. Another type of resonant device, this time 'in line' with the main gas flow, is also shown in Figure 9.15(d). This can also be compared to a Helmholtz resonator in which the air in the interconnecting pipe acts as the mass and the air in the two sections of the chamber acts as the spring. The resonant frequency depends on the length and cross-sectional area of the interconnecting pipe, and on the total volume of the chamber.

A plenum chamber in ventilating systems is an example of a device which incorporates dissipative and reactive sound-reducing mechanisms (see Figure 9.15(e)). Part of the attenuation, mainly at low frequencies, is produced by the expansion chamber principle, but the overall performance is greatly improved if the inside of the chamber is lined with sound-absorbing material and if baffles are used in the chamber to prevent sound travelling directly without reflection from the inlet to the outlet of the chamber.

9.12 Jets and exhausts

A jet is a flow of fast moving gas or air moving from the end of a tube or pipe into free air. The structure of a jet is shown in Figure 9.16. The noise producing mechanism is the turbulence created by the viscous forces which operate between adjacent layers of air in the boundary layer between the jet and the surrounding air, where there is a very high velocity gradient. The sound power level produced depends on the cross-sectional area of the jet and the efflux velocity. The level increases very sharply with increasing velocity, so that any velocity reduction which can be achieved will produce significant noise reductions, more than 20 dB reduction if the velocity is halved. The noise has a broadband spectrum with a peak at a frequency which depends on the jet velocity, v (in m/s) and diameter, d (in mm) according to the formula:

$$f = 200 \times v/d$$

Thus small diameter high velocity jets will have a high frequency spectrum, whereas much larger diameter and lower velocities in exhausts produce a much lower frequency spectrum. The radiation is directional and so some benefit can be obtained by directing outdoor jets vertically upwards.

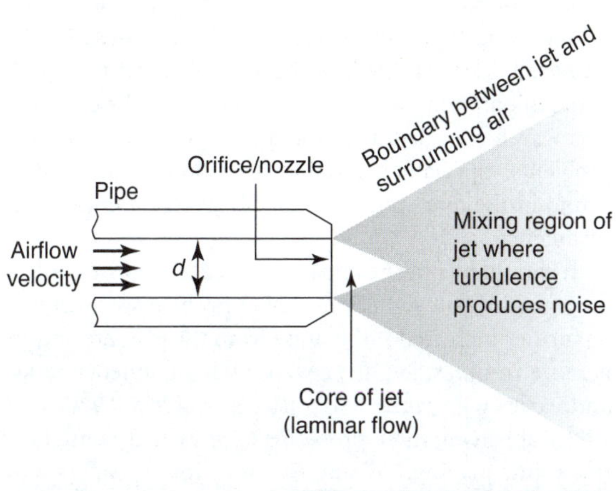

Figure 9.16 The features of a jet

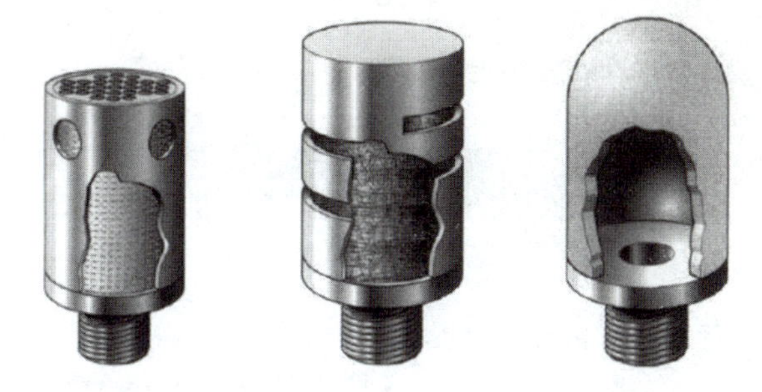

Figure 9.17 Jet silencers

Jet noise is produced in exhausts, vents, valves and other pressure release devices where air, gas or steam at high pressures and velocities is released into the atmosphere, by the exhausts of pneumatically operated actuators and hand tools, and by 'working' air jets used to eject work-pieces and for cleaning and cooling purposes. Compressed air hoses are among the most universal sources of high noise levels in industry.

The principles of diffusion and absorption are used to reduce noise from exhaust jets. Figure 9.17 shows the main features of an exhaust silencer suitable for process exhausts, in which the initial high flow velocity is forced through a perforated diffuser. This has the effect of reducing flow velocity, promoting frictional effects, and also converting one single jet, which for a large diameter process exhaust system may be predominantly low frequency noise, to many smaller high frequency jets which are more easily attenuated by the sound absorbing material lining the silencer. Exhaust silencers working on the diffuser principle, which can easily be screwed onto pneumatic exhausts, are relatively cheap and readily available (see Figure 9.17), and can produce reductions of up to 25 dBA. The diffusing medium may be sintered plastic, compacted wire wool or layers of woven wire cloth. In addition to the noise reduction to be achieved, the selection of the most appropriate type will need to take into account the effects of any back pressure caused by the silencer, and arrangements for cleaning, maintenance or replacement in order to avoid blockage of the diffuser by debris in the airflow. An alternative to fitting a diffusing silencer is to lead the exhaust away through a length of flexible tubing to exhaust in a less noise sensitive area.

Where the air jet is required to do some useful work a variety of 'low noise' nozzles are available. Most of these work on the 'air entrainment' or 'induced flow' principle, in which a flow of air, of intermediate velocity, is introduced into the boundary layer between the fast moving jet stream and the stationary surrounding air, in order to reduce the velocity gradient, and so to reduce turbulence and noise generation (see Figure 9.18).

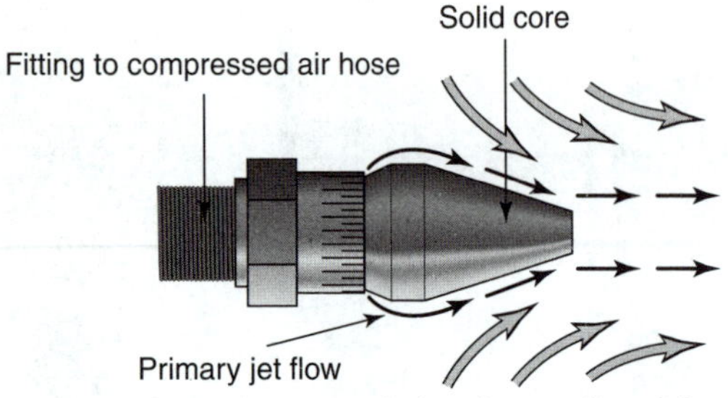

Figure 9.18 A low-noise (induced flow) air nozzle

More detailed information on reducing noise from pneumatic air jet is given in HSE Guidance Note PM56 *Noise from Pneumatic Systems.*

9.13 Active noise control

Active noise control systems apply the principle of destructive interference between waves to reduce noise. They use one or more loudspeakers to radiate sound which is equal in amplitude but opposite in phase to the noise waveform. The term active refers to the use of a source of energy, from the loudspeaker, in the noise reduction process, as compared to the so-called passive noise reduction measures, such as splitter silencers, which rely on sound absorbing materials to convert sound energy to heat.

Although the idea of cancelling the noise with antiphase signals (colloquially called anti-noise) is simple in principle, it is difficult to carry out in practice. If significant noise reduction is to be obtained, the cancellation of the primary waveform (the noise) by the secondary waveform, produced by the loudspeaker, must be achieved to a high degree of accuracy. A 20 dB noise reduction, for example, requires a reduction of sound intensity to one hundredth of its original value; this gives an indication of the degree of exactness with which the inverted secondary waveform must match the primary waveform it is designed to cancel.

The first active noise control system was patented by Lueg in the 1930s. Practical difficulties meant that progress was slow until the 1970s and 1980s, when advances in electronics and signal processing techniques accelerated development.

When explaining the principles and methods involved in active noise control, it is convenient to refer to the cancellation of noise in a duct, a situation which has been the subject of much research. If the sound wavelength is more than twice the maximum duct cross-sectional dimension, the situation is simplified to a one-dimensional problem because only plane waves will travel along the duct. The elements of a simple active noise control system are shown in Figure 9.19. A microphone downstream of the noise source, perhaps a fan, is connected via the signal processor to a loudspeaker, called the secondary source, which is a further distance L downstream.

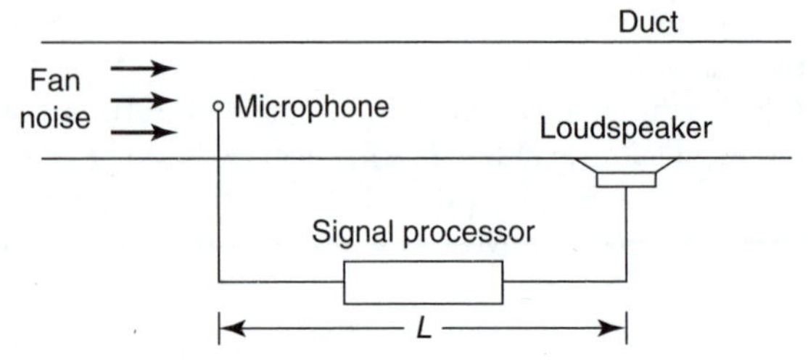

Figure 9.19 Active attenuation of fan noise in a duct

The condition for complete cancellation of noise to occur is that the secondary source must produce a waveform that is equal in amplitude but opposite in phase to the waveform received by the microphone at time L/c earlier, to allow for the travel time of sound upstream from microphone to loudspeaker. This assumes that there are no absorption losses in the duct, and it also means the signal detected at the microphone must be delayed by an amount corresponding to the acoustic transit time minus the time taken to process the signal.

If the primary signal were a pure tone, cancellation could in principle be achieved without any delay by fixing the distance between the microphone and loudspeaker to be half a wavelength. Even such a simple system would be sensitive to changes in temperature, which would affect the sound speed, and therefore the wavelength, or to changes in the phase response of the electronics and the transducers.

Another difficulty arises because the loudspeaker will radiate downstream towards the microphone and the fan as well as upstream towards the end of the duct. Therefore feedback from the loudspeaker will affect the signal at the microphone which is used to provide the cancellation waveform. Airflow in the duct will cause additional problems, especially if variable; it will affect the speed of sound in the duct and will possibly generate noise from turbulence.

It should by now be apparent that active noise control presents difficult signal processing problems, which will be further increased if the noise from the primary source is more complex, i.e. if it covers a wide frequency range and varies with time. Early attempts, in the 1970s and 1980s, at solving these problems used various configurations and placings of one or more loudspeakers and microphones in combination with various fixed signal processing systems. More recent research has successfully

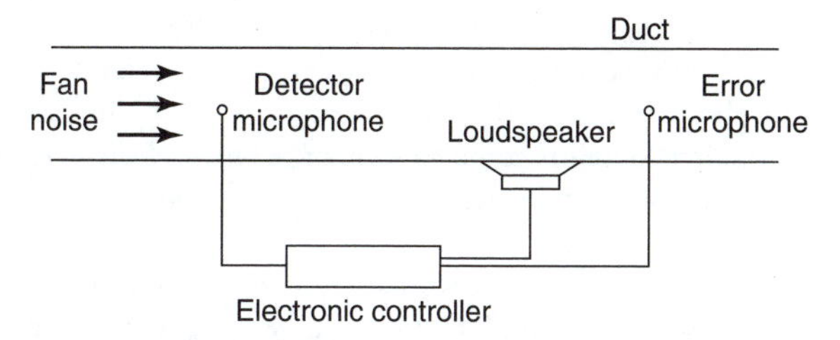

Figure 9.20 Use of an error microphone in the active attenuation of fan noise in a duct

applied the techniques of adaptive digital filtering to produce active noise control systems which can, within limits, adapt to changes in operating conditions, such as changes in temperature, flow rate and characteristics of electronic components and transducers.

One such system (see Figure 9.20) involves the use of an error microphone downstream of the loudspeaker to detect imperfections in the cancelling process. The signal from the error microphone is fed back to the electronic controller, which adjusts the signal to the loudspeaker accordingly.

Active noise control has been successfully applied to a wide variety of noise sources, including industrial fans, air conditioning systems in buildings, engines, generators, transformers, vehicle exhaust noise and noise from turbine exhaust stacks. The principle has been applied to the cancellation of noise in small enclosed spaces, such as aircraft cockpits and inside sports cars, as well as to hearing protectors and communication headsets. Noise reductions of up to 30 dB have been achieved at frequencies of up to 500 Hz. Some problems remain in developing loudspeaker and amplifier systems to provide the high levels of sound power that are sometimes required to cancel low frequency noise from some industrial sources, and to withstand long exposures to hostile industrial environments. Further developments in signal processing hardware and software will lead to improvements in the performance of future systems.

Active noise control systems are most effective at low frequencies: because the sound wavelengths are long, the sound field changes only gradually with position. It is in the low frequency region that active noise control methods are particularly advantageous, compared to passive systems, which are large and expensive at low frequencies. The use of active noise control can also lead to considerable savings in energy because the alternative passive silencers cause high flow resistance, which has to be overcome using increased fan power. Hybrid systems, which use active methods at low frequencies and passive methods at high frequencies, optimize the advantages of both methods of noise control.

Active vibration control

Exactly the same principles and techniques used in active noise control can be applied to the reduction of vibration in machinery and structures.

Vibration waveforms detected using accelerometers, or other vibration transducers, are fed in anti-phase to one or more vibrators in an attempt to cancel the effect of the primary source of vibration.

9.14 The specification of noise from machinery, sound pressure level and sound power level

An important aspect of noise control involves a policy of replacing old machines with quieter versions. To do this it is necessary to have reliable methods by which the noise emission of machinery plant and equipment may be measured and specified.

In principle the sound pressure level produced by a sound source at any distance can be predicted from its sound power level, although the calculations may, in general, be more complicated than for the simple free field and reverberant field situations.

Therefore the output from a sound source may be described either by its sound power level, or by specifying the sound pressure level at a certain distance from the source. There are advantages and disadvantages to each method.

The main advantage of using sound power levels is simplicity – sound power level is a single-figure number which depends only on the sound source, not on any other factor such as distance or details of the acoustic environment. This is the idea behind various EC machinery noise directives which require certain classes of machinery to be labelled with their sound power level (L_W) values, so that users and purchasers can easily compare the noise produced by different types of machine.

But there are several disadvantages. The measurement of sound power level requires specialist test rooms or equipment. Also, the L_W value itself has no significance, subjectively, unlike the sound pressure level, which can be related to what is actually heard. Although, in principle, the sound pressure level in any situation may be predicted from the L_W value, it is in practice only possible to predict far field values; near field values, close to the source, are difficult to predict.

The main disadvantage of specifying the sound pressure level at a particular distance from the source is that it can lead to ambiguity and confusion for users, when trying to compare the noise outputs from machines which are described in different ways, e.g. if different distances are used, or if the tests are carried out in different acoustic environments. But this method does have the advantage that sound pressure levels may be easily measured, and can be related to what is heard.

9.15 Hearing protectors

Hearing protection (sometimes called ear protection) should be seen by managers in industry as a last resort method of protecting people's hearing, after failure to reduce noise levels by other means, and as an interim method until a more acceptable method of controlling their exposure to noise is in place. Nevertheless the effective use of hearing protectors is likely to be the only practicable means available of protecting the hearing of very large numbers of employees for several years to come, and it is therefore very important that a well managed programme is in place to ensure that they are always being used effectively. This needs to cover aspects such as availability, selection, training and monitoring of use. Although the cost of each muff or plug is relatively small, the cost of their continued provision for large numbers of employees over several years, together with the cost of managing their use, including time spent on training and monitoring, can be substantial, and comparable with the costs of investing in noise control measures.

Types of hearing protectors

Earmuffs are fitted over the ears in order to block the path of airborne sound to the ear canal. They consist of four main parts:

- moulded rigid plastic cups which fit over the ears and which provide the basic sound insulation
- sound absorbing foam plastic lining fitted inside each cup to reduce reverberation
- flexible plastic cushions designed to enable the cups to fit closely over the ears and establish a close seal to the wearer's head in order to minimize leakage of sound around the cups to the ear. (The cushions are made out of thin flexible plastic film, and most types are filled with foam plastic, but there are some which are filled with oil.)
- an adjustable elastic, flexible headband which holds the cups in place and helps the cushions to make a good seal to the head.

Although all earmuffs conform to this same basic pattern there is a variety of different cup sizes and weights. Muffs are available which fit onto 'hard hats'.

Earplugs are fitted into the ear canal. There are many different types, made from soft rubber, foam plastic or a special type of mineral fibre (called glass down). (Note that ordinary cotton wool is not effective.) In each case they are designed to completely fill the ear canal and therefore minimize the sound transmitted to the ear drum. It is therefore important that the types selected are a good fit and are inserted properly, according to manufacturer's instructions.

Headband mounted earplugs (sometimes called canal caps) consist of a pair of plastic or rubber plugs connected by a light flexible plastic headband. They are designed to be quick and easy to fit and remove for personnel who frequently move in and out of noisy areas such as managers and supervisors.

Specialist types

Specially developed amplitude sensitive earmuffs are available which allow low amplitude sound, for example from normal conversation, but prevent transmission of high intensities. These are useful in situations where there are sudden unpredictable bursts of high levels of noise.

Other types of plugs are available, which are individually moulded and fitted with an acoustic filter designed to ensure that the device produces more or less the same attenuation at all frequencies. This type of plug is used, by musicians for example, when it is necessary that the shape of the frequency spectrum of the sound being listened to remains unchanged, and so its character is preserved.

Earmuffs fitted with communication headsets are also available so that employees can enjoy protection from high levels of machine noise and also listen to 'piped music' and receive messages from the workplace communications system.

Earmuffs fitted with active sound attenuation are available, i.e. the muffs contain a miniature microphone and loudspeaker which detect the sound waveform and provide a cancelling 'anti-phase' signal.

Noise attenuating performance of hearing protectors

Suppliers of hearing protection with the CE mark are required to satisfy the relevant part of BS EN 352 which sets out basic safety requirements for hearing, e.g. features such as size, weight and durability for:

- earmuffs (BS EN 352-1:2002)
- earplugs (BS EN 352-2:2002)
- helmet-mounted earmuffs (BS EN 352-3:2002).

Hearing protection which complies with BS EN 352 must be supplied with performance information derived from a standard test defined in BS EN 13819-2:2002 (which in turn draws on a method in BS EN 24869-1:1993). The information required is:

- mean and standard deviation attenuation values at each octave band centre frequency from 125 Hz to 8 kHz (63 Hz is optional)
- assumed protection values (APV) at each centre frequency (based on mean minus one standard deviation)
- *H*, *M* and *L* values in accordance with BS EN ISO 4869-2:1995
- *SNR* value in accordance with BS EN ISO 4869-2: 1995.

The *H*, *M*, *L* and *SNR* values are derived from the mean and standard deviation attenuation values, and are explained below.

The first requirement of a hearing protector is that it should provide sufficient noise attenuation in order to reduce the noise at the wearer's ear to a level which will prevent risk of noise induced hearing loss. This obviously depends on the levels and frequency spectrum of the noise.

The sound pressure levels at the ear when hearing protection is worn may be estimated using a number of different methods. The principal three methods for passive hearing protectors are defined in BS EN 4869-2: 1995. They are the octave band method, the HML method and the SNR method.

The first method requires data about the noise level to which the unprotected ear is exposed in octave bands. The two other methods do not require octave band data; the HML method requires both the overall A-weighted and C-weighted noise levels and the SNR method requires only the C-weighted value.

All methods will give similar predictions of sound levels at the ear for general industrial and occupational noise sources. The HML and SNR methods become less accurate when compared with the octave band method where the noise is dominated by noise at single frequencies, particularly where these are at low frequencies.

The octave band (OB) method

The performance of ear protectors is specified as a mean attenuation, measured in each of seven octave bands (from 125 Hz to 8 kHz), according to a test method specified in BS EN 24869-1, according to which the hearing thresholds of test subjects are measured with and without hearing protectors, the attenuation being the difference between the two measurements. The attenuation obtained from individual test subjects will be different, because of differences in quality of fit, and the test specifies that the standard deviation from the mean shall also be specified. If the mean value of the attenuation were to be used in calculations of noise exposure then half the population of wearers would receive a lower value of protection than predicted. Therefore the method given in BS EN 24869-2 requires that an assumed protection value of attenuation of the mean value minus the standard deviation be used in such calculations of noise exposure.

Assuming that the spread of attenuation values follows Gaussian statistics, 68% of the wearing population will receive the assumed level of protection. If a value of mean minus two standard deviations were to be used then 95% of the wearing population would receive the calculated level of protection.

In order to calculate the amount of protection afforded in any particular application it will be necessary to measure the level of noise at the wearer's ear, in octave bands. The assumed protection provided by the protector in each band is then subtracted to obtain the band level at the 'protected' ear. These levels are then A-weighted, and combined, to give the level at the protected ear, in dBA.

The HML method

HML stands for high, medium and low frequency, and the manufacturers provide three different performance figures to be used, depending on the frequency spectrum of the noise. The noise level is measured using both the A and C frequency weighting scales, and the difference between the two readings ($L_A - L_C$) is indicative of the type of spectrum. The *H*, *M* and *L* values are the predicted noise level reduction (PNR) of the hearing protector for noises with values of ($L_A - L_C$) equal to −2 dB, 2 dB and 10 dB respectively. Where the difference ($L_A - L_C$) is different from one of these three values the value of PNR is obtained by interpolation, i.e. by using one of the following formulae:

If ($L_A - L_C$) > 2 dB

$$\text{PNR} = M - [\{(M - L)/8\}(L_A - L_C - 2)]$$

Otherwise:

$$\text{PNR} = M - [\{(H - M)/4\}(L_A - L_C - 2)]$$

The effective A-weighted sound pressure level at the protected ear, L'_A, is then given by:

$$L'_A = L_A - \text{PNR}$$

The SNR method

SNR stands for single number rating and the manufacturers provide an SNR value for each type of hearing protector. The method requires measurement of the C-weighted sound pressure level L_C.

The effective A-weighted sound pressure level at the protected ear, L'_A, is given by subtracting the SNR value for the protector from the C-weighted sound pressure level L_C:

$$L'_A = L_C - \text{SNR}$$

Example 9.3

The sound pressure level close to a machine in a printing works is 104 dBA and the octave band spectrum of the noise is given in the table below. Using the octave band method, the HML method and the SNR method calculate the assumed protected level at the ear of an employee working near to the machine who is wearing ear protectors having the octave band performance given in the table below, and with $H = 25$ dB, $M = 17$ dB, $L = 15$ dB and SNR = 21dB.

Octave band method

The assumed protection, of column 5 is found by subtracting the standard deviation of column 4 from the mean attenuation of column 3. The A-weighted protected octave band level of column 8 is obtained by subtracting the assumed protection value of column 5 from the band level of column 2, to give column 6, and then adding the A-weighting value of column 7. The overall protected level is obtained by combining the levels in column 8, using the rules of decibel arithmetic (see Chapter 1), to give an assumed protected level of 84 dBA.

Thus for this particular application the overall protection provided, in reducing the level at the ear from 104 dBA to 84 dBA, is 20 dBA. If an assumed protection of the mean minus two standard deviations had been used then the predicted level at the ear would be 88 dBA, i.e. 95% of wearers would be exposed to 88 dBA or below.

HML method

$L_A = 104$ dB $L_C = 104$ dB

Therefore $L_A - L_C = 0$

and

$$\text{PNR} = M - [\{(H - M)/4\}(L_A - L_C - 2)]$$
$$= 17 - [\{(25 - 17)/4\}(0 - 2)]$$
$$= 17 - (8/4)(-2) = 21 \text{ dB}$$

and

$$L'_A = L_A - PNR = 104 - 21 = 83 \text{ dB}$$

SNR method

$$L'_A = L_C - \text{SNR} = 104 - 21 = 83 \text{ dB}$$

The optimum range for the sound pressure level at the protected ear (L'_A)

Ideally the hearing protection should reduce the sound pressure level at the ear to well below 85 dBA and preferably in the range 75 to 80 dBA.

However, it must be remembered that it is the exposure level, which depends upon duration as well as sound level, which is important, and so higher levels may be acceptable for shorter periods. The Control of Noise at Work Regulations require that the protectors should provide sufficient attenuation to reduce exposure levels to

(1)	(2)	(3)	(4)	(5)	(6)	(7)	(8)
OB (Hz)	OB SPL (dB)	Mean attenuation (dB)	Standard deviation (dB)	Assumed protection (dB)	OB APL (dB)	OB A-wtg (dB)	A-wtd OB APL (dB)
63	87	20	4	16	71	−26	45
125	90	19	3	16	74	−16	58
250	95	18	5	13	82	−9	73
500	90	18	6	12	78	−3	75
1000	99	21	4	17	82	0	82
2000	100	31	4	27	73	+1	74
4000	95	37	4	33	63	+1	63
8000	94	38	6	32	62	−1	61

below the upper exposure action value (85 dBA $L_{EP,d}$), although the lower exposure action value level would be a better, i.e. safer, objective.

Accounting for 'real-world' protection

Research has shown that in real use the protection provided can be less than predicted by manufacturer's data. To give a realistic estimate, allowing for the imperfect fitting and condition of hearing protectors in the working environment, it is recommended that a real-world factor of 4 dB is applied, i.e. that 4 dB is added to the estimated sound levels at the protected ear calculated by any of the above three methods.

Dual protection

Sometimes employees may be exposed to very high noise levels, and the above calculations may show that neither plugs or muffs will reduce the noise level at the ear to a low enough level. This problem is likely where the daily noise exposure is above 110 dB or the peak sound pressure level exceeds 150 dB, especially if there is substantial noise at frequencies less than about 500 Hz. If both muffs and plugs are worn together then this may provide an increased amount of protection, but only by a few extra decibels, and it is not easy to predict the combined performance. The amount of protection will depend on the particular earmuff and plug combination. In general, the most useful combination is a high performance plug with a moderate performance earmuff (a high performance earmuff adds a little extra protection but is likely to be less comfortable).

If dual protection is used, test data should be obtained for the particular combination of plug and earmuff (and helmet, if used) if available. In practice, the increase in attenuation to be expected from dual protection will be no more than 6 dB over that of the better of the individual protectors.

Under these circumstances it is strongly recommended that alternative working procedures are found, and that exposure times at these levels are reduced to an absolute minimum, and are strictly monitored, even with ear protectors being worn.

Overprotection

Hearing protectors that reduce the level at the ear to below 70 dB should be avoided, since this overprotection may cause difficulties with communication and hearing warning signals. Users may become isolated from their environment, leading to safety risks, and generally may have a tendency to remove the hearing protection and therefore risk damage to their hearing.

Selection of hearing protectors

The noise level of 104 dBA in the above example is very high, so that without hearing protectors the lower exposure action value would be reached within a few minutes (see Chapter 4). In such a case it is certainly necessary to use one of the three methods described above in order to check that the protectors being used provide sufficient protection. At lower levels of exposure, say 92 or 93 dBA for example, any reputable plug or muff with the CE mark indicating compliance with the relevant part of BS EN 352 should provide ample protection if in good condition and fitted properly. In these situations emphasis should be placed on providing a selection of types for the wearer to choose from, in order to reduce discomfort and inconvenience to a minimum, and therefore to maximize the chances that the protectors will be worn for 100% of the time for which they should be worn.

Apart from performance, other factors influencing selection are considerations of comfort, convenience and hygiene. Earmuffs are easier to fit than earplugs, and they can be worn by almost everyone, whereas some people may have a medical condition which prevents them from wearing plugs. Earmuffs generally, but not invariably, provide higher levels of attenuation than plugs (depending upon the particular types) and this attenuation is less variable because less skill is required in fitting them properly. Earmuffs can be inconvenient, uncomfortable, hot and heavy, particularly when they have to be worn for several hours in warm and restricted spaces, and plugs will be lighter and more comfortable in such circumstances. Good hygiene is essential when earplugs are used, and hands should always be washed before inserting or removing the plugs into the ear canal, and earmuffs are therefore preferable in dirty areas where adequate washing facilities are not available, or where they have to be frequently taken on and off because the wearer is moving in and out of ear protection zones. People who wear glasses or earrings, or who have long hair, may receive a reduced amount of protection from earmuffs if the quality of fit of the seals of the muffs to the head is reduced, and there can be problems if other safety equipment has to be worn such as safety glasses, face masks or other respiratory equipment, although special types of muffs are available to deal with some of these situations, for example muffs attached to safety helmets.

Difficulties and limitations of hearing protectors

The difficulties are that they may not be worn, and even if they are worn they may not be giving the estimated degree of protection, either because they have not been fitted correctly or because they are defective in some way.

Many employees understandably do not like having to use ear protectors particularly if they have to be worn for several hours each day. Apart from the discomfort and inconvenience several other reasons for concern are often given, including that they impair the wearer's ability to detect changes in the noise emitted by the machines they are operating, or to hear alarm signals, or communications from fellow workers or from the public address system. In fact the protectors reduce both the level of the background noise and the level of the wanted sound signal, and it has been shown that for wearers with good hearing their detection of these signals is undiminished, or even increased, but there are problems for wearers who already have hearing impairment. Sometimes those already suffering from noise induced hearing loss put forward objections to wearing ear protectors, either on the basis that they are not needed because they have 'got used to the noise', or alternatively that 'it is too late' since they already have hearing impairment. The guiding principle in such cases should be to safeguard all the more carefully the hearing which remains. It is important to counter and overcome all such concerns with a programme of training which explains the need to wear ear protectors and the consequences of not doing so.

The importance of the need for the protection to be worn for 100% of the time exposed to high levels of noise cannot be overestimated. No matter how good the assumed protection provided, and this can be in the region of 20 of 30 dB, it will be reduced to only 3 dB if they are worn for only 50% of the time, and even if they are worn for 90% of the time the degree of protection can be only 10 dB. This is particularly important when the noise levels are very high because the exposure level increases rapidly even in a short time. At a level of 102 dBA for example, a personal daily noise exposure level of 85 dBA (the upper exposure action level of the Control of Noise at Work Regulations) is reached after 30 minutes, and this falls to 15 minutes at 105 dBA and 7.5 minutes at 108 dBA. In these circumstances it is obvious that failure to wear protectors even for a few minutes will be harmful. It is important to explain and emphasize this degree of urgency in training sessions.

The attenuation figures provided by manufacturers are obtained from tests under standardized measurement conditions. They represent the maximum attenuation likely to be met under laboratory conditions, when protectors are new and always fitted properly. Much lower levels of protection are obtained if the protectors are not fitted properly. This is particularly important with plugs, which must always be inserted exactly according to manufacturer's instructions. Training is needed to demonstrate the proper procedure and to ensure that each employee is able to fit the plugs properly, and even then they may work loose in time. Even with earmuffs, which are much easier to fit properly, it is necessary to sweep back the hair, ensure that the cups are completely surrounding the ears, and adjust headband tension to ensure a good seal with the side of the head. It has been shown that a few minutes spent demonstrating and explaining these points leads to higher levels of attenuation for the user.

After prolonged use plastic plugs may become inflexible, the seals of muffs may become worn or hard and therefore ineffective, headbands may lose their elasticity or the cups may become damaged. In all cases the degree of protection will be drastically reduced until the protectors are replaced. Training must therefore cover inspection of protectors to alert wearers to these defects.

Thus it can be seen that if a company's hearing protection policy relies heavily on the use of ear protectors, particularly if there are high levels of noise, then for all of the reasons described above there will be some risk that employees may not be receiving the amount of protection predicted on the basis of assumed protection calculations. It is for this reason that health surveillance in the form of regular audiometric monitoring is used as a check on the effectiveness of the hearing protection policy, as required by the Control of Noise at Work Regulations.

More detailed information about hearing protectors is available in Part 5 of the Health and Safety Executive Guidance to the Regulations, *Controlling Noise at Work.*

9.16 The Control of Noise at Work Regulations 2005: a summary

These regulations follow the requirements of EC Directive 10/2003. They replace the Noise at Work Regulations 1989 which had been in force since 1990.

Action and limit values

- Lower exposure action value:
  - personal daily (or weekly) noise exposure level of 80 dBA
  - peak sound pressure level of 135 dBC

- Upper exposure action value:
  - personal daily (or weekly) noise exposure level of 85 dBA
  - peak sound pressure of 137 dBC
- Exposure limit value:
  - personal daily (or weekly) noise exposure level of 87 dBA
  - peak sound pressure level of 140 dBC.

Duties of employers

- Assess risks to employees.
- Take action to reduce noise exposure.
- Provide employees with hearing protection.
- Make sure legal limits on noise exposure are not exceeded.
- Provide employees with information, instruction and training.
- Carry out health surveillance.

Duties of employees

- To comply with measures put in place by employers to reduce noise exposures.
- To inform management if any such measures (including hearing protection) are in need of maintenance or replacement.
- To wear hearing protection when noise exposure levels are above the upper exposure action values.

Comparison with the 1989 Noise at Work Regulations

The main differences are:

- The two action values for daily noise exposure have been reduced by 5 dB to 85 and 80 dBA.
- There are now two action values for peak noise at 135 dBC and 137 dBC.
- There are new exposure limit values of 87 dB (for daily exposure) and 140 dB (for peak noise) which must not be exceeded, but which may take into account the effect of wearing hearing protection.
- There is a specific requirement to provide health surveillance where there is a risk to health.
- More emphasis on risk, selective use of hearing protectors, action plans, re-emphasis on priority of noise control over hearing protection.

What does 80 dBA sound like?

Significantly noisier than the sounds of everyday life such as:

- busy street
- typical vacuum cleaner
- crowded restaurant
- able to hold a conversation but noise is intrusive.

Simple test to see if noise risk assessment is needed

Test	Probable noise level	A risk assessment will be needed if the noise level is like this for more than:
Noise is intrusive but normal conversation is possible	80 dB	6 hours
Have to shout to talk to someone 2 m away	85 dB	2 hours
Have to shout to talk to someone 1 m away	90 dB	45 minutes

What noise measurements are required?

To determine personal noise exposures:

- for all noise: A-weighted noise average level, $L_{Aeq,T}$
- for impulsive and impact noise: C-weighted peak sound pressure level, L_{Cpeak}.

For selecting ear protectors and noise control measures:

- octave band noise levels, C-weighted noise average levels, $L_{Ceq,T}$.

Noise measurements: sampling strategies

- Sample the activity
- Sample the person (using dosemeters)
- Sample the workplace
- Samples should be representative
- Recognize and allow for uncertainties in your estimate of exposure
- Has the worst case situation been adequately taken into account?

Assessing exposure to noise

- Measurement of noise levels (for each activity)
- Estimation of duration of exposure (for each activity)
- Determine daily noise exposure level
- Compare with action values and exposure limits.

Skills and knowledge needed to carry out a noise exposure assessment

To carry out a competent assessment requires skills and knowledge about:

- the workplace – work patterns including variations
- noise and how it behaves, and how it is measured
- the Regulations.

Examples of processes producing high peak noise levels

- General mechanical handling (bangs and clatters)
- Impact tools
- Hammering (general metal working)
- Punch presses
- Jolt squeezing moulding machine
- Drop hammer
- Drop forge
- Nail gun/nailer
- Explosives, fireworks, gunfire

Managing noise risks

- Stage 1: Is there a risk due to noise?
- Stage 2: Who is at risk and how?
- Stage 3: Evaluate risks and develop a plan to control them.
- Stage 4: Record the findings of the risk assessment and take appropriate actions.
- Stage 5: Review the risk assessment.

Employees who may be particularly vulnerable to risks from noise exposure

- Those with a pre-existing hearing condition, or with a family history of deafness
- Pregnant women
- Young people
- Those also exposed to high levels of hand-transmitted vibration
- Those exposed at work to certain ototoxic substances, particularly solvents.

What is a suitable and sufficient assessment?

- Carried out by a competent person
- Based on advice and information from competent sources
- Identifies where risk occurs and who is affected
- Contains reliable estimates of employees noise exposures and compares these with exposure action and limit values
- Identifies measures needed to eliminate risks and exposures or reduces them as far as is reasonably practicable, (i.e. to comply with the Regulations)
- Identifies employees requiring health surveillance and those at particular risk.

Recording the risk assessment

The record should contain:

- the scope of the assessment: workplaces, areas, jobs, people and activities included
- the date(s) that the assessment was made
- daily personal exposures (and peak exposure levels) of employees, or groups of employees included
- the information used to determine noise exposures
- details of noise measurements and procedures, including name of person responsible
- any further information used to evaluate risks
- the name of the person who made the risk assessment
- action plan to control noise risks.

The action plan

The action plan should contain:

- what has been done to tackle immediate risks
- actions being considered regarding general duty to reduce risks
- plans to develop a programme of noise reduction measures
- arrangements for provision of hearing protection and hearing protection zones
- arrangements for providing information, instruction and training
- arrangements for providing health surveillance
- realistic timescales for the above to be carried out
- people or post holders to be responsible for the above various tasks
- person with overall responsibility to ensure that the plan is competently carried out.

When to review the risk assessment

- If work patterns, processes, procedures, machinery have changed significantly
- If new technologies (new ways of working, or noise control techniques) have become available
- If new noise control measures have been introduced
- If health surveillance shows the need for review
- If measures previously unjustifiable become reasonably practicable
- If two years have elapsed since the last assessment.

Information, instruction and training

What should be included?

- Likely noise exposure levels and risk to hearing
- What is being done to control risks and exposures
- Where and how employees can obtain hearing protection
- How to report defects in hearing protection and noise control equipment
- Duties of employees under the Regulations
- What employees should do to minimize risk, e.g. how to use, store and maintain hearing protectors and other noise control equipment
- What health surveillance system is being provided
- How to detect symptoms of hearing loss and how and to whom to report them.

Reduction of noise exposure by organizational control

- Plan and organise the work to reduce noise exposure
- Job design
- Job rotation
- Different ways of working
  - change of process
  - change of machine
  - change of activity
- Workplace design to minimize noise
- Purchasing policy for new quieter tools and machinery in future.

Reduction of noise exposure by technical/ engineering control

- Maximize distance between source and receiver
- Ensure good maintenance
- Minimize airflow (turbulence) noise
- Avoid impacts
- Machine enclosures
- Screens and barriers
- Noise refuges
- Standard noise control methods:
  - damping
  - isolation
  - use of silencers
  - active noise control.

List of Regulations

- Regulation 1: Citation and commencement
- Regulation 2: Interpretation
- Regulation 3: Application
- Regulation 4: Exposure limits and action values
- Regulation 5: Assessment of the risk to health and safety created by exposure to noise in the workplace
- Regulation 6: Elimination or control of exposure to noise at the workplace
- Regulation 7: Hearing protection
- Regulation 8: Maintenance and use of equipment
- Regulation 9: Health surveillance
- Regulation 10: Information, instruction and training
- Regulation 11: Exemption certificates from hearing protection
- Regulation 12: Exemption certificates for emergency services
- Regulation 13: Exemptions relating to the Ministry of Defence
- Regulation 14: Extension outside Great Britain
- Regulation 15: Revocations, amendments and savings

More detailed information about the Regulations is available in the Health and Safety Executive Guidance to the Regulations, *Controlling Noise at Work*, issued in 2005 (HSE Books).

Appendix 9.1 Derivation of enclosure insertion loss formula

L_{BEFORE}

Sound power level of source L_{WS}

Sound absorption in room containing the enclosure A_R

Therefore

$$L_{BEFORE} = L_{WS} + 10\log(4/A_R)$$
$$= L_{WS} - 10\log(A_R) + 6$$

L_{AFTER}

Step 1

Sound level inside enclosure.

Sound absorption inside enclosure A_{ENCL}

Sound level inside enclosure:

$$L_{INSIDE} = L_{WS} + 10\log(4/A_{ENCL})$$
$$= L_{WS} - 10\log(A_{ENCL}) + 6$$

Step 2

Sound transmission through enclosure into the room containing the enclosure.

Sound reduction index of enclosure wall R_{ENCL}

Sound level outside enclosure, close to enclosure wall:

$$L_{OUTSIDE} = L_{INSIDE} - R_{ENCL} - 6$$

Substituting the expression for L_{INSIDE}:

$$L_{OUTSIDE} = (L_{WS} - 10\log(A_{ENCL}) + 6) - R_{ENCL} - 6$$
$$= L_{WS} - 10\log(A_{ENCL}) - R_{ENCL}$$

Area of enclosure walls and roof = S_{ENCL}

Sound power level radiated by enclosure walls into room containing the enclosure:

$$L_{WofENCL} = L_{OUTSIDE} + 10\log(S_{ENCL})$$
$$= (L_{WS} - 10\log(A_{ENCL}) - R_{ENCL}) + 10\log(S_{ENCL})$$

Step 3

Sound level inside room containing the enclosure = L_{AFTER}

$$L_{AFTER} = L_{WofENCL} + 10\log(4/A_R)$$
$$= L_{WofENCL} - 10\log(A_R) + 6$$

$$L_{AFTER} = (L_{WS} - 10\log(A_{ENCL}) - R_{ENCL}) + 10\log(S_{ENCL}) - 10\log(A_R) + 6$$

Step 4

Insertion loss IL

$$IL = L_{BEFORE} - L_{AFTER}$$

$$IL = (L_{WS} - 10\log(A_R) + 6) - [(L_{WS} - 10\log(A_{ENCL}) - R_{ENCL}) + 10\log(S_{ENCL}) - 10\log(A_R) + 6]$$

$$IL = R_{ENCL} - 10\log(S_{ENCL}) + 10\log(A_{ENCL})$$

Questions

1 Briefly distinguish between: (i) 'noise control at source' and (ii) 'noise control in the transmission path'. Illustrate your answer with an example of each. Which is preferable, and why?

2 Explain why the performance of any noise control device, such as an acoustic enclosure, silencer, barrier etc., should always be specified in octave bands rather than as a single dBA value.

3 Explain carefully the difference between the following methods used to reduce noise: (i) sound insulation, (ii) sound absorption, (iii) vibration isolation. In each case describe the type of materials used to provide effective noise control, and define the parameters used to specify their performance.

4 It is proposed to site a gas turbine-electricity generator plant on the boundary of a residential area (see the figure below).

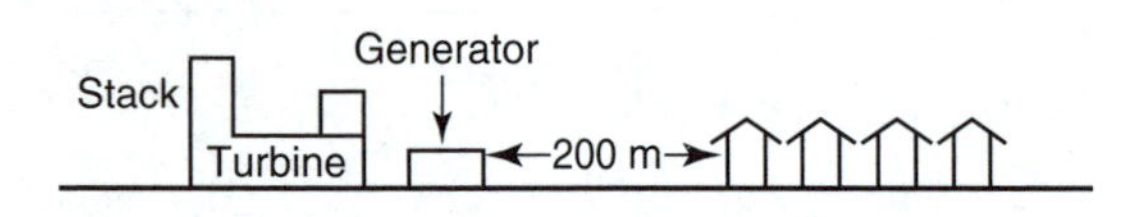

Describe in detail the data calculation procedures and measurements which would be necessary to assess the following:

(i) the potential noise level generated by the plant in the residential area

(ii) the degree, and likely forms, of noise control necessary, and

(iii) the acceptability of the completed installation.

(IOA 1981)

5 The occupants of a first-floor flat are complaining of the noise caused by the ventilation and extract system from the restaurant below. The restaurant kitchen is a flat-roofed extension built onto the restaurant. The extract fan unit is situated on the flat roof with the extract ductwork running across the roof and up the external wall of the flat (see the figure below). There is a window in the flat which overlooks the flat roof and the fan unit, and the ductwork passes close to this window and terminates at roof level.

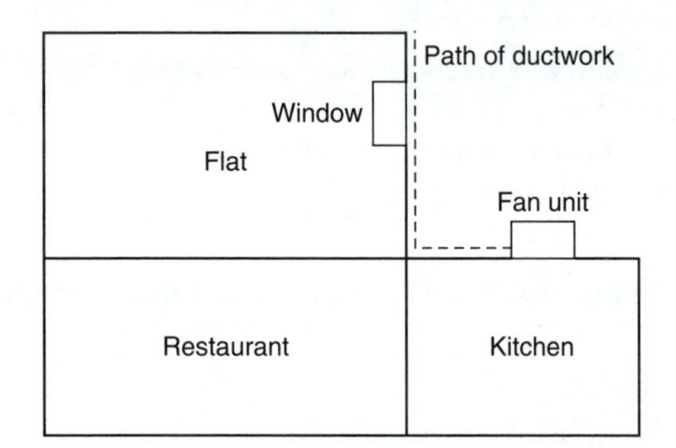

Describe the various paths by which noise caused by the extract system arrives in the flat, and suggest possible noise control remedies in each case.

6 The manager of a factory which contains a large number of different sorts of machinery is aware that there are noise problems in the factory. He receives advertising literature from a firm producing sound absorbing materials, showing examples of their use in reducing factory noise. He asks you to assess and advise on the suitability of this form of noise control for his particular factory.

Describe how you would carry out the assessment. Say what information you would need to obtain, what measurements you would make, and what calculations, if any, you would perform. What alternative noise control treatments might you recommend?

7 Discuss the possible methods for reducing the noise from a vibrating machine panel. Explain why stiffening such panels is not always effective.

8 What are the main mechanisms of noise generation by impacts. How may impact noise be reduced?

9 Explain how the noise control principles of absorption, insulation and isolation are demonstrated in the design of an effective acoustic enclosure. Discuss the advantages and disadvantages of using acoustic enclosures to reduce machinery noise.

10 Distinguish between the techniques of sound absorption, sound insulation and vibration isolation used in noise control, with reference to the design of earmuffs, acoustic enclosures and double-leaf partitions.

11 (a) Explain precisely what is meant by the damping of a flexible structure, distinguishing it clearly from its stiffness. Mention some of the mechanisms of damping which commonly contribute to damping of a structure.
(b) Under what conditions can the application of damping treatment to a structure, which is subject to continuous vibratory or impact forces, constitute an effective means of noise and vibration control?
(c) Describe the various forms of damping treatment available to the noise control engineer, and discuss their practical limitations. (IOA)

12 With reference to duct-borne noise in ventilation systems, discuss the contribution to sound attenuation made by:
(i) rectangular straight ducts (sheet metal);
(ii) circular straight ducts (sheet metal);
(iii) rectangular straight ducts, internally lagged;
(iv) elbows;
(v) elbow with guide vanes; and
(vi) elbow internally lagged. (IOA (part question))

13 A machine used to cut large sheets of material using an abrasive cutting wheel is very noisy. The sheets are fed into the machine by the machine operator. The machine holds the material in place during the cutting process with pneumatically operated clamps. When the cutting is complete the clamps are released, with the compressed air released directly to atmosphere, and the cut components fall into a sheet metal container via a mild steel chute. There is a dust extract system to remove airborne dust but the machine also needs regular cleaning to remove deposits of dust which collect inside the machine.

(a) Describe the various noise sources and noise producing mechanisms involved in the operation of this machine, and in each case, suggest a suitable noise control treatment.
(b) It is proposed to fit an enclosure around this machine in order to reduce the noise exposure of the machine operator. Discuss some of the practical difficulties and limitations associated with this noise control solution. (IOA)

14 It is suspected that the most important source of noise in a workshop arises from radiation from a particular sheet steel panel attached to a particular machine.

Explain what diagnostic tests you would carry out to confirm or refute this suspicion. Assuming that your tests did confirm the importance of noise radiated by the machine panel, describe what methods are available in principle to reduce the noise.

15 Briefly describe the mechanisms of noise generation/transmission in the following situations, and describe the conventional methods of noise control for each case:

(a) beating of steel panels using a hand-held hammer
(b) a compressed air jet used for cleaning machine parts
(c) a rotating shaft attached by bearings to a steel box
(d) noise entering a ventilated room via an air conditioning grille in the ceiling
(e) noise from a train in an underground tunnel to houses above
(f) impact of stones falling from a conveyor belt onto a welded sheet metal hopper
(g) the release of compressed air from a pneumatic cylinder
(h) airborne noise from a fan
(i) a small vibrating electric motor bolted to a steel baseplate. (IOA (adapted))

16 An office is located directly above a plant room in a steel-framed building with concrete walls and floors. The plant room contains several large pumps, which require air cooling which is supplied via a duct system from outside the plant room. The pumps are all attached rigidly to the floor of the plant room, and the air supply ductwork and the various pipes connected to the pumps are rigidly attached to the walls and ceiling of the plant room. The noise level in the offices is unacceptably high, and needs to be reduced.

(a) Describe the different routes by which noise could be transmitted from the plant room to the office.
(b) Explain carefully how the noise transmission by each route might be reduced.
(c) Describe how you would decide which of the proposed solutions to implement first, indicating the measurement you would carry out and the information you would require, and what calculations you would perform in order to reach your decision.

(IOA)

17 Noise from a motor located in a plant room on the flat roof of an office block is causing disturbance in the room immediately below. Describe the investigations which you would carry out in order to find out whether the problem is caused by airborne or structure-borne sound transmission, and describe noise control measures which would be appropriate in each case.

18 A pump running at 1800 rpm is located in a plant room above a conference room in which a pronounced low frequency hum causes disturbance. The sound pressure level in the conference room (L_{p2}) is measured and is found to exceed the required NR 30 (see the table below). The room measures 5 m × 5 m × 2.4 m high and has a reverberation time of 0.5 s (ideal for speech intelligibility). The reverberant sound pressure level inside the plant room (L_{p1}) and the sound reduction index of the party floor (SRI) are given in the table below.

(a) Calculate the sound pressure level in the conference room due to breakout of noise from the plant room floor. Diagnose the cause of the problem by comparing this level to the measured level (L_{p2}).
(b) Recommend a means of noise control.
(c) If it is desired to achieve NR 30, are the necessary noise reductions practicable to achieve? Quantify your answers.

Frequency (Hz)	31.5	63	125	250	500	1000
Location						
Plant room L_{p1}	100	90	88	84	81	80
Floor SRI	33	39	41	45	49	57
Measured L_{p2}	85	55	48	40	33	24
NR 30	76	59	48	40	34	30

(IOA)

19 A fan is located in a plant room next door to a conference room (see the figure below), ducting air into the conference room via a 10 m long duct. The plant room and the conference room share a 10 m^2 party wall.

(a) Using the information in the table, calculate the reverberant sound pressure level in the conference room for:
   (i) duct-borne noise only
   (ii) breakout noise through the plant room wall only
   (iii) both contributions combined.
(b) What insertion loss would be required from a silencer in order to achieve NR 27 in the conference room?

Assume that:

- half the sound power from the fan goes into the duct and half into the plant room
- flow generated noise is insignificant compared with the direct fan noise, and
- the average absorption coefficient in the conference room is much less than 1.

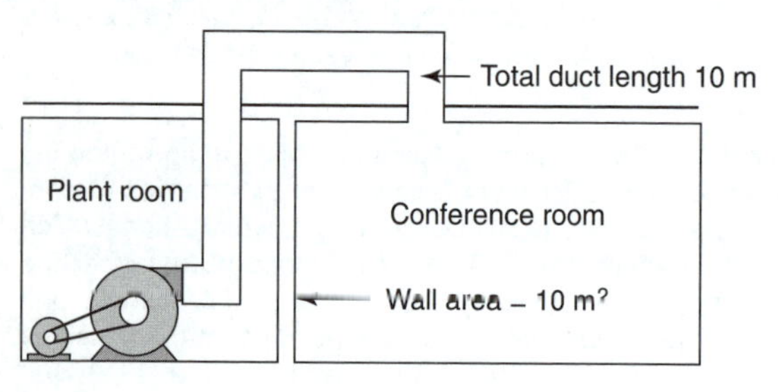

Octave band centre frequency (Hz)	125	250	500	1000
Fan total L_W	88	86	85	81
Attenuation of a bend (dB)	0	1	8	6
Duct attenuation (dB per m)	0.2	0.15	0.1	0.07
End loss (dB)	6	2	0	0
Wall sound reduction Index (dB)	38	42	48	52
Room constant (conference room) (m^2)	24	30	36	36
Room constant (plant room) (m^2)	4	4	5	5
NR 27 octave band L_p (dB)	44	37	31	27

(IOA 2008)

20 An air cooler with a 5 m diameter, four-bladed fan is mounted 25 m from the boundary fence of a factory site. An electric motor running at 1000 rpm drives the fan via a 3:1 reduction toothed belt pulley system.

(a) Fan noise sound power level, L_W can be predicted by: $L_W = 60\log V_t + 10\log N$ where: V_t = fan blade tip speed N = number of blades.
(b) Using this, determine the sound pressure level due to the fan at the boundary.
(c) Calculate the blade passing frequency.
(d) The local planning authority requires the factory to meet a boundary sound pressure level of 65 dBA. How could this result be achieved?
(e) What modifications could be made to the fan to reduce the noise without reducing the cooling performance? (IOA 2007)

21 (a) Briefly describe two main sources of natural attenuation in a ducted ventilation system.
(b) A ducted ventilation system, containing a fan and two attenuators, is as shown in the sketch below.

Fan L_W = 78 dB

Attenuator A insertion loss = 25 dB
Attenuator A airflow regenerated L_W = 55 dB
Attenuator B insertion loss = 15 dB

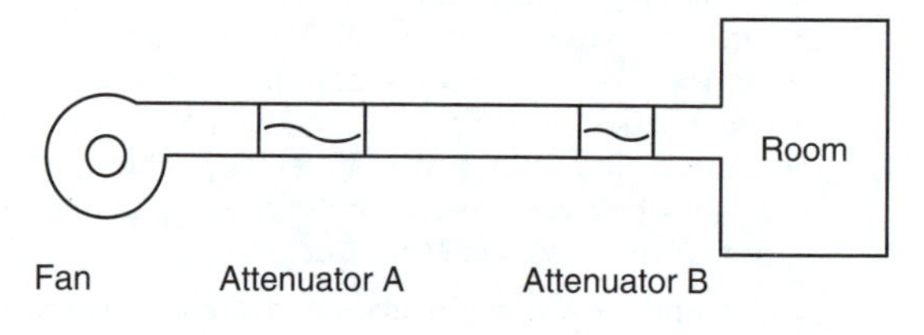

(c) The maximum noise level in the rectangular room (10 m × 6 m × 4 m) at a distance of 1.5 m from the duct opening is 35 dB. Ignoring duct attenuation and end effects, calculate the maximum permissible airflow regenerated noise level for attenuator B. You may assume the following:

Room reverberation time = 0.9 seconds
Directivity of duct opening $Q = 2$

(d) The attenuator proposed by the contractor for position B exceeds the permissible airflow regenerated noise level. Briefly explain how this could be reduced. (IOA)

22 The following equation is often used to estimate the insertion loss of an enclosure used in a reverberant sound field:

$$IL = R - 10\log(1/\alpha) \text{ dB}$$

where R is the sound reduction index of the enclosure and α is the average sound absorption coefficient of the inside surfaces.

(a) Explain why the average absorption, α, is important in determining the transmission performance of the enclosure.

(b) Explain why this equation would be inaccurate if:
   (i) the levels were measured close to the enclosure sides;
   (ii) the enclosure were placed near to a reflecting wall.

(c) An air compressor operates in a factory and generates a reverberant sound pressure level of 105 dB in the 1 kHz octave band. As a noise control measure, it is planned to cover the compressor with an enclosure of 1.5 m × 2.5 m × 1.5 m high.

The enclosure is to be lined with 50 mm thick mineral fibre absorber having an absorption coefficient of 0.8 at 1 kHz. The floor of the enclosure is concrete having an absorption coefficient of 0.04 at 1 kHz.

Calculate the sound reduction index of the enclosure necessary to reduce the reverberant level to below 85 dB. (IOA)

Chapter 10 The law relating to noise

This chapter is a general guide to noise and the law, so many topics have been dealt with in outline only. Some have been dealt with more fully, e.g. the Environmental Protection Act 1990, the Noise and Statutory Nuisance Act 1993, the Noise Act 1996 and the Control of Pollution Act 1974 because of their direct applicability to noise control.

Technological and scientific knowledge today indicates the undesirability of unwanted sound, i.e. noise. But this knowledge has been gained over a relatively short period of history, for although it was recognized long ago that a noisy workplace could cause deafness or discomfort, it was another matter to prove it. That is why the law relating to noise has grown up bit by bit, protecting different interests with different remedies.

This chapter is set out in a series of sections. Sometimes more than one of the topics will be applicable, and wherever possible this has been indicated. But treating the subject in only one chapter has required considerable brevity. In order to help the reader with no legal knowledge, the following outline of English law may be useful.

10.1 English law

The law found in the following pages relates only to the law laid down in England and Wales. Within the political state of the United Kingdom of Great Britain and Northern Ireland and its dependencies, there are a number of different legal systems with different histories. Scotland, Northern Ireland, the Isle of Man and the Channel Islands all have their own. Often, however, Acts of Parliament legislate for all the United Kingdom. But, because each system has its own methods of putting law into practice, sometimes separate Acts have to be passed to take account of these differences. In 1998 additional powers to make law were devolved to Scotland, Northern Ireland and, to a more limited extent, to Wales.

Categories of English law

There are a number of categories of law within the English legal system. Firstly, law can be divided into two types, public and private. Public law is for the benefit of society in general and is designed to prevent the breakdown of law and order. For this reason, any such rules, if broken, are punished by agents of the state. The obvious example is criminal law. If someone murders or steals, he is first apprehended by the police and then brought before the criminal courts and prosecuted by the Crown Prosecution Service. He or she can be punished in many different ways such as imprisonment and fines, disqualification in the case of driving offences, and community service and anti-social behaviour orders for certain examples of anti-social behaviour such as vandalism and public disorder. In today's sophisticated society many types of bad behaviour are dealt with by the criminal law, even if this does not appear to be wholly appropriate at first. For example, driving offences are dealt with in the criminal courts. Breaches of the Environmental Protection Act 1990, the Noise and Statutory Nuisance Act 1993, the Noise Act 1996, the Control of Pollution Act 1974, the Health and Safety at Work, etc., Act 1974 and local government by-laws, are also dealt with by the criminal courts. In the first four examples the law is enforced by environmental health officers (EHOs); the fifth example is enforced by EHOs and the Health and Safety Inspectorate instead of the Crown Prosecution Service. However, serious breaches of safety legislation will be dealt with in the Crown Court and be prosecuted by the Crown Prosecution Service.

Private or civil law is for the benefit of individuals and thus only protects individual interests. The person who has suffered from noise-induced illness caused by an employer's disregard for the Health and Safety at Work Act and the Control of Noise at Work Regulations 2005, cares little for the prosecution of the employer. He or she would prefer to recover compensation in the form of damages by suing the employer in the civil courts for the tort of negligence or the tort of breach of statutory duty (see pages 302 and 308). Private law covers many different types of law. For us, the most important is the law of tort. A tort is a civil wrong which is generally not bad enough to be treated as a crime. Examples include negligence, nuisance, trespass to land, person or goods, libel and slander. Some torts may also be dealt with as crimes in the criminal courts, e.g. assault and battery, which are types of trespass to the person. This is why one may hear of people taking private actions when the police have failed to prosecute for some reason. Torts particularly relevant to noise are dealt with in section 10.3. Other, less relevant, forms of private law include contract, family law, company law, land law and succession on death.

The courts

As there are different forms of law, so there are appropriate courts which are part of a hierarchical system, with the most important court at the top of the ladder, the Supreme Court of the United Kingdom, and many different magistrates' courts dealing with the least serious criminal matters at the bottom. The Supreme Court of the United Kingdom replaced the judicial function of the House of Lords with effect from 1 October 2009, ensuring for the first time a distinct separation, physically and institutionally, between the legislature and the judiciary. Decisions of the House of Lords prior to that date nevertheless continue to be significant in terms of the precedents they have set, until such time as they are overruled by future Supreme Court decisions.

Figure 10.1 shows a simplified version of the court system.

The courts most relevant to us are the magistrates' courts, which deal with minor criminal offences under the Environmental Protection Act 1990, the Control of Pollution Act 1974, the Health and Safety at Work Act 1974 and the Noise Act 1996 (which also uses fixed penalty notices). The Crown Court deals with more serious offences under those Acts.

If someone has committed a tort, then they can be sued in either the Queen's Bench Division of the High Court or the county court. Generally, claims for damages for more than £50,000 or injunctions will be dealt with in the Queen's Bench Division and for cases under £25,000 in the county court. Following amendments to the High and County Courts Jurisdiction Order 1991, S.I. 1991/724, and with effect from 6 April 2009, claims where the county courts have general jurisdiction may go to the High Court *only* if their value is more than £25,000 (High Court and County Court Jurisdiction (Amendment) Order 2009 (S.I. 2009/577) Whether such cases will go the High Court will depend on certain factors such as whether matters of public interest are involved, whether the facts, legal issues, remedies or procedures involved are particularly complex and whether transfer is likely to result in a speedier trial. Those factors could also be used to recommend transfer of an action worth more than £50,000 to the county court.

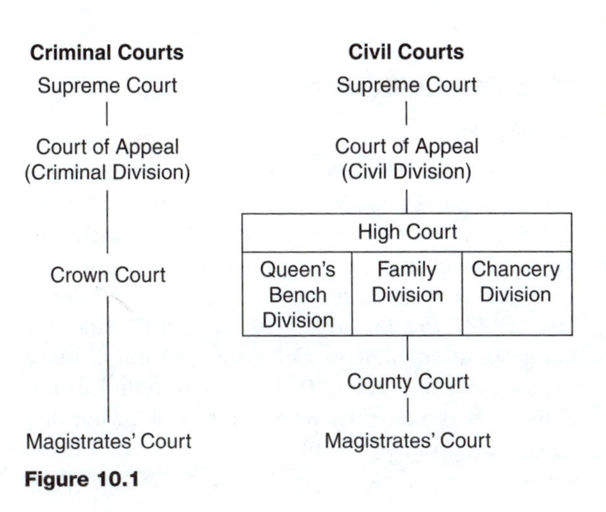

Figure 10.1

The sources of English law

Like a river, English law is created from a number of sources. The oldest source has been that of *customary law*, where people have exercised some right for centuries and eventually it has been recognized in a court of law, e.g. a right of way. Nowadays, only local customs become law and then only if they satisfy certain criteria.

More important to us is law created by *judicial precedent*. Here the judges in court cases apply existing principles of law to new situations as and when they arise. Their decisions on points of law are called precedents; if they are made or confirmed by courts high up the ladder, they bind all lower courts. Lower courts must obey the ruling on that point unless the later case can be distinguished on the facts. One of the advantages of the precedent system is that it only makes law for situations which have actually arisen, whereas the next source, *legislation*, legislates for situations which may never arise.

Legislation is law created by a body specially set up for that purpose. In England and Wales this is the Houses of Parliament. The law is primarily found in Acts of Parliament, often referred to as primary legislation. It is also known as *direct legislation*, being directly enforceable if all the rules relating to a particular aspect are contained in the Act itself. The Environmental Protection Act, Noise and Statutory Nuisance Act, Noise Act and Control of Pollution Act are good examples, as all the basic law is found within the Acts themselves. If the Act merely enables somebody else to make the rules, such as a government minister in consultation with others, then the Act is known as an *enabling Act* and the law made thereunder, as *secondary*, *delegated*, *indirect* or *subordinate* legislation. Such laws are usually called *rules*, *regulations*, orders-in-council or *by-laws*. Thus, the Health and Safety at Work Act is an enabling Act in relation to the Control of Noise at Work Regulations 2005. Government ministers make such regulations and orders by *statutory instrument*, each piece of delegated legislation having a reference starting 'S.I.' followed by the date and reference number. Thus the reference for the Control of Noise at Work Regulations 2005 is S.I. 2005 No. 1643 (often shown as S.I. 2005/1643). Acts of Parliament can also enable local authorities to make delegated legislation in the form of by-laws. A local by-law forbidding the sounding of motor-car horns outside a hospital would be made under the Local Government Act 1972, as amended (see section 10.16).

Following devolution of certain powers to the Welsh Assembly in 1998, the trend has been for recent statutory instruments to apply to England only, with Welsh versions being issued separately.

Acts of Parliament and statutory instruments can be accessed at *www.legislation.gov.uk*. Since 1999, many new Acts of Parliament come with an Explanatory Note, which is not part of the Act itself but provides a useful context, background and summary. From 2004, the Explanatory Memoranda accompanying statutory instruments can also be accessed there.

A relatively new source of law is now found in the *European Union*, see section 10.2 below.

Common law and equity

In English law we also talk of *common law* and *equity*. The common law is that law that has been and is still being created by custom and precedent (see above). It was originally described as common by the Norman kings following the conquest of 1066. In their desire to unite the differing peoples of England under one king, visiting judges were sent around on circuit throughout the kingdom, trying cases using, wherever possible, the king's law and gradually, by absorbing and eliminating undesirable local customary law, a homogeneous law was applied, common to all the land. Thus, wherever one lived, one had, in theory, the opportunity to use the common law and could avoid the corrupt local courts.

Unfortunately, because of power struggles between the Crown and the barons, the developing common law was dealt a serious blow by a number of procedural restrictions. It became very rigid and inflexible. The expressions 'going by the book', 'keeping to the letter of the law' or 'too much red tape' could all have been applied to the common law at that time. As a result, many people were unable to go to law and would have had to fall back on the local courts if a new development had not occurred. The king was regarded as the 'fountain of all justice' and as such it was possible to ask him directly for a remedy where no others existed. Travelling to see the king in the Middle Ages was not easy and sometimes it even involved going behind enemy lines abroad. Even so, many found this method of getting justice so successful that the sheer numbers overwhelmed successive kings. Gradually, this right to hear direct appeals from his subjects was passed by the king to the Lord Chancellor. He was the king's priest and confessor and gradually a new approach to the law and new remedies evolved. This new body of rules was called equity. The Lord Chancellor was more concerned with doing justice and achieving equality than going by the letter of the law. Eventually, he was given his own court, the Court of Chancery. Other judges were appointed and equity developed in a different way from the common law, with its own remedies such as injunctions. Equity offered *not* an alternative system but a supplement 'to stop the gaps' in the common law relating to private legal matters.

Equity did not use precedents (see above) but it did use maxims which even today are at the root of equitable decisions. An example is 'He who comes to equity must come with clean hands.' This means that a claimant (formerly called a 'plaintiff') asking for an equitable remedy or for relief in equity in some way, must have behaved fairly throughout the affair and not merely complied with his or her legal duties.

Another equitable maxim is 'Delay defeats equity.' You cannot expect the court to help you in equity if you delayed seeking that help as it may have adversely affected the opposition's position.

Today the two systems of rules are applied in all the courts, but if there is a conflict between a legal rule and an equitable rule, the equitable rule prevails. More recently, under the Constitutional Reform Act 2005, the Lord Chancellor is no longer entitled to sit as a judge and his judicial role has been transferred to the Lord Chief Justice, although he remains a member of the House of Lords.

10.2 European Union law

Origins

European Union law is the newest source of English law. The European Economic Community (EEC) was set up in 1957 by the first Treaty of Rome, which was signed by the original six member states. Two other communities were also set up, the European Coal and Steel Community (ECSC) by the Treaty of Paris 1951 and the European Atomic Energy Community (Euratom) by a second Treaty of Rome 1957. The administrative aspects of the three communities were merged in 1967 but they remained separate legal entities. The first Treaty of Rome was amended by the Single European Act 1986. As a result the EEC was from then on referred to as the European Community (EC) and the first Treaty of Rome as the EC Treaty (following the Lisbon Treaty, this EC Treaty is now the Treaty on the Functioning of the European Union). By the Single European Act environmental issues were given proper Treaty authority which they had hitherto lacked. The EC thus is the source of environmental policy and law. The Treaty on European Union (the Maastricht Treaty) 1991 was finally adopted by all the member states in 1993. The Maastricht Treaty amended the EC Treaty and now the collective communities are referred to as the *European Union* (EU). Three further treaties have been adopted, the Treaty of Amsterdam 1997, the Treaty of Nice 2000 and the Lisbon Treaty, signed by EU leaders in December 2007 and finally becoming law on 1 December 2009 after ratification by all 27 member states.

Objects

Originally formed primarily for economic and trade purposes (with the ultimate goal of political union), the Treaty of Rome 1957 made no provision for reference to environmental issues/concerns. Now, environmental issues are clearly on the agenda and constitutionally incorporated, which indicates the interconnectedness of economic and environmental affairs and the growth of environmental awareness among the population of EU Member States.

By Article 2 of the first Treaty of Rome (the Treaty) as amended, lastly by the Treaty of Amsterdam 1997, the main object of the Community was to 'promote throughout the Community a harmonious, balanced and sustainable development of economic activities, a high level of employment and of social protection, equality between men and women, sustainable and non-inflationary growth, a high degree of competitiveness and convergence of economic performance, a high level of protection and improvement in the quality of the environment, the raising of the standard of living and quality of life, and economic and social cohesion and solidarity among Member States'. Following the Lisbon Treaty, Article 2 has been repealed. A replacement is provided by what is now an amended Article 3 in the consolidated version of the Treaty on European Union as a result of the Lisbon Treaty, paragraph 3 of which provides:

> The Union shall establish an internal market. It shall work for the sustainable development of Europe based on balanced economic growth and price stability, a highly competitive social market economy, aiming at full employment and social progress, and a high level of protection and improvement of the quality of the environment. It shall promote scientific and technological advance.
>
> It shall combat social exclusion and discrimination, and shall promote social justice and protection, equality between women and men, solidarity between generations and protection of the rights of the child.
>
> It shall promote economic, social and territorial cohesion, and solidarity among Member States.
>
> It shall respect its rich cultural and linguistic diversity, and shall ensure that Europe's cultural heritage is safeguarded and enhanced.

Article 11 of the Treaty on the Functioning of the European Union (formerly the EC Treaty) provides: 'Environmental protection requirements must be integrated into the definition and implementation of the Union policies and activities, in particular with a view to promoting sustainable development.' Thus the environment and sustainable development remain EU priorities following the treaty changes.

In terms of context setting, it is helpful to track the history to these changes in some more detail. In 1985 a step was taken by all member states signing the Single European Act 1986 in order to create the Single European Market on 1 January 1993. With this move trade barriers were abolished, border controls for Community nationals were eliminated and employees were permitted to move freely between the states. Article 130r of the EC Treaty inserted by the 1986 Act stated that the objectives of the Community's environmental action were to be:

(i) To preserve, protect and improve the quality of the environment.
(ii) To contribute towards protecting human health.
(iii) To ensure a prudent and rational utilization of natural resources.

These objectives are now set out in Article 191 of the Treaty of the Functioning of the European Union (the former EC Treaty).

The Treaty on European Union (the Maastricht Treaty) 1991 also emphasized the importance of environmental issues when implementing the main aims of the EU. Thus continued prosperity in Western Europe was encouraged but with an environmental dimension. The Fifth Environmental Action programme 'Towards Sustainability' (1993–2000) reinforced the EU's priorities for action, the previous four programmes all having instigated much environmental law. The Treaty of Amsterdam 1997, Article 2, set out above, further reinforced the importance of protection and improvement in the quality of the environment. The Sixth Environmental Action Programme, adopted in July 2002, provides a strategic framework for the environmental policy of the EU until 2012. It identifies four priority areas: climate change, nature and biodiversity, environment and health, natural resources and waste. Further details are available at *http://ec.europa.eu/environment/newprg/index.htm.*

Accession by the United Kingdom

In 1971 the United Kingdom, Ireland and Denmark signed the Brussels Treaty of Accession by which they agreed to join the three communities. Like all treaties, this would have remained a mere political agreement had the UK's Parliament not passed the European Communities Act 1972 which came into effect on 1 January 1973. The Act ratified the Treaty and incorporated the three treaties as part of the law of the United Kingdom (note not just England and Wales) as well as all the case law from the European Court of Justice. At one fell swoop there was a new source of English, Scottish and Northern Irish law. The Single European Act 1986 and the Maastricht Treaty were ratified in the UK by the European Communities (Amendment) Act 1986 and the European Communities (Amendment) Act 1993 respectively. The European Communities (Amendment) Acts 1998 and 2002 adopted the Treaties of Amsterdam (1997) and Nice (2000) respectively. The European Union (Amendment) Act 2008 enabled UK ratification of the Treaty of Lisbon. Thus, the English courts must apply EU law to a case before them, if it is applicable. If there is conflict, EU law prevails.

EU institutions

The EU is run by four main institutions: the Council of Ministers of the European Union, the European Parliament, the European Commission of the European Union and the European Court of Justice. Each member state has voluntarily given up to these bodies some of its own rights to make law (sovereignty). This is because the communities are *supranational*, having authority to make decisions for all the member states on certain community matters, e.g. the Common Agricultural Policy.

The *Council of Ministers* of the European Union which sits at Brussels is the principal executive body of the EU. The Council of Ministers contains one appropriate representative from the government of each member state. Thus, if an environmental matter is to be discussed then an environment representative is sent. The Council of Ministers has the final decision on proposals put forward by the Commission, and legislation may be passed with a qualified majority (achieved by weighting the votes among the states) but complete unanimity is required in foreign and security policy, justice and home affairs, after consultation with the European Parliament. The Council also comprises a number of working groups of officials from the member states.

The European *Commission* is the executive of the EU. It sits in Brussels and consists of 27 independent members appointed by the member states, one from each (though it is likely that the maximum number of commissioners will be limited in future, appointed by rotation on the basis of nationality, with safeguards in place to ensure fair and balanced representation). The president is appointed by common accord between governments after consulting the European Parliament, but there is a power of veto which applies to commissioners as well as to the president. Commissioners are nominated by governments in consultation with the new president and the resultant list is then sent for approval to the European Parliament.

The European Commission implements EU policy, initiates and draws up proposals for legislation for the Council to approve. It also polices and enforces EU law and has extensive investigative powers. It can also start proceedings before the European Court of Justice. The European Commission draws up the environment action programme and drafts proposed EU legislation. Decisions are reached by majority. Member states must inform the Commission of proposed domestic law. It is divided into a number of directorates. Directorate General XI deals with environment, consumer protection and nuclear safety. The Commission has developed a new framework for noise policy, building on its Green Paper (COM(96) 540) published in 1996. It identified environmental noise from traffic, and from industrial and recreational activities as 'one of the main local environmental problems'. A significant development from the new policy framework is the Environmental Noise Directive (2002/49/EC), requiring member states to produce noise action plans based on data obtained from strategic noise mapping. This is explored in more detail at the end of this section.

The *European Parliament* sits at Strasbourg. It has evolved from a mere advisory assembly into a directly elected representative parliament having equal rights to the European Council on budgetary matters. Its powers *vis-à-vis* the introduction of new legislation have been extended by the Single European Act and the Maastricht Treaty so that there is now a conciliation committee designed, where possible, to produce a joint statement from the Commission and Parliament. It can set up inquiry committees to investigate maladministration, and any citizen of the EU can directly petition Parliament on community matters that affect him or her. He or she can do this either in a private capacity or as a member of an organization, including companies. It has been suggested that this right is not used enough.

The Lisbon Treaty creates two new political posts: President of the European Council and High Representative of the Union for Foreign Affairs and Security Policy.

The member states

Since 1 January 2007 there are twenty-seven member states: Spain, Portugal, United Kingdom, Denmark, Belgium, Italy, France, Finland, Ireland, Netherlands, Germany, Luxembourg, Austria, Greece, Sweden, Hungary, Poland, Estonia, Czech Republic, Slovenia, Cyprus, Slovak Republic, Latvia, Lithuania, Bulgaria, Malta and Romania.

European Union law

The first three treaties are the main source of Community law, which the Single European Act 1986, and the Maastricht, Amsterdam, Nice and Lisbon treaties amend and supplement. The European Commission and Council are allowed to make three types of law called regulations, directives and decisions.

Regulations

These are rules which, once made, are immediately binding on all the member states, without reference to their legal systems.

Directives

These are orders or requirements directed to all or some of the member states, the results of which are binding but the means to achieve them are left to each member state. This is useful because of the difficulties in translating legal ideas between different legal systems. Because of its nature, most environmental law will be created in this way. An example of this is Directive 2003/10/EC of the European Parliament and of the Council on the minimum health and safety requirements regarding the exposure of workers to the risks arising from physical agents (noise) which in the UK has been implemented under the Health and Safety at Work, etc., Act 1974, by the Control of Noise at Work Regulations 2005 (S.I. 2005/1643). Other examples include Directive 86/594/EEC on airborne noise emitted by household appliances, implemented nationally by the Household Appliances (Noise Emission Regulations 1990 (S.I. 1990/161) and Directive 2000/14/EC on noise from outdoor equipment, transposed into national law by the Noise Emission in the Environment by Equipment for Use Outdoors Regulations 2001 (S.I. 2001/1701).

Proposals for directives may come from the Council of Ministers or the Commission. Preparation of drafts of directives is undertaken by the Commission who may set up technical working groups where necessary. Directives must be approved by the Council of Ministers. In some cases ministers may have the power of veto but in others only a majority agreement is required.

Implementation of a directive is normally within a specified time, usually two years. Once the domestic law has changed there has been formal compliance.

Decisions

These are decisions on some aspect of community law which are binding on the person or body to whom they are addressed.

European Court of Justice

This court, situated at Luxembourg, is a court specially set up to deal with problems of Community law. Each state sends at least one judge. The Court rules on the meaning of the treaties, and under the Treaty of Rome if there is any doubt as to the meaning of the treaties a national case must be sent to the Court for clarification of the Community law involved (Art. 267, formerly Art. 234 EC Treaty). The Court also has jurisdiction over disputes involving aspects of Community law between states, corporations and even individuals. It is therefore another source of judicial precedent, binding national courts by its decisions. The ECJ itself is not bound by its own previous decisions.

There have been many pieces of legislation introduced in relation to noise as a result of the EU. Examples include the Lawnmowers (Harmonisation of Noise Emission Standards) Regulations 1992, the Household Appliances (Noise Emission) Regulations 1990, the Town and Country Planning (Assessment of Environmental Effects) Regulations 1999 (previously 1988) and the Control of Noise at Work Regulations 2005 (previously 1989) and Environmental Noise (England) Regulations 2006.

EU noise policy

Nationally, the Department for Environment, Food and Rural Affairs (DEFRA) has been undertaking strategic noise mapping to implement *Directive 2002/49/EC relating to the assessment and management of environmental noise* (the Environmental Noise Directive – END). The Directive's broad objectives, set out in *Article 1*, are to limit exposure to noise by the implementation of action plans for noise management, prevention and reduction based on the mapping results, and to ensure that environmental information on noise and its effects is made available to the public. The Directive focuses on noise from major sources: road and rail vehicles and infrastructure, aircraft, outdoor and industrial equipment and mobile machinery (*Art.1(2)*).

Article 7 and *Annex 4* of the Directive deal more specifically with noise mapping, and *Article 5* and *Annex 1* specify the applicable noise indicators L_{den} (day-evening-night equivalent level) and L_{night} (night equivalent level) for mapping and review. *Article 6* and *Annexes 2 and 3* set out assessment methods, and *Article 8* makes provision for action plans, with Annex 5 setting out their minimum requirements. Action plans, as well as noise maps are subject to review and revision (at least every five years after their preparation/approval) (*Art. 7(5)* and *Art. 8(5)*).

Timescales: mapping

The Directive required member states to ensure that strategic noise maps showing the situation in the preceding calendar year had been made and approved by 30 June 2007 for major roads with more than 6 million vehicle passages a year, railways with more than 60,000 train passages a year, major airports and agglomerations with more than 250,000 inhabitants (*Art. 7(1)*). The deadline for the second round of noise maps for all other major roads, railways and agglomerations is 30 June 2012 (*Art. 7(2)*). An agglomeration is defined in *Article 3 (k)* as 'part of a territory, delimited by the Member State, having a population in excess of 100,000 persons and a population density such that the Member State considers it to be an urbanised area.'

Timescales: action plans

Action plans must have been drawn up by 18 July 2008 for places near the major roads which have more than 6 million vehicle passages a year, major railways which have more than 60,000 train passages per year and major airports and for agglomerations with more than 250,000 inhabitants. For other major roads, railways and agglomerations, the deadline is 18 July 2013. Major roads are those with more than 3 million vehicle passages a year (*Art. 3(n)*); a major railway has more than 30,000 train passages per year (*Art. 3(o)*); and a 'major airport' is a civil airport, which has more than 50,000 movements (a take-off or a landing) per year *Art. 3(p)*).

The Directive has been transposed into national law by the *Environmental Noise (England) Regulations 2006* (S.I. 2006/2238), amended by the Environmental Noise (England) (Amendment) Regulations 2008 (S.I. 2008/375) and by the Environmental Noise (England) (Amendment) Regulations 2009 (S.I. 2009/1610) with equivalent provisions for Wales.

Information on the development of national noise action plans is available at *http://www.defra.gov.uk/environment/quality/noise/mapping/index.htm* and from the Noise Mapping England website: *http://services.defra.gov.uk/wps/portal/noise*

Further information from the EU on noise mapping is available at: *http://ec.europa.eu/environment/noise/mapping.htm*

For a full list of EU directives relating to noise sources, go to: *http://ec.europa.eu/environment/noise/sources.htm* and for further information on EU noise policy in general see: *http://ec.europa.eu/environment/noise/home.htm*

EU legislation is accessible via *http://eur-lex.europa.eu/en/index.htm*

10.3 The eradication or reduction of noise or vibration at common law

The common law has treated noise and vibration to be within a category of behaviour known as *nuisance*. A nuisance may be so bad as to constitute a crime or it can be treated as a *tort*, i.e. a civil wrong.

Nuisance

There are three types of nuisance. A *public* nuisance is a crime created by the common law, i.e. it is considered to be so harmful that people breaking the law should be punished by the state for the good of the people. A *private* nuisance, on the other hand, is a tort. The person suffering may thus sue the wrongdoer, known as the tortfeasor, and may be awarded a civil remedy such as damages or an injunction. *Statutory* nuisances are nuisances created by Acts of Parliament and are dealt with in the criminal courts (see below and section 10.4).

Public nuisance

Public nuisance has been defined as 'an act or omission which materially affects the reasonable comfort and convenience of life of a class of Her Majesty's subjects' (*AG v. PYA Quarries* [1957] 2 QB 169.

Because of its ancient beginnings, it has embraced different types of behaviour which one would not normally consider to be nuisances. Examples include obstructing the highway, keeping a brothel, polluting a public water supply, causing dust, noise and vibration by quarrying activities and causing dirt, noise and an unreasonable amount of traffic by holding a pop festival. A doctor has been prosecuted for performing operations while suffering from an infectious form of hepatitis. In *R v. Holme* [1984] CLY 2471 an eccentric but sane man was found guilty of committing a public nuisance. Apart from threatening people, banging on car roofs and blocking the public highway, he also imitated an ape, provoked dogs into barking in the early hours of the morning, played one chord on the piano through the night, played a radio at top volume all day and night while it was hanging from a rope out of his window, kicked a dog up the street and assaulted people. There seems to be little in common with each example, other than the element of annoyance or inconvenience to the public. Inconvenience and annoyance are also common to private nuisance as we shall see later. In some cases of public nuisance, Parliament has enacted specific legislation enabling action to be taken, for example under the statutory nuisance provisions under Part III, Environmental Protection Act 1990, as amended, or under the Noise Act 1996 in respect of noise emanating from dwellings at night. The more general Anti-Social Behaviour Act 2003 can also be used to tackle cases of noise nuisance, see section 10.4 below.

As public nuisance is technically a crime, it could be tried on indictment in the Crown Court like other serious crimes, e.g. murder. But it is more usual for the Attorney-General, who is the principal law officer of the Crown and head of the Bar, to start the action in the High Court on behalf of a sufficient portion of the public. This is called a relator action. An application has to be made to the Attorney-General to which he may or may not agree. Similarly, a local authority may take action in the same way on behalf of a community within its area under s. 222 Local Government Act 1972. Section 222 states:

> (1) Where a local authority consider it expedient for the promotion or protection of the interests of the inhabitants of their area – (a) they may prosecute or defend or appear in any legal proceedings and, in the case of civil proceedings, may institute them in their own name, and (b) they may, in their own name, make representations in the interests of the inhabitants at any public inquiry held by or on behalf of any Minister or public body under any enactment.

How many constitutes a sufficient portion of the public for it to be dealt with as a public nuisance depends on the circumstances of each case. In *R. v. Lloyd* (1802) the occupants of only three barristers' chambers in Cliffords Inn in London affected by noise was held not to be a sufficient number for a public nuisance. In *AG v. PYA Quarries* (above) Denning LJ, referring to quarrying activities affecting a wide area, stated that 'a public nuisance is a nuisance which is so widespread in its range or so indiscriminate in its effect that it would not be reasonable to expect one person to take proceedings on his own . . . but that it should be taken on the responsibility of the community at large'. Thus, a noise nuisance affecting a whole village or a number of streets could probably be dealt with as a public nuisance, whereas if only a few houses are affected then it seems not.

If a member of the public has been particularly affected by the commission of a public nuisance, provided the person can prove he or she has suffered damage over and above that suffered by others in the neighbourhood, then he or she may sue in tort. Such damage could take the form of personal injury, loss of business or physical damage to property. Such action was taken in *Halsey v. Esso Petroleum Co. Ltd.* [1961] 2 All ER 145 (see below). The damage must have been a foreseeable consequence of the nuisance, e.g. loss of sleep due to vehicles being driven during the night, as in Halsey's case. The person suffering 'extra' damage does not have to have an interest in land nor can the person only sue in relation to the effect of the nuisance on his or her land, as in private nuisance, which is a tort designed to protect use of land only (see below).

As public nuisances are essentially crimes, it is no defence to say there has been consent to the nuisance, as one can never consent to a crime. But one can always consent to commission of a tort (see page 282).

Finally there does not have to be any indirect invasion of private land, as there has to be in private nuisance.

In *East Dorset DC v. Eaglebeam Ltd* [2006] EWHC 2378 (QBD) the High Court was satisfied that both a public and a statutory nuisance existed as a result of 'over-noisy' motorcross activity in a public park and granted an injunction. If one or more owners of the nearby houses affected had decided to take action, they would have been able to bring an action in private nuisance since there was an unreasonable interference with their use and enjoyment of their own land. Rights available in private nuisance are not affected because others (in this case other non-participating park users) suffer similar harm.

Private nuisance

The tort of private nuisance has been defined by Professor Winfield as an unlawful interference or annoyance which causes damage to an occupier or owner in respect of his or her use and enjoyment of his or her land, or of certain rights over or in connection with land. These 'certain rights' are easements – legal rights attaching to land in respect of another's land over which the owner of the easement has, for example, the right to light, water or a right of way.

Private nuisance therefore protects only interests in *land*, not personal interests as in negligence (see below). Thus, damages for personal injuries cannot be recovered under this tort (cf. public nuisance), see *Hunter v. Canary Wharf* [1997] 2 WLR 684. 'Land' does, however, include *buildings* on the land.

Behaviour amounting to a private nuisance may take the form of noise, preventing certain uses of the property such as for sleeping or use of the garden. Vibrations of course can actually physically harm buildings. Nuisance can also take other forms, such as that created by smoke, water, gas, smells, fumes, roots undermining walls and other un-neighbourly behaviour. The courts agree that other types of nuisance may occur in the future.

The behaviour amounting to nuisance may be intentional, negligent or perhaps even unintentional and not negligent.

Criteria for suing in nuisance

Unlike negligence (see section 10.8) there is only one essential element in order to sue for nuisance. But many other factors have to be taken into account when deciding whether a nuisance has or has not been committed.

(i) *Damage or harm must have been caused*

This is the essential element in nuisance (which negligence also requires). There must be some damage or harm suffered on the part of the claimant in relation to his or her land. This may take the form of physical damage to the property, such as cracks caused by vibrations. If there is physical damage, the court looks no further at whether damage has been caused and is less likely to investigate its triviality.

If, on the other hand, the damage is only an interference with the use of land or its enjoyment, then it must not be trivial. The interference must be substantial. In *Andreae v. Selfridge* [1938] 3 All ER 255 it was said by the Master of the Rolls that the loss of only one night's sleep through excessive noise was not trivial and no injury to health need be proved.

The damage that results in these situations is that the claimant is unable to use a room or a garden in the normal way because of the nuisance. This could be quantified by a valuation in respect of the reduction in value of the house. Imagine buying a house, part of which you could not use because of the noise or vibration from next door. You would expect to receive a reduction in price. In the case of the hotel in the above case, the proprietors could by reference to their bookings and accounts show the loss in business caused by the noise.

(ii) *The behaviour must be unusual, excessive or unreasonable*

The courts take the view that there should be 'give and take' between neighbours. So an occasional noisy party or do-it-yourself session would not be actionable. Weekly parties or continuous drilling would almost inevitably amount to nuisance. Normal, reasonable and moderate use of property cannot be actionable. In *Halsey v. Esso Petroleum Co. Ltd.* (1961) there was an excess of noise, smell and acid smuts caused by the defendant's business. In *Andreae v. Selfridge & Co. Ltd* (1938) demolition contractors caused an unreasonable amount of dust and noise which affected the claimant's hotel trade to its detriment. In *Dunton v. Dover District Council* (1977) 76 LGR 87 the noise from a playground adjacent to a private hotel was found to be excessive and in *De Kuyser's Royal Hotel Ltd v. Spicer Bros* (1914) 30 TLR 257, night-time pile-driving was held to be a nuisance. More recently, in *Watson v. Croft Promo-Sport Ltd* [2009] JPL 1178, the Court of Appeal awarded an injunction to restrain a nuisance resulting from the defendant using a former aerodrome for motor racing, even though planning permission had been granted for the activity. By contrast, the Court of Appeal dismissed the claimants' appeal in *Murdoch v. Glacier Metal Co Ltd* [1998] Env LR 732 where they had claimed that their sleep was disturbed by fluctuating noise levels and glare at night from the defendant's factory. That the noise levels were just above the World Health Organization (WHO) levels was not solely determinative of the existence of a nuisance and location of the claimant's premises in a mixed-use area was considered to be significant. The Court of Appeal approved the test applied by the County Court:

> In this case the issue is not whether or not there was any noise and glare from the factory. Both are accepted by Glacier Metal. The issue is whether according to the standards of the average person and taking into account the character of the neighbourhood the noise and the glare were sufficiently serious to constitute a nuisance.

This latter aspect is discussed further in (v) below and also in section 10.11 on spatial planning. The test for common law nuisance reaffirmed in *Murdoch* has been applied in cases of statutory nuisance, see for example *Godfrey v. Conwy CBC* [2001] Env LR 38 (section 10.4 below).

In *Dennis v. MoD* [2003] Env LR 34 the High Court held that noise from low flying military aircraft affecting the claimants' property was an unreasonable interference and not only constituted a nuisance but also an interference with human rights (see section 10.17 below).

In *Toff v. McDowell* (1993) 25 HLR 650, however, the court found that *ordinary* use of premises did, partly in this case, give the right to successfully sue in nuisance where the parties, living in converted flats, shared the responsibility of maintaining an adjoining floor/ceiling, which experts showed contained no form of sound insulation. This decision is quite unusual and may relate only to the particular facts involved. The House of Lords' decision in *Baxter v. Camden LBC (No. 2)* [2000] Env LR 112 (HL) indicates that this is indeed the case, Lord Hoffmann stating:

> I do not think that the normal use of a residential flat can possibly be a nuisance to the neighbours. If it were, we would have the absurd position that each, behaving normally and reasonably, was a nuisance to the other.

Nevertheless *Baxter* does not overrule *Toff* and in *Stannard v. Charles Pitcher Ltd* [2003] Env LR10, the High Court held that the noise penetration from the defendant's flat into the claimant's flat was unreasonable and unacceptable and constituted an actionable nuisance. In that case the defendants had reconfigured the internal arrangements of their flat so that the vertical distribution of the accommodation was 'incompatible with the minimisation of noise invasion' into the claimant's flat: the kitchen and principal bathroom being directly above the claimant's drawing room and a new bathroom installed directly above the claimant's bedroom. To make matters worse, the defendants had also removed carpets and replaced them with marble and ceramic tiles. The court awarded a mandatory injunction requiring the necessary sound installation. A useful analysis is provided by Patrick Bishop in his article, 'Inadequate sound insulation: does the law of nuisance provide an effective remedy?', Env LR 2005 7(4) 238–252.

(iii) *The behaviour must have gone on for some time*

Generally, in relation to noise nuisance, isolated acts are not actionable. But they may give rise to another type of tort, e.g. an explosion could be caused by negligence.

(iv) *The nuisance must be caused by another person on neighbouring property and not on the claimant's own premises*

This is partially self-explanatory. If the nuisance was actually on the claimant's land he or she should obviously take action to stop the anti-social behaviour as he or she is in charge of that land. The land does not have to be adjacent to the claimant's land but obviously it will not be far away for a noise nuisance.

It seems that the nuisance could be committed on a highway and not necessarily on the defendant's own land. In *AG v. Gastonia Coaches* (1977) *The Times*, 12 Nov, a coach business operated from a house and yard in a residential area. The business expanded and the defendants were unable to park all their vehicles on their own premises. They began to park their coaches on the roadway outside and they also carried out maintenance and repairs. The coach drivers also parked their own cars there. The local residents complained of the noise and fumes from the vehicles and also about the obstruction of the road. In a relator action they claimed for both public and private nuisances and asked for damages and an injunction. The court held that only the obstruction was a *public nuisance*. The neighbour living next to the premises established that the obstruction to access to his own home by parking on the highway and the noise and fumes from the vehicles using the highway did amount to a private nuisance. However, the noise of the maintenance did not amount to a private nuisance.

(v) *Character of the neighbourhood*

Thesiger LJ is often quoted from his judgment in *Sturges v. Bridgeman* (1879) 11 Ch D 852 in which he said, 'what would be a nuisance in Belgrave Square would not necessarily be so in Bermondsey.' This is not strictly true, however, as one cannot create nuisances merely because one is in an industrial or urban area. In *Rushmer v. Polsue and Alfieri Ltd* [1907] AC 121, for example, the defendants moved some new printing equipment into their works in Fleet Street, then a very noisy area. The noise from the equipment was over and above what would have been normal in that area. Thus they were liable in nuisance. Nevertheless, it is obvious that one cannot expect the quietness of a rural community in the centre of a city. In *Murdoch v. Glacier Metal Co Ltd* (above), the overall character of the neighbourhood including the presence of a by-pass, together with the lack of complaints from other residents, led the court to conclude that there was no actionable nuisance. In *Watson v. Croft Promo-Sport Ltd* (above), the Court of Appeal held that the grant of planning permissions for motor racing at a former aerodrome had not changed the nature and locality of the area, which remained predominantly rural. Consequently the activities of the defendant were an actionable nuisance. For the influence of planning permission on the locality of an area, see section 10.11 below.

If the area is in a noise abatement zone (see below) then this should obviously influence court decisions in actions for private nuisance. Unfortunately, at present, the small numbers of noise abatement zones that have been set up under the Control of Pollution Act 1974 (the relevant part has not been repealed by the Environmental Protection Act) means that in the majority of cases the court continues to listen to expert evidence and actual evidence of the noise itself in order to decide whether the noise is excessive for that area.

(vi) *Abnormally sensitive claimants*

The law can only protect the average person with average sensitivity to noise. If the claimant is unusually sensitive, he or she will not be able to succeed in an action, unless the noise is over and above what would be acceptable to the average person. For example, in an unreported case in 1977 the ringing of church bells did not constitute a legal nuisance to the neighbours. Their objections indicated that they were unusually sensitive. But if a nuisance is established, the claimant can recover damages for interference with a special use or in relation to his or her sensitivity.

(vii) *The defendant's conduct*

Cases show that behaviour amounting to a nuisance may be intentional, negligent or occasionally unintentional and even non-negligent. The unthinking noise producer is probably most common, but after being given notice of the nuisance by the sufferer, those people who continue to commit noise nuisance are probably intentional or negligent. Malice need not be shown to succeed in nuisance, unlike some other torts, such as malicious prosecution, for which actual malice must be proved. But if it can be shown that the defendant intended to annoy people, this will better support the claimant's case. In *Hollywood Silver Fox Farm Ltd. v. Emmett* [1936] 2 KB 468 the defendant encouraged his son to shoot his gun with the intention of interfering with the breeding of foxes on the claimant's farm. In *Christie v. Davey* [1893] 1 Ch 316 the defendant's neighbour, annoyed by the music lessons being given by the claimant, banged trays and made other noises with the intention to annoy. In *Fraser v. Booth* (1948) 50 SR (NSW) 113 a neighbour got his son to let off fireworks in order to discourage the claimant's homing pigeons from returning to their loft.

(viii) *The behaviour must be indirect in nature*

Nuisance by its very nature is an indirect tort; the bad behaviour is on the defendant's own land or possibly on the highway (see above). If it were direct, e.g. throwing things onto the claimant's land, this would amount to trespass to land, and a different basis for action would have to be used.

(ix) *Could the nuisance have been prevented easily?*

If it could have been prevented for a relatively small cost, the nuisance should have been stopped and the court will take this into account. Switching off record-players costs nothing. This does not mean to say that the court will not find that a nuisance exists if the cost of its eradication is huge.

What sort of claimant may sue?

As the tort only protects interests in land, the claimant must prove that he or she has a right to enjoy the land itself or an easement over that land. The House of Lords reaffirmed this requirement in *Hunter v. Canary Wharf* [1997] 2 WLR 684. Thus, the claimant must normally be in possession of the land, unless the damage is physical and affects the property in some permanent way, in which case persons out of possession may sue. Thus anyone in legal or equitable occupation of the land may sue, e.g. an owner-occupier, tenant in possession, licensees in possession and lessors (but the last only if their reversionary interest has been damaged, e.g. by cracks in the property). (Lessors or landlords lease or rent their property to lessees or tenants, who at the end of the allotted time must deliver up possession of the land back to the landlords. While they are out of possession, the landlords are said to have *reversionary* interests as the land will revert back to them.) Otherwise, as noise is of such a transient nature, people such as lessors, who are not in possession of the land, are generally unable to sue for nuisance.

Who can be sued?

(i) *The creator of the nuisance*

In nuisance generally, the creator will always be liable, even if he or she gives up occupation of the land. This seems to be inappropriate in the case of noise, as noise is normally created by persons currently occupying land.

(ii) *The occupier*

Obviously the occupier should be primarily liable as he or she is in control of the premises and can tell people what they must or must not do. The occupier will therefore be responsible for his or her own and his or her family's acts. By the principle of vicarious liability, an employer is also responsible for the torts committed by any employees during the course of their employment. Thus, a discotheque owner would be liable for the nuisance created by a disc jockey or band unless he or she expressly forbade the particular behaviour causing the nuisance. But even in these cases, the courts often take the view that the prohibition must be so strong as to put the act outside the scope of the employee's employment. A disc jockey playing music loudly would thus be doing what he is employed to do, but in a wrongful way, making his employer vicariously or indirectly liable in tort.

The occupier may also be liable for his or her independent contractors, i.e. those people with whom he or she has a contract, not a contract of service like an employee, but a contract for services such as window-cleaners, building contractors, etc. However, the occupier will be responsible only if he or she has a great deal of control over the independent contractor. Otherwise, the claimant would have to sue the independent contractor as the creator of the nuisance.

The occupier may also be liable for nuisances created by invited guests, such as young cousins addicted to loud music while sunbathing and persons who have licences to enter his or her property. They have no legal or equitable rights to the land, but they have been given a licence by the occupier to enter the land for a particular reason. Anglers may enter for fishing by right of a written fishing licence, or someone such as a neighbour may have been asked to feed the cat and water the plants while the occupier is on holiday. But in such cases, by attempting to abate the nuisance, the occupier may reduce or eliminate his or her liability.

Liability for the nuisance of a trespasser only occurs if the occupier has been negligent in discovering the existence of the nuisance.

An occupier who carries on or adopts a nuisance created by a former occupier will obviously be liable for nuisance, *Sedleigh Denfield v. O'Callaghan* [1940] 3 All ER 349. The making of noise is a more active example of nuisance, whereby the noise-maker will be sued as the creator rather than the adopter even if, for example, he or she is merely taking over a noisy business (see prescription below).

However, in *Sampson v. Hodson-Pressinger* [1981] 3 All ER 710, (1984) 12 HLR 40 a new landlord of a building, previously converted into flats, was liable in nuisance because, at the time he was assigned the freehold reversion, he knew that a roof terrace allowed noise to penetrate to the flat below. He was effectively authorizing the nuisance (see below).

(iii) *The person who authorizes the nuisance*

If a person lets land for a purpose which by its very nature entails the commission of a nuisance, then that person may be liable in tort. This applies even if it is only foreseeable that the nuisance *may* be committed by the occupiers. In *Tetley v. Chitty* [1986] 1 All ER 663 a local authority gave planning permission to a go-kart club on a site leased to them by the council specifically for go-karting. The court held that the council was liable in nuisance as the noise was an ordinary, natural and necessary consequence of the letting and as such the council had given express or implied consent. But in *Smith v. Scott* [1972] 3 WLR 783 a local authority let property to a known 'problem' family who, it could have been foreseen, would create nuisances by causing damage and noise. By including in the tenancy agreement a clause prohibiting such behaviour, the authority was held not to have authorized the nuisance. More recently the Court of Appeal applied this principle in *Hussain v. Lancaster CC* [1999] 2 WLR 1142 and in *Mowan v. Wandsworth LBC* (2001) 33 HLR 56.

Any landlord who retains the power to inspect premises let by him or her without a prohibition clause may be open to action even though he or she has not in fact authorized the nuisance, likewise where a landlord has covenanted to repair or has the right to enter and repair. A landlord is not liable for the ordinary noise of other tenants, see *Southwark LBC v. Mills, Baxter v. Camden LBC (No. 2)* [1999] 3 WLR 939 (*Southwark LBC v. Mills* was heard by the House of Lords together with *Baxter* case, considered above).

Defences

(i) *Prescription*

Generally, as a defence to a nuisance action, one may gain the legal right to continue to commit a nuisance by prescription (long user). However, in relation to noise nuisances this seems to be less likely as the behaviour amounting to a nuisance would have to be capable of existing as an easement. Easements are legal rights attaching to land, such as rights of light, way and water, and it is open to conjecture whether a prescriptive right can be obtained in relation to noise, especially in the light of the Environmental Protection Act 1990. A right to commit a public nuisance which is a crime can never be obtained by prescription.

For a prescriptive right to be gained, the owner of the land must prove that he or she has been committing the nuisance for 20 years at least. He or she must not have committed the nuisance with the other person's permission, nor must it have been done secretly (usually impossible in any event in relation to noise), nor must force have been used to exercise it (*nec vi, nec clam, nec precario* – without force, secrecy or permission). If the neighbour disputes the nuisance then this, too, would negate the prescriptive right.

The behaviour must have amounted to a *nuisance* for the period of 20 years. In *Sturges v. Bridgeman* (1879) LR 11 Ch D 852 a confectioner operated at the rear of property abutting onto a garden belonging to a doctor in Wimpole Street. The doctor moved his consulting rooms from the main house to the bottom of the garden, where the noise from the confectioner became a nuisance and the doctor sued. The confectioner, although able to prove that he had operated his business for over 20 years in the same manner, was not able to show that it amounted to a *nuisance* for those years, so the doctor was successful in his action. In other words, the noise only became a nuisance when the doctor moved.

(ii) *'Coming to the nuisance'*

What happens if someone knows that a nuisance is being committed in relation to property he or she is proposing to occupy? It is sometimes argued by the defence that where the behaviour amounting to a nuisance was being committed before the claimants arrived and now continues, the claimants cannot then successfully pursue a claim in nuisance. This is *no* defence. But there have been cases where the court in such situations, when deciding whether or not to grant an injunction, has weighed up the general interest of the public, who may have been benefiting from the nuisance behaviour, as opposed to the person coming to the nuisance in full knowledge of its existence; the court has refused to grant the injunction. (The court, however, could not refuse to grant damages once the tort has been proved, see 'Remedies' below) *Miller v. Jackson* [1977] QB 966 involved the nuisance of cricket balls from a village cricket green landing in an abutting garden. Although a nuisance was proved to exist, the injunction requested was refused because the claimants came to the nuisance and indeed were probably attracted by the closeness of the cricket ground. A similar decision was reached in a county court case in 1994 when it was decided that no nuisance had been committed in the first place, even though cricket balls had landed on a number of occasions in the claimant's garden.

(iii) *Consent*

It is usually a good defence if a claimant consents to the nuisance being committed. Even if not a defence, it may still have the effect of defeating a plea for an injunction. There have been unreported cases where builders, for example, have informed local people of the likelihood of building noise and occasionally made payments to them or even paid for holidays while the worst of the work has been carried out. Consent could be given gratuitously or could be paid for. But before refusing to grant a remedy, the courts would look carefully at the nature of the consent and could take the view that the claimants may not have realized to what extent of noise they had consented. Thus, to be successful as a defence, one would probably have to prove that there was consent and secondly that the claimant knew exactly to what he or she was consenting. If there was a written contract between the parties, this would make it more difficult for the claimant's case, especially if the claimant in accepting payment specifically stated that he or she gave up his or her right to sue for nuisance.

(iv) *Mitigating the nuisance*

In law it is always the duty of a claimant to mitigate his or her loss. This means that, whatever the damage suffered, he or she has tried to reduce it accordingly. But only reasonable steps need be taken, and if such action requires spending money, the defendant would not be able to say that the claimant had not mitigated his or her loss.

(v) *Statutory authority*

If an Act of Parliament or a relevant rule or regulation imposes a *duty* on someone or somebody to do something which inevitably will cause a nuisance, the authority will be a defence, provided that no negligence was attached to the action. It will be up to the defendant to prove that he or she took care to avoid negligence, even if the Act expressly made the defendant liable for nuisances.

There are slightly different rules in relation to *powers.* Duties *must* be carried out; powers *may* be carried out. Once again there must be no negligence on the part of the defendant, but this time the statute must either expressly exclude liability for nuisance or it must make no mention of the matter. In *Allen v. Gulf Oil Refinery Ltd.* [1981] 2 WLR 188 powers given by the Gulf Oil Refinery Act 1965 permitted the building of an oil refinery. Local villagers complained about noxious odours, vibrations and excessive noise. The House of Lords held that the section of the Act relied on gave the company immunity from an action for nuisance which might be the inevitable result of building such a refinery.

Remedies

(i) *At common law: damages*

On proving his or her case, the claimant is entitled to compensation (damages) for actual damage caused to his or her land, such as depreciation or loss of business. Entitlement is as of right because it is a *legal* remedy, and even if the claimant has behaved inequitably (see below), the court cannot refuse damages if he or she satisfactorily proved the case and there are no adequate defences. The court will take into account and offset the loss caused by reasonable behaviour, when the job which caused the nuisance was necessary. For example, in *Andreae v. Selfridge & Co. Ltd* [1938] 3 All ER 255 there would in any event have been loss of custom due to work, but not as much as was actually caused.

(ii) *In equity: injunctions*

An injunction is a court order which orders someone to do or not to do some specific thing. In relation to noise nuisance it usually orders the defendant to stop the work or behaviour which amounts to a nuisance. Such injunctions are called *prohibitory* or *restrictive* injunctions. They may be awarded before the case actually comes to court for trial (*interlocutory* or *interim*); if they are awarded at the trial they are called *perpetual.* However, this does not mean that the injunction necessarily carries on in perpetuity. The case of *Stannard v. Charles Pitcher Ltd* (above), provides an example of the court awarding a *mandatory* injunction, requiring the defendant to install acoustic barriers to prevent excessive noise entering the flat below.

Because an injunction is an equitable remedy, it is *discretionary* and will only be awarded if damages are inadequate on their own. If there is no likelihood of a repeat of the nuisance, an injunction would not be the appropriate remedy.

In using the court's discretion, the equitable maxims 'He who seeks equity must do equity' and 'Delay defeats equity' will be taken into account by the court. The first means that the claimant must have behaved equitably or fairly if he or she is going to be availed of an equitable remedy, e.g. if he or she had told a neighbour that he or she didn't mind the senior citizens' tap-dancing class practising every week next door, he or she may be refused an injunction. (But if he or she proves a nuisance exists, he or she can still get damages by right.)

The second maxim means that if one requires an equitable remedy the claimant must not delay; he or she must go to court as soon as possible.

A request for an injunction often accompanies a request for damages. In *Kennaway v. Thompson and Another* [1980] All ER 329 the Court of Appeal reversed the decision of the trial judge, thus allowing an injunction where previously only damages had been granted for past and future nuisance. The nuisance caused by a motor-boat club was restrained only as to behaviour, which did amount to a nuisance. Thus, the club members could still enjoy their water sports but not at the expense of the claimant. The court provided an agenda for a season of events with suitable gaps in between. Furthermore, noise limits were set for boats not taking part in national or international events, and no more than six boats were to be used at any one time. The injunction may be permanent or it may request modification of behaviour, e.g. operations to be carried out during specified hours only. An injunction may also be granted but suspended in order to give the defendant the opportunity of modifying his or her behaviour. See also *Miller v. Jackson* (above).

Quia timet injunctions may be granted even though the nuisance is not actual, only threatened. But in order to obtain such an injunction, one would have to show that there was imminent danger of irreparable damage (in nuisance, damage to land). One cannot really imagine a situation in relation to pure noise nuisance which would give such a result. However, the prospect of an unauthorized pop festival, perhaps resulting in many different types of nuisance, could possibly give rise to a *quia timet* injunction.

Historically, injunctions were only granted by the Court of Chancery (the court applying equity) and the court had no power to grant damages, as this was a legal remedy only available in the common law courts. However, the Chancery Amendment Act 1858 (Lord Cairns Act) changed the law and permitted the award of damages in lieu of, or in addition to, an injunction, e.g. where only an injunction had been asked for. The remedy is now found in s. 50 Senior Courts Act 1981 (formerly the Supreme Court Act 1981 but re-titled in consequence of the Constitutional Reform Act 2005 that created (from 2009) the Supreme Court). (Damages under this heading could be granted even in lieu of a *quia timet* injunction, but this would be extremely unusual as it would allow the award of damages even though there had been no tort committed and thus no possibility of damages being awarded at common law.)

The leading case, *Shelfer v. City of London Electric Lighting Co.* [1895] 1 Ch 287, involved nuisance caused by noise and vibration. It stated that damages in lieu of injunctions would only be awarded if (i) there was only a small injury; (ii) the injury could be estimated in money terms; (iii) the injury could be adequately compensated by a small amount of damages, and (iv) the award of an injunction would be oppressive. But each case must be looked at individually and in the light of the equitable maxims.

If someone carries on with acts in breach of terms of the injunction, they will be in contempt of court and could be fined or sent to prison. Anyone knowingly aiding them in breach could also be dealt with for contempt, even if not party to the original action. If a company or other type of corporation is in breach, its property may be sequestrated and its officers committed for contempt.

(iii) *Non-judicial action: abatement of nuisance*

Self-help is not usually advisable. The person suffering from the nuisance, after giving the other party notice, enters the land of the person committing the nuisance and does whatever necessary to abate it. He or she must not do anything which causes unnecessary damage and if there are alternative methods of abatement, then he or she must choose one that causes least harm, e.g. a person infuriated with the noise of a neighbour's defective burglar alarm entering the property and cutting wires. However, in so doing he or she is committing the torts of trespass to land and property. Fortunately, abatement is a defence to trespass. But this type of approach is still not advisable because one could possibly make oneself liable for negligence if there were a subsequent burglary. The burglary might have been prevented had the defective alarm been noticed and mended by keyholders. If someone does exercise his or her right to abate, he or she cannot then sue for damages in nuisance. (There are now better remedies under legislation see below.)

10.4 Statute law: nuisances

Environmental Protection Act 1990, Noise and Statutory Nuisance Act 1993

Preliminary

The common law is undoubtedly adequate in awarding compensation or injunctions when a nuisance has been committed. But the civil procedures in the county court, or High Court if an injunction is sought, are quite complex and relatively slow, requiring in nearly every case the services of a solicitor. If the case is uncertain, a barrister may be asked to give 'counsel's opinion' on its merits; if the case actually goes to the High Court, a barrister or an advocate solicitor must be briefed to act as advocate. All this adds to the expense and time involved.

In the main, common law actions can only be made *after* the event, and it has been left to Parliament and local authorities to lay down rules preventing noise or restricting it. This legislation is cheaper to enforce, relying on the criminal court structure. Noise control under legislation, whether under the following legislation or under by-laws made by local authorities, is also quicker and easier, and does not necessarily require the services of lawyers.

The Control of Pollution Act 1974 (the 1974 Act) made great strides in noise control and part of this Act remains in force in relation to noise abatement zones (see section 10.7) and construction sites (see section 10.5 below).

The Environmental Protection Act 1990, which for the most part supersedes the 1974 Act, specifically allows recourse to the civil courts for relief, e.g. by way of injunction, if the Act falls short, as in *Hammersmith London Borough Council v. Magnum Automated Forecourts* [1978] 1 WLR 50 and *London Borough of Lewisham v. Saunders* (unreported). The important case of *Barnes (NE) Ltd v. Newcastle upon Tyne City Council* [2006] Env LR 25 regarding applications for injunctions is considered in the statutory nuisance procedure section below.

Statutory nuisance and related matters

The main legislation regarding noise control is now the Environmental Protection Act 1990 (the 1990 Act or EPA) Part III ss. 79–82 as amended and supplemented by the Noise and Statutory Nuisance Act 1993 (the 1993 Act) and the Noise Act 1996 (the 1996 Act), as amended. Part III of the 1990 Act is concerned with statutory nuisances in general as amended by the 1993 Act. The 1996 Act deals with residential night-time noise and has itself been amended by the Anti-Social Behaviour Act 2003 (see section 10.6 below).

Section 79 EPA defines certain matters which constitute 'statutory nuisances'. For this book, however, only those relating to noise will be considered. The subsections of importance to us are therefore:

s. 79(1)(a) Any premises in such a state as to be prejudicial to health or a nuisance.

s. 79(1)(f) Any animal kept in such a place or manner as to be prejudicial to health or a nuisance.

s. 79(1)(g) Noise emitted from premises so as to be prejudicial to health or a nuisance.

s. 79(1)(ga) Noise that is prejudicial to health or a nuisance and is emitted from or caused by a vehicle, machinery or equipment in a street. (This was introduced by the 1993 Act.)

Thus, in each case a statutory nuisance is one which is *either* prejudicial to health *or* a nuisance. As far as noise control in general is concerned, this is a big improvement because the previous legislation was only concerned with nuisances. Now, in applying the EPA to a situation, the local authority's environmental health officer (as enforcing agent) has two approaches in each statutory nuisance situation.

Section 79(7) specifically defines 'prejudicial to health' as 'injurious, or likely to cause injury to health'. The test for whether the circumstances or activities specified in s. 79 are prejudicial to health is an *objective* one, so that where a claimant is hypersensitive, it will be necessary also to provide evidence that an 'ordinary' person would be similarly affected, see *Cunningham v. Birmingham City Council* [1998] Env LR 1 and *R (Anne) v. Test Valley BC* [2002] Env LR 553.

The word *nuisance*, however, is not defined. It is and has been implicit in noise legislation that the word *nuisance* is to be read in terms of both public and private nuisance, and no understanding of the present legislation is possible without this; the reader is referred to pages 276–283 above.

Before looking at each specific statutory nuisance, it should be noted that the word *premises* as defined by s. 79(7) includes land and any vessel other than one powered by steam reciprocating machinery (s. 79(12)). Thus, premises do not have to be indoors, like a house or factory, but will include gardens, fields, and railway premises, etc.

Furthermore, by s. 79(7), the definition clause has been extended by the 1993 Act to define the word *equipment* to include a musical instrument. Also the word *street* means 'a highway and any other road, footway, square or court that is for the time being open to the public'. *Noise* includes vibration. Each of the relevant categories of statutory nuisance is considered further below.

Categories of statutory nuisance

Section 79(1)(a) Any premises in such a state as to be prejudicial to health or a nuisance

Section 79(1)(a) is a re-enactment of s. 91(1)(a) of the Public Health Act 1936 and the similarities between the 1990 Act and the old Public Health Acts leads one to believe that much of the old case law will continue to be of relevance in the future. This section is not intended to cover *noise-making* activities. Instead it is concerned with the passive state of the premises which by their very nature are either prejudicial to health or causing a nuisance. This would include, e.g. condensation, mould and dampness. As to *noise*, landlords have been held liable for statutory nuisance caused by lack of adequate soundproofing, once they are aware of the noise nuisance: *Network Housing Association v. Westminster City Council* (1994) 27 HLR 189, but the *Baxter* case, a decision in the *tort* of nuisance, considered above, held that landlords are not usually liable for the ordinary noise of other tenants. This approach was subsequently adopted in a statutory nuisance case, *Vella v. Lambeth LBC* [2006] Env LR 33 where it was held that the premises were not a statutory nuisance: in themselves they were not in such a state as to be injurious or likely to cause injury to health because of the inadequate sound insulation. The noise from the flat above and from the communal hallway and stairs from the ordinary activities of other tenants was insufficient to establish liability. The claimant had started proceedings under s. 82 but the court held that the local authority was entitled not to serve an abatement notice under s. 79(1)(a).

As regards exposure to traffic noise, in *London Borough of Southwark v. Ince and Another* (1989) 21 HLR 504, a case brought under a similar section of the Public Health Act 1936, the local authority was taken to court as owners of premises which were inadequately insulated against *noise from adjacent railway lines and roads*. The noise was making the occupants ill, and the court held that premises in such a state as to admit noise could be prejudicial to health. This case has come under subsequent judicial scrutiny and criticism, most recently in *Vella v. Lambeth LBC* (above) and while not expressly overruled, may not be followed today, particularly in view of the Noise and Statutory Nuisance Act 1993. A contrasting case to *Ince*, coming after the changes introduced by the 1993 Act is *LB Haringey v. Jowett* (2000) 32 HLR 308, where a council tenant's claim on similar grounds was dismissed on appeal as an attempt to get around the traffic exemption applicable to s. 79(1)(ga) (noise in streets, see further page 297 below and section 10.13 on road traffic below).

Section 79(1)(f) Any animal kept in such a place or manner as to be prejudicial to health or a nuisance

The barking of dogs or the sound of other animals, such as cockerels, has long been a ground for neighbourhood noise disputes. The problem with the 1974 Act's approach was that one could only deal with the situation if the barking, etc., amounted to a nuisance in common law terms. Also prosecutions under the Public Health Act 1936, s. 92(1)b, an identical section to s. 79(1)(f), have cast doubt on whether one could apply it to animals who were *only* noisy. See *Galer v. Morrissey* [1955] 1 All ER 380, where it was held not to apply, and *Coventry City Council v. Cartwright* [1975 1 WLR 845, where doubt was cast on this viewpoint. It remains to be seen whether action can be taken, under this section and/or under s. 79(1)(g), if it can be proved that the noise is *only* prejudicial to health, e.g. if the noise is only occasional. Probably, if there is only noise involved, subsection (g) may be more appropriate, as indicated in more recent cases such as *Lowe & Watson v. South Somerset DC* [1998] Env LR 143 (cockerels and waterfowl) and *Budd v. Colchester BC* [1999] Env LR 739 (greyhounds). (Remember that the keeping of animals can also be dealt with under other legislation such as Town and Country Planning, see pages 321–325, and local by-laws, see page 330.) If cruelty is involved, the police and RSPCA can also deal with this under the Protection of Animals Act 1911. DEFRA Guidance, *Constant Barking Can Be Avoided*, was published July 2005. The Clean Neighbourhoods and Environment Act 2005, Part 6 makes provision for, among other things, dog control orders.

Section 79(1)(g) Noise emitted from premises so as to be prejudicial to health or a nuisance

This replaces and extends the previous law found in Part III of the Control of Pollution Act 1974. It should be emphasized that this subsection allows action to be taken whether the noise is or is likely to be harmful to health, or if it is a nuisance. 'Noise' includes vibration. This subsection does not apply to premises occupied on behalf of the Crown for Navy, Army or Air Force purposes, for the Department of Defence or for the purposes of a visiting force as defined by the Visiting Forces Act 1952 s. 79(2). Furthermore, no action can be taken under these provisions in relation to noise caused by aircraft in general, except for model aircraft, which are subject to a code of practice (see page 302). Remember, however, that a common law action in the tort of nuisance may nevertheless be available. In *Dennis v. MoD* (2003), the claimant's action at common law for noise nuisance in respect of RAF fighter jets flying over his land resulted in his being awarded substantial damages.

Under this section the noise must be emitted from *premises*. In *Tower Hamlets London Borough Council v. Manzoni & Walder* [1984] JPL 437 the word *premises* was held not to cover noise created in streets or public places. Here demonstrators used megaphones to protest against the sale of animals at a street market. This continued for a number of weeks and the council served a noise nuisance notice under s. 58 of the 1974 Act. On appeal to the Crown Court against the notice, the court held that the section should be read in the light of the whole of Part III of the 1974 Act and s. 58 should therefore be construed to apply to noise emanating from premises.

As a result of the limitations of this subsection, the 1993 Act introduced s. 79(1)(ga): noise that is prejudicial to health or a nuisance and is emitted from or caused by a vehicle, machinery or equipment in a street.

This subsection does not apply to noise made by (a) traffic, (b) any Navy, Army or Air Force of the Crown or by a visiting force as defined previously, or (c) by a political demonstration or a demonstration supporting or opposing a cause or campaign (s. 79(6A)). Thus the court allowed the council's appeal in *LB Haringey v. Jowett* (see above). The tenant argued that the intrusion of traffic noise into his inadequately soundproofed flat was a statutory nuisance under s. 79(1)(a), as it was prejudicial to health. The authority argued that their tenant could not rely on s. 79(1)(a) because s. 79(1)(ga) specifically dealt with noise in the street and the s. 79(6A) exemption provided that s. 79(1)(ga) did not apply to noise from traffic (see also section 10.13 below).

Car radios, cars being worked on or revved up, outdoor do-it-yourself work on the roadside and car alarms could all fall within the scope of this subsection. Defective exhausts will probably come within its scope, too. *Westminster CC v. French Connection Retail Ltd* [2005] EWHC 933 (a prosecution under s. 62(1)(b) Control of Pollution Act 1974), indicates that 'in the street' is likely to be broadly construed under this provision also. The Divisional Court held that an offence under the 1974 Act was committed when sound equipment attached on the inside of a shop window broadcast music into the street. The equipment made the window a loudspeaker and the outer face of that window was in the street. Street and neighbourhood noise in general is dealt with in section 10.6 below.

In total the 1990 Act, as amended, sets out 12 categories of statutory nuisance, two of which concern noise specifically and two others which may do so. Noise is, however, the most common complaint to environmental health officers (Stookes, 2009: 19.47 – see Appendix 10.3 below). In 2008/9, based on returns from 46% of local authorities, the CIEH reported that the number of incidents confirmed as a statutory nuisance was over 25,000. We can now move on to examine more closely the duty imposed on a local authority and relevant procedures. The statutory duty, which is set out below, cannot be avoided.

Statutory nuisance procedure

The local authority's duty

Section 79(1) continues after defining the different matters that constitute 'statutory nuisances' by stating:

> and it shall be the duty of every local authority to cause its area to be inspected from time to time to detect any statutory nuisances which ought to be dealt with under section 80 below and, where a complaint of a statutory nuisance is made to it by a person living within its area, to take such steps as are reasonably practicable to investigate the complaint.

Duty

In law if a *duty* is imposed on someone or somebody then that duty must be carried out. It is not a mere *power* which may or may not be carried out. This does indeed mean that local authorities are in somewhat of a dilemma. They must carry out this duty but may not have sufficient money to fund it. What could happen if they do not make their investigation? Schedule 3 para. 4 of the 1990 Act specifically deals with this duty. If the Secretary of State is satisfied that the local authority has failed, in any respect, to discharge this function, he or she may make an order declaring the authority to be in default. Such an order may direct the defaulting authority to perform the function specified in the order and may specify the manner of its performance and the time(s) when it should be performed. Should the authority fail to comply with the order, the Secretary of State, instead of enforcing the order by a mandatory order, may make an order transferring to himself or herself the function of that authority. (A mandatory order, formerly called 'Mandamus' meaning 'we command', is a High Court order used to order the performance of a public duty, often by a local authority.) Any expenses incurred by the Secretary of State must be reimbursed by the defaulting authority. The Secretary of State can vary or revoke any such order. It is argued that, as the Act has specifically enacted these procedures, it is unlikely any other remedies for maladministration are available.

Local authorities *must* therefore inspect for statutory nuisances. Local authorities may contract out their enforcement authority or have civic wardens for this purpose.

Complaints made to the local authority

This reflects what has usually always happened in practice. If anyone living in the local authority's area complains to the local authority of a statutory nuisance, the local authority must take 'such steps as are reasonably practicable to investigate the complaint.' It is much more likely that claims for maladministration in this respect would arise where the local authority ignores the complaint.

Following the investigation

Section 80 abatement notices

Section 80(1) states that where a local authority is satisfied that a statutory nuisance *exists*, or is likely to *occur* or *recur*, in the area of the authority, the local authority shall serve a notice ('an abatement notice') imposing all or any of the following requirements:

(a) requiring the abatement of the nuisance or prohibiting or restricting its occurrence or recurrence
(b) requiring the execution of such works and the taking of such other steps as may be necessary for any of those purposes;

and the notice shall specify the time or times within which the requirements of the notice are to be complied with.

This section imposes a *duty* on the authority. It *must* serve an abatement notice if a statutory nuisance:

- exists
- is likely to occur
- is likely to recur.

Subsections (2A) to (2E) of s. 80 (added by s. 86 Clean Neighbourhoods and Environment Act 2005) allow the local authority to defer its duty to serve an abatement notice for up to seven days in the case of a statutory nuisance under category (g).

Compare this with a civil action for nuisance, where the sufferer can only sue when there has been nuisance behaviour on more than one occasion (see page 279). Normally, civil action can never be used for behaviour which may occur in the future.

The abatement notice must impose certain requirements. It can require:

1. abatement of the nuisance, e.g. asking someone to reduce the sound from their television
2. prohibition of a nuisance likely to occur or recur, e.g. a forthcoming pay party that will probably cause a nuisance
3. restriction of such nuisances, e.g. imposing days and times during which the nuisance behaviour must be stopped
4. execution of such works and pursuance of such other steps as may be necessary for those purposes.

(Mnemonic, PEAR: prohibition, execution, abatement and restriction).

The notice must specify the time(s) for compliance. Normally this should be a reasonable time, but cases under s. 58 of the 1974 Act had held that no time need be specified; indeed in certain cases, a very short time would be sufficient, depending on the facts. In *Strathclyde Regional Council v. Tudhope* [1983] JPL 536 the City of Glasgow Council served a s. 58 notice on the Strathclyde Authority in respect of noisy roadworking operations, requiring that all pneumatic drills should be fitted with effective exhaust silencers and dampened tool-bits. No time was stated for compliance; the court held that the notice should come into effect at midnight following the date of service and that this was not unreasonable. However, to avoid appeals being made in relation to the reasonableness of time limits imposed by abatement notices, it is much more sensible to state a reasonable time at the outset. If someone is being asked to build something or modify machinery, tasks which require expert advice, a longer time should be given.

Should the abatement notice contain a prohibition on the occurrence or recurrence of the nuisance then this prohibition continues indefinitely. In *Wellingborough Borough Council v. Gordon* (1991) the council had served a prohibition notice in 1985 banning the recurrence of a noise nuisance. The recipient subsequently held a birthday party to which all his neighbours were invited and to which there were no complaints. The council took the recipient to the magistrates' court for breach of the notice. The magistrates' court agreed with the recipient that this was an isolated incident and this was a reasonable excuse, especially since there had been no complaints. This was overturned on appeal. No appeal had been made against the notice in the first place and the reasonable excuse defence should not have been allowed by the magistrates. The loud music at the party was exactly the sort of noise the notice had been designed to prevent in the first place.

A House of Lords decision settled a vexed question once and for all. In *Aitken v. South Hams District Council* [1994] 3 WLR 333, even though it could be argued that s. 58 of the Control of Pollution Act 1974 seemed to have been repealed completely, without any provision for the continued validity of the notices, the Lords held that the notice did, in law, continue to be effective. They reversed the decision taken in the Divisional Court. The case concerned the appellant who appealed against a conviction for contravening a s. 58 notice served in November 1983, during the period between August and October 1991, by allowing his dogs to bark so as to cause a nuisance. The legislation involved both the Environmental Protection Act 1990 and s. 16(1) of the Interpretation Act 1978.

Who must the notice be served upon?

Section 80(2) states that subject to s. 80A(1), below, the abatement notice must be served as follows:

(a) On the person responsible for the nuisance.
(b) On the owner of the premises if the nuisance arises from any defect of a structural character (as in the *Ince* case, see page 285).
(c) Where the person responsible for the nuisance cannot be found or the nuisance has not yet occurred, on the owner or occupier of the premises.

The 'person responsible' is defined in s. 79(7) as the person to whose *act, default* or *sufferance* the nuisance is attributable. This shows that the person may be responsible in a positive way, by actually doing the act, or in a negative way, by not doing something he or she should be doing or by allowing a state of affairs to continue.

Trespassers who are by definition in occupation, even though they are not the legal owners of the land, may be served with notices if they create the nuisance or as occupiers.

'Where the person responsible . . . cannot be found' may cover situations where the local authority would only be able to ascertain the person truly responsible for the nuisance after extensive and costly research, and the court should be told what and how extensive the enquiries were.

The 1993 Act inserted a new section, s. 80(A), which applies to statutory nuisances under s. 79(1)(ga). In such cases, where the nuisance (a) has *not* yet occurred or (b) arises from noise emitted from or caused by an unattended vehicle or unattended machinery or equipment, the abatement notice shall be served in accordance with s. 80(A)(1).

Section 80(A)(2) states that the notice shall be served:

(a) Where the person responsible for the vehicle, machinery or equipment *can* be found, on *that* person.
(b) Where the person cannot be found or where the local authority determines that the paragraph should apply, by fixing the notice to the vehicle, machinery or equipment.

Section 80(A)(3) states that where:

(a) an abatement notice is served in accordance with s. 2(b) above by virtue of a determination of the local authority, and
(b) the person responsible for the vehicle, etc., *can* be found and served with a copy of the notice within an hour of the notice being fixed to the vehicle, etc., a copy of the notice shall be served on that person accordingly.

This would cover the situation where someone parks his or her car with the radio playing while visiting a shop. The notice is attached to the windscreen before the driver returns within the hour. A copy of the notice is then served on the driver. If anyone without authority removes the notice, it is a criminal offence liable on summary conviction (s. 3(7) of the 1993 Act).

By s. 80(A)(4) if an abatement notice is served under s. 2(b), i.e. the local authority determines that the person responsible cannot be found, the notice affixed to the vehicle, etc., must state that the time specified in the notice for compliance with requirements will be extended by a further specified time if a subsequent copy notice is served under s. 2(3).

Thus s. 80 gives the local authority wide-ranging powers. Not only may it give notice to stop or reduce existing noise, but it may also restrict the emission of noise to certain parts of the day or week. It can prohibit noise nuisance which has not yet occurred or nuisance that may happen again, and it can order works to be carried out. The owner or occupier of premises may be held responsible.

Execution of works

The local authority need not necessarily specify in the abatement notice what works must be carried out to comply with it but can, in appropriate cases leave the choice of the means of abatement to the perpetrator (*Lambie v. Thanet DC* [2001] Env LR 397). The authority must only 'require' the execution of such works and the taking of such steps as may be necessary to fulfil the terms of the abatement notice. These words allow the authority to show that work must be done without having to specify what exactly has to be done. It has happened in the past that the specified works have been carried out yet the nuisance remained. In *Budd v. Colchester BC* [1999] Env LR 739 the Court of Appeal held that it was unnecessary to specify the steps to be taken in the case of barking dogs. Where, however, the local authority chooses to specify the necessary works or steps required to abate the nuisance, then they must be specified with sufficient clarity, i.e. so that the recipient of the notice is reasonably certain what he is required to do. In *R v. Falmouth & Truro PHA ex p South West Water Ltd* [2000] Env LR 658, Simon Brown LJ stated, '. . . in all cases the local authority can if it wishes leave the choice of means of abatement to the perpetrator of the nuisance. If, however, the means of abatement *are* required by the local authority, then they must be specified.'

It should be noted that by s. 81(3) if an abatement notice has not been complied with, the local authority may abate the nuisance itself and do whatever may be necessary in execution of the notice. This is so even if the authority does or does not decide to take proceedings in court for an offence under s. 80(4), see below.

Furthermore, by s. 81(4) any expenses reasonably incurred by a local authority under s. 81(3) may be recovered from the person who caused the nuisance, and if that person is the owner of the premises, from *any* person who is for the time being the owner. Presumably this means it is possible to claim against people besides the person who was an owner of the property at the time the notice was served, thus avoiding a hiatus if the property is sold while the proceedings are carried on. Under this subsection the court may also apportion the expenses between persons whose acts or default have caused the nuisance, in such a manner as it thinks fair and reasonable.

Co-responsibility for nuisances

What happens if a number of people have contributed to a statutory nuisance but individually their behaviour would not necessarily amount to a nuisance? Section 79(1) clearly states that s. 80 will apply to each person in such situations. This has been added to by the 1993 Act. An additional s. 81(1A) states that in the case of a s. 79(1)(ga) statutory nuisance where more than one person is responsible (whether or not their individual behaviour would amount to a nuisance), s. 80(2)(a) shall apply with the substitution of 'any one of the persons' for 'the person'.

Nuisance created outside the local authority's area

By the very nature of noise nuisances, it could happen that the point of immission, where the noise nuisance actually has its effect, is different from the point of emission, where it has been created. Indeed, the two sites could occur across local authority borders. But s. 81(2) allows that in such situations the local authority may still take action under s. 80 as if the act or default was wholly in its area, but any appeal made against the notice must be heard by the magistrates' court having jurisdiction where the act or default is alleged to have taken place. Liaison between the authorities to obviate reliance on this section is obviously desirable, so that reciprocal arrangements are made whereby the authority from whose area the nuisance is emitted takes action.

Appeals

By s. 80(3) the person served with the notice may appeal against the notice to a magistrates' court within the period of 21 days beginning with the date on which he or she was *served* with the notice. Under Schedule 3 to the Act the appeal is made *by way of complaint for an order* and the Magistrates' Courts Act 1980 applies (para. 1(2)).

Any appeal *from* the magistrates' court in relation to an appeal under s. 80(3) will be made to the Crown Court at the instance of *any* party to the proceedings in which the decision was given (Schedule 3, para. 1(3)).

The Secretary of State has made regulations in relation to such appeals. These are the Statutory Nuisance (Appeals) Regulations 1995 (S.I. 1995/2644) made under the 1990 Act, as amended. Regulation 2(2) sets out the following grounds for appeal under s. 80:

(a) That the abatement notice is not justified by the terms of s. 80.
(b) That there has been some informality, defect or error in or in connection with the abatement notice or in, or in connection with, any copy of the abatement notice served under s. 80A(3).
(c) That the authority has refused unreasonably to accept compliance with alternative requirements, or that the requirements of the abatement notice are otherwise unreasonable in character or extent, or are unnecessary.
(d) That the time, or times within which the requirements of the abatement notice are to be complied with is not reasonably sufficient for the purpose.
(e) Where the noise to which the notice relates is a nuisance falling within categories (a), (f), (g) of the Act and arises on industrial, trade or business premises; or category (ga) noise emitted from or caused by a vehicle, machinery or equipment in a street used for industrial, trade or business purposes, that the best practicable means have been used to prevent or to counteract the effect of the noise.
(f) That in the case of category (g) (or ga) the requirements imposed by the abatement notice by virtue of s. 80(1)(a) are more onerous than the requirements for the time being in force, in relation to the noise to which the notice relates, of:
    (i) any notice served under s. 60 or s. 66 of the Control of Pollution Act 1974 (control of noise on construction sites and from certain premises),
    (ii) any consent given under s. 61 or s. 65 of the Control of Pollution Act 1974 (consent for work on construction sites and consent for noise to exceed registered level in a noise abatement zone),
    (iii) any determination made under s. 67 of the Control of Pollution Act 1974 (noise control of new buildings).

(g) That, in the case of a nuisance under (ga) (noise emitted from or caused by vehicles, machinery or equipment in a street), the requirements imposed by the abatement notice by virtue of s. 80(1)(a) of the Act are more onerous than the requirements for the time being in force, in relation to the noise to which the notice relates, of any condition of a consent given under para. 1 of Schedule 2 to the 1993 Act (loudspeakers in streets or roads).
(h) That the abatement notice *should* have been served on some person instead of the appellant, being:
  (i) the person responsible for the noise, or
  (ii) the person responsible for the vehicle, machinery or equipment, or
  (iii) in the case of a nuisance arising from any defect of a structural character, the owner of the premises, or
  (iv) in the case where the person responsible for the nuisance cannot be found or the nuisance has not yet occurred, the owner or occupier of the premises.
(i) That the abatement notice *might* lawfully have been served on some person instead of the appellant, being:
  (i) in the case where the appellant is the *owner* of the premises, the *occupier* of the premises,
  (ii) in the case where the appellant is the *occupier* of the premises, the *owner* of the premises,

  and that it would have been equitable for it to have been so served.
(j) That the abatement notice might lawfully have been served on some person in addition to the appellant, being:
  (i) a person also responsible for the nuisance
  (ii) a person who is also an owner of the premises
  (iii) a person who is also an occupier of the premises
  (iv) a person who is also the person responsible for the vehicle, machinery or equipment,

  and that it would have been equitable for it to have been so served.

If the court is satisfied that there was an informality, defect or error in the abatement notice but that it was not material, then the court shall dismiss the appeal (reg. 2(3)).

If the grounds for appeal include one specified in para. 2(i) or (j), the appellant must serve a copy of the notice of appeal on any other person referred to, or any other person having an estate or interest in the premises, vehicle, machinery or equipment in question.

On hearing the appeal, the court may quash the notice, vary it in favour of the appellant or dismiss the appeal.

In *Sfi Group Plc (Formerly Surrey Free Inns Plc) v. Gosport Borough Council* [1999] Env LR 1, the Court of Appeal held that when considering an appeal, the time for assessing the noise nuisance is at the time the abatement notice is served, not at the time the appeal against the notice is heard. The Court held that the case of *Johnsons News of London v. Ealing London Borough Council* (1989) 154 JP 33 on this point was wrongly decided.

Regulation 3 specifies two situations in which a notice will be *suspended* pending appeal:

(a) When the notice requires expenditure on works before the appeal has been heard.
(b) In the case of category g (or ga), when the notice to which the noise relates is caused in the course of the performance of a legal duty, e.g. by a statutory undertaker such as British Gas.

However, the notice shall not be suspended pending appeal in the above situations if:

1. the noise is injurious to health,

or

2. the noise is of such limited duration that suspension would render the notice of no practical effect, or the expenditure incurred on the works before the hearing would not be disproportionate to the public benefit.

In such cases the notice must include a statement that it shall not be suspended pending appeal. Such a statement was included in the leading case *Hammersmith London Borough Council v. Magnum Automated Forecourts Ltd.* [1978] 1 WLR 50.

Powers of entry

Schedule 3 para. 2 of the Act provides that any person authorized by a local authority may, on production of his or her authority (if required), enter any premises at any reasonable time:

(a) in order to ascertain whether or not a statutory nuisance exists,

or

(b) for the purpose of taking any action, or executing any work, authorized or required by Part III of the Act.

Paragraph 3(2) further provides that in the case of wholly or mainly residential premises, normally 24 hours' notice of the intended entry under the previous subparagraph must be given to the occupier, unless there is an emergency. Paragraph 2(3) deals with refusal of admission and other difficulties in gaining entry and makes provision for the magistrates' court to authorize a warrant of entry:

> Should it be shown to the satisfaction of a justice of the peace in a sworn information in writing
> (a) that admission to any premises has been refused, or that refusal is apprehended, or the premises are unoccupied, or the occupier is temporarily absent or that the case is one of emergency or that an application for admission would defeat the object of the entry; and
> (b) that there is reasonable ground for entry into the premises for the purpose for which entry is required, the justice may by warrant under his hand authorise the local authority by any authorised person to enter the premises, if need be by force. (para. 2(3))

When entering the premises under any of the above circumstances, the authorized person may:

(a) take other people with him and such equipment as may be necessary
(b) carry out such inspections, measurements and tests as he considers necessary for that purpose (para. 2(4)).

After such authorized entries, the person must then leave the premises *as effectively secured against trespassers as he or she found them* (para. 2(5)).

> Any warrant issued under sub-para. 3 shall continue in force until the purpose for which the entry is required, is satisfied (para. 2(6)).
>
> An emergency is to be interpreted as a situation where the authorized person has reasonable cause to believe that there is immediate danger to life or health and that immediate entry is necessary to verify the facts or to ascertain their cause and to effect a remedy (para. 2(7)).

The 1993 Act has extended the powers in Schedule 3 to provide any person authorized by a local authority with powers to:

(a) enter or open a vehicle, machinery or equipment, if necessary by force,

or

(b) remove a vehicle, machinery or equipment from a street to a secure place.

Both powers are for the purpose of taking any action, or executing any work, authorized by or required under Part III in relation to a statutory nuisance within s. 79(1)(ga).

Once again, on leaving the unattended vehicle, etc., it must be left properly secured as effectively as it was found according to s. 4(2) of the 1993 Act. Otherwise it should be immobilized or removed from the street (s. 4(3)). No more damage should be committed than necessary (s. 4(4)). Before any of the above action is taken, the police should be notified; they should also be notified if the vehicle, etc., is removed and should be informed of its new location.

Offences relating to entry

Paragraph 3(1) of Schedule 3 to the Act provides that if a person *wilfully obstructs* someone exercising the above powers he or she shall be liable on summary conviction (in the magistrates' court) to a fine not exceeding level 3 on the standard scale.

Trade secrets

Someone carrying out the powers conferred under para. 2 must be careful that he or she does not disclose a trade secret discovered while exercising those powers, otherwise he or she will be liable on summary conviction to a fine not exceeding level 5 on the standard scale. This is so unless the disclosure was made in the performance of his or her duty, or with the consent of the person having the right to disclose (para. 3(2)).

Breach of notice

By s. 80(4), as amended, if a recipient of an abatement notice contravenes any requirement contained in the notice without *reasonable excuse*, he or she commits an offence against this part of the Act. Summary proceedings will then proceed in the magistrates' court.

If the offence is committed on industrial, trade or business premises, the guilty person will be liable to a fine on summary conviction not exceeding £20,000, according to s. 80(6). But if the offence is committed elsewhere, the fine will be not more than level 5 on the standard scale (currently £5,000), with a further fine equal to one-tenth of that level for each day the offence continues.

In *Cooke v. Adatia* (1989) it was held that, in proving there has indeed been a breach of the notice, it is not necessary to show that a particular occupier of premises has suffered from the nuisance (cf. an action for private nuisance). Thus, if it is impossible for one reason or another to produce admissible evidence from a particular occupier, then noise meter evidence or other evidence of the noise can be produced by the environmental health officer. There is no specified type of evidence that need be produced under the Act. More recent cases confirm this view. *Lewisham LBC v. Hall* [2003] Env LR 4 held that merely because there was no evidence of acoustic measurement before the magistrates, they should not refuse to convict a defendant charged with an alleged breach of an abatement notice. And where expert evidence is produced, the Divisional Court in *R (Hackney LBC) v. Rottenberg* [2007] Env LR 24 held, conversely, that there was no obligation to accept that evidence uncritically: judgment does indeed rest with the court. See also *Roper v. Tussauds*

Theme Parks Ltd [2007] Env LR 31 on this issue, considered in section 10.11 below.

While the *categories* of statutory nuisance are clearly set out in the legislation, the question of the *extent* to which an activity or set of circumstances constitutes a nuisance, is determined by applying the principles and criteria for suing in nuisance at common law. *Godfrey v. Conwy CBC* [2001] Env LR 38, held that where the noise causes an unreasonable interference with the claimant, it can constitute a statutory nuisance if it has qualities that are irritating, without necessarily exceeding background noise levels and being injurious to health (remember that abnormally sensitive individuals will not succeed in their nuisance action unless the average person would be similarly affected). Each case will turn on its particular facts. A contrasting case is *Lewisham LBC v. Fenner* (1995) ENDS Report 248 where it was held that there was no statutory nuisance despite complaints, since the noise levels from a humidifier at a swimming pool fell below the recommended threshold for intrusive noise in BS 8233. This is not to say, however, that compliance with recommended standards, licences and other authorizations, e.g. planning permission, is a defence to statutory nuisance proceedings: *Cambridge City Council v. Douglas* [2001] Env LR 715 (noise in compliance with a public entertainments licence held to be a statutory nuisance).

Defences

(i) *Reasonable excuse*

Firstly, it is a defence if the recipient had 'reasonable excuse' to breach the abatement notice (s. 80(4)). This will depend on the facts of the situation but would probably cover emergency situations.

In *Wellingborough Borough Council v. Gordon* [1991] JPL 874 reasonable excuse was not held to be a proper defence in those circumstances, although the recipient's behaviour could be pleaded in mitigation. In *Lambert (A) Flat Management v. Lomas* [1981] 2 All ER 280 the 'reasonable excuse' defence was said to have been provided in the 1974 Act (Control of Pollution Act) only as a defence to a criminal charge; it was not to provide an opportunity to challenge the notice, which should properly have been done by appealing against its issue. (Read the grounds for appeal against notice; reasonable excuse does not figure among them.)

(ii) *Best practicable means*

If the noise was caused in the course of a trade or business, then it is a defence to prove that the 'best practicable means' have been used to counteract the effects of the nuisance (s. 80(7)). This defence is available in the case of all the possible noise-related statutory nuisance situations, s. 79(1)(a), (f), (g) and (ga), but only if the premises or vehicle, machinery or equipment is used for trade or business.

Section 79(9) gives a statutory definition of what is meant by best practicable means. The word *practicable* means reasonably practicable having regard, among other things, to local conditions and circumstances, to the current state of technical knowledge and to the financial implications. Although a court may have sympathy with a factory vital to a community, it will not necessarily acquit the owner because of financial implications. In *Wivenhoe Port v. Colchester Borough Council* [1985] JPL 175, a case concerned with dust nuisance, the court held that the defence was not maintained by proving that the business would require extra money to operate or even would become unprofitable. The *means to be employed* include the design, installation, maintenance and manner and periods of operation of plant and machinery, and the design, construction and maintenance of buildings and structures.

The test, according to the section, is to apply only so far as is compatible with any duty imposed by law. Thus, if there are legal duties imposed at common law in nuisance, in negligence or in statute in the same situation, this test must be compatible with those duties. Finally the test is to apply only so far as compatible with safety and safe working conditions, and with the exigencies of any emergency or unforeseeable circumstances. If there is a code of practice under s. 71 of the Control of Pollution Act (noise minimization), this must be regarded when applying the best practicable means defence.

(iii) *Noise nuisance special defences*

The above defences apply on the whole to all types of statutory nuisances. There are additional special defences relating to s. 79(1)(g) noise nuisances, introduced by the 1974 Act, and to s. 79(1)(ga), introduced by the 1993 Act.

By s. 80(9)(a) it will be a defence if the alleged offence under s. 80(4) was covered by a notice served under s. 60 of the 1974 Act, or consent had been given under s. 61 or s. 65 of the 1974 Act. These sections relate to noise on construction sites and noise level registers (see below).

There will also be a good defence if a s. 66 Control of Pollution Act noise reduction notice was in force at the time of the alleged offence and the noise was not in contravention of that limit (s. 80(9)(b)).

It will be a defence if a noise reduction notice had been made in relation to a particular day or time and the alleged offence was committed outside the period for which the limit was laid down (s. 80(9)(b)).

Finally, if a s. 67 Control of Pollution Act noise level (new buildings liable to an abatement *order*, not an abatement notice) had been fixed and the noise of the alleged offence did not exceed that level, then this would be a good defence (s. 80(9)(c)).

In the last three paragraphs, the defences apply whether or not the relevant notice was subject to appeal at the time when the offence was alleged to have been committed (s. 80(10)).

The right to take High Court action

Section 81(5) allows the local authority to take proceedings in the High Court, if it feels that proceedings for an offence under s. 80(4) would be an inadequate remedy. The intention would be to secure the abatement, prohibition or restriction of the nuisance, even though the local authority has suffered no damage from the nuisance.

This may be instead of, or as is possibly more usual, following failure of summary proceedings in the magistrates' court.

This section can only be relied on if the nuisance complained of is a statutory nuisance. If the nuisance is only a private nuisance, the local authority or the person suffering only have the right to proceed in the High Court if they have suffered interference with their own land. If the nuisance is a public nuisance, the local authority would have to use s. 222 of the Local Government Act 1972 (see page 272).

Compare the use of s. 222 in respect of breach of a construction abatement notice under s. 60 of the 1974 Act, where no statutory right was given within the Act itself, unlike s. 81(5) above (see the *Bovis* case below, in section 10.5).

An example of a situation where this section was used is the leading case of *Hammersmith London Borough Council v. Magnum Automated Forecourts Ltd.* (1978) which sorted out the confusion over jurisdiction when using the equivalent section under the old legislation. In 1976, the defendant company erected a new building in a quiet street and used it for a 'taxi care centre'. This provided 24-hour service for taxis, selling fuel, washing facilities and vending-machines. People living nearby complained about the noise, particularly in the early hours of the morning. On being satisfied that a noise amounting to a nuisance existed, the local authority served notice under the then s. 58(1) of the 1974 Act, requiring Magnum Ltd. to stop all activities on the premises between 11:00 pm and 7:00 am. In the notice the authority had stated that the expenditure necessary to comply with the notice was not disproportionate to the public benefit. Thus, the notice was not suspended pending any appeal. The defendants continued to operate as before. They also appealed against the notice. The local authority therefore decided to apply to the High Court for an injunction. When the appeal came before the magistrates' court, the magistrates said that the High Court must now deal with the matter. When the case was heard in the High Court, it decided that the magistrates must determine the case. Before the magistrates' appeal was heard, the appeal from the High Court was heard by the Court of Appeal. It decided that the present situation was adequately covered by s. 58(8) and that recourse to the High Court for a civil remedy was thus available. The notice had specifically stated, as required, that it would not be suspended pending an appeal. When the nuisance continued, in contravention of the notice, action had to be taken, and if the authority was of the opinion that proceedings under s. 58(4) would be inadequate, there would be no alternative but to seek an injunction. The Court of Appeal thus granted an interlocutory injunction (i.e. pending the outcome of the action) pending determination of the matter in the magistrates' court.

If the local authority felt at the outset that there would be no response at all to an abatement notice, it could proceed immediately under s. 81(5) in the High Court and the whole matter would be dealt with in the High Court, without any reference to the magistrates' court.

In *Barnes (NE) Ltd v. Newcastle upon Tyne City Council* [2006] Env LR 25 the Court of Appeal held that there had to be prior service of an abatement notice by a local authority under s. 80(1) before seeking injunctive relief under s. 81(5) EPA 1990.

If such proceedings are brought under s. 81(5), in relation to a statutory noise nuisance s. 79(1)(g) or (ga) only, then it will be a good defence to prove that the noise was authorized by a notice under s. 60 or a consent under s. 61 (construction sites) of the Control of Pollution Act 1974.

Aggrieved persons' rights (s. 82)

In recessionary times the local authority's ability to take action under s. 80 can often be reduced through lack of money. Night-time activities, such as pay parties, cannot be properly monitored if there are no environmental health officers on duty. Irregular nuisances are also extremely difficult to monitor. So, with all the will in the world, it may be necessary to turn away a person suffering from a noise nuisance because the local authority is unable, for whatever reason, to take action. Fortunately, s. 82 is available to help in such situations. (Remember that this section can be used *against* a local authority, especially in respect of category (a) statutory nuisances.)

This section is a considerable improvement on its old equivalent, s. 59 of the 1974 Act, which only gave *occupiers of property* rights to complain to a magistrates' court. Furthermore, the 1993 Act has greatly extended this section by covering category (ga) cases, too.

Now a magistrate's court may act on a complaint made by *any person* that is *aggrieved by the existence of a statutory nuisance.* Who is a person aggrieved? It appears from cases that the aggrieved person should not be 'a mere busybody who is interfering in things that do not concern him' as per Lord Denning in *Att.-Gen. (Gambia) v. N'Jie* [1961] 2 WLR 845. There should be some connection between the aggrieved person and the statutory nuisance complained of. Thus, if someone's health is being affected or someone is suffering from a nuisance directly, he or she should be able to make a complaint. Similarly, if his or her family is being affected, he or she will be able to complain. (As proceedings under s. 82 are criminal, not civil, the correct procedure is for the individual affected to lay an information before the court.)

The magistrates may, if satisfied that the alleged nuisance exists or that, although abated, it is likely to recur on the same premises (or in the case of a category (ga) nuisance, the same street), make a nuisance order for either or both of the following purposes:

(a) requiring the defendant to abate the nuisance, within a time specified in the order, and to execute any works necessary for that purpose;
(b) prohibiting a recurrence of the nuisance and requiring the defendant, within a time specified in the order, to execute any works necessary to prevent the recurrence.

The court may also impose a fine not exceeding level 5 on the standard scale.

This section does not allow action to be taken if there is only a likelihood of a statutory nuisance occurring. One must either *exist* or there must be a likelihood of its recurrence (cf. the local authority's duties).

The date on which it must be decided whether a nuisance exists or, if abated, is likely to recur is the date of the hearing of the complaint.

By s. 82(3) if the magistrates are satisfied that the alleged nuisance exists and in their opinion renders the premises unfit for human habitation, they may, by order under subsection (2), prohibit the use of the premises for human habitation until the premises are to the satisfaction of the court rendered fit.

The proceedings should be brought against the same people as in s. 80 and the co-responsibility section also applies (s. 82(4) and (5)).

The local authority's rights are different from those of the aggrieved person, in that magistrates' court action would only occur following the service of an abatement notice by the authority. However, before instituting proceedings under s. 82, the aggrieved person must now give notice in writing of his or her intention to bring proceedings, and the notice must specify the matter complained of. In relation to category (g) noise nuisances, at least three days' notice must be given. For all other categories, a minimum of 21 days' notice is necessary.

The order must be unambiguous and need not necessarily specify what works must be done to abate the nuisance, unless it would otherwise be unclear what is required. Compare a simple noise nuisance situation with a category (a) situation, which could require the soundproofing of walls and floors; the order must then be served as soon as possible on the defendant.

If, after receiving the court order, the defendant, without reasonable excuse, contravenes any requirement or prohibition imposed by that order, he or she will be guilty of an offence and liable to summary conviction in the magistrates' courts to a fine not exceeding level 5 on the standard scale together with a further fine of an amount equal to one-tenth for each day the offence continues after the conviction (s. 82(8)).

A recent example of the use of the s. 82 procedure is *Roper v. Tussauds Theme Park Ltd* [2007] Env LR 31 where residents living near Alton Towers claimed that the noise from Alton Towers park amounted to a statutory nuisance. This case is discussed further in section 10.11 below.

Defences

The 'best practicable means' defence is available in relation to category (a), (f) and (g) (and ga) statutory nuisance, but only if the nuisance is committed on industrial, trade or business premises. It is not available if the nuisance is such as to render the premises unfit for human habitation (s. 82(9) and (10)).

Local authority works

By s. 82(11), if someone is convicted of an offence under s. 82(8), the magistrates, *after* giving the local authority the opportunity of being heard on the subject, may direct the authority to do anything which the convicted person was required to do by the order to which the conviction relates.

Also by s. 82(13), if the potential defendants cannot be found, then after listening to representations by the local authority, the magistrates may direct the authority to do anything which the court would have ordered those people to do.

Compensation

Section 82(12) gives the complainant an opportunity to be compensated by the defendant. If it is proved that the nuisance existed at the date of the complaint, whether or not it is shown that the nuisance still exists or is likely to recur, the courts shall order the defendant(s) to pay the complainant a reasonably sufficient sum to compensate him or her for any expenses properly incurred by him or her in the proceedings. Expenses are not the same as damages in a civil case, where the claimant would be compensated for the interference with the use of his or her land. Nevertheless, it is an improvement on the previous law, although there was a similar provision under the Public Health Act 1936 for non-noise statutory nuisances.

10.5 Construction sites

Cf. cases in section 10.3 above

The Control of Pollution Act 1974 applies here. As previously mentioned the Act was not completely repealed by the Environmental Protection Act 1990.

One of the most difficult areas involving noise control relates to construction sites, which by virtue of the type of work involved, often create noise nuisance. The Act lays down special rules peculiar to construction site work. Such work is defined by s. 60(1) as:

(a) the erection, construction, alteration, repair or maintenance of buildings, structures or roads
(b) breaking up, opening or boring under any road or adjacent land in connection with the construction, inspection, maintenance or removal of works
(c) demolition or dredging work
(d) any work of engineering construction.

Section 60 notices

If it appears to a local authority that any of the above work is to be carried out or is being carried out, it may serve a notice imposing requirements as to the way the work is to be carried out and, if necessary, publish the notice of the requirements in such a way as the authority thinks appropriate.

Such a notice may specify:

(a) plant or machinery which is or is not to be used
(b) times of operation
(c) levels of noise for:
  (i) emission from the premises
  (ii) emission from any particular part of the premises during specified hours.

It may also provide for change in circumstances (s. 60(3)).

According to s. 60(4)(a), the local authority must have regard to any codes of practice issued under this part of the Act. The Control of Noise (Codes of Practice for Construction and Open Sites) (England) Order 2002/461, and the Control of Noise (Codes of Practice for Construction and Open Sites) (Wales) Order 2002/1795 have adopted the British Standards Institution Code of Practice BS 5228: *Noise and Vibration Control on Construction and Open Sites* (2009). The 2009 version is published in two volumes, for noise and vibration respectively. The 2009 Code replaces the 1997 version.

They must also ensure that the 'best practicable means' are employed to minimize noise (s. 60(4)(b)).

Under s. 60(4)(c) the local authority must, before specifying a particular method, plant or machinery, have regard to the desirability of specifying other methods, plant or machinery which would be almost as effective in minimizing noise and more acceptable to builders, in the interests of the recipients of notices. It seems from this subsection that it is important to look at the work from the viewpoint of the builder, with noise reduction achievable using acceptable alternative equipment.

The local authority must have regard to the need to protect persons from the effects of noise in the locality in which the premises are situated (s. 60(4)(d)). In view of the complaints that the local authority receives in relation to construction work, one could argue that this is a most important consideration.

The notice must be served on the person who is going to carry out the works and on such persons as have control over the operations, as the local authority thinks fit (s. 60(5)).

Under s. 60(6) the notice *may* specify the time within which the notice must be complied with, and it may require works to be executed, or the taking of steps as may be necessary or specified, to comply with the notice.

By s. 60(7) appeals may be made to the magistrates' court within 21 days and the Control of Noise (Appeals) Regulations 1995/2644 apply. The following are the grounds of appeal:

(a) That the notice is not justified by the terms of s. 60.
(b) That there has been some informality, defect or error in, or in connection with the notice.
(c) That the authority has refused unreasonably to accept compliance with alternative requirements, or that the requirements of the notice are otherwise unreasonable in character or extent, or are unnecessary.
(d) That the time – or where more than one time is specified, any of the times – is not reasonably sufficient for the purpose.
(e) That the notice should have been served on some person instead of the appellant, being a person who is carrying out the works or is going to carry out the

works, or a person who is responsible for, or has control over, the carrying out of the works.

(f) That the notice might lawfully have been served on some person in addition to the appellant, being a person who is carrying out the works or is going to carry out the works, or a person who is responsible for, or has control over, the carrying out of the works. And that it would have been equitable for it to have been so served.
(g) That the authority has not had regard to some or all of the provisions of s. 60(4).

If the person served with a notice fails to do anything required within the specified time and without reasonable excuse, he or she commits an automatic offence (s. 60(8)). But if consent has been given under s. 61, it will be a defence.

In *Walter Lilly & Co. v. Westminster City Council* (1994) 158 JP 805 building contractors were served with s. 60 notices; they were also asked under s. 93 of the Act to give details of the work to be carried out and the proposed time for completion. Noise nuisances were committed over a weekend period but they were in relation to a second contract, not specified in the original works to which the s. 60 notice related (nor contemplated originally when the s. 93 details had been supplied). The council then took magistrates' court action against the builders. The contractors appealed to the Divisional Court, stating that the notice should only relate to the works being carried out at the time of the notice. The builders could have appealed against the notice in the first instance, but the problem works, the second contract, had not been contemplated at that time. The court upheld the contractors' appeal. Thus, it appears from this case that an s. 60 notice can only be used in relation to the work being carried out at the time it is served. A statement providing for changes in circumstances is only meant in relation to the works in hand or contemplated at that time. It seems as if this decision could cause problems in the future, although the law has been logically applied.

It should be noted that, like proceedings under s. 82 of the Environmental Protection Act 1990, if recipients of notices give the appearance that they will continue to ignore the notice, recourse can still be made to the High Court. The Control of Pollution Act 1974 does not specifically state this, so either s. 222 of the Local Government Act must be used or there must be an unlawful interference with the local authority's land to enable them to invoke private nuisance. Indeed, *City of London Corporation v. Bovis Construction Limited* (1988) 86 LGR 660 particularly looked at situations where proceedings under s. 60 of the Control of Pollution Act 1974 had commenced but an injunction was being sought. In that case, the court said, injunctions were to be used with caution. Only if the recipient was obviously going to ignore the notice and there was behaviour over and above a breach of the criminal law should an injunction be sought. In that case s. 222 was used to instigate the course of action.

Prior consent on construction sites, s. 61 amended by 1990 Act

Section 61 gives those people responsible for construction work, and the local authorities, an opportunity to settle any problems relating to the potential noise before the work starts. Advice is given in the BSI Code of Practice 5228 that the noise requirements of a local authority should be ascertained before the tender documents are sent out, so that they can be incorporated into those documents. This is only right as one of the main reasons why builders ignore s. 60 notices is that the financial implications of fixed damage clauses in the building contract make it imperative that they finish work on time. To restrict their working hours to the notice requirements reduces the time available to finish the work before incurring fixed damages (the erroneously named penalty clauses). Had the local authority told the employer *before* the work was contracted out that they would expect work to be carried out only between certain hours and without using certain equipment, then the tender documents could have alerted the potential contractors to the problem in advance and allowed them to make a sensible judgement as to the amount of time needed to complete the work.

Application may be made before the work begins and consent may be given to the applicant. The parliamentary draftpersons probably envisaged that the contractor would make such an application. It has been suggested, however, that the consulting engineer would be a more suitable person and that the consent should be sought before the tender is made, in order to take account of any requirements made by the local authority.

By s. 61, if building regulation approval is required, the application must be made at the same time or later than the request for approval. The application must contain the following particulars:

(a) the works and the method by which they are to be carried out
(b) noise minimization steps (s. 61(3)).

According to s. 60(4), (5) and (6), if the local authority considers that sufficient information has been given and that, if the works were carried out in accordance with the application, it would *not* serve a notice under the preceding section in respect of those works, the authority shall grant its consent within 28 days and must not serve a notice under s. 60, but in so doing may:

(a) attach conditions
(b) limit consent where there is a change in circumstances
(c) limit duration of consent.

If there is a contravention of any of these items, an offence will have been committed (s. 60(5)). The consent may be published if the authority thinks fit (s. 60(6)). Where the local authority does not give its consent or gives its consent but subject to conditions, the applicant may appeal to a magistrates' court within 21 days (s. 60(7)). Should proceedings be brought under s. 60(8), it would be a good defence to prove that the alleged contravention amounted to the carrying out of the works in accordance with consent given under this section.

Any consent given does not itself constitute a defence to proceedings under s. 82 of the 1990 Act (an aggrieved person's right to take proceedings, s. 61(9)).

Where consent had been obtained by someone other than the site worker, e.g. an employer, an architect or a consulting engineer, they must bring the consent to the notice of the site worker, otherwise the applicant will be guilty of an offence (s. 61(10)).

10.6 Noise in streets and neighbourhoods

Section 62 Control of Pollution Act 1974 amended by the Environmental Protection Act 1990 and the Noise and Statutory Nuisance Act 1993, the Noise Act 1996, Licensing Act 2003, Anti-Social Behaviour Act 2003, Clean Neighbourhoods and Environment Act 2005

Loudspeakers

Section 62 of the 1974 Act is concerned not with general noise in streets, but specifically with the use of loudspeakers. By s. 62 a street is defined for this section as a highway and any other road, footway, square or court which is for the time being open to the public. A highway has at common law been defined as a right of way over which the public have the right to pass and repass, thus a street probably does not have to be metalled.

Under s. 62(1) a loudspeaker in a street shall not be operated:

(a) between 9:00 pm and 8:00 am the following morning for any purpose (s. 62(1)(a))
(b) at *any* other time, for the purpose of *advertising* any entertainment, trade or business (s. 62(1)(b)).

This section has been amended by the 1993 Act (ss. 7 and 8), and Schedule 2. Section 62(1A) and (1B) has been inserted after s. 62(1). This states that 'Subject to subsection (1B) of this section the Secretary of State may by order amend the times specified in subsection (1)(a) of this section.' Furthermore, by s. 62(1B) any order under the new subsection shall *not* amend the times so as to *permit* the operation of a loudspeaker in a street at any time between the hours of 9:00 pm and 8:00 am.

Any person contravening the above is guilty of an offence and liable to a fine of up to £5000, with a continuing daily fine of £50 for each day after conviction the offence continues.

Westminster CC v. French Connection Retail (above, section 10.4) provides an example. From their shop in Regent Street, French Connection Retail Ltd were broadcasting music and advertisements for their business via audio pucks attached to the inside of the shop window, which had the additional effect of broadcasting it into the street. The Divisional Court held that the pucks on the inside turned the window itself into a loudspeaker which was itself 'in the street' by virtue of its outer face.

Subsection (b) would obviously affect the trade of traditional travelling food vendors, so s. 62(3) states that it shall not apply to the operation of a loudspeaker between 12:00 noon and 7:00 pm of the same day, provided *all* the following apply:

(a) The speaker is fixed to a vehicle being used to sell perishable foods (for humans).
(b) The speaker is operated solely for informing members of the public that the commodity is for sale from the vehicle.
(c) The speaker is operated so as not to give reasonable cause for annoyance to people in the vicinity.

Note: The Control of Noise (Code of Practice on Noise from Ice-Cream Van Chimes, Etc.) Order 1981/1828 made under s. 71 is also appropriate here (see page 302).

Exceptions to the general prohibition under s. 62

Obviously there have to be exceptions. By s. 62(2) the general prohibition does not apply to:

(a) Operation of loudspeakers by the police, fire brigade, ambulances, Environment Agency, water undertakers, sewerage undertakers exercising their functions, or a local authority within its area.
(b) Operation of loudspeakers by persons for communicating with someone on a vessel in order to direct its movement or the movement of another vessel.

(c) Operation of loudspeakers if they are part of a public telephone system.
(d) Operation of loudspeakers if they are in or fixed to a vehicle and they are operated either for the entertainment of the driver or passengers (such as a radio or television), or they are horns or other devices for warning traffic *and* they are operated so as not to cause reasonable annoyance to persons in the vicinity.
(e) Operation of loudspeakers not on a highway, when in connection with a transport undertaking used by the public, such as British Rail or a bus company, when making announcements to passengers or employees.
(f) Operation of loudspeakers by travelling showmen on land being used for a fair.
(g) Operation of loudspeakers in cases of emergency.

The 1993 Act has inserted s. 3(A), which states that the prohibition section of the Act (banning loudspeakers between 9:00 pm and 8:00 am) shall not apply to loudspeakers operating in accordance with a consent granted by a local authority under Schedule 2 to the 1993 Act.

Schedule 2 consents

By s. 8(1) of the 1993 Act, a local authority *may* (not shall) resolve that Schedule 2 is to apply to its area. Such a resolution must be published for two consecutive weeks in a local newspaper in circulation in that area. The notice must state, in addition to the fact that the resolution has been passed, the general effect of Schedule 2 and the procedure to apply for a consent under the schedule.

Schedule 2 para. 1 states that any person may apply to the local authority for consent to operate a loudspeaker in contravention of s. 62(1) of the 1974 Act in its area. Paragraph 2 states that a consent shall not be given to the operation of a loudspeaker in connection with any election or for the purpose of advertising any entertainment, trade or business. The consent may be granted subject to such conditions as the local authority considers appropriate.

Procedure

The application must be made in writing giving sufficient information as may be reasonably required by the local authority. From the date of receipt, the application must be considered and the decision notified within 21 days. If consent is granted subject to conditions, the conditions must be specified. The authority may publish details in a local newspaper.

Burglar alarms

Part 7 and Schedule 5 of the Clean Neighbourhoods and Environment Act 2005 (CNEA 2005) introduced new controls over audible intruder alarms, replacing s. 9 and Schedule 3 of the 1993 Act. Under s. 69 a local authority may designate all or any part of its area as an alarm notification area. The person responsible for premises installed with an alarm in a designated area must nominate a key holder and notify the local authority of the key holder's contact details within 28 days. Failure to comply is an offence under s. 71(4), with a fine not exceeding level 3 on the standard scale (s. 71(5)). Section 73 makes provision for fixed penalty notices.

Entry to premises

Section 77 gives the much needed statutory authority to turn off the alarm. If the alarm is sounding continuously for more than 20 minutes or intermittently for more than one hour after it was activated and the noise is such as to give people nearby *reasonable cause for annoyance*, and reasonable steps have been taken to get the nominated key holder to silence the alarm, an authorized local authority officer may enter the premises to turn off the alarm. No force may be used under this section.

But under s. 78, the officer may apply to a justice of the peace for a warrant to enter the premises by force if need be. The JP must be satisfied:

(a) that the alarm has been operating for 20 minutes or one hour, as explained above
(b) that the audible operation of the alarm is such as to give persons living or working nearby reasonable cause for annoyance
(c) that reasonable steps have been taken to get the nominated key holder to silence the alarm
(d) that the officer has been unable to gain entry without using force.

By s. 78(3) before applying for a warrant, the officer shall leave a notice at the premises stating that the alarm is operating so as to cause annoyance to those close by and that an application is being made to a JP for a warrant to enter and turn off the alarm.

The warrant shall continue in force until the alarm has been turned off and the officer has complied with s. 79(5) if that subsection applies (s. 79(8)) (see below).

The officer may take with him anyone, such as a locksmith, and such equipment as may be necessary (s. 79(3)).

Anyone entering premises by virtue of s. 79(3) must not cause more damage or disturbance than necessary for the purpose of silencing the alarm (s. 79(4)). (Should they do so, they would be liable in tort for negligence or trespass to land or property.)

By s. 79(5) an officer entering premises which are unoccupied, or from which the occupier is temporarily absent, shall:

(a) leave a notice giving details of action taken under this section or under s. 77 or s. 78
(b) leave the premises, so far as is reasonably practicable, as effectually secured against entry as he or she found them.

But the officer is not required to re-set the alarm (s. 79(6)).

The local authority may recover expenses reasonably incurred (s. 79(7)).

Finally, provided anything done by the officers or anyone authorized is done in good faith for the purposes of ss. 77–79, they and the local authority are protected from any action, liability, claim or demand.

DEFRA has published guidance on ss. 69–81 CNEA 2005 (the power to deal with intruder alarms), accessible at *http://www.defra.gov.uk/environment/quality/noise/guidance/index.htm*

Night-time noise: the noise Act 1996 as amended

Separately from statutory nuisance provisions discussed in section 10.4 above, the Noise Act 1996 (the 1996 Act), as amended by s. 42 of the Anti-Social Behaviour Act 2003, deals specifically with *night-time* noise from, and affecting, residential premises. Where a local authority receives a complaint from an individual present in a dwelling during night hours (for the purpose of the 1996 Act defined as being from 11.00 pm to 7.00 am) that excessive noise is being emitted from *another dwelling*, the local authority must take reasonable steps to investigate the complaint (s. 2(1) and (2)). If the local authority's investigating officer is satisfied that the noise being emitted from the offending dwelling would or might exceed the permitted level if measured from within the complainant's dwelling, he may serve a warning notice under s. 3. The notice warns that any person who is responsible for the noise in question may be guilty of an offence. Where a warning notice has been served and the noise does exceed the permitted level, the person responsible is guilty of an offence and liable on summary conviction to a fine not exceeding level 3 on the standard scale (s. 4(1) and (3)). Section 4(2) provides a defence of reasonable excuse. Section 8 of the Act provides for fixed penalty notices where the local authority's officer has reason to believe that a person is committing or has just committed an offence under s. 4, he may give that person a fixed penalty notice of £100. Section 10 confers powers of entry and seizure, so that an officer of the local authority can enter the dwelling in question and seize and remove any equipment which it appears to him is being or has been used in the emission of the noise.

The current guidance to local authorities in England from the Department for Environment, Food and Rural Affairs (DEFRA) on the 1996 Act, its procedures and what may constitute excessive noise was published in March 2008 and is entitled *The Noise Act 1996 as amended by the Anti-Social Behaviour Act 2003 and the Clean Neighbourhoods and Environment Act 2005*, accessible at *www.defra.gov.uk*. The permitted level of noise is set at 34 dBA if the underlying level of noise is no more than 24 dBA, or 10 dBA above the underlying level of noise where this exceeds 24 dBA. Paragraph 32 of the guidance explains that if the investigating officer is satisfied that a statutory nuisance is occurring and the permitted level of noise is being exceeded, the mandatory duty to serve an abatement notice under s. 80(1) EPA 1990 will apply. The duty to serve an abatement notice may, however, be deferred for seven days under s. 80(2A)–(2E) while other appropriate steps are taken to abate the noise nuisance. Deferral of the duty to serve an abatement notice was introduced by s. 86 Clean Neighbourhoods and Environment Act 2005, amending the 1990 Act. Guidance on fixed penalty notices and on the deferral of the duty to serve an abatement notice, entitled *Fixed Penalty Notices* and *Noise* respectively, was issued by DEFRA in 2006, also accessible at the DEFRA website.

The Clean Neighbourhoods and Environment Act 2005 extends the 1996 Act to *licensed premises*, more specifically to 'any premises in respect of which a premises licence or temporary event notice has effect' (s. 84 and Schedule 1). This includes pubs, nightclubs and music venues.

Licensed premises: general controls

Local authorities derive their power to issue licences from a number of statutes, including the Private Places of Entertainment (Licensing) Act 1967, the Local Government (Miscellaneous Provisions) Act 1982 and the Licensing Act 2003. Licences will frequently contain conditions intended to control noise which if breached could result in a prosecution in the magistrates' court and a fine up to a maximum of £20,000.

Section 4 Licensing Act 2003 sets out the general duties of licensing authorities:

1. A licensing authority must carry out its functions under this Act ('licensing functions') with a view to promoting the licensing objectives.
2. The licensing objectives are—
   (a) the prevention of crime and disorder;
   (b) public safety;
   (c) the prevention of public nuisance; and
   (d) the protection of children from harm.

3. In carrying out its licensing functions, a licensing authority must also have regard to—
   (a) its licensing statement published under section 5, and
   (b) any guidance issued by the Secretary of State under section 182.

Closure orders. Both the Licensing Act 2003 and the Anti-Social Behaviour Act 2003 confer powers on the police and local authorities respectively to close premises, licensed or otherwise, for a period of 24 hours. Section 161(1) Licensing Act 2003 provides:

> a senior police officer may make a closure order in relation to any relevant premises if he reasonably believes that—
> (a) there is, or is likely imminently to be, disorder on, or in the vicinity of and related to, the premises and their closure is necessary in the interests of public safety, or
> (b) a public nuisance is being caused by noise coming from the premises and the closure of the premises is necessary to prevent that nuisance.

Sections 162 and 163 contain respectively provisions for closure orders' extension and cancellation. Section 40 Anti-Social Behaviour Act 2003 provides:

1. The chief executive officer of the relevant local authority may make a closure order in relation to premises to which this section applies if he reasonably believes that—
   (a) a public nuisance is being caused by noise coming from the premises, and
   (b) the closure of the premises is necessary to prevent that nuisance.
2. This section applies to premises if—
   (a) a premises licence has effect in respect of them, or
   (b) a temporary event notice has effect in respect of them.

A person commits an offence if without reasonable excuse he permits premises to be open in contravention of closure order (s. 40(4)) and is liable on summary conviction to a fine not exceeding £20,000 or a maximum term of 3 months' imprisonment or both (s. 40(5)). Section 161(6) and (7) Licensing Act 2003 contains equivalent provisions.

It is worth noting that ss. 63–66 Criminal Justice and Public Order Act 1994 provide powers for the police to deal with illegal raves, including powers to remove persons attending or preparing for a rave and to enter and seize equipment. Section 58 Anti-Social Behaviour Act 2003 extends the power of the police to deal with raves.

10.7 Noise abatement zones

Section 63 Control of Pollution Act 1974

By the remaining part of s. 57 of the Control of Pollution Act 1974 a local authority is under a statutory duty to cause its area to be inspected from time to time to decide how to exercise its powers concerning noise abatement zones. There is no need, however, for the local authority to make an inspection before making an order (*Morganite Special Carbons v. Secretary of State for the Environment* (1980)).

By s. 63 a local authority may, by order, designate all or any part of its area a noise abatement zone. Its purpose is to prevent a deterioration in environmental noise levels and, wherever practicable, to achieve reduction in noise levels.

The order must specify the classes of premises to which it applies (s. 63(2)). Examples include public utility installations, such as waterworks, power stations, gasworks, launderettes, and entertainment halls. The control of noise that will be achieved is limited to controlling noise from such individual premises and not by laying down a standard maximum noise level for the zone. The classes of premises specifically exclude domestic premises, as it was felt that the other methods of controlling noise, formerly under the Control of Pollution Act itself and now s. 80 and s. 81 of the Environmental Protection Act 1990, would be more appropriate. The order designating the zone is called a noise abatement order. Do *not* confuse it with the noise abatement *notice* under s. 80 EPA.

The order may subsequently be revoked or varied by another order (s. 63(3)).

Register of noise levels, s. 64

Where a local authority has designated its area or any part of it a noise abatement zone, it must measure the level of noise emanating from the premises falling within the particular category noted in the order, e.g. commercial premises (s. 64(1)). Which category is noted will depend on whether it is considered to be a problem in that area.

The measurements shall then be recorded in a noise level register to be kept by the local authority (s. 64(2)). A copy of the record shall be served upon the owner and occupier of the affected premises and he or she may appeal to the Secretary of State within 28 days, who may then give such directions as he or she thinks fit (s. 64(3) and (5)). The Control of Noise (Appeals) Regulations (S.I. 1975/2116) are the appropriate regulations for such an appeal. Otherwise, once the levels have been recorded, the record cannot be properly challenged, except as to whether it was properly served. The register shall be open for inspection and copies may be obtained (s. 64(7)). The Secretary of State, by virtue of s. 64(8), has made regulations giving the methods by which the measurements are to be made, the Control of Noise (Measurement and

Registers) Regulations 1976 (SI 1976/37); the measuring device should comply with BS 61672-1: 2003.

Exceeding the registered level, s. 65

Once a noise level has been determined, it must not be exceeded, unless written consent has been given previously (s. 65(1)). Such prior consent must have been entered on the register (s. 65(2)). If the level or the higher permitted level has been exceeded, then an offence will have been committed (s. 65(5)). The magistrates' court convicting a person of an offence may, if satisfied that the offence is likely to recur, make an order requiring the execution of any works necessary to prevent it continuing or recurring. Failure to carry out the works is also an offence (s. 65(6)). The local authority may be heard on the matter and may be empowered to do any such work and recover the cost from the offender (s. 65(7) and s. 69).

The consent mentioned above must be applied for if the owner wishes to be able to emit noise above the registered level. Unless a decision is made in writing within two months, it will be deemed refused by the local authority unless a further period is agreed in writing (s. 65(3)). The failed applicant may appeal to the Secretary of State (s. 65(4)). The Control of Noise (Appeals) Regulations 1995 apply. Such consents may be made subject to such conditions as necessary, e.g. the amount by which the noise levels may be increased or the periods during which the level may be increased (s. 65(2)). All these particulars must be recorded on the noise level register. The consent will contain a statement to the effect that the consent does not of itself constitute any ground of defence against proceedings instituted under s. 82 (aggrieved person's rights) of the 1990 Act (s. 65(8)). It may, however, be a defence to an offence under s. 80 (notice from local authority).

Reduction of noise levels, s. 66

If it appears to the local authority that the level of noise emanating from any premises under a noise abatement order is not acceptable, having regard to the purposes for which the order was made, and that reduction is practicable at reasonable cost and would serve a public benefit, then the authority may serve a notice on the person responsible (s. 66(1)). Such a notice is known as a noise reduction notice. Do *not* confuse it with the noise *abatement* notice under s. 80 EPA.

According to s. 66(2), the notice shall require:

(a) reduction of noise to a specified level
(b) prohibition of an increase without consent
(c) such steps to be taken so as to achieve those purposes.

Such steps and the abatement of the noise must be taken within a period of not less than six months from the date of service (s. 66(3)). Contravention of the requirements amounts to an offence (s. 66(8)). In such proceedings, if the offence was committed in the course of a trade or business, the best practicable means defence can be used (see page 292).

The notice may specify times, dates and different noise levels (s. 66(4)). Section 66 noise reduction notices *override* the consent given under s. 65 (s. 66(5)). The local authority must record the details of the noise reduction notice in the noise level register (s. 66(6)).

The recipient has three months in which to appeal to a magistrates' court (s. 66(7)) and the 1975 Control of Noise (Appeals) Regulations apply. The following are the grounds for an appeal:

(a) that the notice is not justified by the terms of s. 66
(b) that there has been some informality, defect or error in, or in connection with, the notice
(c) that the authority has refused unreasonably to accept compliance with alternative requirements, or that the requirements of the notice are otherwise unreasonable in character or extent, or are unnecessary
(d) that the time – or where more than one time is specified, any of the times – within which the requirements of the notice are to be complied with is not reasonably sufficient for the purpose
(e) where the noise to which the notice relates is noise caused in the course of a trade or business, that the best practicable means have been used for preventing, the noise or for counteracting the effect of the noise
(f) that the noise should have been served on some person instead of the appellant, being the person responsible for the noise
(g) that the notice might lawfully have been served on some person in addition to the appellant, being a person also responsible for the noise, and that it would have been equitable for it to have been so served.

New buildings, s. 67

Where a building is being constructed within a noise abatement zone or works are being done to a building, so that they fall within a class specified within the noise abatement order, the local authority may determine the level of noise which will be acceptable as that emanating from the premises (s. 67(1)). This does not apply to private houses. The level must be placed on the noise level register (s. 66(2)). This must be done on the authority's own initiative or in response to a request by the owner or occupier of the premises, or even a developer who is in the process of acquiring the premises. The applicant or the recipient of the notice stating the determination of the level has a right of appeal within three months to the Secretary of State (s. 66(3)). The Control of Noise (Appeals) Regulations 1975 apply. The usual two-month refusal period applies where an application has been made to determine the acceptable noise level (s. 66(4)).

If a noise abatement order comes into force and later any premises become premises to which the order would apply, as a result of construction work (from scratch) or works carried out on the building, but no noise level has been determined, then s. 66 applies but the recipient of a determination of noise level has six months in which to appeal (s. 66(5)).

Noise from plant or machinery, s. 68

Regulations may be made under s. 68 in relation to the reduction of noise caused by plant and machinery by the use of devices or arrangements. Also, regulations may be made to limit noise from plant or machinery used on construction work or by machinery operating in a factory, as defined by the Factories Act 1961 (s. 175). Standards, specifications, methods of testing and descriptions not forming part of the regulations may be used. Before these regulations are made, the Secretary of State must consult with relevant persons and bodies representing producers and users in order to prevent rules being made which are impracticable or involve unreasonable expense. Contravention means the commission of an offence, but the best practicable means defence may be used.

Codes of practice, s. 71

In order to provide guidance on methods of minimizing noise, the Secretary of State may, by order, issue codes of practice or approve codes of practice made by other bodies. Current orders relating to construction sites are the Control of Noise (Code of Practice for Construction and Open Sites) (England) Order 2002 (S.I. 2002/461) and the corresponding order for Wales (S.I. 2002/1795) (see section 10.5 above). Other orders include the Control of Noise (Code of Practice on Noise from Ice-Cream Van Chimes, Etc.) Order 1981 (S.I. 1981/1828), Control of Noise (Code of Practice on Noise from Audible Intruder Alarms) Order 1981, Control of Noise (Code of Practice on Noise from Model Aircraft) Order 1981 (the code itself itself issued by the DoE in 1982).

Other codes of practice have been issued by a variety of bodies and cover such matters as audible bird scarers, and recreational equipment and activities such as powerboats, clay pigeon shooting and off-road motorcycle sport. They have been issued by the National Farmers' Union (bird scarers) and the Chartered Institute of Environmental Health (CIEH) (clay target shooting, 2003), the British Water Skiing Federation (powerboats, 1999), the Noise Council (environmental noise at concerts, 1995, currently under review), the Noise Council in association with others (organized off-road motorcycle sport, (1994), and oval motor racing circuits, 1997).

In August 2002 DEFRA issued *Guidance on preparing Codes of Practice for minimising noise in England*, and in September 2006 jointly produced with the CIEH a Management Guide *Neighbourhood Noise, Policies and Practice for Local Authorities* accessible at *http://www.defra. gov.uk/environment/quality/noise/guidance/index.htm*

To this should now be added the Noise Policy Statement for England (NPSE), published March 2010, accessible at *http://www.defra.gov.uk/environment/quality/noise/policy/documents/noise-policy.pdf*

Thus it can be seen that the Acts provide a comprehensive treatment of most noise problems; enforcement is achieved at local authority level. Many of the problems with the legislation are caused by lack of funding, so that local authorities are unable to monitor noise nuisances properly. Other legislation that can only be noted here for reasons of space is the Fireworks Act 2003 and accompanying Fireworks Regulations 2004 (S.I. 2004/1836).

10.8 Noise and personal health and welfare

As we have already seen, the tort of nuisance, the Environmental Protection Act, the Noise and Statutory Nuisance Act, etc., try to provide a quieter society and, in so doing, indirectly reduce the effects of noise on health. However, only in exceptional circumstances will compensation be awarded for illnesses or deafness caused by noise. Until fairly recently there was not even statutory guidance on reducing the effects of noise in the workplace.

The law deals with personal health and welfare in a piecemeal way, through the tort of negligence, social security legislation, the Health and Safety at Work, etc., Act 1974 and the Control of Noise at Work Regulations 2005 (S.I. 2005/1643). Compensation may be awarded in some cases; state pensions may be given in others. The Court of Appeal decision *Baker v. Quantum Clothing Group* [2009] EWCA Civ 499 is a significant case on employers' liability for hearing impairment at work and is considered in detail below. Wrongdoers, under health and safety legislation, unmindful of others' safety, may be dealt with in the criminal courts.

Negligence

The tort of negligence is a tort which protects personal interests and not just interests in land (cf. nuisance). If someone causes damage to person or property by being negligent, he or she may be sued for damages. (For criteria as to whether the case should be started in the county court or Queen's Bench Division of the High Court, see page 271.) Most damage caused by noise will be to health, but vibrations or an explosion could damage property.

Requirements

In order to sue successfully in negligence, the claimant has to prove that *all* the following elements exist:

1. That a duty of care was owed in law by the defendant to the claimant.
2. That this duty has been breached.
3. That the actual damage was caused as a direct result of the breach.

1. *The duty of care*

In law there can be no action taken unless there are laws giving people rights and duties. Often there are statutory rights and duties contained in Acts or statutory instruments. The problem with common law rights and duties is that one often has to test whether they existed *after* the event. We cannot predict what is going to happen to us in the future. Fortunately, in one respect involving employer/employee situations, the existence of this elusive duty of care is predetermined. If the person suffering from noise-induced illness is an employee and wishes to sue his employer, then it has long been recognized that in common law a duty of care is automatically owed by employers to their employees while they are still at work or affected by work.

In other situations the matter is not so simple and has been made more complex by a series of legal decisions in recent years. The history of negligence as a tort in its own right, not dependent on mere precedents, began with the famous case of *Donoghue v. Stevenson* [1932] AC 562. This case was important in that the Law Lords put forward a formula for determining whether or not a duty of care existed in a given case. This was called the *neighbour principle* and stated: 'You must take reasonable care to avoid acts or omissions which you can reasonably foresee would be likely to injure your neighbour.' Who, then, in law, is my neighbour? The answer seems to be 'persons who are so closely and directly affected by my acts, that I ought reasonably to have them in contemplation as being so affected, when I am directing my mind to the acts or omissions which are called in question', as per Lord Atkin. Over the years this formula was considered too wide in its application, but many cases have been decided on this basis and would still provide a precedent in 'same facts' situations. An updated version was provided in 1977 with *Anns v. London Borough of Merton* [1977] 2 WLR 1024, Lord Wilberforce said that whether a duty of care exists could be approached in two stages:

> First, one has to ask whether as between the alleged wrongdoer and the person who has suffered damage, there is sufficient relationship of proximity or neighbourhood such that in the reasonable contemplation of the former, carelessness on his part may be likely to cause damage to the latter, in which case a prima facie duty of care arises.
>
> Secondly, if the first question is answered affirmatively, it is necessary to consider whether there are any considerations which ought to negative or to reduce or limit the scope of the duty or the class of person to whom it is owed or the damages to which a breach of it may give rise.

Over the years both formulae were considered to be too wide. In *Caparo Industries Plc v. Dickman* [1990] 2 WLR 358, a case which has been confirmed by subsequent cases, a formula was arrived at incorporating facets of the other two decisions reducing the scope of the duty of care. (*Anns v. Merton LBC* was also overruled on its particular facts in another decision.) The Lords decided that three requirements were necessary:

(a) There must be reasonable foreseeability of the relevant loss.
(b) It must be just and reasonable that a duty should exist.
(c) There must exist a sufficient relationship of proximity between defendant and claimant.

This test is more stringent than the earlier ones. Let us apply it to a hypothetical situation. A lady lives next door to a factory operating noisy machinery. She now suffers deafness and tinnitus. No medical explanation can be given other than it has been caused by the noisy machinery. Is she owed a duty of care? Is there reasonable foreseeability of the relevant loss? Could that factory owner have reasonably imagined that a neighbour could be affected by that work and suffer from deafness, especially if he has provided his workers with ear defenders in accordance with the Control of Noise at Work Regulations? Secondly, is it just and reasonable that she, a neighbour, should be owed such a duty? Probably, yes. Thirdly, is there a sufficient relationship of proximity between the parties? In all three cases one could argue in favour of a duty being owed.

2. *Breach of the duty*

Having established that a duty of care was owed in the circumstances, the claimant must then prove to the court that there was a breach of that duty, i.e. the defendant was negligent in fact and law.

This is done in two stages. First, taking the circumstances of the case, the court decides on the standard of care, i.e. what should have been done by the defendant. Second, the defendant's actual behaviour is measured against the established standard. If it falls short, the defendant is negligent; if it matches or is above the standard, the defendant has committed no breach.

(i) *The standard of care.* The standard of care is not of superman nor that of a fool. The court 'fixes' the standard by examining evidence which would show what the *average reasonable* person should have done when placed in the circumstances of the case. Experts may be called to give evidence of trade practice. So, the average factory owner has to meet the standard of the average factory owner doing that particular sort of work. The same factory owner driving his or her car home at night has to meet the standard of care of an average driver. If he or she decides to repair his television set, the standard of care is that of the average layperson. It makes no difference whether he or she is extremely intelligent, habitually lazy or unusually careless because this introduces a further subjective element, which has been rejected by the courts.

Furthermore, the knowledge which one is supposed to possess depends on the current state of knowledge. So, at present, we are probably doing many things which in years to come will prove to have been dangerous. Once knowledge becomes available to society, any failure to use it appropriately could amount to negligence, even if individual defendants are ignorant of that knowledge.

(ii) *Other factors which may raise or lower the basic standard of care – (a) Is the risk reasonably likely?* The courts are not going to be impressed with a defendant who failed to avoid an obvious risk. Once again, the reasonable foresight test is used. Thus failure to ensure that ear defenders are being worn in noisy factories or on building sites will amount to negligence because of the extreme likelihood of impairing one's hearing.

Even if the risk is not likely but is reasonably foreseeable, it should be taken into account when determining the standard of care. This follows dicta in the *Wagon Mound No. 2* [1967] 1 AC 617, a Privy Council shipping case which involved an unusual chain of events. For example, in *Haley v. London Electricity Board* [1965] 3 WLR 479, the electricity board's employees had dug a hole in a pavement and, although they erected adequate barriers to prevent a sighted person from falling down the hole, they were inadequate for a blind person. Although the risk of harm being caused by their action was slight, it was reasonably foreseeable that a blind person could have fallen down the hole.

If the risk is a 'fantastic possibility' then probably there would be no need to do more than take the usual precautions.

If you are dealing with a known hazardous situation, such as working in a noisy industry or with noisy equipment, the courts take the view that you must take extra care of those people affected by it.

(b) Are the claimants at particular risk? Having decided that certain categories of people are owed a duty of care, if it then appears they would be at particular risk because of age, infirmity or other special reason, such as being women of child-bearing age, then the standard of care is much higher than normal. So schoolteachers owe a high standard of care towards their pupils. Employers of handicapped people must see that such people are safe, e.g. they must make sure that emergency warning alarms are easily seen by deaf workers, as they cannot hear normal sirens. Apprentices or anyone doing a new job is at particular risk until they understand what is required of them. They should be adequately supervised; warnings and advice should be given and seen to be understood. If they do not speak English, translators and notices in their own language should be provided. This need is increasing with increased mobility of workers in the European Union, where sub- and co-contractors may be from other countries, and it exists in areas where there are many immigrants who may not necessarily speak English.

In *Paris v. Stepney Borough Council* [1951] 1 KB 320, the employing council was held to be negligent in failing to give goggles to a one-eyed workman, although it would not have been negligent (then) had he been fully sighted. This seems an odd decision, but the court specifically mentioned the high degree of risk to a partially sighted person. Obviously, failure to provide earmuffs in situations where they are required by the Control of Noise at Work Regulations 2005 would amount to negligence (and be grounds for suing for another tort, breach of statutory duty, see page 308). By analogy with the *Paris* case, it is possible that failure to provide earmuffs for a partially deaf person, working in relatively noisy circumstances, would amount to negligence, but not if he were of good hearing although such a proposition must now be considered in the light of the decision in the *Baker* case (below).

(c) Could the risk of injury or harm have been eradicated by a small outlay? No court will be impressed with the defendant arguing that he could only have avoided the risk of harm at great expense (unless coupled with the argument that the risk was a fantastic possibility). But, if the harm could have been avoided at small cost and the defendant failed to take avoiding action, the courts are more likely to find that the standard of care was higher than normal and there would be a case of negligence.

(d) Is the defendant an expert? If someone falls down a pothole, smashing their leg so the only way to free it is by amputation, then the standard of care to be shown by an

unqualified rescuer is that of the average rescuer in an emergency. But if the injured person's leg is amputated in hospital, the surgeon is expected to show such skill and experience as is appropriate to his position, qualifications and experience. Thus, in the first case the standard is quite low, and in the second extremely high. In *McCafferty v. Metropolitan Police District Receiver* [1977] 2 All ER 756, the claimant, a ballistics expert, sued his employers for damage to his hearing. The employers were distinguished scientists who knew more of the effect of such work than the ballistics expert (see below).

Amateurs doing work which could be better done by experts should show reasonable care. This is because their standard of care would be that of the average amateur doing a particular job. In determining such a standard, the courts assume that all people have a certain degree of common sense and knowledge.

(e) Are there regulations, codes of practice, etc., to follow? If there are statutory regulations laying down conditions to be observed (such as the Control of Noise at Work Regulations 2005, see page 313), they should be complied with; then the prime cause of action will be for breach of statutory duty (see page 308). If it is impossible to sue in this way, due to lack of one of the essential preconditions, the claimant can sue in negligence and plead that the regulations established a particular standard of care which was not met. But the claimant will usually sue for both negligence and breach of statutory duty as it would be impossible to add claims at a later date if one claim failed for some reason. (If there has been a breach of the regulations, there will probably be a criminal prosecution first; this will provide good evidence for the subsequent civil action.)

Any statutory codes of practice are not themselves absolute legal requirements, but they will usually establish a set standard of care.

If there are house or company rules or regulations, they will indicate a certain level of care, which may or may not be what is acceptable in the circumstances. For example, some company rules may be far too stringent; their breach would perhaps be grounds for dismissal under a contract of employment but would not amount to negligence.

(iii) *Circumstances which may lower the required standard of care – (a) Was there an emergency?* Whether or not an emergency lowers the standard of care depends on the circumstances of each case. If, for example, there is a fire, rules should already exist so as to avoid worsening the situation. If these rules are not followed, there may be a case of negligence. But, if there are no rules and the defendant does his best under the circumstances, then he will have acted as the average reasonable person; there can be no negligence. Ear defenders may have been removed during an emergency in order to follow instructions; the resulting damage to hearing may not be actionable.

(b) Is the claimant an expert? If the defendant is working with a claimant who is an expert, this may reduce the standard of care, so as to negative any claim of negligence against the defendant, provided that the expert knew what was being done. Even if negligence is proved, the defendant could claim that the claimant was contributorily negligent or even that there was consent (see pages 306–7).

(c) What did the defendant do? Having determined the standard of care which should have been exercised in the given situation, the claimant must then prove that the defendant did not conform with that standard. If car drivers should normally be sober, then a defendant causing an accident while inebriated will be found to be negligent.

Normally the burden of proving the breach will fall on the shoulders of the claimant. It is not for the defendant to show that he or she was not negligent. Evidence must be produced to satisfy the court that, on a balance of probabilities, the defendant was negligent in the circumstances. Consequently, all evidence must be properly produced and preserved. The type of evidence will depend on the situation; it could include photographs, pieces of equipment or machines, medical evidence, witness accounts, expert evidence, accident reports, letters, contracts, drawings and plans. It is up to the claimant to present sufficient evidence and sufficiently persuasive evidence so the court will agree that the defendant was negligent.

A conviction in criminal proceedings will be good evidence in subsequent civil actions.

If, from the outset, the evidence is overwhelmingly in favour of the claimant, it is highly unlikely that the case would ever actually go to a court hearing. The lawyers would then settle out of court. But sometimes, even though the defendants have admitted their liability, they still dispute the amount of damages, known as the quantum. This is especially true of personal injuries or fatalities because the courts have a highly complex approach to arriving at a final global amount. Such cases will go to court, but merely for the judge to decide on the amount of damages to be awarded; no evidence need be produced.

3. *Damage must result from the breach*

(i) *Damage must be caused.* First, there must be damage caused. If there is no damage, there can be no action, even if one can prove the first two elements. This is because negligence is *not* actionable *per se*, like nuisance, but unlike trespass, where no damage need be proved. Damage may be caused to people – injuries – or to buildings. In noise cases the injuries may be occupational deafness or tinnitus; in vibration cases the injuries may be white finger or carpal tunnel syndrome, caused by vibrating tools. Physical damage to buildings could also be caused by noise and vibration.

(ii) *The damage must not be too remote from the negligent act in fact.* If there is damage, it must not be too far removed from the negligent act. There has to be a causal link between the initial damage and the resultant damage. If no audiometric tests are undertaken at the commencement of a worker's employment, how can she later prove that any occupational deafness is as a result of her noisy work conditions? Conversely, if the employer is trying to defend an action, having performed no tests at the outset, how can he claim that the worker was already deaf before being taken on? It is obviously in the interests of truth that tests are undertaken in noisy industries. Medical evidence will have to be produced in these cases, but is not always conclusive.

(iii) *The damage must not be too remote in law.* This is the reasonable foresight approach again. To limit the liability for all the direct *factual* consequences of negligence, the courts now recognize that one should only be liable for those consequences which could have been *reasonably foreseen* at the time of the negligence. The test was approved by the Privy Council in *Wagon Mound No. 1* [1961] AC 388. Thus, it is reasonably foreseeable that excess noise can cause a number of symptoms. If, on the other hand, noise also causes conditions that are currently unknown to society, the employer could not be sued in the future because the damage would not have been reasonably foreseen at the time of the alleged negligent acts. Once a hazard becomes known, one must guard against that hazard.

It is never a defence in law to claim that you did not realize what you were doing was legally wrong. Ignorance is no defence. Thus, employers in business have a particularly difficult task in taking note of all modern developments affecting their work.

Defences

The defendant in court does not sit back with fingers crossed, hoping the claimant will fail to prove a case. He will attempt to raise appropriate defences which could cancel out any negligence. The burden of proving a defence obviously falls on the defendant, and he must produce evidence to support this defence.

Necessity

It is occasionally possible that the defendant may claim that the tort was committed necessarily to prevent greater harm occurring. However, most of the cases involving necessity as a defence are concerned with other torts, where the act was deliberate.

Volenti non fit inuria

Volenti non fit inuria is consent, literally 'there is no injury to a willing person'. This is an absolute defence and, if proved, cancels out the negligence, thus defeating the claimant's claim. It arises where the claimant:

- knew that the defendant was being negligent
- knew of the risk
- continued to cooperate with the defendant
- suffered injury of the type one would expect to result from such negligence.

Nowadays the courts are very reluctant to find a case of consent existing in all but the most open-and-shut cases. Consequently, its application adheres to strict guidelines.

(i) *The claimant must know that the defendant is being negligent.* There can be no valid consent if the claimant did not know there was any negligence. Thus, the defendant must prove there was either express or implied consent *to the negligence.* Express consent will be either oral (and therefore witnessed in order to stand up in court) or written, which may be witnessed. But under the Unfair Contract Terms Act 1977, liability for what amounts to negligence that causes death or personal injury cannot be excluded *at all*, by contract or notice, in the course of a *business.* Consent may be implied where the claimant continues to cooperate despite the negligence.

Consent concerns the negligence, not merely the hazards of a job. People who do dangerous work may be paid danger money and they consent to run the risk of the ordinary hazards of the work (and possibly sign an agreement to that effect), but they are *not* consenting to actual negligence. So if an employee arrived at work to find the safety equipment vandalized and the employer unwilling to do anything about the situation, this unwillingness will amount to negligence. Furthermore, if the employee carries on working, it could be argued that he is willing to run the risk of the employer's negligence.

(ii) *The claimant must have known of the risks which could be caused by the negligence.* Someone who consents to run the risk of a negligent act will not lose his or her action for negligence if he or she was injured in some other way.

(iii) *Consent must be freely given.* The defence will fail if the worker is under duress from his or her employer to carry on work regardless of the negligence.

Contributory negligence

Contributory negligence is not a complete defence, but it acknowledges that, although negligence may have been proved against the defendant, the claimant may at the same time, by his own fault, have contributed to his own injuries or loss, e.g. by failing to wear earmuffs. The Law Reform (Contributory Negligence) Act 1945 introduced this defence and by s. 1(1) 'the damages recoverable . . . shall be reduced to such an extent as the court thinks fit and equitable, having regard to the claimants' share in the responsibility for the damage'.

The wording of the Act allows the defendant to plead this without having to prove *negligence* on the part of the claimant. All he or she needs to do is to demonstrate the claimant was at fault in some way which contributed to the damage. The claimant should have reasonably foreseen that his or her behaviour would contribute to his or her injury, otherwise the defence will fail.

As a result, the court works out the damages in the usual way then reduces them proportionately, taking account of the blameworthiness of the claimant, e.g. that by failing to wear his or her earmuffs the claimant is 50% to blame and therefore gets only 50% damages.Obviously, it is in the worker's own interests to do all he or she can to prevent greater personal injury by complying with all safety requirements, such as wearing safety clothing.

Limitation Act 1980

It is unfair on potential defendants if claimants do not take advantage of their right to sue within a reasonable period. The Act lays down various periods in which the action must be brought. For most torts this is a period of six years. The exceptions are actions brought for damages for negligence, nuisance or breach of duty, when the damages claimed include damages for personal injuries and when the period is three years. The limitation period begins to run from the date on which the cause of action accrued. This is difficult to calculate in relation to noise-induced illnesses, as such illnesses usually start imperceptibly and progressively get worse. Deafness caused by an explosion would be easier to deal with.

By s. 2(A) and (B) of the Limitation Act 1980, the time now runs either from the date of the accrual of the cause of action or from the date of the claimant's knowledge. The knowledge is that the injury is significant, that the injury is due to the alleged negligence of the defendant and the identity of the defendant. Furthermore, the court has the right to override the time limits. These points were raised in *McCafferty v. Metropolitan Police District Receiver* [1977] 2 All ER 756 (above). The claimant was a senior experimental officer in charge of a ballistics section of a police laboratory. Although he possessed considerable knowledge of ballistics, he had no scientific or medical qualifications. Part of his job involved shooting rounds of ammunition. From 1965 this took place in a room 22 feet by 6 feet which had no sound absorption. On the advice of his superiors, who were distinguished scientists, the claimant protected his hearing by putting cotton wool in his ears. By 1967 he was suffering from tinnitus. On consulting a specialist, he was told to get earmuffs immediately. Earmuffs were supplied by his employers, but evidence was produced at the trial that they would only be effective in a room which had special acoustic protection. He continued to use the earmuffs and noticed no deterioration, although his tinnitus continued. In 1973 he underwent hearing tests, which showed that he was suffering from severe acoustic trauma and, on advice, he quit his job. In 1974 he started proceedings to claim damages for negligence resulting in damage to his hearing by 1967 and the premature termination of his job, caused by the defendants' failure to protect his hearing.

The defendants pleaded contributory negligence, claiming that the claimant had been told to find out about precautions and, furthermore, that the claim was statute-barred by being outside the limitation period.

The Court of Appeal held that there had been no contributory negligence because, despite the delegation of responsibility regarding precautions, he was not the proper person to do this. He was an expert, but not in medical and acoustic matters. Also he had always complied with instructions given him.

As far as his claim about his hearing was concerned, the date of knowledge that the injury was significant, for the purposes of s. 2(A), was in 1968, and by then he realized the gravity of the matter. Thus, the action was prima facie statute-barred. Nevertheless, the court could exercise its discretion and override the three-year period. They took into account such matters as the reluctance of the claimant to start an action in order to keep good relations with his employers and his obvious interest in his work.

Vicarious liability

If a worker is negligent and causes injury to someone, it is often the case that his or her employer will be sued either with the worker or instead. This is because the employer is vicariously or indirectly liable for the torts of his or her servant, committed during the course of that servant's employment, even though that servant is obviously working badly. If the tort was committed while the worker was on a 'frolic' of his or her own, i.e. doing something totally outside his or her duties, the employer will not be responsible. Whether the worker is doing the act in the course of his or her employment is a question of fact in each case.

Employers' insurance

By the Employers' Liability (Compulsory Insurance) Act 1969 (and by various statutory instruments itemized below), every employer in Great Britain, with certain exceptions, must insure and maintain proper insurance against liability for bodily injury and disease sustained by his employees arising out of their employment. It has the effect that in most cases of negligence caused by employees, or by the employer, the employer's insurance company will make payment directly to the sufferer. Only where the offer is too low, or where the insurers refuse to make any payment, will the sufferer need to go to court.

The 1969 legislation must be read in conjunction with a number of amending and exemption regulations, first issued in 1971. The current main regulations are the Employers' Liability (Compulsory Insurance) Regulations 1998 (S.I. 1998/2573), the Employers' Liability (Compulsory Insurance) (Amendment) Regulations 2004 (S.I. 2004/2882), and the Employers' Liability (Compulsory Insurance)(Amendment) Regulations 2008 (S.I. 2008/1765). There is a helpful guide for employers published by the Heath and Safety Executive (HSE), November 2008 and one for employees and their representatives, published September 2009, accessible respectively at *http://www.hse.gov.uk/pubns/hse40.pdf* and *http://www.hse.gov.uk/pubns/hse39.pdf*

Breach of statutory duty

Certain Acts of Parliament and regulations impose duties on employers and employees to do or not to do specific things, e.g. the Control of Noise at Work Regulations 2005.

A breach of such duties is primarily punished by criminal proceedings. Whether the relevant Act or regulations allow someone to be able to sue in *civil law* initially depends on what was intended by the criminal Act. Thus there is no right to sue for breach of one of the general duties under the Health and Safety at Work, etc., Act 1974 because this is specifically stated in the Act (s. 47). Following a prosecution under the Act, an action would merely have to be brought in ordinary negligence. But, if there has been a breach of a specific duty in a set of regulations, such as the Control of Noise at Work Regulations, action could be brought for the tort of breach of statutory duty.

What must be proved to be able to sue?

There are five things which must be proved:

1. That the regulation or Act imposes a duty to do something.
2. That the particular regulation has been broken.
3. That the claimant is within the class which the regulation was designed to protect, e.g. a worker in a shipyard.
4. That the Act or regulation conferred a right to sue on the claimant.
5. That the claimant is suffering from an injury which the regulation was designed to protect against, even if the injury did not occur in the precise way anticipated.

There are similar defences to those of negligence and some may be specified in the Act or the regulations themselves.

The case of *Baker v. Quantum Clothing Group and others* [2009] EWCA Civ 499 is a significant judgment from the Court of Appeal regarding the liability of employers and the rights of their employees to recover compensation for hearing loss due to exposure to noise and merits some detailed consideration here. While the focus in this chapter is on the issue of employers' liability both under common law and under s. 29 Factories Act 1961, the judgment of Smith LJ (with which the rest of the court agreed) provides a clear and succinct account of the historical background and context for the evolution of noise measurement methods and corresponding policy development and legislation to which the reader is directed.

Mrs Stephanie Baker had been employed in the knitting industry by Quantum Clothing Ltd (and its predecessors) from 1971 to 1991. Her claim for noise-induced hearing loss caused by exposure to noise at work was among a group of seven heard originally and dismissed by the judge in the High Court. The seven claimants had been employed by four different employers in knitting factories and all of them had been exposed to noise in excess of 80 dBAL_{epd} but less than 90 dBAL_{epd} (dBA being the measurement of noise levels by reference to the weighted average for all frequencies and dBAL_{epd} the measurement for assessing the equivalent noise exposure over an eight-hour working day). For six of the claimants, the judge held that they had failed to prove that they were suffering from noise-induced hearing loss. In Mrs Baker's case, he held that she did have noise-induced hearing loss but that her employer was not in breach of duty either at common law (negligence as discussed above) or under s. 29 Factories Act 1961 which deals with safe means of access and safe place of employment. Section 29(1) provides:

> There shall, so far as is reasonably practicable, be provided and maintained safe means of access to every place at which any person has at any time to work, and every such place shall, so far as is reasonably practicable, be made and kept safe for any person working there.

The judge held that the standard of safety would be to exposure to 90dBAL_{epd} and that the duty under s. 29 did not add materially to the common law duty.

The ambient noise at Mrs Baker's various workplaces was about 85 to 86 dBAL_{epd}. In 1989 her employer provided her with ear protectors which she wore. Nevertheless for a period of 18 years or so, she had been exposed to noise at or above 85 dBAL_{epd}. On appeal she contended that her employer had been under a duty to provide her with ear protectors from about 1972 or if not then, at least before 1989. The Court of Appeal therefore had to consider: *when* employers should have known about the risk of harm from exposure to noise between 85 and 90 dBAL_{epd}; whether the judge at first instance had been right to equate employers' duty at common law with the duty under s. 29, or whether s. 29 imposed a higher duty (and thus a higher standard of safety); and whether, for the purposes of s. 29, an employer could reasonably have thought the degree of risk of harm was acceptable in considering whether a place of work was safe, i.e. whether the obligation under s. 29 to ensure that the place of work was safe was absolute or was qualified by the need to show reasonable foreseeability of harm.

After considering the authorities, the Court of Appeal held that it was bound by the decision in *Larner v. British Steel* [1993] 4 All ER 102 and that the obligation under s. 29 to provide a safe place of work was absolute. The safety of a place of work is to be judged objectively and not by reference to what was reasonably foreseeable at the time. Thus s. 29 does impose a higher duty on employers than does the common law. Smith LJ at para. 76 stated:

> If the safety of a place of work is to be judged objectively without reference to the reasonable foresight of injury, it must follow *a fortiori* that it must be considered without reference to what society might at that time have thought was an acceptable degree of danger for their employees to have to face.

On that basis he held that the High Court had erred in adopting a test of safety based on reasonable foresight of harm (the common law test) and equating the duty under s. 29 with that of the common law. The learned judge continued:

> [W]hat is objectively safe cannot change with time. If 85 dB(A)L_{epd} causes deafness to a particular Claimant, that Claimant's place of work was not safe for him or her. It might have been safe for another person working alongside. But for the susceptible worker who has in fact been damaged, it can be demonstrated, without more, that his or her place of work was not safe. Looking at matters from the point of view of the workforce generally, it is known that a minority of people will suffer appreciable harm as the result of prolonged exposure to 85 dB(A)L_{epd}. Therefore it can be said that the place of work is not safe for the workforce because there is a risk of injury to all of them. Some of them will be injured but not all; no-one knows which of them will not be injured. They all face some risk, save for those workers who are already known to be of average susceptibility. For those reasons . . . on the evidence before the judge, which was not controversial, the places of work where the ambient noise levels were 85 dB(A)L_{epd} or above were not safe.

The court noted that imposing a strict duty on employers is not unfair because the duty is qualified by the defence of reasonable practicability. Under s. 29, the burden of proof for the claimant is to prove that the place was unsafe; the employer then has to plead and prove the defence that he had done all that was reasonably practicable to keep the place safe (where considerations of knowledge and foresight would be relevant). To avoid liability, the defendant has to show that the burden of eliminating the risk substantially outweighed the 'quantum of risk', i.e. the gravity of the harm which might occur if the risk came to pass as well as the likelihood of its doing so. The court could not see how the provision of ear protectors was disproportionate to the risk, since provision was 'neither difficult nor expensive'.

Smith LJ went on to explain that:

> [A]t common law, the burden of proof remains with the claimant throughout and he must show that the employer has failed to take reasonable care to avoid the risks of harm which he ought reasonably to have foreseen might arise in the circumstances. The hallmark of liability at common law is that the employer must be shown not to have acted reasonably. Reasonableness pervades the whole concept of common law liability. If the employer has acted reasonably he will avoid common law liability.

The position under statute was different:

> [T]he adjective 'reasonably' serves only to qualify the concept of practicability. Reasonableness of conduct does not stand as the hallmark by which statutory liability is avoided as it does at common law. . . . Under the statute, the employer must first consider whether the employee's place of work is safe. If the place of work is not safe (even though the danger is not of grave injury or the risk very likely to occur) the employer's duty is to do what is reasonably practicable to eliminate it.

As to whether it was reasonably practicable for employers in the knitting industry to eliminate the risk of hearing loss from exposure to noise between 85 and 90 dBAL_{epd}, the court considered that from early 1977, the defendants and any other employer of average size in the knitting industry who exposed their employees to 85 dBAL_{epd} or more without protection were in breach of their duty under s. 29 Factories Act 1961. The court allowed six to nine months for the provision of ear protectors once the decision had been taken that they should be provided and fixed the date by which action should have been taken at 1 January 1978.

The Court of Appeal therefore allowed Mrs Baker's appeal: she should have been provided with ear protectors in early 1977 or at least before 1 January 1978. Her employers were therefore liable for damage from that date. Mrs Baker was therefore exposed from 1971 to 1989, an 18-year period. The exposure in breach of duty was for about 12 years. The court awarded damages of two thirds of £5000, i.e. £3334.

In consequence of this case, people whose hearing was damaged by noise levels under 90 dBAL_{epd} would be able to claim compensation. Compare the exposure limit value and the upper and lower exposure action values in the Control of Noise at Work Regulations 2005, considered below. Remember that potential claimants will still have to prove causation, i.e. that exposure to noise at work caused their deafness. The judgment will only apply to claimants whose symptoms appeared within the limitation period available to bring a claim.

HSE statistics on noise induced deafness can be accessed at *http://www.hse.gov.uk/statistics/causdis/deafness/index.htm*

HSE guidance: *Noise at Work: Guidance for Employers on the Control of Noise at Work Regulations 2005*, accessible at *http://www.hse.gov.uk/pubns/indg362.pdf*

State compensation for accidents at work or diseases or conditions contracted during the course of employment

The law governing this area is found principally under the Social Security Contributions and Benefits Act 1992 Part V, the Social Security Administration Act 1992 and the Social Security (Consequential Provisions) Act 1992, as amended by the Social Security Act 1998.

In order to claim for a benefit as a result of an accident, the accident must have occurred:

- at work or in the course of the employee's employment
- as a result of no one's fault
- as a result of another's misconduct or act of God.

In addition, the employee must not have caused the accident himself or herself, either directly or indirectly.

Prescribed diseases

The insured must be shown to be suffering from one of the prescribed diseases or conditions set out in the Social Security (Industrial Injuries) (Prescribed Diseases) Regulations 1985 (S.I. 1985/967) as amended by S.I. 1993/862 which added carpal tunnel syndrome (CTS)), and later by S.I. 1996/425 which amended the entries for occupational deafness and CTS, and the 2005 Regulations (S.I. 2005/324). There are four categories, A–D. Occupational deafness, white finger and carpal tunnel syndrome caused by using hand-held vibrating tools (added to the list in 1993) come within category A in relation to conditions due to physical agents, and are reproduced here in full.

A10 Occupational deafness. Sensorineural hearing loss amounting to at least 50 dB in each ear, being the average of hearing losses at 1, 2 and 3 kHz frequencies and being due in the case of at least one ear to occupational noise. The types of job which must have caused the condition are:

(a) any job involving the use of, or work wholly or mainly in the immediate vicinity of a band saw, circular saw or cutting disc to cut metal in the metal founding or forging industries, circular saw to cut products in the manufacture of steel, powered (other than hand powered) grinding tool on metal (other than sheet metal or plate metal), pneumatic percussive tool on metal, pressurised air arc tool to gouge metal, burner or torch to cut or dress steel-based products, skid transfer bank, knock out and shake out grid in a foundry, machine (other than a power press machine) to forge metal including a machine used to drop stamp metal by means of closed or open dies or drop hammers, machine to cut or shape or clean metal nails, or plasma spray gun to spray molten metal; or

(b) any job involving the use of, or work wholly or mainly in the immediate vicinity of a pneumatic percussive tool to drill rock in a quarry, on stone in a quarry works, underground, for mining coal, for sinking a shaft, or for tunnelling in civil engineering works; or

(c) any job involving the use of, or work wholly or mainly in the immediate vicinity of a vibrating metal moulding box in the concrete products industry, or circular saw to cut concrete masonry blocks; or

(d) any job involving the use of, or work wholly or mainly in the immediate vicinity of a machine in the manufacture of textiles for weaving man-made or natural fibres (including mineral fibres), high speed false twisting of fibres, or the mechanical cleaning of bobbins; or

(e) any job involving the use of, or work wholly or mainly in the immediate vicinity of a multi-cutter moulding machine on wood, planing machine on wood, automatic or semi-automatic lathe on wood, multiple cross-cut machine on wood, automatic shaping machine on wood, vertical spindle moulding machine (including a high speed routing machine) on wood, edge banding machine on wood, bandsawing machine (with a blade width of not less than 75 millimetres) on wood, circular sawing machine on wood including one operated by moving the blade towards the material being cut, or chain saw on wood; or

(f) any job involving the use of, or work wholly or mainly in the immediate vicinity of a jet of water (or mixture of water and abrasive material) at a pressure above 680 bar, or jet channelling process to burn stone in a quarry;

(g) any job involving the use of, or work wholly or mainly in the immediate vicinity of a machine in a ship's engine room, or gas turbine for performance testing on a test bed, installation testing of a replacement engine in an aircraft, or acceptance testing of an Armed Service fixed wing combat aircraft; or

(h) any job involving the use of, or work wholly or mainly in the immediate vicinity of a machine in the manufacture of glass containers or hollow ware for automatic moulding, automatic blow moulding, or automatic glass pressing and forming; or

(i) any job involving the use of, or work wholly or mainly in the immediate vicinity of a spinning machine using compressed air to produce glass wool or mineral wool; or

(j) any job involving the use of, or work wholly or mainly in the immediate vicinity of a continuous glass toughening furnace; or

(k) any job involving the use of, or work wholly or mainly in the immediate vicinity of a firearm by a police firearms training officer; or

(l) any job involving the use of, or work wholly or mainly in the immediate vicinity of a shot-blaster to carry abrasives in air for cleaning.

A11 Vibration white finger.

(a) intense blanching of the skin, with a sharp demarcation line between affected and non-affected skin, where the blanching is cold-induced, episodic, occurs throughout the year and affects the skin of the distal with the middle and proximal phalanges, or distal with the middle phalanx (or in the case of a thumb the distal with the proximal phalanx), of
   (i) in the case of a person with five fingers (including thumb) on one hand, any three of those fingers; or
   (ii) in the case of a person with only four such fingers, any two of those fingers; or
   (iii) in the case of a person with less than four such fingers, any one of those fingers or, as the case may be, the one remaining finger;

   where none of the person's fingers was subject to any degree of cold-induced, episodic blanching of the skin prior to the person's employment in an occupation described below in relation to this paragraph, or

(b) significant, demonstrable reduction in both sensory perception and manipulative dexterity with continuous numbness or continuous tingling all present at the same time in the same distal phalanx of any finger (including thumb) where none of the person's fingers was subject to any degree of reduction in sensory perception, manipulative dexterity, numbness or tingling prior to the person's employment in an occupation described below, where the symptoms in paragraph a) or b) were caused by vibration.

The types of job that must have caused this condition are any job involving:

(a) the use of hand-held chain-saws on wood; or

(b) the use of hand-held rotary tools in grinding or in the sanding or polishing of metal, or the holding of material being ground or metal being sanded or polished by rotary tools; or

(c) the use of hand-held percussive metalworking tools, or the holding of metal being worked upon by percussive tools, in riveting, caulking, chipping, hammering, fettling or swaging; or

(d) the use of hand-held powered percussive drills or hand-held powered percussive hammers in mining, quarrying, demolition, or on roads or footpaths, including road construction; or

(e) the holding of material being worked upon by pounding machines in shoe manufacture.

A12 Carpal tunnel syndrome. The types of job that must have caused this condition are any job involving:

(a) the use, at the time the symptoms first develop, of hand-held powered tools whose internal parts vibrate so as to transmit that vibration to the hand, but excluding those tools which are solely powered by hand; or

(b) repeated palmar flexion and dorsiflexion of the wrist for at least 20 hours per week for a period or periods amounting in aggregate to at least 12months in the 24 months prior to the onset of symptoms. We use 'repeated' to mean once or more often in every thirty seconds.

Unfortunately, for occupational deafness the claimant must have worked in any of the occupations listed in A10 for a period or periods of not less than 10 years.

Industrial Injuries Disablement Benefit

Disablement benefit takes the form of a pension or gratuity paid after a period of 90 days, excluding Sundays. The period begins at the time of the accident or the onset of the industrial disease or condition. The money is paid where there remains a loss of physical or mental faculty when entitlement to other benefits has ceased. It is based on medically assessing the person's loss of faculty as a result of his or her incapacity as compared with another person of the same age and sex. Besides this pension, if the person is severely disabled or has particular difficulties, he or she may also claim other benefits, e.g. constant attendance allowance, but this is unlikely in relation to occupational deafness. It is also unlikely that occupational deafness would give rise to 100% disablement but, if it did, the 2009 benefit would be £143.60 per week if aged over 18.

Claims are made to the appropriate regional office of the Department for Work and Pensions, of which there are five, formally known as Industrial Injuries Disablement Benefit Delivery Centres. After filling in the appropriate claim form (there is a special form for occupational deafness), a claimant is then referred to an otologist, who will provide a report to a special medical board. These boards are appointed under the Social Security Act to determine whether the claimant actually suffers from occupational deafness, or any other prescribed disease, and to what extent his or her hearing has been affected.

The decision is first given by the insurance officer. But if the claimant is dissatisfied, he or she has the right to go to the local medical tribunal. Appeals go from there to the National Insurance Commissioner and may be made by the claimant, an employees' association such as a trade union, or by the insurance officer. The Commissioner's decision is final.

Health and Safety at Work, etc., Act 1974

Although the majority of accidents occur in the home, unfortunately, it would be impossible to monitor domestic safety conditions. Legislation has therefore concentrated on trying to control accidents at work.

Legislation was originally introduced in a piecemeal way, initially covering only the most dangerous of occupations and later covering the most common – factories, offices and shops. This piecemeal approach to legislation has now been tied together by the Health and Safety at Work, etc., Act 1974, which covers all work situations. Under its umbrella come all the old Acts and regulations, which remain in force until they are gradually replaced by new regulations under the 1974 Act.

The 1974 Act does *not* set out to enable compensation to be recovered from wrongdoers, although compensation may be awarded if there is a breach of a *specific* duty under the Act or regulations made thereunder (s. 47). In relation to noise, they would be the Control of Noise at Work Regulations 2005 (see pages 313–16). The Act's prime responsibility, therefore, is to impose *criminal* liability on those breaching the Act or regulations made under it.

Reasons for enactment

The 1974 Act came into effect in 1975 and was designed to provide, for the first time, health, safety and welfare protection for *all* people at work (except domestic servants), *and* for people *affected* by the dangers of work.

The Act thus provides a framework on which complex technical and scientific regulations can be formulated to cover all work processes, but which will be administered and enforced in a standardized way.

The Health and Safety Commission

The Health and Safety Commission has corporate status; it was set up by the 1974 Act with the purpose of achieving better health and safety by undertaking research and training, formulating policy, providing advice and information and acting in close liaison with appropriate government departments. Its present address is Rose Court, 2 Southwark Bridge, London SE1 9HS. The Commission may also issue codes of practice; they do not actually form part of the legislation or general law, but they do provide good evidence in court if any recommended practice has not been followed.

The Commission may also approve existing codes of practice issued by other specialist bodies or government departments. It must have the consent of the Secretary of State before approving a code and it must first consult appropriate bodies.

The Commission may also hold informal investigations and formal inquiries into accidents or situations when it thinks necessary.

The Health and Safety Executive

Before the 1974 Act there were a number of different inspectorates having similar responsibilities but whose authority stemmed from their controlling Acts, e.g. the factory inspector was empowered by the Factories Act 1961. All the inspectorates have now been brought under the control of the executive, which is the health and safety enforcement agency.

Some of the inspectors' powers may be transferred to other enforcement agencies. For example, activities involving catering, offices and the sale or storage of goods in shops, warehouses and launderettes are dealt with by local authority environmental health departments. They have exactly the same powers (under the 1974 Act) as other health and safety inspectors.

In brief, under s. 2 of the 1974 Act an employer must ensure, so far as is reasonably practicable, the health, safety and welfare of his or her employees. This duty is expanded by s. 2(2), which specifies five areas of responsibility:

1. The employer must provide and maintain plant and systems of work which are, as far as is reasonably practicable, safe and without risks to health.

2. He must ensure the safety and absence of risks to health in the use, storage, handling and transport of substances and articles.
3. He must provide sufficient information, instruction, training and supervision in order to ensure the health and safety of his employees.
4. He must maintain the place of work in a condition which is safe and without risk to health and ensure safe access and egress to it.
5. He must provide a safe working environment with adequate welfare facilities.

Furthermore, by s. 3 every employer and every self-employed person is under a duty to ensure so far as is reasonably practicable that persons *not* in his or her employment are not exposed to risks to their health or safety by the conduct of the business. Thus passers-by, visitors and neighbours are owed a duty under s. 3.

The term *reasonably practicable*, which qualifies many of the duties, means that the employers are not required to go to unreasonable lengths of trouble and expense to eliminate a small risk of accident.

Section 6 imposes duties on manufacturers, designers, importers and suppliers of articles for use at work or in fairground equipment. They must ensure that the article is so designed that it will be safe and without risks to health at all times during their use at work. Also, by s. 6, arrangements must be made for testing and examining the articles or equipment and the employees should be provided with such information as is necessary to secure their health when using it.

Section 7 imposes two duties on every employee at work:

(a) To take reasonable care for the health and safety of himself or herself and of other persons who may be affected by his or her acts or omissions at work.
(b) As regards any duty or requirement imposed on his or her employer, to cooperate with that employer so far as is necessary to enable the duty or requirement to be performed or complied with.

Failure to abide by these duties amounts to a criminal offence.

By s. 8 a person who wilfully or recklessly interferes or misuses anything provided in pursuance of a relevant statutory provision, such as safety apparatus, is guilty of an offence.

In order to supplement the 1974 Act, regulations have been enacted, such as the Control of Noise at Work Regulations 2005 (see below). Codes of practice may also be issued or approved by the Commission. Once approved they may then be used in evidence in court to show that someone has infringed the Act.

Powers of the inspectorate

The inspectors have wide-ranging investigatory powers; they can enter premises, examine and take photographs, take samples and measurements, dismantle machinery, take possession of anything relevant and require anyone to give information. If an inspector finds that there has been a breach of an Act or regulations, he or she may take the following enforcement steps. By s. 21 the inspector can issue an improvement notice requiring the situation to be remedied within a specified period. The inspector can issue a prohibition notice if there is a risk of serious personal injury, ordering the work to cease immediately or to be suspended for a specific period. Appeals against both notices may be made to an industrial tribunal. The inspector can also seize, render harmless or destroy dangerous articles. Finally, the inspector can prosecute either in the magistrates' court for less serious offences or in the Crown Court for serious breaches. Fines and imprisonment can be ordered. Anyone can be prosecuted, including corporations and companies.

An offence under the 1974 Act would provide good evidence in any civil action taken by an employee for compensation, but it does not in itself allow for damages to be paid. Action could, however, be taken for breach of statutory duty if there are relevant grounds.

The Control of Noise at Work Regulations 2005 (S.I. 2005/1643)

These regulations came into force on 6 April 2006, with the exception of the music and entertainment sectors, where they apply from 6 April 2008. They were made under the Health and Safety at Work, etc., Act 1974 and replace the Noise at Work Regulations 1989. They were issued to implement in the UK the contents of EU Directive 2003/10/EC on the minimum health and safety requirements regarding the exposure of workers to the risks arising from physical agents (noise). This is the seventeenth directive made under the Directive 89/391/EEC, the health and safety Framework Directive on the introduction of measures to encourage improvements to the health and safety of workers at work. The Health and Safety (Miscellaneous and Revocations) Regulations 2009 (S.I. 2009/693) amend reg. 7 on hearing protection so as to require personal hearing protectors made available or provided by the employer to comply with any requirement applicable to them under the Personal Protective Equipment Regulations 2002 (S.I. 2002/1144). The 2002 Regulations themselves maintain the implementation of

Directive 89/656 on the minimum health and safety requirements for the use by workers of personal protective equipment in the workplace.

Under reg. 3, the provisions apply to *all employees and the self-employed* in workplaces, excluding for the most part those on ships (reg. 3(4)) and, by analogy with the Act, all those at home. Where the Regulations place a duty on an employer in respect of his employees, the employer shall, so far as is reasonably practicable, be under a duty in respect of any *other* person at work who may be affected by the work carried out by the employer: reg. 3(2) (cf. s. 3 of HASWA 1974, above). This would include, for example, trainees, people on work experience, and, in the case of educational institutions and hospitals, students and patients respectively. The Regulations also cover the music and entertainment sectors, so while bars, pubs, restaurants, theatres, concert halls and clubs are workplaces for many, patrons of those establishments are also owed a duty. There are two exceptions to this. The health surveillance duty (reg. 9) does not extend to persons who are not his employees, nor does reg. 10 (information, instruction and training) unless those persons are present at the workplace where the work is being carried out.

Regulation 2 is an interpretation clause:

- *Daily personal noise exposure (dpne)* means the level of dpne of an employee as ascertained in accordance with Schedule 1 Part 1 of the Regulations, taking account of the level of noise and duration of exposure and covering all noise. This is calculated by reference to the formula,

$$L_{\mathrm{EP,d}} = L_{\mathrm{Aeq},T_e} + 10\log_{10}\left(\frac{T_e}{T_0}\right)$$

where:
T_e is the duration of the person's working day, in seconds
T_0 is 28,800 seconds (8 hours) and
L_{Aeq,T_e} is the equivalent continuous A-weighted sound pressure level, as defined in ISO 1999:1990 clause 3.5, in decibels, that represents the sound the person is exposed to during the working day.

- *Exposure limit value* means the level of daily or weekly personal noise exposure or of peak sound pressure set out in reg. 4 which must not be exceeded.
- *Lower exposure action value* means the lower of the two levels of daily or weekly personal noise exposure or of peak sound pressure set out in reg. 4 which, if reached or exceeded, requires specific action to be taken to reduce risk.
- *Upper exposure action value* means the higher of the two levels of daily or weekly personal noise exposure or of peak sound pressure set out in reg. 4 which, if reached or exceeded, requires specific action to be taken to reduce risk.
- *Peak sound pressure* means the maximum sound pressure to which an employee is exposed, ascertained in accordance with Schedule 2.
- *Weekly personal noise exposure* means the level of weekly personal noise exposure as ascertained in accordance with Schedule 1 Part 2, taking account of the level of noise and the duration of exposure and covering all noise.
- *Noise* means any audible sound.

By reg. 2(2), a reference to an employee being exposed to noise is a reference to an the exposure of that employee to noise which arises while he is at work, or arises out of or in connection with his work.

By reg. 4, the lower exposure action values are: a daily or weekly personal noise exposure of 80 dB (A-weighted); and a peak sound pressure of 135 dB (C-weighted). The upper exposure action values are: a daily or weekly personal noise exposure of 85 dB (A-weighted); and a peak sound pressure of 137 dB (C-weighted). The exposure limit values are: a daily or weekly personal noise exposure of 87 dB (A-weighted); and a peak sound pressure of 140 dB (C-weighted). An employer may use weekly personal noise exposure instead of daily personal noise exposure where an employee's exposure to noise varies markedly from day to day (reg. 7(4)). In applying the exposure limit values, account shall be taken of the protection given to the employee by any personal hearing protectors the employer has provided in accordance with reg. 7(2). Such protection is not taken into account in applying the lower and upper exposure action values (reg. 7(5)).

Exposure limit values and upper and lower exposure action values replace the concepts of peak action level and first and second action levels in the old (1989) Regulations. The second action level meant a dpne of 90 dBA and had been criticized as being too high.

The employer's general duty: elimination or control of exposure to noise in the workplace

By reg. 6, every employer shall ensure that risk from the exposure of his employees to noise is either eliminated at source or, where this is not reasonably practicable, reduced to as low a level as is possible. If any employee is likely to be exposed to noise at or above an upper exposure action value, the employer must reduce the exposure to as low a level as reasonably practicable by establishing and implementing a programme of organizational and technical measures, *excluding* the provision of personal hearing protectors, which is appropriate to the activity (s. 6(2)).

Section 6(3) sets out matters which the employer should take into consideration to comply with his duty and provides that the actions taken must be based on the general principles of prevention set out in Schedule 1 to the Management of Health and Safety Regulations 1999 (S.I. 1999/3242, as amended).

Every employer must ensure that their employees are not exposed to noise above an exposure limit value or if an exposure limit value is exceeded, must immediately reduce exposure to noise below that value, identify the reason for the exposure limit value being exceeded and modify the technical and organizational measures to prevent it being exceeded again (s. 6(4)).

Risk assessment

By reg. 5, where work is liable to expose any employees to noise at or above a lower exposure action value, employers must make a suitable and sufficient assessment of the risk from that noise to the health and safety of those employees. The assessment must identify the action to be taken, must be reviewed regularly and immediately if there has been a significant change in the work or a suspicion that the assessment is no longer valid. Where necessary, changes must be made (s. 5(1),(4)). The employer must record the significant findings of the risk assessment and the measures taken and intended to be taken to meet the requirements of regs. 6, 7 and 10.

Provision of hearing protectors

Regulation 7(1) requires that an employer who carries out work which is likely to expose any employees to noise at or above a lower exposure action value shall make personal hearing protectors available on request to any employee who is so exposed. Additionally s. 7(2) provides that if an employer cannot by other means reduce the levels of noise to which an employee is likely to be exposed to below an upper exposure action value, he must provide personal heating protectors to any employee who is so exposed.

By reg. 7(4), as amended (see above) personal ear protectors provided under this regulation must comply with any requirement of the Personal Protective Equipment Regulations 2002 (S.I. 2002/1144) which is applicable to ear protectors. Regulation 7 also requires employers to consult with the employees concerned or their representatives prior to selection of personal hearing protectors.

Hearing protection zones, reg. 7(3)

Every employer must demarcate as a hearing protection zone (HPZ) those areas of his workplace where there is likely to be exposure to noise at or above an upper exposure action value. Hearing protection zones are demarcated and identified by a sign (see BS 5499-5: 2002) as specified in the Safety Signs and Signals Regulations 1996 (S.I. 1996/341). The sign should indicate that the area is an HPZ and that protectors must be worn. Where practicable and the risk from exposure justifies it, the employer must ensure that access to the area is restricted. The employer must ensure, so far as reasonably practicable, that no employee enters the HPZ unless that employee is wearing personal ear protectors.

Maintenance and use of equipment, reg. 8(1)

Employers must ensure, so far as is reasonably practicable, that anything provided by them in compliance with their duties, other than personal hearing protectors provided under reg. 7(1) is fully and properly used. Equipment should be maintained in an efficient state, in efficient working order and in good repair.

Employees' duties, reg. 8(2)

Every employee must make full and proper use of personal hearing protectors the employer has supplied under reg. 7(2) and of any other control measures provided by his employer in compliance with his duties under these Regulations and must report any defects to the employer as soon as is practicable.

Health surveillance

Employers must ensure that their employees are placed under suitable health surveillance, including testing of their hearing, if the risk assessment indicates there is a risk to the health of employees who are, or are liable to be, exposed to noise (reg. 9(1)). A health record of each employee under health surveillance must be made and maintained, and on reasonable notice being given, the employer must allow an employee access to his personal health record. The employer shall provide the enforcing authority (i.e. the Health and Safety Executive or local authority) with copies of such health records as it may require (reg. 9(2) and (3)). An employee to whom this regulation applies shall, when required by his employer, and at the cost of his employer, present himself during his working hours for such health surveillance procedures as may be required (reg. 9(5)). Where he is found to have identifiable hearing damage, the employer must ensure that the employee is examined by a doctor and if the doctor considers that the damage is likely to be the result of exposure to noise, the employer has further duties. He must ensure that a suitably qualified person informs the employee accordingly; review the risk assessment; review any measures taken to comply with regs. 6, 7 and 8; consider assigning the employee to alternative work where there is no risk from further exposure to noise; and ensure continued health surveillance (reg. 9(4)). The revoked 1989 Regulations contained no equivalent duty.

Information, instruction and training, reg. 10

Where his employees are exposed to noise which is likely to be at or above a lower exposure action value, the employer shall provide those employees and their representatives with suitable and sufficient (formerly 'adequate' under the 1989 Regulations) information, instruction and training including (reg. 10(2)):

(a) the nature of risks from exposure to noise
(b) the organizational and technical measures taken in order to comply with the requirements of reg. 6
(c) the exposure limit values and upper and lower exposure action values set out in reg. 4
(d) the significant findings of the risk assessment, including any measurements taken, with an explanation of those findings
(e) the availability and provision of personal hearing protectors under reg. 7 and their correct use in accordance with reg. 8(2)
(f) why and how to detect and report signs of hearing damage
(g) the entitlement to health surveillance under reg. 9 and its purposes
(h) safe working practices to minimise exposure to noise
(i) the collective results of any health surveillance undertaken under reg. 9 in a form calculated to prevent those results from being identified as relating to a particular person.

Under reg. 10(4) the employer has a duty to ensure that any person, whether or not his employee, who carries out work in connection with the employer's duties under these Regulations has suitable and sufficient information and training. Regulation 10 duties are considerably extended from the information duty under the 1989 Regulations.

Exemptions: regs. 11, 12 and 13

Regulation 11 permits the Health and Safety Executive to grant exemptions from reg. 6(4) and reg. 7(1) and (2) requirements where because of the nature of the work the full and proper use of personal hearing protectors would be likely to cause greater risk to health or safety than not using such protectors. Nevertheless the resulting risks must be reduced to as low a level as is reasonably practicable. Exemptions may be granted subject to conditions and may be revoked at any time. Exemptions cannot be granted without prior consultation with appropriate persons, including employers, employees and their representatives and all employees concerned must be subject to increased health surveillance.

Regulation 12 permits the Health and Safety Executive to grant exemptions from reg. 6(4) and reg. 7(1) to (3) in respect of activities carried out by the emergency services and reg. 13 enables the Secretary of State for Defence to grant exemptions under reg. 6(4) and reg. 7(1) to (3) in respect of activities carried out in the interests of national security which conflict with any of those provisions. Both reg. 12 and reg. 13 require the exempting authority to be satisfied that the health and safety of the employees concerned is ensured as far as possible in the light of the objectives of these Regulations.

The Control of Vibration at Work Regulations 2005 (S.I. 2005/1093)

Since noise includes vibration, it should be noted that the Control of Vibration at Work Regulations 2005 (S.I. 2005/1093) came into in force on 6 July 2005. They were issued under the 1974 Act to implement Directive 2002/44/EC on minimum health and safety requirements regarding exposure to vibration at work, the sixteenth directive made under the health and safety Framework Directive 89/391/EEC. Like the Noise Regulations, these regulations impose duties on employers to ensure the elimination or control of exposure to vibration in the workplace (reg. 6), to carry out risk assessment (reg. 5), implement health surveillance (reg. 7) and provide information, instruction and training (reg. 8). Regulation 4 sets out the exposure limit values and action values for hand-arm vibration and whole body vibration.

10.9 The Land Compensation Act 1973: compensation for noise caused by public works

There are many Acts empowering government departments and local authorities to compulsorily purchase property and give compensation, where property is adversely affected by public works, but the Land Compensation Act extends these rights. Part I of the Act confers a right to compensation for depreciation in the value of land caused by public works. Part II, as amended by the Highways Act 1980 and the Local Government, Planning and Land Act 1980, confers on public authorities power to acquire land and to do works to reduce the harmful effects of public work on their surroundings.

Part I: the right to compensation

By s. 1, certain specified landowners in certain specified circumstances may be paid compensation in respect of the depreciation in value of their land as a result of certain physical factors, including noise, vibration, smell, fumes, smoke and artificial lighting, caused by the use of a new or altered *highway*, *aerodrome* or *other public works* provided in exercise of statutory powers.

By s. 9(7), altering or changing the use of public works does not cover the intensification of use. The compensation will be paid by the appropriate highway authority or else by the person or body managing the works. The source of the physical factors must be situated on or in the public works, or in relation to aerodromes, it must be caused by the use of the aerodrome as such or by aircraft departing or arriving.

In the case of public works, according to s. 1(6), the authorities must have statutory immunity against actions for nuisance, otherwise compensation will not be awarded, e.g. immunity under the Civil Aviation Act 1982 (see page 325). If they do not have immunity then the appropriate course of action for a person suffering is to sue in tort for nuisance. In *Vickers v. Dover District Council* (1993) 20 EG 132 the council exercised its power under the Road Traffic Regulation Act 1984 to provide a car park. There was nothing in the Act giving immunity from action, and the claimant could and should proceed in nuisance. The Lands Tribunal said that there was no right to compensation under the Land Compensation Act.

By s. 2, where the claim relates to a dwelling-house, the claimant must own the legal fee simple, i.e. the freehold, or must have a tenancy of at least three years to run and be in occupation of the property.

Claims

By s. 3 the claim is made by serving the responsible authority with a notice containing *inter alia* particulars of the land, the interest held, the public works to which the claim relates and the amount of compensation sought. No claim shall be made before the expiration of 12 months from the relevant date. The authority may enter the land and survey it for valuation purposes. If there are any disputes, they must be determined by the Lands Tribunal.

O'Connor v. Wiltshire CC [2007] EWCA Civ 426, provides a recent example of a claim for noise compensation. The noise came from the construction of a distributor road, which the Court of Appeal, reversing the decision of the Lands Tribunal, held had been constructed by the local highway authority, Wiltshire County Council, albeit via an acceleration agreement with developers. Consequently the works were 'public works' and the appellants, who were owners and occupiers of a house nearby, were entitled to claim compensation.

Part II: mitigation of injurious effect of public works

By s. 20(1) the Secretary of State may make regulations requiring authorities responsible for public works to insulate buildings against noise from construction or use of public works, or to make grants towards the cost of such insulation. The regulations are the Noise Insulation Regulations (S.I. 1975/1763) as amended by the Noise Insulation (Amendment) Regulations (S.I. 1988/2000), the Noise Insulation (Railways and Other Guided Transport Systems) Regulations (S.I. 1996/428), with amending Regulations (S.I. 1998/1701). The Highways Noise Payment and Movable Homes (England) Regulations (S.I. 2000/2887) extend noise payments to occupiers of caravans and mobile homes who meet the qualifying conditions.

These regulations apply to buildings, affected by the construction and use of new highways and additional carriageways to existing highways, railways, tramways and other guided transport systems subject to certain time limits, and impose duties or give powers to insulate against noise or give grants to do such works. Buildings which are eligible under the regulations are dwellings or other buildings used for residential purposes not being more than 300 metres from the nearest point of the carriageway. Certain types of buildings are specifically excluded. The regulations apply to caravans and mobile homes in the case of highways only.

By reg. 3(1) of the 1975 Regulations, the appropriate highway authority is under a duty to carry out or make a grant in respect of the cost of carrying out insulation work on an eligible building, where the *use* of a relevant highway causes or is expected to cause noise at a level not less than the 'specified level', i.e. L_{10} (18-hour) of 68 dBA. For the purpose of this regulation, the use of the highway will cause noise at a level not less than the specified level if the 'relevant noise level', i.e. the level of noise expressed as a level of L_{10} (18-hour) 1 metre in front of the most exposed of any windows and doors in a façade of a building caused by the traffic using or expected to be using the highway is greater by at least 1 dBA than the 'prevailing noise level' (measured in much the same way as the relevant noise level but before the works for the construction of the road began) and is not less than the specified level. Additionally, the noise caused or expected to be caused by traffic using or expected to be using that highway makes an effective contribution to the relevant noise level of at least 1 dBA.

In the above case, the highway or additional carriageway must have been first open to the public after 16 October 1972.

Regulation 4(1) imposes a similar duty on the 'responsible authority' in respect of railways, tramways and other guided transport systems.

The noise levels must be calculated in accordance with memoranda published by HMSO: *Calculation of Road Traffic Noise* 1975 and 1988, *Calculation of Railway Noise* 1995, with Supplement 1.

Acquisition of land in connection with highways

Section 246 of the Highways Act 1980 empowers a highway authority to acquire, compulsorily or by agreement, land for the purpose of reducing any adverse effect caused by any highway, either existing or proposed.

By s. 26 of the Land Compensation Act 1973, an authority may acquire land by agreement in order to mitigate any adverse effect which the public works will have on the surrounding area. Such land could be seriously affected by the construction, alteration or use of the public works. By s. 27 of the same Act, an authority having s. 26 powers to acquire land may carry out works to reduce the injurious effect on land it already owns or the land it has acquired under the previous section. Such work may include planting trees, shrubs or plants, or laying out the area as grassland.

Thus the Land Compensation Act provides a number of remedies for people who are affected by works which in the opinion of the state are important to the rest of the community.

10.10 Insulation requirements under the Building Regulations

One way of reducing sound at source is to provide sufficient insulation when constructing buildings. The Building Regulations (S.I. 2000/2531), as amended, and made under the Building Act 1984 provide a framework of control of building standards supplemented by practical guidance documents, called approved documents, on different aspects of construction work. The approved documents provide the relevant detail. These documents are approved or issued by the Secretary of State. In relation to noise, the relevant document is Approved Document E *Resistance to the Passage of Sound*, 2003 edition, incorporating 2004 amendments (published 2006).

The 2000 Regulations replace, but largely take forward, the 1991 Regulations which themselves had been subject to amendment over the years. The 2000 Regulations have themselves been amended substantially by the Building (Amendment) (No. 2) Regulations (S.I. 2002/2871) (see below). Failure to work to Building Regulation standards does not necessarily mean there has been a breach. Failure to comply with the guidance documents does not in itself render a person liable to civil or criminal proceedings, but it would provide good evidence of a general failure to abide by the regulations (s. 7 of the 1984 Act). If the builder has followed the guidance in the approved document, it will be evidence tending to show that he or she has complied with the regulations, and vice versa. If he or she chooses other methods of building, the onus falls on the builder to show that he or she has satisfied the requirements of the regulations. The builder can do this by showing that he or she has followed an appropriate British Standard, or an equivalent national technical specification of any member state of the European Union; he or she has used a product bearing an EC mark in accordance with Construction Products Directive 89/106/EEC, as amended by Directive 93/68/EC; he or she has followed an appropriate technical specification as defined by the Directive or by the British Board of Agrément Certificate.

The purpose of the provisions is to control noise from *other* parts of the building and adjoining buildings, not to control noise *entering* through external walls.

The regulations apply to any 'building work' or 'material change in use' of a building. Regulation 3 defines *building work* to include *inter alia* the erection or extension of a building and the material alteration of a building. The rest of the definition applies to sanitary fittings, bathroom fittings, sewers, etc. A material change covers changes to dwellings where not previously so used; similarly, flats, hotels or institutions, public buildings and rooms for residential purposes (reg. 5, as amended). Where building work is carried out, the requirements of resistance to the passage of sound are required (reg. 4), likewise where there is material change in use of a building to a dwelling, or so that it contains a flat, or is used as a hotel or boarding house, or, where it contains at least one dwelling, it now contains a greater or lesser number of dwellings than it did previously, or the building now contains a room for residential purposes, where previously it did not, or the building, which contains at least one room for residential purposes, contains a greater or lesser number of such rooms than it did previously (reg. 6, as amended). A 'room for residential purposes' is defined as, 'a room, or suite of rooms, which is not a dwelling-house or a flat and which is used by one or more persons to live and sleep and includes a room in a hostel, an hotel, a boarding house, a hall of residence or a residential home, whether or not the room is separated from or arranged in a cluster group with other rooms, but does not include a room in a hospital, or other similar establishment, used for patient accommodation and, for the purposes of this definition a "cluster" is a group of rooms for residential purposes which is – (a) separated from the rest of the building in which it is situated by a door which is designed to be locked; and (b) not designed to be occupied by a single household' (reg. 2, as amended).

Where a building now consists of, or contains a school, it must comply with the requirements of acoustic conditions in schools (E4, below) (reg. 6, as amended).

By reg. 7: 'Building work shall be carried out –

(a) with adequate and proper materials which –
  (i) are appropriate for the circumstances in which they are used,
  (ii) are adequately mixed or prepared, and
  (iii) are applied, used or fixed so as adequately to perform the functions for which they are designed; and
(b) in a workmanlike manner.'

It is this regulation in particular that the approved documents are intended to support.

Note that Part E (along with Part M – access and facilities for disabled people) is *excluded* from reg. 8, as amended, which generally does not require anything to be done, except for the purpose of securing reasonable standards of health and safety for persons in or about the building, because it addresses the welfare and convenience of building users.

A new Part E of Schedule 1 to the Building Regulations 2000 was substituted by the Building (Amendment) (No. 2) Regulations (S.I. 2002/2871). Part E (resistance to the passage of sound) refers to protection against sound from other adjoining buildings and parts of buildings, protection against sound within a dwelling-house, reverberation in common internal parts of property with rooms for residential purposes and acoustic conditions in schools (reg. 2(9)). It is set out in full below.

Protection against sound from other parts of the building and adjoining buildings

E1. Dwelling-houses, flats and rooms for residential purposes shall be designed and constructed in such a way that they provide reasonable resistance to sound from other parts of the same building and from adjoining buildings. (It is interesting to note that the 1991 Regulations, as originally promulgated, used the phrase 'shall resist' rather than 'provide reasonable resistance to'.)

Protection against sound within a dwellinghouse etc

E2. Dwelling-houses, flats and rooms for residential purposes shall be designed and constructed in such a way that –

(a) internal walls between a bedroom or room containing a water closet, and other rooms; and
(b) internal floors,

provide reasonable resistance to sound. Requirement E2 does not apply to –

(a) an internal wall which contains a door;
(b) an internal wall which separates an en suite toilet from the associated bedroom;
(c) existing walls and floors in a building which is subject to a material change of use.

Reverberation in common internal parts of buildings containing flats or rooms or residential purposes

E3. The common internal parts of buildings which contain flats or rooms for residential purposes shall be designed and constructed in such a way as to prevent more reverberation around the common parts than is reasonable.

Requirement E3 only applies to corridors, stairwells, hallways and entrance halls which give access to the flat or room for residential purposes.

Acoustic conditions in schools

E4. (1) Each room or other space in a school building shall be designed and constructed in such a way that it has the acoustic conditions and the insulation against disturbance by noise appropriate to its intended use.

(2) For the purpose of this part – 'school' has the same meaning as in section 4 of the Education Act 1996 and 'school building' means any building forming a school or part of a school.

These are the four requirements in relation to sound. The Approved Document describes how to satisfy them.

Pre-completion testing

The 2002 Amendment Regulations introduced a requirement for sound insulation testing for building work in relation to which paragraph E1 (above) applies (reg. 2(7), inserting a new s. 20A into the Building Regulations 2000). Appropriate testing must be carried out in accordance with an approved procedure and a copy of the test results given to the local authority (reg. 20A(2)). Pre-completion testing applies to rooms for residential purposes and house and flat conversions from 1 July 2003 and to new houses and flats from 1 July 2004. Annex B of Approved Document E contains guidance on the testing procedure which must be done in accordance with the relevant British Standards for field and laboratory measurements, relating to airborne sound insulation between rooms and of building elements (BS EN ISO 140-4:1998 and BS EN ISO 140-3:1998), impact sound insulation of floors (BS EN ISO 140-7:1998 and BS EN ISO 140-6: 1998), and the reduction of transmitted impact noise by floor coverings on a heavyweight standard floor (BS EN ISO 140-8:1998). British Standard BS EN ISO 354:2003 deals with measurement of sound absorption in a reverberation room and BS EN 29052-1:1992 sets out the method for the determination of dynamic stiffness: materials used under floating floors in dwellings. Other British Standards relate to the rating of airborne sound insulation and impact sound insulation in buildings and of building elements (BS EN ISO 717-1:1997 and BS EN ISO 717-2:1997) and the rating of sound absorption in buildings (BS EN ISO 11654:1997). Annex D of the Approved Document provides full details of all standards, key guidance and legislation.

Use of 'robust details'

An alternative method for demonstrating compliance in the case of new houses and flats is the use of robust details, described as follows in the Approved Document:

> Robust details are high performance separating wall and floor constructions (with associated construction details) that are expected to be sufficiently reliable not to need the check provided by pre-completion testing.

Design details approved by Robust Details Ltd (a non-profit distributing company, set up by the house building industry) are set out in Annex E of the Approved Document. A handbook containing robust design details can be obtained from the company at PO Box 7289, Milton Keynes, MK14 6ZQ, telephone 0870 240 8210, fax 0870 240 8203, email *administration@robustdetails.com*, website *www.robustdetails.com*. This alternative approach was introduced into the Building Regulations by the Building (Amendment) Regulations (S.I. 2004/1465). Regulation 2(5) inserts reg. 20A(4) into the 2000 Regulations, which sets out the substantive and procedural requirements, which are also set out in Annex E:

> (a) the building work consists of a new dwelling-house or building containing flats;
> (b) the person carrying out the work must notify the building control body (usually the local authority) that he is using one or more specified design details approved by Robust Details Ltd;
> (c) the notification specifies the unique number or numbers issued by Robust Details Ltd in respect of the specified use of the design details;
> (d) the building work is carried out in accordance with those details.
>
> Late notification or lack of detail specificity will render the building subject to sound insulation testing under Regulation 20A. Regulation 20A will also apply where notification is timely but the work is not carried out in accordance with the specified design details.

The Secretary of State's view, set out in s. 0 of the Approved Document, is that the requirements of E1 are met if the relevant parts of the building are designed and built in such a way that they achieve the sound insulation values set out in Table 1a and the values for rooms for residential purposes set out in Table 1b in the Approved Document. Preferably the testing body should have UKAS accreditation (or European equivalent). The use of robust details is an alternative for new residential developments. The normal way of satisfying requirement E2 will be to use the constructions that provide the laboratory sound insulation values set out in Table 2. Satisfying requirement E3 will normally be by applying the sound absorption measures described in s. 7 of the Approved Document, and for E4 it will be by meeting the relevant requirements set out in s. 1 of Building Bulletin 93, *The Acoustic Design of Schools*, 2003, DfES/TSO.

Work must be supervised either by the local authority's building control officers or an 'approved inspector', i.e. one approved by the Secretary of State. The builder must provide the local authority with a building notice (for small and minor works) or deposit full plans. The local authority can ask for more information. If full plans are deposited, they must be passed or rejected by the authority within five weeks. Provided the plans show that the work will be carried out in accordance with the Building Regulations, the local authority must approve them. During the course of the work, notification must be made to the local authority at various times. Firstly, at least 48 hours' notice of commencement. Secondly, at least 24 hours before covering up of any excavation for a foundation or a damp-proof course or any concrete or other material laid over a site material. Thirdly, at least 24 hours' notice before covering up drains and finally within seven days of completion of the work.

Breaching the regulations is an offence triable in the magistrates' court under s. 35 of the Building Act 1984. Each day the offence continues will incur a further fine after summary conviction. The local authority could also serve notice on the owner, requiring demolition or removal of the works, or order the owner to make such alterations as are necessary (s. 36(1)(a) and (b)). Similarly, if work is carried out and no plans were deposited (and they should have been) or the plans were rejected or the work was carried out otherwise than in accordance with any requirements made by the local authority, then the local authority can require the demolition, removal or alteration as before (s. 36(2)). The local authority may also do the works themselves and charge the owner (s. 36(3)). No s. 36 notice can be served after the expiration of 12 months from the date of completion of the works (s. 36(4)). By s. 36(6) there remains the right of the local authority, the Attorney-General or any other person to apply for an injunction for the removal or alteration of any work on the ground that it contravenes any regulation or provision of the Act. However, if plans were deposited and they were passed by the local authority, or notice of their rejection was not given in the relevant period and the work was executed in accordance with the plans, then the court, on granting an injunction, has power to order the local authority to pay compensation as the court thinks just and can join the local authority as party to the action, if it is not one already.

10.11 Spatial planning (town and country planning)

Spatial planning can obviously play a large part in the control of noise. It may predetermine what should be located in a particular area by the use of regional spatial strategies and local development frameworks, thus avoiding new problems. And it may refuse planning permission to build potentially noisy premises or lay down specifications or conditions, if planning permission is given.

Planning Policy Guidance Note 24 (PPG24) *Planning and Noise*, issued by the government in 1994, gives advice to local authorities on the use of their planning powers to achieve those aims, indicating sources of material and information and encouraging liaison between authorities affected by the same problems.

The relevant legislation is found in the Town and Country Planning Act 1990 (TCPA) as amended by the Planning and Compensation Act 1991 (PCA), the Planning and Compulsory Purchase Act 2004 (PCPA 2004) and the Planning Act 2008. At the base of all planning control are development plans, the main plans being Regional Strategies (replacing regional spatial strategies, but themselves to be abolished by the Localisation Bill) and Local Development Frameworks (LDF). The Regional Strategy covered a 15-year period and was prepared by the responsible regional authority. While there is always the prospect of the reintroduction of regional planning, this section can merely note its passing. The LDF (now the main development plan) is prepared by local planning authorities (district councils, borough councils and unitary authorities). The LDF comprises a local authority's local development plan scheme in a 'portfolio' of documents which collectively delivers the area's spatial planning strategy. The portfolio of local development documents (LDDs) includes: development plan documents (which have development plan status), supplementary planning documents (these do not have development plan status but will still be subject to community involvement), a statement of community involvement and a local development scheme (details of timescale and arrangements for the production of LDDs). DPDs are required to contain a core strategy, action area plans (where needed), site specific allocations of land and a proposals map (ss. 13–37 PCPA 2004).

Development plans are based on a survey of the area, taking into account, *inter alia*, the principal physical, economic, social and environmental characteristics of the authority's area, and effect of neighbouring areas, the population, communications, transport systems and traffic, plus anything else of relevance. All plans must be kept under review and during their formative stages adequate publicity must be made of the draft so that ordinary people and other relevant interests may make representations.

The plans lay down aims and objectives, policies and goals but do not necessarily have to be followed: s. 38(6) PCPA 2004 provides: 'If regard is to be had to the development plan for the purpose of any determination to be made under the planning Acts the determination must be made in accordance with the plan unless material considerations indicate otherwise.' Section 39 makes sustainable development a statutory objective in planning.

Planning permission

The Act can only control 'development' which is defined by s. 55(1) of the 1990 Act as 'the carrying out of building, engineering, mining or other operations in, on, over or under land, or the making of any material change in the use of any buildings or other land'. Section 1A inserted by the 1991 Act now specifically includes in the definition of 'building operations' (a) demolition of buildings; (b) rebuilding; (c) structural alterations of or additions to buildings; and (d) other operations normally undertaken by a person carrying on business as a builder. Section 1A has put an end to doubt in this area and has particular relevance to noise because demolition work is often just as noisy as straightforward building work.

Thus, any operation falling outside the definition will not require planning permission.

By s. 57(1) planning permission is required for the carrying out of any development of land.

If there is any doubt whether a project may require planning permission then, under s. 192 of the 1990 Act, a person may apply to the local planning authority as to whether the work to be carried out or the proposed change of use is lawful. If, on an application under this section, the local planning authority is supplied with such information as would satisfy it, the local planning authority shall issue a certificate to that effect and in any other case it shall refuse the application (s. 192(2)). If a certificate is issued, by s. 192(4) there is a conclusive presumption of lawfulness of any use or operations unless there is a material change before the specified use or work has started. Thus, this gives the local planning authority's seal of approval to work which would not require permission. If a certificate is refused then planning permission must be applied for.

There is no development if there is a change of use within the same use class, as specified in the Town and Country Planning (Use Classes) Order 1987 and amended by the Town and Country Planning (Use Classes) (Amendment) Orders 1991 and 2005. But a change from one use class to another use class does require permission where that change is material. Certain changes between use classes constitute permitted development (see below) under the GPDO 1995, Part 3 as amended by the GPDO (Amendment) (England) Order 2005 (S.I. 2005/85). No classes in the order include use as a theatre, amusement arcade, centre or funfair; the sale of fuel for motor vehicles or the sale or display of motor vehicles; a scrapyard or a yard for storing or distributing minerals (like coal) or breaking motor vehicles, work registrable under the Alkali, etc., Works Regulation Act 1906, hostel, waste disposal installation for the incineration, chemical treatment or landfill of waste to which Directive 91/689 applies, retail warehouse club, nightclub, and casino. They are potentially nuisance-causing and would require specific planning permission.

Here are some examples of the sorts of premises included in the order:

- *Class A1* includes shops with certain specific exceptions, e.g. take-away hot food which comes under *Class A5*.
- *Class A2* is concerned with financial and professional services covering banks and building societies.
- *Class B1* covers offices (but not A2), including those connected with research and industrial processes, being those uses which could be carried on in any residential area without detriment to the amenity of that area by reason of noise, vibration, smell, fumes, smoke, soot, dust, ash or grit.
- *Class B2* is general industrial for industrial processes not falling within class B1.
- *Class B8* covers storage and distribution centres.
- *Class C1* covers hotels and hostels.
- *Class C3* covers houses used by single people or families or those with not more than six residents living together as a household.
- *Class D2* covers assembly and leisure arenas, cinemas, concert halls, bingo halls, dance-halls, swimming-baths, skating-rinks, gymnasia or places for other indoor or outdoor sports that do not involve motorized vehicles or firearms.

Another order, the General Permitted Development Order 1995 (GPDO 1995), as amended, sets out many classes of development for which planning permission is automatic, i.e. granted by the Order itself. Such development is known as 'permitted development'. The GPDO 1995 is frequently amended, see e.g. S.I. 2005/2935, S.I. 2008/675, and S.I. 2008/2362, accessible via *www.legislation.gov.uk*. A local planning authority or the Secretary of State may make a Direction removing any development from the provisions of the GPDO.

The Order is extremely detailed, each category having strict limitations and exclusions. Obviously, the development proposed is of a relatively minor nature and such that, had it been necessary, planning permission would have been granted. This also prevents waste of time and money in the local planning authority. Any planning permission granted by virtue of this order is subject to any relevant limitations or conditions specified in Schedule 2. Application still has to be made to the relevant local planning authority.

Local development orders, introduced into the planning system by the PCPA 2004, enable LPAs to make local development orders so as to implement policies in the relevant Development Plan Document. The Town and Country Planning (General Development Procedure) (Amendment) Order 2006 (S.I. 2006/1062), Art. 5, refers.

Applying for planning permission

Section 62 of the 1990 Act, the Town and Country Planning (Applications) Regulations 1988 (S.I. 1988/1812) and the General Development Procedure Order (S.I. 1995/419), as amended, lay down the requirements for applying for planning permission. Applications are made on forms supplied by the local planning authority, to which all necessary information, site plans and drawings are attached. By s. 70 the local planning authority may grant permission, either unconditionally or subject to such conditions as it thinks fit, or it may refuse permission altogether. Model conditions are contained in PPG 24, Annex 4, with further guidance in Circular 11/95, *The Use of Conditions in Planning Permissions.* Case law has shown that conditions must be relevant to planning and to the development to be permitted; they must also be reasonable. The government has also said in its circular that 'conditions should be necessary, precise and enforceable. The key test is whether planning permission would have to be refused if the condition were not imposed. If not, then such a condition needs special and precise justification.' There should be close liaison with the environmental health department in order to get expert advice on noise as well as to consider the imposition and monitoring of compliance with the conditions. Here are some examples of relevant conditions suggested in PPG24 *Planning and Noise*:

1. Construction work shall not begin until a scheme for protecting the proposed [noise-sensitive development]

from noise from the ______ has been submitted to and approved by the local planning authority; all works which form part of the scheme shall be completed before [any part of] the [noise-sensitive development] is occupied.

In relation to aerodromes examples are:

3. The total number of movements shall not exceed ( ) per [period of time], except in an emergency.
7. The runways shall not be used by [class of aircraft], except in an emergency.

In relation to noise emitted from industrial or commercial buildings and sites:

11. Before the use commences, the [specified buildings] shall be insulated in accordance with a scheme agreed with the local planning authority.
18. No [specified machinery] shall be operated on the premises before [time in the morning] on weekdays and [time in the morning] on Saturdays nor after [time in the morning] on weekdays and [time in the evening] on Saturdays, nor at any time on Sundays or Bank Holidays.

In dealing with applications the local planning authority's decision must be in accordance with the development plan, unless material considerations indicate otherwise (s. 38(6) PCPA 2004).

Nowadays, one material consideration is the environmental impact assessment, made under the Town and Country Planning (Environmental Impact Assessment) (England and Wales) Regulations 1999 (S.I. 1999/293), as amended. The original 1988 regulations were made as a result of EEC Council Directive 85/337/EEC, itself amended by Directive 97/11. Under these regulations, in certain situations before development has begun, an assessment must be made of the likely effects of the works on the environment. By Annex 1 of the Directive in some cases there *must* be an assessment, e.g. for motorways, airports and long-distance railways. In other cases (Annex 2 development) an assessment is required if the development is *likely* to have a significant environmental effect because of its size, nature or location. Guidance is found in Circular 2/99.

In other cases there are many regulations on whether an assessment is necessary. The assessment could be voluntary, ordered by the local planning authority or ordered by the Secretary of State. If required, an environmental statement must be prepared by the developer, who must identify those objects which may be affected by the works – humans, climate, animals, etc. – how the works affect them and the proposed methods of reducing or avoiding the harmful effects of the works. The local planning authority (if the appropriate body) will then prepare the assessment which must be taken into account before permitting the development.

Other material considerations include overlap with certain areas of statutory control, perhaps in relation to noise clashes with building regulation control and noise control under the Environmental Protection Act 1990 and the Pollution Prevention and Control Act 1999 (environmental permitting).

Directive 2001/42/EC introduced strategic environment assessment (SEA), i.e. environmental assessment of plans and programmes, as opposed to development projects. The Directive was brought into force by the Environmental Assessment of Plans and Programmes Regulations 2004 (S.I. 2004/1633). In national planning terms this applies to Regional Strategies and Local Development Frameworks, considered above. The objectives of SEA are set out in Article 1 of the Directive:

> [T]o provide for a high level of protection of the environment and to contribute to the integration of environmental considerations into the preparation and adoption of plans and programmes with a view to promoting sustainable development, by ensuring that, in accordance with this Directive, an environmental assessment is carried out of certain plans and programmes which are likely to have significant effects on the environment.

An interesting decision was given in *Gillingham Borough Council v. Medway (Chatham) Dock Co. Ltd.* [1992] 3 All ER 923. Here a local authority granted planning permission to develop part of the former Chatham Royal Dockyard into a commercial dockyard. Initially it was clear that the new development would be operating 24 hours a day. Between 300 and 400 lorries would use the local access roads, all residential streets. But after the yard began operating, complaints came from local residents and the local authority applied for an injunction under s. 222 of the Local Government Act 1972 on the basis that the noise and other discomfort amounted to a public nuisance (see page 276). The local authority was refused an injunction, so it appealed to the High Court. The High Court turned down the appeal principally, it said, because the nature of the locality had changed, a change authorized by the local authority when it granted planning permission in the first place. The usage of the neighbourhood was now characteristic of the environs of a commercial port. It could even be argued that changes in the development plan could have a similar effect on an area. It was in some respect analogous to the defence of statutory authority (see page 282). When granting the permission, great account had been taken of the financial advantages for the area, which had suffered greatly as a result of the closure of the dockyard. No environmental assessment had been undertaken (there was no need at the time). It may be that this decision could be used as a precedent in non-planning cases because it relates to the nature of the neighbourhood. Local planning authorities must therefore be even more wary of granting permissions in environmentally difficult situations and the average person must be more vigilant in voicing objections before permissions are granted.

This case should not, however, be construed as authority for the proposition that planning permission will provide immunity from liability in nuisance in every case. In *Wheeler v. J.J. Saunders* [1995] 3 WLR 466, the defendants had planning permission to extend their pig farm which was near to the claimant's holiday cottages. The Court of Appeal held that the smell from the pig farm was an actionable nuisance, despite the grant of planning permission. In *Watson v. Croft Promo-Sport Ltd* [2009] JPL 1178 the judge awarded compensation to the claimant, on the basis that planning permission for a motor-racing circuit operated on a disused aerodrome and 300 metres from the claimant's home had not changed the nature and character of the locality, which remained essentially rural, such that the activities, which generated excessive noise, constituted an actionable nuisance. The Court of Appeal (2009) went further and granted an injunction to the claimant, who, in the court's opinion, had suffered a substantial injury with respect to the enjoyment of his property. Only in very special circumstances should damages in lieu of an injunction be awarded and in this instance the judge had erred in not restricting the number of days when the defendants could use their facilities. Thus planning permission does not affect private law rights, save for those cases, where the permission changes the nature and locality as a whole. Such cases are likely to be relatively rare.

Mid Beds Model Aircraft Club v. The Secretary of State and Bedford Borough Council [2009] EWCA 681 (Admin), provides a useful example of noise disturbance being the key consideration in deciding a planning application and related appeal. The local authority had refused the claimant's planning application for a temporary change of use of land from agriculture to model aircraft flying on the grounds that the resulting noise from the flying activities 'would cause an unacceptable impact on the amenities of the occupiers of nearby residential premises.' As such the proposal was contrary to development plan policies. The claimant's appeal to the Planning Inspector was dismissed primarily for the same reason. The claimant's challenge on points of law and procedure to the Inspector's decision was dismissed by the High Court.

Enforcing the law

By s. 171A, TCPA 1990, as amended, if no permission has been given and the developer has carried out the work or there has been a contravention of a condition, the local planning authority can take the following steps.

Firstly, it can issue an enforcement notice under s. 172. A copy must be served on the owner and the occupier of the land and any other person with an interest in the land. By s. 173 the enforcement notice must specify the breach and whether the breach is caused by unauthorized development or by failure to comply with a condition or limitation under s. 171A; it must also be clear to the recipient what this means.

The notice must also specify the steps which the authority requires to be taken or the activities which the authority requires to be stopped in order to achieve the remedying of the breach. Thus, according to s. 173(5), the enforcement notice may require:

(a) alteration or removal of works
(b) carrying out of building or other operations
(c) any activity not to be carried on, except as limited by the terms of the notice.

When specifying the steps to be taken they must not be too vague, otherwise the courts may regard the notice as a nullity. Thus mere directions to 'install satisfactory soundproofing' is not sufficiently precise (*Metallic Protectives Ltd. v. Secretary of State for Environment*, 1976).

The recipient of the notice may appeal to the Secretary of State against it (s. 174).

If the notice is ignored, the local planning authority has the right to enter the land and carry out the required work at the expense of the developer (s. 178).

By s. 179, if at the end of the period for compliance with the enforcement notice, the required steps have not been carried out or work directed to cease has not ceased, then the *owner* of the land is in breach and is guilty of an offence. The local planning authority can then prosecute and, on finding the defendant guilty, will be liable on summary conviction to a fine not exceeding £20,000 and on conviction on indictment to an unlimited fine.

Where a condition has been imposed by s. 187A, the local planning authority may serve a breach of condition notice, requiring the steps to be taken. Ignoring the notice amounts to an offence.

Finally, the local planning authority also has the power to serve a stop notice, prohibiting work carried out in contravention of planning conditions, if it considers it expedient that a relevant activity should cease before the expiry of the period for compliance with an enforcement notice. The stop notice ceases to have effect once the enforcement notice has been quashed or withdrawn at an appeal or when the period for compliance has expired. There are no rights of appeal; failure to comply with the notice constitutes an offence as above.

Section 187B introduced by the 1991 Act has introduced a similar right to take High Court action, as we have already seen, under the Environmental Protection Act. Thus, if a local planning authority considers it 'necessary or expedient for any actual or apprehended breach of planning control' to be restrained by injunction, it may apply to the court for an injunction, whether or not it has exercised or is proposing to exercise any of its other powers under the Act.

In *Barnet LBC v. Alder* [2009] EWHC 2012 (QBD) the High Court granted the local authority's application for an injunction to restrain a breach of planning control. A semi-detached residential property was being used as a school. The local authority refused to grant retrospective planning permission because the increased activity and noise consequent upon this unauthorized use caused significant harm to neighbouring properties. The council issued an enforcement notice requiring closure of the school within 11 months. The first respondent, whose property it was, appealed unsuccessfully against the enforcement notice, the Planning Inspector finding that although the school would provide a particular community benefit (it provided a strict Orthodox Jewish religious education alongside the national curriculum), it would cause unacceptable levels of noise and disturbance to nearby properties. Continuing the use as a school would also be detrimental to the supply of local housing. Despite the Inspector's finding that an 11-month period was reasonable, three years elapsed during which time the school tried and failed to find alternative premises. The council refused a second application for retrospective planning permission and while the appeal was pending sought an injunction to restrain the continuing use of the school which was granted but suspended for five months so that pupils would not be unreasonably interrupted in their education. The prospects of success on appeal were slight: in the first appeal, the Inspector considered the breach of planning law was serious. As to alternative arrangements for the pupils, Walker J stated:

> [I]t seems to me that the pupils and their parents must recognise that they have only had the benefit of places at [the school] because [the school] contravened planning law. If [the school] cannot find a location that complies with planning law, then in the circumstances of the present case it is just and proportionate to require them to accept that the position will have to revert to what it was before [the school] was established.

10.12 Civil aviation

The problems of intensification in air travel are obvious in the field of noise control. People suffering the most are those living close to airports, such as Gatwick and Heathrow, especially when further runways are threatened. How are these problems being tackled? Because of the different and localized problems, special laws have been enacted. Many Acts have been passed and will be passed to try to diminish the effects of what is an essential part of a modern transport system.

The relevant legislation is mainly found in the Civil Aviation Act 1982 (CAA), as amended by the Civil Aviation Act 2006, ss. 1–4 and all subordinate legislation made thereunder.

Statutory immunity

The early air industry was considered to require protection from litigation in case large payouts crippled this new mode of transport. As early as 1920 legislation was passed to prevent lawsuits.

Thus, in normal flying situations there is immunity from legal action. By s. 76(1) of the Civil Aviation Act 1982:

> No action shall lie in respect of trespass or . . . nuisance by reason only of the flight of an aircraft over any property at a height above the ground which having regard to wind, weather and all the circumstances of the case is reasonable, or the ordinary incidents of such flight, so long as the provisions of any Air Navigation Order and of any orders under section 62 above have been duly complied with and that there has been no breach of section 81.

Section 62 orders concerned control of navigation in time of war or emergency and have been repealed by the Transport Act 2000. Section 81 concerns dangerous flying – a criminal offence, punishable by a fine not exceeding level 4 on the standard scale and/or imprisonment not exceeding six months.

In normal circumstances, even if a landowner is troubled by incessant noise from overflying aircraft, he or she will not be able to sue for noise nuisance.

Furthermore, by s. 77(1):

> An Air Navigation Order may provide for regulating the conditions under which noise and vibration may be caused by aircraft on aerodromes and may provide that subsection (2) below shall apply to any aerodrome as respects which provisions as to noise and vibration caused by aircraft is so made.

Section 77(2) states:

> No action shall lie in respect of nuisance by reasons only of the noise and vibration caused by aircraft on an aerodrome

> to which this subsection applies by virtue of an Air Navigation Order as long as the provisions of any such Order are duly complied with.

Similarly, if our landowner finds that the nearby aerodrome is the source of noise nuisance while the aircraft are on the aerodrome, no action can be taken, providing they comply with any Air Navigation Orders (see below).

The 1989 Air Navigation Order (S.I. 1989/2004) provides in Art. 83 of the order, that the Secretary of State may prescribe the conditions under which noise and vibrations *may be caused* by aircraft in virtually all types of aircraft and on virtually all types of aerodromes.

The Secretary of State has issued the Air Navigation (General) Regulations 1993 (S.I. 1993/1622). Regulation 13 states:

> For the purpose of Article 83 the conditions under which noise and vibration *may be caused* by aircraft (including military aircraft) on Government aerodromes, aerodromes owned or managed by the Civil Aviation Authority, licensed aerodromes or aerodromes at which the manufacture, repair or maintenance of aircraft is carried out by persons carrying on a business as manufacturers or repairers of aircraft shall be as follows:
>
> (a) the aircraft is taking off or landing; or
> (b) the aircraft is moving on the ground or water; or
> (c) the engines are being operated in the aircraft:
> (i) for the purpose of ensuring their satisfactory performance; or
> (ii) for the purpose of bringing them to a proper temperature in preparation for, or at the end of, a flight; or
> (iii) for the purpose of ensuring that the instruments, accessories or other components of the aircraft are in a satisfactory condition.

In other words, Regulation 13 lays down the situations in which aircraft may make noise and vibrations without there being any legal comeback from people suffering as a result.

These sections therefore cut down the possibility of litigation in the majority of cases, but they do not rule out an action being brought, where the noise or vibration was being caused outside the protection of a section, e.g. by ignoring regulations, flying too low or where there is another cause of action in addition to nuisance (or trespass). Many groups over the years have criticized this special protection and have pointed out that in other countries, such as the United States, which has no similar immunity, civil aviation has continued to flourish. The Batho Report (Report of the Noise Review Working Party 1990, DoE) also considered that s. 76 is no longer appropriate. The report especially differentiated between commercial aircraft and private and leisure aircraft, the latter being less deserving of protection. The report indicated there were about 280 airfields in the United Kingdom with approximately 7000 aircraft registered in Britain alone, not counting the foreign-registered aircraft; it added that most of the 7000 aircraft were used for private flying.

The protection afforded by this legislation does not extend to situations where persons or property are damaged by the aircraft itself or things falling from the aircraft, including people. In such cases the owner of the aircraft is strictly liable, i.e. without proof of negligence or intention, as it is obviously in the interests of public safety that bits don't fall off aeroplanes. Thus by s. 76(2):

> Where material loss or damage is caused to any person or property on land or water by or by a person in or an article, animal or person falling from, an aircraft while in flight, taking off or landing, then unless the loss or damage was caused or contributed to by the negligence of the person by whom it was suffered, damages in respect of the loss or damage shall be recoverable without proof of negligence or intention or other cause of action, as if the loss or damage had been caused by the wilful act, neglect or default of the owner of the aircraft.

Such loss or damage includes that caused by noise or vibration. Thus, the Act differentiates between mere nuisance and direct damage caused by aircraft.

Legal control of aircraft noise

Legal control of aircraft noise is achieved in a number of ways. Firstly, by the making of regulations under the Act to control noise and vibration caused in the operation of aircraft. Secondly, by phasing out noisy aircraft through aircraft certification. Thirdly, by providing compensation under the Land Compensation Act 1973 as amended (see page 316) for diminution in value of property, and in certain areas, by providing grants towards the insulation of buildings affected by aircraft noise. There exist other practical measures to reduce noise, but they are not the concern of the law.

Regulations

By s. 78(1) the Secretary of State may by notice impose a duty on aircraft operators taking off or landing at a designated aerodrome to secure compliance with any requirements specified in the notice. At present Heathrow, Gatwick and Stansted have been designated. The requirements of the notice will be for the purpose of limiting or mitigating the effect of noise and vibration at landing and take-off. By s. 78(2) if such persons do not comply with the requirements, then after being given the opportunity of making representations to the Secretary of State, the manager of the aerodrome may be directed to withhold airport facilities from the operator and his or her servants. The manager is under a duty to comply. Such notices include requirements for minimum noise routes and night levels of noise on take-off.

By s. 78(3) the Secretary of State may also make an order to reduce the effects of noise and vibration by limiting the number of occasions on which aircraft may take off or land during certain periods. To that end the Secretary of State may specify the *maximum* number of take-offs or landings during these periods in relation to particular types of aircraft. Furthermore, the Secretary of State can determine which aircraft operators may take off and land within these periods and the number of occasions particular types of aircraft belonging to the operators may take off and land. It is the responsibility of the aerodrome manager to see that these rules are complied with.

In *R v. Secretary of State for Transport ex p Richmond-upon-Thames London Borough Council* (1993) *The Times*, 12 Oct, an application for judicial review of a decision of the Secretary of State was made by various local authorities. The Secretary under s. 78(3)(b) had issued a press announcement that he intended to introduce a new system of night-flying restrictions at Heathrow, Stansted and Gatwick. The numbers would no longer be *fixed*; they would *vary* according to the noise produced by the aircraft. The court granted the application. The section specifically states that the requirements are to state the maximum number of movements 'the linchpin of any order made' as per Laws J. Thus, the decision was not authorized by the section and was invalid.

If it appears to the Secretary of State that an aircraft is about to take off in contravention of limitations imposed, then without prejudicing the powers of the airport manager, any person with the Secretary's authority may detain the aircraft for such period as necessary and may enter on any land for that purpose (s. 78(5)(c) and (d)).

The Secretary of State by s. 78(6) may give the manager of the aerodrome directions in order to limit or mitigate the effect of noise and vibration and the manager must comply. Section 78(6A) (inserted by the Civil Aviation Act 2006) provides that directions under s. 78(6) may be given for the purpose of avoiding, limiting or mitigating the effect of such noise and vibration either generally or in any particular area or areas.

By s. 78(7) the Secretary of State may also, after consulting with the manager of the aerodrome, order the manager to provide, maintain and operate in a particular area, at a specified time, such equipment for measuring noise in the vicinity of the area as is specified. The manager must then make reports and permit any authorized person to inspect the equipment.

Failure to comply with any of the duties will make the person defaulting guilty of an offence liable to summary conviction, and continued non-compliance will lead to the commission of separate offences for each day on which the default occurs (s. 78(9)). Section 78(A) (added by the Civil Aviation Act 2006) grants power for aerodromes to establish penalty schemes (and see the Civil Aviation Act 2006 (Commencement No. 1) Order 2007 (S.I. 2007/598).

Section 63(2)(b) of the Airports Act 1986 authorizes the power to make byelaws for controlling the operations of aircraft within or directly above the airport for the purpose of limiting or mitigating the effect of noise, vibration and atmospheric pollution caused by aircraft using the airport.

Section 38 of the Civil Aviation Act 1982, as substituted by s. 1 Civil Aviation Act 2006 fixes, by reference to noise and emissions factors, the charges that may be levied for using licensed aerodromes. An aerodrome authority may fix its charges in respect of an aircraft or class of aircraft by reference to the amount of noise caused by the aircraft or the extent or nature of any inconvenience resulting from such noise for the purposes of encouraging the use of quieter aircraft and reducing inconvenience from aircraft noise. Section 38(1) and (2) contain further provisions for aerodrome charges to be fixed by reference to the level of noise and noise and emissions requirements for the respective purposes of controlling the level of noise or atmospheric pollution and promoting compliance with noise or emissions requirements.

Section 38A (added by the CAA 2006) gives power for aerodromes to establish noise control schemes. Section 38C makes provision for breaches of noise control schemes, when payment of a penalty may be required.

By s. 35 the Secretary of State may, by order, require the management of any specified aerodrome to provide adequate facilities for consultation regarding the management of aerodromes to the following parties where their interests may be affected: the users of the aerodrome, interested local authorities and any organizations representing the interests of persons concerned with the locality in which the aerodrome is situated. All national and regional airports have been designated in this way (as well as some general aviation aerodromes). In 1994 Heathrow, Gatwick and Stansted airports took over the responsibility for aircraft complaints from the Department of Transport.

By s. 6 there is a power enabling the Secretary of State to direct the Civil Aviation Authority to take action to prevent noise pollution caused by civil aviation, providing it already has powers to act. Following the completion of the M25, London's orbital motorway, s. 6 powers were used to prevent a temporary helicopter link between Heathrow and Gatwick from continuing in operation beyond its original permission.

Finally, by s. 5 the Secretary of State is empowered to require the Civil Aviation Authority to consider environmental matters when licensing or renewing a licence for an aerodrome but not those designated under s. 78. This power appears never to have been exercised.

The Aerodromes (Noise Restrictions) (Rules and Procedures) Regulations 2003 (S.I. 2003/1742 set out the rules and procedures applicable at all public airports with over 50,000 aircraft movements a year, including noise-related operating instructions. The Regulations transpose the requirements of Directive 2002/30/EC.

Noise certification

Noise certification is a concept which lays down rules prescribing noise limits for aircraft calculated in relation to their maximum certificated weight. The authority for relevant legislation is found in s. 60 of the Act and EC Council Directive 89/629.

By s. 60, 'Her Majesty the Queen may by Order in [the Privy] Council (called here an Air Navigation Order) make such provisions as authorised in subsections (2) and (3).'

> [s. 60(3)] . . . an Air Navigation Order may contain provisions . . .
>
> (r) for prohibiting aircraft from taking off or landing in the United Kingdom unless there are in force in respect of these aircraft such certificates of compliance with standards of noise as may be specified in the Order and except upon compliance with the conditions of those certificates and
>
> (s) for regulating or prohibiting the flight of aircraft over the United Kingdom at speeds in excess of Mach 1.

Section 60 also empowers the implementation into English law of the provisions of Annex 16 of the 1944 Chicago Convention on International Civil Aviation. (Remember that international law found in conventions such as this, or accords, or treaties, does not become part of an internal legal system until ratified in some way.)

The relevant order is the Air Navigation (Environmental Standards) Order 2002 (S.I. 2002/798). The order applies (subject to the right of the Civil Aviation Authority, after consultation with the Secretary of State to make exceptions) to the following cases:

(a) certain supersonic aeroplanes
(b) every microlight aeroplane
(c) every helicopter (of certain specifications).

By this order no relevant aircraft may take off or land in the United Kingdom unless the Civil Aviation Authority, or other competent authority of the state where the aircraft has been registered, has issued a noise certificate and the conditions have been complied with. Military aircraft and visiting foreign military aircraft are exempt. The certificates in this country are issued by the Civil Aviation Authority, and the applicant must supply all necessary information and satisfy all the relevant tests to his or her aircraft. The certificate has no time limit but will end if the aircraft is modified in such a way as to affect its noise emission.

Council Directive 92/14/EEC, issued in 1992 by the European Commission, was incorporated into English law by the Aeroplane Noise (Limitation on Operations of Aeroplane) Regulations 1993 (S.I. 1993/1409). These regulations have been revoked by the Aeroplane Noise Regulations 1999 (S.I. 1999/1452) which set out the current standards for aircraft and certification requirements for registered propeller driven planes and civil subsonic jet aeroplanes.

These regulations apply, with certain exceptions, to every civil subsonic jet aeroplane with a maximum take-off mass greater than or equal to 34,000 kg and with more than 19 passenger seats. In such cases no aircraft is permitted to take off or land within the United Kingdom unless it is carrying a noise certificate, as required by its country of registration, issued to certain specified standards.

Noise insulation

Despite all attempts to reduce noise at source, it is impossible to eradicate the problem for those living close to airports. By s. 79 of the Act, the Secretary of State may make schemes requiring the person for the time being managing the aerodrome to make grants to certain people occupying certain types of buildings close to designated aerodromes towards the cost of insulating such buildings. Heathrow and Gatwick both have schemes but some other airports have schemes authorized by private Acts of Parliament.

The insulation required involves double glazing of windows and the installation of specified ventilation systems.

Compensation may be also awarded under the Land Compensation Act 1973 (see page 316).

10.13 Road traffic noise

Road traffic noise is created in a number of ways, from operating hooters, noisy exhausts, car radios, revving up, to mere high density of traffic. The problems can be dealt with in many ways, some of which we have already seen. Under the Land Compensation Act 1973, land can be purchased or compensation be given in certain circumstances; under the Noise Insulation Regulations 1975 (S.I. 1995/1763) as amended, works to buildings may be carried out or grants given (the counterpart for *railway* noise being the Noise Insulation (Railways and Other Guided Transport Systems) Regulations 1996 (S.I. 1996/428), as amended); in some situations the Environmental Protection Act 1990 or Noise and Statutory Nuisances Act 1993 may be implemented and local by-laws may be of particular use.

Also regulations have been made under the Road Traffic Act 1972 and s. 41 Road Traffic Act 1988 to control the use and construction of certain motor vehicles, i.e. the Road Vehicles (Construction and Use) Regulations 1986 (S.I. 1986/1078) much amended, not least to comply with a number of EU Directives. These regulations cover a number of different aspects of noise from manufacturing to use. Breach of any regulation is an offence.

Thus, a vehicle propelled by an internal combustion engine must be fitted with an exhaust system including a silencer and the exhaust gases from the engine must not escape into the atmosphere without first passing through the silencer (reg. 54(1)). All exhaust systems and silencers must be maintained in good and efficient working order and must not be altered after manufacture so as to increase the noise made by the escape of exhaust gases (reg. 54(2)).

Regulation 55 applies to any wheeled motor vehicle with three or more wheels, such as motor cars, first used after 1 October 1983, subject to certain exceptions such as motorcycles, tractors or road rollers. These vehicles must be constructed so that they comply with the requirements set out in the Table made under reg. 55(3) and which comply with certain European Union Directives so that the sound level from the specified vehicle does not exceed the set limits as measured using the specified methods and apparatus. Regulation 55A applies to all motor vehicles with less than four wheels, first used on or after 1 October 1996. Regulations 56, 57 and 57A and 57B cover similar matters in relation to agricultural motor vehicles and industrial tractors, motorcycle construction, motorcycle exhaust systems, and motorcycle maintenance respectively. Regulation 59 excepts vehicles from the above requirements if they are proceeding to a place where noise emission levels are going to be measured in order to determine whether they comply with the Regulations or if they are going to be adjusted or modified so that they do comply with the Regulations. The Motor Cycle Noise Act 1987 prohibits the supply of motorcycle exhaust systems and silencers likely to result in the emission of excessive noise, i.e. those not complying with the requirements of the relevant regulations. Anyone contravening the Act is guilty of an offence and liable to a fine or imprisonment.

Apart from the technical control of noise emission in the construction and use of motor vehicles, the Regulations also lay down rules restricting the driver in the way he or she drives. Regulation 97 lays down a general prohibition on noisy use. 'No motor vehicle shall be used on a road in such manner as to cause any excessive noise which could have been avoided by the exercise of reasonable care on the part of the driver.'

By reg. 98 the driver of a vehicle must stop the engine when stationary unless he or she is stuck in traffic or it is necessary to keep the engine running in order to examine or repair it.

Regulation 99 deals with the problem of warning devices such as a car horn. Thus, no person (note not just the driver) must sound or cause to be sounded any audible warning instrument such as a horn, gong, bell or siren fitted or carried on the vehicle, if the vehicle is stationary *at any time* unless to warn of danger to another moving vehicle on or near the road. Nor must a person sound such devices while moving on restricted roads between 23.30 pm and 7.00 am the following day. Those provisions do not apply to reversing warning devices or boarding aid alarm provided they are on certain specified types of large vehicles such as goods vehicles or buses. In addition, subject to certain exceptions, no one must sound or cause to sound a gong, bell, siren or two-tone horn fitted to or carried on a vehicle whether it is stationary or not. The exceptions are for emergency services when it is necessary to warn other road users of the urgency of the situation or to warn them of the vehicle's presence; theft alarms (but this does not include two-tone horns); bus alarms to summon help for the bus personnel; an apparatus designed to inform the public that the vehicle is conveying goods for sale but only between 12.00 midday and 19.00 pm and subject to s. 62 Control of Pollution Act 1974 (see page 297).

Spatial planning (town and country planning) can also play an important part in reducing noise caused by traffic. This can be done by predetermining what should be located in a particular area by the use of the development plan for the area and thus preventing new problems. Also, planning permission may be refused if the development would increase the traffic inordinately.

Finally, the reader is reminded of EU noise policy and legislation outlined in section 10.2 earlier.

10.14 Covenants in a lease

A covenant is a promise contained in a document, in this case, a lease. When someone acquires the lease of a house or flat, the agreement usually contains many covenants with which the lessee must comply, e.g. to pay the rent, to repair and not to sublet the property without permission of the lessor. Often such leases also contain particular covenants dealing with the prevention of nuisances, e.g. not to operate a washing-machine or other domestic appliance between the hours of 11:00 pm and 8:30 am the following day so as to cause a nuisance or annoyance to the neighbours. These sorts of covenants are especially important in leases of flats within a block, as noise can be potentially contentious. Where there is a breach of the covenant then the lessor may be able to forfeit the lease, or in less extreme cases, take other forms of action such as obtaining an injunction.

10.15 Restrictive covenants

Restrictive covenants are different from covenants in leases; they must not be confused as there are complicated rules relating to whether or not they are enforceable. The Law Commission has recommended the reform of restrictive covenants, to modernize and simplify the law, and a draft Bill is anticipated in Spring 2011. Until such time as the law is changed, an outline of the current nature and scope of restrictive covenants is necessary.

Restrictive covenants are in essence covenants of a negative nature, e.g. not to commit a nuisance or carry on a trade or business on particular land, or even not to build on a particular piece of property, and they can apply to leasehold or freehold land. They are usually imposed by a person selling part only of his or her land for building purposes. Obviously he or she wishes to keep the remaining land free from nuisances or generally to maintain the original character of the neighbourhood and thus he or she imposes the restrictive covenant on the purchaser. It is clear that the original parties to the covenants must be bound by what they agreed. But what if the seller sells the land he or she retained, or the purchaser sells the land he or she bought, which was subject to the covenant? Equity allows both the benefit and the burden of the covenant to pass to subsequent purchasers of either the retained land or the part sold, and appropriate action may be taken where there is a breach. Where a restrictive covenant no longer benefits the original property, e.g. because the character of the district has gone down, then the Lands Tribunal may, on application, discharge or modify it. It may also merely fall into disuse. It is probable nowadays that action under the Environmental Protection Act 1990 is more useful to a person affected by noise. However, if the noise would not amount to a nuisance, but is in breach of a restrictive covenant, then the action taken would be in relation to that.

10.16 By-laws

Like restrictive covenants, by-laws can be effectively used against a particular type of noise which may not perhaps amount to a nuisance. By-laws are local laws made by the local authority with power delegated to it by Act of Parliament, i.e. in most cases by s. 235 of the Local Government Act 1972, and confirmed by the appropriate authority, the Secretary of State. The Local Government and Public Involvement in Health Act 2007 (ss. 129–135) amends the 1972 Act by enabling the Secretary of State to make regulations setting up simplified procedures for local authorities when making by-laws so that they can be enacted without confirmation by the Secretary of State. This part of the 2007 Act came into force on 27 January 2010 (S.I. 2010/112 refers). (Note also all the by-laws made in relation to civil aviation, above.)

In order to maintain some sort of uniformity, sets of model by-laws have been issued by the Department for Communities and Local Government covering such matters as the control of power-driven model aircraft and excessive noise from singing, shouting, playing on a musical instrument, or by operating any radio, amplifier, tape recorder or similar device in pleasure grounds, public walks and open spaces. Breach of any by-law will result in a summary conviction and a fine not exceeding such sum as may be fixed by the enactment conferring the power to make the by-laws. The 2007 Act provides for the enforcement of by-laws through the issuing of fixed penalty notices instead of through enforcement proceedings in the magistrates' courts.

10.17 Human rights

Neither the European Convention on Human Rights (ECHR) nor the Human Rights Act 1998 grants expressly rights to a decent or quiet environment. At best they play a subsidiary role in the legislative control of noise and anyone seeking a remedy should first consider the options discussed elsewhere in this chapter. Nevertheless case law from national courts and from the European Court of Human Rights (ECtHR) indicates that exposure to excessive noise may in an appropriate case constitute a breach of human rights, hence it merits brief discussion here.

Origins

Although the UK was a party to the ECHR from its inception in 1951, the Convention was only incorporated into domestic law by the Human Rights Act 1998 (HRA). The HRA provides that:

- UK legislation must be interpreted so as to be compatible with the ECHR (s. 3)

- public authorities must not act in such a way that is incompatible with the ECHR (s. 6)
- courts and tribunals must take into account Convention jurisprudence (s. 2).

Case law developments

As far back as 1994, the ECtHR stated that '. . . severe environmental pollution may affect individuals' well-being and prevent them from enjoying their homes in such a way as to affect their private and family life adversely without . . . seriously endangering their health': *Lopez Ostra v. Spain* (1995) 20 EHRR 277. The Court in that case held unanimously that there had been a breach of Article 8 ECHR: *Everyone has the right to respect for his private and family life, his home and correspondence.*

The *Lopez Ostra* case was an extreme example of air pollution from fumes and smells. In the context of noise pollution, the sources of noise complained of in the context of human rights include low-flying fighter jets, road traffic, night flights at Heathrow and nightclubs in Valencia, Spain. It is important to note that Article 8 (like most of the other Articles in the ECHR) is not absolute: a state can interfere with this right and noise pollution claims have been successful only in extreme factual circumstances (thus the noise pollution would have to be very serious). State interference is permissible only if what is done has its basis in law; and is done to secure a permissible aim set out in the relevant Article; and is necessary in a democratic society, i.e. must fulfil a pressing social need, pursue a legitimate aim and be proportionate to the aims being pursued. Article 8(2) provides that:

> There shall be no interference by a public authority with the exercise of this right except such as is in accordance with the law and is necessary in a democratic society in the interests of national security, public safety or the economic well-being of the country, for the prevention of disorder or crime, for the protection of health or morals, or for the protection of the rights and freedoms of others.

Thus in *Hatton and Others v. UK* (2003) 37 EHRR 101 (the Heathrow Airport night flights case), the Grand Chamber of the ECtHR held (overturning the decision of the ECtHR) that the government's scheme to regulate night flight noise (and allegedly increase flights) did not breach Article 8: the UK had a 'margin of appreciation' to assess the varying competing interests (e.g. national, local, individual, economic, social, environmental) and accord them appropriate weight so that a 'fair balance' was struck between the interests of the wider public and the interests of the persons directly affected by the noise. In essence, the country's economic interests in keeping Heathrow operating in that way outweighed the interests of the applicant, who lived nearby. The availability of mitigating measures against the effects of the interference is relevant in assessing whether an appropriate balance has been struck.

A contrasting case in the High Court is *Dennis v. Ministry of Defence* [2003] EWHC 793, where the noise from low-flying Harrier jet fighters from RAF Wittering on training flights over the claimant's home was held not only to be a nuisance, with no defence of public benefit, but also a violation of Article 8 as well as of Protocol 1, Article 1 ECHR:

> Every natural or legal person is entitled to the peaceful enjoyment of his possessions. No-one shall be deprived of his possessions except in the public interest and subject to the conditions provided for by law and by the general principles of international law.

Damages of £950,000 were awarded and the case described by the judge as 'exceptional'.

In *Moreno Gomez v. Spain* (2005) 41 EHRR 40, the applicant's complaint was upheld by the ECtHR. The court held unanimously that the local authority's failure to tackle night-time noise disturbances from nightclubs, bars and restaurants near the applicant's home was a breach of Article 8. The applicant lived in what had been formally designated by the local authority as an 'acoustically saturated zone' which under the relevant by-law meant an area where local residents were exposed to high noise levels causing serious disturbance. The case also illustrates that an omission to act can constitute a breach of Article 8, since although the local authority had made the designation under the by-law, it had subsequently tolerated repeated flouting of the rules and taken no steps to prevent the night-time disturbances. The Court awarded damages, to include the cost of double glazing the claimant's home.

Finally, in *Andrews v. Reading BC (No. 2)* [2005] EWHC 256 (QBD), the claimant alleged excessive traffic noise as the result of a traffic regulation order made by the defendant authority and claimed compensation for the cost of the consequent noise insulation work to his home. The increase in traffic noise was, he contended, a breach of his rights under Article 8 and the court agreed, awarding him £2000. The court held that the increase in noise levels was substantial and that the Noise Insulation Regulations 1975 (see section 10.13 above) did not preclude a claim under the HRA. No fair balance had been struck, since while the traffic regulation order conferred benefits on the community at large that outweighed its disadvantages to the individual, the defendant council had not considered the availability of mitigating measures, i.e. compensating the claimant for the adverse impact of the order on him.

Appendix 10.1 Industrial noise control: the integrated pollution prevention and control regime

In addition to the sectoral and locational control and regulation of noise discussed elsewhere in this chapter, the Integrated Pollution and Prevention Control (IPPC) regime makes provision for cross-cutting noise control across all industry sectors. The main legislation is the Pollution Prevention and Control Act 1999, which sets out the broad framework for control, leaving detailed control to regulations. The current regulations are the Environmental Permitting (England and Wales) Regulations 2007 (S.I. 2007 No 3538) which came into force on 6 April 2008. The 2007 regulations are significant in that they introduce a new system of 'environmental permitting' for all IPPC processes. The system has since been extended by the Environmental Permitting (England and Wales) Regulations 2010 (S.I. 2010 No 675).

The Pollution Prevention and Control Act 1999 implemented EU IPPC Directive 96/61. The Directive was modelled on the Environmental Protection Act 1990 Part 1 but dealt with wider environmental impacts, redefining 'pollution' to include noise and vibration and focusing on activities at prescribed installations, not merely emissions from industrial processes. Control frameworks were based on broad principles and procedures, rather than specified standards. The provisions of Directive 96/61 have been carried forward and replaced by EU Directive 2008/1 on IPPC.

The IPPC/environmental permitting system requires an environmental permit for activities at prescribed industrial installations (fuel and power, metal production and processing, mineral industries, chemical industries, waste disposal and recycling and 'other'). The emissions and pollutants that the system seeks, among other things, to control, includes noise and vibration from industrial sites. The permit may be a standard or 'tailor-made' permit but in either case will be issued subject to conditions, non-compliance with which is a criminal offence. Permit conditions are based on Best Available Techniques (BAT). BAT encompasses not only the use of certain technology or equipment but also the provision of adequate personnel, premises, operational and management systems. Guidance notes on IPPC are issued by the Environment Agency (see below), including guidance on the BAT framework. Industrial installations are categorized as either A1, and controlled by the Environment Agency, or as A2, in which case they come under the control of the local authority.

The Environment Agency's Horizontal Guidance Note for Noise (IPPC H3) – Part 1, *Regulation and Permitting* was issued in 2002. Its guidance is based on the Pollution Prevention and Control (England and Wales) Regulations 2000 (S.I. 2000/1973), which have now been replaced by the Environmental Permitting Regulations 2007 (see above). The current version of Part 2, *Noise Assessment and Control*, was published in 2004. The documents are accessible at *www.environment-agency.gov.uk*. The Note is 'horizontal' as it provides general advice for all industry sectors. Sector-specific guidance notes have also been issued which should be consulted as appropriate. Following the 2010 Environmental Permitting Regulations (see above), the Department for Environment, Food and Rural Affairs issued updated general guidance on environmental permitting, accessible at *www.defra.gov.uk*.

Applicants for permits must carry out a noise risk assessment and permit conditions can cover such aspects as the provision of noise insulation and noise barriers and on-going monitoring of emission levels.

Appendix 10.2 Case law: significance, operation and access – a short guide for non-lawyers

Significance and operation

Case law is a vital component of the English legal system and is a rich source of law in action. To the non-lawyer, however, the system of case law and judicial precedent can at first appear confusing and complex. The purpose of this note is to clarify the role of case law, outline the operation of judicial precedent, and explain law reporting and how to access law reports electronically through legal databases.

Case law is judge-made law. It relies on litigation to grow, i.e. someone must be willing to bring a case to court for a ruling. There are three different aspects to case law. First, interpreting the meaning of legislation in cases brought before the court. Second, developing principles of common law which are entirely judge-made. While legal frameworks for the prevention of noise pollution are statutory, i.e. contained in Acts of Parliament and Regulations (Statutory Instruments), common law principles in environmental law, e.g. the tort of nuisance, provide remedies for individuals and organizations suffering environmental harm, and common law principles in land law, e.g. the rules relating to restrictive covenants, provide them with enforceable rights without legislative intervention. Third, its judicial review function, i.e. the exercise of the High Court's inherent supervisory jurisdiction over the administrative actions and decisions of public bodies, e.g. local authorities and the Secretary of State.

Judicial precedent

Case law is also known as judicial precedent, a key principle in English law. The principle is that once a court has stated the legal position in a given situation, then the same decision will be reached in any future case where the material facts are the same. The hierarchy of the courts determines the significance of decisions between one court and another: all courts stand in a distinct relationship to one another and precedents can be binding or persuasive. Chapter 10 contains an outline of the court hierarchy. A court is bound by the decisions of a court above itself and often by a court of equal standing. Superior courts have the power to overrule or reverse decisions of inferior courts. Decisions of the Supreme Court of the UK (formerly the Judicial Committee of the House of Lords) deal with points of law of general public importance and bind all lower courts. Most cases that go on appeal stop at the Court of Appeal, therefore in terms of precedent, decisions of the Court of Appeal and the Supreme Court (and the House of Lords before it) are the most significant.

Law reports

The doctrine of binding precedent is dependent on the law reports which contain full details of decided cases. They contain the facts of a case, the judge's decision and his reason for that decision. They may also summarize counsels' argument. Law reports are published by a number of organizations. The reports by the Incorporated Council of Law Reporting for England and Wales are called the 'Law Reports'. The Council also provides a weekly publication service in the form of the Weekly Law Reports. Other series of law reports published by private organizations include the All England Law Reports, the Environmental Law Reports, the Planning Property and Compensation Reports, the Planning Law Reports, and the Housing Law Reports. Law reports also appear in legal journals and in the newspapers. High Court and Court of Appeal decisions are available online through the Court Service website: *www.hmcourtservice.gov.uk*; and those of the House of Lords from 1996 to 30 July 2009 on the Lords' section of Parliament's website: *www.parliament.uk*. Decided cases from the Supreme Court of the UK (which replaced the Appellate Committee of the House of Lords with effect from 1 October 2009) are accessible at *www.supremecourt.gov.uk*

Electronic legal databases available on subscription include Westlaw, LexisNexis and Lawtel. LexisNexis is a major law database and contains decisions from the Crown Court, High Court, Court of Appeal and House of Lords/Supreme Court. Westlaw UK has links to case law, legislation, legal journals and is particularly useful for cases on environmental law and planning law cases.

Citation of the Law Reports and other series

1. The name of the party bringing the court action comes first, the defendant's name comes second. In a civil case, the party bringing the action is called the claimant (formerly 'the plaintiff'). In a criminal case he will be the prosecutor.
2. After the parties' names appear the year of the report, the volume number, if necessary, the abbreviated title of the report and the page number.

   Thus the High Court case of *New Windsor Corporation v. Mellor* [1974] 1 WLR 1504 can be found in volume one of the 1974 Weekly Law Reports at page 1504. The Court of Appeal decision reference in this case is [1975] 3 WLR 25.
3. A case will often be reported in more than one series of reports. The All England Reports references for the High Court and Court of Appeal decisions in the above case are respectively [1974] 2 All ER 510 and [1975] 3 All ER 44.
4. The 'Law Reports' series classifies cases by the court in which they were heard. There are separate volumes for each High Court division (reference QB, Ch and Fam) which also includes Court of Appeal decisions and there is a further volume for House of Lords' and Privy Council cases (reference AC].

   The Law Reports reference for the Court of Appeal decision in the above case is [1975] Ch 380 and it will therefore be found in the 1975 Chancery Reports at page 380.

Neutral citations

From 2001 High Court (Administrative Court) and Court of Appeal judgments are numbered and have numbered paragraphs, not pages. These citations make it easy to access law reports online. (See Practice Direction *Judgments: Form and Citation*, 2001, LCJ.) Judgments are numbered as follows:

Court of Appeal (Civil Division) [2001] EWCA Civ 1, 2, 3 etc.
Court of Appeal (Criminal Division) [2001] EWCA Crim 1, 2, 3 etc.
High Court (Administrative Court) [2001] EWHC Admin 1, 2, 3 etc.

Thus paragraph 49 in *Smith v. Jones*, the 8th numbered judgment in the Civil Division of the Court of Appeal in 2002 would be cited: *Smith v. Jones* [2002] EWCA Civ 8 at [49].

With effect from 14 January 2002, the practice of neutral citation was extended to all judgments given by judges in the High Court in London.

The neutral citation for the House of Lords is [year] UKHL 1, 2, 3 etc.

Thus the House of Lords' decision in *R (on the application of Holding and Barnes plc) v. Secretary of State for the Environment, Transport and the Regions* [2001] UKHL 23 was the 23rd judgment in 2001. The case is also reported in the Weekly Law Reports at [2001] 2 WLR 1389.

The neutral citation for the Supreme Court is [year] UKSC 1, 2, 3 etc.

More details about neutral citations can be found at: *www.lawreports.co.uk/WLRD/2001PracticeDirection.htm*

Abbreviations of other law report titles include

Journal of Planning and Environment Law – JPL
Environmental Law Reports – Env LR
Journal of Environmental Law – JEL
Planning Law Reports – PLR
Property and Compensation Reports – P&CR (renamed the Planning Property and Compensation Reports)
Estates Gazette Case Summaries – EGCS
Estates Gazette Law Reports – EGLR
Housing Law Reports – HLR.

Appendix 10.3 General textbooks on environmental law (including law on noise)

Bell, S., and McGillivray, D., 2008, *Environmental Law*, Oxford University Press, 7th edition

Salter, I., and Robinson, P., 2006, *Property Transactions: Planning and the Environment*, Sweet and Maxwell

Seggar, M., and Khalfan, A., 2004, *Sustainable Development Law: Principles, Practices and Prospects*, Oxford University Press

Stallworthy, M., 2008, *Understanding Environmental Law*, Sweet and Maxwell

Stookes, P., 2009, *A Practical Approach to Environmental Law*, Oxford University Press

Wolf, S., and Stanley, N., 2010, *Wolf and Stanley on Environmental Law*, Routledge, 5th edition

Websites

www.communities.gov.uk
www.decc.gov.uk
www.defra.gov.uk
www.environment-agency.gov.uk
www.legislation.gov.uk
www.parliament.uk
www.rcep.gov.uk
www.sustainable-development.gov.uk

Legal databases

Lawtel
LexisNexis
Westlaw

Questions

1 Control of neighbourhood noise from construction sites may be effected by use of special or general provisions laid down in the Control of Pollution Act, 1974. State what these are, and give an example of each situation in which you would expect one procedure to be used more appropriately than any of the others.

Describe briefly those supporting regulations and codes of practice which would need to be considered during application of these procedures.

2 An application is received by the planning authority from a company which intends to make reinforced concrete fencing posts and panels. The site in question is only a few yards from good quality housing which surrounds it, but it has been used for general industrial purposes before without causing complaint. This manufacturing process, however, will be carried out mainly in the open air and will involve the use of noisy machinery. Experience indicates that noise nuisance will be created if the development proceeds: there is a local unemployment problem which may influence the authority's decision.

What powers are available to the local planning authority to restrict or control noise from development of this type, what alternative action may be available to the local authority to control noise levels thought likely to be generated, and what advice would you give to the planning officer?

3 The licensee of the public house known as the 'Pheasant and Owl' has, over the last two years, been regularly hiring out a room in the premises, known as the 'Partridge Room', to various bands. The bands perform during the evenings three or four times every week. A relevant licence covering musical and dancing entertainment has been in force for many years.

The public house is situated in Conduit Street, a narrow street in a densely populated residential area. Pamela bought 8 Conduit Street in 1959 and has lived there continuously ever since. She has recently written a letter to the licensee complaining about the intolerable noise arising from the premises.

Pamela has received a letter from the licensee's solicitor stating that as Pamela has not complained about noise from music on any previous occasion, the 'Pheasant and Owl' has acquired a prescriptive right to commit a noise nuisance. The letter is brought to the notice of the environmental health officer. Discuss the position.

4 Mr Smith is buying his house on a mortgage and finds that a new motorway is to be constructed quite close to his house. How will it be determined whether Mr Smith's house is eligible for improved insulation against traffic noise?

What are the main components of the standard package which will be installed if his house is eligible? What other compensation might Mr Smith be able to claim? How would you advise Mr Smith on the procedure for claiming this compensation?

5 Distinguish between private, public and statutory nuisance.

Section 80 of the Environmental Protection Act 1990 gives a local authority powers to deal with nuisance caused by noise but does not define *nuisance*. By reference to case law indicate what principles you would need to consider before taking action under s. 58.

6 You have been asked to prepare a proposal for your district council's first noise abatement zone. What factors would you need to take into account in making your recommendations?

What information must be included in the noise level register? Is there any additional information which you consider it important to record?

7 Describe the procedure for producing noise legislation in the United Kingdom with particular reference to the Environmental Protection Act, Part III.

Briefly outline the procedure for the production of EU directives.

8 Outline the options available to a local authority for controlling noise from construction sites. Discuss the advantages and disadvantages of different options.

9 As an environmental health officer, you have received a complaint about high levels of amplified music from an adjoining flat. Describe how you would investigate such a complaint to determine whether the amplified music caused a statutory nuisance and what action would be appropriate.

If you decided that it would not be appropriate for the local authority to take action, what legal action could the complainant take if he/she wished to pursue the matter?

10 Distinguish between private, public and statutory nuisance.

Discuss the concept of assessment, as opposed to measurement, in the context of noise nuisance investigation.

11 In the context of initiatives on noise control originating from the European Union, what are (i) the Council of Ministers, (ii) a directive, (iii) a derogation, (iv) a regulation?

Using the example of the Control of Noise at Work Regulations 2005, explain the process by which EU requirements affect UK law.

12 A developer receives conditional planning permission from the local planning authority in respect of a development proposal. You are consulted to assess the action of the planning authority. Explain the features which should be used to test appropriateness of a planning condition.

13 (i) Describe the main provisions of the Control of Noise at Work Regulations 2005.
(ii) Do the Control of Noise at Work Regulations 2005 comply fully with the requirements of Directive 2003/10/EC? Discuss.

14 Breaches under planning law may be controlled by a variety of notices.

Explain:

(i) Who is empowered to serve these notices.
(ii) Which types of breaches can be controlled by these notices.
(iii) The main provisions related to each notice.

15 Case law has established a number of factors which are relevant to assessing the existence of a nuisance. One of them is the *Standard of comfort*.

(i) What are the other factors? Briefly explain each.
(ii) Discuss the concept of Standard of comfort in relation to noise nuisance.

16 Outline the main provisions of the Building Regulations covering sound transmission in dwellings.

What differences exist between the way the regulations deal with conversions compared to new-build? (Students in Scotland or Northern Ireland should specify which regulations they are addressing.)

17 You are asked to investigate, on behalf of a local authority, an alleged noise nuisance occurring between two adjoining flats. Describe how you would determine your opinion on the following:

(i) Whether or not the noise constitutes a nuisance.
(ii) Whether the nuisance has arisen from inadequate sound insulation, anti-social behaviour, or both.
(iii) Who is responsible for the nuisance.
(iv) What statutory controls are available.

18 Under what circumstances may a local authority apply for an injunction in respect of an alleged noise nuisance? Illustrate your answer with a practical example.

Outline the relevant legislation.

What is an interlocutory injunction?

19 Discuss the meaning of *precise* in relation to (a) noise nuisance notices and (b) planning conditions.

In the case of planning conditions, outline all appeal provisions which are available.

Appendix 1 Glossary of acoustical terms

This glossary is mainly related to the scope of the text. More specific glossaries may be found, for example in major reports, guidelines and codes of practice documents. There are also standards giving definitions of terms, and various 'dictionaries of acoustics'.

A-weighting a frequency weighting devised to attempt to take into account the fact that human response to sound is not equally sensitive to all frequencies; it consists of an electronic filter in a sound level meter, which attempts to build this variability into the indicated noise level reading so that it will correlate, approximately, with human response (defined in BS EN ISO 61672-1)

absorption see under sound absorption

absorption coefficient see under sound absorption coefficient

acceleration rate of change of velocity (in m/s^2)

accelerometer a transducer which measures acceleration

acoustic calibrator a device for producing an accurately known sound pressure level; used for the calibration of sound level meters

acoustic enclosure a structure built around a machine to reduce noise

acoustic impedance of a surface or acoustic source; the (complex) ratio of the sound pressure averaged over the surface to the volume velocity through it; the volume velocity is the product of the surface area and acoustic particle velocity; see also under characteristic, mechanical and specific acoustic impedance

acoustic lagging materials applied externally to the surface of pipes and ducts to reduce the radiation of noise; not to be confused with thermal lagging

acoustic model A device for predicting acoustic parameters such as sound pressure level or reverberation time. Often used in connection with predicting acoustic performance of rooms such as concert halls and other performance spaces. Scale modelling involves creating a physical model of 1/*N* full scale and modelling the acoustic characteristics of sound sources, microphones and material surfaces at *N* times the frequencies to be encountered in the full scale version of the space. Computer modelling involves creating a virtual three dimensional model of the space and predicting its acoustic performance using software packages which use ray tracing, beam (or cone) tracing or source / image techniques (or combination of these)

acoustic particle velocity the velocity of a vibrating particle in an acoustic wave

acoustic reactance the imaginary part of the complex acoustic impedance

acoustic resistance the real part of the complex acoustic impedance

acoustic trauma sudden permanent hearing damage caused by exposure to a burst of high level noise

acoustics (1) the science of sound; (2) of a room: those factors which determine its character with respect to the quality of the received sound

action value a noise exposure level in the workplace above which certain actions are required under the 2005 Control of Noise at Work Regulations

active filter a filter which contains transistors, integrated circuits or other components requiring a power supply

active noise control a noise control system which uses antiphase signals from loudspeakers to reduce noise by destructive interference

agglomeration An area having a population in excess of 100,000 persons and a population density equal to or greater than 500 people per km^2 and which is considered to be urbanized (a term used in noise mapping and defined in the EU Noise Directive)

AI see articulation index

airborne sound sound or noise radiated directly from a source, such as a loudspeaker or machine, into the surrounding air (in contrast to structure-borne sound)

airborne sound insulation the reduction or attenuation of airborne sound by a solid partition between source and receiver; this may be a building partition, e.g. a floor, wall or ceiling, a screen or barrier or an acoustic enclosure

aliasing introduction of false spectral lines (aliases) into a spectrum by having the maximum frequency of the signal greater than one-half the digital sampling frequency

ambient noise the totally encompassing noise in a given situation at a given time; it is usually composed of noise from many sources, near and far (defined in BS 4142)

amplitude the maximum value of a sinusoidally varying quantity

analogue signal an analogue signal is one in which continuous variations in an electrical signal faithfully represents (i.e. is analogous to) the variation in some physical variable such as temperature or sound pressure

analogue-to-digital converter (A/D converter or ADC), a device which samples and digitizes analogue signals, preparatory for digital signal processing; the continuously varying analogue signal is converted into a finite number of discrete steps or levels then represented as a series of numbers

ANCON UK civil aircraft noise computer model developed by the Environmental Research and Consultancy Department of the Civil Aviation Authority (CAA). ANCON calculates contours from data describing aircraft movements, routes, noise generation and sound propagation

angle of view a term used in the prediction of road traffic or railway noise giving the angle of view of the road or railway line subtended at the receiving point.

anechoic literally 'without echo', i.e. without any sound reflections. An anechoic room is one in which all the interior surfaces (walls, floor and ceiling) are lined with sound absorbing materials so that there are no reflections; it provides a standard environment for acoustic tests

angular frequency the product 2π times frequency; symbol ω; measured in radians per second

annoyance noise annoyance is a feeling of displeasure caused by noise

anti-aliasing filter a low pass filter inserted in an instrument, before the ADC, in order to prevent aliasing

antinode a point, line or surface of an interference pattern at which the amplitude of the sound pressure or particle velocity is at a maximum

antivibration (AV) mounts springs or other resilient materials used to reduce vibration (and noise) by isolating the source from its surroundings

apparent sound reduction index (R′) a term relating to the sound insulation performance of partitions, defined in BS EN ISO 140-4, measured in octave or third octave frequency bands

articulation index (AI) a measure of the intelligibility of speech; the percentage of words correctly heard and recorded in an intelligibility test, or predicted from the levels of speech, background noise and reverberant sound in a room. The value of AI varies between 0 and 1 (representing 100% score or perfect intelligibility)

attenuation a general term used to indicate the reduction of noise or vibration, by whatever method or for whatever reason, and the amount, usually in decibels, by which it is reduced

attenuator a device introduced into air or gas flow systems in order to reduce noise; absorptive types contain sound absorbing materials; reactive types are designed to tune out noise at particular frequencies

audibility the ability of a sound to be heard; the concept of audibility has been used as a criterion for setting limits to noise levels, particularly from amplified music; it is a subjective criterion, i.e. one which can only be determined by the ear of the listener, not by measurement of sound levels; also used as a criterion to determine the degree of privacy between rooms (e.g. offices)

audibility threshold the minimum sound pressure which can just be heard at a particular frequency by people with normal hearing; usually taken to be 20 μPa at 1000 Hz

audible range frequencies from 20 Hz to 20 kHz (approx.); sound pressures from 20 mPa to 100 Pa (approx.)

audiogram a chart or graph of hearing level against frequency

audiometer an instrument which measures hearing sensitivity

audiometry the measurement of hearing

auditory cortex the region of the brain which receives signals from the ear

aural of or relating to hearing or the hearing mechanism

axial mode the room modes associated with each pair of parallel surfaces

background noise Ambient noise which remains at a given site when occasional and transient bursts of higher level ambient noise levels have subsided to typically low levels; the noise normally present for most of the time at a given site. It is usually described by the L_{A90} level

background noise level defined in BS 4142 as the value of the A-weighted residual noise at the assessment position that is exceeded for 90% of a given time interval, T, (i.e. $L_{A90,T}$) measured using time weighting, F, and quoted to the nearest whole number of decibels (also see under residual noise)

band pass filter a filter which provides zero attenuation to all frequencies within a certain band but which attenuates completely all other frequencies

band sound pressure level the sound pressure level of the sound signal within a certain frequency band

bandwidth the range of frequencies contained within a signal, passed by a filter, or transmitted by a structure or device

basic noise level a term used in connection with the prediction of environmental noise such as road traffic or train noise, where, as a first stage, factors relating to the noise source are used to predict the noise level at some arbitrary point close to the source prior to using other factors relating to the propagation of the sound to the reception point(s).

basilar membrane a membrane inside the cochlea of the inner ear which vibrates in response to sound, thus exciting the hair cells

beats periodic variations which are heard when two pure tones of slightly differing frequencies are superimposed

bel 10 decibels; a unit of level on a logarithmic scale which is based on the ratio of two powers, or of power-related quantities such as sound intensity or the square of sound pressure

bending or flexural waves elastic waves in plates, panels, beams etc., which are a combination of compression and shear waves and which are responsible for the transmission of structure-borne sound in buildings and other structures

binaural relating to hearing using both ears, e.g. binaural localization; the use of both ears to locate the direction of sounds

bit abbreviation of binary digit; the smallest possible unit of information in binary form, i.e. on or off, yes or no, 0 or 1

broadband containing a wide range of frequencies

byte a binary word or group of bits

C; C_{tr} spectral adaptation terms used in connection with the measurement and assessment of airborne sound insulation and defined in BS EN ISO 7172-1

C-weighting one of the frequency weightings defined in BS EN ISO 61672-1; it corresponds to the 100 Phon contour and is the closest to the linear or unweighted value

capacitor one of the basic elements of an electrical circuit consisting of two conducting plates separated by a gap containing an insulator, or dielectric; it has the property of capacitance, measured in farads (F), microfarads (μF) or picofarads (pF)

centre frequency the centre of a band of frequencies; in the cases of octave or one-third octave it is the geometric mean of the upper and lower limiting frequencies of the band

characteristic acoustic impedance (of a medium) the ratio of sound pressure to acoustic particle velocity at a point in the medium during the transmission of a plane wave; it is the product (ρc) of the speed of sound (c) in the medium and its density (ρ), measured in rayls (Nsm^{-3}, Pas/m or kg/m^{2}s)

charge amplifier a type of preamplifier suitable for use with piezoelectric accelerometers; it gives an output which is proportional to the electric charge present in the input signal

cochlea a coiled, snail-shaped structure in the inner ear; it is fluid-filled and contains a complex arrangement of membranes and hair cells which convert mechanical vibrations of the fluid into electrical impulses transmitted to the brain

coherence, coherent two sounds are coherent if there is a constant phase difference between their two waveforms (usually because they have originated from the same source)

coincidence effect an effect which leads to increased transmission of sound by panels and partitions when the speed (and wavelength) of flexural waves in the panel coincide with the speed (and wavelength) of the sound waves exciting the panel

colouration some change to a sound from the original version, detectable to a listener, caused for example by the sound reproduction system or by the room in which the sound is produced.

community noise defined by the World Health Organization as 'Noise emitted from all noise sources except noise at the industrial workplace'

compressional wave an elastic wave in a fluid or solid in which the elements of the medium are subjected to deformations which are purely compression, i.e. which do not contain any element of rotation or shear, and of which sound waves in air are an example

condenser see under capacitor

conductive deafness hearing loss which is caused by some defect or fault in the outer or middle ear

continuous descent approach (CDA) a noise abatement technique for arriving aircraft in which the pilot, when given clearance by air traffic control (ATC), will descend along a path which as far as is practicable corresponds to a continuous descent (rather than periods of level flight in between periods of descent). The idea is to remain as high as possible for as long as possible during descent. CDA has benefits in terms of both reduced noise and improved fuel economy

continuous equivalent noise level, L_{Aeq} of a time varying noise; the steady noise level (usually in dBA) which, over the period of time under consideration, contains the same amount of (A-weighted) sound energy as the time varying noise, over the same period of time; also called time averaged sound level

continuous spectrum a sound or vibration spectrum whose components are continuously distributed over the particular frequency range, for example random noise; contrast with a line spectrum from a harmonic sound

cortex see under auditory cortex

coulomb damping a form of damping in which the damping force is constant, independent of either displacement or velocity (also called dry friction damping)

crest factor of a signal; the ratio of the peak to the root mean square (RMS) value

criterion the basis on which a noise or vibration is to be judged, e.g. damage to hearing, interference with speech, annoyance

critical band in human hearing, only those frequency components within a narrow band, called the critical band, will mask a given tone. Critical bandwidth varies with frequency but is usually between 1/6 and 1/3 octaves.

critical damping the amount of viscous damping in a system which will allow the system to return to its equilibrium position, in the minimum time, without overshoot, i.e. without oscillation; the boundary between overdamping and underdamping

critical frequency the lowest frequency at which the coincidence effect takes place for a particular panel or partition, and above which the sound insulation performance starts to deteriorate

CRN (Calculation of Railway Noise) the UK method for predicting rail (i.e. train) noise, in terms of L_{Aeq} and L_{den}. May be replaced by integrated Europe-wide methods such as Harmonoise.

crosstalk a signal from one track, channel or circuit which is transmitted, unwanted, into another track, channel or circuit

CRTN (Calculation of Road Traffic Noise) the UK method for predicting road traffic noise, in terms of $L_{A10,18hour}$. First published in 1968 and revised in 1998 it was developed as a method for determining eligibility for sound insulation under the Noise Insulation of the Land Compensation Act, but is more widely used for Noise Impact Assessment Regulations and for planning applications. May be replaced by integrated Europe-wide methods such as Harmonoise

Curie point the temperature above which a piezoelectric material becomes polarized, and loses its piezoelectric properties

cycle of a periodically varying quantity; the complete sequence of variations of the quantity which occurs during one period

cycle per second unit of frequency; one cycle per second is one hertz (Hz)

***D* level difference** the difference in sound levels in two spaces separated by a partition, measured as part of a sound insulation test according to BS EN ISO 140

D_{nc}, D_{ne}, D_{nf}, standardized level difference values for small elements (e.g. ventilators), D_{ne}, suspended ceilings, D_{nc}, and raised floors, D_{nf}, defined in parts 10, 12 and 9 respectively of BS EN ISO 140

D_n normalized level difference a measurement of airborne sound insulation, corrected according to BS EN ISO 140-4 for receiving room characteristics (sound absorption); a complete set of measurements consists of 16 third octave band values, from 100 Hz to 3150 Hz

D_{nT} standardized level difference a measurement of airborne sound insulation, corrected according to BS EN ISO 140-4 for receiving room characteristics (reverberation times); a complete set of measurements consists of 16 third octave band values, from 100 Hz to 3150 Hz

$D_{nT,w}$ weighted standardized level difference a single-figure value of airborne sound insulation performance, derived according to procedures in BS EN ISO 717-1 used for rating and comparing partitions and based on the values of D at different frequencies; values of $D_{nT,w}$ are specified in the Building Regulations

damping a process whereby vibrational energy is converted into heat through some frictional mechanism, thus causing the level of vibration to decrease

damping ratio the ratio of the amount of damping in a vibrating system to the amount of damping when critical

day–evening–night level, L_{den} the L_{Aeq} over the period 00.00–24.00, but with the evening values (19.00–23.00) weighted by the addition of 5 dBA, and the night values (23.00–07.00) weighted by the addition of 10 dBA

day–night level L_{dn} an index of environmental noise which is a 24 h L_{eq}, but with a 10 dB weighting added to the night-time noise levels (22.00 to 07.00) to allow for increased sensitivity to noise during the night-time

dBA the A-weighted sound pressure level; see under A-weighting

decade a range of ten to one, e.g. from 100 Hz to 1000 Hz

decibel (dB) the decibel scale is a scale for comparing the ratios of two powers, or of quantities related to power, such as sound intensity; on the decibel scale the difference in level between two powers, W_1 and W_2 is N dB, where $N = 10\log_{10}(W_1/W_2)$; the decibel scale may also be used to compare quantities, whose squared values may be related to powers, including sound pressure, vibration displacement, velocity or acceleration, voltage and microphone sensitivity; in these cases the difference in level between two signals, of magnitude S_1 and S_2, is given by $N = 20\log_{10}(S_2/S_1)$; the decibel scale may be used to measure absolute levels of quantities by specifying reference values which fix one point in the scale (0 dB) in absolute terms; a decibel is one tenth of a bel

degrees of freedom the number of degrees of freedom of a mass–spring model of a vibrating system is the minimum number of coordinates required to specify all the different possible modes of vibration of the system

deterministic a deterministic signal is one whose value can be predicted with certainty from a knowledge of its behaviour at previous times, as opposed to a random signal, where this is not possible

dielectric a material which is an electrical non-conductor or insulator; it is used between the plates of a capacitor

diffraction the process whereby an acoustic wave is disturbed and its energy redistributed in space as a result of an obstacle in its path; the relative sizes of the sound wavelength and the object are always important in diffraction; reflection may be considered to be a special case of diffraction when the size of the obstacle is very large compared to the wavelength; the combined effects of diffraction from an irregular array of objects in the path of the sound is also known as scattering; diffraction theory deals with all aspects of the interactions between matter (i.e. obstacles) and waves, so it also determines the directional patterns of sound radiation from vibrating objects

diffuse reflection reflection of sound at a rough irregular profiled surface which scatters sound in different directions (as opposed to specular reflection)

diffuse sound field a sound field of statistically uniform energy density in which the directions of propagation of waves are random from point to point

diffuser an object or surface profile designed to scatter sound in random directions and so to minimize specular reflection (according to the law of reflection)

diffusion the scattering of sound wave in many different directions by an object or a surface

digital signal a signal having a discrete number of values, which can be represented as a sequence of numbers; see also analogue-to-digital converter and digital-to-analogue converter

digital audio tape recorder (DAT) a tape recorder which includes an ADC (and a DAC) and which records analogue signals on tape in coded digital form

digital-to-analogue converter (DAC) an electronic device which converts digital signals into analogue signals

dipole A sound source which has the characteristics of two monopole sources of equal amplitude but opposite phase a short distance apart. An example is the sound radiated by an unbaffled loudspeaker cone which approximates to a dipole source

directivity factor the ratio of the sound intensity at a given distance from the source, in a specified direction, to the average intensity over all directions, at the same distance

directivity index the directivity factor (DF) of a source, expressed in decibels, i.e. $10\log_{10}$ (DF)

direct sound sound which arrives at the receiver having travelled directly from the source, without reflection

direct sound field that part of the sound field produced by the source where the effects of reflections may be neglected

distortion a lack of faithfulness in a signal, such as the introduction of harmonics into the frequency spectrum, introduced, for example, because of non-linearity or of overload of some component of the measurement system

disturbance an effect of noise which may be indicated by some change of behaviour, e.g. closing windows, interruption to speech, moving bedrooms; the objectively measureable effect of noise on the performance of a task or activity, such as listening to speech, or getting to sleep (as compared with annoyance which can only be measured by asking people questions, e.g. via a questionnaire)

Doppler effect the change in the observed frequency of a wave caused by relative motion between source and receiver

dose–response the relationship between the human response to noise or vibration and the received exposure (or dose) of noise or vibration received.

dynamic magnification factor (*Q* factor) a quantity which is a measure of the sharpness of resonance of an oscillating system (either mechanical or electrical); it is related to the amount of damping in the system

dynamic range the range of magnitudes of a signal which a measuring system, or component of a system, can faithfully record, process or measure, from highest to lowest; usually expressed in decibels

dynamic stiffness the ratio of change of force to change of displacement in a vibrating system; it may be different from the static stiffness of the system

ear defenders or ear protectors earmuffs or earplugs worn to provide attenuation of sounds reaching the ear and reduce the risk of noise induced hearing loss

early sound reflections (or early sound) reflections of sound from surfaces in a room which arrive at the receiver within a certain time after the direct sound. This is usually considered to be about 50 ms for speech or about 80 ms for music. Early sound is treated by the listener's ear/brain system as being helpful in reinforcing the direct sound signal and therefore acoustic designers will seek to encourage the production of early sound reflections while discouraging late reflections which produce an unfavourable response in the listener

echo a sound reflection whose magnitude and time delay are such that it is perceived as a separate, distinguishable repetition of the direct sound, as opposed to reverberation which is perceived as part of the original sound

electret or prepolarized microphone a type of condenser microphone in which a prepolarized layer of electret polymer is used as a dielectric between the diaphragm and backing-plate which form the condenser

electrostatic actuator a device which fits over a microphone, close to the diaphragm, and is used for remote calibration

emission a measure of sound energy produced by a source of sound and radiated into the environment (see also immisssion)

EPNdB (effective perceived noise decibel) a metric used internationally in the noise certification of aircraft

equal loudness contours a standardized set of curves which show how the loudness of pure tone sounds varies with frequency at various sound pressure levels

equivalent continuous noise level see under continuous equivalent noise level, L_{Aeq}

environmental noise the European Environmental Noise Directive 2002/49/EC defines environmental noise as 'unwanted or harmful outdoor sound created by human activities, including noise emitted by means of transport, road traffic, rail traffic, air traffic, and from sites of industrial activity'

Eustachian tube the passage from the middle ear to the back of the throat which serves to equalize the pressure across the eardrum

eVDV estimated vibration dose value; a measure of a cumulative amount of vibration based upon weighted RMS acceleration values and durations; for signals of limited crest factor, the eVDV approximates to the vibration dose value, VDV; see also under root mean quad and vibration dose value

Eyring's formula a modified version of Sabine's formula, for reverberation time, which takes into account the discrete nature of sound reflections; also known as the Norris–Eyring formula

exposure limit value a noise exposure level defined in the 2005 Control of Noise at Work Regulations requiring action from employers (a personal daily (or weekly) noise exposure level of 87 dBA or peak sound pressure level of 140 dBC)

F (fast) time weighting an averaging time used in sound level meters, and defined in BS EN ISO 61672-1

façade noise level a noise level measured close to (less than 3 m away) from the façade of a building, which contains a contribution arising from reflection of sound at the façade. The difference between the façade level and the free field level (in the absence of the façade) is called the façade correction factor

far field of a sound source; that part of the sound field of the source where the sound pressure and acoustic particle velocity are substantially in phase, and the sound intensity is inversely proportional to the square of the distance from the source

fast Fourier transform (FFT) an algorithm or calculation procedure for the rapid evaluation of Fourier transforms; an FFT analyser is a device which uses FFTs to convert digitized waveform signals into frequency spectra, and vice versa

field measurements measurements carried out on-site, away from controlled laboratory conditions; the results of field tests of sound insulation may include the effects of flanking paths as well as direct sound transmission, which would not be the case for laboratory tests

filter a device which transmits signals within a certain band of frequencies but attenuates all other frequencies; filters may be electrical, mechanical or acoustical

flanking transmission the transmission of airborne sound between two adjacent rooms by paths other than via the separating partition between the rooms, e.g. via floors, ceilings and flanking walls

flutter echo a series of repeating echoes caused by parallel reflecting surfaces

forced vibration steady-state vibration of a system caused by a continuous external force

form factor the ratio of the RMS value of a signal to the mean value between two successive zero-crossings

Fourier analysis/series/spectrum Fourier's theorem shows that any periodic function may be broken down (or analysed) into a series of discrete harmonically related frequency components which may be represented as a line spectrum

Fourier transform a mathematical process which transforms a non-periodic function of time into a continuous function of frequency, and vice versa (in the case of the inverse transform)

fractional dose a fractional component of a total noise exposure, defined in connection with assessment of workplace noise exposure in connection with the Noise at Work Regulations

free field conditions a situation in which the radiation from a sound source is completely unaffected by the presence of any reflecting boundaries; see also under anechoic

frequency of a sinusoidally varying quantity such as sound pressure or vibration displacement; the repetition rate of the cycle, i.e. the reciprocal of the period of the cycle, the number of cycles per second; measured in hertz (Hz)

frequency analysis the separation and measurement of a signal into frequency bands

frequency response of measurement system or component of such a system, e.g. a sound level meter or microphone; the variation in performance, e.g. sensitivity, with change of frequency

frequency spectrum a graph resulting from a frequency analysis and showing the different levels of the signal in the various frequency bands

frequency weighting an electronic filter built into a sound level meter according with BS EN ISO 61672-1; see also under A- and C-weighting

fundamental frequency the lowest natural frequency of a vibrating system; the repetition rate of a harmonic waveform

ground attenuation an attenuation of sound at a distance from a receiver caused by interference between sound waves travelling directly from source to receiver and sound arriving at receiver after reflection at the ground

Haas effect a psycho-acoustic phenomenon in which precedence is given to the direction of the first arrival of direct sound in attributing the direction from which the sound is coming

hair cells biological cells in the cochlea of the inner ear where vibration is turned into a neural signal which is transmitted to the brain

hard ground ground such as concrete, most other paving materials, and water which is considered to be an acoustically reflecting surface, in contrast with 'soft ground' which is considered to be sound absorbing

harmonic a signal having a repetitive pattern

Harmonoise an integrated method for predicting environmental noise throughout Europe which may replace individual national methods such as CRTN and CRN

hearing level a measured threshold of hearing, expressed in decibels relative to a specified standard threshold for normal hearing

hearing loss any increase of an individual's hearing levels above the specified standard of normal hearing

Helmholtz resonator a vibrating system having a single degree of freedom; it consists of an air-filled enclosure connected to the open air by a narrow column; the air in the enclosure acts as the spring and the air in the column acts as the mass

henry (H) the unit of electrical inductance

hertz (Hz) the unit of frequency; the number of cycles per second

high pass filter a filter which transmits frequency components of a signal that are higher than a certain cut-off frequency but which attenuates those below the cut-off

HML method a method for estimating the A-weighted sound pressure level at the ear when hearing protector is being worn based on the attenuation provided by the protector at high (H), medium (M) and low (L) frequencies

hyperacusis unusual sensitivity and discomfort caused by sounds which are usually tolerable to other listeners. The condition can sometimes be associated with hearing loss

hysteresis damping a type of damping that occurs within materials as a result of phase changes which occur between stress and strain during the vibration cycle

immission a measure of sound energy received at a particular location in the environment

impact noise sound resulting from the impact between colliding bodies

impact sound insulation the resistance of a floor to the transmission of impact sound; measured according to BS EN ISO 140-7

impedance see under acoustic impedance

impedance matching the use of a device to act as a buffer between a system, or component of a system, with a high output impedance and a system, or component of a system, with a low input impedance. An example of electrical impedance matching is the use of a preamplifier between a condenser microphone and the signal processing electronics in a sound level meter. An example of acoustics impedance matching is the function of the middle ear to match the very different acoustic impedances of air (in the outer ear) and cochlear fluid (in the inner ear): see also under input and output impedance

impulse a transient signal of short duration; impulsive noise is often described by words such as bang, thump, clatter

impulse response a sound pressure versus time measurement showing how a device or room responds to an impulse

incus or anvil, the middle of the three bones in the middle ear

inductance the property of an electrical coil, or inductor, associated with the rate of change of magnetic field; measured in henrys (H)

inertia base a concrete slab used under antivibration mounts to provide additional mass, rigidity and stability

infrasound acoustic waves with frequencies below the audible range, i.e. below about 20 Hz

initial noise ambient noise prevailing in an area before any modification of the existing situation (used in BS 7445)

input and output impedance output impedance is an important property of any device which delivers a signal, e.g. a battery, a microphone, accelerometer or an amplifier. Devices with a low output impedance can deliver higher electrical currents (and more energy) than those with high output impedance. Input impedance is an important property of any device which receives a signal (i.e. of any 'load'), such as a loudspeaker, a level recorder or an amplifier. Devices with a low input impedance draw a higher current (and more energy) from the source than those with a high input impedance. If a source device with a very high output impedance is connected to a receiving device with a low input impedance the output voltage will be much reduced because the source will be unable to deliver the current (or energy) demanded by the receiver. This situation may be improved by the use of an impedance matching device interposed between source and receiver: see also under impedance matching

insertion loss a measure of the effectiveness of noise control devices such as silencers and enclosures; the insertion loss of a device is the difference, in dB, between the noise level with and without the device present

insulation see under sound insulation

integrated impulse response method a method for determining room acoustics parameters such as reverberation time, based on measurement of the impulse response of the room and defined in BS EN ISO 3382 and BS EN ISO 354

Integrated Pollution Prevention and Control (IPPC) Regulations introduced to comply with European Commission Directive 96/61 whereby a permit system is used for controlling pollution from industrial activities

integrating circuit an electrical circuit which converts an acceleration signal into a velocity or displacement signal

integrating sound level meter a sound level meter which electrically integrates sound pressure signals to measure the equivalent continuous sound level, L_{Aeq}

intelligibility of speech signals; the degree to which each individual syllable of speech can be identified and understood

intensity see under sound intensity

interference (1) the principle of interference governs how waves interact; the combined wave disturbance is the algebraic sum of the individual wave disturbances, leading to the possibility of constructive and destructive interference; (2) the disturbing effect of unwanted signals on the wanted signal, often electrical in nature

IPPC see under Integrated Pollution Prevention and Control Regulations

ISO International Organization for Standardization

isolation see under vibration isolation

isolation efficiency a measure of the effectiveness of a vibration isolation; isolation efficiency = $(1-T) \times 100\%$, where T is the transmissibility of the system; see also under transmissibility

jerk the rate of change of acceleration

joule (J) unit of energy: the amount of energy used in 1 second by a source of energy delivering power at a rate of 1 watt

just noticeable difference (JND) a concept used in psycho-acoustic measurement; the difference between two (acoustic) stimuli which is just noticeable in some defined condition

***L* (level)** sound pressure level, SPL; in general, it implies the use of decibels related to the ratio of powers, or power-related quantities such as sound intensity or sound pressure

L_A see under A-weighted sound pressure level

L_{AE} see under sound exposure level, SEL

$L_{Aeq,16hour}$ the L_{Aeq} over the period 07.00–23.00, local time (for strategic noise mapping this is an annual average)

$L_{Aeq,T}$ see under continuous equivalent sound level

L_{Amax} the maximum RMS A-weighted sound pressure level occurring within a specified time period; the time weighting, Fast or Slow, is usually specified

$L_{AN,T}$ percentile level, i.e. the sound pressure level in dBA which is exceeded for N% of the time interval T, e.g. in L_{A10} and L_{A90}

L_{day} the L_{Aeq} over the period 07.00–19.00, local time (for strategic noise mapping this is an annual average)

L_{den} (day–evening–night level) the L_{Aeq} over the period 00.00–240.00, but with the evening values (19.00–23.00) weighted by the addition of 5 dBA, and the night values (23.00–07.00) weighted by the addition of 10 dBA

$L_{EP,d}$ see under personal daily noise exposure level

$L_{evening}$ the L_{Aeq} over the period 19.00–23.00, local time (for strategic noise mapping this is an annual average)

L_{night} the L_{Aeq} over the period 23.00–07.00, local time (for strategic noise mapping this is an annual average)

L'_{nT} see under standardized impact sound pressure level

$L'_{nT,w}$ see under weighted standardized impact sound pressure level

L_{peak} see under peak sound pressure level

L_W see under sound power level

late sound reflections (or late sound) reflections of sound from surfaces in a room which arrive at the receiver within 50 ms of the arrival of the direct sound signal (for speech) or within 80 ms for music. Late sound is treated by the listener's ear/brain system as being unpleasant and discordant and therefore acoustic designers will seek to discourage the production of late sound reflections while encouraging early reflections which are considered to be helpful in reinforcing the direct sound signal response in the listener

level difference, *D* BS EN ISO 140-4 uses the difference in level between two rooms as the basic measure of airborne sound insulation

level recorder an instrument for registering and measuring the variation of signals, such as sound pressures, with time

linear a measurement device is linear if its output is directly proportional to its input; in the case of a microphone, for example, this means that the sensitivity is constant and does not change with sound pressure level; linear SPL means unweighted

linearity the degree to which a device is linear

logarithmic decrement (δ) a measure of the amount of damping in a vibrating system, based on the rate of decay of natural vibrations of the system

Lombard effect an effect whereby a speaker will often raise the level of his/her voice when the level of background noise increases

longitudinal wave a wave in which the vibratory movement of the particles in the medium is parallel to the direction in which the wave is travelling; compressional waves in a fluid medium are longitudinal

long-term average rating level average over the long-term time interval of the rating levels for a series of reference time intervals, carried out as described in BS 7445-2 (ISO 1996-2)

long-term average sound level average over the long-term time interval of the equivalent continuous A-weighted sound pressure levels for a series of reference time intervals comprised within the long-term time interval, carried out as described in BS 7445-2 (ISO 1996-2)

long-term sound level, long-term average rating level, long-term time interval terms used in connection with the description and measurement of environmental noise, and defined in BS 7445

long-term time interval specified time interval for which the results of the noise measurement are representative. Note that the long-term time interval consists of a series of reference time intervals and is determined for the purpose of describing the environmental noise and is generally designated by competent authorities

loss factor a term used to describe the amount of damping in a system or material; it is twice the damping ratio

loudness the measure of the subjective impression of the magnitude or strength of a sound

loudness level the loudness level of a sound is the sound pressure level of a standard pure tone, of specified frequency, which is equally loud, according to the assessment of a panel of normal observers

lower exposure action value a noise exposure level defined in the 2005 Control of Noise at Work Regulations requiring action from employers and employees (a personal daily (or weekly) noise exposure level of 80 dBA, or peak sound pressure level of 135 dBC)

low frequency noise a term generally used to refer to sound below a frequency of about 100 to 150 Hz. It is much less well attenuated during transmission both outdoors and indoors than sounds of higher frequencies, and is therefore often heard at considerable distances from its source particularly late at night when other background noise from other sounds has decreased. In such cases it can often give rise to annoyance because it is often tonal in nature

low pass filter a filter which transmits signals at frequencies below a certain cut-off frequency and attenuates all higher frequencies

lumped parameter model a model of a vibrating system in which mass, stiffness and damping are represented as discrete elements

magnetic tape recorder a device for capturing, storing and replaying analogue signals onto a tape medium containing ferromagnetic metal oxide particles

malleus or hammer; one of the three bones of the middle ear, connected to the eardrum

masking the process whereby the threshold of hearing for one sound is raised due to the presence of another, thus rendering the first sound inaudible

mass law an approximate relationship for predicting the sound reduction index of panels and partitions, based only on the surface density of the panel and the frequency of the sound

mean free path a term used in the statistical treatment of sound in rooms, relating to the average distance between reflections

measurement time interval a term used in standards on the measurement and rating of environmental noise (BS 4142 and BS 7445) to indicate the total time over which measurements should be taken or predicted , for assessment purposes. Note that this may consist of the sum of a number of non-contiguous, short-term measurement time intervals

measurement uncertainty see under uncertainty

mechanical filter a resilient pad or layer which prevents the transmission of high frequency vibration acts as a low pass, mechanical filter

mechanical impedance the (complex) ratio of force to velocity at a point in a vibrating system

mel a unit of pitch; the pitch of any sound judged by listeners to be *n* times that of a 1 mel tone is *n* mels; 1000 mels is the pitch of a 1000 Hz tone at a sensation level of 40 dB

micron (μm) one thousandth of a millimetre or one millionth of a metre

microphone a transducer which converts acoustic signals into electrical (voltage) signals

middle ear an air-filled space which connects the eardrum of the outer ear to the oval window of the inner ear by three small bones, called ossicles

milli- a standard metric prefix meaning one thousandth

MLS (maximum length sequence) a method for measuring room acoustics parameters featuring the use of maximum length sequence, pseudo-random but deterministic signals.

modal analysis a method for investigating modes of vibration of structures involving mapping of the amplitude and phase of the vibration when the structure is excited into vibration

mode of vibration a pattern of vibration of a vibrating system, characterized by a series of nodes and antinodes

mode shape the shape of a particular mode of vibration is usually represented as the maximum displacement of the system from its mean or equilibrium position

modulus of elasticity the stress divided by the strain for an elastic medium; an important factor in determining the speed of elastic waves in the medium; there are different types of elastic modulus, e.g. shear modulus, compression or bulk modulus and torsional modulus, for the different types of elastic deformation

monopole a model or idealized point source of sound which radiates spherical waves

music noise level (MNL) the A-weighted continuous equivalent sound level of the music noise measured at a particular location. Defined in the 1995 Noise Council Code of Practice on Environmental Noise Control at Concerts

nano- a standard metric prefix meaning one thousand millionth (i.e. 10^{-9})

narrowband filter a band pass filter with a small bandwidth, i.e. less than one third octave

natural frequency the frequency of free or natural vibrations of a system

near field of a sound source; the region of space surrounding the source where sound pressure and acoustic particle velocity are not in phase, and the sound pressure varies with position in a complex way

neighbour or neighbourhood noise noise from domestic premises: household appliances, radios, televisions, music systems, noisy pets, DIY activities, intruder alarms, parties or similar events.

newton (N) the SI unit of force; the force required to produce an acceleration of 1 m/s^2 in a mass of 1 kg

node a point, line or surface in a standing-wave pattern where some characteristic of the vibration, e.g. the displacement, is zero

noise unwanted sound or unwanted signal (usually electrical) in a measurement or instrumentation system

noise and track keeping (NTK) system a computerized system used at some major airports whereby radar is used to monitor flightpaths (tracks) of arriving and departing aircraft in order to identify aircraft which are not keeping within required limits, and to match noise events from nearby noise monitoring terminals with aircraft

noise criteria (NC) curves a method devised by Beranek in the 1940s for rating or assessing internal (mainly office) noise. It consists of a set of curves relating octave band sound pressure level to octave band centre frequencies; each curve is given an NC number, which is numerically equal to its value at 1000 Hz. The NC value of a noise is obtained by plotting the octave band spectrum against the family of curves. In order to meet a particular NC specification the noise level must be either below or equal to the SPL in each octave band

noise dose an amount of noise energy, usually A-weighted, received by a person, resulting from a combination of sound pressure level and exposure time; see also under personal daily noise exposure level, $L_{EP,d}$

noise exposure category a term used in Planning Policy Guidance Note 24 *Planning and Noise*

noise exposure forecast (NEF) a noise index used mainly in the United States for aircraft noise

noise immission the amount of noise exposure received at a particular location

noise index a method of evaluating or rating a noise, usually by assigning a single number to it, based on some combination of its physical characteristics (sound pressure level, frequency, duration) and other factors such as time of day, tonal characteristics and impulsive characteristics

noise limit a maximum or minimum value imposed on a noise index, e.g. for some legal purpose or to determine eligibility for some benefit

noise mapping the production of computer software generated maps showing how the predicted levels of outdoor noise levels vary with location, e.g. from street to street in an area. They show 'sound (or noise) immission contours' (see also noise immission) and may be used (according to the EU Noise Directive) to help generate noise action plans

noise nuisance has been defined by the World Health Organization as 'a feeling of displeasure evoked by noise'. Statutory nuisance has a more specific meaning and is subject to legal action under the Environmental Protection Act 1990

noise pollution level, L_{PN} an index devised in the 1960s for assessing environmental noise, based on a combination of its L_{Aeq} value and its variability, expressed in terms of its standard deviation; it is now rarely used

noise rating (NR) curves a method of rating noise which is similar to the NC system but intended to be applicable to a wider range of situations; the method was defined in ISO R1996, now withdrawn, but the NR system continues to be used, particularly for offices and is used in BS 8233

noise reduction coefficient a single-figure number sometimes used to describe the performance of sound absorbing materials, based on a combination of its absorption coefficient at various frequencies

noise zone region where the long-term average rating level lies between two specified levels such as, for example, between 65 and 70 dB. The noise zone number for this example is 65–70 dB (defined in BS 7445)

non-linear in general there is a non-linear relationship between two quantities if they are not directly proportional to each other; if in measurement and instrumentation systems the input exceeds the linear range, then non-linearity results in a distorted output

normalized corrected or standardized in some way, as in normalized level difference, D_n, defined in BS EN ISO 140-4, where the measured level difference is corrected on the basis of the amount of sound absorption in the receiving room.

normal mode a natural mode of a vibrating system

normal threshold of hearing the modal value of the thresholds of hearing of a large number of otologically normal observers between 18 and 25 years of age

Noy a unit of noisiness related to the perceived noise level in PNdB by the formula: $\text{PNdB} = 40 + 10 \log_2 (\text{Noy})$

NPRs (noise preferential routes) specified aircraft departure routes at some airports designed to minimize noise exposure to major centres of population near to the airport. Compliance with NPRs is monitored by noise and track keeping (NTK) systems

Nyquist frequency the frequency which corresponds to half the sampling rate of digitized data, above which aliasing occurs

octave the range between two frequencies whose ratio is 2 : 1

organ of Corti a complex structure in the cochlea of the middle ear, supported by the basilar membrane and containing the hair cells

oscillation a to-and-fro motion; a fluctuation of a quantity or value about a mean

oscilloscope a device for displaying oscillatory signals on a cathode-ray screen

ossicles the three small bones of the middle ear which connect the eardrum with the oval window in the cochlea

otitis media inflammation and /or infection of the middle ear leading to build-up of fluid and blockage of the eustachian tube and may result in temporary or permanent hearing loss (also sometimes called 'glue ear')

otoacoustic emission sounds emitted by the eardrum and detected by a miniature microphone in the ear canal arising from activity in the cochlea in response to the stimulus of external sound

outer ear the outer part of the hearing mechanism which collects and guides airborne sound down the ear canal to the eardrum

output impedance the impedance of a device measured at its output

oval window diaphragm connecting the cochlea to the middle ear

overdamping (1) an amount of damping, in excess of critical, which is sufficient to prevent oscillation in a mass–spring system; (2) producing a damping ratio greater than one

overload a situation in which a component or system is used beyond its range of linearity

overload indicator a device which indicates when an instrument is likely to read incorrectly because it is being overloaded

overtone a higher (i.e. not the lowest) harmonic or natural frequency of a vibrating system

P-wave a longitudinal compression wave in an elastic medium

particle velocity see under acoustic particle velocity

pascal a unit of pressure equal to 1 N/m^2

pass band a band of frequencies which are transmitted by a band pass filter

passive a device which does not require a source of power for its operation, e.g. a passive filter or a passive noise control (cf. active)

peak the maximum deviation of a signal from its mean value within a specified time interval

peak-to-peak the algebraic difference between the extreme values of a signal occurring within a specified time interval

perceived noise level of a sound; the sound pressure level of a reference sound which is assessed by normal observers as being equally noisy; the reference sound consists of a band of random noise centred on 1000 Hz

percentile level, $L_{AN,T}$ the sound level, in dBA which is exceeded for N% of the time interval T, for example in L_{A10} and L_{A90}. The time weighting (F or S) should always be specified, e.g. L_{AF10} or L_{AS10}. Note that percentile levels determined over a certain time interval cannot generally be extrapolated to other time intervals

period of a repetitive signal; the time for one cycle

periodic signal one which repeats itself exactly

permanent threshold shift the component of threshold shift which shows no progressive reduction with the passage of time when the apparent cause has been removed

personal daily noise exposure level, $L_{EP,d}$ that steady or constant level which, over eight hours, contains the same amount of A-weighted sound energy as is received by the subject during the working day

phase of a sinusoidal signal; an angle whose value determines the point in the cycle, i.e. the magnitude of the signal, at some reference time

phase difference the difference between the phase angles (of two sinusoidal signals of the same frequency)

Phon the unit of loudness level; the loudness level of a sound, in Phons, is the sound pressure level of a 1000 Hz pure tone judged by the average listener to be equally loud

piezoelectric the behaviour of certain crystalline materials whereby a deformation of the material (caused by force or stress) results in the production of electric charge on the stressed faces, and a voltage difference between them

pink noise a random broadband signal which has equal power per percentage bandwidth and therefore has a flat, i.e. horizontal, frequency spectrum when plotted on a logarithmic frequency scale (cf. white noise)

pinna the external part of the ear leading to the ear canal

pitch that attribute of auditory sensation in terms of which sounds may be ordered on a scale related primarily to frequency; the unit of pitch is the mel

plane wave a wave in which the wavefronts are plane and parallel everywhere, so the sound energy does not diverge with increasing distance from the source

plenum a chamber or space used to collect air prior to its distribution via a duct system

PNdB the unit of perceived noise level

point source an idealized concept of an acoustic source which radiates spherical waves

polarization a property of transverse waves but not longitudinal waves; it relates to the direction of the particle displacement in the plane normal to the direction of propagation

power spectral density the amount of power per unit of frequency in a signal. It is measured in W/Hz, or more generally for a voltage signal in V^2/Hz (since power is proportional to V^2)

ppv, peak particle velocity a measure of vibration, usually measured in mm/s, used for assessing likelihood of damage to buildings from vibration

preamplifier a circuit which acts as an electrical impedance matching device between a transducer with a high output impedance, such as a microphone or accelerometer, and the signal processing circuits of the sound level or vibration meter

precedence effect see under Haas effect

pre-completion testing (PCT) sound insulation tests required of newly built (and some converted) dwellings, in order to test compliance with the performance requirements of the 2002 Building Regulations (unless built in accordance with Robust Standard details)

preferred speech interference level the arithmetic average of the sound pressure levels in the three octave bands 500 Hz, 1000 Hz and 2000 Hz

prepolarized microphone see under electret microphone

presbycusis hearing loss, mainly of high frequencies, that occurs with advancing age

progressive wave a wave that travels outwards, from its source, and is not being reflected

psycho-acoustics the study of the relationship between the physical parameters of a sound and its human perception

pure tone a sound for which the waveform is a sine wave, i.e. for which the sound pressure varies sinusoidally with time

pure tone audiometer an instrument for measuring hearing acuity to pure tones by determination of hearing levels

***Q* factor** a quantity which measures the sharpness of the resonance of a single degree of freedom mechanical or electrical vibrating system; in a mechanical system it is related to the damping ratio, the amplification produced at resonance and the shape of the resonance peak

R* and *R′ sound reduction index (R) and apparent sound reduction index (R'). Terms relating to the sound insulation performance of partitions defined in BS EN ISO 140-4, measured in octave or third octave frequency bands

R_w and R'_w single figure frequency weighted values of R and R' defined in BS EN ISO 717-1

random noise/vibration/signal a noise, vibration or signal which has a random waveform, with no periodicity

RASTI rapid analysis speech transmission index; a measurement parameter for assessing the speech intelligibility in a room. It has a value between 0 and 1: 1 representing perfect speech intelligibility and 0 representing zero speech intelligibility. RASTI is a shortened version of Speech Transmission Index, STI

rating level $L_{Ar,Tr}$ a noise index defined in BS 4142 and BS 7445; the specific noise level plus any adjustment for the characteristic features of the noise (e.g. tonal and impulsive features) during a specified time (see under specific noise level)

ray a straight line representing the direction in which a sound is travelling, used in situations where the size of reflecting surfaces is large compared to the sound wavelength

rayl the unit of specific acoustic impedance

Rayleigh wave a type of elastic wave which propagates close to the surface of a solid

Raynaud's disease a disorder affecting the blood vessels, nerves, connective tissues and bones of the fingers; one of its causes is prolonged exposure to high levels of vibration

ray tracing a method of modelling room acoustic performance using computer software package (also called beam tracing)

reactance the complex component of impedance associated with energy being stored and converted from one form to another (e.g. from potential to kinetic, or from electrostatic to electromagnetic) rather than being converted to heat

reactive silencer a silencer which reduces sound levels by using changes in impedance instead of sound absorbing materials

real time, in quickly enough to observe changes in a situation as they happen

real-time analyser a device which is capable of analysing signals (usually in the frequency domain) in real time

real world protection the degree of sound level attenuation provided by hearing protectors under realistic working conditions, as opposed to that from laboratory tests; an allowance to be deducted from manufacturer's test data to allow for real world conditions

recruitment an aspect of certain forms of perceptive deafness; an abnormally rapid increase in the sensation of loudness with increasing sound pressure level

reference time interval, Tr the specified interval over which a noise index such as equivalent continuous A-weighted sound pressure level is determined for assessment purposes e.g. in BS 7445 and BS 4142. The value of Tr is 1 hour during the day and 5 minutes at night.

reference value standardized values used as the basis for decibel scales of sound pressure, sound intensity, sound power, vibration acceleration, velocity and displacement

reflection the redirection of waves which occurs at a boundary between media when the size of the boundary interface is large compared with the wavelength (see also diffuse reflection and specular reflection)

refraction the change in direction of waves caused by changes in the wave velocity in the medium

repeatability the variability of measurements when repeated under the same measurement conditions

reproducibility the variability of measurements when repeated under different measurement conditions

residual noise The ambient noise remaining at a given position in a given situation when the specific noise source is suppressed to a degree such that it does not contribute to the ambient noise (defined in BS 4142 and BS 7445; see also specific noise and ambient noise)

residual noise level, $L_{Aeq,T}$ the equivalent continuous A-weighted sound pressure level of the residual noise

resonance the situation in which the amplitude of forced vibration of a system reaches a maximum at a certain forcing frequency (called the resonance frequency)

resonance frequency the frequency at which resonance occurs, i.e. at which the forced vibration amplitude in response to a force of constant amplitude is a maximum; for an undamped system the resonance frequency is the same as the natural frequency of the system; for a damped system the resonance frequency is slightly reduced

reverberant room a standard acoustic test environment designed to produce diffuse sound conditions throughout the space

reverberant sound/reverberation the sound in an enclosed space which results from repeated reflections at the boundaries

reverberant sound field the region in an enclosed space in which the reverberant sound is the major contributor to the total sound pressure level

reverberation time the time required for the steady sound pressure level in an enclosed space to decay by 60 dB, measured from the moment the sound source is switched off

ringing transient free vibration of bodies caused by impact

Robust Standard details a method of building new dwellings according to prescribed details which guarantees sound insulation sufficient to meet the requirements of the 2002 Building Regulations and so avoid the need for pre-completion sound insulation testing

room constant, R_c a constant used in the calculation of reverberant sound pressure level in a room: $R_c = S\alpha_{AVG}/(1 - \alpha_{AVG})$, where S = total area of room surfaces and α_{AVG} = average absorption of room surfaces

room mode a three-dimensional standing-wave sound pressure pattern, i.e. a mode shape, associated with one of the natural frequencies of a room

root mean quad (RMQ) the RMQ value of a set of numbers is the fourth root of the average of the fourth powers of the numbers; for a vibration waveform the RMQ value over a given time period is the fourth root of the average value of the fourth power of the waveform over that time period

root mean square (RMS) the RMS value of a set of numbers is the square root of the average of their squares; for a sound or vibration waveform the RMS value over a given time period is the square root of the average of the square of the waveform over that time period

round window a diaphragm or membrane at the end of the cochlea which connects with the middle ear

S (slow) time weighting one of the standard averaging times for sound level meter displays, defined in BS EN ISO 61672-1

sabin unit of sound absorption; one sabin is the amount of absorption equivalent to one square metre of perfect absorber

Sabine's formula a formula for predicting reverberation times of rooms

sampling frequency of a digitized signal; the number of samples per second (see Shannon's sampling theorem)

sampling interval the time interval between samples

scalar a quantity that may be completely defined by its magnitude alone, i.e. it has no direction

semi-anechoic a room with anechoic walls and ceiling, but with a sound reflecting floor

semi-reverberant a room which is neither completely anechoic nor reverberant, but somewhere in between

sensation value of a specified sound; the sound pressure level when the reference sound pressure corresponds to the threshold of hearing for the sound

sensitivity of a transducer; the ratio of output to input, e.g. for a microphone the sensitivity is measured as output voltage (V)/input pressure (Pa)

Shannon's sampling theorem an important result of signal processing theory that states that a minimum of two samples per hertz is needed adequately to represent the highest frequency component in a signal. Therefore a sample rate of at least 40,000 samples per second would be needed adequately to represent a signal with frequency components of up to 20,000 Hz. It is important that this condition is met otherwise false frequencies, called aliases, start to appear in the analysis

shear wave a transverse wave of shear stress propagating in an elastic medium

shock a sudden transient disturbance to a vibrating system caused by a rapid change in force, displacement, velocity or acceleration

signal-to-noise ratio a measure of the strength of a signal, indicating its magnitude relative to the background electrical noise in the measurement system; usually expressed in decibels

silencer a device for reducing noise in air and gas flow systems; silencers are either absorptive or reactive; also called attenuators or mufflers

simple harmonic motion a single frequency vibration, i.e. one in which the displacement varies sinusoidally with time

simple source an idealized model of an acoustic source, which radiates spherical waves, under free field conditions; see also under point source, monopole

sine wave the graph of a sinusoidal function, which indicates the simplest possible repeating waveform, characterized by a single frequency and constant amplitude

single degree of freedom system a vibrating system consisting of only one mass, one spring and one dashpot (damper); such a system has one natural frequency and one mode of vibration; its motion can be completely described by one variable

sinusoidal relating to a sine wave

SNR (single noise rating) a single figure method for evaluating the attenuation performance of hearing protectors

snubber a device used to restrict the maximum displacement of a vibrating system, e.g. at resonance

sociocusis hearing loss arising from everyday activities

soft ground ground such as grassland, cultivated land which is considered to be an acoustically absorbing surface, in contrast with 'hard ground' which is considered to be sound reflecting

Sone the unit of loudness; the tone scale is devised to give numbers which are approximately proportional to the loudness; it is related to the Phon scale as follows: $P = 40 + 10 \log_2 S$, where P represents Phons and S represents Sones

sound (1) pressure fluctuations in a fluid medium within the (audible) range of amplitudes and frequencies which excite the sensation of hearing; (2) the sensation of hearing produced by such pressure fluctuations

sound absorbing material material designed and used to maximize the absorption of sound by promoting frictional processes; the most commonly used materials are porous, such as mineral fibre materials or certain types of open-cell foam polymer materials

sound absorption (1) the process whereby sound energy is converted into heat, leading to a reduction in sound pressure level; (2) the property of a material which allows it to absorb sound energy

sound absorption coefficient a measure of the effectiveness of materials as sound absorbers; it is the ratio of the sound energy absorbed or transmitted (i.e. not reflected) by a surface to the total sound energy incident upon that surface; the value of the coefficient varies from 0 (for very poor absorbers and good reflectors) to 1 (for very good absorbers and poor reflectors)

sound exposure level, SEL (L_{AE}) a measure of A-weighted sound energy used to describe noise events such as the passing of a train or aircraft; it is the A-weighted sound pressure level which, if occurring over a period of one second, would contain the same amount of A-weighted sound energy as the event

sound insulating material material designed and used as partitions in order to minimize the transmission of sound; the best materials are those which are dense and solid, such as wood, metal or brick, although lightweight panels can also be effective when in the form of double-skin constructions

sound insulation the reduction or attenuation of airborne sound by a solid partition between source and receiver; this may be a building partition (e.g. a floor, wall or ceiling), a screen or barrier, or an acoustic enclosure

sound intensity the sound power flowing per unit area, in a given direction, measured over an area perpendicular to the direction of flow; its units are W/m^2

sound intensity level, L_I sound intensity measured on a decibel scale: $L_I = 10 \log_{10}(I/I_0)$, where I_0 is the reference value of sound intensity, $10^{-12} W/m^2$

sound level a frequency-weighted sound pressure level, such as the A-weighted value

sound level meter an instrument for measuring sound pressure levels

sound power the sound energy radiated per unit time by a sound source, measured in watts (W)

sound power level, L_W sound power measured on a decibel scale: $L_W = 10 \log_{10}(W/W_0)$, where W_0 is the reference value of sound power, 10^{-12}W

sound pressure the fluctuations in air pressure, from the steady atmospheric pressure, created by sound, measured in pascals (Pa)

sound pressure level, SPL (L_p) sound pressure measured on a decibel scale: $L_p = 20 \log_{10}(p/p_0)$, where p_0 is the reference sound pressure, 20×10^{-6}Pa

sound propagation the transmission or transfer of sound energy from one point to another

sound reduction index, R a measure of the airborne sound insulating properties, in a particular frequency band, of a material in the form of a panel or partition, or of a building element such as a wall, window or floor; it is measured in decibels: $R = 10 \log_{10}(1/t)$, where t is the sound transmission coefficient; it is measured under laboratory conditions according to BS EN ISO 140-4; also known as transmission loss

soundscape the total sound environment at a particular location, implying much more than can be described just in terms of sound level. A soundscape approach to improving the urban acoustic environment focuses on improving the acoustic quality of the environment, i.e. how it may be made more pleasing to the ear, rather than simply on noise reduction

sound transmission the transfer of sound energy across a boundary from one medium to another

sound transmission coefficient the ratio of the sound energy transmitted by a partition, or across a boundary, to the sound energy incident upon the partition or the boundary

sound wave a pressure wave in a fluid which transmits sound energy through the medium by virtue of the inertial, elastic and damping properties of the medium

specific acoustic impedance at a point in a sound field; the complex ratio of sound pressure to the acoustic particle velocity. In terms of the fundamental units m, kg and s, the units of specific acoustic impedance are kgm/s^2, but sometimes the quantity is expressed in terms of newtons (N) or pascals (Pa), as either Nsm^{-3} or as Pas/m, and also as the rayl

specific noise the particular component of the ambient noise which is under consideration or investigation, e.g.

in connection with a planning application or noise complaint; defined in BS 4142

specific noise source The noise source under investigation for assessing the likelihood of complaints (defined in BS 4142)

specific noise level, $L_{Aeq,Tr}$ the equivalent continuous A-weighted sound pressure level at the assessment position produced by the specific noise source over a given reference time interval

spectral adaptation terms C, Crt; terms used in connection with the measurement and assessment of airborne sound insulation and defined in BS EN ISO 7172-1

spectrum a frequency spectrum is a graph showing variation of sound pressure level (or other quantity) with frequency

specular reflection sound reflection which obeys the law of reflection that angle of incidence equals angle of reflection and which occurs at surfaces which are smooth on a scale comparable with the wavelength of the sound (as opposed to **diffuse reflection**)

speech intelligibility the ability of speech to be understood; the ability of a listener to hear and correctly interpret verbal messages. The concept of intelligibility is used as a criterion to determine the degree of acoustic privacy between rooms

speech interference level a measure of the ambient noise level in offices which gives an indication of the degree to which speech will be intelligible; it is based on the arithmetic mean of the octave band sound pressure levels 500 Hz, 1000 Hz and 2000 Hz, which are most significant for good speech intelligibility

speech privacy the degree to which speech is unintelligible between offices. Three ratings are used: confidential, normal (non-obtrusive) and minimal. It is inversely related to speech intelligibility, e.g. good speech intelligibility leads to poor acoustic privacy, and vice versa

speech transmission index (STI) a measurement parameter for assessing the speech intelligibility in a room. It has a value between 0 and 1: 1 representing perfect speech intelligibility and 0 representing zero speech intelligibility

spherical waves an idealized model of how sound propagates in free field conditions, and used as the basis of certain sound level prediction methods

standard deviation a measure of the deviation or scatter of a set of values (e.g. sound pressure level measurements) from the mean value

standardized impact sound pressure level, L'_{nT} a measurement of impact sound insulation, corrected according to BS EN ISO 140-7 for room characteristics; a complete set of measurements consists of 16 values, one for each third octave frequency band from 100 Hz to 3150 Hz

standardized level difference, D_{nT} a measurement of airborne sound insulation, corrected according to BS EN ISO 140-4 for receiving room characteristics; a complete set of measurements consists of 16 values, one for each third octave frequency band from 100 Hz to 3150 Hz

standing waves a wave system characterized by a stationary pattern of amplitude distribution in space arising from the interference of progressive waves; also called stationary waves

stapes or stirrup; one of the three bones of the middle ear, connected to the oval window of the inner ear

static deflection the deflection produced in the spring of a mass–spring system by the weight of the mass; it is related to the natural frequency of the system and is used to specify the stiffness of springs for vibration isolation

stationary waves see under standing waves

steady noise noise for which the fluctuations in time are small enough to permit measurement of average sound pressure level to be made satisfactorily without the need to measure L_{Aeq} using an integrating sound level meter; defined in BS 4142

STI see speech transmission index

STIPA a version of STI for public address systems

STITEL a version of STI for telecommunications systems

strain the fractional change in shape due to an elastic deformation in a material caused by an applied stress

stress force per unit area, measured in N/m^2; stress applied to elastic materials causes strain

structure-borne sound sound which reaches the receiver after travelling from the source via a building or machine structure; structure-borne sound travels very efficiently in buildings, and is more difficult to predict than airborne sound

subjective depending upon the response of the individual

superposition according to the principle of superposition, the wave disturbances in a medium caused by different sources may be combined algebraically

tangential mode a room mode which involves reflections between two pairs of parallel surfaces (e.g. walls)

temporary threshold shift the component of threshold shift which shows progressive reduction with the passage of time, when the apparent source has been removed

threshold of hearing for a given listener the lowest sound pressure level of a particular sound that can be heard under specified measurement conditions, assuming the sound reaching the ears from other sources is negligible

threshold of pain for a given listener the minimum sound pressure level of a specified sound which will produce the sensation of pain in the ear

threshold shift the deviation, in decibels, of a measured hearing level from one previously established

timbre the quality of a sound which is related to its harmonic structure

time averaged sound level see under continuous equivalent sound level, $L_{\text{Aeq},T}$

time constant of a process or quantity which decays exponentially with time; the time required for the value to reduce by a factor of 1/e, where e is the exponential number 2.7183 . . .

time weighting one of the standard averaging times (F, S, I) used for the measurement of RMS sound pressure level in sound level meters, specified in BS EN ISO 61671-1

tinnitus a subjective sense of noises in the head or ringing in the ears for which there is no observable cause

tonality the degree to which a noise contains audible pure tones: broadband noise is generally less annoying than noise with identifiable tones

tone a sound which produces the sensation of pitch; see also under pure tone

traffic noise index an index used for the assessment of environmental noise in the 1960s, based on a combination of the L_{A10} value and the L_{A90} value; it is now rarely used

transducer a device for converting signals from one form to another; frequently, the requirement is to convert changes in some physical variable, such as temperature or sound pressure, into analogous changes in electrical voltage or charge

transfer function of a vibrating system; the ratio of the output or response of the system to the input excitation, usually expressed as a complex function of frequency

transfer standard a calibrated noise source designed to fit over a microphone

transient a noise or vibration signal which is not continuous but which decreases to zero then remains zero

transmissibility of a vibrating system; the non-dimensional ratio of vibration amplitude at two points in the system; frequently, the two points are on either side of springs used as antivibration mounts, and the transmissibility is used as an indicator of the effectiveness of the isolation

transmission coefficient see under sound transmission coefficient

transmission loss see under sound reduction index

transverse sensitivity of an accelerometer; the sensitivity to vibration in a direction perpendicular to the axis of the accelerometer

transverse wave a wave in which the direction of vibration of the particles of the medium is perpendicular to the direction of wave travel; an example is a shear wave in a solid medium

triboelectric effect the production of electric charge as a result of vibration in an accelerometer cable, leading to electrical noise in the accelerometer signal, unless the cable is secured to prevent movement

ultrasonics the study of ultrasound

ultrasound acoustic waves with frequencies which are too high to be heard by human ears

uncertainty an estimate of the degree of variability or dispersion associated with the result of a measurement

unweighted sound pressure level a sound pressure level which has not been frequency weighted, sometimes known as the linear sound pressure level; symbol L_p (see also Z-weighting)

upper exposure action value a noise exposure level defined in the 2005 Control of Noise at Work Regulations requiring action from employers and employees (a personal daily (or weekly) noise exposure level of 85 dBA, or peak sound pressure level of 137 dBC)

vector a quantity which has a direction as well as a magnitude, e.g. force, displacement, velocity and acceleration

velocity the rate of change of displacement, measured in m/s or mm/s

vibration a to-and-fro motion; a motion which oscillates about a fixed equilibrium position

vibration dose value (VDV) a measure of vibration exposure; the fourth root of the integral, over the measurement period, of the fourth power of the frequency-weighted time-varying acceleration; see also under eVDV and root mean quad

vibration isolation the reduction of vibration and structure-borne sound by the use of resilient materials inserted in the transmission path between source and receiver

vibration white finger blanching of the fingers and other symptoms caused by exposure to hand-transmitted vibration; see also under Raynaud's disease

viscous damping damping of the sort which occurs in viscous fluid layers, in which the damping force is proportional to the velocity of the fluid element

volt (V) the unit of electrical potential; the difference in electrical potential between two points on an electric conductor which is carrying a constant electric current of one ampere (A) when the power dissipated between the points is one watt (W)

watt (W) the unit of power; the power dissipated when one joule of energy is expended in one second

wave in an elastic medium; a mechanism whereby a disturbance, and the energy associated with it, is propagated through an elastic medium; the disturbance results in vibrations of the particles of the medium, vibrations transmitted to nearby regions as a result of the elastic and inertial nature of the medium, resulting in a disturbance which is a function of both position and time

waveform a graph showing how a variable at one point in a wave (e.g. sound pressure or particle velocity) or vibration varies with time

wavefront the leading edge of a progressive wave, along which the vibration of the particles of the medium are in phase

wavelength the minimum distance between two points that are in phase within a medium transmitting a progressive wave

Weber–Fechner law a law of psychology which states that the change of subjective response to a physical stimulus is proportional to the logarithm of the stimulus

weighted sound reduction index, R_w a single figure value of sound reduction index, derived according to procedures given in BS EN ISO 717-1, used for rating and comparing partitions and based on the values of sound reduction index at different frequencies

weighted standardized impact sound pressure level, $L'_{nT,w}$ a single-figure value of impact sound insulation performance, derived according to procedures in BS EN ISO 717-2, used for comparing and rating floors and based on the values of L'_{nT} at different frequencies; values of $L'_{nT,W}$ are specified in the Building Regulations

weighted standardized level difference, $D_{nT,w}$ a single-figure value of airborne sound insulation performance, derived according to procedures given in BS EN ISO 717-1, used for rating and comparing partitions and based on the values of D_{nT} at different frequencies; sound insulation performance requirements in the Building Regulations are specified in terms of values of $D_{nT,w} + C_{tr}$

weighting see under frequency weighting and time weighting

white finger see under vibration white finger

white noise a random broadband noise which contains equal power per unit bandwidth, so it has a flat, i.e. horizontal, frequency spectrum when plotted on a linear frequency scale (cf. pink noise)

whole-body vibration vibration transmitted to the body as a whole

Z-weighting a (zero) frequency weighting defined in BS EN ISO 61672-1

Appendix 2 List of formulae

Chapter 1

Relationship between frequency, period, sound velocity and wavelength:

$f = 1/T \qquad c = f\lambda$

One-dimensional progressive wave:

$p = A\sin(\omega t - kx) + B\cos(\omega t - kx)$

Sound intensity pressure and particle velocity relationships:

$p = zv \quad I = pv \quad I = p^2/z \quad I = zv^2 \quad I = p^2$

Relationship between upper, lower and centre octave bands:

$f_C = \sqrt{(f_L \times f_U)}\ f_U = 2f_L\ f_U = (\sqrt{2})f_C\ f_L = f_C/\sqrt{2}$

For 1/3 octaves:

$f_U = f_L \times (2)^{1/3} \quad f_U = f_C \times (2)^{1/6} \quad f_L = f_C/(2)^{1/6}$

Peak and RMS sound pressure:

$p_{RMS} = p_{peak}/\sqrt{2} = 0.707 \times p_{peak}$

Decibels:

$N = 10\log(I_2/I_1)$

$N = 10\log(W_2/W_1)$

$N = 10\log(p_2/p_1)^2 = 20\log(p_2/p_1)$

$L_p = 20\log(p/p_0)$

$L_I = 10\log(I/I_0)$

$L_W = 10\log(W/W_0)$

Combining and averaging decibels:

$$L_T = 10\log[10^{L_1/10} + 10^{L_2/10} + 10^{L_3/10} + 10^{L_4/10} + \cdots + 10^{L_N/10}]$$

$$L_{AVGE} = 10\log[(10^{L_1/10} + 10^{L_2/10} + 10^{L_3/10} + 10^{L_4/10} + \cdots + 10^{L_N/10})/N)]$$

Subtracting decibels:

$L_D = 10\log(10^{L_1/10} - 10^{L_2/10})$

Velocity of elastic waves:

$c = \sqrt{(K/\rho)}$

Elastic modulus = stress/strain

Speed of sound in air:

$c = \sqrt{(\gamma P/\rho)}$

Variation with temperature:

$c = \sqrt{(\gamma RT)}$

Reflection of sound at a boundary:

$R = [(z_1 - z_2)/(z_1 + z_2)]^2$

Beats:

$f_B = f_1 - f_2$

One-dimensional standing waves:

$p = A\cos(kx)\sin(\omega t)$, or $p = A\cos(\omega t)\sin(kx)$

Snell's law of refraction:

$\sin(i)/\sin(r) = c_1/c_2$

Chapter 2

Inverse square law, variation of sound intensity and pressure with distance:

$I = W/S \qquad I = W/4\pi r^2 \qquad I = 1/r^2$

$(I_2/I_1) = (r_1/r_2)^2$

$p = 1/r \quad (p_2/p_1) = (r_1/r_2)$

Inverse square law in decibels:

$L_p = L_W - 20\log r - 11 \qquad L_1 - L_2 = 20\log(r_2/r_1)$

Directivity factor and index:

$Q = I/I_{avge} = (p/p_{avge})^2 \quad D = L_\theta - L_{avge} \quad D = 10\log Q$

and $Q = 10^{D/10} \quad L_p = L_W - 20\log r - 11 + D$

$L_p = L_W - 20\log r - 11 + D - A_{excess}$

Barriers:

$\delta = (a + b) - c \quad N = 2\delta/\lambda$

Attenuation $= 10\log(3 + 20N)$ dB

Outdoor sound pressure level:

$$L_p = L_W - 20\log r - 11 + D - A_{air} - A_{ground} - A_{turbulence} - A_{refraction} - A_{barrier} - \cdots$$

Doppler effect:

moving source $f' = [c/(c - v)]f$
moving receiver $f' = [(c + v)/c]f$

Chapter 3

Voice level for intelligibility:

$VL_A \geq 1.33\,(\text{SIL} + 20\log r) - 36$

Loudness:

$\text{Sones} = 2^{(\text{Phons}-40)/10}$ $\quad \text{Phons} = 40 + 33\log(\text{Sones})$

$S_{tot} = S_{i,\max} + 0.3((\sum S_i) - S_{i,\max})$

Noisiness and perceived noise level:

$\text{Noys} = 2^{(L_{PN}-40)/10}$ $\quad L_{PN} = 40 + 33\log(\text{Noys})$

Noise and Number Index:

$\text{NNI} = L_{PN,\max} + 15\log N - 80$

Chapter 4

$L_{Aeq,T} = 10\log[(t_1 \times 10^{L1/10} + t_2 \times 10^{L2/10} + t_3 \times 10^{L3/10} + \cdots + t_N \times 10^{LN/10})/T]$

$L_{Aeq,T} = 10\log[(\int^T p^2(t)dt/T)]$

$L_{Aeq,T_2} - L_{Aeq,T_2} = 10\log(T_2/T_1)$

$L_{AE} = L_{Aeq,T} + 10\log T$

$L_{Aeq,T} = L_{AE} + 10\log N - 10\log T$

$L_{Aeq,T} = 10\log[\{(N_1 \times 10^{LAE1/10}) + (N_2 \times 10^{LAE2/10}) + (N_3 \times 10^{LAE3/10}) + \cdots + (N_n \times 10^{LAEn/10})\}/T]$

$L_{dn} = 10\log((15 \times 10^{(Ld/10)} + 9 \times 10^{(Ln+10)/10})/24)$

$L_{den} = 10\log((12 \times 10^{(Lday/10)} + 4 \times 10^{(Leven+5)/10} + 8 \times 10^{(Lnight+10)/10})/24)$

CRTN shortened measurement procedure:

$L_{10(18\text{ hour})} = L_{10(3\text{ hour})} - 1\text{ dBA}$

CRTN approximate relationships:

$L_{Aeq,16h} = L_{A10,18h} - 2\text{ dB}$

$L_{Aeq,1hour} = 0.94 \times L_{A10,1hour} + 0.77\text{ dB}$

$L_{Aeq,1hour} = 0.57 \times L_{A10,1hour} + 24.46\text{ dB}$

$L_{den} = 0.92 \times L_{A10,18h} + 4.20$ for non-motorway roads and $L_{den} = 0.90 \times L_{A10,18h} + 9.69$ for motorways

$\text{TNI} = 4 \times L_{A10} - 3 \times L_{A90} - 30$

$L_{NP} = L_{Aeq} + 2.56\sigma = L_{Aeq} + (L_{A10} - L_{A90})$

Chapter 5

Reflection, transmission and absorption coefficients:

$I_i = I_r + I_t + I_a$

$r = I_r/I_t$ and $t = I_t/I_i$ $\quad r = (p_r^2/p_i^2) = (p_r/p_i)^2$ and $t = (p_t^2/p_i^2) = (p_t/p_i)^2$

$\alpha = (I_a + I_t)/I_i$ and $\alpha = 1 - r$

Sound absorption:

$A = S\alpha$ and $\alpha = A/S$

$A_{TOTAL} = A_1S_1 + A_2S_2 + A_3S_3 + \cdots$

$\alpha_{AVGE} = A_{TOTAL}/S_{TOTAL}$

$S_{TOTAL} = 2(LB + LH + BH)$

Panel resonance frequency: $f = 60/\sqrt{(m \times d)}$

Helmholtz resonator frequency:

$f = (c/2\pi)\sqrt{(S/(L'V))}$ and $L' = L + 1.5 \times r$ (approximately)

Sabine:

$T = 0.16 \times V/A$

Eyring:

$T = 0.16 \times V/[-S \times \alpha_{AVGE} \times \ln(1 - \alpha_{AVGE})$

Effect of air absorption on RT:

$T = 0.16V/\{(S \times \alpha_{AVGE})\} + kV\}$

α using reverberant room:

$\alpha = (0.16V/S)[(1/T_1) - (1/T_2)]$

Direct and reverberant sound:

$L_{REV} = L_W + 10\log(4/R_C)$

$R_C = S \times \alpha_{AVGE}/(1 - \alpha_{AVGE})$

$L_{DIRECT} = L_W + 10\log(Q/4\pi r^2) = L_W - 20\log r - 11$

$L_{TOTAL} = L_W + 10\log[(Q/4\pi r^2) + (4/R_C)]$

$R = \sqrt{[QR_C/16\pi]}$

Use of absorption for reducing reverberant sound:

$L_1 - L_2 = 10\log(R_{C2}/R_{C1})$ and $L_1 - L_2 = 10\log(A_2/A_1)$

α using standing wave tube:

$\alpha = 4n/(n^2 + 2n + 1)$

Room modes:

$f = (c/2)\sqrt{[(N_L/L)^2 + (N_W/W)^2 + (N_H/H)^2]}$

Energy density:

$\varepsilon = p/\rho c^2 = I/c$ and $I_{EFF} = p^2/4\rho c = \varepsilon c/4$

Transmission from inside to outside:

$L_{IN} = L_{OUT} - 6$

Chapter 6

Sound reduction index:

$R = 10\log(1/t)$ dB $\quad t = 1/10^{R/10} = 10^{-R/10}$

Composite sound reduction index:

$t_{AVGE} = (t_1S_1 + t_1S_1 + t_1S_1 + \cdots + t_NS_N)/S_{TOTAL}$
$R_{AVGE} = 10\log(1/t_{AVGE})$

Maximum allowable glazing area:

$S = S_{TOTAL} \times [t_{AVGE} - t_W]/[t_G - t_W]$

Room to room sound transmission:

$L_1 - L_2 = R - 10\log S + 10\log A$ and
$R = L_1 - L_2 + 10\log S - 10\log A$

Inside to outside sound transmission:

$L_{OUT} = L_{IN} - R - 6 \quad L_W = L_{OUT} + 10\log S$
$L_r = L_W - 20\log r - 11 + D$

$L_r = L_{IN} + 10\log S - 20\log r - 14$ and
$L_{2m} = L_{FF} + 3$

Outside to inside sound transmission:

$L_{IN} = L_{2m} - R + 10\log S - 10\log A$ or
$L_{IN} = L_{FF} - R + 10\log S - 10\log A + 3$

$L_{2m} = L_{FF} + 3$ and $L_{CLOSE} = L_{FF} + 6 = L_{2m} + 3$

$D_n = D - 10\log(A/A_0)$

$D_{nT} = D + 10\log(T/0.5) = L_1 - L_2 + 10\log(T/0.5)$

Apparent sound reduction index:

$R' = D + 10\log(S/A) = D + 10\log S - 10\log A$

Normalized impact sound pressure level:

$L_n = L_i + 10\log(A/A_0)$

Standardized impact sound pressure level:

$L'_n = L_i - 10\log(T/T_0)$

Small element, ceiling and floor insulation:

$D_{NE} = L_1 - L_2 + 10\log(A_0/A)$ and
$L_1 - L_2 = D_{NE} + 10\log(A/A_0)$

$D_{nC} = L_1 - L_2 + 10\log(A_0/A)$

$D_{nF} = L_1 - L_2 + 10\log(A_0/A)$

Spectral adaptation terms:

$X_{A1} = 10\log\sum(10(L_{i1} - R_i)/10)$
$X_{A2} = 10\log\sum(10(L_{i2} - R_i)/10)$

$C = X_{A1} - D_{nT,w} \quad C_{tr} = X_{A2} - D_{nT,w}$

Mass law:

$R = 20\log(fm) - 48$ where $m = \rho t$

Mass–spring–mass resonance:

$f_0 = 60\sqrt{\{(m_1 + m_2)/m_1m_2d)\}}$

Chapter 7

Relationships between vibration, displacement, velocity and acceleration:

Displacement, $x = X\sin(2\pi ft) = X\sin(\omega t)$
Velocity, $v = V\cos(2\pi ft) = V\cos(\omega t)$

Acceleration, $a = -A\sin(2\pi ft) = -A\sin(\omega t)$

where $\omega = 2\pi f$ = angular frequency, in radians per second

Relationships between displacement, velocity and acceleration amplitudes:

$$V = 2\pi fX$$
$$A = 2\pi fV$$
$$A = 4\pi^2f^2X$$

Expressing vibration magnitudes in decibels:
$N\,(\text{dB}) = 10\log(A/A_0)^2 = 20\log(A/A_0)$ and
$A = A_0 \times 10^{(N/20)}$

Natural frequency of a mass–spring system:

$f_0 = (1/2\pi)\sqrt{(k/m)}$

Combining spring stiffnesses:

$k_T = k_1 + k_2 + k_3 + \cdots + k_N$

Natural frequency and static deflection:

$f_0 = 15.8\sqrt{(1/X_S)}$

Critical damping:

$r_C = 2\sqrt{(k/m)}$

Damped natural frequency:

$f_d = f_0\sqrt{(1 - \xi^2)}$

Logarithmic decrement:

$\delta = (1/n)\ln(x_1/x_n)$ and $\delta = 2\pi\xi$

Q factor:

$Q = f_0/(f_2 - f_1)$ and $Q = 1/(2 \times \xi)$

Definition of transmissibility:

$T_F = F_t/F_i$ and $T_X = X_t/X_i$

Isolation efficiency:

$\eta = (1 - T) \times 100\%$

Transmissibility in terms of dB reduction:

$N = 10\log(1/T)^2 = 20\log(1/T)$

Transmissibility:

$$T = \sqrt{\frac{1 + 4\xi^2(f/f_0)^2}{(1 - (f/f_0)^2)^2 + 4\xi^2(f/f_0)^2}}$$

and $T = 1/[(f/f_0)^2 - 1]$ for $\xi = 0, f > f_0$

Vibration dose value:

$$\text{RMS value of } a_W(t) = \left[\frac{1}{T}\int_0^T a_W^2(t)dt\right]^{0.5}$$

$$\text{RMQ value of } a_W(t) = \left[\frac{1}{T}\int_0^T a_W^4(t)dt\right]^{0.25}$$

$$VDV = \left[\int_0^T a_W^4(t)dt\right]^{0.25}$$

$V_T = [V_1{}^4 + V_2{}^4 + \cdots + V_N{}^4]^{0.25}$

$V_T = [NV^4]^{0.25} = N^{0.25}V$

$\text{eVDV} = 1.4a_{\text{rms}}T^{0.25}$

Hand-arm vibration:

$a_{hv} = \sqrt{(a^2_{hwx} + a^2_{hwy} + a^2_{hwx})}$

$$a_{hv(eq,8h)} = A(8) = \sqrt{\{[(a_{hv1}{}^2 \times t_1) + (a_{hv2}{}^2 \times t_2) + \cdots + (a_{hvN}{}^2 \times t_N)]/8\}}$$

Motion sickness:

$\text{MSDV} = a_W T^{0.5}$

Sound radiation from a vibrating surface:

$W = \sigma S z V_{\text{avge}}{}^2$

$L_W = 10\log(V_{\text{avge}}{}^2) + 10\log S + 10\log\sigma + 146$

$L_W = L_V + 10\log S + 10\log\sigma - 34$

Chapter 8

Microphone sensitivity in decibels:

$N = 20\log(S_2/S_1)$

FFT relationships:

number of samples $= N$

interval between successive samples $= \Delta T$ seconds
total duration of sample $= T = N\Delta T$ seconds

sampling rate $= N/T = 1/\Delta T$ samples per second

line spacing, $\Delta f = 1/T$ Hz

number of frequency lines $= N/2$
maximum frequency, $f_{\max} = (N/2)(1/T)$ Hz

Time Bandwidth Product Theorem:

$\varepsilon = 1/(2\sqrt{(BT)})$

Power spectral density:

PSD = mean square value/bandwidth
Mean square value = PSD × bandwidth

Sound intensity measurement:

sound pressure gradient $= (\Delta p/\Delta x) = -(p_1 - p_2)/\Delta x$

Sound power measurement:

$L_W = L_{p,\text{AVGE}} + 10\log S$
$L_W = L_I + 10\log S$
$L_W = L_p + 10\log 10\ V - 10\log 10\ T - 14$
$L_{W\ \text{Test source}} - L_{W\ \text{Standard source}} = L_{p\ \text{Test source}} - L_{p\ \text{Standard source}}$

charge sensitivity = voltage sensitivity × capacitance
[picocoulombs per m/s^2 = volts per m/s^2 × picofarads]

Chapter 9

Noise enclosure performance:

$\text{NR} = L_{\text{IN}} - L_{\text{OUT}} = R_E - 10\log S_E + 10\log A_R$
$\text{IL} = L_{\text{BEFORE}} - L_{\text{AFTER}} = R_E - 10\log S_E + 10\log A_E$

Fan noise:

$W \propto V^N$
$L_{W1} - L_{W2} = 10\log(V_2/V_1)^N = N \times 10\log[(V_2/V_1)]$
$f = \text{RPM} \times N/60$ Hz

Jet noise:

$f = 200 \times v/d$

Hearing protectors:

The HML method:

If $(L_A - L_C) > 2$ dB

$\text{PNR} = M - [\{(M - L)/8\}(L_A - L_C - 2)$

Otherwise

$\text{PNR} = M - [\{(H - M)/4\}(L_A - L_C - 2)$

$L'_A = L_A - \text{PNR}$

The SNR method:

$L'_A = L_C - \text{SNR}$

Appendix 3 Suggested list of experiments, tests and observations

Doing seeing and listening will always enhance your understanding and enjoyment of any subject and this is particularly so in the study of acoustics.

A list of possible experiments investigations and tests is given below. In some cases these may be adapted as necessary to suite the equipment available. Detailed procedures are given in this text for some of them, or may be found in British or ISO standards. In some cases demonstrations will be available on the Internet.

With just a basic sound level meter capable of measuring only dBA it will be possible to measure simple steady sound levels (e.g. of sounds from sources in the home) and background noise levels, and try to relate dBA levels to loudness, and to investigate the difference between maximum levels of impulsive sounds measured using Fast and Slow time weightings. If octave bands are also available it will be possible to investigate frequency spectra of different sources and indoor environments and measure NR and NC values and estimate loudness of some common sounds. Finally if the sound level meter has the ability to measure L_{Aeq}, L_{AN} and L_{Amax} it will be possible to carry out a much wider range of environmental noise measurements, e.g. from road traffic, trains and aircraft. A simple but thought provoking experiment is to fit a calibrator over the microphone and observe how the values of the various parameters (L_{Aeq}, L_{A10} L_{A50} L_{A90} change with time (use a stopwatch) after the calibrator has been switched off (leave the calibrator in position because removing will create noise at the microphone). Try to explain how the various levels change over time.

List of possible experiments

- Investigation of sound level with distance from a noise source, both indoors and outdoors. Indoors investigate rooms of different size and furnishings (e.g. 'live' and 'dead' rooms), and along corridors. (Plot noise levels against the log of distance to produce a 'straight line' drop off with distance, and check against the inverse square law prediction.)
- Measurement of airborne sound insulation (listen to investigate flanking paths, investigate of effect of opening doors and windows).
- Measurement of impact sound insulation. Effect of different floor coverings.
- Measurement of reverberation time (interrupted source and impulse source methods) (select different rooms).
- Measurement of absorption coefficient by reverberant room and standing wave tube methods.
- Speech intelligibility using lists of words (investigate the effect of increasing background noise e.g. white or pink noise from a loudspeaker. Repeat experiment in live and dead spaces.)
- Sound power level of a source by various methods.
- Investigation of room modes, particularly if you have access to a reverberant room. Play pure tones from a loudspeaker, tune to a room resonance, and then simply walk about and listen. (Room mode frequencies may be estimated (see Chapter 5) but may only be approximate for non-rectangular shaped rooms.)
- Measurement of road traffic noise. Investigate variation with distance from road.
- Subjective loudness tests. Attempt to compare loudness of pure tones at different levels and frequencies, Listen to a number of everyday sounds and rank their loudness and then compare with simple dBA measurements.
- Measurement of vibration and of vibration isolation.
- Insertion loss of a barrier and an enclosure (use octave band measurements if possible).
- Carry out an audiometric test. Investigate the repeatability of the test. If audiometric test equipment is not available the various websites offer tests but without proper calibration these can only be approximate.
- Investigate frequency spectrum and directionality of a loudspeaker and of human voice (male and female).
- Seek out audio demonstrations on the internet e.g. audio range of pure tones, white and pink noise, octave bands of noise, sounds played in anechoic and reverberant rooms, simulations of hearing loss.
- Seek out animated demonstrations of: how the ear works, standing waves in pipes and plates, of the Doppler effect, of properties of waves (reflection, diffraction, refraction and interference).

Other tests and exercises

- Investigate how the noise from a common source (e.g. vacuum cleaner, washing machine, hair dryer, electric drill) is created and transmitted to the receiver, and how it may be controlled.
- Investigate how sound is generated by a particular musical instrument of your choice.
- Listen to: the drop off in sound level as you walk away from a noise source of sound and to the directionality of the source.
- Listen to the effect of screening of noise by a barrier or as you walk around a corner out of line of sight of a sound source.
- Just listen, and wonder!

Appendix 4 Some electrical principles

These notes are intended for the reader with little of no knowledge of electrical terminology and principles; they aim to introduce some of the technical terms and ideas met in Chapter 9. They offer only the very briefest introduction to the subject, and readers requiring a more comprehensive treatment should consult an appropriate textbook on electrical or physical science.

A4.1 Ohm's law and resistors: volts, amps and ohms

Figure A4.1 shows the simplest possible electrical circuit; it consists of a cell or battery, as found in a car or a transistor radio, connected by copper wire to a **resistor**. It is possible to draw analogies with a simple water circuit in which water is pumped via copper pipes through a radiator. The pump provides a difference in water pressure, which drives the water through the circuit. The radiator provides resistance to the flow of water, and if more radiators are added, the flow rate decreases, The cell is the equivalent of the pump and provides a difference in electrical pressure or **electric potential**; it is measured in **volts** (V). The electric current is related to the rate at which electrons flow through the circuit; it is measured in **amperes** (A). The resistor provides a **resistance** to the current flow; it is measured in **ohms** (Ω). Resistors commonly found in electrical circuits; (e.g. in a radio or television) can vary from a few ohms to millions of ohms. The resistance of the copper wire in the circuit is probably a few hundredths of an ohm and is taken as negligible by comparison with the resistor. **Ohm's law** relates the current (I in amps) flowing through a resistor (R in ohms) to the potential difference (V in volts) across it

$$V = IR$$

In the case of the simple circuit in Figure A4.1, the voltage drop across the resistor is the voltage provided by the cell. Suppose it is 12 V, as in a car battery, and that the resistance is 3 Ω, then $12 = I \times 3$, from which the current is 4 A.

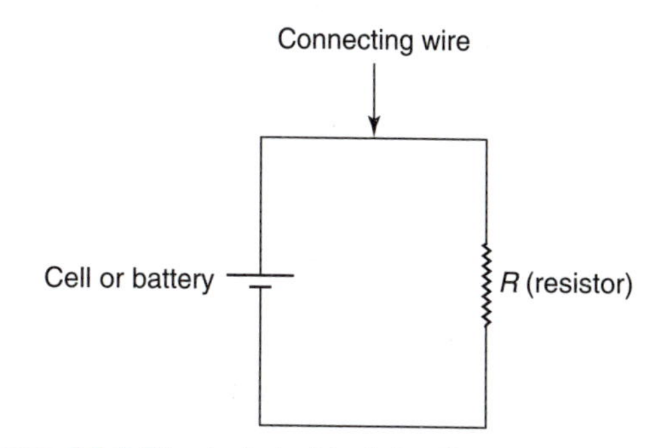

Figure A4.1 Simplest electrical circuit

A4.2 Resistors in series and parallel: attenuator chains

A second resistor has been added in Figure A4.2. The same current flows through each resistor in turn. The resistors are said to be **in series**. An alternative arrangement (Figure A4.3) shows the two resistors **in parallel**. In this arrangement the current divides and a different portion flows through each resistor. Returning to the series arrangement, it can be shown that the total effective resistance is simply $R_1 + R_2$, the sum of the individual resistances. If, for example, both resistances were equal to 3 Ω, the total resistance would be 6 Ω and the current, according to Ohm's law, would be 2 A. Another point to notice about this two-resistor circuit is that the potential or voltage drop across each resistor is 6 V, making a total drop around the circuit of 12 V as before.

Figure A4.4 shows the same idea extended to a chain or ladder of 10 identical resistors in series. The voltage drop across each individual resistor is exactly one tenth

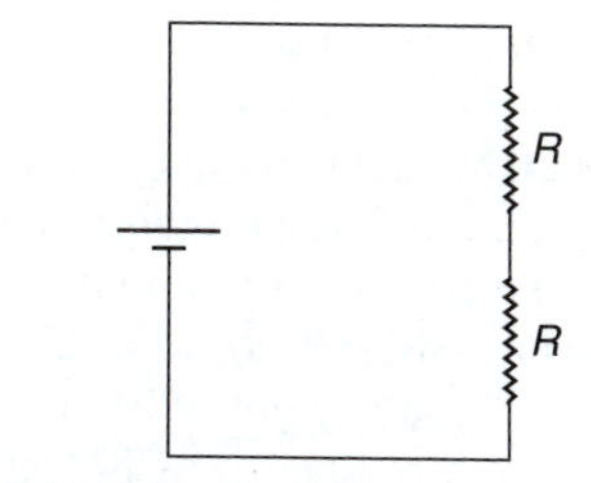

Figure A4.2 Resistors in series

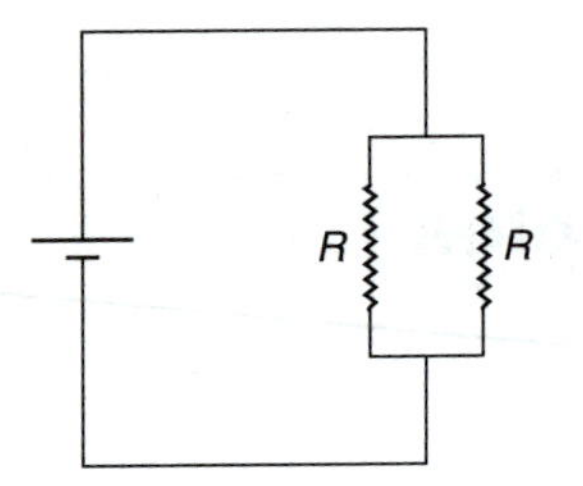

Figure A4.3 Resistors in parallel

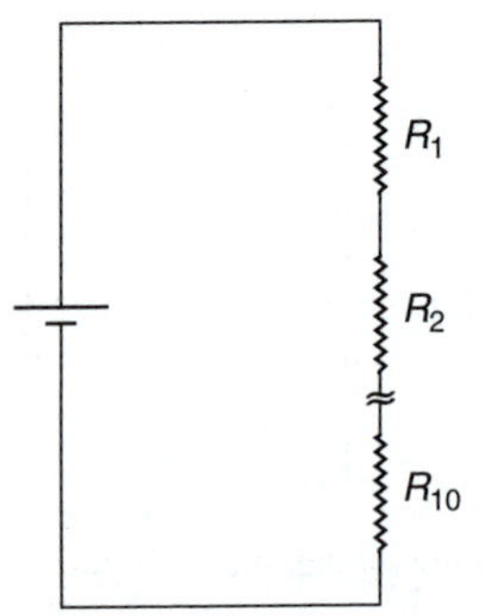

Figure A4.4 Attenuator: a chain of resistors in series

of the voltage provided by the cell or battery. Chains of resistors can therefore be used to divide an electrical voltage signal. This is the principle of the **attenuator**, used in sound level meters, to provide changes in range of exactly 10 dB, 20 dB, 30 dB, etc. Attenuators can be made with great accuracy and stability because they do not rely on components such as transistors whose performance can vary with age and with variations in supply (i.e. battery) voltage.

A4.3 Electrical loads: input and output resistance

Figure A4.5 shows a slightly more complex but more realistic version of Figure A4.2. The cell is connected to an external device which has an effective resistance R_L. This device, which draws current from the cell, is called the **load**. However, this is not the only resistance in the circuit because the cell itself provides a certain amount of resistance to current flow. This is called the internal resistance of the cell and is shown as r in Figure A4.5. The load R_L behaves in the circuit as though it were a simple single resistor, but it could be a complicated device such as a transistor radio circuit or the amplifiers in a sound level meter. R_L is the resistance as seen by the cell looking into the load; it is called the **input resistance** of the load. The cell itself presents a resistance r to the load; it is called the **output resistance** of the cell. In this particularly

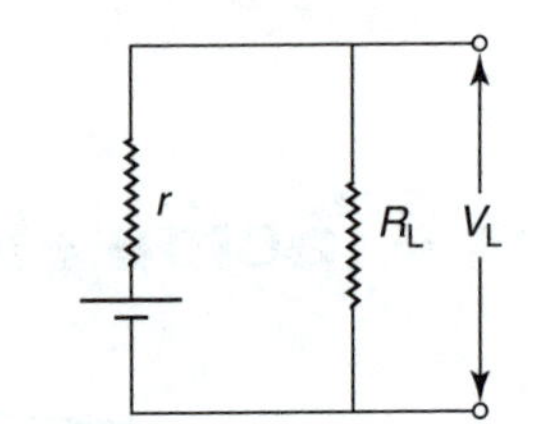

Figure A4.5 Series circuit including internal resistance of the cell

simple case the output resistance of the cell is the same as its internal resistance.

The maximum potential difference provided by the battery is V_B and the total resistance in the circuit is $r + R_L$. The current in the circuit is therefore given by $I = V_B/(R + r)$. The voltage drop across the load, V_L, is given by Ohms law:

$$V_L = IR_L = V_B \frac{R_L}{R + r}$$

Let us consider two extreme cases. First, suppose the internal resistance r is negligible compared to the load. The potential drop across the load, and across the terminals of the cell, will be the maximum possible, i.e. V_B, and the current through the load (V_B/R_L) will depend on the value of the load itself. **The smaller the value of the load resistance, the higher the current drawn from the cell.** Now suppose the cell has a high internal resistance compared to the load. The potential drop across the load, and across the terminals of the cell, will be given, approximately, by $V_L = V_B R_L/r$ and will be much less than V_B. The current through the load will be given by $I = V_B/r$ and will be limited by the internal resistance of the cell. **The larger the output resistance of the voltage source, the lower the current it can deliver.**

To illustrate the two cases, consider two 12 V cells: a lead-acid accumulator (i.e. a car battery) and a dry cell (or bank of cells) such as used in a transistor radio. Both cells identically show 12 V when a voltmeter is connected across their terminals. Both cells are capable of providing the current to operate the transistor radio, and in both cases the voltmeter across the terminals remains unchanged at 12 V when the radio is switched on. But when the cells are connected to a car headlamp, there is a different story. The accumulator lights up the bulb and there is very little drop in the voltmeter reading when the lamp is switched on. The dry cells are not capable of providing enough current to light up the headlamp, so the voltmeter reading drops rapidly from the 12 V mark when the lamp is switched on. The difference in the performance of the two cells arises from the difference in their output resistances. The accumulator has a very low internal resistance, about 1 Ω, so it is capable of delivering a current of several amps, sufficient to operate the lamp. The dry cells have a much higher internal resistance, so they can only deliver a maximum current of a few thousandths of an amp, sufficient for the transistor radio but not for the lamp.

Although we must not push the analogy too far, the condenser microphone is rather like the dry cell whose output voltage drops drastically under the heavy load of the headlamp bulb. The microphone acts as a cell (a source of alternating voltage) and, without a suitable preamplifier, its output signal may be drastically reduced by the loading effect of the sound level meter circuits.

A4.4 Alternating current

So far we have considered circuits in which the electric current flows in one direction only, i.e. direct current (DC). Circuits involving only resistors will operate in exactly the same way if the cell is replaced by a source which generates an alternating potential difference. The mains supply of electricity provides a voltage which varies sinusoidally with a frequency of 50 Hz and a voltage amplitude of 240 V. This produces an alternating current (AC) in devices connected to it. In a circuit containing only resistors, the voltage and current will always be in phase. There are, however, two other circuit components, inductors and capacitors, which have to be considered in AC circuits; their voltage–current relationships are more complicated.

A4.5 Inductors

The simplest *inductor* is a piece of wire wound into the form of a coil. The first law of electromagnetism says that a piece of wire carrying an electric current generates a magnetic field around itself. In a long, straight wire this is not very significant. But when the wire is wound into a coil, the magnetic field is concentrated in and around the coil. This is the principle of the electromagnet, and the coil behaves like a bar magnet while the current is switched on.

Changes in current through the coil cause changes in the magnetic field it produces, so when an alternating voltage is applied to the coil, the magnetic field has to go through a cycle of building up, decreasing, then building up again in the opposite direction, and so on. A magnetic field takes its energy from the electrical energy of the voltage source; therefore, an inductor or coil in a circuit will hamper or impede the flow of current. This is rather similar to the effect of a resistance in a DC circuit, except that the electrical energy needed to overcome resistance in a circuit is converted to heat instead of magnetic field energy. The electrical *impedance* (Z) provided by the inductor is given by the expression:

$$Z = 2\pi f L$$

where f is the frequency of the alternating voltage and L is the *inductance* of the coil, which depends on its physical dimensions and the number of turns. Inductance is measured in henrys (H). Electrical impedance is measured in ohms, like electrical resistance. In fact resistance is just one special form of impedance. Unlike inductance, resistance does not depend on the frequency of the current. Another difference is that the current through an inductor and the voltage across it are out of phase by 90°. Ohm's law also applies to inductors (between current and voltage amplitudes); impedance (Z) replaces resistance (R) in $V = IR$.

A4.6 Input and output impedance

Input and output resistance have already been introduced. In general they may be replaced by input impedance and output impedance. The *output impedance* is an important property of any device which *delivers* a signal, e.g. a battery, a microphone or an amplifier. Devices with a low output impedance can deliver higher currents than those with high output impedance. The *input impedance* is an important property of any device which *receives* a signal (i.e. of any load), such as a loudspeaker, a level recorder or an amplifier. Devices with a low input impedance draw a higher current from the source than those with a high input impedance.

A4.7 The electrodynamic loudspeaker and electrodynamic microphone

The operation of electrodynamic loudspeakers and microphones is based upon the behaviour of a coil in a magnetic field. The loudspeaker operates on the same principle as the electric motor. The loudspeaker cone, which radiates the sound, is attached to a coil that is free to move in the magnetic field produced by a permanent magnet. When an alternating current signal is supplied to the coil, it becomes a small electromagnet. The coil is alternately attracted and repelled by the permanent magnet as the polarity of the coil's magnetic field changes with the alternating current of the signal. As the coil vibrates, it vibrates the attached loudspeaker cone, radiating a sound pressure waveform similar to the alternating current.

The microphone operates on the reverse principle, which is also the basis of the generator or dynamo. Sound waves strike the microphone diaphragm, causing it to vibrate to and fro. The vibratory motion is transferred to the coil attached to the diaphragm and situated in the field of a permanent magnet. According to the principle of electromagnetic induction (Faraday's law), a current is induced to flow in a coil moving in a magnetic field. This induced current is the microphone signal, and if the operation is faithful, it has the same waveform as the incoming sound pressure wave.

A number of vibration transducers are coil-operated and work in a similar way to the electrodynamic microphone. They produce an electrical signal (a voltage) which is proportional to the vibration velocity.

A4.8 Capacitors

A **capacitor**, also called a condenser, consists of two conducting surfaces, often parallel plates, separated by a non-conductor such as air. A continuous direct current cannot flow through the non-conductor, called a **dielectric**, but if the plates are connected to a DC battery, a momentary or transient current flows from the cell to the plates until they reach the same electrical potential as the cell terminals. Once the potentials are equal, the flow of current ceases and the capacitor is fully charged. The capacitor can be thought of as a device which stores electric charge. The amount of charge it can store depends on the voltage difference across its plates and upon its **capacitance.** The capacitance depends on the physical dimensions of the capacitor, i.e. the area of the plates, the separation of the plates and the dielectric material between them.

Capacitance is measured in farads (F), but one farad is a very very large unit, so capacitors used in electronic circuits usually have values measured in microfarads ($1\,\mu F = 10^{-6}$ F) or in picofarads ($1\,pF = 10^{-12}$ F). A value of $1.0\,\mu F$ represents a fairly large capacitance, whereas 1 pF is a very small capacitance indeed. A typical value for the capacitance of a one-inch condenser microphone is about 60 pF.

If an alternating voltage is applied to a circuit containing a condenser, an alternating electric current flows in the circuit even though electric charge cannot pass across the gap between the plates. Electric charge is continually moving to and from the plates in response to the cyclic variations in the voltage. The capacitance presents an impedance to the flow of current; it is measured in ohms and is given by:

$$Z = \frac{1}{2\pi f C}$$

where C is the capacitance in farads. For a condenser microphone of 60 pF operating at 1000 Hz:

$$Z = \frac{1}{2 \times \pi \times 1000 \times 60 \times 10^{-12}} = 2.65 \times 10^{6}\ \Omega$$

$$= 2.65\ M\Omega$$

A4.9 Resistance–capacitance (RC) circuits

Figure A4.6 shows a DC battery, a resistor and a capacitor connected in series; a switch is also in the circuit. When the switch is closed, the voltage at point A rises instantaneously to the battery voltage. The voltage also rises at point B, i.e. the voltage rises on the plate of the capacitor as the capacitor becomes charged, but it rises more gradually than at A. The rate of increase of the voltage at B is in fact exponential; it depends on the **time constant** of the circuit, which is given in seconds by RC, i.e. by the production of R in ohms and C in farads. As the voltage at B reaches the battery voltage, it rises more and more

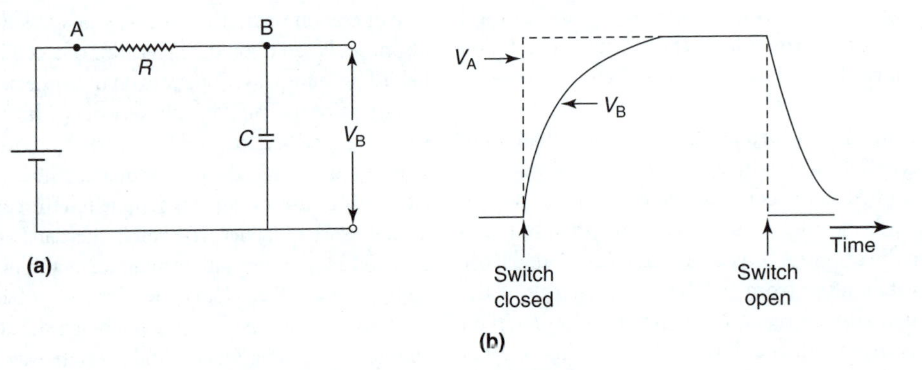

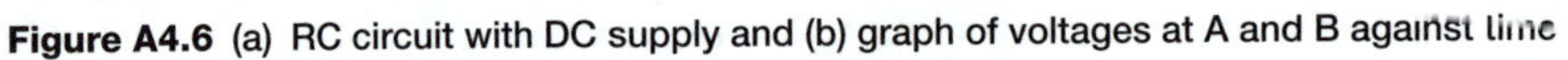
Figure A4.6 (a) RC circuit with DC supply and (b) graph of voltages at A and B against time

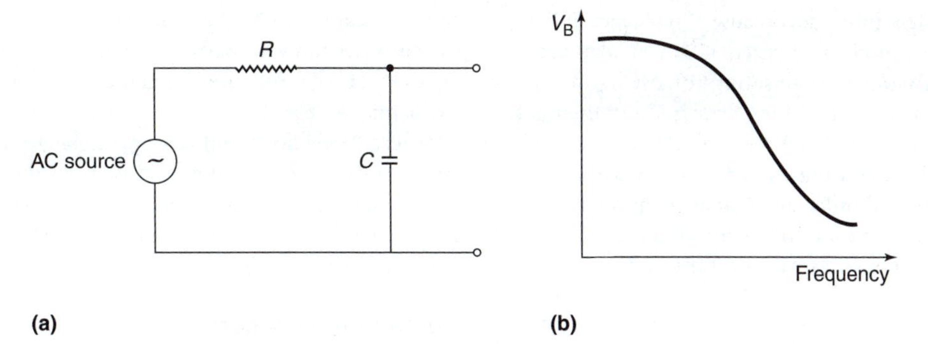

Figure A4.7 (a) RC circuit with AC supply and (b) graph of V_B against supply frequency. This configuration is a low pass filter

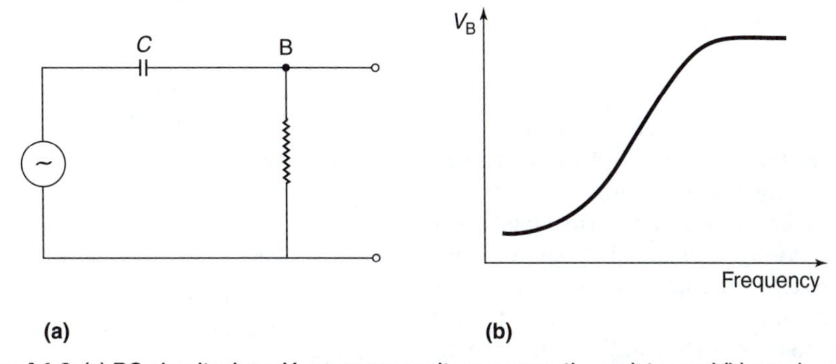

Figure A4.8 (a) RC circuit where V_B measures voltage across the resistor and (b) graph of V_B against supply frequency. This configuration is a high pass filter

slowly and the flow of current becomes less and less. If the voltage at A suddenly drops to zero, e.g. if the battery is suddenly shorted out, the voltage at B also falls, but more slowly (exponentially in fact) as the capacitor discharges. **A capacitor in a circuit can have the effect of smoothing out or slowing down rapid changes in voltage signals.** This is similar to the way a large reservoir smooths out variations of flow in air or water systems. The same property of capacitors can be used to smooth out or average time-varying signals; this is the basis for the averaging circuits in sound level meters. The averaging time (which is 1.0 s for slow mode and 0.125 s for fast mode) refers to the time constant of the circuit.

If the battery is replaced by an AC voltage source (Figure A4.7), a steady current flows through the circuit. The voltage amplitude at B, at the capacitor plate, depends on the relative magnitudes of the impedances of C and R. The situation is similar to that of Figure A4.5, but with two impedances instead of the two resistances R and r. At low frequencies the impedance of the capacitor will be far greater than the impedance of the resistor, so the voltage across the capacitor will be almost the source voltage. The voltage across the resistor will be small. As the frequency increases, the voltage across the capacitor decreases and the voltage across the resistor increases. If we consider the voltage at A as an input to the RC circuit and the voltage at B as the output, then the circuit is acting as a very crude **low pass filter**. If R and C are interchanged (Figure A4.8), similar reasoning shows that the circuit now behaves as a crude **high pass filter**. The filters used in sound measuring equipment are much more complex and have a much sharper cut-off rate, but some of them are based on multi-stage RC networks.

A4.10 Oscillating electrical circuits and analogy with mechanical oscillators

Inductors store energy as magnetic field energy. Capacitors store energy in the electric field between their two plates. In a circuit containing an inductor and a capacitor, the energy continually flows from one to the other and back again, resulting in an oscillating or alternating current flow. This electrical oscillation or vibration is analogous to the vibration of a mechanical oscillator, i.e. a mass on a spring. In both cases there will be a loss of energy. In the electrical system this arises because of resistance in the circuit and in the mechanical system it is due to damping.

The behaviour of a vibrating (or oscillating) mass spring system is discussed in Chapter 7 and the mathematical theory derived in Appendix 7.1.

The motion of the undamped system was described in terms of a second order differential equation, known as the equation of motion for the displacement (x) of the mass (m) attached to a spring of stiffness k:

$$m\ddot{x} + kx = 0$$

The analogous differential equation for an inductor and capacitor joined together (with no resistance) is expressed in terms of the electric charge, q, moving around the circuit as an electric current:

$$q/C + L\ddot{q} = 0$$

The first term (q/C) in this equation represents the voltage generated across the capacitor as a result of charge q stored on its plates. The voltage induced across the inductor as a result of the rate of flow of current (and hence of electric charge) passing through it, is given, according to Faraday's law of electromagnetic induction as $-L\ddot{q}$. In the case of free or natural electrical oscillations there is no applied voltage to the capacitor and inductance in the above equation simply expresses the fact that these two voltages must total to zero. (Note that electric current is the rate of flow of charge i.e. $\frac{dq}{dt}$ or $\dot{q}$, and so rate of change of current is $\frac{d^2q}{dt^2}$ or $\ddot{q}$.)

Following exactly the same argument as in Appendix 7.1 the solution of this differential equation will be a sinusoidal variation of electric charge with time:

$$q = \mathrm{Q} \sin \omega t$$

where q is the instantaneous amount of charge on the capacitor plates at time t, ω is the angular frequency ($= 2\pi f$) where f is the frequency of oscillation or vibration, and Q is a constant representing the initial conditions, i.e. the amount of charge at time = 0.

The equation indicates that electrical oscillations occur, leading to a sinusoidally varying current through the circuit and a sinusoidally varying voltage across each of its two components.

Comparing the two differential equations it can be seen that mass (m) in the mechanical oscillator corresponds to inductance (L) in the electrical oscillator, and the spring stiffness (k) corresponds to the reciprocal of capacitance (i.e. to $1/C$).

Using the analogy between the electrical and mechanical cases the natural frequency of the electrical oscillator can be obtained from the formula $f_0 = (1/2\pi)\sqrt{(k/m)}$ for the mechanical oscillator, which results in: $f_0 = (1/2\pi)\sqrt{(1/LC)}$ for the frequency of the electrical oscillator.

Alternatively this formula for f_0 can be derived exactly as in Appendix 7.1, by substituting the solution of the differential equation ($q = \mathrm{Q}\sin \omega t$) back into the equation, which involves differentiating twice to obtain the mean expression for $\ddot{q}$.

Electrical impedance

Electrical impedance tells us how much electrical current, I, will flow through an inductor, capacitor or resistor in response to an applied voltage, V:

$$Z_E = \text{voltage/current} = V/I$$

The electrical impedance for each of the three components of an electric oscillator are:

Inductor: $Z_L = 2\pi f L = \omega L$

Capacitor: $Z_C = 1/2\pi f C = 1/\omega C$

Resistor: $Z_R = R$

For an electrical circuit in which the three components – inductance, capacitor and resistor – are in series the total impedance Z_T is given by:

$$Z_T = Z_L + Z_C + Z_R$$

The electric current and voltage are alternating, i.e. varying sinusoidally with time. The above relationships only tell us about the magnitude of the current and voltage, but we also have to consider the phase between the voltage and the current. For a resistor voltage and current will be in phase, but for the inductor and capacitor the current will be 90°(quarter of a cycle) out of phase with the voltage; in one case current will lag behind the voltage and in the other the current will lead the voltage. This means of course that the currents in the inductor and capacitor will, for the same voltage, be 180°, or half a cycle, out of phase, i.e. in opposite directions. The situation may be represented by showing the three impedances as vector quantities, as shown below:

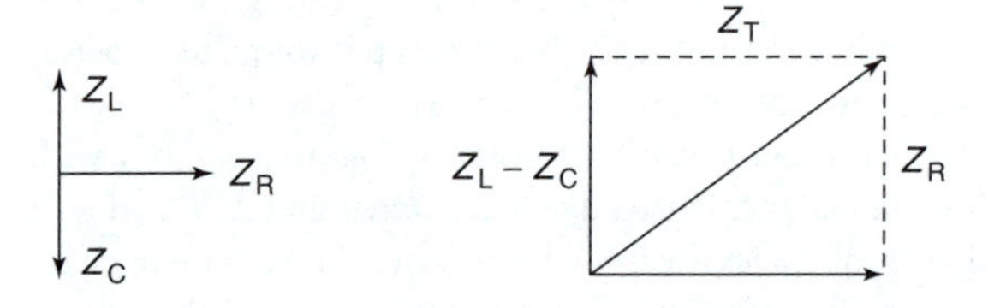

Using Pythagoras' theorem: $Z_T = [(Z_L - Z_C)^2 + Z_R^2]^{1/2}$

Both Z_L and Z_C vary with frequency but Z_R does not. The magnitude of the total impedance, Z_T, varies with frequency as shown in the sketch graph below. The sketch showing how the current, I, varies with frequency is analogous to the resonance curve for a mass–spring system (mechanical oscillator) shown in Figure 7.9.

The total impedance will be a minimum, and hence current will be a maximum, at the frequency at which Z_L and Z_C cancel, leaving only the resistance component, i.e. $Z_T = Z_R$. The frequency at which this occurs is called the resonance frequency, f_0, or angular frequency ω_0 so that:

$$\omega_0 L = 1/\omega_0 C$$

and $\omega_0^2 = 1/LC$, so that $\omega_0 = \sqrt{(1/LC)}$, and $f_0 = (1/2\pi)\sqrt{(1/LC)}$

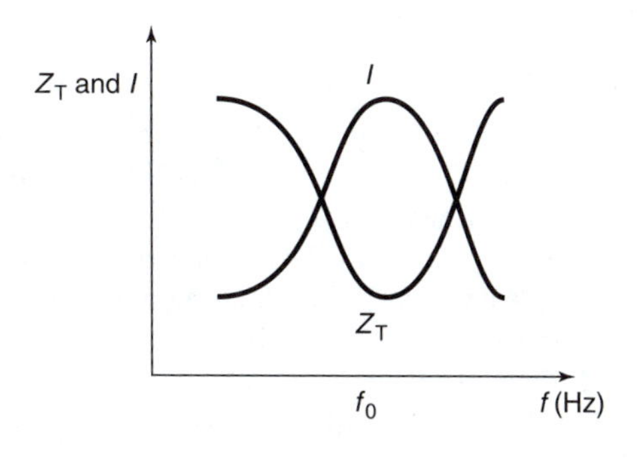

Complex number representation

The different components of the impedance are usually represented as a complex number, with the real part, on the horizontal axis, representing the 'in phase' resistive component and the vertical axis representing the 'out of phase' components Z_L and Z_C usually called the reactive components:

$$Z_T = Z_R + j(Z_L - Z_C)$$

The overall magnitude is given, as before by:

$$Z_T = [(Z_L - Z_C)^2 + Z_R^2]^{1/2}$$

Mechanical impedance

Mechanical impedance tells us the vibration velocity, V, that will be produced by the application of a vibratory force, F, to the component of the vibrating system:

$$Z = \text{voltage/current} = V/F$$

A similar approach may be adopted for a mass–spring–damper system where the mechanical impedance of the three components are given by:

Mass: $Z_m = 2\pi fm = \omega m$

Stiffness: $Z_K = 1/2\pi fC = 1/\omega C$

Damper: $Z_R = r$

For a damper the alternating force and the alternating velocity it produces will be in phase, but for the mass and stiffness the velocity will be 90° (quarter of a cycle) out of phase with the force; in one case velocity will lag behind the force, and in the other case velocity will lead the force. This means that the impedance of the mass and spring will be 180°, or half a cycle, out of phase, i.e. in opposite directions. The situation may be represented by showing the three impedances as vector quantities, as for the electrical case, and the overall magnitude of the impedance Z_T will be given by:

$$Z_T = [(Z_m - Z_K)^2 + Z_r^2]^{1/2}$$

As with the electrical case, the mechanical impedance will be a minimum, and the vibration velocity amplitude will be a maximum, at the angular frequency (ω_0) and frequency (f_0), at which the impedances of the mass and stiffness are equal in magnitude, and therefore, because they are opposite in phase, they cancel:

$$\omega_0 m = k/\omega_0$$

and $\omega_0^2 = k/m$, so that $\omega_0 = \sqrt{(k/m)}$, and $f_0 = (1/2\pi)\sqrt{(k/m)}$

which is the result obtained from solving the differential equation of motion as shown in Appendix 7.1.

Answers

Chapter 1

2. 70 to 77 dB
3. 1.2×10^{-5} W, 4.3 J
5. 270°; 180°
6. (1): 2.41×10^{-5} m/s, 2.41×10^{-7} W/m^2; (2): 0.415 Pa, 4.15×10^{-4} W/m^2; (3): 0.20Pa, 4.91×10^{-4} m/s; (4): 4.82×10^{-4} m/s, 9.64×10^{-4} W/m^2; (5): 0.83 Pa, 1.66×10^{-3} W/m^2; (6): 0.091 Pa, 2.2×10^{-4} m/s
7. (a) 2.24 times, 5 times, 7 dB, 7 dB; (b) 3.87 times, 15 times, 12 dB, 12 dB (c) 5 times, 25 times, 14 dB, 14 dB
8. (a) 25 times, 14 dB, 14 dB; (b) 225 times, 23.5 dB, 23.5 dB; (c) 625 times, 28 dB, 28 dB
9. 3.16 times, 10 times, 31.6 times
10. 94 dB, 114 dB, 52 dB, 26 dB, 0 dB, −6 dB
11. 2.0×10^{-3} Pa; 1.4×10^{-3} Pa, 1.0×10^{-3} Pa; 2.0 Pa; 1.4 Pa; 0.2 Pa
12. 96.6 dB
13. 85.8 dBA, 92.6 dBZ
14. 87 dBA, 85 dBA
15. 26 dB
16. 81.9 dB
17. 75 dB
18. 86.9 dBA
19. 4 machines
20. 7 machines
22. (a) 34 Hz, 68 Hz, 102 Hz; (b) 17 Hz, 51 Hz, 85 Hz; all frequencies will increase by 15.8%
23. 200 Hz, 400 Hz, 600 Hz; all frequencies will increase by 41.4%
24. 0.65 m
25. 707 Hz, 354 Hz; 449 Hz, 356 Hz; 561 Hz, 445 Hz; 707 Hz, 561 Hz
27. (c) (i) 500 Hz, (ii) 0.69 m, (iii) 343 m/s, (iv) 0.71
28. (a) 256 Hz, 1.33 m, 90 dB; (b) 80.6 dBA, 84.9 dBA, 79.5 dBA
29. 0.41 m, 0.15 Pa, 50 Hz
31. (a) 5×10^{-6} m, 1000 Hz, 340 m/s

Chapter 2

7. 44.7 dBA
10. 143.5 dB (assume audible when siren exceeds background level)
11. 4.8 dB; 13.7 dB
12. −5 dB, 6 dB, −7 dB, −20 dB
13. 79 dB, 56 dB; estimated barrier attenuation = 12.7 dB, so yes, just
15. 52.5 dB, 5 dB
19. 1.04
21. 12 dB, 80 dB, 69 dB, 11 dB
22. (d) (ii) about 3 dB

Chapter 3

2. A outer ear or pinna, B ossicles (malleus or hammer), C auditory nerve, D cochlea, E eardrum (tympanic membrane)
3. A basilar membrane, B tectorial membrane, C arch or rods of Corti, D hair cells, E auditory nerve
6. 34 Sones, 91 Phons, 77.5 dBA
13. 35 Phons, 91 PNdB
14. 109 Phons, 104 PNdB

Chapter 4

6. (to nearest dB): L_{A10} = 81 dBA, L_{A50} = 75 dBA, and L_{A90} = 69 dBA
8. 66.6 dBA, 68.0, dBA, 71.5 dBA
9. 63.4 dBA, 64.6 dBA
10. 53.9 dBA, 46.1 dBA
13. (a) little or no difference; (b) rating level is 21dBA, complaints likely
15. an increase of 1.8 dB which is feasible if speeds are slow so that tyre–surface noise is negligible
16. 7 old trains
17. existing L_{Aeq16h} = 58.4 dBA, future L_{Aeq16h} = 62.4 dBA, therefore noticeable impact, owner is wrong
18. 53.8, 51.4, 56, 53, 50 dBA
19. 63.2 dBA, 62.4 dBA, 66.6 dBA
20. survey 2 with situation C, survey 4 with situation D, survey 1 with situation A, survey 5 with situation B, survey 1 with situation E

Chapter 5

1. 0.28
2. 50.8 m^2
3. 0.27, 0.32, 0.36, 0.39, 0.37

4. 40.5 dB, 32.2 dB
5. 954 m^2
6. 92.6 dBA
7. 99 dB
8. 6.31, 0.47
12. 49 people
13. 0.59 s, 0.51 s
14. 0.75 s
15. 111.8 dB, 111 dB, 109.5 dB, 108 dB
16. (a) 96.7 m^2; (b) 0.46 s, 139 m^2; (c) 36.2 dB
17. for RT of 1 second: 6.25 m
19. (a) 43 Hz, 61 Hz, 105 Hz; (c) 84 dB re. 10^{-12} W
21. 18.8 dB per diffuser
22. 93 dBA, 86 dBA, 93 dBA, 84 dBA

Chapter 6

1. 1.9 m^2
2. 5 dB
3. 36.5 dB
4. 27.5 dB
5. 5 kg/m^2
6. 42 dB
7. 46 dBA
8. 28.5 dB, 26 dB
9. 38 dB
10. 51.4 dB
11. 27 dB
12. 35.8 dB, 0.52
13. 54.3 dB
14. 41.1 dB, 3 dB, 33.1 dB
15. 32.2 dB, 270.8 m^2
16. 74 dB, 67 dB
17. 30 dB, 6 dB
18. 43.8 dB
19. 52.5 dB
20. 17 dB
21. 78 dB, 72 dB and 60 dB using Rathe method (79, 71 and 59 dB using point source method)
22. 20 dB
23. (b) 39 dB
24. for plant room 37.3 dB, so 38 dB door required; for computer room, 32.6 dB, so 34 dB door required

Chapter 7

1. 1.6 mm/s, 8.4 microns
2. 0.63 mm/s, 0.2 m/s^2
3. 159 microns, 0.628 m/s^2
4. 22.0 m/s^2, 5.52 m/s^2, 147 dB, 135 dB
5. 1.58 m/s^2
6. 0.14 mm/s
7. 930 Hz
8. 93.5 dB
9. 0.184 W, 131 dB
10. 16.1 Hz
11. 20 dB, $1.7 \times 10^7 Nm^{-1}$, 0.0011 m, 43%
12. $111 \times 10^3 Nm^{-1}$
13. $2.64 \times 10^6 Nm^{-1}$ per spring
14. above 36 Hz
15. 0.89
16. 2.83 m/s^2, 8 h 40 min
17. 4.3 m/s^2, 2.6 m/s^2

Chapter 8

11. 30.3 dB
12. 100 mV
13. 1.22 mV
14. 24.8 mms^{-1}
15. 40 dB to 150 dB (assumed to be 10 dB above noise level to 10 dB below distortion level)
16. (a) (i) 4.0 m/s^2, (ii) 12.6 mm/s, 40.1 μm; (b) 1122 Hz, 891 Hz
17. 1.6 V, 9 m/s^2
19. 44.7 microvolts, assuming that 38 dB is 5 dB above the noise floor of the instrument, 60 dB to 162 dB
20. not real time

Chapter 9

19. 50 dB, 19 dB, 50 dB, 24 dB
20. 83.4 dB, 22.2 Hz
21. 36 dB
22. 22 dB

Bibliography

Books

Association of Noise Consultants, *Measurement and Assessment of Groundborne Noise and Vibration*, ANC Guidelines, 2001

Attenborough, K., Li, K. M. and Horoshenkov, K., *Predicting Outdoor Sound*, Taylor and Francis, 2007

Barron, M., *Auditorium Acoustics and Architectural Design*, Taylor and Francis, 2000

Beranek, L., *Concert Halls and Opera Houses: Music, Acoustics, and Architecture*, 2nd edtion, Springer Verlag, 2003

Beranek, L. L., *Acoustics*, Acoustical Society of America through the American Institute of Acoustics, 1993

Beranek, L. L. and Ver, I. (eds), *Noise and Vibration Control Engineering*, Wiley, 1992

Bies, D. A. and Hansen, C. H., *Engineering Noise Control*, 4th edition, Spon Press, 2009

Broch, J. T., *Mechanical Vibration and Shock Measurements*, Bruel and Kjaer, 1984

Cox, T. J. and D'Antonio, P., *Acoustic Absorbers and Diffusers*, Taylor and Francis, 2009

Everest, F. A., *Master Handbook of Acoustics*, McGraw-Hill, 2001

Fahy, F., *Foundations of Engineering Acoustics*, Academic Press, 2001

Fahy, F. and Walker, J. G. (eds), *Advanced Applications in Acoustics, Noise and Vibration*, Taylor and Francis, 2004

Fahy, F. J. and Walker, J. G. (eds), *Fundamentals of Noise and Vibration*, Taylor and Francis, 1998

Griffin, M. J., *Handbook of Human Vibration*, Academic Press, 1990

Hall, D. E., *Basic Acoustics*, Wiley, 1987

Hassal, J. R. and Zaveri, K., *Acoustic Noise Measurements*, 5th edition, Bruel and Kjaer, 1988

Haughton, P., *Acoustics for Audiologists*, Academic Press, 2002

Hopkins, C., *Sound Insulation: Theory into Practice*, Elsevier Science and Technology, 2007

Kang, J., *Urban Sound Environment*, Taylor and Francis, 2006

Kinsler, L. E., Frey, A. R., Coppens, A. B. and Sanders, J. V., *Fundamentals of Acoustics*, 3rd edition, Wiley, 1982

Kryter, K. D., *The Effects of Noise on Man*, 2nd edition, Academic Press, 1985

Kuttruff, H. *Acoustics: An Introduction*, Taylor and Francis, 2006

Kuttruff, H., *Room Acoustics*, 4th edition, Elsevier Applied Science, 1991

Lord, P. and Templeton, D., *Detailing for Acoustics*, Taylor and Francis, 1995

Nelson, P. M. (ed), *Transportation Noise Reference Book*, Butterworth, 1987

Parkin, P. H., Humphreys, H. R. and Cowell, J. R. *Acoustics, Noise and Buildings*, 4th edition Faber and Faber, 1979

Porges, G., *Applied Acoustics*, Edward Arnold, 1977

Sharland, I. J., *Woods Practical Guide to Noise Control*, Woods of Colchester, 1972 (reissued by Flakt)

Sound Research Laboratories, *Noise Control in Mechanical Services*, 1976

Templeton, D., *Acoustics in the Built Environment: Advice for the Design Team*, 2nd edition, Architectural Press, 1997

Watson, R. and Downey, O., *The Little Red Book of Acoustics*, 2nd edition, Blue Tree Acoustics, 2008

Webb, J. D. (ed.), *Noise Control in Industry*, Sound Research Laboratories, 1976

Reports and codes of practice

BRE, *The 1999/2000 National Survey of Attitudes to Environmental Noise – Volume 2: Trends in England and Wales*, Building Research Establishment/DEFRA, 2002

BRE, *The National Noise Incidence Study 2000 (England and Wales)*, Building Research Establishment/DEFRA, 2002

BRE/CIRIA, *Sound Control for Homes*, Building Research Establishment/Construction Industry Research and Information Association, 1995

CAA, DR Report 8402, *United Kingdom Aircraft Noise Index Study: Main Report*, Civil Aviation Authority, 1985

CIBSE, *Guide A: Environmental Design*, Section 1.9, The acoustic environment, Chartered Institute of Building Services Engineers (London), 2006

Department for Education and Skills, Building Bulletin 93, *Acoustic Design of Schools*, The Stationery Office, 2003

Department of the Environment, *Report of the Noise Review Working Party* (Batho Report), HMSO, 1990

Department of the Environment, Planning Policy Guidance Note (PPG) 24, *Planning and Noise*, 1994

Department of the Environment, Welsh Office, *Calculation of Road Traffic Noise*, HMSO, 1975 and 1988.

Department of Health, Health Technical Memorandum HTM 08-01, *Acoustic Performance*, 2008

Department of Transport, *Railway Noise and the Insulation of Dwellings*, HMSO, 1991

Department of Transport, *The Calculation of Railway Noise*, HMSO, 1995

Department of Transport, *Attitudes to Noise from Aviation Sources in England* (ANASE study), HMSO, 2007

Health and Safety Executive, *Controlling Noise at Work*, The Control of Noise at Work Regulations 2005, Guidance on Regulations, L108, HSE Books, 2005

Health and Safety Executive, *Hand-Arm Vibrations*, The Control of Vibration at Work Regulations 2005, Guidance on Regulations, L140, HSE Books, 2005

Health and Safety Executive, *Whole-Body Vibration*, The Control of Vibration at Work Regulations 2005, Guidance on Regulations, L141, HSE Books, 2005

Highways Agency, *Design Manual for Roads and Bridges*, Volume 11 section 3 part 7, Traffic noise and vibration, HA 231.08, Highways Agency, 2008

IEH, *Report on the Non-auditory Effects of Noise*, Report R10, Institute of Environmental Health/Medical Research Council, 1997

IOA/IEMA, *Guidelines on Noise Impact Assessment* (Draft), Institute of Acoustics/Institute of Environmental Management and Assessment

Noise, Final Report on the Problem of Noise (Wilson Report), HMSO, 1963

Noise Council, *Code of Practice on Environmental Noise Control at Concerts*, The Noise Council, 1995

Resistance to the Passage of Sound, Approved Document E, The Building Regulations 2000, ODPM, 2003 edition

Smith, A. P. and Broadbent, D. E., *Non-auditory Effects of Noise at Work: A Review of the Literature*, HSE, 1992

Steffens, R. J., *Structural Vibration and Damage*, Building Research Establishment Report, HMSO, 1974 (reissued 1992)

WHO, *Guidelines for Community Noise*, World Health Organization, 2000

Standards

Lists of standards on acoustics and noise control may be found by interrogating BSI and ISO websites. The following list is not comprehensive and relates mainly to material covered in the text.

General

BS 8233:1999 Code of practice for sound insulation and noise reduction for buildings

BS EN 12354 Building acoustics. Estimation of acoustic performance in buildings from the performance of elements:

Part 1, 2000: Airborne sound insulation between rooms

Part 2, 2000: Impact sound insulation between rooms

Part 3, 2000: Airborne sound insulation against outdoor sound

Part 4, 2000: Transmission of indoor sound to the outside

Part 5, 2009: Sound levels due to the service equipment

Part 6, 2003: Sound absorption in enclosed spaces

BS EN ISO 1683:2008 Acoustics. Preferred reference values for acoustical and vibratory levels

BS EN ISO 80000-8:2007 Quantities and units. Acoustics

ISO 266 Acoustics – Preferred frequencies

ISO 9613 Acoustics – Attenuation of sound during propagation outdoors – Part 1 Calculation of sound absorption by the atmosphere (1993); Part 2 General method of calculation (1996)

Relating to Chapter 3

BS EN 60645-1:2001 Audiometers. Pure-tone audiometers

BS EN ISO 389-3:1999 Acoustics. Reference zero for the calibration of audiometric equipment. Reference equivalent threshold force levels for pure tones and bone vibrators

BS EN ISO 389-4:1999 Acoustics. Reference zero for the calibration of audiometric equipment. Reference levels for narrow-band masking noise

BS EN ISO 389-6:2007 Acoustics. Reference zero for the calibration of audiometric equipment. Reference threshold of hearing for test signals of short duration

BS EN ISO 389-7:2005 Acoustics. Reference zero for the calibration of audiometric equipment. Reference threshold of hearing under free-field and diffuse-field listening conditions

BS EN ISO 7029:2000 Acoustics. Statistical distribution of hearing thresholds as a function of age

BS EN ISO 8253-1:1998 Acoustics. Audiometric test methods. Basic pure tone air and bone conduction threshold audiometry

BS EN ISO 8253-2:1998 Acoustics. Audiometric test methods. Sound field audiometry with pure tone and narrow-band test signals

BS EN ISO 8253-3:1998 Acoustics. Audiometric test methods. Speech audiometry

BS ISO 226:2003 Acoustics. Normal equal-loudness-level contours

BS ISO 12124:2001 Acoustics. Procedures for the measurement of real-ear acoustical characteristics of hearing aids

EN 26189:1991 Specification for pure tone air conduction threshold audiometry for hearing conservation purposes

ISO 1999:1990 Determination of occupational noise exposure and of noise induced hearing impairment.

Relating to Chapter 4

BS 4142:1997 Method for Rating industrial noise affecting mixed residential and industrial areas

BS 5228-1:2009 Code of practice for noise and vibration control on construction and open sites – Part 1: Noise, Part 2: Vibration

BS 7445-1:2003 Description and measurement of environmental noise. Guide to quantities and procedures

BS 7445-2:1991 (ISO 1996-2:1987) Description and measurement of environmental noise – Part 2: Guide to the acquisition of data pertinent to land use

BS 7445-3:1991 (ISO 1996-3:1987) Description and measurement of environmental noise – Part 3: Guide to application to noise limits

BS 8233:1999 Code of practice for sound insulation and noise reduction for buildings

BS ISO 362-1:2007 Acoustics – Measurement of noise emitted by accelerating road vehicles. Engineering method

BS ISO 5130:2007 Acoustics – Measurements of sound pressure level emitted by stationary road vehicles

ISO 1996-1:2003 Acoustics – Description, measurement and assessment of environmental noise – Part 1: Basic quantities and assessment procedures

ISO 1996-2:2007 Acoustics – Description, assessment and measurement of environmental noise – Part 2: Determination of environmental noise levels

ISO 3891 Acoustics – Procedure for describing aircraft noise heard on the ground.

ISO 13325:2003. Tyres – Coast-by methods for measurement of tyre-to-road sound emission.

Relating to Chapter 5

BS 8233:1999 Code of practice for sound insulation and noise reduction for buildings

BS EN ISO 354:2003 acoustics. Measurement of sound absorption in a reverberation room

BS EN ISO 3382-2:2008 Acoustics. Measurement of room acoustic parameters. Reverberation time in ordinary rooms

BS EN ISO 9921:2003 Assessment of Speech Communication

BS EN ISO 10534-1:2001 acoustics. Determination of sound absorption coefficient and impedance in impedances tubes. Method using standing wave ratio

BS EN ISO 10534-2:2001 acoustics. Determination of sound absorption coefficient and impedance in impedance tubes. Transfer-function method

BS EN ISO 11654:1997. Acoustics. Sound absorbers for use in buildings. Rating of sound absorption

BS EN ISO 18233:2006 acoustics. Application of new measurement methods in building and room acoustics

BS EN 60268-16:2003 Sound system equipment. Objective rating of speech intelligibility by speech transmission index

BS EN 60849:1998 IEC 60849:1998 Sound systems for emergency purposes

ISO 3382:1997 Acoustics – Measurement of the reverberation time of rooms with reference to other acoustical parameters

ISO 3382-2:2008 Acoustics – Measurement of room acoustic parameters – Part 2: Reverberation time in ordinary rooms

Relating to Chapter 6

BS 8233:1999 Code of practice for sound insulation and noise reduction for buildings

BS EN ISO 140 Acoustics – Measurement of sound insulation in buildings and of building elements:

Part 1: Requirements for laboratory test facilities with suppressed flanking transmission

Part 2: Determination, verification and application of precision data

Part 3: Laboratory measurements of airborne sound insulation of building elements

Part 4: Field measurements of airborne sound insulation between rooms

Part 5: Field measurements of airborne sound insulation of façade elements and facades

Part 6: Laboratory measurements of impact sound insulation of floors

Part 7: Field measurements of impact sound insulation of floors

Part 8: Laboratory measurements of the reduction of transmitted impact noise by floor coverings on a heavyweight standard floor

Part 9: Laboratory measurements of room-to-room airborne sound insulation of a suspended ceiling with a plenum above it

Part 10: Laboratory measurements of airborne sound insulation of small building elements

Part 11: Laboratory measurements of the reduction of transmitted impact noise by floor coverings on lightweight reference floors

Part 12: Laboratory measurements of room-to-room airborne and impact sound insulation of an access floor

Part 13: Guidelines

Part 14: 2004 Guidelines for special situations in the field

Part 16: Laboratory measurement of the sound reduction index improvement by additional linings (2006)

Part 18: Laboratory measurement of sound generated by rainfall on building elements

BS EN ISO 717-1:1997 acoustics. Rating of sound insulation in buildings and of building elements. Airborne sound insulation

BS EN ISO 717-2:1997 acoustics. Rating of sound insulation in buildings and of building elements. Impact sound insulation

ISO 717-1:2004 Acoustics–Rating of sound insulation in buildings and of building elements, Part 1: Airborne sound insulation

ISO 717-2:2004 Rating of sound insulation in buildings and of building elements Part 2: Impact sound insulation

Relating to Chapter 7

BS 3015:1991 ISO 2041:1990 Glossary of Terms relating to mechanical vibration and shock

BS 5228-2:2009 Code of practice for noise and vibration control on construction and open sites – Part 2: Vibration

BS 6472-1:2008 Guide to evaluation of human exposure to vibration in buildings – Part 1: Vibration sources other than blasting

BS 6472-2:2008 Guide to evaluation of human exposure to vibration in buildings Part 2: Blast-induced vibration

BS 6841:1987 Guide to Measurement and evaluation of human exposure to whole-body mechanical vibration and repeated shock

BS 7385-1:1990 (ISO 4866:1990) Evaluation and measurement for vibration in buildings – Part 1: Guide for measurement of vibrations and evaluation of their effects on buildings

BS 7385-2:1993 Evaluation and measurement for vibration in buildings – Part 2: Guide to damage levels from groundborne vibration

BS EN 1032:2003 Mechanical vibration. Testing of mobile machinery in order to determine the vibration emission value British Standards Institution

BS EN ISO 5349-1:2001 Mechanical vibration – Measurement and evaluation of human exposure to hand-transmitted vibration – Part 1: General requirements

BS EN ISO 5349-2:2002 Mechanical vibration – Measurement and evaluation of human exposure to hand-transmitted vibration – Part 2: Practical guidance for measurement at the workplace

BS EN ISO 8041:2005 Human response to vibration – Measuring instrumentation

BS EN ISO 12096:1997 Mechanical vibration. Declaration and verification of vibration emission values British Standards Institution

BS EN 13490:2002 Mechanical vibration. Industrial trucks. Laboratory evaluation and specification of operator seat vibration British Standards Institution

BS EN 14253:2003 + A1:2007 Mechanical vibration – Measurement and calculation of occupational exposure to whole-body vibration with reference to health – Practical guidance

BS ISO 2631- 4:2001 Mechanical vibration and shock – Evaluation of human exposure to whole-body vibration: Part 4: Guidelines for the evaluation of the effects of vibration and rotational motion on passenger and crew comfort in fixed guide-way transport systems

BS ISO 5348:1998 Mechanical vibration and shock – Mechanical mounting of accelerometers

ISO 2631 Mechanical vibration and shock – Evaluation of human exposure to whole-body vibration: Part 1: General requirements

ISO 2631 Mechanical vibration and shock – Evaluation of human exposure to whole-body vibration: Part 2: Vibration in Buildings (1 Hz to 8 Hz)

ISO 8569:1996 Mechanical vibration and shock – Measurement and evaluation of shock and vibration effects on sensitive equipment in buildings

Relating to Chapter 8

BS EN 60942:2003 Electroacoustics. Sound calibrators

BS EN 61252:1997 Electroacoustics. Specifications for personal sound exposure meters

BS EN 61672 Electroacoustics – Sound level meters – Part 1: Specifications; Part 2: Pattern evaluation tests; Part 3: Periodic testing

IEC 61260 Electroacoustics – Octave-band and fractional-octave-band filters

Relating to Chapter 9

BS EN ISO 15667:2000 Acoustics. Guidelines for noise control by enclosures and cabins

ISO 3740 series on sound power level measurement:

ISO 3740:2000 Acoustics – Determination of sound power levels of noise sources – Guidelines for the use of basic standards

ISO 3741:1999 Acoustics – Determination of sound power levels of noise sources using sound pressure – Precision methods for reverberation rooms

ISO 3743-1:1994 Acoustics – Determination of sound power levels of noise sources using sound pressure – Engineering methods for small, movable sources in reverberant fields – Part 1: Comparison method for hard-walled test rooms

ISO 3743-2:1994 Acoustics – Determination of sound power levels of noise sources using sound pressure – Engineering methods for small, movable sources in reverberant fields – Part 2: Methods for special reverberation test rooms

ISO 3744:1994 Acoustics – Determination of sound power levels of noise sources using sound pressure – Engineering method in an essentially free field over a reflecting plane

ISO 3745:2003 Acoustics – Determination of sound power levels of noise sources using sound pressure – Precision methods for anechoic and hemi- anechoic rooms

ISO 3746:1995 Acoustics – Determination of sound power levels of noise sources using sound pressure – Survey method using an enveloping measurement surface over a reflecting plane

BS EN ISO 3747:2000 Acoustics – Determination of sound power levels of noise sources using sound pressure – Comparison method in situ

BS EN ISO 9614 series on – sound intensity methods:

BS EN ISO 9614-1:1993 Acoustics – Determination of sound power levels of noise sources using sound - intensity – Part 1: Measurement at discrete points

BS EN ISO 9614-2:1997 Acoustics – Determination of sound power levels of noise sources using sound intensity – Part 2: Measurement by scanning

BS EN ISO 1680:2000 Acoustics. Test code for the measurement of airborne noise emitted by rotating electrical machinery

BS EN ISO 5136:2003 Acoustics. Determination of sound power radiated into a duct by fans and other air-moving devices. In-duct method

Standards relating to hearing protectors:

BS EN 352:2002 Hearing protectors. Safety requirements and testing.

Part 1: Ear-muffs

Part 2: Ear-plugs

Part 3: Ear-muffs attached to an industrial safety helmet

Part 4: Level-dependent ear-muffs

Part 5: Active noise reduction ear-muffs

Part 6: Ear-muffs with electrical audio input

Part 7: Level-dependent ear-plugs

BS EN 13819-2:2002 Hearing protectors. Testing. Acoustic test methods

BS EN 24869-1:1993 Acoustics. Hearing protectors. Sound attenuation of hearing protectors. Subjective method of measurement

BS EN ISO 4869-2:1995 Acoustics. Hearing protectors. Estimation of effective A-weighted sound pressure levels when hearing protectors are worn

Index

WITHDRAWN FROM STOCK QMUL LIBRARY

QM LIBRARY (MILE END)